Digital Signal Processing

Digital Signal Processing
A Computer Science Perspective

Jonathan (Y) Stein

A Wiley-Interscience Publication

JOHN WILEY & SONS, INC.

New York • Chichester • Weinheim • Brisbane • Singapore • Toronto

Copyright © 2000 by John Wiley & Sons, Inc.

All rights reserved. Published simultaneously in Canada.

For ordering and customer service, call 1-800-CALL-WILEY.

Library of Congress Cataloging in Publication Data

Stein, Jonathan Y.
  Digital signal processing : a computer science perspective / Jonathan Y. Stein
    p. cm.
  "A Wiley-Interscience publication."
  Includes bibliographical references and index.
  ISBN 0-471-29546-9 (cloth : alk. paper)
  1. Signal processing—Digital techniques.   I. Title.

  TK5102.9. S745 2000
  621.382'2—dc21   00-035905

Printed in the United States of America.

10 9 8 7 6 5 4 3 2 1

To Ethel, Hanna and Noga

Contents

Preface   xv

1   **Introductions**   **1**
   1.1   Prehistory of DSP 2
   1.2   Some Applications of Signal Processing 4
   1.3   Analog Signal Processing 7
   1.4   Digital Signal Processing 10

Part I   **Signal Analysis**

2   **Signals**   **15**
   2.1   *Signal* Defined 15
   2.2   The Simplest Signals 20
   2.3   Characteristics of Signals 30
   2.4   Signal Arithmetic 33
   2.5   The Vector Space of All Possible Signals 40
   2.6   *Time* and *Frequency* Domains 44
   2.7   Analog and Digital Domains 47
   2.8   Sampling 49
   2.9   Digitization 57
   2.10   Antialiasing and Reconstruction Filters 62
   2.11   Practical Analog to Digital Conversion 64

3   **The Spectrum of Periodic Signals**   **71**
   3.1   Newton's Discovery 72
   3.2   Frequency Components 74
   3.3   Fourier's Discovery 77
   3.4   Representation by Fourier Series 80
   3.5   Gibbs Phenomenon 86
   3.6   Complex FS and Negative Frequencies 90

3.7   Properties of Fourier Series 94
3.8   The Fourier Series of Rectangular Wave 96

4   The Frequency Domain   103
4.1   From Fourier Series to Fourier Transform 103
4.2   Fourier Transform Examples 110
4.3   FT Properties 113
4.4   The Uncertainty Theorem 117
4.5   Power Spectrum 122
4.6   Short Time Fourier Transform (STFT) 126
4.7   The Discrete Fourier Transform (DFT) 132
4.8   DFT Properties 135
4.9   Further Insights into the DFT 141
4.10 The z Transform 143
4.11 More on the z Transform 151
4.12 The Other Meaning of Frequency 155

5   Noise   161
5.1   Unpredictable Signals 162
5.2   A Naive View of Noise 164
5.3   Noise Reduction by Averaging 171
5.4   Pseudorandom Signals 174
5.5   Chaotic Signals 180
5.6   Stochastic Signals 192
5.7   Spectrum of Random Signals 198
5.8   Stochastic Approximation Methods 202
5.9   Probabilistic Algorithms 203

Part II   Signal Processing Systems

6   Systems   207
6.1   *System* Defined 208
6.2   The Simplest Systems 209
6.3   The Simplest Systems with Memory 213
6.4   Characteristics of Systems 221
6.5   Filters 226
6.6   Moving Averages in the Time Domain 228
6.7   Moving Averages in the Frequency Domain 231
6.8   Why Convolve? 237

6.9   Purely Recursive Systems 241
6.10 Difference Equations 245
6.11 The Sinusoid's Equation 249
6.12 System Identification—The Easy Case 252
6.13 System Identification—The Hard Case 259
6.14 System Identification in the z Domain 265

7 **Filters**   **271**
7.1   Filter Specification 272
7.2   Phase and Group Delay 275
7.3   Special Filters 279
7.4   Feedback 289
7.5   The ARMA Transfer Function 293
7.6   Pole-Zero Plots 298
7.7   Classical Filter Design 303
7.8   Digital Filter Design 309
7.9   Spatial Filtering 315

8 **Nonfilters**   **321**
8.1   Nonlinearities 322
8.2   Clippers and Slicers 324
8.3   Median Filters 326
8.4   Multilayer Nonlinear Systems 329
8.5   Mixers 332
8.6   Phase-Locked Loops 338
8.7   Time Warping 343

9 **Correlation**   **349**
9.1   Signal Comparison and Detection 350
9.2   Crosscorrelation and Autocorrelation 354
9.3   The Wiener-Khintchine Theorem 357
9.4   The Frequency Domain Signal Detector 359
9.5   Correlation and Convolution 361
9.6   Application to Radar 362
9.7   The Wiener Filter 365
9.8   Correlation and Prediction 369
9.9   Linear Predictive Coding 371
9.10 The Levinson-Durbin Recursion 376
9.11 Line Spectral Pairs 383
9.12 Higher-Order Signal Processing 386

10   Adaptation   **393**
   10.1 Adaptive Noise Cancellation 394
   10.2 Adaptive Echo Cancellation 400
   10.3 Adaptive Equalization 404
   10.4 Weight Space 408
   10.5 The LMS Algorithm 413
   10.6 Other Adaptive Algorithms 420

11   Biological Signal Processing   **427**
   11.1 Weber's Discovery 428
   11.2 The Birth of Psychophysics 430
   11.3 Speech Production 435
   11.4 Speech Perception 439
   11.5 Brains and Neurons 442
   11.6 The Essential Neural Network 446
   11.7 The Simplest Model Neuron 448
   11.8 Man vs. Machine 452

Part III   Architectures and Algorithms

12   Graphical Techniques   **461**
   12.1 Graph Theory 462
   12.2 DSP Flow Graphs 467
   12.3 DSP Graph Manipulation 476
   12.4 RAX Externals 481
   12.5 RAX Internals 487

13   Spectral Analysis   **495**
   13.1 Zero Crossings 496
   13.2 Bank of Filters 498
   13.3 The Periodogram 502
   13.4 Windows 506
   13.5 Finding a Sinusoid in Noise 512
   13.6 Finding Sinusoids in Noise 515
   13.7 IIR Methods 520
   13.8 Walsh Functions 523
   13.9 Wavelets 526

14   The Fast Fourier Transform   **531**
   14.1 Complexity of the DFT 532
   14.2 Two Preliminary Examples 536
   14.3 Derivation of the DIT FFT 539
   14.4 Other Common FFT Algorithms 546
   14.5 The Matrix Interpretation of the FFT 552
   14.6 Practical Matters 554
   14.7 Special Cases 558
   14.8 Goertzel's Algorithm 561
   14.9 FIFO Fourier Transform 565

15   Digital Filter Implementation   **569**
   15.1 Computation of Convolutions 570
   15.2 FIR Filtering in the Frequency Domain 573
   15.3 FIR Structures 579
   15.4 Polyphase Filters 584
   15.5 Fixed Point Computation 590
   15.6 IIR Structures 595
   15.7 FIR vs. IIR 602

16   Function Evaluation Algorithms   **605**
   16.1 Sine and Cosine Generation 606
   16.2 Arctangent 609
   16.3 Logarithm 610
   16.4 Square Root and Pythagorean Addition 611
   16.5 CORDIC Algorithms 613

17   Digital Signal Processors   **619**
   17.1 Multiply-and-Accumulate (MAC) 620
   17.2 Memory Architecture 623
   17.3 Pipelines 627
   17.4 Interrupts, Ports 631
   17.5 Fixed and Floating Point 633
   17.6 A Real-Time Filter 635
   17.7 DSP Programming Projects 639
   17.8 DSP Development Teams 641

Part IV   Applications

18   Communications Signal Processing   **647**
    18.1 History of Communications 648
    18.2 Analog Modulation Types 652
    18.3 AM 655
    18.4 FM and PM 659
    18.5 Data Communications 664
    18.6 Information Theory 666
    18.7 Communications Theory 670
    18.8 Channel Capacity 674
    18.9 Error Correcting Codes 680
    18.10 Block Codes 683
    18.11 Convolutional Codes 690
    18.12 PAM and FSK 698
    18.13 PSK 704
    18.14 Modem Spectra 708
    18.15 Timing Recovery 710
    18.16 Equalization 714
    18.17 QAM 716
    18.18 QAM Slicers 720
    18.19 Trellis Coding 723
    18.20 Telephone-Grade Modems 729
    18.21 Beyond the Shannon Limit 733

19   Speech Signal Processing   **739**
    19.1 LPC Speech Synthesis 740
    19.2 LPC Speech Analysis 742
    19.3 Cepstrum 744
    19.4 Other Features 747
    19.5 Pitch Tracking and Voicing Determination 750
    19.6 Speech Compression 753
    19.7 PCM 757
    19.8 DPCM, DM, and ADPCM 760
    19.9 Vector Quantization 765
    19.10 SBC 768
    19.11 LPC Speech Compression 770
    19.12 CELP Coders 771
    19.13 Telephone-Grade Speech Coding 775

A   Whirlwind Exposition of Mathematics   **781**

A.1   Numbers 781

A.2   Integers 782

A.3   Real Numbers 784

A.4   Complex Numbers 785

A.5   Abstract Algebra 788

A.6   Functions and Polynomials 791

A.7   Elementary Functions 793

A.8   Trigonometric (and Similar) Functions 795

A.9   Analysis 800

A.10 Differential Equations 803

A.11 The Dirac Delta 808

A.12 Approximation by Polynomials 809

A.13 Probability Theory 815

A.14 Linear Algebra 819

A.15 Matrices 821

A.16 Solution of Linear Algebraic Equations 826

Bibliography   829

Index   849

Preface

I know what you are asking yourself—'there are a lot of books available about DSP, is this book the one for *me*?' Well that depends on who *you* are. If

- you are interested in doing research and development in one of the many state-of-the-art applications of DSP, such as speech compression, speech recognition, or modem design,
- your main proficiency is in computer science, abstract mathematics, or science rather than electronics or electrical engineering,
- your mathematical background is relatively strong (flip back now to the appendix—you should be comfortable with about half of what you see there),

then you are definitely in the *target group* of this book. If in addition

- you don't mind a challenge and maybe even enjoy tackling brain-teasers,
- you're looking for one comprehensive text in all aspects of DSP (even if you don't intend reading all of it now) and don't want to have to study several different books with inconsistent notations, in order to become competent in the subject,
- you enjoy and learn more from texts with a light style (such as have become common for computer science texts) rather than formal, dry tomes that introduce principles and thereafter endlessly derive corollaries thereof,

then this is probably the book you have been waiting for.

This book is the direct result of a chain of events, the first link of which took place in mid-1995. I had been working at a high-tech company in Tel

Aviv that was a subsidiary of a New York company. In Tel Aviv it was relatively easy to locate and hire people knowledgeable in all aspects of DSP, including speech processing, digital communications, biomedical applications, and digital signal processor programming. Then, in 1995, I relocated to a different subsidiary of the same company, located on Long Island, New York. One of my first priorities was to locate and hire competent DSP software personnel, for work on speech and modem signal processing.

A year-long search turned up next to no-one. Assignment agencies were uncertain as to what *DSP* was, advertisements in major New York area newspapers brought irrelevant responses (digital design engineers, database programmers), and, for some inexplicable reason, attempts to persuade more appropriate people from Silicon Valley to leave the California climate, for one of the worst winters New York has ever seen, failed.

It struck me as rather odd that there was no indigenous DSP population to speak of, in an area noted for its multitude of universities and diversity of high-tech industries. I soon found out that DSP was not taught at undergraduate level at the local universities, and that even graduate-level courses were not universally available. Courses that *were* offered were Electrical Engineering courses, with Computer Science students never learning about the subject at all. Since I was searching for people with algorithm development and coding experience, preferably strong enough in software engineering to be able to work on large, complex software systems, CS graduates seemed to be more appropriate than EEs. The ideal candidate would be knowledgeable in DSP and would in the *target group* mentioned above.

Soon after my move to New York I had started teaching graduate level courses, in Artificial Intelligence and Neural Networks, at the Computer and Informations Sciences department of Polytechnic University. I inquired of the department head as to why a DSP course was not offered to Computer Science undergraduates (it *was* being offered as an elective to Electrical Engineering graduate students). He replied that the main reason was lack of a suitable teacher, a deficiency that could be easily remedied by my volunteering.

I thus found myself 'volunteered' to teach a new Computer Science undergraduate elective course in DSP. My first task was to decide on course goals and to flesh out a syllabus. It was clear to me that there would be little overlap between the CS undergraduate course and the EE graduate-level course. I tried to visualize the ideal candidate for the positions I needed to fill at my company, and set the course objectives in order to train the perfect candidate. The objectives were thus:

- to give the student a basic understanding of the theory and practice of DSP, at a level sufficient for reading journal articles and conference papers,

- to cover the fundamental algorithms and structures used in DSP computation, in order to enable the student to correctly design and efficiently code DSP applications in a high-level language,

- to explain the principles of digital signal processors and the differences between them and conventional CPUs, laying the framework for the later in-depth study of assembly languages of specific processors,

- to review the background and special algorithms used in several important areas of state-of-the-art DSP research and development, including speech compression/recognition, and digital communications,

- to enable the student who completes the course to easily fit in and contribute to a high-tech R&D team.

Objectives defined, the next task was to choose a textbook for the course. I perused web sites, visited libraries, spoke with publisher representatives at conferences, and ordered new books. I discovered that the extant DSP texts fall into three, almost mutually exclusive, categories.

About 75% of the available texts target the EE student. These books assume familiarity with advanced calculus (including complex variables and ordinary differential equations), linear system theory, and perhaps even stochastic processes. The major part of such a text deals with semirigorous proofs of theorems, and the flavor and terminology of these texts would certainly completely alienate most of my target group. The CS student, for example, has a good basic understanding of derivatives and integrals, knows a little linear algebra and probably a bit of probability, but has little need for long, involved proofs, is singularly uninterested in poles in the complex plane, and is apt to view too many integral signs as just so many snakes, and flee in terror from them.

In addition, these *type-one* texts ignore those very aspects of the subject that most interest our target students, namely algorithm design, computational efficiency and special computational architectures, and advanced applications. The MAC instruction and Harvard architecture, arguably the defining features of digital signal processors, are generally not even mentioned in passing. Generally only the FFT, and perhaps the Levinson-Durbin recursion, are presented as algorithms, and even here the terminology is often alien to the CS student's ear, with no attention paid to their relation with other problems well known to the computer scientist. The exercises generally involve extending proofs or dealing with simplistic signals that can be handled analytically; computer assignments are rare.

Finally, due perhaps to the depth of their coverage, the *type-one* texts tend to cover only the most basic theory, and no applications. In other words, these books finish before getting to the really interesting topics. Some cover the rudiments of speech processing, e.g. LPC and cepstral coefficients, but all consider speech compression and modem design beyond their scope. More advanced or specific texts are thus absolutely necessary before real-world applications can be tackled. These texts thus do not achieve our goal of preparing the student for participation in a real R&D team.

The next category, counting for about 20% of the texts, *do* target people who are more at home with the computer. *Type-two* texts tend to be 'recipe books', often accompanied by a diskette or CD. The newer trend is to replace the book with interactive instruction and experimentation software. These books usually contain between fifty and one hundred black box routines that can be called from a high-level language (e.g. C or MATLAB). The bulk of the text consists of instructions for calling these routines, with discussion of the underlying theory kept to a minimum.

While very useful for the computer professional who on odd occasions has need for some DSP procedures, these books do not instill a deep unified comprehension of the subject. Admittedly these books often explain algorithms in greater depth than *type-one* texts, but our target readers would benefit even more from a combination of *type-one* depth with *type-two* emphasis.

Of course there is nothing wrong with obtaining a well tested program or routine that fulfills the purpose at hand. Indeed it would not be prudent for the implementor to reinvent wheels in places where tire shops abound. However, due to their generality, library routines are often inefficient and may even be impractical for specific purposes. I wanted to enable my students to meet specific DSP needs by evaluating existing programs and library routines, or by writing original, tailored DSP code as required. The reader should also be able to port libraries to a new platform, understanding both the algorithm and the platform idiosyncrasies.

Finally, there are *type-three* texts, often written by DSP processor manufacturers. They emphasize the architecture, programming language, and programming tools of the manufacturer's line of digital signal processors, and while they may explain some theory, they mostly assume prior knowledge or claim that such knowledge is not really required for the comprehension of the subject matter. The programming techniques developed, usually in lengthy detail, may be applicable to some extent to other manufacturers' processors, but considerable adaptation would normally be required. *Type-three* texts tend to stress FIR and IIR filter structures, the radix 2 FFT algorithms, the LMS and perhaps Viterbi algorithms, and often describe various practical applications of these in great depth.

Due to their lack of mathematical sophistication, these books do not attempt to seriously treat DSP theory. Such critical topics as the sampling theorem, filtering, and adaptive systems are only trivially covered; true explanation of noise, filtering, and Fourier transforms are replaced by historical accounts, and algorithms are displayed in pseudocode *fait accompli* rather than derived. On the other hand, the manufacturers apparently feel that the typical reader will be lacking in CS background, and thus overly stress such obvious features as loops and numeric representation.

I thus reached the conclusion that none of the available DSP texts was truly suitable for the course, and was compelled to create my own course materials. These became the corner-stone of the present book. Often I found myself rethinking my own understanding of the subject matter, and frequently connections with other computer science subjects would only become clear during lecture preparation, or even during the lecture itself. I also found that the elimination of the conventional mathematical apparatus and rigorous proofs not only did not deplete the subject matter of meaning, but actually enriched it.

The topics included in this text may, at first, surprise the reader who is used to more conventional DSP texts. Subjects such as the matched filters, adaptive algorithms, the CORDIC algorithm, the Viterbi algorithm, speech compression, and modern modem theory are normally considered too complex and specialized for presentation at this level. I have found that these *advanced* topics are no more difficult for the newcomer to grasp than filter design or limit cycles, and perhaps more interesting and relevant. However, in order to keep the book size moderate, some of the more classical subjects had to be curtailed. These subjects are adequately covered in traditional texts, which may be consulted to supplement the present one.

Even so, the present book contains more material than can be actually taught in a single-semester course. A first course in DSP could cover most of the material in the early chapters, with the instructor then selecting algorithms and applications according to personal preference. The remaining subjects may be relegated to a more advanced course, or be assigned as self-study topics. My initial course went through the basic theory at break-neck speed, in order to rapidly get to speech compression and recognition. A second attempt emphasized modems and DSP for data communications.

Every section ends with a number of exercises that are designed to be entertaining and enriching. Some of these should not be difficult for the reader who understands the section, being designed to reinforce basic understanding of the material. Many are somewhat challenging, complementing the text, extending the theory, or presenting actual applications of the subject studied. Some are only loosely defined; for these one can give a quick answer, or develop them into a term project. Others introduce new material that will ease the understanding of the following sections, as well as widening the reader's DSP horizons.

I purposely avoid taking sides on the divisive issue of programming language and environment for algorithm design and test on general-purpose computers. Realizing that C, MATLAB, SPW, Mathematica and the like will all have their staunch supporters, and all have their strengths and weaknesses, I leave it to the student or instructor to select that language with which they are the most comfortable. Every seasoned programmer is most effective in his or her native language, and although some languages are obviously better DSP 'environments' than others, the difference can be minimized by the use of appropriate libraries.

Although the book was written to serve as a course textbook, it may be used by non-students as well. DSP practitioners are like master craftsmen; when they are called upon to construct some object they must exploit their box of tools. Novices have only a few such tools, and even these may not be sufficiently sharp. With time more tools are acquired, but almost all craftsmen tend to continue using those tools with which they have the most experience. The purpose of this book is to fill the toolbox with tools, and to help the DSP professional become more proficient in their proper use. Even people working in the field several years will probably find here new tools and new ways of using tools already acquired.

I would like to thank my students, who had to suffer through courses with no textbook and with continually changing syllabus, for their comments; my colleagues, particularly Yair Karelic, Mauro Caputi, and Tony Grgas, for their conscientious proofreading and insights; my wife Ethel for her encouragement (even allowing me untold late-night sessions banging away at the keyboard, although she had long ago banished all computers from the house); and our two girls, Hanna and Noga, who (now that this book is complete) will have their father back.

Jonathan (Y) Stein
Jerusalem, Israel
31 December 1999

1

Introductions

The reader is already an expert in signal processing, although possibly unaware of it. We are all remarkably complex signal processing systems, adaptively processing intricate audio and video signals every moment of our lives. While awake we input intricate signals from our environment, extract high-level representations of information carried by these signals, make decisions based on this information, record some of the information for later recall and processing, and produce new signals to change our environment in *real time*. Even while sleeping, although most of the input has been removed, we unconsciously continue the processing *off-line*; we reintroduce recently input signals in order to correlate them with previously stored signals, decide which signals should be stored for long periods of time, and generally perfect our signal processing performance. Due to this signal processing we are extremely good at understanding speech and immediately reacting based on what has been said. We scarcely think about our ability to recognize faces and greet (or avoid) their owners. We take our proficiency at reading handwriting for granted, except when admiring the pharmacist's even greater competency when presented with a physician's scrawl.

It is therefore extremely frustrating to discover how difficult it is to design artificial devices that can perform as well. After decades of research, devices that can understand unconstrained human speech are still extremely primitive, and even speech synthesis is still to be considered a nontrivial problem. Machine recognition of human faces is possible only in severely restricted environments, and even our limited capabilities are not yet commonplace due to being prohibitively expensive. While optical character recognition of high quality printed fonts has been perfected, acceptable machine reading of handwriting has yet to be attained.

These three examples—speech understanding, face recognition, and reading of handwriting—are typical of a long list of tasks which we find almost trivial, but which have turned out to be extremely difficult for machines. It is only due to our meager attempts at designing machines to perform

1

these functions that we have come to grasp their extreme complexity. Due to this inherent complexity, researchers and implementors attempting to mechanize these functions have turned to the strongest and most intelligent devices available. The most sophisticated and capable invention humankind has devised to date is the digital computer. For this reason it is natural that much of the state-of-the-art signal processing is performed digitally. In this, the first chapter, we introduce *signal processing*, and more specifically Digital Signal Processing, which from now on we shall call **DSP**.

In order to acquaint the reader with the concept of using digital technology in order to process signals, we will first trace the early history of signal processing. We then jump ahead to a survey of state-of-the-art applications, in order to convince the reader that the problem is still alive and interesting. Next we introduce the concept of signal processing by demonstrating analog signal processing on a simple example. Finally we present the basic ideas behind the use of computers in signal processing.

1.1   Prehistory of DSP

The first major accomplishments of humankind involved mastering the processing of *material* objects, and indeed humankind is often defined as the animal who fashions tools. Lower forms of animals do not, in general, change naturally occurring objects in order to adapt them to their needs. When humankind discovered that one could take stones and bones and by relatively simple processing convert them into arrows, knives, needles, fire-making implements, and the like, the species transcended all those that had come before it. More and more complex processing algorithms were then developed. For example, humans learned to till the soil, plant wheat seeds, water and fertilize them, harvest the wheat, separate the chaff from the grain, ground the grain into flour, mix the flour with water and yeast, and bake the dough to make bread. This represents a highly developed culture of material object processing.

The next stage in humankind's development involved the processing of *signals*. Signals, like materials, are real physical objects, but are intangible. Humankind learned to adapt a mouth (originally developed for eating) and an ear (designed for hearing predators approach) into a highly flexible acoustic communications system. The medium for this communications exchange was pressure waves in air, and to some extent visual clues conveyed via light. Primitive peoples also developed techniques for communications over distances, such as tom-tom drums and smoke signals. Then came the development of the telegraph and telephone, which used the electrical signals, and radio, which used electromagnetic waves. The objects being manipulated remain physically existing quantities, although they became less and less tangible.

The final stage (so far) in humankind's development entailed learning to process *information*. Unlike material objects and signals, information is entirely abstract and cannot really be said to exist in the physical world. Information is like ideas, and while it can be quantified it is not held back by physical limitations. The seeds of information-processing were sown with the invention of writing and arithmetic, philosophy and algebra, art and logic, but were brought to full fruition with the invention of the digital computer. The computer can transcend nature by predicting physical phenomena before they occur, simulating worlds that cannot exist, and creating new information where none was before.

The marriage of the last two developments in mankind's history, i.e., utilizing digital computation for the purpose of processing of signals in the real world, is the objective of DSP. While perhaps not a major milestone in the history of humankind, DSP is a significant enough endeavor to warrant study by all interested in manipulating their world using information-processing techniques.

EXERCISES

1.1.1 Does *Digital Signal Processing* mean 'the processing of digital signals' or 'the digital processing of signals'?

1.1.2 What possible relationships might there be between DSP and the following *computer science* fields?
    1. Numerical Analysis
    2. Compiler Design
    3. Operating Systems
    4. Database Programming
    5. Artificial Intelligence

1.1.3 Listen to an extremely weak station on an AM radio. Can you understand what is being said? Would you be able to understand were the language spoken to be one in which you are not completely proficient? What happens if there are interference and whistles? Other radio stations? Does the same happen with an FM radio station? Repeat the above experiment with a shortwave radio. Find stations using SSB modulation. What happens if you do not tune the signal in properly? Sit in a cocktail party where many groups of people are talking. Focus on one conversation after another. How well can you separate out voices? What have you learned from this exercise?

1.2   Some Applications of Signal Processing

So what exactly *is* signal processing and why do we want to do it? Signal processing is the discipline of detecting, manipulating, and extracting information from physical quantities that vary in time (*signals*).

The only way to really understand what we mean by this definition is to consider examples of signal processing applications.

Voice communications, processing, and store-and-forward. The main means of communications between humans is speech. One human broadcasts information as an acoustic signal that can be detected by other humans. When the persons desiring to converse are not colocated, we must provide a mechanism to transfer the signal from place to place. When they are not available simultaneously, we need to record this acoustic signal for later playback. Digital forwarding and recording of speech have certain advantages, as we shall discuss later. In order to use digital transfer and storage we require a method for making a digital representation of the acoustic signal, as well as algorithms for **A**utomatic **G**ain **C**ontrol (AGC), **V**oice **A**ctivity **D**etection (VAD), and perhaps compressing the digital representation in order to preserve disk space or communications bandwidth. Additional processing entails enhancing the quality of speech in noise, and acceleration/deceleration of the playback speed without distortion. More complex processing algorithms are required for separation of one voice from others (cocktail-party effect), machine-synthesized speech (text to speech), speech recognition and speaker identification.

Music synthesis, recording, and playback. Much of what we said about speech holds for 'wider bandwidth' acoustic signals, such as music. Here the emphasis is on high-quality transfer (e.g., broadcast), compression (e.g., MPEG files), storage (e.g., compact disks), and noise reduction, (for example, restoration of old recordings). However, there are also processes specific to music such as accurate recreation of the original sound in different acoustic environments (equalization), digital simulation of musical instruments (synthesizers, keyboard organs, MIDI), and special effects (mixing, echo, reverberation).

Data communications on voice-grade channels. Another extension of voice processing is the adding of data bearing signals to channels originally designed for voice use. *Touch-tone* dialing (technically known as DTMF) has become almost universal for dialing and for menu selection. Facsimile machines that transmit documents over public telephone circuitry have also become commonplace, and high speed modems enable computers to interconnect over this medium. It is also useful to convert the audio itself to digital form, for example, to enable several conversations to share one telephone line, or for the purposes of secure communications (encryption).

Automobile Industry. A conventional muffler reduces noise by passing exhaust gases through a series of baffles that reduce their velocity. Unfortunately this same process requires the engine to waste energy forcing these gases through the muffler, energy that would otherwise have been used to increase horsepower and fuel efficiency. The electronic muffler uses *active noise cancellation* instead; the noise is sensed by a microphone, and identical noise is added 180° out of phase. This same technique can be utilized to add out-of-phase vibration to the mounts of the engine on the chassis. Acoustic DSP can also be used to diagnose and control engine faults.

Industrial Applications. The measurement of vibrational modes and the discovery of their underlying causes and mechanical structural problems they may indicate is a well-known industrial application of signal processing. Chemical process control relies heavily on instrumentation that employs advanced signal processing. Robots on assembly lines receive signals from sensors and adaptively act upon them by moving their mechanical appendages. Other applications include the diagnosis of electric motor faults from current signatures, the rapid and precise measurement of fluid flow, the control of welding and smelting apparatus, and pump wear monitoring.

Biomedical engineering. The human brain is a massively parallel computer containing about 10^{10} processing units called *neurons*. These neurons fire electric impulses that are not externally observable, but by placing electrodes at various positions on the scalp, voltages that represent sums of many neurons are detectable. These recordings are known as electroencephalograms (EEG) and after proper processing they can be used for diagnosis of sleep disorders, epilepsy, and brain disease. The electric activity of the heart can also be monitored, using the electrocardiogram (ECG). Processing this signal aids the physician in diagnosing heart problems. Monitoring during labor involves continual display of fetal heart rate as well as uterine muscular activity. These signals require removal of hum introduced from the electric power source and extensive real-time preprocessing.

Radar and sonar processing. The purpose of radar and sonar is to locate bodies in space and optionally to determine their speeds. Well-known radar applications include air traffic control, aircraft radar, smart-missiles, weather satellite radar, and police speed traps. The distance determination relies on the sensitive detection and accurate timing of return signals; electromagnetic signals for radar and acoustic signals in water for sonar. This processing relies on *matched filtering* and high resolution spectral analysis. Doppler radar speed measurement requires precise frequency measurement.

Radar signals usually have very high bandwidths, and consequently require very fast processing rates. Sonar bandwidths are much lower than those of radar, but the processing power required is high due to the interference being stronger, and the return signals being weaker and more distorted. Multipath reception complicates the location effort and often arrays of sensors are employed and *beamforming* used. Electronic intelligence (ELINT) and electronic warfare (EW) exploit interception of radar signals in order to detect/identify and to deceive/defeat the radar system, respectively.

Seismology. Seismic signal analysis is used by the oil and gas industries in the exploration of subsurface hydrocarbon reserves; by government agencies for nuclear detonation detection; and by long-term planning authorities for investigation of subsurface geological formations and their significance to architecture and urban development. Signals passively collected during naturally occurring seismic events such as earthquakes and volcanic eruptions may aid in their detection, epicenter location, and prediction. During active exploration such seismic disturbances must be initiated, for example, by setting off high-energy charges (although environmental considerations may mandate the use of lower energy sources such as acoustic speakers). The seismic waves are scattered by interfaces between different geological strata, and collected at the earth's surface by an array of seismometers. Thus multiple seismic signals must be digitized and processed to lead to source location or mapping of the geological strata.

EXERCISES

1.2.1 What other areas utilize signal processing? List several applications not on the above list. Research at a library or search the Internet.

1.2.2 What areas may potentially benefit from signal processing, but are not yet using it? Write up a detailed description and submit to the patent office.

1.2.3 Consider a mobile robot able only to avoid obstacles, and to locate an electric outlet when its batteries are low. What technologies would be needed to implement such a robot? Where is DSP needed? Now give the robot the ability to receive verbal commands, to retrieve objects, and to keep its owner informed. What DSP is needed now?

1.2.4 Dual Tone Multi Frequency (DTMF) tones consists of two frequencies. The same is true for *dial tone* and *ring-back tone*. What simple tunes can you play recognizably using DTMF? Who answers the phone when you do? Why are two different frequencies used—wouldn't it be easier to use only one? (Hint: Take the expense of incorrectly routed calls into account.)

1.2.5 Using a computer with multimedia capabilities, record some naturally spoken speech and observe its waveform graphically. Does this picture contain all the information we can obtain from listening to the speech? You can easily find long silences and tell the difference between whispering and screaming. Try to tell where the words begin and end. Can you differentiate between male and female voices? Can you guess what is being said? Assuming you answered in the affirmative to the first question, where exactly is the information?

1.2.6 You are given the job of saving the several megabytes of information from an old computer, about to be discarded. The computer has no serial output ports or modem, but *does* have an analog output that can produce 256 different voltage levels. What is the simplest encoding method for outputting information? How can you decode and store the information (you can use any readily available computer or peripheral)? How fast can you go? Do you think it can be decoded this fast? What are the real limitations? What happens if background noise is recorded along with the signal? This is the basic idea behind the download path of the so-called *PCM modem* that achieves 56 Kb/s over telephone lines.

1.2.7 Same problem but this time the computer has an internal speaker and can generate tones of different frequencies (all of the same amplitude). You may decide to convert the data to be saved, byte by byte, into one of 256 tones, and to record the tones onto an audio cassette. Design a transmitter (*modulator*) for this case (try writing a program). What do you need to decode this information (*demodulator*)? How fast can you go? Perhaps you decide to convert the data to be saved, bit by bit, into one of only two tones. What do the modulator and demodulator look like now? This is the FSK modem, capable of 300 b/s on phone lines.

1.3 Analog Signal Processing

Signal processing is the discipline of detecting, manipulating and extracting information from physical quantities that vary in time. Now that we have seen *why* we want to do it, we can begin to discuss *how* to do it. DSP processes signals *digitally*, that is, by programming, rather than by building *analog* electronic circuits. However, before we jump into digital processing, a brief discussion of analog processing is in order.

It is clear that signals can be processed using analog circuits such as amplifiers and filters. These devices take analog signals as inputs and return analog signals as outputs. Electronic engineers know how to design these circuits to obtain specific processing characteristics (obtaining certain voltage levels, amplifying certain frequency ranges while eliminating others, etc.). Quite complex systems can be designed, for example, receivers that are sensitive only to very specific waveforms. In-depth explanation of the techniques that have been developed in this field is beyond the scope of our book, and for our purposes it is sufficient to analyze a simple example.

Assume that wish to input a sine wave of arbitrary frequency, offset and amplitude (within bounds of course) and output a train of narrow pulses of equal frequency. One can observe the input and desired output as 'X' and 'Y' in Figure 1.2. Why would one want to perform this operation? There may be a number of reasons. For example, one may want to measure the frequency of the sine wave using a digital counter that increments upon receiving a narrow pulse. Or one may need the pulse as a synchronization signal for some process that should be locked in time with the sine wave. It could be that we need to generate a pulse for triggering an oscilloscope or some other instrument.

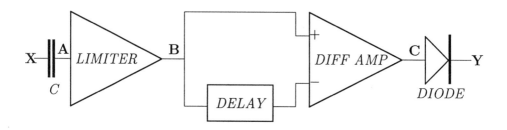

Figure 1.1: Diagram of an analog sine to pulse converter.

One way of producing the desired effect is depicted in Figure 1.1, with the input, intermediate signals, and output drawn in Figure 1.2. The first step is to pass the signal (waveform X) through a *capacitor* that acts as a *DC blocker*. This ensures that the signal values are centered around zero voltage (waveform A). Next we put the signal through a *hard limiter*. This is an amplifier driven to its maximum amplification, so that its output will be $+V_{max}$ for any positive input, and $-V_{max}$ for any negative input (waveform B). Next we split the signal so that it traverses two paths, one slightly delayed with respect to the other (this delay determines the width of the pulse to be obtained). The delayed signal is now subtracted from the nondelayed version, producing an output that is almost always zero (waveform C). The subtraction is performed, once again, using an amplifier, this time a *differential amplifier* that has noninverting and inverting inputs. The amplifier's output is nonzero and positive at the leading edge of the square wave (since

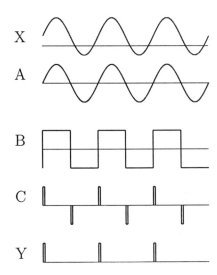

Figure 1.2: Analog signals from the sine to pulse converter.

there we have $+V_{max} - -V_{max}$) and nonzero and negative one half-cycle later. This latter artifact is eliminated by the use of a *half-wave rectifier*, a component that only passes positive voltages, suppressing negative ones. The final result (waveform Y) is a narrow positive pulse locked to the leading edge of the original sine, as required.

EXERCISES

1.3.1 The above example assumes the existence of a *delay* element, which may be quite difficult to implement. For high-frequency signals, a long piece of cable may be used, relying on the finite speed of propagation of the signal through the cable to introduce the time delay. For low frequencies, even extremely long lengths of cable introduce delays that are insignificant fractions of the period. Assume you have an analog differentiator, a device whose output is the derivative of its input. How would you use it in our sine to pulse converter? What would the output pulse look like?

1.3.2 The device we described above is basically a *zero crossing detector*, a device that determines when the signal goes through zero voltage. We can avoid the need for a rectifier if we employ a *peak picker*, which outputs pulses at the maxima of the input signal. How can a peak picker be implemented given a differential amplifier and a reference voltage source?

1.3.3 How can an *integrator* (a device whose output is the integral of its input) be used to solve our problem?

1.3.4 Assume you have a digital representation of the input; that is, a sequence of voltage measurements uniformly spaced in time. Write software routines for the zero crossing detector and the peak picker, assuming that the sine wave is sampled very densely (many equally-spaced samples per cycle). Will your routines work if the sampling rate is much lower, for example, eight samples per cycle? Four samples per cycle?

1.4 Digital Signal Processing

In the previous section we saw an example of how signals can be processed using analog circuits. How can we similarly process analog signals digitally? A very general scheme is depicted in Figure 1.3.

Figure 1.3: Generic DSP scenario.

The purpose of the filters will only become clear later on (see Section 2.10). The blocks marked **A/D** and **D/A** represent devices that convert **A**nalog signals into **D**igital ones, and vice versa. These devices allow us to translate signals in the physical world into sequences of numbers that computers can accept as input and process, and to convert sequences of numbers output by computers back into physical signals.

The heart of the system is the digital signal processor, which we shall usually just call the *DSP*. (This double use of the acronym DSP should not be confusing, with the differentiation between the *processor* and *processing* being easily understood from context.) You may think of the DSP as basically a computer that performs the needed computation. It may be a general-purpose computer, such as a desktop workstation, readily available and easily programmed. Or it may be special purpose digital hardware designed specifically for the task at hand. Intermediate between these extremes is a general-purpose programmable digital signal processor. These DSP *chips* are similar to microprocessors, with arithmetic capabilities, memory access, input and output ports, etc. However, as we shall discuss in Chapter 17, they are augmented with special commands and architecture extensions in order to make them particularly efficient for computations of the type most prevalent in DSP applications. While programming DSPs in high-level languages is becoming popular, their special architectures can be best exploited by low-level (assembly) programming.

At this point you are probably asking yourself whether DSP is truly superior to analog signal processing. Why should we replace a handful of electronic components with two filters, an A/D and a D/A, and an expensive and hard-to-program DSP? The main reasons to favor digital techniques over analog ones, are:

- greater functionality,

- accuracy and reproducibility,

- modularity and flexibility,

- increased price/performance, and

- reduced time-to-market.

The greater functionality derives from the possibility of implementing processes that would be extremely difficult and/or expensive to build in analog circuitry. In particular, arbitrary time delays, noncausal response, linear-phase (see Chapter 6), and adaptivity (see Chapter 10) are simple to implement in DSP, while practically impossible in analog.

Accuracy and reproducibility are characteristics of digital numbers in contrast to analog voltages. Precision is a function of the number of bits used in computation, and digital numbers can be protected against inaccuracy by error-correcting codes. A copy of a copy of a copy of a digital recording is identical to the original, with no added noise and no 'drift' caused by temperature or aging.

The modularity and flexibility are byproducts of programmability; DSP code can readily be reused and modified. DSP code, like all software, can be made generic and placed into libraries with little sacrifice. Last minute changes are a hardware engineer's worst nightmare, while field debugging is commonplace in the software arena.

In recent years significant advances have been achieved in digital technology, including the development of smaller, faster, more power efficient and less expensive digital processors. For these reasons digital technology is finding its way into almost every facet of our lives. Once the process *can* be performed digitally, it usually takes only a very short time until it is profitable to do so.

There are, admittedly, a few drawbacks associated with the use of DSP. The most notable of these are:

- limited speed of general-purpose DSPs,

- finite word-length problems, compounding of round-off errors, and 'stability',   as well as

- the need for specialized algorithms and programming.

As a result of the first problem many applications, for example, those dealing with real-time processing of high bandwidth signals, cannot yet be handled digitally. The second shortcoming is more a hindrance than a true impediment, compelling us to analyze our numeric algorithms more cautiously. The third drawback is actually a favorable opportunity for students of DSP. It ensures a steady demand for competent DSP personnel for many years to come.

EXERCISES

1.4.1 'Some **DSP** practitioners rarely deal with **DSP** theory at all, rather are experts at programming **DSPs** for control and general algorithmic applications, rather than as a **DSP**.' The acronym **DSP** appears four times in this sentence. Explain which of the various meanings (processing, block diagram function, programmable processor) best matches each.

1.4.2 Figure 1.3 depicts a situation with exactly one input signal and one output signal. Describe an application with no inputs and one analog output. Two analog inputs and one output. One input and two outputs. Can there be useful applications with no outputs?

1.4.3 An *amplifier* increases the magnitude of a signal, while an *attenuator* decreases the magnitude. An *inverter* inverts the polarity of a signal, while a *clipper* limits the magnitude of a signal. A *DC blocker* shifts the average to zero. What mathematical functions are performed by these components? Code a routine for each.

1.4.4 What differences do you expect to find between DSPs and conventional CPUs?

1.4.5 Are there functions that can be performed in analog electronics, but cannot be performed in DSP?

1.4.6 Compare digital **C**ompact **D**isc (CD) technology with the older **L**ong **P**laying (LP) records. Explain why CD technology has totally replaced LPs by considering sound quality, playing duration, noise, stereo separation, the effect of aging media, and the ability to make copies.

Part I

Signal Analysis

Signals

We are now ready to commence our study of signals and signal processing systems, the former to be treated in Part I of this book and the latter in Part II. Part III extends the knowledge thus gained by presentation of specific algorithms and computational architectures, and Part IV applies all we will have learned to communications and speech signal processing.

At times one wants to emphasize signals as basic entities, and to consider systems as devices to manipulate them or to measure their parameters. The resulting discipline may then be called *signal analysis*. At other times it is more natural to consider systems as the more fundamental ingredients, with signals merely inputs and outputs to such systems. The consequence of this viewpoint is called *signal processing*. This term is also most commonly used when it is not clear which aspect one wishes to stress.

In this chapter we introduce the concept of a signal. We will see that there are analog signals and digital signals, and that under certain conditions we can convert one type into the other. We will learn that signals can be described in terms of either their time or frequency characteristics, and that here too there are ways to transform one description into the other. We present some of the simplest signals, and discover that arbitrary signals can be represented in terms of simple ones. On the way we learn how to perform arithmetic on signals, and about the connection between signals and vectors.

2.1   *Signal* Defined

The first question we must ask when approaching the subject of signal analysis is 'What exactly do we mean by *signal*?' The reader may understand intuitively that a signal is some function of time that is derived from the physical world. However, in scientific and technological disciplines it is customary to provide formal mathematical definitions for the main concepts, and it would be foolish to oppose this tradition. In order to answer the question satisfactorily, we must differentiate between analog and digital signals.

Definition: signal

An *analog signal s* is a finite real-valued function $s(t)$ of a continuous variable t (called time), defined for all times on the interval $-\infty < t < +\infty$. A *digital signal s* is a bounded discrete-valued sequence s_n with a single index n (called discrete time), defined for all times $n = -\infty \ldots + \infty$.   ∎

The requirement that analog signals be *real-valued*, rather than integer or complex, has its origin in the notion that *real-world* signals, such as speeds, voltages, and acoustic pressures, are simple continuous variables. Complex numbers are usually considered purely mathematical inventions that can never appear in nature. Digital signals are constrained more by the requirement of representability in a digital computer than by physical realizability. What we mean here by 'discrete' is that the possible values are *quantized* to discrete values, such as integers or all multiples of 2^{-b}. 'Bounded' means that there are only a finite number of possible signal values. Bounded discrete values are exactly the kinds of numbers represented by computer words with some finite number of bits.

Finiteness is another *physical* requirement, and comes in three varieties, namely finite signal value, finite energy, and finite bandwidth. Finite-valuedness simply means that the function desiring to be a signal must never diverge or become mathematically singular. We are quite confident that true physical quantities never become infinite since such behavior would require infinite energy or force or expense of one type or another. Digital signals are necessarily bounded in order to be representable, and so are always finite valued. The range over which a signal varies is called its *dynamic range*. Finite energy and finite bandwidth constraints are similarly grounded, but the concepts of energy and bandwidth require a little more explanation for the uninitiated.

Energy is a measure of the *size* of a signal, invented to enable the analyst to compare the infinitely many possible signals. One way to define such a measure might be to use the highest value the signal attains (and thus finite energy would imply finite signal value). This would be unsatisfactory because a generally small signal that attains a high value at one isolated point in time would be regarded as larger than a second signal that is almost always higher than the first. We would certainly prefer a measure that takes all times into account. Were signals to have only positive values we could possibly use the average signal value, but since they are not the average is ineffectual as many seemingly large signals (e.g., $A\sin(\omega t)$ with large A) have zero average due to positive and negative contributions cancelling. The simplest satisfactory measure is given by the following definition.

Definition: energy

The energy of an analog or digital signal s is defined to be

$$E_s = \int_{-\infty}^{\infty} s^2(t)dt \quad \mathbf{A} \;\Big|\; \mathbf{D} \quad E_s = \sum_{n=-\infty}^{\infty} s_n^2 \tag{2.1}$$

the sum (or integral for the analog case) of the signal's values squared. ∎

This measure is analogous to the squared length of multidimensional vectors, and is proportional to the physical quantity known as energy when the signal is a velocity, voltage, or current. The energy we have just defined is also directly related to the expense involved in producing the signal; this being the basis for the physical requirement of finite energy. The square root of the energy defines a kind of average signal value, called the **R**oot **M**ean **S**quared (RMS) value.

Bandwidth is a measure not of size but of speed, the full discussion of which we must postpone until after the notion of spectrum has been properly introduced. A signal that fluctuates rapidly has higher bandwidth than one that only varies slowly. Requiring finite bandwidth imposes a smoothness constraint, disallowing sudden jump discontinuities and sharp corners. Once again such functions violate what we believe nature considers good taste. Physical bodies do not disappear from one place and appear in another without traveling through all points in between. A vehicle's velocity does not go from zero to some large value without smoothly accelerating through intermediate speeds. Even seemingly instantaneous ricochets are not truly discontinuous; filming such an event with a high-speed camera would reveal intermediate speeds and directions.

Finally, the provision *for all times* really means *for all times of interest*, and is imposed in order to disallow various pathological cases. Certainly a body no longer has a velocity once destroyed, and a voltage is meaningless once the experimental apparatus is taken apart and stored. However, we want the experimental values to *settle down* before we start observing, and wish our phenomena to exist for a reasonable amount of time after we stop tending to them.

Now that we fully understand the definition of *signal*, we perceive that it is quite precise, and seemingly inoffensive. It gives us clear-cut criteria for determining which functions or sequences are signals and which are not, all such criteria being simple physical requirements that we would not wish to forgo. Alas this definition is more honored in the breach than the observance. We shall often relax its injunctions in the interests of mathematical simplicity, and we permit ourselves to transgress its decrees knowing full well that the 'signals' we employ could never really exist.

For example, although the definition requires signals to be *real-valued functions*, we often use complex values in order to simplify the algebra. What we really mean is that the 'real' signal is the real part of this complex signal. This use of an 'imaginary' complex signal doesn't overly bother us for we know that we *could* reach the same conclusions using real values, but it would take us longer and we would be more apt to make mistakes. We even allow entities that aren't actually functions at all, when it saves us a few lines of proof text or program code!

Our definition relies on the existence of a *time* variable. At times the above definition is extended to functions of other time-like independent variables, and even to functions of more than one variable. In particular, *image processing*, that deals with functions of two spatial coördinates, invokes many signal processing concepts. However, in most of this book we will not consider image processing to be part of signal processing. Although certain basic ideas, notably filtering and spectral analysis, *are* common to both image and signal processing, the truly strong techniques of each are actually quite different.

We tend to scoff at the requirement for finite-valuedness and smoothness, routinely utilizing such nonphysical constructs as tangents and square waves, that possess an infinite number of discontinuities! Once again the reader should understand that real-world signals can only approximate such behavior, and that such refractory functions are introduced as mathematical scaffolding.

Of course signals are defined over an infinite range of times, and consequently for a signal's energy to be finite the signal must be zero over most times, or at least decay to zero sufficiently rapidly. Strictly requiring finite energy would rule out such useful signals as constants and periodic functions. Accordingly this requirement too is usually relaxed, with the understanding that outside the interval of time we observe the signal, it may well be set to zero. Alternatively, we may allow signals to be nonzero over infinite times, but to have finite *power*. Power is the energy per time

$$P_s(\tau) = \lim_{T \to 0} \frac{1}{T} \int_{\tau - \frac{T}{2}}^{\tau + \frac{T}{2}} s^2(t)\, dt \qquad \mathbf{A} \,\Big|\, \mathbf{D} \qquad P_{s_\nu} = \lim_{N \to 0} \frac{1}{N} \sum_{n = \nu - \frac{N}{2}}^{\nu + \frac{N}{2}} s_n^2 \quad (2.2)$$

which is time-dependent in general.

Hence although the definition we gave for *signal* is of good intent, its dictates go unheeded; there is scarcely a single clause in the definition that we shan't violate at some time or other. In practice entities are more often considered signals because of the utility in so doing, rather than based on their obeying the requirements of this definition (or any other).

In addition to all its possibly ignorable requirements, our definition also leaves something out. It is quiet about any possible connection between analog and digital signals. It turns out that a digital signal can be obtained from an analog signal by **A**nalog to **D**igital conversion (the 'A/D' of Figure 1.3) also known as *sampling* and *digitizing*. When the sampling is properly carried out, the digital signal is somehow equivalent to the analog one. An analog signal can be obtained from a digital signal by **D**igital to **A**nalog conversion (the 'D/A' block), that surprisingly suffers from a dearth of alternative names. Similar remarks can be made about equivalence. A/D and D/A conversion will be considered more fully in Section 2.7.

EXERCISES

2.1.1 Which of the following are *signals*? Explain which requirement of the definition is possibly violated and why it is acceptable or unacceptable to do so.

1. the height of Mount Everest
2. $\left(e^{it} + e^{-it} \right)$
3. the price of a slice of pizza
4. the 'sinc' function $\frac{\sin(t)}{t}$
5. Euler's totient function $\phi(n)$, the number of positive integers less than n having no proper divisors in common with n
6. the water level in a toilet's holding tank
7. $\lfloor t \rfloor$ the greatest integer not exceeding t
8. the position of the tip of a mosquito's wing
9. $\sqrt{t}$
10. the Dow Jones Industrial Average
11. $\sin(\frac{1}{t})$
12. the size of water drops from a leaky faucet
13. the sequence of values x_n in the interval $[0 \ldots 1]$ defined by the *logistics recursion* $x_{n+1} = \lambda x_n (1 - x_n)$ for $0 \le \lambda \le 4$

2.1.2 What is the power of $s(t) = A\sin(\omega t)$? The RMS value?

2.1.3 A signal's *peak factor* is defined to be the ratio between its highest value and its RMS value. What is the peak factor for $s(t) = A\sin(\omega t)$? The sum of N sinusoids of different frequencies?

2.1.4 Define a size measure M for signals *different* from the energy (or RMS value). This measure should have the following properties.

- The zero signal must have zero measure $M_0 = 0$, and no other signal should have zero measure.
- If signal y is identical to signal x shifted in time then $M_y = M_x$.
- If $y_n = \alpha x_n$ for all times, then $M_y > M_x$ if $\alpha > 1$ and $M_y < M_x$ if $\alpha < 1$.
- If $y_n > x_n$ almost all of the time, then $M_y > M_x$.

What advantages and disadvantages does your measure have in comparison with the energy?

2.2   The Simplest Signals

Let us now present a few signals, ones that will be useful throughout our studies. The simplest signal is the *unit constant*, that is, $s(t) = 1$ in analog time or $s_n = 1$ in digital time.

$$s(t) = 1 \quad \mathbf{A} \mid \mathbf{D} \quad s_n = 1 \qquad (2.3)$$

Although this is the simplest signal we can imagine, it has infinite energy, and therefore violates one of the finiteness constraints. Hence technically it isn't really a signal at all! Arbitrary constant signals can be obtained by multiplying the unit constant signal by appropriate values. The constant signal, depicted in Figure 2.1, although admittedly trivial, can still be useful. We will often call it **Direct Current (DC)**, one of the many electronics

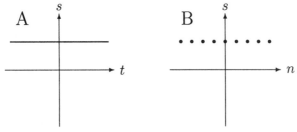

Figure 2.1: The constant signal. In (A) we depict the analog constant and in (B) the digital constant.

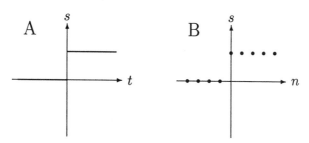

Figure 2.2: The unit step signal. In (A) we depict the analog step $u(t)$ and in (B) the digital step u_n.

terms imported into signal processing. The gist is that a battery's voltage is constant, $v(t) = V_0$, and consequently induces a current that always flows in one direction. In contrast the voltage from a wall outlet is sinusoidal, $v(t) = V_0 \sin(\omega t)$, and induces an **A**lternating **C**urrent (AC).

We cannot learn much more from this signal, which although technically a 'function of time' in reality is not time dependent at all. Arguably the simplest time-dependent signal is the *unit step*, which changes value at only one point in time (see Figure 2.2). Mathematically, the analog and digital unit step signals are:

$$u(t) = \Theta(t) = \begin{cases} 0 & t < 0 \\ 1 & t > 0 \end{cases} \quad \mathbf{A}\,\Big|\,\mathbf{D} \quad u_n = \Theta(n) = \begin{cases} 0 & n < 0 \\ 1 & n \geq 0 \end{cases} \quad (2.4)$$

respectively. In some of the literature the step function is called *Heaviside's step function*. Once again the finite energy requirement is unheeded, and in the analog version we have a jump discontinuity as well. Here we have set our clocks by this discontinuity, that is, we arranged for the change to occur at time zero. It is a simple matter to translate the transition to any other time; $u(t - T)$ has its discontinuity at $t = T$ and u_{n-N} has its step at $n = N$. It is also not difficult to make step functions of different sizes $Au(t)$ and Au_n, and even with any two levels $Au(t) + B$ and $Au_n + B$. The unit step is often used to model phenomena that are 'switched on' at some specific time.

By subtracting a digital unit step shifted one to the right from the unshifted digital unit step we obtain the digital unit impulse. This signal, depicted in Figure 2.3.B, is zero everywhere except at time zero, where it is unity. This is our first true signal, conforming to all the requirements of our definition. In Chapter 6 we will see that the unit impulse is an invaluable tool in the study of systems. Rather than invent a new mathematical

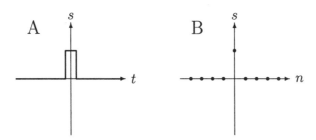

Figure 2.3: The unit impulse. In (A) we depict an analog impulse of unity width. In (B) the digital unit impulse $\delta_{\hat{n},0}$.

symbol for this signal, we utilize one known as the *Kronecker delta* $\delta_{n,m}$. This doubly indexed entity is defined to be one, if and only if its indices are equal; otherwise it is zero. In terms of the Kronecker delta, the digital unit impulse is $s_n = \delta_{n,0}$.

The full Kronecker delta corresponds to a **Shifted Unit Impulse** (SUI)

$$s_n = \delta_{n,m} \tag{2.5}$$

that is zero for all times except for time $n = m$, when it equals one. The importance of the set of all SUIs will become clear in Section 2.5.

One might similarly define an analog unit impulse by subtracting analog unit steps, obtaining the Figure 2.3.A. This analog signal flagrantly displays two jump discontinuities, but by now that should not make us feel uncomfortable. However, this is not the signal usually referred to as the analog unit impulse. There is no profound meaning to the width of this signal, since in the analog world the meaning of a unit time interval depends on the time units! What *is* meaningful is the *energy* of the impulse, which is its amplitude squared times its width. There are good reasons to expect that once the width is small enough (i.e., small compared to all significant times in the problem) all impulses with the same energy will have basically the same effect on systems. Accordingly, when one speaks of a 'unit impulse' in the analog domain, conventionally this alludes to a 'unit energy' impulse. Of course the unit width impulse in Figure 2.3 is a unit impulse in this sense; but so are all the others in Figure 2.4.

The unit energy impulses in the figure are given by:

$$I(t) = \begin{cases} 0 & |t| > T \\ \frac{1}{\sqrt{2T}} & |t| < T \end{cases}$$

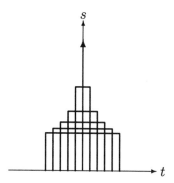

Figure 2.4: Analog unit energy impulses. Since all of these signals have the same energy, the height increases as the width decreases. The vertical arrow is a symbolic way of designating Dirac's delta function.

where T is the width. In the limit $T \to 0$ we obtain a mathematical entity called Dirac's delta function $\delta(t)$, first used by P.A.M. Dirac in his mathematical description of quantum physics. The name *delta* is purposely utilized to emphasize that this is the 'analog analog' of Kronecker's delta. The word *function* is a misnomer, since Dirac's delta is not a true function at all. Indeed, Dirac's delta is defined by the two properties:

- $\delta(t)$ is zero everywhere except at the origin $t = 0$
- the integral of the delta function is unity $\int_{-\infty}^{\infty} \delta(t)dt = 1$

and clearly there can be no such function! However, Dirac's delta is such an extremely useful abstraction, and since its use *can* be justified mathematically, we shall accept it without further question. Indeed, Dirac's delta is *so* useful, that when one refers without further qualification to the analog unit impulse, one normally means $\delta(t)$.

$$s(t) = \delta(t) \qquad \mathbf{A} \;\Big|\; \mathbf{D} \qquad s_n = \delta_{n,0} \qquad (2.6)$$

The next signal we wish to discuss is the square wave $\square(t)$, depicted in Figure 2.5.A. It takes on only two values, ± 1, but switches back and forth between these values periodically. One mathematical definition of the analog square wave is

$$\square(t) = \begin{cases} 1 & \lfloor t \rfloor \text{ is even} \\ -1 & \lfloor t \rfloor \text{ is odd} \end{cases} \qquad (2.7)$$

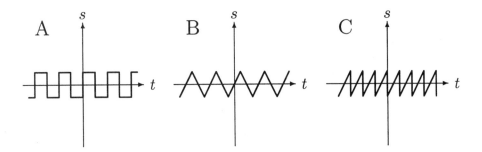

Figure 2.5: Three periodic analog signals. In (A) we depict the square wave, in (B) the triangle wave and in (C) the sawtooth.

where $\lfloor t \rfloor$ (pronounced 'floor' of t) is the greatest integer less than or equal to the real number t. We have already mentioned that this signal has an infinite number of jump discontinuities, and it has infinite energy as well! Once again we can stretch and offset this signal to obtain any two levels, and we can also change its *period* from unity to T by employing $\Box(t/T)$. We can further generalize the square wave to a rectangular wave by having it spend more time in one state than the other. In this case the percentage of the time in the higher level is called the *duty cycle*, the standard square wave having a 50% duty cycle. For digital signals the minimal duty cycle signal that is not a constant has a single high sample and all the rest low. This is the periodic unit impulse

$$d_n = \sum_m \delta_{n,mP} \tag{2.8}$$

where the period is P samples.

Similarly we can define the analog triangle wave $\triangle(t)$ of Figure 2.5.B and the sawtooth $\mathcal{T}(t)$ of Figure 2.5.C. Both, although continuous, have slope discontinuities. We leave the mathematical definitions of these, as well as the plotting of their digital versions, to the reader. These signals pop up again and again in applications. The square wave and its close brethren are useful for triggering comparators and counters, the triangle is utilized when constant slope is required, and the sawtooth is vital as the 'time base' of oscilloscopes and the 'raster scan' in television. Equipment known as 'function generators' are used to generate these signals.

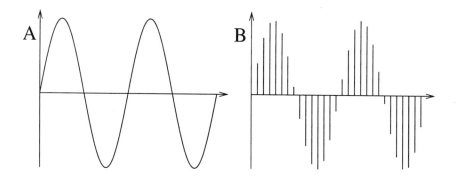

Figure 2.6: Sinusoidal signals. In (A) we depict the analog sinusoid with given amplitude, frequency and phase. In (B) the digital sinusoid is shown.

Of course the most famous periodic signal is none of these, but the sine and cosine functions, either of which we call a *sinusoid*.

$$s(t) = \sin(2\pi f t) \quad \mathbf{A} \,\Big|\, \mathbf{D} \quad s_n = \sin(2\pi f_d\, n) \qquad (2.9)$$

The connection between the *frequency f* of an analog sinusoid and its period T can be made clear by recalling that the sine function completes a full cycle after 2π radians. Accordingly, the frequency is the reciprocal of the period

$$f = \frac{1}{T}$$

and its units must be *full cycles per second*, also known as Hertz or Hz. The period represents the number of seconds per cycle while the frequency in Hz describes the number of full cycles per second. Since discrete time n carries no units, the *digital frequency* f_d will be essentially a pure number. The periodicity of digital sinusoids will be discussed later.

In order to avoid factors of 2π we often rewrite equation 2.9 as follows.

$$s(t) = \sin(\omega t) \quad \mathbf{A} \,\Big|\, \mathbf{D} \quad s_n = \sin(\omega_d\, n) \qquad (2.10)$$

Since the argument of a trigonometric function must be in radians (or degrees), the units of the *angular frequency* $\omega = 2\pi f$ must be *radians per second*, and those of the *digital angular frequency* $\omega_d = 2\pi f_d$ simply *radians*.

In many respects $\sin(t)$ is very similar to $\square(t)$ or $\triangle(t)$, but it possesses a major benefit, its smoothness. Sinusoids have neither jump nor slope discontinuities, elegantly oscillating back and forth (see Figure 2.6.A). More general sinusoids can be obtained by appropriate mathematical manipulation

$$A\sin(\omega t + \phi) + B$$

where A is called the *amplitude*, ω the frequency, ϕ the *phase*, and B the *DC* component. Sines of infinite time duration have infinite energy, but are otherwise eminent members of the signal community. Sinusoidal signals are used extensively in all facets of signal processing; communications are carried by them, music is modeled as combinations of them, mechanical vibrations are analyzed in terms of them, clocks are set by comparing to them, and so forth.

Although the signals $\sin(\omega t)$ and $\cos(\omega t)$ look exactly the same when viewed separately, when several signals are involved the relative phases become critical. For example, adding the signal $\sin(\omega t)$ to another $\sin(\omega t)$ produces $2\sin(\omega t)$; adding $\sin(\omega t)$ to $\cos(\omega t)$ creates $\sqrt{2}\sin(\omega t + \frac{\pi}{4})$; but adding $\sin(\omega t)$ to $\sin(\omega t + \pi) = -\sin(\omega t)$ results in zero. We can conclude that when adding sinusoids $1+1$ doesn't necessarily equal 2; rather it can be anything between 0 and 2 depending on the phases. This addition operation is analogous to the addition of vectors in the plane, and many authors define *phasors* in order to reduce sinusoid summation to the more easily visualized vector addition. We will not need to do so, but instead caution the reader to take phase into account whenever more than one signal is present.

Another basic mathematical function with a free parameter that is commonly employed in signal processing is the exponential signal

$$s(t) = e^{\Lambda t} \qquad \mathbf{A} \mid \mathbf{D} \qquad s_n = e^{\Lambda_d n}$$

depicted in Figure 2.7 for negative Λ. For positive Λ and any finite time this function is finite, and so technically it is a well-behaved signal. In practice the function explodes violently for even moderately sized negative times, and unless somehow restricted does not correspond to anything we actually see in nature. Mathematically the exponent has unique qualities that make it ideal for studying signal processing systems.

We shall now do something new; for the first time we will allow complex-valued functions. We do this by allowing the constant in the argument of the exponential to be a pure imaginary number $\Lambda = i\omega$, thus radically chang-

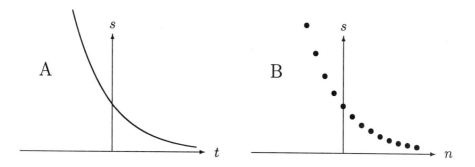

Figure 2.7: Exponentially decreasing signals. In (A) we depict the analog exponential, in (B) the digital.

ing the character of the signal. Recalling the remarkable identity (equation (A.7))

$$e^{i\varphi} = \cos(\varphi) + i\sin(\varphi)$$

we see that exponentials with imaginary coefficients are complex sinusoids.

$$A e^{i\omega t} = A\cos(\omega t) + iA\sin(\omega t)$$

When we deal with complex signals like $\mathbf{s}(t) = A e^{i\omega t}$, what we really mean is that the real-world signal is the real part

$$s(t) = \Re\,\mathbf{s}(t) = A\cos(\omega t)$$

while the imaginary part is just that—imaginary. Since the imaginary part is 90° (one quarter of a cycle) out of phase with the real signal, it is called the *quadrature component*. Hence the complex signal is composed of in-phase (real) and quadrature (imaginary) components.

At first it would seem that using complex signals makes things more *complex* but often the opposite is the case. To demonstrate this, consider what happens when we multiply two sinusoidal signals $s_1(t) = \sin(\omega_1 t)$ and $s_2(t) = \sin(\omega_2 t)$. The resulting signal is

$$s(t) = s_1(t)s_2(t) = \sin(\omega_1 t)\cos(\omega_2 t) + \cos(\omega_1 t)\sin(\omega_2 t)$$

which is somewhat bewildering. Were we to use complex signals, the product would be easy

$$\mathbf{s}(t) = \mathbf{s_1}(t)\mathbf{s_2}(t) = e^{i\omega_1 t} e^{i\omega_2 t} = e^{i(\omega_1 + \omega_2)t}$$

due to the symmetries of the exponential function. The apparent contradiction between these two results is taken up in the exercises.

A further variation on the exponential is to allow the constant in the argument of the exponential to be a complex number with both real and imaginary parts $\Lambda = \lambda + i\omega$. This results in

$$\mathbf{s}(t) = Ae^{(\lambda+i\omega)t} = Ae^{\lambda t}\cos(\omega t) + iAe^{\lambda t}\sin(\omega t) \qquad (2.11)$$

corresponding to the real signal

$$s(t) = Ae^{\lambda t}\cos(\omega t) \qquad (2.12)$$

which combines the exponential with the sinusoid. For negative λ, this is a damped sinusoid, while for positive λ it is an exponentially growing one.

Summarizing, we have seen the following archetypical simple signals:

unit constant	$s(t) = 1$	$s_n = 1$
unit step	$s(t) = u(t)$	$s_n = u_n$
unit impulse	$s(t) = \delta(t)$	$s_n = \delta_{n,0}$
square wave	$s(t) = \square(\omega t)$	$s_n = \square(\omega_d n)$
sinusoid	$s(t) = A\sin(\omega t + \phi)$	$s_n = A\sin(\omega_d n + \phi)$
damped sinusoid	$s(t) = Ae^{-\lambda t}\sin(\omega t + \phi)$	$s_n = A\alpha^{-n}\sin(\omega_d n + \phi)$
real exponential	$s(t) = Ae^{\lambda t}$	$s_n = \alpha^n$
complex sinusoid	$s(t) = Ae^{i(\omega t + \phi)}$	$s_n = Ae^{i(\omega_d n + \phi)}$
damped complex sinusoid	$s(t) = Ae^{(\lambda+i\omega)t}$	$s_n = Ae^{(\lambda+i\omega_d)n}$

EXERCISES

2.2.1 Thomas Alva Edison didn't believe that AC electricity was useful, since the current first went one way and then returned. It was Nikola Tesla who claimed that AC was actually better than DC. Why was Edison wrong (hint: energy) and Tesla right (hint: 'transformers')?

2.2.2 In the text we depicted digital signals graphically by placing dots at signal values. We will usually use such *dot graphs*, but other formats are prevalent as well. A *comb graph* uses lines from the time axis to the signal point; a *slint graph* (straight line **int**erpolation) simply connects successive signal values; comb-dot and slint-dot combinations are useful when the signal takes on zero values. These formats are depicted in Figure 2.8. Write general routines for plotting digital signals in these formats in whatever computer programming language you usually use. Depending on your programming language you may first have to prepare low-level primitives. Plot the digital sinusoidal signal $s_n = \sin(\omega_d n)$ for various frequencies ω in all of these formats. Decide which you like the best. You may use this format from now on.

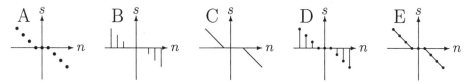

Figure 2.8: Different formats for graphical representation of digital signals. In (A) we depict a signal using our usual *dot graph*. In (B) the same signal is plotted as a *comb graph*. In (C) it is graphed as a *slint graph*. (D) and (E) are *comb-dot* and *slint-dot* representations respectively.

2.2.3 Give mathematical definitions for the analog triangle signal $\triangle(t)$ of Figure 2.5.B and for the analog sawtooth saw(t) of Figure 2.5.C.

2.2.4 What is the integral of the square wave signal? What is its derivative?

2.2.5 Using your favorite graphic format plot the digital square wave, triangle wave and sawtooth, for various periods.

2.2.6 Perform the following experiment (you will need an assistant). Darken the room and have your assistant turn on a pen-flashlight and draw large circles in the air. Observe the light from the side, so that you see a point of light moving up and down. Now have the assistant start walking while still drawing circles. Concentrate on the vertical and horizontal motion of the point of light, disregarding the depth sensation. You should see a sinusoidal signal. Prove this. What happens when you rotate your hand in the opposite direction? What can you infer regarding negative frequency sinusoids?

2.2.7 Dirac's delta function can be obtained as the limit of sequences of functions other than those depicted in Figure 2.4. For example,

$$\text{asymmetric unit impulses} \quad \mathcal{I}_T(t) \quad = \begin{cases} 0 & t < 0 \\ \frac{1}{\sqrt{T}} & 0 < t < T \\ 0 & t > T \end{cases}$$

$$\text{Gaussian functions} \quad G_\sigma(t) \quad = \frac{1}{\sqrt{2\pi}\sigma} e^{\frac{t^2}{\sigma^2}}$$

$$\text{Sinc functions} \quad \tfrac{1}{\pi}\text{sinc}_\omega(t) \quad = \frac{\sin(\omega t)}{\pi t}$$

$$\text{Lorentzian functions} \quad \mathcal{L}(t) \quad = \frac{1}{\pi} \frac{\epsilon}{\epsilon^2 + t^2}$$

Graph these functions for decreasing T, ϵ and increasing σ, ω, graphically showing the appearance of the Dirac delta. What new features appear? Show that in the proper limit these functions approach zero for all nonzero times.

2.2.8 The integral of the analog impulse $\delta(t)$ is the unit step $u(t)$, and conversely the derivative of $u(t)$ is $\delta(t)$. Explain these facts and depict graphically.

2.2.9 Explain the following representation of Dirac's delta.

$$\delta(t) = \tfrac{1}{2}\frac{d}{dx}|x|$$

2.2.10 Show that

$$\int_{-\infty}^{\infty} f(t)\delta(t-t')dt = f(t')$$

both graphically and by using basic calculus. From this result show that $\delta(t)$ must be zero for all nonzero arguments. Compare the above relation with the Fourier identity

$$f(t') = \frac{1}{2\pi}\int_{-\infty}^{\infty} du \int_{-\infty}^{\infty} dt f(t)e^{iu(t-t')}$$

and derive an integral representation for the Dirac delta. What meaning can be given to the derivative of the Dirac delta?

2.2.11 Plot the analog complex exponential. You will need to simultaneously plot two sinusoids in such fashion that one is able to differentiate between them. Extend the routines you wrote in the previous exercise to handle the digital complex exponential.

2.2.12 Explain *why* the real signal corresponding to the product of two complex exponentials is *not* the same as the product of the two real sinusoids.

2.3   Characteristics of Signals

Now that we have some experience with signals, let us discuss some general characteristic signals can have. Signals are characterized as being:

- deterministic or stochastic
- if deterministic: periodic or nonperiodic
- if stochastic: stationary or nonstationary
- of finite or infinite time duration
- of finite bandwidth or of full spectrum

Perhaps the most significant characteristic of a signal is whether it is deterministic or stochastic. Deterministic signals are those that are generated by some nonprobabilistic algorithm. They are thus reproducible, predictable (at least over short time scales—but see Section 5.5) and well-behaved mathematically. Stochastic signals are generated by systems that contain randomness (see Section 5.6). At any particular time the signal is a *random variable*, (see Appendix A.13), which may have well defined average and variance, but is not completely defined in value. Any particular sequence of measurements of the signal's values at various times captures a specific instantiation of the stochastic signal, but different sequence of measurements under the same conditions would retrieve somewhat different values.

In practice we never see a purely deterministic signal, since even the purest of deterministic signals will inevitably become contaminated with *noise*. 'Pure noise' is the name we give to a quintessential stochastic signal, one that has only probabilistic elements and no deterministic ones. When a deterministic signal becomes contaminated with additive noise, as depicted in Figure 2.9,

$$y(t) = x(t) + n(t)$$

we can quantify its 'noisiness' by the **S**ignal to **N**oise **R**atio (SNR). The SNR is defined as the ratio of the signal energy to the noise energy, and is normally measured in *dB*. (equation (A.16))

$$\text{SNR(dB)} = 10 \log_{10} \frac{E_x}{E_n} = 10 \left(\log_{10} E_x - \log_{10} E_n \right) \qquad (2.13)$$

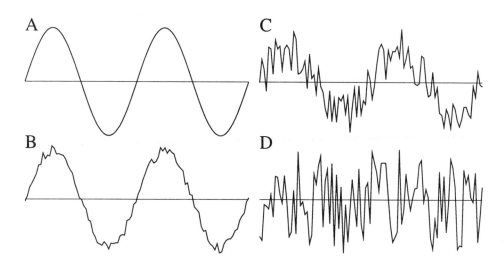

Figure 2.9: Deterministic signal (simple sine) with gradually increasing additive noise. In (A) the deterministic signal is much stronger than the noise, while in (D) the opposite is the case.

When measuring in, we usually talk about the signal as being *above* the noise by SNR(dB).

Not all the signals we encounter are stochastic due solely to contamination by additive noise. Some signals, for example speech, are inherently stochastic. Were we to pronounce a single vowel for an extended period of time the acoustic signal would be roughly deterministic; but true speech is random because of its changing content. Speech is also stochastic for another reason. Unvoiced sounds such as **s** and **f** are made by constricting air passages at the teeth and lips and are close to being pure noise. The **h** sound starts as noise produced in the throat, but is subsequently filtered by the mouth cavity; it is therefore partially random and partially deterministic.

Deterministic signals can be *periodic*, meaning that they *exactly* repeat themselves after a time known as the *period*. The falling exponential is not periodic, while the analog sine $A\sin(2\pi ft)$, as we discussed above, is periodic with period $T = \frac{1}{f}$. The electric voltage supplied to our houses and the acoustic pressure waves from a flute are both nearly perfect sinusoids and hence periodic. The frequency of the AC supplied by the electric company is 60 Hz (sixty cycles per second) in the United States, and 50 Hz (fifty cycles per second) in Europe; the periods are thus $16\frac{2}{3}$ and 20 milliseconds respectively. The transverse flutes used in orchestral music can produce frequencies from middle C (524 Hz) to about three and a half octaves, or over ten times, higher!

While the analog sinusoid is always periodic the digital counterpart is not. Consider an analog signal with a period of 2 seconds. If we create a digital sinusoid by 'sampling' it 10 times per second, the digital signal will be periodic with digital period 20. However, if we sample at 10.5 times per second, after 2 seconds we are a half-second out of phase; only after four seconds, (i.e., 21 samples) does the digital signal coincide with its previous values. Were we to sample at some other rate it would take even longer for the digital version to precisely duplicate itself; and if ratio of the period to the sampling interval is not rational this precise duplication will never occur.

Stochastic signals may be stationary, which means that their probabilistic description does not change with time. This implies that all the signal's statistics, such as the mean and variance, are constant. If a stochastic signal gets stronger or weaker or somehow noisier with time, it is not stationary. For example, speech is a stochastic signal that is highly nonstationary; indeed it is by changing the statistics that we convey information. However, over short enough time intervals, say 30 milliseconds, speech seems stationary because we can't move our mouth and tongue this fast.

A signal, analog or digital, can be of infinite or finite time duration. We required that signals be *defined* for all times $-\infty < t < \infty$ or $n = -\infty, \infty$, but not that they be nonzero for all times. Real physical signals are of finite energy, and hence are often zero for times much before or after their peak.

In like fashion, signals, analog or digital, can be of infinite or finite bandwidth. According to our original definition an analog signal should be finite bandwidth, but noise and signals with discontinuities are *full spectrum*. The interpretation of this concept for digital signals must be postponed until after clarification of the *sampling theorem*, in the Section 2.8.

EXERCISES

2.3.1 Look closely at the graphs of the digital sinusoid $s_n = \sin(\omega n)$ that you prepared in exercise 2.2.2. When is the digital sinusoid periodic? Under what conditions is the period the same as that of the analog sinusoid? Verify the statement in the text regarding nonperiodic digital sinusoids.

2.3.2 The purpose of this exercise is to examine the periodicity of the sum of two analog sines. For example, the sum of a sine of period 4 seconds and one of period 6 seconds is periodic with period 12 seconds. This is due to the first sine completing three full periods while the second competes two full periods in 12 seconds. Give an example of a sum that is *not* periodic. Give a general rule for the periodicity. What *can* be said about cases when the sum is not exactly periodic?

2.3.3 Plot analog signals composed of the sum of two sinusoids with identical amplitudes and frequencies f_1 and f_2. Note that when the frequencies are close the resultant seems to have two periods, one short and one long. What are the frequencies corresponding to these periods? Prove your assertion using the trigonometric identities.

2.4   Signal Arithmetic

Some of the requirements in our definition of *signal* were constraints on signal values $s(t)$ or s_n, while some dealt with the signal as a whole. For example, finite valuedness is a constraint on every signal value separately, while finite energy and finite bandwidth requirements mix all the signal values together into one inequality. However, even the former type of requirement is most concisely viewed as a single requirement on the signal s, rather than an infinite number of requirements on the values.

This is one of the economies of notation that make it advantageous to define signals in the first place. This is similar to what is done when one defines complex numbers or n-dimensional vectors (n-vectors); in one concise equation one represents two or even n equations. With a similar motivation of economy we define arithmetic operations on signals, thus enabling us to write single equations rather than a (possibly nondenumerable) infinite number! Hence in some ways signals are just like n-vectors of infinite dimension.

First let us define the multiplication of a signal by a real number

$$
\begin{array}{c|c}
\begin{array}{c} y = ax \\ \text{means} \\ y(t) = ax(t) \quad \forall t \end{array}
\quad \mathbf{A} \;\bigg|\; \mathbf{D} \quad
\begin{array}{c} y = ax \\ \text{means} \\ y_n = ax_n \quad \forall n \end{array}
\end{array}
\qquad (2.14)
$$

that is, we individually multiply every signal value by the real number. It might seem overly trivial even to define this operation, but it really is important to do so. A signal is not merely a large collection of values, it is an entity in its own right. Think of a vector in three-dimensional space (a 3-vector). Of course it is composed of three real numbers and accordingly doubling its size can be accomplished by multiplying each of these numbers by two; yet the effect is that of creating a new 3-vector whose direction is the same as the original vector but whose length is doubled. We can visualize this operation as stretching the 3-vector along its own direction, without thinking of the individual components. In a similar fashion amplification of the signal should be visualized as a transformation of the signal as a whole, even though we may accomplish this by multiplying each signal value separately.

We already know that multiplication of a signal by a real number can represent an amplification or an attenuation. It can also perform an *inversion*

$$
\begin{array}{c|c}
\begin{array}{c} y = -x \\ \text{means} \\ y(t) = -x(t) \quad \forall t \end{array}
\quad \mathbf{A} \;\bigg|\; \mathbf{D} \quad
\begin{array}{c} y = -x \\ \text{means} \\ y_n = -x_n \quad \forall n \end{array}
\end{array}
\qquad (2.15)
$$

if we take the real number to be $a = -1$ Here the minus sign is an 'operator', transforming a signal into another, related, signal. The inverted signal has the same energy and bandwidth as the original, and we shall see later on has the same *power spectrum*. Nevertheless, every time the original signal increases, the inverted one decreases; when the signal attains its maximum, the inverted signal attains its minimum.

There is another way to make a signal of the same energy and power spectrum as the original, but somehow backwards. We can *reverse* a signal using the operator Rev

$$
\begin{array}{c|c}
\begin{array}{c} y = \text{Rev}\, x \\ \text{means} \\ y(t) = x(-t) \quad \forall t \end{array} & \mathbf{A} \;\bigg|\; \mathbf{D} \quad \begin{array}{c} y = \text{Rev}\, x \\ \text{means} \\ y_n = x_{-n} \quad \forall n \end{array}
\end{array} \tag{2.16}
$$

which makes it run backwards in time. If you whistle a constant note it will sound the same when reversed, but if you whistle with ascending pitch the reversed signal will have descending pitch. This operation has no counterpart for n-vectors.

Frequently we will need to add two signals,

$$
\begin{array}{c|c}
\begin{array}{c} z = x + y \\ \text{means} \\ z(t) = x(t) + y(t) \quad \forall t \end{array} & \mathbf{A} \;\bigg|\; \mathbf{D} \quad \begin{array}{c} z = x + y \\ \text{means} \\ z_n = x_n + y_n \quad \forall n \end{array}
\end{array} \tag{2.17}
$$

one simply adds the values. This is the familiar addition of two n-vectors, and is the similar to the addition of complex numbers as well. Signal addition is commutative $(x + y = y + x)$ and associative $(x + (y + z) = (x + y) + z)$ and adding a signal to its inversion yields the zero signal. Hence signals, like real numbers, complex numbers, and n-vectors, obey all the normal rules of arithmetic.

We will also need to multiply two signals, and you have probably already guessed that

$$
\begin{array}{c|c}
\begin{array}{c} z = x\,y \\ \text{means} \\ z(t) = x(t)\,y(t) \quad \forall t \end{array} & \mathbf{A} \;\bigg|\; \mathbf{D} \quad \begin{array}{c} z = x\,y \\ \text{means} \\ z_n = x_n\,y_n \quad \forall n \end{array}
\end{array} \tag{2.18}
$$

one simply multiplies *value by value*. Multiplication of a signal by a number is consistent with this definition of multiplication—just think of the number as a constant signal. However, this multiplication is different from multiplication of 3-vectors or complex numbers. The usual 'dot product' multiplication of two 3-vectors yields a scalar and not a 3-vector. There *is* a *cross* or *vector product* kind of multiplication that yields a vector, but it doesn't generalize to n-vectors and it isn't even commutative. Multiplication of complex numbers yields a complex number, but there

$$
z = x\,y \quad \text{does not mean} \quad \Re z = \Re x\, \Re y \quad \text{and} \quad \Im z = \Im x\, \Im y
$$

which is quite different from *value by value* multiplication of signals.

Although value by value multiplication of signals can be very useful, for instance in 'mixing' of signals (see Section 8.5), there is another type of multiplication, known as *dot product*, that is more important yet. This product *is* analogous to the usual scalar product of n-vectors, and it yields a real number that depends on the entire signal.

$$
\begin{array}{c|c}
\begin{array}{c}
r = x \cdot y \\
\text{means} \\
r = \int_{-\infty}^{\infty} x(t)y(t)dt
\end{array}
\quad \mathbf{A} \ \Big| \ \mathbf{D} \quad
\begin{array}{c}
r = x \cdot y \\
\text{means} \\
r = \sum_{n=-\infty}^{\infty} x_n y_n
\end{array}
\end{array}
\qquad (2.19)
$$

This is the proper definition for *real* signals; although it can be extended for complex signals. The energy of a signal is the dot product of the signal with itself, while the dot product of two different signals measures their similarity (see Chapter 9). Signals for which the dot product vanishes are said to be *orthogonal*, while those for which it is large are said to be strongly correlated.

For digital signals there is another operator known as the *time advance operator* z,

$$
y = z\,x \qquad \text{means} \qquad y_n = x_{n+1} \quad \forall n \qquad (2.20)
$$

which would certainly be meaningless for vectors in space. What meaning could there possibly be for an operator that transforms the x coördinate of a vector into the y coördinate? However, signals are not static vectors; they are dynamic entities. The time variable is not a dummy variable or index; it is physical time. We can always renumber the axes of a vector, thus scrambling the order of elements, and still understand that the same physical vector is described. For signals such an action is unthinkable. This is the reason that $\text{Rev}(x)$ had no vector counterpart. This is the reason that our original definition of *signal* emphasized that the independent variable or index was *time*.

You can think of z as the 'just wait a little while and see what happens' operator. For digital signals the natural amount of time to wait is one unit, from n to $n + 1$. If we wish to peek further forward in time, we can do so. For example, we can jump forward two units of time by first advancing one unit and then one more

$$
y = zz\,x = z^2 x \qquad \text{means} \qquad y_n = x_{n+2} \quad \forall n
$$

and so on.

We may also wish to go backwards in time. This doesn't require us to invent a time machine, it just means that we wish to recall the value a signal had a moment ago. A little reflection leads us to define the *time delay operator* z^{-1}

$$
y = z^{-1} x \qquad \text{means} \qquad y_n = x_{n-1} \quad \forall n \qquad (2.21)
$$

so that $z\,z^{-1}\,x = z^{-1}\,z\,x = x$. The operator z^{-1} will turn out to be even more useful than z, since it is usually easier to remember what just happened than to predict what is about to occur. The standard method for implementing the digital delay of L units of time is through a FIFO buffer of length L. A signal value that enters the FIFO at time n exits at time $n + L$, and so the output of the FIFO is delayed exactly L time units with respect to its input. When used in this fashion the FIFO is called a *delay line*.

We can make these operators more concrete with a simple example. In exercise 2.1.1.13 we introduced a family of recursively defined signals, often called the *logistics signals*

$$x_{n+1} = \lambda x_n(1 - x_n) \tag{2.22}$$

where the x_n are all in the range $0 \leq x_n \leq 1$. In order to enforce this last restriction we must restrict λ to be in the range $0 \leq \lambda \leq 4$. A particular signal in this family is determined by giving x_0 and λ. It is most instructive to generate and plot values for various x_0 and λ, and the reader will be requested to do so as an exercise. In this case the operation of the time advance operator can be simply specified

$$zx = \lambda x(1 - x)$$

which should be understood as an equation in signals. This stands for an infinite number of equations of the form (2.22), one for each n. However, we needn't return to these equations to understand it. We start with $1-x$, which really means $1 + (-x)$. $(-x)$ is the inversion of the signal x; we add to it the signal 1 that is the constant signal whose value is 1 for all times. Addition between signals is value by value of course. Next we multiply this signal by the original signal, using signal multiplication, value by value. Finally we multiply this resulting signal by a real number λ. So for this special case, the time advance operator can be specified in terms of simple signal arithmetic.

Operators can be combined to create new operators. The finite difference operator Δ is defined as

$$\Delta \equiv 1 - z^{-1} \tag{2.23}$$

that is, for any digital signal s, the following holds for all time n.

$$\Delta s_n = s_n - s_{n-1}$$

The finite difference operator for digital signals is vaguely similar to the differentiation operator for continuous signals. Common characteristics include linearity and the fact that they are identically zero only for a constant. Δ is a linear operator since for any two signals x and y, $\Delta(x+y) = \Delta x + \Delta y$ and for any number c and signal x, $\Delta cx = c\Delta x$. $\Delta s = 0$ (the zero signal) if and only if the signal is constant. In other ways finite differences are similar to, but not identical to derivatives. For example, $\Delta(xy) = x\Delta y + \Delta x\, z^{-1}y$. In some things finite differences are completely different, e.g., $\Delta\alpha^n = \alpha^n(1 - \alpha^{-1})$.

This last example leads us to an important property of the time delay operator. For the exponential signal $s_n = e^{\Lambda n}$ it is easy to see that

$$s_{n-1} = e^{\Lambda(n-1)} = e^{-\Lambda}e^{\Lambda n} = e^{-\Lambda}s_n$$

so that

$$z^{-1}s = e^{-\Lambda}s$$

i.e., the operation of time delay on the exponential signal is equivalent to multiplication of the signal by a number. In linear algebra when the effect of an operator on a vector is to multiply it by a scalar, we call that vector an 'eigenvector' of the operator. Similarly we can say that the exponential signal is an *eigensignal* of the time delay operator, with eigenvalue $e^{-\Lambda}$.

The fact that the exponential is an eigensignal of the time delay operator will turn out to be very useful. It would have been even nicer were the sinusoid to have been an eigensignal of time delay, but alas equation (A.23) tells us that

$$s_{n-1} = \sin\left(\omega(n-1)\right) = \sin(\omega n)\cos(\omega) - \cos(\omega n)\sin(\omega)$$

which mixes in phase-shifted versions of the original signal. The sinusoid *is* the eigensignal of a more complex operator, one that contains two time delays; this derives from the fact that sinusoids obey second-order differential equations rather than first-order ones like the exponential. Nonetheless, there is a trick that saves the day, one that we have mentioned before. We simply work with complex exponentials, which *are* eigensignals of time delay, remembering at the end to take the real part. This tactic is perhaps the main reason for the use of complex signals in DSP.

EXERCISES

2.4.1 Show that the exponential signal $s_n = Ae^{\Lambda n}$ is an eigensignal of the time *advance* operator. What is its eigenvalue? The real sinusoid $s_n = A\sin(\omega n + \phi)$ is the eigensignal of an operator that contains z^{-1} and z^{-2}. Can you find this operator?

2.4.2 What is the effect of the time advance operator on the unit impulse? Express the general SUI $\delta_{n,m}$ in terms of $\delta_{n,0}$ and the time delay operator.

2.4.3 Compare the energy of a time delayed, advanced, or reversed signal with that of the original signal. What is the energy of $y = ax$ in terms of that of x? What can be said about the energy of the sum of two signals? For example, consider summing two sinusoids of the same frequency but different amplitudes and phases. What about two sinusoids of different frequencies? Why is there a difference between these two cases?

2.4.4 Plot the logistics signal of equation (2.22) using several different x_0 for each λ. Try $\lambda = 0.75$ and various x_0—what happens after a while? Next try $\lambda = 1.5, 2.0$, and 2.75. How is the long time behavior different? Can you predict the behavior as a function of λ? Are there any starting points where the previous behavior is still observed? Next try $\lambda = 3.2, 3.5, 3.55, 3.5675$, and 3.75. What is the asymptotic behavior (for almost all x_0)?

2.4.5 Using the program from the previous exercise try $\lambda = 3.826, 3.625$ and 3.7373. What is the asymptotic behavior? Try $\lambda = 4$. How is this different?

2.4.6 Canons are musical compositions composed of several related *voices* heard together. The 'canonical' relations require the voices to repeat the theme of the first voice:

time offset: after a time delay,

key shift: in a different key,

diminution: at twice normal speed,

augmentation: at half normal speed,

inversion: with high and low tones interchanged,

crab order: time reversed,

or with combinations of these. Describe the signal processing operators that transform the basic theme into the various voices. In order for the resulting canon to sound pleasing, at (almost) every instant of time the voices must be harmonically related. Can you write a program that composes canons?

2.4.7 In the text we discussed the usefulness of considering a signal as a single entity. This exercise deals with the usefulness of considering a signal as a collection of values. A *streaming signal* is a digital signal that is made available as time progresses. When the signal is not being streamed one must wait for the signal to be completely prepared and placed into a file before processing. Explain the usefulness of streaming digital audio. In certain computer languages a *stream* is defined to be a sequentially accessed file. Compare this use of 'stream' with the previous one.

2.5   The Vector Space of All Possible Signals

In Section 2.2 we presented the simplest of signals; in this section we are going to introduce you to all the rest. Of course there are an infinite number of different signals, but that doesn't mean that it will take a long time to introduce them all. How can this be? Well, there are an infinite number of points in the plane, but we can concisely describe every one using just two real numbers, the x and y coördinates. There are an infinite number of places on earth, but all can be located using longitude and latitude. Similarly there are an infinite number of different colors, but three numbers suffice to describe them all; for example, in the RGB system we give red, green, and blue components. All events that have already taken place or will ever take place in the entire universe can be located using just four numbers (three spatial coördinates and the time). These concise descriptions are made possible by identifying *basis elements*, and describing all others as weighted sums of these. When we do so we have introduced a *vector space* (see Appendix A.14). The points in the plane and in space are well known to be two-dimensional and three-dimensional vector spaces, respectively.

In the case of places on earth, it is conventional to start at the point where the equator meets the prime meridian, and describe how to reach any point by traveling first north and then east. However, we could just as well travel west first and then south, or northeast and then southwest. The choice of basic directions is arbitrary, as long as the second is not the same as the first or its reverse. Similarly the choice of x and y directions in the plane is arbitrary; instead of RGB we can use CMY (cyan, magenta, and yellow), or HSV (hue, saturation, and value); and it is up to us to choose the directions in space to arrive at any point in the universe (although the direction in time is *not* arbitrary).

Can all possible signals be described in terms of some set of basic signals? We will now convince you that the answer is affirmative by introducing the vector space of signals. It might seem strange to you that signals form a vector space; they don't seem to be magnitudes and directions like the vectors you may be used to. However, the colors also form a vector space, and they aren't obviously magnitudes and directions either. The proper way to dispel our skepticism is to verify that signals obey the basic axioms of vector spaces (presented in Appendix A.14). We will now show that not only do signals (both the analog and digital types) form a vector space, but this space has an inner product and norm as well! The fact that signals form a vector space gives them algebraic structure that will enable us to efficiently describe them.

Addition: Signal addition $s = s_1 + s_2$ according to equation (2.17),

Zero: The constant signal $s_n = 0$ for all times n,

Inverse: The inversion $-s$ according to equation (2.15),

Multiplication: Multiplication by a real number as in equation (2.14),

Inner Product: The dot product of equation (2.19),

Norm: The energy as defined in equation (2.1),

Metric: The energy of the difference signal obeys all the requirements.

Since signals form a vector space, the theorems of linear algebra guarantee that there is a basis $\{v_k\}$, i.e., a set of signals in terms of which *any* signal s can be expanded.

$$s = \sum_k c_k v_k \qquad (2.24)$$

The use of the summation sigma assumes that there are a finite or denumerable number of basis signals; when a nondenumerable infinity of basis signals is required the sum must be replaced by integration.

$$s = \int c(k)v(k)\,dk \qquad (2.25)$$

From linear algebra we can show that every vector space has a basis, but in general this basis is not unique. For example, in two-dimensional space we have the natural basis of unit vectors along the horizontal 'x' axis and vertical 'y' axis; but we could have easily chosen any two perpendicular directions. In fact we can use any two nonparallel vectors, although orthonormal vectors have advantages (equation (A.85)). Similarly, for the vector space of signals there is a lot of flexibility in the choice of basis; the most common choices are based on signals we have already met, namely the SUIs and the sinusoids. When we represent a signal by expanding it in the basis of SUIs we say that the signal is in the *time domain*; when we the basis of sinusoids is used we say that the signal is in the *frequency domain*.

We are not yet ready to prove that the sinusoids are a basis; this will be shown in Chapters 3 and 4. In this section we demonstrate that the SUIs are a basis, i.e., that arbitrary signals can be uniquely constructed from SUIs. We start with an example, depicted in Figure 2.10, of a digital signal that is nonzero only between times $n = 0$ and $n = 8$. We build up this signal by first taking the unit impulse $\delta_{n,0}$, multiplying it by the first signal value s_0, thereby obtaining a signal that conforms with the desired signal at time

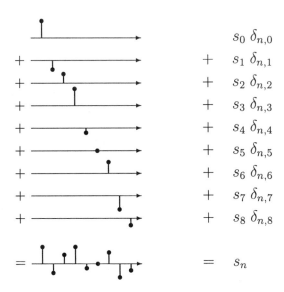

Figure 2.10: Comb-dot graph depicting building up a digital signal from shifted unit impulses.

$n = 0$ but which is zero elsewhere. Next we take the shifted unit impulse $\delta_{n,1}$, which is nonzero only for $n = 1$, and multiply it by s_1, thus obtaining a signal that agrees with s_n for $n = 1$ but is otherwise zero. Adding together these two signals we obtain a signal that is identical to the desired signal both at time $n = 0$ and at time $n = 1$ but otherwise zero. We proceed in a similar fashion to build up a signal that is identical to the desired signal for all times.

In a similar fashion we can expand *any* digital signal in terms of SUIs

$$s_n = \sum_{n=-\infty}^{\infty} s_m \delta_{n,m} \tag{2.26}$$

thus proving that these signals span the entire space. Now, it is obvious that no two SUIs overlap, and so the SUIs are orthogonal and linearly independent (no $\delta_{n,m}$ can be expanded in terms of others). Therefore the SUIs are a linearly independent set that spans the entire space, and so they are a basis.

Hence we see that the SUIs form a basis of the vector space of digital signals. Since there are (denumerably) infinite signals in the basis, we see that the vector space of signals is of infinite dimension. Similar statements are true for analog signals as well. In Figure 2.11 we demonstrate approximating a function using shifted unit width analog impulses. We leave it for the reader to complete the argument to show that any analog signal can be

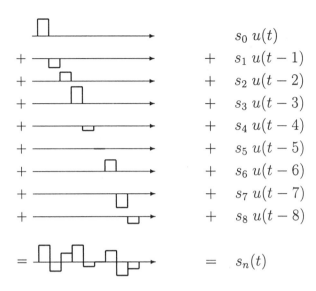

	$s_0\ u(t)$
$+$	$+\quad s_1\ u(t-1)$
$+$	$+\quad s_2\ u(t-2)$
$+$	$+\quad s_3\ u(t-3)$
$+$	$+\quad s_4\ u(t-4)$
$+$	$+\quad s_5\ u(t-5)$
$+$	$+\quad s_6\ u(t-6)$
$+$	$+\quad s_7\ u(t-7)$
$+$	$+\quad s_8\ u(t-8)$
$=$	$=\quad s_n(t)$

Figure 2.11: Building up an analog signal from shifted unit width impulses.

expanded in terms of shifted Dirac deltas. Dirac deltas are consequently a basis of a vector space of (nondenumerably) infinite dimension. The deltas (whether Kronecker or Dirac) form a basis that induces the time domain representation of the signal.

EXERCISES

2.5.1 Show that the triangle inequality is obeyed for signals.

$$\sum (s_{1_n} - s_{3_n})^2 \geq \left(\sum s_{1_n} - s_{2_n}\right)^2 + \left(\sum s_{2_n} - s_{3_n}\right)^2$$

2.5.2 Show that the set of digital signals of finite time duration is a finite dimension vector space.

2.5.3 Express a general digital signal x_n as a sum involving only the impulse at time zero and time delay operators.

2.5.4 Let's try to approximate the 3-vector $v = (v_x, v_y, v_z)$ by a vector parallel to the x axis $\alpha_x \hat{x}$. The best such approximation requires that the error vector $\epsilon = v - \alpha_x \hat{x}$ be of minimal squared length. Show that this criterion leads to $\alpha_x = v_x$ and that the error vector lies entirely in the y-z plane. Similarly, show that best approximation of v by a vector in the x-y plane $\alpha_x \hat{x} + \alpha_y \hat{y}$, requires $\alpha_x = v_x$ and $\alpha_y = v_y$, and for the error vector must be parallel to the z axis. When can the error become zero?

2.5.5 The previous exercise leads us to define the coefficients v_i as those real numbers that minimize the approximation error. Use this same approach to find the expansion of a given signal $s(t)$ in terms of a set of normalized signals $v_k(t)$, by requiring the error signal to be of minimal energy. Show that this approach demystifies the use of equation (2.19) as the dot product for signals.

2.5.6 Show how to expand analog signals in terms of shifted Dirac delta functions, by starting with Figure 2.11 and sending the impulse width to zero.

2.5.7 Explain why the set of all analog signals forms a vector space. What new features are there? What is the dimensionality of this vector space? In what sense are there more analog signals than digital ones?

2.5.8 Show that the set of all analog periodic signals with the same period is a vector space. Is it denumerably or nondenumerably infinite in dimension?

2.6   *Time* and *Frequency* Domains

According to our definition a signal is a function of a signal variable, or a singly-indexed sequence. Doesn't that mean that digital signal processing is some subset of mathematics, similar to analysis (calculus)?

Technically *yes*, of course, but in a deeper sense *not at all*. The first requirement for a signal was for it to be a physical quantity; a requirement that imparts a special flavor to signal processing, quite distinct from the seasonings with which mathematical treatments of analysis are spiced.

The differential calculus was originally invented to help in the abstract mathematical treatment of the kinematics of ideal bodies. As such, the emphasis is on derivatives and the basic functions used are polynomials. Consider the kinematical quantity $s = s_0 + v_0 t + \frac{1}{2}at^2$—this function is not a physically plausible signal as it stands, since although continuous, for large times it diverges! Physically realizable functions should remain bounded for all times, which rules out all polynomials except constants.

The fundamental law of differential calculus states that any function (well not *any* function, but we won't worry about that now) can be described in the following way. First pick some time of interest, which we will call t_0. Find the value of the function at that point, $f(t_0)$. Close enough to t_0 the function is always approximately $f(t_0)$ due to continuity constraints. To go a little further away from t_0 we need the first derivative. The first derivative describes what the function looks like close enough to t_0 since all well-behaved functions are approximately linear over a small enough interval

$f(t) \approx f(t_0) + \frac{df}{dt}|t_0 (t - t_0)$. If you want to know what the function does even further away, find the second derivative evaluated at t_0, and then the third derivative, etc. Higher and higher derivatives allow one to stray further and further from the original point in time. Knowing all derivatives at any one point in time is equivalent to knowing the function's values at all times. This law is called *Taylor's Theorem* and is the very fabric of the classical analysis way of looking at functions. It approximates functions using polynomials as the basis for the vector space of functions.

The fundamental law of signal processing proclaims a different way of representing signals. 'Real-world' signals have finite energy and occupy some finite bandwidth. Hence polynomials are not a natural basis for describing them. The signal processing approximation is global rather that local, i.e., for any finite order is about as good (or bad) simultaneously for all times $-\infty < t < +\infty$. Rather than using derivatives and polynomials, the signal processing way of looking at the world emphasizes *spectrum* and its basic signals are sinusoids. The signal processing law (the *Fourier transform*) states that all signals can be approximated by summing together basic sinusoids.

Because of this unique way of representing signals, signal processing tends to be quite schizophrenic. One has to continuously jump back and forth between the *time domain* representation, which gives the value of the signal for all times, and the *frequency domain* representation, where the harmonic content of the signal at every frequency is given.

Spectrum is simply a shorter way of saying 'frequency domain representation', and the idea is probably not new to you. You surely realize that the operation of a prism on white light consists of its decomposition into different frequencies (colors). You certainly have tuned in a station on the radio by changing the center frequency being demodulated. You may even have an audio system with a graphic equalizer enables amplifying certain component acoustic frequencies more than others.

The spectrum of a signal that consists of a pure sine wave has a single line at the frequency of this sine. The sum of two sines corresponds to two lines in the frequency domain. If the sum is weighted the relative heights of these lines will reflect this. In general, any signal that can be constructed by weighted combination of a finite number of sines will have a discrete spectrum with lines corresponding to all the frequencies and weights.

Not all signals have spectra comprised of discrete lines. For example, the analog unit width impulse has a sinc-shaped spectrum, where the sinc function

$$\text{sinc}(f) \equiv \frac{\sin(f)}{f}$$

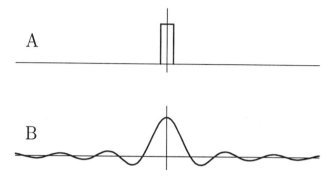

Figure 2.12: The unit width analog impulse and its spectrum. In (A) we depict the unit width impulse in the time domain, and in (B) its (sinc-function) frequency domain representation. The latter is the raw spectrum including negative frequencies.

(see Figure 2.12). The meaning of negative spectral values and negative frequencies will become clear later on. The spectrum has a strong DC component because the impulse is nonnegative. In order to make the infinitesimally sharp corners of the impulse, an infinite range of frequencies is required. So although this spectrum decreases with increasing frequency, it never becomes zero. Its bandwidth, defined as the spectral width wherein *most* of the energy is contained, is finite.

Signal processing stresses the dual nature of signals—signals have time domain and frequency domain (spectral) characteristics. Although the signal (time domain) and its Fourier transform (frequency domain) contain exactly the same information, and indeed either can be constructed from the other, some signal processing algorithms are more natural in one domain than in the other. This dual way of looking at signals is what makes signal processing different from mathematical analysis.

EXERCISES

2.6.1 Experiment with plotting signals composed of several sinusoids with various frequencies and amplitudes. Can you recognize the original frequencies in the resulting waveform? What do you observe when one sinusoid is much stronger than the others? When all the frequencies are multiples of a common frequency? When the frequencies are very close together? When they are well separated? When does the signal seem unpredictable?

2.6.2 Taylor expand a sine wave (you can do this by hand since you only need to know the derivatives of sinusoids). Fourier expand a parabola (it will probably be easiest to use numeric Fourier transform software). What can you say about the compactness of these descriptions?

2.6.3 The Taylor expansion can be interpreted as the expansion of arbitrary continuous functions in a basis of polynomials. Are the functions $f_0(x) = 1, f_1(x) = x, f_2(x) = x^2, f_3(x) = x^3, \ldots$ a basis? Are they an orthonormal basis?

2.6.4 Let's examine a more complex signal with a discrete line spectrum. The V.34 probe signal is composed of 21 sinusoids $\sin(2\pi ft + \phi)$ with frequencies f that are multiples of 150 Hz, and phases ϕ given in the following table.

f(Hz.)	ϕ(deg)	f(Hz.)	ϕ(deg)	f(Hz.)	ϕ(deg)
150	0	1500	0	2850	0
300	180	1650	180	3000	180
450	0	1950	0	3150	180
600	0	2100	0	3300	180
750	0	2250	180	3450	180
1050	0	2550	0	3600	0
1350	0	2700	180	3750	0

Plot a representative portion of the final signal. What is special about the phases in the table? (Hint: Try altering a few phases and replotting. Observe the maximum absolute value of the signal.)

2.7 Analog and Digital Domains

At the end of Section 2.1 we mentioned that one can go back and forth between analog and digital signals. A device that converts an analog signal into a digital one is aptly named an **A**nalog to **D**igital converter or *A/D* (pronounced *A to D*) for short. The reverse device is obviously a **D**igital to **A**nalog converter or *D/A* (*D to A*). You will encounter many other names, such as sampler, digitizer and codec, but we shall see that these are not entirely interchangeable. In this and the next two sections we will explain that A/D and D/A devices *can* work, leaving the details of *how* they work for the following two sections.

In explaining the function of an A/D there are two issues to be addressed, corresponding to the two axes on the graph of the analog signal in Figure 2.13. You can think of the A/D as being composed of two quantizers, the *sampler* and the *digitizer*. The sampler samples the signals at discrete times while the digitizer converts the signal values at these times to a digital representation.

Converting a continuously varying function into a discrete time sequence requires sampling the former at specific time instants. This may lead to a loss

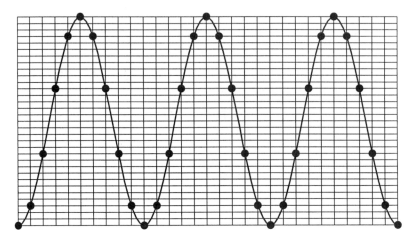

Figure 2.13: Conversion of an analog signal into a corresponding digital one involves quantizing both axes, sampling time and digitizing signal value. In the figure we see the original analog signal overlaid with the sampled time and digitized signal value grid. The resulting digital signal is depicted by the dots.

of information, since many different continuous functions can correspond to the same sampled sequence, but under certain conditions there is no such loss. The key to understanding this surprising result is the *sampling theorem.* This theorem tells us what happens when we create a discrete time signal by sampling an analog signal at a uniform rate. The sampling theorem will be discussed in the next section.

Converting the continuous real values of the analog signal into bounded digital ones requires rounding them to the nearest allowed level. This will inevitably lead to a loss of precision, which can be interpreted as adding (real-valued) noise to each value $a_n = d_n + \nu_n$, where ν_n can never exceed one half the distance to nearby quantization levels. The effect of this noise is to degrade the **S**ignal to **N**oise **R**atio (SNR) of the signal, a degradation that decreases in magnitude when the number of available levels is increased.

Digital signals obtained from analog ones are sometimes called PCM streams. Let's understand this terminology. Imagine wiping out (zeroing) the analog signal at all times that are not to be sampled. This amounts to replacing the original continuously varying signal by a sequence of pulses of varying amplitudes. We could have reached this same result in a slightly different way. We start with a train of pulses of constant amplitude. We then vary the amplitude of each incoming pulse in order to reflect the amplitude of the analog signal to be digitized. The amplitude changes of the original signal are now reflected in the varying heights of the pulses. The process of varying some aspect of a signal in order to carry information is called *modulation*. In this case we have modulated the amplitudes of the pulse stream, and so have produced **Pulse Amplitude Modulation** (PAM). Other aspects of the pulse stream could have been varied as well, resulting in **Pulse Width Modulation** (PWM), and **Pulse Position Modulation** (PPM). We now wish to digitally record the amplitude of each pulse, which we do by giving each a *code*, e.g. the binary representation of the closest quantization level. From this code we can accurately (but not necessarily precisely) reconstruct the amplitude of the pulse, and ultimately of the original signal. The resulting sequence of numbers is called a **Pulse Code Modulation** (PCM) stream.

EXERCISES

2.7.1 It would seem that sampling always gives rise to some loss of information, since it always produces gaps between the sampled time instants; but sometimes we can accurately guess how to fill in these gaps. Plot a few cycles of a sinusoid by connecting a finite number of points by straight lines (linear interpolation). How many samples per cycle are required for the plot to look natural, i.e., for linear interpolation to accurately predict the missing data? How many samples per cycles are required for the maximum error to be less than 5%? Less than 1%?

2.7.2 Drastically reduce the number of samples per cycle in the previous exercise, but generate intermediate samples using quadratic interpolation. How many true samples per cycle are required for the predictions to be reasonably accurate?

2.7.3 The sampling theorem gives a more accurate method of interpolation than the linear or quadratic interpolation of the previous exercises. However, even this method breaks down at some point. At what number of samples per cycle can *no* method of interpolation work?

2.8 Sampling

We will generally sample the analog signal at a uniform rate, corresponding to a *sampling frequency* f_s. This means that we select a signal value every $t_s = \frac{1}{f_s}$ seconds. How does t_s influence the resulting digital signal? The main effects can be observed in Figures 2.14–2.17.

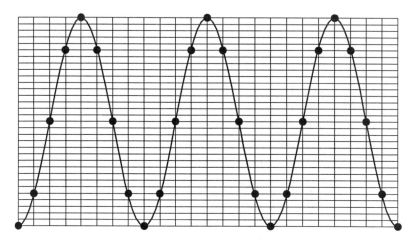

Figure 2.14: Conversion of an analog signal into the corresponding digital one with a lower sampling rate. As in the previous figure, the original analog signal has been overlaid with the sampled time and digitized signal value grid. However, the time interval between samples t_s is longer.

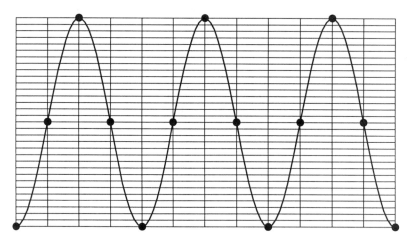

Figure 2.15: Conversion of an analog signal into the corresponding digital one with yet a lower sampling rate. Once again the original analog signal has been overlaid with the sampled time and digitized signal value grid. Although there are only four samples per cycle, the original signal is still somewhat recognizable.

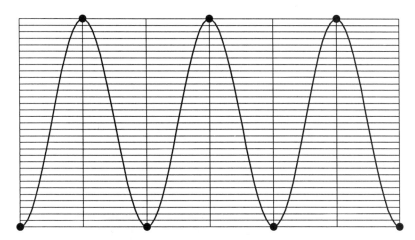

Figure 2.16: Conversion of an analog signal into a digital one at the minimal sampling rate. Once again the original analog signal has been overlaid with the sampled time and digitized signal value grid. Although there are only two samples per cycle, the frequency of the original sine wave is still retrievable.

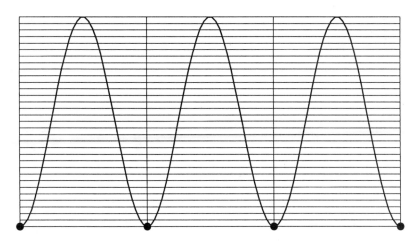

Figure 2.17: Conversion of an analog signal into a digital one at too low a sampling rate. Once again the original analog signal has been overlaid with the sampled time and digitized signal value grid. With only one sample per cycle, all information is lost.

In Figures 2.14 and 2.15 the sampling rate is eight and four samples per cycle respectively, which is high enough for the detailed shape of the signal to be clearly seen (it is a simple sinusoid). At these sampling rates even simple linear interpolation (connecting the sample points with straight lines) is not a bad approximation, although peaks will usually be somewhat truncated. In Figure 2.16, with only two samples per cycle, we can no longer make out the detailed form of the signal, but the basic frequency is discernible. With only a single sample per cycle, as in Figure 2.17, even this basic frequency is lost and the signal masquerades as DC.

Have you ever watched the wagon wheels in an old *western*? When the wagon starts to move the wheels start turning as they should; but then at some speed they anomalously seem to stand still and then start to spin backwards! Then when the coach is going faster yet they straighten out for a while. What is happening? Each second of the moving picture is composed of some number (say 25) still pictures, called frames, played in rapid succession. When the wheel is rotating slowly we can follow one spoke advancing smoothly around the axle, from frame to frame. But when the wheel is rotating somewhat faster the spoke advances so far between one frame and the next that it seems to be the next spoke, only somewhat behind. This gives the impression of retrograde rotation. When the wheel rotates exactly the speed for one spoke to move to the next spoke's position, the wheel appears to stand still.

This phenomenon, whereby sampling causes one frequency to look like a different one, is called *aliasing*. The sampled pictures are consistent with different interpretations of the continuous world, the real one now going under the alias of the apparent one. Hence in this case the sampling caused a loss of information, irreversibly distorting the signal. This is a general phenomenon. Sampling causes many analog signals to be mapped into the same digital signal. This is because the digitized signal only records the values of the continuous signal at particular times $t = nt_s$; all analog signals that agree at these points in time, but differ in between them, are aliased together to the same digital signal.

Since sampling always maps many analog signals into the same digital signal, the question arises—are there conditions under which A/D does *not* cause irreparable damage? That is, is there any way to guarantee that we will be able to recover the value of the analog signal *at all times* based on the sampled signal alone? We expect the answer to be negative. Surely the analog signal can take on arbitrary values at times not corresponding to sample periods, and therefore many different analog signals correspond to the same digital one. An affirmative answer would imply a one-to-one correspondence between analog signals obeying these conditions and the digital signals obtained by sampling them.

Surprisingly the answer *is* affirmative; but what stipulation can confound our simple logic? What restrictions can ensure that we incur no loss of information when representing a continuous function at discrete points only? What conditions on the signal will allow us to correctly guess the value of a function between two times separated by t_s where it is known? The answer is *finite bandwidth*.

Theorem: The Sampling Theorem

Assume that the analog signal $s(t)$ is sampled with a sampling frequency $f_s = 1/t_s$ producing the digital signal $s_n = s(nt_s)$.

A. If the sampling frequency is over twice that of the highest frequency component of the signal $f_s > f_{max}$, then the analog signal can be reconstructed for any desired time.

B. The reconstructed value of the analog signal at time t

$$s(t) = \sum_{n=-\infty}^{\infty} s_n \operatorname{sinc}\left(\pi f_s(t - nt_s)\right) \qquad (2.27)$$

is a linear combination of the digital signal values with $\operatorname{sinc}(t) \equiv \sin(t)/t$ weighting.   ∎

At first sight the sampling theorem seems counterintuitive. We specify the values of a signal at certain discrete instants and claim to be able to exactly predict its value at other instants. Surely the signal should be able to oscillate arbitrarily in between sampling instants, and thus be unpredictable. The explanation of this paradox is made clear by the conditions of the sampling theorem. The bandwidth limitation restricts the possible oscillations of the analog signal between the sample instants. The signal cannot do more than smoothly interpolate between these times, for to do so would require higher frequency components than it possesses.

The minimal sampling frequency (a little more than twice the highest frequency component) is called the Nyquist frequency $f_N \equiv 2f_{max}$ in honor of Harry Nyquist, the engineer who first published the requirement in 1928. It wasn't until 1949 that mathematician Claude Shannon published a formal proof of the sampling theorem and the reconstruction formula. An inaccurate, but easy to remember, formulation of the contributions of these two men is that Nyquist specified *when* an A/D can work, and Shannon dictated *how* a D/A should work.

To better understand the Nyquist criterion consider the simple case of a single sinusoid. Here the minimum sampling frequency is twice per cycle. One of these sample instants will usually be in the positive half-cycle and the in the negative one. It is just this observation of positive and negative half-cycles that makes the sampling theorem work. It is intuitively obvious that sampling at a lesser rate could not possibly be sufficient, since entire half cycles will be lost. Actually even sampling precisely twice per cycle is not sufficient, since sampling at precisely the zero or peaks conceals the half-cycles, which is what happened in Figure 2.17. This is why the sampling theorem requires us to sample at a strictly higher rate.

The catastrophe of Figure 2.17 is a special case of the more general phenomenon of *aliasing*. What the sampling theorem tells us is that discrete time signals with sampling rate f_s uniquely correspond to continuous time signals with frequency components less than $\frac{f_s}{2}$. Sampling any continuous time signal with higher-frequency components still provides a discrete time signal, but one that uniquely corresponds to another, simpler signal, called the *alias*. Figure 2.18 demonstrates how a high-frequency sinusoidal signal is aliased to a lower frequency one by sampling. The two signals agree at the sample points, but the simpler interpretation of these points is the lower-frequency signal.

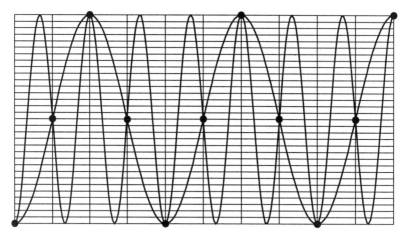

Figure 2.18: Aliasing of high-frequency analog signal into lower-frequency one. The high-frequency signal has only a sample every one and a half cycles, i.e., it corresponds to a digital frequency of $\frac{3}{4}$. The lower-frequency sinusoid is sampled at four samples per cycle, i.e., $\varphi = \frac{1}{4}$.

It is conventional to define a *digital frequency* in the following way

$$\varphi \equiv \frac{f}{f_s}$$

and the sampling theorem tells us that we must have $\varphi < \frac{1}{2}$. Consistently using this digital frequency frees us from having to think about real (analog) frequencies and aliasing. All the DSP will be exactly the same if a 2 Hz signal is sampled at 10 Hz or a 2 MHz signal is sampled at 10 MHz.

Before continuing we should mention that the sampling theorem we have been discussing is not the final word on this subject. Technically it is only the 'low-pass sampling theorem for uniform time intervals'. If the signals of interest have small bandwidth but are centered on some high frequency, it is certainly sufficient to sample at over twice the highest frequency component, but only necessary to sample at about twice the bandwidth. This is the content of the *band-pass sampling theorem*. It is also feasible in some instances to sample nonuniformly in time, for example, at times $0, \frac{1}{2}, 2, 2\frac{1}{2}, 4, \ldots$. For such cases there are 'nonuniform sampling theorems'.

Now that we understand the first half of the sampling theorem, we are ready to study the reconstruction formula in the second half. We can rewrite equation (2.27) as

$$s(t) = \sum_{n=-\infty}^{\infty} s_n h(t - nt_s) \tag{2.28}$$

where $h(t) \equiv \text{sinc}(\pi f_s t)$ is called the *sampling kernel*. As a consequence the reconstruction operation consists of placing a sampling kernel at every sample time nt_s, weighting it by the sampled value there s_n, and adding up all the contributions (see Figure 2.19). The sine in the numerator of the sinc is zero for all sample times nt_s, and hence the sampling kernel obeys $h(nt_s) = \delta_{n,0}$. From this we immediately conclude $s(nt_s) = s_n$ as required. Consequently, the reconstruction formula guarantees consistency at sample times by allowing only the correct digital signal value to contribute there. At no other times are the sampling kernels are truly zero, and the analog signal value is composed of an infinite number of contributions.

In order for the reconstruction formula to be used in practice we must somehow limit the sum in (2.28) to a finite number of contributions. Noting that the kernel $h(t)$ decays as $\frac{1}{\pi f_s t}$ we can approximate the sum by restricting the duration in time of each sample's contribution. Specifically, if we wish to take into account only terms larger than some fraction p, we should limit each sample's contributions to $\pm\frac{1}{\pi p}$ samples from its center. Conversely this restriction implies that each point in time to be interpolated will only receive

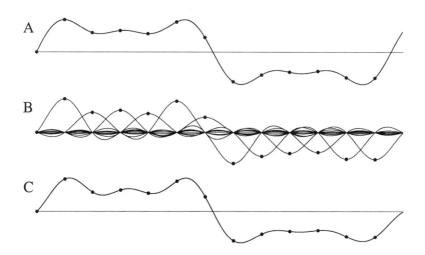

Figure 2.19: The reconstruction formula depicted graphically. In (A) we see an analog signal and the samples digitized slightly higher than twice the highest frequency component. (B) shows the sinc kernels weighted by the sample value placed at each sample time; note that at sample times all other sincs contribute zero. In (C) we sum the contributions from all kernels in the area and reconstruct the original analog signal.

a finite number of contributions (from those sample instants no further than $\frac{1}{\pi p}$ away).

Proceeding in this fashion we obtain the following algorithm:

```
Given:      a sampled signal xₙ,
            its sampling interval tₛ,
            a desired time t, and
            a cut-off fraction p
```

$w \leftarrow \text{Round}(\frac{1}{\pi p})$

Initialize $i \leftarrow 0$

$n_{mid} \leftarrow \frac{t}{t_s}$

$n_{lo} \leftarrow n_{mid} - w$

$n_{hi} \leftarrow n_{mid} + w$

$x \leftarrow 0$

for n $\leftarrow n_{lo}$ to n_{hi}

   $x \leftarrow x + x_n \, \text{sinc}(\pi \frac{t - n t_s}{t_s})$

EXERCISES

2.8.1 The wagon wheel introduced in the text demonstrates the principle of aliasing in a popular context. What is the observed frequency as a function of intended frequency.

2.8.2 Redraw Figures 2.13–2.17 with sample times at different phases of the sinusoid. Is a sine wave sampled at exactly twice per cycle (as in Figure 2.16) always recoverable?

2.8.3 Redraw Figures 2.13–2.17 with a noninteger number of samples per cycle. What new effects are observed? Are there any advantages to such sampling? Doesn't this contradict the sampling theorem?

2.8.4 Plot an analog signal composed of several sinusoids at ten times the Nyquist frequency (vastly oversampled). Overlay this plot with the plots obtained for slightly above and slightly below Nyquist. What do you observe?

2.8.5 Write a program for sampling rate conversion based on the algorithm for reconstruction of the analog signal at arbitrary times.

2.9 Digitization

Now we return to the issue of signal value quantization. For this problem, unfortunately, there is no panacea; there is no critical number of bits above which no information is lost. The more bits we allocate per sample the less noise we add to the signal. Decreasing the number of bits monotonically reduces the SNR.

Even more critical is the matching of the spacing of the quantization levels to the signal's dynamic range. Were the spacing set such that the signal resided entirely between two levels, the signal would effectively disappear upon digitizing. Assuming there are only a finite number of quantization levels, were the signal to vary over a much larger range than that occupied by the quantization levels, once again the digital representation would be close to meaningless. For the time being we will assume that the digitizer range is set to match the dynamic range of the signal (in practice the signal is usually amplified to match the range of the digitizer).

For the sake of our discussion we further assume that the analog signal is linearly digitized, corresponding to b bits. This means that we select the signal level $l = -(2^{b-1} - 1) \ldots + 2^{b-1}$ that is closest to $s(t_n)$. How does b influence the resulting digital signal? The main effects can be observed in Figures 2.20–2.24.

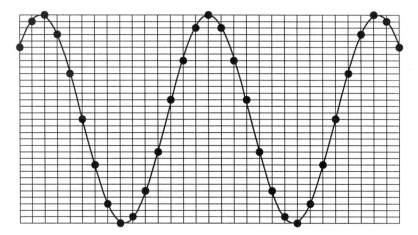

Figure 2.20: Conversion of an analog signal into a corresponding digital one involves quantizing both axes, sampling time and digitizing signal value. In the figure we see the original analog signal overlaid with the sampled time and digitized signal value grid. The resulting digital signal is depicted by the dots.

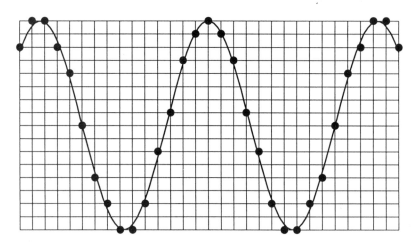

Figure 2.21: Conversion of an analog signal into the corresponding digital one with fewer digitizing levels. As in the previous figure the original analog signal has been overlaid with the sampled time and digitized signal value grid. However, here only 17 levels (about four bits) are used to represent the signal.

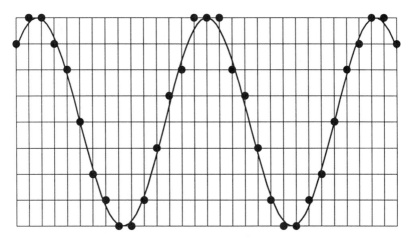

Figure 2.22: Conversion of an analog signal into the corresponding digital one with fewer digitizing levels. Once again the original analog signal has been overlaid with the sampled time and digitized signal value grid. Here only nine levels (a little more than three bits) are used to represent the signal.

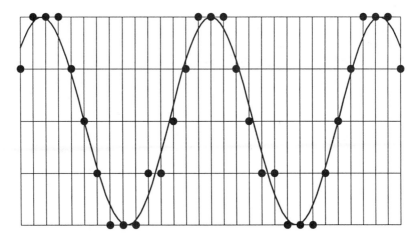

Figure 2.23: Conversion of an analog signal into the corresponding digital one with fewer digitizing levels. Once again the original analog signal has been overlaid with the sampled time and digitized signal value grid. Here only five levels (about two bits) are used to represent the signal.

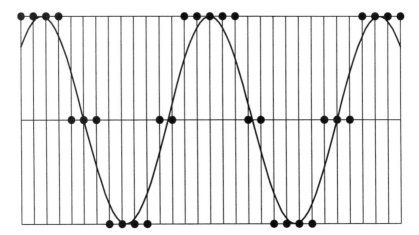

Figure 2.24: Conversion of an analog signal into the corresponding digital one with the minimum number of digitizing levels. Once again the original analog signal has been overlaid with the sampled time and digitized signal value grid. Here only three levels (one and a half bits) are used to represent the signal.

Reflect upon the discrete time signal before signal value quantization (the pulses before coding). This sequence of real numbers can be viewed as the sum of two parts

$$a_n = d_n + \nu_n \qquad \text{where} \qquad d_n \equiv \text{Round}(a_n)$$

and so d_n are integers and $|\nu_n| \leq \frac{1}{2}$. Assuming a_n to be within the range of our digitizer the result of coding is to replace a_n with d_n, thus introducing an error ν_n (see Figure 2.25). Were we to immediately reconvert the digital signal to an analog one with a D/A converter, we would obtain a signal similar to the original one, but with this noise added to the signal.

The proper way of quantifying the amount of quantization noise is to compare the signal energy with the noise energy and compute the SNR from equation (2.13). For a given analog signal strength, as the quantization levels become closer together, the relative amount of noise decreases. Alternatively, from a digital point of view, the quantization noise is always a constant $\pm \frac{1}{2}$ levels, while increasing the number of bits in the digital representation increases the digital signal value. Since each new bit doubles the number of levels and hence the digital signal value

$$\text{SNR(dB)} \approx 10 \left(\log_{10}(2^b)^2 - \log_{10} 1^2 \right) = 20b \log_{10} 2 \approx 6b \qquad (2.29)$$

that is, each bit contributes about 6 dB to the SNR. The exact relation will be derived in exercise 2.9.2.

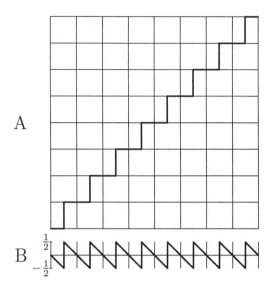

Figure 2.25: Noise created by digitizing an analog signal. In (A) we see the output of a digitizer as a function of its input. In (B) the noise is the rounding error, i.e., the output minus the input.

We have been tacitly assuming a digitizer of infinite range. In practice all digitizers have a maximum number of bits and thus a minimum and maximum level. The interval of analog signal values that are translated into valid digital values is called the *dynamic range* of the digitizer. Analog signal values outside the allowed range are clipped to the maximum or minimum permitted levels. Most digitizers have a fixed number of bits and a fixed dynamic range; in order to minimize the quantization noise the analog signal should be amplified (or attenuated) until it optimally exploits the dynamic range of the digitizer. Exceeding the dynamic range of the digitizer should be avoided as much as possible. Although moderate amounts of saturation are not usually harmful to the digitizer hardware, signal clipping is introduced. For a signal with high **P**eak to **A**verage **R**atio (PAR), one must trade off the cost of occasional clipping with the additional quantization noise.

Signal to noise ratios only have significance when the 'noise' is truly random and uncorrelated with the signal. Otherwise we could divide a noiseless signal into two equal signals and claim that one is the true signal, the other noise, and the SNR is 0 dB! We have been tacitly assuming here that the quantization noise is truly noise-like and independent of the signal, although this is clearly not the case. What is the character of this 'noise'?

Imagine continuously increasing the input to a perfect digitizer from the minimum to the maximum possible input. The output will only take quantized values, essentially rounding each input to the closest output level. Hence the output as a function of the input will produce a graph that looks like a staircase, as in Figure 2.25.A. Accordingly the rounding error, the output minus the input, will look like a sawtooth, as in Figure 2.25.B. Thus the quantization 'noise' is predictable and strongly correlated with the signal, not random and uncorrelated as we tacitly assumed. This result seems contradictory—if the noise signal is predictable, then it isn't *noise* at all. Were the error to be truly predictable, then one could always compensate for it, and digitizing would not harm the signal at all. The resolution of this paradox is simple. The noise signal is indeed correlated to the analog signal, but independent of the digitized signal. After digitizing the analog signal is unavailable, and the noise becomes, in general, unpredictable.

EXERCISES

2.9.1 Dither noise is an analog noise signal that can be added to the analog signal before digitizing in order to lessen perceived artifacts of round-off error. The dither must be strong enough to effectively eliminate spurious square wave signals, but weak enough not to overly damage the SNR. How much dither should be used? When is dither needed?

2.9.2 Refine equation (2.29) and derive SNR $= (2\log_{10} 2b + 1.8)$dB by exploiting the statistical uniformity of the error, and the definition of standard deviation.

2.9.3 Plot the round-off error as a function of time for sinusoids of amplitude 15, and frequencies 1000, 2000, 3000, 1100, 1300, 2225, and 3141.5 Hz, when sampled at 8000 samples per second and digitized to integer levels (-15, -14, ..., 0, ..., 14, 15). Does the error look noise-like?

2.10   Antialiasing and Reconstruction Filters

Recall that in Figure 1.3 there were two filters marked *antialiasing filter* and *reconstruction filter* that we avoided discussing at the time. Their purpose should now be clear. The antialiasing filter should guarantee that no frequencies over Nyquist may pass. Of course no filter is perfect, and the best we can hope for is adequate attenuation of illegal frequencies with minimal distortion of the legal ones. The reconstruction filter needs to smooth out the D/A output, which is properly defined only at the sampling instants, and recreate the proper behavior at all times. In this section we will briefly discuss these filters.

Assume that the highest frequency of importance in the signal to be sampled is f_{max}. Strictly speaking the sampling theorem allows us to sample at any frequency above the Nyquist frequency $f_N = 2f_{max}$, but in practice we can only sample this way if there is absolutely nothing above f_{max}. If there are components of the signal (albeit unimportant ones) or other signals, or even just background noise, these will fold back onto the desired signal after sampling unless removed by the antialiasing filter. Only an *ideal* antialiasing filter, one that passes perfectly all signals of frequency less than f_{max} and blocks completely all frequencies greater than f_{max}, would be able to completely remove the undesired signals; and unfortunately, as we shall learn in Section 7.1, such an ideal filter cannot be built in practice.

Realizable antialiasing filters pass low frequencies, start attenuating at some frequency f_1, and attenuate more and more strongly for higher and higher frequencies, until they effectively block all frequencies above some f_2. We must be sure that the spectral areas of interest are below f_1 since above that they will become attenuated and distorted; however, we can't use $2f_1$ as our sampling frequency since aliasing will occur. Thus in order to utilize realizable filters we must sample at a frequency $2f_2$, higher than the sampling theorem strictly requires. Typically sampling frequencies between 20% and 100% higher ($1.2f_N \leq f_s \leq 2f_N$) are used. The extra spectral 'real-estate' included in the range below $\frac{f_s}{2}$ is called a *guard band*.

The D/A reconstruction filter's purpose is slightly less obvious than that of the antialiasing filter. The output of the D/A must jump to the required digital value at the sampling time, but what should it do until the next sampling time? Since we have no information about what the analog signal does, the easiest thing to do is to stay constant until the next sampling time. Doing this we obtain a piecewise constant or 'boxcar' signal that doesn't approximate the original analog signal very well. Alternatively, we might wish to linearly interpolate between sampling points, but there are two difficulties with this tactic. First, the linear interpolation, although perhaps better looking than the boxcar, is not the proper type of interpolation from the signal processing point of view. Second, and more importantly, interpolation of any kind is noncausal, that is, requires us to know the next sample value before its time. This is impossible to implement in real-time hardware. What we *can* do is create the boxcar signal, and then filter it with an analog filter to smooth the sharp transitions and eliminate unwanted frequencies.

The antialiasing and reconstruction filters may be external circuits that the designer must supply, or may be integral to the A/D and D/A devices themselves. They may have fixed cutoff frequencies, or may be switchable, or completely programmable. Frequently DSP software must set up these filters along with initialization and setting sampling frequency of the A/D and D/A. So although we shall not mention them again, when designing, building, or programming a DSP system, don't forget your filters!

EXERCISES

2.10.1 Simulate aliasing by adding sinusoids with frequencies above Nyquist to properly sampled sinusoidal signals. (You can perform this experiment using analog signals or entirely on the computer.) Make the aliases much weaker than the desired signals. Plot the resulting signals.

2.10.2 If you have a DSP board with A/D and D/A determine how the filters are implemented. Are there filters at all or are you supposed to supply them externally? Perhaps you have a 'sigma-delta' converter that effectively has the filter built into the A/D. Is there a single compromise filter, several filters, or a programmable filter? Can you control the filters using software? Measure the antialiasing filter's response by injecting a series of sine waves of equal amplitude and increasing frequency.

2.10.3 What does speech sound like when the antialiasing filter is turned off? What about music?

2.11   Practical Analog to Digital Conversion

Although in this book we do not usually dwell on hardware topics, we will briefly discuss circuitry for A/D and D/A in this section. We have two reasons for doing this. First, the specifications of the analog hardware *are* of great important to the DSP software engineer. The DSP programmer understand what is meant by such terms as 'one-bit sampling' and 'effective bits' in order to properly design and debug software systems. Also, although we all love designing and coding advanced signal processing algorithms, much of the day-to-day DSP programming has to do with interfacing to the outside world, often by directly communicating with A/D and D/A devices. Such communication involves initializing, setting parameter values, checking status, and sending/receiving data from specific hardware components that the programmer must understand well. In addition, it is a fact of life that A/D

components occasionally fail, especially special-purpose fast A/D converters. The DSP software professional should know how to read the signs of a
failing A/D, and how to test for deficiencies and to evaluate performance.

Perhaps the simplest A/D to start with is the so-called *flash converter*,
the block diagram of which is given in Figure 2.26. The triangles marked
'comp' are *comparators* that output 'one' when the voltage applied to the *in*
input is higher than that applied to the reference input *ref*, and 'zero' otherwise. For a b bit A/D converter we require 2^b such comparators (including
the highest one to indicate an *overflow* condition). The reference inputs to
the comparators must be as precise as possible, and for this reason are often
derived from a single voltage source.

Every sampling time a voltage x is applied to the input of the digitizer.
All the comparators whose reference voltages are less than x will fire, while
those with higher references will not. This behavior reminds one of a mercury
thermometer, where the line of mercury reaches from the bottom up to a
line corresponding to the correct temperature, and therefore this encoding is
called a *thermometer code*. The thermometer code requires 2^b bits to encode
2^b values, while standard binary encoding requires only b bits. It would
accordingly be not only nonstandard but also extremely inefficient to use it
directly. The function of the block marked 'thermometer to binary decoder'
in the diagram is to convert thermometer code into standard binary. It is
left as an exercise to efficiently implement this decoder.

The main drawback of the flash converter is its excessive cost when a
large number of bits is desired. A straightforward implementation for 16-
bit resolution would require 2^{16} reference voltages and comparators and a
2^{16} by 16 decoder! We could save about half of these, at the expense of
increasing the time required to measure each voltage, by using the following
tactic. As a first step we use a single comparator to determine whether
the incoming voltage is above or below half-scale. If it is below half-scale,
we then determine its exact value by applying it to a bank of $\frac{1}{2}2^b = 2^{b-1}$
comparators. If it is above half-scale we first shift up the reference voltages
to all of these 2^{b-1} comparators by the voltage corresponding to half-scale,
and only then apply the input voltage. This method amounts to separately
determining the MSB, and requires only $2^{b-1} + 1$ comparators.

Why should we stop with determining the MSB separately? Once it
has been determined we could easily add another step to our algorithm to
determine the second most significant bit, thus reducing to $2^{b-2} + 3$ the
number of comparators needed. Continuing recursively in this fashion we
find that we now require only b stages, in each of which we find one bit,
and only b comparators in all. Of course other compromises are possible,

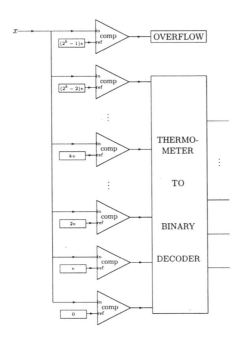

Figure 2.26: Schematic diagram of a flash converter A/D.

for example, n most significant bits can be determined by a coarse flash converter, and then the remaining $b - n$ bits by an appropriately shifted fine converter. These methods go by the name *serial-parallel* or *half-flash* converters.

In order to use such a device we would have to ensure that the input voltage remains constant during the various stages of the conversion. The time taken to measure the voltage is known as the *aperture time*. Were the voltage to fluctuate faster than the aperture time, the result would be meaningless. In order to guarantee constancy of the input for a sufficient interval a *sample and hold* circuit is used. The word 'hold' is quite descriptive of the circuit's function, that of converting the continuously varying analog signal into a piecewise constant, boxcar signal.

When a sample and hold circuit is employed, we can even reduce the number of comparators employed to one, at the expense of yet a further increase in aperture time. We simply vary the reference voltage through all the voltages in the desired range. We could discretely step the voltage through 2^b discrete levels, while at the same time incrementing a counter; the desired level is the value of this counter when the reference voltage first passes the sample and hold voltage. Stepping through 2^b levels can be a complex and time-intensive job, and can be replaced by a continuously increasing ramp. The counter is replaced by a mechanism that measures the time until the comparator triggers. A sawtooth waveform is usually utilized in order to quickly return to the starting point. This class of A/D converters is called a *counting converter* or a *slope converter.*

High-precision counting converters are by their very nature extremely slow. *Successive-approximation* converters are faster for the same reason that half-flash are faster than flash converters. The principle is to start with a steep slope, thus quickly determining a rough approximation to the input voltage. Once the reference passes the input it is reduced one level and further increased at a slower rate. This process continues until the desired number of bits has been obtained.

Now that we understand some of the principles behind the operation of real-world A/D devices, we can discuss their performance specifications. Obviously the device chosen must be able to operate at the required sampling rate, with as many bits of accuracy as further processing requires. However, bits are not always bits. Imagine a less-than-ethical hardware engineer, whose design fails to implement the require number of bits. This engineer could simply add a few more pins to his A/D chip, not connecting them to anything in particular, and claim that they are the least significant bits of the converter. Of course they turn out to be totally uncorrelated to the input signal, but that may be claimed to be a sign of noise. Conversely, if a noisy input amplifier reduces the SNR below that given by equation (2.29) we can eliminate LSBs without losing any signal-related information. A/D specifications often talk about the number of *effective bits* as distinct from the number of output bits. Effective bits are bits that one can trust, the number of bits that are truly input-signal correlated. We can find this number by reversing the use of equation (2.29).

The number of effective bits will usually decrease with the frequency of the input signal. Let's understand why this is the case. Recall that the A/D must actually observe the signal over some finite interval, known as the aperture time, in order to determine its value. For a low-frequency signal this is not problematic since the signal is essentially constant during this entire time. However, the higher the frequency the more the signal will change during this interval, giving rise to *aperture uncertainty*. Consider a pure sine wave near where it crosses the axis. The sine wave is approximately linear in this vicinity, and its slope (derivative) is proportional to the frequency.

From these considerations it is easy to see that

$$fT_{aperture} \leq 2^{-b} \qquad (2.30)$$

so that the effective bits decrease with increasing frequency.

The sigma-delta, or one-bit, digitizer is a fundamentally different kind of A/D device. Although the principles have been known for a long time, sigma-delta digitizing has become fashionable only in the last few years. This is because its implementation has only become practical (read *inexpensive*) with recent developments in VLSI practice.

With *delta-PCM* one records the differences ('delta's) between successive signal values rather than the values themselves. It is clear that given the initial value and a sequence of such differences the original signal may be recovered. Hence delta-PCM carries information equivalent to the original PCM. The desirability of this encoding is realized when the signal does not vary too rapidly from sample to sample. In this case these differences will be smaller in absolute value (and consequently require fewer bits to capture) than the signal values themselves. This principle is often exploited to compress speech, which as we shall see in Section 19.8 contains more energy at low frequencies.

When the signal does vary too much from sample to sample we will constantly overflow the number of bits we have allotted to encode the difference. To reduce the possibility of this happening we can increase the sampling rate. Each doubling of the sampling rate should reduce the absolute value of the maximum difference by a factor of two and accordingly decrease the number of bits required to encode it by one. We therefore see a trade-off between sampling frequency and bits; we can sample at Nyquist with many bits, or *oversample* with fewer bits. Considering only the number of bits produced, slower is always better; but recalling that the number of comparators required in a flash converter increases exponentially in the number of bits encoded, faster may be cheaper and more reliable. In addition there is another factor that makes an oversampled design desirable. Since we are oversampling, we can implement the antialiasing filter digitally, making it more dependable and flexible.

It would seem that we have just made our A/D more complex by requiring digital computation to be performed. However, reconstructing the original signal from its delta encoding requires digital computation in any case, and the antialiasing filter can be combined with the reconstruction. The overall computation is a summing (represented mathematically by the letter sigma) of weighted differences (deltas) and consequently these designs are called *sigma-delta* converters.

Carried to its logical extreme delta encoding can be limited to a one-bit representation of the analog signal, an encoding designated *delta modulation*. As in a conventional A/D we observe the signal at uniformly spaced intervals, but now we record only whether the signal has increased or decreased as compared to the last sampling interval. When the signal is sufficiently oversampled, and now we may require extremely high sampling frequencies, we can still recover the original signal. This is the principle behind what is advertised as *one-bit* sampling.

Before leaving our discussion of hardware for moving between analog and digital domains, we should mention D/A designs. D/A devices are in general similar to A/D ones. The first stage of the D/A is the antidigitizer (a device that converts the digital representation into an appropriate analog voltage). In principle there need be no error in such a device, since all digitized levels are certainly available in the continuous world. Next comes the antisampler, which must output the antidigitized values at the appropriate clock times. Once again this can, in principle, be done perfectly. The only quandary is what to output in between sampling instants. We could output zero, but this would require expensive quickly responding circuits, and the resulting analog signal would not really resemble the original signal at all. The easiest compromise is to output a boxcar (piecewise constant) signal, a sort of anti-sample-and-hold! The signal thus created still has a lot of 'corners' and accordingly is full of high-frequency components, and must be smoothed by an appropriate low-pass filter. This 'anti-antialiasing filter' is what we called the 'reconstruction filter' in Figure 1.3. It goes by yet a third name as well, the *sinc* filter, a name that may be understood from equation (2.27).

EXERCISES

2.11.1 Design a thermometer to binary converter circuit for an eight level digitizer (one with eight inputs and three outputs). You may only use logical gates, devices that perform the logical NOT, AND, OR, and XOR of their inputs.

2.11.2 A useful diagnostic tool for testing A/D circuits is the *level histogram*. One inputs a known signal that optimally occupies the input range and counts the number of times each level is attained. What level histogram is expected for a white noise signal? What about a sinusoid? Write a program and find the histograms for various sounds (e.g., speech, musical instruments).

2.11.3 An A/D is said to have *bad transitions* when certain levels *hog* more of the input range than they should. An A/D is said to have a *stuck bit* when an output bit is constant, not dependent on the input signal. Discuss using sawtooth and sinusoidal inputs to test for these malfunctions.

2.11.4 A signal that is too weak to be digitized can sometimes be captured using a technique known as dithering whereby a small amount of random noise is added before digitizing. Explain and demonstrate how dithering works.

2.11.5 Delta encoding is often allocated fewer bits than actually needed. In this cases we must round the differences to the nearest available level. Assuming uniform spacing of quantization levels, how much noise is introduced as a function of the number of bits. Write a program to simulate this case and try it on a speech signal. It is often the case that smaller differences are more probable than larger ones. How can we exploit this to reduce the quantization error?

2.11.6 Fixed step size delta modulation encodes only the sign of the difference between successive signal values, $d_n = \text{sgn}(s_n - s_{n-1})$, but can afford to oversample by b, the number of bits in the original digitized signal. Reconstruction of the signal involves adding or subtracting a fixed δ, according to $\hat{s}_n = \hat{s}_{n-1} + d_n\delta$. What problems arise when δ is too small or too large? Invent a method for fixes these problems and implement it.

2.11.7 Prove equation (2.30).

Bibliographical Notes

The material in this chapter is treated to some extent in all of the elementary books on DSP. Probably the first book devoted to DSP was the 1969 text by Gold and Rader [79]. The author, like many others, first learned DSP from the classical text by Oppenheim and Schafer [185] that has been updated and reissued as [186]. A more introductory text co-authored by Oppenheim is [187]. Another comprehensive textbook with a similar 'engineering approach' is by Proakis and Manolakis [200]. A very comprehensive but condensed source for almost everything related to DSP is the handbook edited by Mitra and Kaiser [241].

More accessible to non-engineers, but at a much more elementary level and covering much less ground is the book by Marven and Ewers [159]. Steiglitz has written a short but informative introductory book [252]. Finally, Mclellan, Schafer and Yoder have compiled a course for first year engineering students that includes demos and labs on a CD [167].

The Spectrum of Periodic Signals

Signals dwell both in the time and frequency domains; we can equally accurately think of them as values changing in time (time domain), or as blendings of fundamental frequencies (spectral domain). The method for determining these fundamental frequencies from the time variations is called *Fourier* or *spectral* analysis. Similar techniques allow returning to the time domain representation from the frequency domain description.

It is hard to believe that 300 years ago the very idea of *spectrum* didn't even exist, that less than 200 years ago the basic mechanism for its calculation was still controversial, and that as recently as 1965 the algorithm that made its digital computation practical almost went unpublished due to lack of interest. Fourier analysis is used so widely today that even passing mention of its most important applications is a lengthy endeavor. Fourier analysis is used in quantum physics to uncover the structure of matter on the smallest of scales, and in cosmology to study the universe as a whole. Spectroscopy and X-ray crystallography rely on Fourier analysis to analyze the chemical composition and physical structure from minute quantities of materials, and spectral analysis of light from stars tells us of the composition and temperature of bodies separated from us by light years. Engineers routinely compute Fourier transforms in the analysis of mechanical vibrations, in the acoustical design of concert halls, and in the building of aircraft and bridges. In medicine Fourier techniques are called upon to reconstruct body organs from CAT scans and MRI, to detect heart malfunctions and sleep disorders. Watson and Crick discovered the double-helix nature of DNA from data obtained using Fourier analysis. Fourier techniques can help us differentiate musical instruments made by masters from inferior copies, can assist in bringing back to life deteriorated audio recordings of great vocalists, and can help in verifying a speaker's true identity.

In this chapter we focus on the concepts of spectrum and frequency, but only for periodic signals where they are easiest to grasp. We feel that several brief historical accounts will assist in placing the basic ideas in proper context. We derive the Fourier series (FS) of a periodic signal, find the FS for various signals, and see how it can be utilized in radar signal processing. We briefly discuss its convergence and properties, as well as its major drawback, the Gibbs phenomenon. We also introduce a new notation that uses complex numbers and negative frequencies, in order to set the stage for the use of Fourier techniques in the analysis of nonperiodic signals in the next chapter.

3.1   Newton's Discovery

Isaac Newton went over to the window and shuttered it, completely darkening the room. He returned to his lab bench, eager to get on with the experiment. Although he was completely sure of the outcome, he had been waiting to complete this experiment for a long time.

The year was 1669 and Newton had just taken over the prestigious Lucasian chair at Cambridge. He had decided that the first subject of his researches and lectures would be optics, postponing his further development of the theory of fluxions (which we now call the differential calculus). During the years 1665 and 1666 Newton had been forced to live at his family's farm in Lincolnshire for months at time, due to the College being closed on account of the plague. While at home he had worked out his theory of fluxions, but he had also done something else. He had perfected a new method of grinding lenses.

While working with these lenses he had found that when white light passed through lenses it always produced colors. He finally gave up on trying to eliminate this 'chromatic aberration' and concluded (incorrectly) that the only way to make a truly good telescope was with a parabolic mirror instead of a lens. He had just built what we now call a Newtonian reflector telescope proving his theory. However, he was not pleased with the theoretical aspects of the problem. He had managed to avoid the chromatic aberration, but had not yet explained the source of the problem. Where did the colors come from?

His own theory was that white light was actually composed of all possible colors mixed together. The lenses were not creating the colors, they were simply decomposing the light into its constituents. His critics on this matter were many, and he could not risk publishing this result without iron clad proof; and this present experiment would vindicate his ideas.

He looked over the experimental setup. There were two prisms, one to break the white light into its constituent colors, and one that would hopefully combine those colors back into white light again. He had worked hard in polishing these prisms, knowing that if the experiment failed it would be because of imperfections in the glass. He carefully lit up his light source and positioned the prisms. After a little experimentation he saw what he had expected; in between the prisms was a rainbow of colors, but after the second prism the light was perfectly white. He tried blocking off various colors and observed the recomposed light's color, putting back more and more colors until the light was white again. Yes, even his most vehement detractors at the Royal society would not be able to argue with this proof.

Newton realized that the white light had all the colors in it. He thought of these colors as *ghosts* which could not normally be seen, and in his Latin write-up he actually used the word *specter*. Later generations would adopt this word into other languages as *spectrum*, meaning all of the colors of the rainbow.

Newton's next step in understanding these components of white light should have been the realization that the different colors he observed corresponded to different frequencies of radiation. Unfortunately, Newton, the greatest scientist of his era, could not make that step, due to his firm belief that light was not composed of waves. His years of experimentation with lenses led him to refute such a wave theory as proposed by others, and to assert a corpuscular theory, that light was composed of small particles. Only in the twentieth century was more of the truth finally known; light is both waves and particles, combined in a way that seventeenth-century science could not have imagined. Thus, paradoxically, Newton discovered the spectrum of light, without being able to admit that *frequency* was involved.

EXERCISES

3.1.1 Each of the colors of the rainbow is characterized by a single frequency, while artists and computer screens combine *three* basic colors. Reconcile the one-dimensional *physical* concept of frequency with the three-dimensional *psychological* concept of color.

3.1.2 Wavepackets are *particle-like* waves, that is, waves that are localized in space. For example, you can create a wavepacket by multiplying a sine wave by a Gaussian

$$s(t) = e^{\frac{(x-\mu)^2}{2\sigma^2}} \sin(\omega t)$$

where μ is the approximate location. Plot the signal in space for a given time, and in time for a given location. What is the uncertainty in the location of the 'particle'? If one wishes the 'particle' to travel at a speed v, one can substitute $\mu = vt$. What happens to the space plot now? How accurately can the velocity be measured?

3.2   Frequency Components

Consider a simple analog sinusoid. This signal may represent monochromatic light (despite Newton's prejudices), or a single tone of sound, or a simple radio wave. This signal is obviously periodic, and its *basic period* T is the time it takes to complete one cycle. The reciprocal of the basic period, $f = \frac{1}{T}$, the number of cycles it completes in a second, is called the *frequency*. Periods are usually measured in seconds per cycle and frequencies in cycles per second, or Hertz (Hz). When the period is a millisecond the frequency is a kilohertz (KHz) and a microsecond leads to a megahertz (MHz).

Why did we need the qualifier *basic* in 'basic period'? Well, a signal which is periodic with basic period T, is necessarily also periodic with period $2T$, $3T$, and all other multiples of the basic period. All we need for periodicity with period P is for $s(t + P)$ to equal $s(t)$ for all t, and this is obviously the case for periods P which contain any whole number of cycles. Hence if a sinusoid of frequency f is periodic with period P, the sinusoid with double that frequency is also periodic with period P. In general, sinusoids with period nf (where n is any integer) will all be periodic with period P. Frequencies that are related in this fashion are called *harmonics*.

A pure sine is completely specified by its frequency (or basic period), its amplitude, and its phase at time $t = 0$. For more complex periodic signals the frequency alone does not completely specify the signal; one has to specify the content of each cycle as well. There are several ways of doing this. The most straightforward would seem to require full specification of the *waveform*, that is the values of the signal in the basic period. This is feasible for digital signals, while for analog signals this would require an infinite number of values to be specified. A more sophisticated way is to recognize that complex periodic signals have, in addition to their main frequency, many other component frequencies. Specification of the contributions of all these components determines the signal. This specification is called the signal's *spectrum*.

What do we mean by frequency components? Note the following facts.

- The multiplication of a periodic signal by a number, and the addition of a constant signal, do not affect the periodicity.

- Sinusoids with period nf (where n is any integer) are all periodic with period $P = \frac{1}{f}$. These are harmonics of the basic frequency sinusoid.

- The sum of any number of signals all of which are periodic with period T, is also periodic with the same period.

From all of these facts together we can conclude that a signal that results from weighted summing of sinusoidal signals with frequencies nf, and possibly addition of a constant signal, is itself periodic with period $P = \frac{1}{f}$. Such a trigonometric series is no longer sinusoidal, indeed it can look like just about anything, but it *is* periodic. You can think of the spectrum as a recipe for preparing an arbitrary signal; the frequencies needed are the ingredients, and the weights indicate how much of each ingredient is required.

The wealth of waveforms that can be created in this fashion can be demonstrated with a few examples. In Figure 3.1 we start with a simple sine, and progressively add harmonics, each with decreased amplitude (the sine of frequency kf having amplitude $\frac{1}{k}$). On the left side we see the harmonics themselves, while the partial sums of all harmonics up to that point appear on the right. It would seem that the sum tends to a periodic sawtooth signal,

$$\sum_{k=0}^{K} \frac{\sin(k\omega t)}{k} \xrightarrow{K \to \infty} -\mathcal{T}(t) \tag{3.1}$$

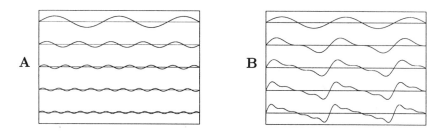

Figure 3.1: Building up a periodic sawtooth signal $-\mathcal{T}(t)$ from a sine and its harmonics. In (A) are the component sinusoids, and in (B) the composite signal.

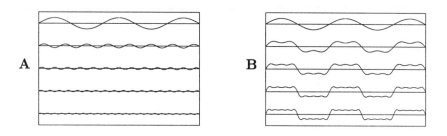

Figure 3.2: Building up a periodic square wave signal from a sine and its odd harmonics. In (A) are the component sinusoids, and in (B) the composite signal.

and this feeling is strengthened when the summation is carried out to higher harmonics. Surprisingly, when we repeat this feat with odd harmonics only we get a square wave

$$\sum_{k=0}^{K} \frac{1}{2k+1} \sin\left((2k+1)\omega t\right) \longrightarrow \Box(t) \tag{3.2}$$

as can be seen in Figure 3.2.

The signal $f(t) = \sin(\omega t)$ is an *odd* function of t, that is $f(-t) = -f(t)$. Since the sum of odd functions is odd, all signals generated by summing only harmonically related sines will be odd as well. If our problem requires an *even* function, one for which $f(-t) = f(t)$, we could sum cosines in a similar way. In order to produce a signal that is neither odd nor even, we need to sum harmonically related sines *and* cosines, which from here on we shall call **H**armonically **R**elated **S**inusoids (HRSs). In this way we can produce a huge array of general periodic signals, since any combination of sines and cosines with frequencies all multiples of some basic frequency will be periodic with that frequency.

In fact, just about anything, as long as it *is* periodic, can be represented as a trigonometric series involving harmonically related sinusoids. Just about anything, as long as it *is* periodic, can be broken down into the weighted sum of sinusoidal signals with frequencies nf, and possibly a constant signal. When first discovered, this statement surprised even the greatest of mathematicians.

EXERCISES

3.2.1 In the text we considered summing all harmonics and all odd harmonics with amplitude decreasing as $\frac{1}{n}$. Why didn't we consider all even harmonics?

3.2.2 When two sinusoids with close frequencies are added *beats* with two observable frequencies result. Explain this in terms of the arguments of this section.

3.2.3 To what waveforms do the following converge?

1. $\frac{1}{2} - \frac{4}{\pi^2}\left(\frac{\cos(x)}{1^2} + \frac{\cos(3x)}{3^2} + \frac{\cos(5x)}{5^2} + \cdots\right)$

2. $\frac{2}{\pi} - \frac{4}{\pi}\left(\frac{\cos(2x)}{1\cdot3} + \frac{\cos(4x)}{3\cdot5} + \frac{\cos(6x)}{5\cdot7} + \cdots\right)$

3. $\frac{1}{\pi} + \frac{1}{2}\sin(x) - \frac{2}{\pi}\left(\frac{\cos(2x)}{1\cdot3} + \frac{\cos(4x)}{3\cdot5} + \frac{\cos(6x)}{5\cdot7} + \cdots\right)$

4. $\frac{1}{3} - \frac{4}{\pi^2}\left(\frac{\cos(x)}{1^2} - \frac{\cos(2x)}{2^2} + \frac{\cos(3x)}{3^2} + \cdots\right)$

3.3 Fourier's Discovery

The idea of constructing complex periodic functions by summing trigonometric functions is very old; indeed it is probable that the ancient Babylonians and Egyptians used it to predict astronomical events. In the mideighteenth century this idea engendered a great deal of excitement due to its possible application to the description of vibrating strings (such as violin strings). The great eighteenth-century Swiss mathematician Leonard Euler realized that the equations for the deflection of a freely vibrating string admit sinusoidal solutions. That is, if we freeze the string's motion, we may observe a sinusoidal pattern. If the string's ends are fixed, the boundary conditions of nondeflecting endpoints requires that there be an even number of half wavelengths, as depicted in Figure 3.3. These different *modes* are accordingly harmonically related. The lowest spatial frequency has one half-wavelength in the string's length L, and so is of spatial frequency $\frac{1}{2L}$ cycles per unit length. The next completes a single cycle in L, and so is of frequency $\frac{1}{L}$. This is followed by three half cycles giving frequency $\frac{3}{2L}$, and so on. The boundary conditions ensure that all sinusoidal deflection patterns have spatial frequency that is a multiple of $\frac{1}{2L}$.

However, since the equations for the deflection of the string are linear, any linear combination of sinusoids that satisfy the boundary conditions is also a possible oscillation pattern. Consequently, a more general transverse deflection trace will be the sum of the basic modes (the sum of HRSs). The

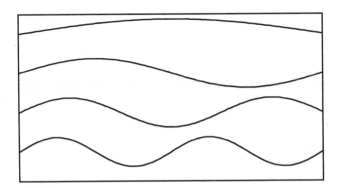

Figure 3.3: The instantaneous deflection of a vibrating string may be sinusoidal, and the boundary conditions restrict the possible frequencies of these sines. The top string contains only half of its wavelength between the string's supports; the next contains a full wavelength, the third three-quarters, etc.

question is whether this is the *most* general pattern of deflection. In the eighteenth and nineteenth century there were good reasons for suspecting the answer to be negative. Not having the benefit of the computer-generated plots of sums of HRSs presented in the previous section, even such great mathematicians as Lagrange believed that all such sums would yield *smooth* curves. However, it was easy to deform the string such that its shape would be noncontinuous (e.g., by pulling it up at its middle point forcing a triangular shape). What would happen the moment such a plucked string was released? Since the initial state was supposedly not representable in terms of the basic sinusoidal modes, there must be other, nonsinusoidal, solutions. This was considered to be a fatal blow to the utility of the theory of trigonometric series. It caused all of the mathematicians of the day to lose interest in them; all except Jean Baptiste Joseph Fourier. In his honor we are more apt today to say 'Fourier series' than 'trigonometric series'.

Although mathematics was Fourier's true interest, his training was for the military and clergy. He was sorely vexed upon reaching his twenty-first birthday without attaining the stature of Newton, but his aspirations had to wait for some time due to his involvement in the French revolution. Fourier (foolishly) openly criticized corrupt practices of officials of Robespierre's government, an act that led to his arrest and incarceration. He would have gone to the guillotine were it not for Robespierre himself having met that fate first. Fourier returned to mathematics for a time, studying at the Ecole Normal in Paris under the greatest mathematicians of the era, Lagrange and Laplace. After that school closed, he began teaching mathematics at the Ecole Polytechnique, and later succeeded Lagrange to the chair of mathematical analysis. He was considered a gifted lecturer, but as yet had made no outstanding contributions to science or mathematics.

Fourier then once again left his dreams of mathematics in order to join Napoleon's army in its invasion of Egypt. After Napoleon's loss to Nelson in the Battle of the Nile, the French troops were trapped in Egypt, and Fourier's responsibilities in the French administration in Cairo included founding of the Institut d'Egypte (of which he was secretary and member of the mathematics division), the overseeing of archaeological explorations, and the cataloging of their finds. When he finally returned to France, he resumed his post as Professor of Analysis at the Ecole Polytechnique, but Napoleon, recalling his administrative abilities, snatched him once again from the university, sending him to Grenoble as Prefect. Although Fourier was a most active Prefect, directing a number of major public works, he neglected neither his Egyptological writing nor his scientific research. His contributions to Egyptology won him election to the French Academy and to the Royal

Society in London. His most significant mathematical work is also from this period. This scientific research eventually led to his being named perpetual secretary of the Paris Academy of Sciences.

Fourier was very interested in the problem of heat propagation in solids, and in his studies derived the partial differential equation

$$\frac{\partial v}{\partial t} = K \frac{\partial^2 v}{\partial x^2}$$

now commonly known as the *diffusion equation*. The solution to such an equation is, in general, difficult, but Fourier noticed that there were solutions of the form $f(t)g(x)$, where $f(t)$ were decreasing exponentials and $g(x)$ were either $\sin(nx)$ or $\cos(nx)$. Fourier claimed that the most general $g(x)$ would therefore be a linear combination of such sinusoids

$$g(x) = \sum_{k=0}^{\infty} \Big(a_k \sin(kx) + b_k \cos(kx) \Big) \qquad (3.3)$$

the expansion known today as the Fourier series. This expansion is more general than that of Euler, allowing both sines and cosines to appear simultaneously. Basically Fourier was claiming that arbitrary functions could be written as weighted sums of the sinusoids $\sin(nx)$ and $\cos(nx)$, a result we now call Fourier's theorem.

Fourier presented his theorem to the Paris Institute in 1807, but his old mentors Lagrange and Laplace criticized it and blocked its publication. Lagrange once again brought up his old arguments based on the inability of producing nonsmooth curves by trigonometric series. Fourier eventually had to write an entire book to answer the criticisms, and only this work was ever published. However, even this book fell short of complete rigorous refutation of Lagrange's claims. The full proof of validity of Fourier's ideas was only established later by the works of mathematicians such as Dirichlet, Riemann, and Lebesgue. Today we know that all functions that obey certain conditions (known as the Dirichlet conditions), even if they have discontinuous derivatives or even if they are themselves discontinuous, have Fourier expansions.

EXERCISES

3.3.1 Consider functions $f(t)$ defined on the interval $-1 \leq t \leq 1$ that are defined by finite weighted sums of the form $\sum_k f_k \cos(\pi k t)$, where k is an integer. What do all these functions have in common? What about weighted sums of $\sin(\pi k t)$?

3.3.2 Show that any function $f(t)$ defined on the interval $-1 \leq t \leq 1$ can be written as the sum of an even function $f_e(t)$ $(f_e(-t) = f_e(-t))$ and an odd function $(f_o(-t) = -f_o(-t))$.

3.3.3 Assume that all even functions can be represented as weighted sums of cosines as in the first exercise, and that all odd functions can be similarly represented as weighted sums of sines. Explain how Fourier came to propose equation (3.3).

3.3.4 How significant is the difference between a parabola and half a period of a sinusoid? To find out, approximate $x(t) = \cos(t)$ for $-\frac{\pi}{2} \leq t \leq \frac{\pi}{2}$ by $y(t) = at^2 + bt + c$. Find the coefficients by requiring $y(-t) = y(t)$, $y(0) = 1$ and $y(\pm\frac{\pi}{2}) = 0$. Plot the cosine and its approximation. What is the maximal error? The cosine has slope 1 at the ends of the interval; what is the slope of the approximation? In order to match the slope at $t = \pm\frac{\pi}{2}$ as well, we need more degrees of freedom, so we can try $y(t) = at^4 + bt^2 + c$. Find the coefficients and the maximum error.

3.4 Representation by Fourier Series

In this section we extend our discussion of the mathematics behind the Fourier series. We will not dwell upon formal issues such as conditions for convergence of the series. Rather, we have two related tasks to perform. First, we must convince ourselves that Fourier was right, that indeed *any* function (including nonsmooth ones) can be uniquely expanded in a **Fourier Series (FS)**. This will demonstrate that the sinusoids, like the SUIs of Section 2.5, form a basis for the vector space of periodic signals with period T. The second task is a practical one. In Section 3.2 we posited a series and graphically determined the periodic signal it represented. Our second task is to find a way to do the converse operation—given the periodic signal to find the series.

In Section 2.5 we saw that *any* digital signal could be expanded in the set of all SUIs. It was left as exercises there to show that the same is true for the analog domain, and in particular for periodic analog signals. The set of all shifted analog impulses (Dirac delta functions) $\delta(t - \tau)$ forms a basis in which all analog signals may be expanded. Now, since we are dealing with periodic signals let us focus on the signal's values in the time interval between time zero and time T. It is clear that it is sufficient to employ shifted impulses for times from zero to T to recreate any waveform in this time interval.

The desired proof of a similar claim for HRSs can rest on our showing that any shifted analog impulse in the required time interval can be built up from such sinusoids. Due to the HRS's periodicity in T, the shifted impulse will automatically be replicated in time to become a periodic 'impulse train'. Consequently the following algorithm finds the HRS expansion of any function of period T.

```
focus on the interval of time from t = 0 to t = T
expand the desired signal in this interval in shifted impulses
for each impulse substitute its HRS expansion
rearrange and sort the HRS terms
consider this to be the desired expansion for all t
```

All that remains is to figure out how to represent an impulse in terms of HRSs. In Section 3.2 we experimented with adding together an infinite number of HRSs, but always with amplitudes that decreased with increasing frequency. What would happen if we used all harmonics equally?

$$b_0 + \cos(t) + \cos(2t) + \cos(3t) + \cos(4t) + \ldots \qquad (3.4)$$

At time zero all the terms contribute unity and so the infinite sum diverges. At all other values the oscillations cancel themselves out. We demonstrate graphically in Figure 3.4 that this sum converges to an impulse centered at time zero. We could similarly make an impulse centered at any desired time by using combinations of sin and cos terms. This completes the demonstration that any analog impulse centered in the basic period, and thus any periodic signal, can be expanded in the infinite set of HRSs.

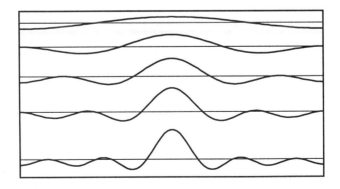

Figure 3.4: Building up an impulse from a cosine and its harmonics.

We are almost done. We have just shown that the HRSs span the vector space of periodic analog signals. In order for this set to be a basis the expansions must be unique. The usual method of proving uniqueness involves showing that there are no extraneous signals in the set, i.e., by showing that the HRSs are linearly independent. Here, however, there is a short-cut; we can show that the HRSs comprise an orthonormal set, and we know from Appendix A.14 that all orthonormal sets are linearly independent.

In Section 2.5 the dot product was shown to be a valid scalar multiplication operation for the vector space of analog signals. For periodic analog signals we needn't integrate over all times, rather the product given by

$$r = x \cdot y \qquad \text{means} \qquad r = \int_0^T x(t)\, y(t)\, dt \qquad (3.5)$$

(where the integration can actually be performed over any whole period) should be as good. Actually it is strictly better since the product over all times of finite-valued periodic signals may be infinite, while the present product always finite. Now it will be useful to try out the dot product on sinusoids.

We will need to know only a few definite integrals, all of which are derivable from equation A.34. First, the integral of any sinusoid over any number of whole periods gives zero

$$\int_0^T \sin\left(\frac{2\pi}{T}t\right) dt = 0 \qquad (3.6)$$

since $\sin(-x) = -\sin(x)$, and so for every positive contribution to the integral there is an equal and opposite negative contribution. Second, the integral of $\sin^2$ (or $\cos^2$) over a single period is

$$\int_0^T \sin^2\left(\frac{2\pi}{T}t\right) dt = \frac{T}{2} \qquad (3.7)$$

which can be derived by realizing that symmetry dictates

$$I = \int_0^T \sin^2\left(\frac{2\pi}{T}t\right) dt = \int_0^T \cos^2\left(\frac{2\pi}{T}t\right) dt$$

and so

$$2I = \int_0^T \left(\sin^2\left(\frac{2\pi}{T}t\right) + \cos^2\left(\frac{2\pi}{T}t\right)\right) dt = \int_0^T 1\, dt = T$$

by identity (A.20). Somewhat harder to guess is the fact that the integral of the product of different harmonics is always zero, i.e.

$$
\begin{aligned}
\int_0^T \sin\left(\frac{2\pi n}{T}t\right) \cos\left(\frac{2\pi m}{T}t\right) dt &= 0 \qquad \forall\, n, m > 0 \\
\int_0^T \sin\left(\frac{2\pi n}{T}t\right) \sin\left(\frac{2\pi m}{T}t\right) dt &= \delta_{n,m}\frac{T}{2} \qquad\qquad (3.8) \\
\int_0^T \cos\left(\frac{2\pi n}{T}t\right) \cos\left(\frac{2\pi m}{T}t\right) dt &= \delta_{n,m}\frac{T}{2}
\end{aligned}
$$

the proof of which is left as an exercise.

These relations tell us that the set of normalized signals $\{v_k\}_{k=1}^\infty$ defined by

$$
\begin{aligned}
v_0(t) &= \sqrt{\frac{1}{T}} \\
v_{2k+1}(t) &= \sqrt{\frac{2}{T}}\cos\left(\frac{2\pi k}{T}t\right) \qquad \forall\, k > 0 \\
v_{2k}(t) &= \sqrt{\frac{2}{T}}\sin\left(\frac{2\pi k}{T}t\right) \qquad \forall\, k > 0
\end{aligned}
$$

forms an orthonormal set of signals. Since we have proven that *any* signal of period T can be expanded in these signals, they are an orthonormal set of signals that span the space of periodic signals, and so an orthonormal basis. The $\{v_k\}$ are precisely the HRSs to within unimportant multiplicative constants, and hence the HRSs are an orthogonal basis of the periodic signals. The Fourier series takes on a new meaning. It is the expansion of an arbitrary periodic signal in terms of the orthogonal basis of HRSs.

We now return to our second task—given a periodic signal $s(t)$, we now know there is an expansion:

$$
s(t) = \sum_{k=1}^\infty c_k\, v_k(t)
$$

How do we find the expansion coefficients c_k? This task is simple due to the basis $\{v_k\}$ being orthonormal. From equation A.85 we know that for an orthonormal basis we need only to *project* the given signal onto each basis signal (using the dot product we defined above).

$$
c_k = s \cdot v = \int_0^T s(t)\, v_k(t)\, dt
$$

This will give us the coefficients for the normalized basis. To return to the usual HRSs

$$
\begin{aligned}
s(t) &= \sum_{k=1}^{\infty} a_k \sin\left(\frac{2\pi k}{T}t\right) + \sum_{k=0}^{\infty} b_k \cos\left(\frac{2\pi k}{T}t\right) \\
&= \sum_{k=1}^{\infty} a_k \sin\left(\frac{2\pi k}{T}t\right) + b_0 + \sum_{k=1}^{\infty} b_k \cos\left(\frac{2\pi k}{T}t\right) \quad (3.9)
\end{aligned}
$$

is not difficult.

$$
\begin{aligned}
a_k &= \frac{2}{T}\int_0^T s(t)\sin\left(\frac{2\pi k}{T}t\right) dt \\
b_0 &= \frac{1}{T}\int_0^T s(t)\, dt \quad\quad (3.10) \\
b_k &= \frac{2}{T}\int_0^T s(t)\cos\left(\frac{2\pi k}{T}t\right) dt
\end{aligned}
$$

This result is most fortunate; were the sinusoids not orthogonal, finding the appropriate coefficients would require solving 'normal equations' (see Appendix A.14). When there are a finite number N of basis functions, this is a set of N equations in N variables; if the basis is infinite we are not even able to write down the equations!

These expressions for the FS coefficients might seem a bit abstract, so let's see how they really work. First let's start with a simple sinusoid $s(t) = A\sin(\omega t) + B$. The basic period is $T = \frac{2\pi}{\omega}$ and so the expansion can contain only sines and cosines with periods that divide this T. The DC term is, using equations (3.6) and (3.7),

$$
b_0 = \frac{1}{T}\int_0^T s(t)\, dt = \frac{1}{T}\int_0^T \left(A\sin\left(\frac{2\pi}{T}t\right) + B \right) dt = \frac{1}{T}BT = B
$$

as expected, while from equations (3.8) all other terms are zero except for one.

$$
\begin{aligned}
a_1 &= \frac{2}{T}\int_0^T s(t)\sin\left(\frac{2\pi k}{T}t\right) dt \\
&= \frac{2}{T}\int_0^T \left(A\sin\left(\frac{2\pi}{T}t\right) + B \right) \sin\left(\frac{2\pi}{T}t\right) dt = \frac{2}{T}A\frac{T}{2} = A
\end{aligned}
$$

This result doesn't surprise us since the expansion of one of basis signals must be exactly that signal!

Slightly more interesting is the case of the square wave $\square(t/T)$. There will be no DC term nor any cosine terms, as can be seen by direct symmetry. To show this mathematically we can exploit a fact we have previously mentioned, that the domain of integration can be over any whole period. In this case it is advantageous to use the interval from $-T/2$ to $T/2$. Since $\square(t/T)$ is an odd function, i.e., $\square(-t/T) = -\square(t/T)$, the contribution from the left half interval exactly cancels out the contribution of the right half interval. This is a manifestation of a general principle; odd functions have only sine terms, while even functions have only DC and cosine term contributions. The main contribution for $\square(t/T)$ will be from the sine of period T, with coefficient

$$
\begin{aligned}
a_1 &= \frac{2}{T} \int_0^T s(t) \sin\left(\frac{2\pi}{T}t\right) dt \\
&= \frac{2}{T} \int_0^{\frac{T}{2}} \sin\left(\frac{2\pi}{T}t\right) dt - \frac{2}{T} \int_{\frac{T}{2}}^T \sin\left(\frac{2\pi}{T}t\right) dt \\
&= 2\frac{2}{T} \int_0^{\frac{T}{2}} \sin\left(\frac{2\pi}{T}t\right) dt = \frac{4}{\pi}
\end{aligned}
$$

while the sine of double this frequency

$$
a_2 = \frac{2}{T} \int_0^T s(t) \sin\left(\frac{4\pi}{T}t\right) dt = \frac{2}{T} \int_0^{\frac{T}{2}} \sin\left(\frac{4\pi}{T}t\right) dt - \frac{2}{T} \int_{\frac{T}{2}}^T \sin\left(\frac{4\pi}{T}t\right) dt = 0
$$

cannot contribute because of the odd problem once again. Therefore only odd harmonic sinusoids can appear, and for them

$$
\begin{aligned}
a_k &= \frac{2}{T} \int_0^T s(t) \sin\left(\frac{2\pi k}{T}t\right) dt \\
&= \frac{2}{T} \int_0^{\frac{T}{2}} \sin\left(\frac{2\pi k}{T}t\right) dt - \frac{2}{T} \int_{\frac{T}{2}}^T \sin\left(\frac{2\pi k}{T}t\right) dt \\
&= 2\frac{2}{T} \int_0^{\frac{T}{2}} \sin\left(\frac{2\pi k}{T}t\right) dt = \frac{4}{\pi k}
\end{aligned}
$$

which is exactly equation (3.2).

EXERCISES

3.4.1 Our proof that the HRSs span the space of periodic signals required the HRSs to be able to reproduce all SUIs, while Figure 3.4 reproduced only an impulse centered at zero. Show how to generate arbitrary SUIs (use a trigonometric sum formula).

3.4.2 Observe the sidelobes in Figure 3.4. What should the constant term b_0 be for the sidelobes to oscillate around zero? In the figure each increase in the number of cosines seems to add another half cycle of oscillation. Research numerically the number and amplitude of these oscillations by plotting the sums of larger numbers of cosines. Do they ever disappear?

3.4.3 Reproduce a graph similar to Figure 3.4 but using sines instead of cosines. Explain the results (remember that sine is an odd function). Why isn't the result simply a shifted version of cosine case?

3.4.4 Find the Fourier series coefficients for the following periodic signals. In order to check your results plot the original signal and the partial sums.

1. Sum of two sines $a_1 \sin(\omega t) + a_2 \sin(2\omega t)$
2. Triangular wave
3. Fully rectified sine $|\sin(x)|$
4. Half wave rectified sine $\sin(x) u(\sin(x))$

3.4.5 We can consider the signal $s(t) = A \sin(\omega t) + B$ to be periodic with period $T = \frac{4\pi}{\omega}$. What is the expansion now? Is there really a difference?

3.4.6 For the two-dimensional plane consider the basis made up of unit vectors along the x axis $A_1 = (1,0)$ and along the 45° diagonal $A = (\frac{1}{\sqrt{2}}, \frac{1}{\sqrt{2}})$. The unit vector of slope $\frac{1}{2}$ is $Y = (\frac{2}{\sqrt{5}}, \frac{1}{\sqrt{5}})$. Find the coefficients of the expansion $Y = \alpha_1 A_1 + \alpha_2 A_2$ by projecting Y on both A_1 and A_2 and solving the resulting equations.

3.4.7 Find explicitly the normal equations for a set of basis signals $A_k(t)$ and estimate the computational complexity of solving these equations.

3.5   Gibbs Phenomenon

Albert Abraham Michelson was the first American to receive a Nobel prize in the sciences. He is justly famous for his measurement of the speed of light and for his part in the 1887 Michelson-Morley experiment that led to the birth of the special theory of relativity. He invented the *interferometer* which allows measurement of extremely small time differences by allowing two light waves to interfere with each other. What is perhaps less known is that just after the Michelson-Morley experiment he built a practical Fourier analysis device providing a sort of physical proof of Fourier's mathematical claims regarding representation of periodic signals in terms of sinusoids. He was quite surprised when he found that the Fourier series for the square wave

$\square(t)$ didn't converge very well. In fact there was significant 'ringing', bothersome oscillations that wouldn't go away with increasing number of terms. Unsure whether he had discovered a new mathematical phenomenon or simply a bug in his analyzer he turned to the eminent American theoretical physicist of the time, Josiah Willard Gibbs. Gibbs realized that the problem was caused by discontinuities. Dirichlet had shown that the Fourier series converged to the midpoint at discontinuities, and that as long as there were a finite number of such discontinuities the series would globally converge; but no one had previously asked what happened near a discontinuity for a finite number of terms. In 1899 Gibbs published in *Nature* his explanation of what has become known as the Gibbs phenomenon.

In Section 3.3 we mentioned the Dirichlet conditions for convergence of the Fourier series.

Theorem: Dirichlet's Convergence Conditions
Given a periodic signal $s(t)$, if

1. $s(t)$ is absolutely integratable, i.e., $\int |s(t)|dt < \infty$, where the integral is over one period,

2. $s(t)$ has at most a finite number of extrema,   and

3. $s(t)$ has at most a finite number of finite discontinuities,

then the Fourier series converges for every time. At discontinuities the series converges to the midpoint. ∎

To rigorously prove Dirichlet's theorem would take us too far afield so we will just give a taste of the mathematics one would need to employ. What is necessary is an analytical expression for the partial sums $S_K(t)$ of the first K terms of the Fourier series. It is useful to define the following sum

$$D_K(t) = \tfrac{1}{2} + \cos(t) + \cos(2t) + \ldots + \cos(Kt) = \tfrac{1}{2} + \sum_{k=1}^{K} \cos(kt) \quad (3.11)$$

and to find for it an explicit expression by using trigonometric identities.

$$D_K(t) = \frac{\sin\left((K + \tfrac{1}{2})t\right)}{2\sin(\tfrac{1}{2}t)} \quad (3.12)$$

It can then be shown that for any signal $s(t)$ the partial sums equal

$$S_K(t) = \frac{2}{T} \int s(t+\tau) D_K\left(\frac{2\pi}{T}\tau\right) d\tau \quad (3.13)$$

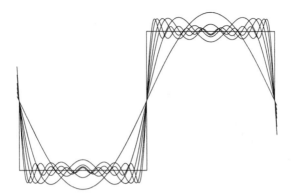

Figure 3.5: Partial sums of the Fourier series of a periodic square wave signal $\square(t)$ for $K = 0, 1, 2, 3, 5$ and 7. Note that although far from the discontinuity the series converges to the square wave, near it the overshoot remains.

(the integration being over one period of duration T) from which Dirichlet's convergence results emerge.

Now you may believe, as everyone did before Gibbs, that Dirichlet's theorem implies that amplitude of the oscillations around the true values decreases as we increase the number of terms in the series. This is the case *except* for the vicinity of a discontinuity, as can be seen in Figure 3.5. We see that close to a discontinuity the partial sums always overshoot their target, and that while the time from the discontinuity to the maximum overshoot decreases with increasing K, the overshoot amplitude does not decrease very much. This behavior does not contradict Dirichlet's theorem since although points close to jump discontinuities may initially be affected by the overshoot, after enough terms have been summed the overshoot will pass them and the error will decay.

For concreteness think of the square wave $\square(t)$. For positive times close to the discontinuity at $t = 0$ equation (3.13) can be approximated by

$$S_K(t) = \frac{2}{\pi}\, \mathrm{sgn}(t)\, \mathrm{Sinc}\,(4\pi K|t|) \tag{3.14}$$

as depicted in Figure 3.6. Sinc is the sine integral.

$$\mathrm{Sinc}(t) = \int_0^t \mathrm{sinc}(\tau)\,d\tau$$

Sinc approaches $\frac{\pi}{2}$ for large arguments, and thus $S_K(t)$ *does* approach unity for large K and/or t. The maximum amplitude of Sinc occurs when its derivative (sinc) is zero, i.e., when its argument is π. It is not hard to find

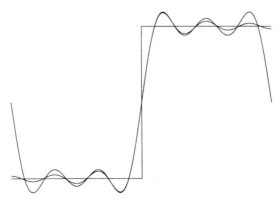

Figure 3.6: Gibbs phenomenon for the discontinuity of the square wave at $t = 0$. Plotted are the square wave, the partial sum with $K = 3$ terms, and the approximation using the sine integral.

numerically that for large K this leads to an overshoot of approximately 0.18, or a little less than 9% of the height of the jump. Also, the sine integral decays to its limiting value like $\frac{1}{t}$; hence with every doubling of distance from the discontinuity the amplitude of the oscillation is halved. We derived these results for the step function, but it is easy to see that they carry over to a general jump discontinuity.

That's what the mathematics says, but what does it mean? The oscillations themselves are not surprising, this is the best way to smoothly approximate a signal—sometimes too high, sometimes too low. As long as these oscillations rapidly die out with increasing number of terms the approximation can be considered good. What do we expect to happen near a discontinuity? The more rapid a change in the signal in the time domain is, the wider the bandwidth will be in the frequency domain. In fact the uncertainty theorem (to be discussed in Section 4.4) tells us that the required bandwidth is inversely proportional to the transition time. A discontinuous jump requires an infinite bandwidth and thus no combination of a finite number of frequencies, no matter how many frequencies are included, can do it justice. Of course the coefficients of the frequency components of the square wave *do* decrease very rapidly with increasing frequency. Hence by including more and more components, that is, by using higher and higher bandwidth, signal values closer and closer to the discontinuity, approach their proper values. However, when we approximate a discontinuity using bandwidth BW, within about $1/BW$ of the discontinuity the approximation cannot possibly approach the true signal.

We can now summarize the Gibbs phenomenon. Whenever a signal has a jump discontinuity its Fourier series converges at the jump time to the midpoint of the jump. The partial sums display oscillations before and after the jump, the number of cycles of oscillation being equal to the number of terms taken in the series. The size of the overshoot decreases somewhat with the number of terms, approaching about 9% of the size of the jump. The amplitude of the oscillations decreases as one moves away from the discontinuity, halving in amplitude with every doubling of distance.

EXERCISES

3.5.1 Numerically integrate sinc(t) and plot Sinc(t). Show that it approaches $\pm\frac{\pi}{2}$ for large absolute values. Find the maximum amplitude. Where does it occur? Verify that the asymptotic behavior of the amplitude is $\frac{1}{t}$.

3.5.2 The following exercises are for the mathematically inclined. Prove equation (3.12) by term-by-term multiplication of the sum in the definition of $D_K(t)$ by $\sin\left(\frac{t}{2}\right)$ and using trigonometric identity (A.32).

3.5.3 Prove equation (3.13) and show Dirichlet's convergence results.

3.5.4 Prove the approximation (3.14).

3.5.5 Lanczos proposed suppressing the Gibbs phenomenon in the partial sum S_K by multiplying the kth Fourier coefficient (except the DC) by sinc$\left(\frac{\pi k}{2K}\right)$. Try this for the square wave. How much does it help? Why does it help?

3.5.6 We concentrated on the Gibbs phenomenon for the square wave. How do we know that other periodic signals with discontinuities act similarly? (Hint: Consider the Fourier series for $s(t) + a\square(t)$ where $s(t)$ is a continuous signal and a a constant.)

3.6 Complex FS and Negative Frequencies

The good news about the Fourier series as we have developed it is that its basis signals are the familiar *sine* and *cosine* functions. The bad news is that its basis signals are the familiar sine *and* cosine functions. The fact that there are two different kinds of basis functions, and that the DC term is somewhat special, makes the FS as we have presented it somewhat clumsy to use. Unfortunately, sines alone span only the subspace composed of all odd signals, while cosines alone span only the subspace of all even signals.

Signals which are neither odd nor even. truly require combinations of both Since the FS in equation (3.9) includes for every frequency both a sine and cosine function (which differ by 90° or a quarter cycle), it is said to be in quadrature form.

The first signal space basis we studied, the SUI basis, required only one functional form. Is there a single set of sinusoidal signals, all of the same type, that forms a basis for the space of periodic signals? Well, for each frequency component ω the FS consists of the sum of two terms $a\cos(\omega t) + b\sin(\omega t)$. Such a sum produces a pure sinusoid of the same frequency, but with some phase offset $d\sin(\omega t + \varphi)$. In fact, it is easy to show that

$$a_k \sin(\omega t) + b_k \cos(\omega t) = d_k \sin(\omega t + \varphi_k) \tag{3.15}$$

as long as

$$d_k = \sqrt{a_k^2 + b_k^2} \qquad \varphi_k = \tan^{-1}(b_k, a_k) \tag{3.16}$$

where the arctangent is the full four-quadrant function, and

$$a_k = d_k \cos \varphi_k \qquad b_k = d_k \sin \varphi_k \tag{3.17}$$

in the other direction.

As a result we *can* expand periodic signals $s(t)$ as

$$s(t) = d_0 + \sum_{k=0}^{\infty} d_k \sin\left(\frac{2\pi k}{T}t + \varphi_k\right) \tag{3.18}$$

with both amplitudes and phases being parameters to be determined.

The amplitude and phase form is intellectually more satisfying than the quadrature one. It represents every periodic signal in terms of harmonic frequency components, each with characteristic amplitude and phase. This is more comprehensible than representing a signal in terms of pairs of sinusoids in quadrature. Also, we are often only interested in the *power spectrum*, which is the amount of energy in each harmonic frequency. This is given by $|d_k|^2$ with the phases ignored.

There *are* drawbacks to the amplitude and phase representation. Chief among them are the lack of symmetry between d_k and φ_k and the lack of simple formulas for these coefficients. In fact, the standard method to calculate d_k and φ_k is to find a_k and b_k and use equations (3.16)!

We therefore return to our original question: Is there a single set of sinusoidal signals, all of the same type, that forms a basis for the space of periodic signals and that can be calculated quickly and with resort to the quadrature representation? The answer turns out to be affirmative.

To find this new representation recall the connection between sinusoids and complex exponentials of equation (A.8).

$$\cos(\omega t) = \frac{1}{2}\left(e^{i\omega t} + e^{-i\omega t}\right) \qquad\qquad \sin(\omega t) = \frac{1}{2i}\left(e^{i\omega t} - e^{-i\omega t}\right) \qquad (3.19)$$

We can think of the exponents with positive $e^{i\omega t}$ and negative $e^{-i\omega t}$ exponents as a single type of exponential $e^{i\omega t}$ with positive and negative frequencies ω. Using only such complex exponentials, although of both positive and negative frequencies, we can produce both the sine and cosine signals of the quadrature representation, and accordingly represent any periodic signal.

$$s(t) = \sum_{k=-\infty1}^{\infty} c_k e^{i\frac{2\pi k}{T}t} \qquad (3.20)$$

We *could* once again derive the expression for the coefficients c_k from those for the quadrature representation, but it is simple enough to derive them from scratch. We need to know only a single integral.

$$\int_0^T e^{i\frac{2\pi n}{T}t}\, e^{-i\frac{2\pi m}{T}t}\, dt = \delta_{m,n}T \qquad (3.21)$$

This shows that the complex exponentials are orthogonal with respect to the dot product for complex signals

$$s_1 \cdot s_2 = \int_0^T s_1(t)\, s_2^*(t)\, dt \qquad (3.22)$$

and that

$$v_k(t) = \frac{1}{\sqrt{T}}\, e^{i\frac{2\pi k}{T}t} \qquad (3.23)$$

form a (complex) orthonormal set. From this it is easy to see that

$$c_k(t) = \frac{1}{T}\int_0^T s(t)\, e^{-i\frac{2\pi k}{T}t}\, dt \qquad (3.24)$$

with a minus sign appearing in the exponent. Thus Fourier's theorem can be stated in a new form: All periodic functions (which obey certain conditions) can be written as weighted sums of complex exponentials.

The complex exponential form of the FS is mathematically the simplest possible. There is only one type of function, one kind of coefficient, and there is strong symmetry between equations (3.20) and (3.24) that makes them easier to remember. The price to pay has been the introduction of mysterious negative frequencies. What do we mean by -100 Hz? How can something cycle *minus 100* times per second?

Physically, negative frequency signals are almost identical to their positive counterparts, since only the real part of a complex signal counts. Recall the pen-flashlight experiment that you were requested to perform in exercise 2.2.6. The complex exponential corresponds to observing the flashlight head-on, while the real sinusoid is observing it from the side. Rotation of the light in clockwise or counterclockwise (corresponding to positive or negative frequencies) produces the same effect on an observer who perceives just the vertical (real) component; only an observer with a full view notices the difference. However, it would be foolhardy to conclude that negative frequencies are of no importance; when more than one signal is present the relative phases are crucial.

We conclude this section with the computation of a simple complex exponential FS—that of a real sinusoid. Let $s(t) = A\cos(\frac{2\pi k}{T}t)$. The period is of course T, and

$$c_k = \frac{1}{T}\int_0^T A\cos(\frac{2\pi}{T}t)\, e^{-i\frac{2\pi k}{T}t}\, dt = \frac{A}{T}\int_0^T \frac{1}{2}\left(e^{i\frac{2\pi}{T}t} + e^{-i\frac{2\pi}{T}t}\right) e^{-i\frac{2\pi k}{T}t}\, dt$$

which after using the orthogonality relation (3.21) leaves two terms.

$$c_k = \frac{A}{2T}\delta_{k,-1} + \frac{A}{2T}\delta_{k,+1}$$

This is exactly what we expected considering equation (3.19). Had we chosen $s(t) = A\sin(\frac{2\pi k}{T}t)$ we would have still found two terms with identical k and amplitudes but with phases shifted by 90°. This is hardly surprising; indeed it is easy to see that all $s(t) = A\cos(\frac{2\pi k}{T}t + \varphi)$ will have the same FS except for phase shifts of φ. Such constant phase shifts are meaningless, there being no meaning to absolute phase, only to changes in phase.

EXERCISES

3.6.1 Plot $\sin(x) + \sin(2x + \varphi)$ with $\varphi = 0, \frac{\pi}{2}, \pi, \frac{2\pi}{2}$. What can you say about the effect of phase? Change the phases in the Fourier series for a square wave. What signals can you make?

3.6.2 Derive all the relations between coefficients of the quadrature, amplitude and phase, and complex exponential representations. In other words, show how to obtain a_k and b_k from c_k and vice versa; a_k and b_k from d_k and vice versa; c_k from d_k and vice versa. In your proofs use only trigonometric identities and equation (A.7).

3.6.3 Prove equation (3.21).

3.6.4 Calculate the complex exponential FS of $s(t) = A\sin(\frac{2\pi k}{T}t)$. How does it differ from that of the cosine?

3.6.5 Consistency requires that substituting equation (3.20) for the FS into equation (3.24) for c_k should bring us to an identity. Show this using (3.21). What new expression for the delta function is implied by the reverse consistency argument?

3.6.6 What transformations can be performed on a signal without effecting its power spectrum $|c_k|^2$? What is the physical meaning of such transformations?

3.7   Properties of Fourier Series

In this section we continue our study of Fourier series. We will exclusively use the complex exponential representation of the FS since it is simplest, and in any case we can always convert to other representations if the need arises.

The first property, which is obvious from the expression for c_k, is linearity. Assume $s_1(t)$ has FS coefficients c_k^1 and $s_2(t)$ has coefficients c_k^2, then $s(t) = As_1(t) + Bs_2(t)$ has as its coefficients $c_k = Ac_k^1 + Bc_k^2$. This property is often useful in simplifying calculations, and indeed we already implicitly used it in our calculation of the FS of $\cos(\omega t) = \frac{1}{2}e^{i\omega t} + \frac{1}{2}e^{-i\omega t}$. As a further example, suppose that we need to find the FS of a constant (DC) term plus a sinusoid. We can immediately conclude that there will be exactly three nonzero c_k terms, c_{-l}, c_0, and c_{+l}.

In addition to its being used as a purely computational ploy, the linearity of c_k has theoretic significance. The world would be a completely different place were the FS not to be linear. Were the FS of $As(t)$ not to be Ac_k then simple amplification would change the observed harmonic content of a signal. Linear operators have various other desirable features. For example, small changes to the input of a linear operator can only cause bounded changes to the output. In our case this means that were one to slightly perturb a signal with known FS, there is a limit to how much c_k can change.

The next property of interest is the effect of time shifts on the FS. By time shift we mean replacing t by $t - \tau$, which is equivalent to resetting our clock to read zero at time τ. Since the time we start our clock is arbitrary such time shifts cannot alter any physical aspects of the signal being studied. Once again going back to the expression for c_k we find that the FS of $s(t-\tau)$

is $e^{-i\frac{2\pi k}{T}} c_k$. The coefficients magnitudes are unchanged, but the phases have been linearly shifted. As we know from exercise 3.6.6 such phase shifts do not change the power spectrum but still may be significant. We see here that phase shifts that are linear in frequency correspond to time shifts.

When a transformation leaves a signal unchanged or changes it in some simple way we call that transformation a *symmetry*. Time shift is one interesting symmetry, and another is time reversal Rev s. Although the import of the latter is less compelling than the former many physical operations are unchanged by time reversal. It is not difficult to show that the effect of time reversal is to reverse the FS to c_{-k}.

The next property of importance was discovered by Parseval and tells us how the energy can be recovered from the FS coefficients.

$$E = \frac{1}{T} \int_0^T |s(t)|^2 \, dt = \sum_{k=-\infty}^{\infty} |c_k|^2 \tag{3.25}$$

What does Parseval's relation mean? The left hand side is the power computed over a single period of the periodic signal. The power of the sum of two signals equals the sum of the powers if and only if the signals are orthogonal.

$$\begin{aligned}
\frac{1}{T} \int_0^T |x(t) + y(t)|^2 \, dt &= \frac{1}{T} \int_0^T \left(x(t) + y(t) \right)^* \left(x(t) + y(t) \right) \, dt \\
&= \frac{1}{T} \int_0^T |x(t)|^2 + |y(t)|^2 + 2 \Re \left(x^*(t) y(t) \right) \, dt
\end{aligned}$$

Since any two different sinusoids are uncorrelated, their powers add, and this can be generalized to the sum of any number of sinusoids. So Parseval's relation is another consequence of the fact that sinusoids are orthogonal.

For complex valued signals $s(t)$ there is a relation between the FS of the signal and that of its complex conjugate $s^*(t)$. The FS of the complex conjugate is c_{-k}^*. For real signals this implies a symmetry of c_k (i.e., $c_{-k} = c_k^*$), which means $|c_{-k}| = |c_k|$ and $\Re(c_{-k}) = \Re(c_k)$ but $\Im(c_{-k}) = -\Im(c_k)$.

There are many more symmetries and relations that can be derived for the FS, e.g., the relationship between the FS of a signal and those of its derivative and integral. There is also an important rule for the FS of the product of two signals, which the reader is not yet ready to digest.

EXERCISES

3.7.1 Show that adding to the argument of a sinusoid a phase that varies linearly with time shifts its frequency by a constant. Relate this to the time shift property of the FS.

3.7.2 Plot the sum of several sinusoids with various phases. Demonstrate that a linear phase shift causes a time shift. Can you tell that all these signals have the same power spectrum?

3.7.3 How does change of time scale $s(\alpha t)$ affect c_k? Prove that the effect of time reversal is to reverse the FS.

3.7.4 Derive Parseval's relation for the FS.

3.7.5 Show that if a signal is symmetric(antisymmetric), i.e., if $s(t + \frac{T}{2}) = \pm s(t)$, then its FS contains only even (odd) harmonics.

3.7.6 The FS of s is c_k; what is the FS of its derivative? Its integral?

3.8   The Fourier Series of Rectangular Wave

Since we have decided to use the complex exponential representation almost exclusively, we really should try it out. First, we want to introduce a slightly different notation. When we are dealing with several signals at a time, say $q(t)$, $r(t)$, and $s(t)$, using c_k for the FS coefficients of all of them, would be confusing to say the least. Since the Fourier coefficients contain exactly the same information as the periodic signal, using the name of the signal, as in q_k, r_k, or S_k, would be justified. There won't be any confusion since $s(t)$ is continuous and S_k is discrete; however, later we will deal with continuous spectra where it wouldn't be clear. So most people prefer to capitalize the Fourier coefficients, i.e., to use Q_k, R_k, and S_K, in order to emphasize the distinction between time and frequency domains. Hence from now on we shall use

$$S_k = \frac{1}{T} \int s(t) \, e^{-i\frac{2\pi k}{T}t} \, dt \tag{3.26}$$

(with the integration over any full period) to go from a signal $s(t)$ to its FS $\{S_k\}_{k=-\infty}^{\infty}$, and

$$s(t) = \sum_{k=-\infty}^{\infty} S_k e^{i\frac{2\pi k}{T}t} \tag{3.27}$$

to get back again.

Now to work. We have already derived the FS of a square wave, at least in the quadrature representation. Here we wish to extend this result to the slightly more general case of a rectangular wave, i.e., a periodic signal that does not necessarily spend half of its time at each level. The fraction of time

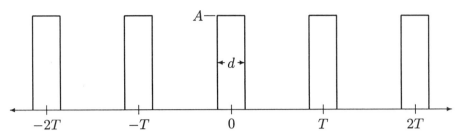

Figure 3.7: The rectangular signal with amplitude A, period T, and duty cycle $\delta = \frac{d}{T}$.

a rectangular wave spends in the higher of its levels is called its *duty cycle* $\delta = \frac{d}{T}$, and a rectangular wave with $\delta = \frac{1}{2}$ duty cycle is a square wave. We also wish to make the amplitude and period explicit, and to have the signal more symmetric in the time domain; we accordingly introduce A, T, and $d = \delta T$, and require the signal to be high from $-\frac{d}{2}$ to $\frac{d}{2}$. Unlike the square wave, a non-50% duty cycle rectangular signal will always have a DC component. There is consequently no reason for keeping the levels symmetric around zero, and we will use 0 and A rather than $\pm A$.

Thus we will study

$$s(t) = A \begin{cases} 1 & |\text{frac}(\frac{t}{T})| < \frac{d}{2} \\ 0 & \frac{d}{2} < |\text{frac}(\frac{t}{T})| < T - \frac{d}{2} \\ 1 & T - \frac{d}{2} < |\text{frac}(\frac{t}{T})| < T \end{cases} \qquad (3.28)$$

(where $\text{frac}(x)$ is the fractional part of x) as depicted in Figure 3.7.

The period is T and therefore the angular frequencies in the Fourier series will all be of the form $\omega_k = \frac{2\pi}{T}k$. We can choose the interval of integration in equation (3.24) as we desire, as long as it encompasses a complete period. The most symmetric choice here is from $-\frac{T}{2}$ to $\frac{T}{2}$, since the signal then becomes simply

$$s(t) = A \begin{cases} 1 & |(\frac{t}{T})| < \frac{d}{2} \\ 0 & \text{else} \end{cases} \qquad (3.29)$$

and as a consequence

$$\begin{aligned} S_k &= \frac{1}{T} \int_{-\frac{T}{2}}^{\frac{T}{2}} s(t)\, e^{-i\frac{2\pi k}{T}t}\, dt \\ &= \frac{A}{T} \int_{-\frac{d}{2}}^{\frac{d}{2}} e^{-i\omega_k t}\, dt \end{aligned}$$

which after change of variable and use of equation (A.8) becomes

$$S_k = A\frac{\sin(\frac{\omega_k d}{2})}{\frac{\omega_k d}{2}} = A\,\mathrm{sinc}\left(\frac{\omega_k d}{2}\right) = A\,\mathrm{sinc}\left(\frac{\pi k d}{T}\right) = A\,\mathrm{sinc}(\pi k\delta) \qquad (3.30)$$

where we have recognized our old friend sinc. The FS is dependent only on the duty cycle, not directly on T. Of course this does not mean that the Fourier series is not dependent on T! The coefficient S_k multiplies the term containing $\omega_k = \frac{2\pi k}{T}$, and consequently the distribution on the frequency axis indeed changes. Taking into account this meaning of S_k we see that the spectral envelope is influenced by the pulse width but not the period.

The main lobe of the sinc function is between $-\pi$ and π, which here means between $\delta k = -1$ and $\delta k = 1$. Hence most of the energy is between $\omega_k = \frac{2\pi k}{T} = \pm\frac{2\pi}{\delta T}$, or otherwise stated, the frequency spread is $\Delta\omega = \frac{4\pi}{\delta T}$. The minimum spacing between two points in time that represent the same point on the periodic signal is obviously $\Delta t = T$. The relationship between the time and frequency spreads can therefore be expressed as

$$\Delta\omega\Delta t = \frac{4\pi}{\delta} \qquad (3.31)$$

which is called the 'time-frequency uncertainty product'. The effect of varying the duty cycle δ at constant period T is demonstrated in Figure 3.8. As δ is decreased the width of the spectrum increases (i.e., the spectral amplitudes become more constant) until finally at zero duty cycle (the signal being a periodic train of impulses) all the amplitudes are equal. If the duty cycle is increased to one (the signal becoming a constant $s(t) = A$), only the DC component remains nonzero.

What happens when the period T is increased, with δ constant? We know that the wider the spacing in the time domain, the narrower the spacing of the frequency components will be. The constancy of the time-frequency uncertainty product tells us that the extent of the sinc function on the frequency axis doesn't change, just the frequency resolution. This is demonstrated in Figure 3.9.

These characteristics of the FS of a rectangular wave are important in the design of pulse radar systems. We will discuss radar in more detail in Section 5.3, for now it is sufficient to assume the following simplistic model. The radar transmits a periodic train of short duration pulses, the period of which is called the **P**ulse **R**epetition **I**nterval (PRI); the reciprocal of the PRI is called the **P**ulse **R**epetition **F**requency (PRF).

This transmitted radar signal is reflected by a target and received back at the radar at this same PRI but offset by the round-trip time. Dividing

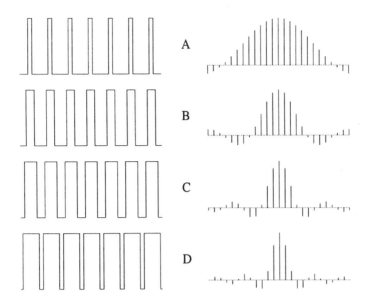

Figure 3.8: The effect of changing the duty cycle at constant period. In these figures we see on the left a periodic rectangular signal, and on the right the absolute squares of its FS amplitudes represented as vertical bars placed at the appropriate frequencies. (A) represents a duty cycle of 20%, (B) 40%, (C) 60% and (D) 80%. Note that when the duty cycle vanishes all amplitudes become equal, while when the signal becomes a constant, only the DC term remains.

the time offset by two and multiplying by the speed of radar waves (the speed of light c) we obtain the distance from radar to target. The round-trip time should be kept lower than the PRI; and echo returning after precisely the PRI is not received since the radar receiver is 'blanked' during transmission; if the round-trip time exceeds the PRI we get aliasing, just as in sampling analog signals. Hence we generally strive to use long PRIs so that the distance to even remote targets can be unambiguously determined. More sophisticated radars vary the PRI from pulse to pulse in order to disambiguate the range while keeping the echo from returning precisely when the next pulse is to be transmitted.

Due to the Doppler effect, the PRF of the reflection from target moving at velocity v is shifted from its nominal value.

$$\Delta \mathrm{PRF} = \mathrm{PRF}\,\frac{v}{c} \qquad (3.32)$$

An approaching target is observed with PRF *higher* than that transmitted, while a receding target has a *lower* PRF. The PRF is conveniently found

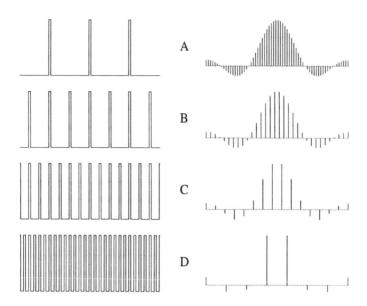

Figure 3.9: The effect of changing the period at constant duty cycle. In these figures we see on the left a periodic rectangular signal, and on the right the absolute squares of its FS amplitudes represented as vertical bars placed at the appropriate frequencies. As we progress from (A) through (D) the period is halved each time. Note that as the period is decreased with constant pulse width the frequency resolution decreases but the underlying sinc is unchanged.

using Fourier analysis techniques, with precise frequency determination favoring high PRF. Since the requirements of unambiguous range (high PRI) and precise velocity (high PRF) are mutually incompatible, simple pulse radars can not provide both simultaneously.

The radar signal is roughly a low duty cycle rectangular wave, and so its FS is approximately that of Figures 3.8 and 3.9. In order to maximize the probability of detecting the echo, we endeavor to transmit as much energy as possible, and thus desire wider pulses and higher duty cycles. Higher duty cycles entail both longer receiver blanking times and narrower sinc functions in the frequency domain. The former problem is easily understood but the latter may be more damaging. In the presence of interfering signals, such as reflections from 'clutter', intentional jamming, and coincidental use of the same spectral region by other services, the loss of significant spectral lines results in reduced target detection capability.

EXERCISES

3.8.1 Show how to regain the Fourier series of the square wave (equation (3.2)), from (3.30) by taking a 50% duty cycle.

3.8.2 We assumed that A in equation (3.28) was constant, independent of T and d. Alternative choices are also of interest. One could demand that the basic rectangle be of unit area $A = \frac{1}{d}$, or of unit energy $A = \frac{1}{\sqrt{d}}$, or that the power (energy per time) be unity $A = \frac{T}{\sqrt{d}}$. Explain the effect of the different choices on the signal and its FS when δ and T are varied.

3.8.3 Show that the FS of a train of impulses $s(t) = \sum \delta(t - kT)$ is a train of impulses in the frequency domain. How does this relate to the calculations of this section? To which choice of A does this correspond?

3.8.4 One technique that radar designers use to disambiguate longer ranges is PRI staggering. Staggering involves alternating between several PRIs. How does staggering help disambiguate? How should the PRIs be chosen to maximize the range? (Hint: Use the Chinese remainder theorem.)

3.8.5 What is the FS of a rectangular wave with stagger two (i.e., alternation between two periods T_1 and T_2)?

Bibliographical Notes

For historical background to the development of the concept of frequency consult [223]. Newton's account of the breaking up of white light into a spectrum of colors can be read in his book *Opticks* [179]. For more information on the colorful life of Fourier consult [83]. Incidentally, Marc Antoine Parseval was a royalist, who had to flee France for a while to avoid arrest by Napoleon. Lord Rayleigh, in his influential 1877 book on the theory of sound (started interestingly enough on a vacation to Egypt where Fourier lived eighty years earlier), was perhaps the first to call the trigonometric series by the name 'Fourier series'. Gibbs' presentation of his phenomenon is [74].

There are many books devoted entirely to Fourier series and transforms. To get more practice in the mechanics of Fourier analysis try [104]. In-depth discussion of the Dirichlet conditions can be found in the mathematical literature on Fourier analysis.

The Frequency Domain

The concept of frequency is clearest for simple sinusoids, but we saw in the previous chapter that it can be useful for nonsinusoidal periodic signals as well. The Fourier series is a useful tool for description of arbitrary periodic signals, describing them in terms of a spectrum of sinusoids, the frequencies of which are multiples of a basic frequency.

It is not immediately obvious that the concepts of spectrum and frequency can be generalized to *nonperiodic* signals. After all, *frequency* is only meaningful if something is periodic! Surprisingly, the concept of spectrum turns out to be quite robust; for nonperiodic signals we simply need a continuum of frequencies rather than harmonically related ones. Thus analog signals can be viewed either as continuous functions of time or as continuous functions of frequency. This leads to a pleasingly symmetric view, whereby the signal can be described in the *time domain* or the *frequency domain.*

The mathematical tool for transforming an analog signal from its time domain representation to the frequency domain, or vice versa, is called the Fourier transform (FT). The name hints at the fact that it is closely related to the Fourier series that we have already discussed. For digital signals we have close relatives, namely the discrete Fourier transform (DFT) and the z transform (zT). In this chapter we introduce all of these, review their properties, and compute them for a few example signals. We also introduce a non-Fourier concept of frequency, the *instantaneous frequency.* The FS, FT, DFT, zT, and instantaneous frequency, each in its own domain of applicability, is in some sense the *proper* definition of frequency.

4.1 From Fourier Series to Fourier Transform

In the previous chapter we learned that the set of harmonically related sinusoids or complex exponentials form a basis for the vector space of periodic signals. We now wish to extend this result to the vector space of all analog signals. The expansion in this basis is the *Fourier transform.*

Looking back at the steps in proving the existence of the Fourier series we see that the periodicity of the signals was not really crucial; in fact the whole periodicity constraint was quite a nuisance! The SUIs form a basis for *all* signals, whether periodic or not. It was only when we introduced the HRSs, sums of which are necessarily periodic, that we had to restrict ourselves to representing periodic signals. It would seem that had we allowed arbitrary frequency sinusoids we would have been able to represent any signal, and indeed this is the case. In fact it would have been just as easy for us to have directly derived the Fourier transform without the annoyance of the Fourier series; however this would have involved a grave break with mathematical tradition that mandates deriving the Fourier transform from the Fourier series.

The basic idea behind this latter derivation is inherent in the FS derived in Section 3.8. There we saw how increasing the period of the signal to be analyzed required decreasing the fundamental frequency of the HRSs. It is a general result that the longer the time duration that we must accurately reproduce, the more frequency resolution is required to do so. Now let us imagine the period going to infinity, so that the signal effectively is no longer periodic. If you find this infinity troublesome just imagine a period longer than the time during which you are willing to wait for the signal to repeat. The required frequency resolution will then become infinitesimal, and at every step of the way the corresponding HRSs form a basis for the signals with this large period. In the limit of aperiodic signals and continuous spectrum we discover that the set of *all* sinusoids forms a basis for the entire vector space of signals. Of course, for our basis signals we can choose to use sinusoids in quadrature $\sin(\omega t)$ and $\cos(\omega t)$, sinusoids with arbitrary phases $\sin(\omega t + \varphi)$, or complex exponentials $e^{\mathrm{i}\omega t}$ with both positive and negative frequencies.

We have neglected an essential technical detail—as long as the fundamental frequency is small, but still finite, there are a denumerably infinite number of basis signals, and so the dimension of the space is $\aleph_0$ and expansions of arbitrary signals are infinite sums. Once the spectrum becomes continuous, there are a nondenumerable infinity of basis functions, and we must replace the infinite sums with integrals. The set of 'coefficients' S_k becomes a single continuous function of frequency $S(\omega)$.

The result is an expression for a signal as an integral over all time of a function of frequency times a complex exponential.

$$S(\omega) = \int_{t=-\infty}^{\infty} s(t)\, e^{-\mathrm{i}\omega t}\, dt \qquad (4.1)$$

This function of frequency is called the Fourier transform (FT). As we shall show, you may think of it as the *spectrum* of a nonperiodic signal. The extension of Fourier's theorem now states that every (not necessarily periodic) function (that obeys certain conditions) can be written as the integral over complex exponentials. The conditions for the convergence of the Fourier transform are almost the same as Dirichlet's conditions for the Fourier series; just remember to increase the region of integration to all times and insist on at most a finite number of extrema and discontinuities in any *finite amount of time.*

Paradoxically, while in normal speech *to transform* usually means to change the form of a quantity without changing its meaning, in mathematics *a transform* is a changing of meaning that does not alter the form. The Fourier transform changes the meaning from time to frequency domain, but the form remains a continuous function. The Fourier series is not a transform since it changes a continuous function into an infinite-dimensional vector of coefficients. We will see later that the discrete Fourier transform translates infinite-dimensional vectors into infinite-dimensional vectors. Specifically, *integral transforms*, like the FT, are representations of continuous functions as

$$F(\omega) = \int f(t) \, K(t, w) \, dt$$

where K is called the *kernel* of the transform.

When dealing with transforms we often use *operator notation*, i.e., we write $S(\omega) = \text{FT}\left(s(t)\right)$ and

$$S(\omega) = \text{FT}\left(s(t)\right) = \int_{t=-\infty}^{\infty} s(t) \, e^{-i\omega t} \, dt \qquad (4.2)$$

and think of FT as an operator that transforms the time domain representation of a signal into the frequency domain representation.

As was the case for periodic signals, the spectrum contains all possible information about the signal, and therefore the signal can be reconstructed from the spectrum alone. Consequently, we can define the inverse Fourier transform (iFT), $s(t) = \text{FT}^{-1}\left(S(\omega)\right)$ where FT^{-1} is the *inverse operator.*

$$s(t) = \text{FT}^{-1}\left(S(\omega)\right) = \frac{1}{2\pi} \int_{\omega=-\infty}^{\infty} S(\omega) \, e^{i\omega t} \, d\omega \qquad (4.3)$$

The form of the iFT is almost identical to that of the transform itself, but it integrates out the frequency variable leaving the time variable. The only differences are the normalization constant (more about that shortly) and the sign of the exponent.

The inverse operator obeys $\mathrm{FT}^{-1}\mathrm{FT} = 1$ where 1 is the identity operator that leaves every signal completely unchanged

$$s(t) = \mathrm{FT}^{-1}\left(S(\omega)\right) = \mathrm{FT}^{-1}\,\mathrm{FT}\left(s(t)\right) \tag{4.4}$$

an identity sometimes called the Fourier Integral Theorem. The two representations related by the FT and FT^{-1} operators are called a *Fourier transform pair*. They are both functions of a single continuous variable and contain exactly the same information about the signal, but in different forms. The function $s(t)$ is the *time domain* representation of the signal, while $S(\omega)$ is its *frequency domain* representation.

Let's prove equation (4.4).

$$
\begin{aligned}
\mathrm{FT}^{-1}\,\mathrm{FT}\left(s(t)\right) &= \frac{1}{2\pi}\int_{\omega=-\infty}^{\infty} S(\omega)\,e^{\mathrm{i}\omega t}\,d\omega \\
&= \frac{1}{2\pi}\int_{\omega=-\infty}^{\infty}\left(\int_{t'=-\infty}^{\infty} s(t')e^{-\mathrm{i}\omega t'}\,dt'\right)e^{\mathrm{i}\omega t}\,d\omega \\
&= \int_{t'=-\infty}^{\infty} s(t')\frac{1}{2\pi}\int_{\omega=-\infty}^{\infty} e^{-\mathrm{i}\omega t'}e^{\mathrm{i}\omega t}\,d\omega\,dt' \\
&= \int_{t'=-\infty}^{\infty} s(t')\delta(t-t')\,dt' = s(t)
\end{aligned}
$$

We now see why the exponents have different signs—it's required to get the needed delta function. Incidentally, we see that instead of the normalization constant $\frac{1}{2\pi}$ in the iFT we could have used any constants in both FT and iFT whose product is $\frac{1}{2\pi}$. For instance, physicists usually define a more symmetric pair

$$S(\omega) = \mathrm{FT}\left(s(t)\right) = \frac{1}{\sqrt{2\pi}}\int_{t=-\infty}^{\infty} s(t)\,e^{-\mathrm{i}\omega t}\,dt$$
$$s(t) = \mathrm{FT}^{-1}\left(S(\omega)\right) = \frac{1}{\sqrt{2\pi}}\int_{\omega=-\infty}^{\infty} S(\omega)\,e^{\mathrm{i}\omega t}\,d\omega \tag{4.5}$$

but any other combination could be used as well. The DSP convention of putting the constant only in the definition of the inverse transform becomes more symmetric when using the frequency f in Hz (cycles per second) rather than the angular frequency ω in radians per second.

$$S(f) = \int_{t=-\infty}^{\infty} s(t)e^{-2\pi\mathrm{i}ft}dt$$
$$s(t) = \int_{f=-\infty}^{\infty} S(f)e^{2\pi\mathrm{i}ft}df \tag{4.6}$$

We have shown that the FT indeed delivers a function of frequency that contains all the information in the signal itself. What we haven't shown is its relationship to the concept of *frequency spectrum* as we understand it. The true spectrum should be prominent at frequencies that are provably significant components of the signal, and should be zero at frequencies not corresponding to any physical aspect of the signal. We *could* show this directly, starting with a single sinusoid such as $s(t) = A\cos(\omega' t)$ and showing that it is

$$
\begin{aligned}
S(\omega) = \mathrm{FT}\left(A\cos(\omega' t)\right) &= \int_{t=-\infty}^{\infty} A\cos(\omega' t)e^{-i\omega t}dt \\
&= \int_{t=-\infty}^{\infty} \frac{A}{2}\left(e^{i\omega' t} + e^{-i\omega' t}\right)e^{-i\omega t}dt \\
&= \frac{A}{2}\left(\int_{t=-\infty}^{\infty} e^{i\omega' t}e^{-i\omega t}dt + \int_{t=-\infty}^{\infty} e^{-i\omega' t}e^{-i\omega t}dt\right) \\
&= \frac{A}{2}\left(\int_{t=-\infty}^{\infty} e^{-i(\omega-\omega')t}dt + \int_{t=-\infty}^{\infty} e^{-i(\omega+\omega')t}dt\right) \\
&= 2\pi\frac{A}{2}\left(\delta(\omega-\omega') + \delta(\omega+\omega')\right)
\end{aligned}
$$

and accordingly has only components at $\pm\omega'$ as expected. Then we would have to invoke the linearity of the FT and claim that for all combinations of sinusoids

$$
\sum_{k=0}^{K} A_k\cos(\omega_k t)
$$

the FT has discrete lines of precisely the expected relative weights. Next we would have to consider the continuous spectra of nonperiodic signals and show that the FT captures the meaning we anticipate. Finally, we would need to show that the FT is zero for unwanted frequencies. This could conceivably involve forcibly *notching out* frequencies from an arbitrary signal, and observing the FT at these frequencies to be zero.

This prescription is perhaps overly ambitious for us at this point, and in any case there is a shrewd way out. All we really need do is to show that the FT is the proper generalization of the FS for possibly nonperiodic signals. This will ensure that all well-known properties of FS spectra will survive in the FT, and all new properties of the FT will be taken to be the definition of what the true spectrum should be.

We start from slightly modified versions of equations (3.26) and (3.27) for the FS of a periodic signal $s(t)$ with period T

$$S_k = \frac{1}{T} \int_{-\frac{T}{2}}^{\frac{T}{2}} s(t) e^{-i\frac{2\pi k}{T} t}$$

$$s(t) = \sum_{k=-\infty}^{\infty} S_k e^{i\frac{2\pi k}{T} t}$$

and define $\omega \equiv \frac{2\pi k}{T}$. We can now think of the FS as S_ω instead of S_k; of course the indices are no longer integers, but there still are a denumerable number of them. They are uniformly spaced with $\Delta\omega = \frac{2\pi}{T}$ between them, and they still run from minus infinity to plus infinity.

$$S_\omega = \frac{\Delta\omega}{2\pi} \int_{-\frac{T}{2}}^{\frac{T}{2}} s(t) e^{-i\omega t} dt$$

$$s(t) = \sum_{\omega=-\infty}^{\infty} S_\omega e^{i\omega t}$$

We next envision increasing the period T without limit $T \to \infty$. As we have already discussed, the frequency spacing Δ will become smaller and smaller $\Delta \to 0$, until the sequence $\{S_\omega\}_{\omega=-\infty}^{\infty}$ becomes a continuous function $S(\omega)$. Unfortunately, this definition of $S(\omega)$ is unsatisfactory. Looking back at the equation for S_ω we see that it is proportional to $\Delta\omega$. Assuming the integral approaches a finite value, S_ω will vanish as $\Delta\omega \to 0$. However, the ratio $\frac{S_\omega}{\Delta\omega}$ will remain finite in this limit, and has the pleasing interpretation of being the density of Fourier components per unit frequency.

We therefore propose defining $S(\omega) \equiv \frac{S_\omega}{\Delta\omega}$, in terms of which

$$S(\omega) = \frac{1}{2\pi} \int_{-\frac{T}{2}}^{\frac{T}{2}} s(t) e^{-i\omega t} dt$$

$$s(t) = \sum_{\omega=-\infty}^{\infty} S(\omega) e^{i\omega t} \Delta\omega$$

In the limit $T \to \infty$ and $\Delta\omega \to 0$ several things happen. The integral over t now runs from $-\infty$ to $+\infty$. The finite difference $\Delta\omega$ becomes the infinitesimal $d\omega$. The sum over the discrete ω index in the formula for $s(t)$ will of course become an integral over the continuous ω variable. Substitution of these brings us to

$$S(\omega) = \frac{1}{2\pi} \int_{-\infty}^{\infty} s(t) e^{-i\omega t} dt$$

$$s(t) = \int_{-\infty}^{\infty} S(\omega) e^{i\omega t} d\omega$$

which is the FT in an unusual but legitimate normalization scheme. Of course had we defined $S(\omega) \equiv 2\pi \frac{S_\omega}{\Delta \omega}$ we would have obtained exactly (4.2) and (4.3), and $S(\omega) \equiv \sqrt{2\pi} \frac{S_\omega}{\Delta \omega}$ would have produced the physicist's (4.5).

In Section 3.4 we interpreted the FS as the expansion of a periodic signal in the basis of sines and cosines. We have just derived the FT by a limiting process starting from the FS, so it is not surprising that we can interpret the FT as the expansion a nonperiodic signal in a basis. Due to the nondenumerably infinite amount of information in a general nonperiodic signal, it is not surprising that we need a nondenumerable number of basis functions, and that the sum in the expansion becomes an integral.

Reiterating what we have accomplished, we have shown that the FT as we have defined it is the natural generalization to nonperiodic signals of Fourier's expansion of periodic signals into sinusoids. The function $S(\omega)$ has a meaningful interpretation as the Fourier spectral density, so that $S(\omega)d\omega$ is the proper extension of the FS component. The FT is therefore seen to truly be the best definition of spectrum (so far).

EXERCISES

4.1.1 Prove the opposite direction of (4.4), namely

$$S(\omega) = \mathrm{FT}\,\mathrm{FT}^{-1}\left(S(\omega)\right)$$

4.1.2 Find the FT of $A\sin(\omega' t)$. How is it different from that of $A\cos(\omega' T)$?

4.1.3 Find the FT of the rectangular wave of Section 3.8. How does it relate to the FS found there? Find the FT of a single rectangle. How does it relate to that of the first part?

4.1.4 Write a routine that computes the value of the FT of a real signal $s(t)$ at frequency $f = \frac{\omega}{2\pi}$. The signal is nonzero only between times $t = 0$ and $t = T$, and is assumed to be reasonably well behaved. You should use numerical Riemann integration with the time resolution Δt variable.

4.1.5 Generate a signal composed of a constant plus a small number of unrelated sinusoids. Using the routines developed in the previous exercise, plot the real and imaginary parts of its FT for a frequency band containing all frequencies of interest. Vary the time resolution. How is the accuracy affected? Vary the frequency resolution. Are the frequencies of the sinusoids exact or is there some width to the lines? Is this width influenced by the time resolution? How much time is needed to compute the entire FT (as a function of time and frequency resolution)?

4.2   Fourier Transform Examples

The time has come to tackle a few examples of FT calculation. Although it *is* instructive to go through the mechanics of integration a few times, that is not our only motivation. We have selected examples that will be truly useful in our later studies.

The simplest signal to try is a constant signal $s(t) = 1$, and for this signal we almost know the answer as well! There can only be a DC (zero frequency) component, but how much DC is there? The integral in (4.2) is

$$S(\omega) = \int_{t=-\infty}^{\infty} e^{-i\omega t} dt = \int_{t=-\infty}^{\infty} (\cos \omega t - i \sin \omega t)\, dt \qquad (4.7)$$

(we simply replaced $s(t)$ by 1). Now we are stuck, since the required definite integrals don't appear in any table of integrals. We can't do the indefinite integral and substitute the values at the endpoints, since $\sin(\pm\infty)$ and $\cos(\pm\infty)$ don't approach a constant value; and don't confuse this integral with equation (3.21) for $m = 1$, since the integral is over the entire t axis. Whenever we're stuck like this, it is best to think about what the integral means. When $\omega = 0$ we are trying to integrate unity over the entire t axis, which obviously diverges. For all other ω we are integrating sinusoids over all time. Over full periods sinusoids are positive just as much as they are negative, and assuming infinity can be considered to be a whole number of periods, the integral should be zero. We have thus deduced a delta function to within a constant $S(\omega) = \gamma \delta(\omega)$. To find γ we need to integrate over ω. We know from (4.4) that

$$\frac{1}{2\pi} \int_{-\infty}^{\infty} S(\omega) dw = \text{FT}^{-1}\left(S(\omega)\right) = s(t) = 1$$

from which we conclude that $\gamma = 2\pi$.

Let's try the other way around. What is the transform of an analog impulse $s(t) = \delta(t)$? Well it's just

$$\text{FT}\left(s(t)\right) = \int_{t=-\infty}^{\infty} \delta(t) e^{-i\omega t} dt = e^0 = 1$$

using property (A.69) of the delta function. So it works the other way as well—the transform of a delta is a constant. With only minimal additional effort we can find the transform of an impulse at any nonzero time τ. In this case we pick out the exponential at some other time

$$\text{FT}\left(\delta(t-\tau)\right) = \int_{t=-\infty}^{\infty} \delta(t-\tau) e^{-i\omega t} dt = e^{-i\omega \tau} \qquad (4.8)$$

which is a complex exponential in the frequency domain. The interpretation of this sinusoid is slightly different from the usual. Remember that here τ is a constant that we are given and ω is the variable. The sinusoidal behavior is as a function of frequency, and the higher τ is, the more compressed the oscillation becomes. So τ plays the role of frequency here, which is not surprising due to the dual nature of time and frequency.

Conversely, a non-DC complex exponential $s(t) = e^{i\Omega t}$ has the transform

$$\mathrm{FT}\left(s(t)\right) = \int_{t=-\infty}^{\infty} e^{i\Omega t}e^{-i\omega t}dt = \int_{t=-\infty}^{\infty} e^{i(\Omega-\omega)t}dt = 2\pi\delta(\omega - \Omega) \quad (4.9)$$

(we could interchange the omegas since the delta function is symmetric). Thus the complex exponential corresponds to a single frequency line, as expected.

What about a real sinusoid $\sin(\Omega t)$ or $\cos(\Omega t)$? Using the linearity of the FT and the expressions (A.8) we can immediately conclude that sine and cosine consist of two delta functions in the frequency domain. One delta is at $+\Omega$ and the other at $-\Omega$.

$$\begin{aligned}\mathrm{FT}\left(\sin(\omega t)\right) &= \frac{\pi}{i}\left(\delta(\omega - \Omega) - \delta(\omega + \Omega)\right)\\ \mathrm{FT}\left(\cos(\omega t)\right) &= \pi\left(\delta(\omega - \Omega) + \delta(\omega + \Omega)\right)\end{aligned} \quad (4.10)$$

The absolute value of the spectrum is symmetric, as it must be for real functions, but sine and cosine differ in the relative phase of the deltas.

The FT decaying exponential can also be useful to know. It is simply

$$\mathrm{FT}\left(e^{-\lambda t}u(t)\right) = \frac{1}{\lambda + i\omega} \quad (4.11)$$

and actually the same transform holds for complex λ, as long as the real part of λ is positive.

Up to now we have treated rather smooth signals and impossibly singular ones (the delta). We will also need to investigate archetypical jump discontinuities, the sgn and step functions. Since sgn is odd, $\mathrm{sgn}(-t) = -\mathrm{sgn}(t)$, we can immediately deduce that the zero frequency component of sgn's FT must be zero. The zero frequency component of $u(t)$ is obviously infinite and so we know that $u(\omega)$ must have a $k\delta(\omega)$ component. The value of k can be determined from the fact that $u(-t) + u(t) = 1$ and from linearity $\mathrm{FT}(u(-t)) + \mathrm{FT}(u(t)) = \mathrm{FT}(1) = 2\pi\delta(\omega)$; so the DC component is simply $\pi\delta(\omega)$.

Trying to find the nonzero frequency components of either sgn or $u(t)$ we stumble upon one of those impossible integrals, like (4.7). For large ω it should go to zero since the integral over an even number of cycles of sinusoids is zero; but for smaller w there is the issue of the end effects. We will be able to prove later that the spectrum decays as $\frac{1}{w}$, i.e., every time we double the frequency the amplitude drops to half its previous value. When displaying the spectrum on a logarithmic scale this translates to a linear drop of 6 dB per octave. Since any signal with a single discontinuity can be considered to be continuous signal plus a step or sgn, all signals with step discontinuities have this 6 dB per octave drop in their spectra.

EXERCISES

4.2.1 Calculate the FT of a complex exponential from those of sin and cos using linearity and equation (A.8).

4.2.2 What is the difference between the FT of sin and cos? Explain the effect of A and φ on the FT of $A\sin(\omega t + \varphi)$.

4.2.3 Find the FT of the single rectangle (equation (3.29)).

4.2.4 Show that $\sum_{n=-\infty}^{\infty} e^{-\mathrm{i}\omega n T} = 0$ when ω is not a multiple of $\frac{2\pi}{T}$.

4.2.5 Formally prove that the FT of the impulse train $s(t) = \sum \delta(t - kT)$ is an impulse train in the frequency domain by finding its Fourier *series* and relating the transform to the series.

4.2.6 Our proof of the universality of the Fourier series in Section 3.4 rested on the expansion of shifted delta functions in the basic period in terms of harmonically related sinusoids. Show how this can be simplified using our results for impulse trains.

4.2.7 Prove that the following are FT pairs:

$u(t)$	$\pi\delta(\omega) + \frac{1}{\mathrm{i}\omega}$
$e^{-\lambda t}u(t)$	$\frac{1}{\lambda + \mathrm{i}\omega}$
$te^{-\lambda t}u(t)$	$\frac{1}{(\lambda + \mathrm{i}\omega)^2}$
$\alpha^{\lvert n \rvert}$	$\frac{1-\alpha^2}{1 - 2\alpha\cos(\omega) + \alpha^2}$
$\lvert t \rvert$	$-\frac{2}{\omega^2}$
$e^{-a\lvert t \rvert}$	$\frac{2a}{a^2 + \omega^2}$

4.3   FT Properties

As we saw in the examples, the Fourier transform of a signal may look like just about anything. It is customary to differentiate between continuous and discrete line spectra. When the FT is a continuous smooth function of frequency a nondenumerable number of frequency components are required to reproduce the signal. A FT composed of some number of sharp discrete lines results from a signal that is the sum of that number of sinusoids. In general, spectra may have both continuous and discrete parts. In fact all signals encountered in practice are noisy and so cannot be precisely periodic, and hence some continuous spectrum contribution is always present.

The question of the 'mathematical existence' of the FT is an important one for mathematicians, but one we will not cover extensively. The Dirichlet conditions for the FT require that the integral over all time of the absolute value of the signal be finite, as well as there being only a finite number of extrema and discontinuities in any finite interval. The FT obviously does not exist in the technical sense for periodic signals such as sinusoids, but by allowing delta functions we bypass this problem.

Although we will not dwell on existence, there are many other characteristics of the FT that we will need. Many times we can find the FT of signals without actually integrating, by exploiting known transforms and some of the following characteristics. These characteristics are often closely related to characteristics of the FS, and so we need not derive them in detail.

First, it is important to restate the Fourier Integral Theorem that the inverse FT given by equation (4.3) is indeed the inverse operation.

$$\mathrm{FT}^{-1}\,\mathrm{FT}\,x = x \qquad\qquad \mathrm{FT}\,\mathrm{FT}^{-1}\,X = X \qquad\qquad (4.12)$$

Next, the FT is linear, i.e.,

$$\mathrm{FT}\left(x(t)+y(t)\right) \;=\; X(\omega)+Y(\omega) \qquad\qquad (4.13)$$
$$\mathrm{FT}\left(as(t)\right) \;=\; aS(\omega)$$

a property already used in our derivation of the FT of real sinusoids.

Speaking of real signals, it is easy to see that the FT of a real signal is Hermitian even,

$$S(-\omega) = S^*(\omega)$$

meaning that $\Re S(\omega)$ is even, $\Im S(\omega)$ is odd, $|S(\omega)|$ is even, and $\angle S(\omega)$ is odd. Conversely the FT of an even signal $(s(-t) = s(t))$ is real, and that of an odd signal is pure imaginary.

There are two properties that deal with changing the clock, namely the time shifting property

$$\text{FT}\left(s(t-\tau))\right) = e^{-i\omega\tau}S(\omega) \qquad (4.14)$$

and the time scaling property.

$$\text{FT}\left(s(ct)\right) = \frac{1}{|c|}S(\frac{\omega}{c}) \qquad (4.15)$$

Conversely, there is a property that deals with shifting the frequency axis

$$\text{FT}\left(s(t)e^{i\Omega t}\right) = S(\omega - \Omega) \qquad (4.16)$$

an operation we usually call *mixing*.

What happens when you differentiate $s(t) = e^{i\omega t}$? You get $i\omega s(t)$. Similarly, integrating it you get $\frac{1}{i\omega}s(t)$. It follows that differentiating or integrating an arbitrary signal affects the FT in a simple way.

$$\text{FT}\left(\frac{ds(t)}{dt}\right) = i\omega S(\omega) \qquad \bigg| \qquad \text{FT}\left(\int_{\tau=-\infty}^{t} s(\tau)d\tau\right) = \frac{1}{i\omega}S(\omega) \qquad (4.17)$$

These are surprising results; we think of differentiation and integration as purely time domain operations, but they are even simpler in the frequency domain! We will see in Section 7.3 that the DSP approach to differentiation and integration in the time domain involves first designing a filter in the frequency domain. Note also that differentiation emphasizes high frequencies, while integration emphasizes lows. This is because derivatives involve subtracting nearby values, while integrals are basically averaging operators.

Linearity told us how to find the spectrum when adding signals; what happens when we multiply them? Since we have never tried this before we will have to actually do the integral.

$$\begin{aligned}
\int_{-\infty}^{\infty} x(t)y(t)e^{-i\omega t}\,dt &= \int_{t=-\infty}^{\infty}\left(\frac{1}{2\pi}\int_{\Omega=-\infty}^{\infty}X(\Omega)e^{i\Omega t}\,d\Omega\right)y(t)e^{-i\omega t}\,dt \\
&= \frac{1}{2\pi}\int_{\Omega=-\infty}^{\infty}X(\Omega)\int_{t=-\infty}^{\infty}y(t)e^{-i(\omega-\Omega)t}\,dt\,d\Omega \\
&= \frac{1}{2\pi}\int_{\Omega=-\infty}^{\infty}X(\Omega)Y(\omega-\Omega)\,d\Omega
\end{aligned}$$

What we did was simply to replace $x(t)$ by its iFT, change the order of integration, and recognize the FT of y. So we have found the following:

$$\text{FT}\left(x(t)y(t)\right) = \frac{1}{2\pi}\int_{\Omega=-\infty}^{\infty}X(\Omega)Y(\omega-\Omega)\,d\Omega \equiv X * Y \qquad (4.18)$$

Now that we have the answer, what does it mean? The FT of the product of two signals in the time domain is the integral of a strange-looking product in the frequency domain. We hide this strangeness by using the symbol $*$, implying a product of some sort. It's a truly unusual product since the integration variable in Y runs in the opposite direction to that of the X variable. If that is not bad enough, repeating the above computation for iFT of a product in the frequency domain, we find

$$\text{FT}^{-1}\left(X(\omega)Y(\omega)\right) = \int_{T=-\infty}^{\infty} x(T)y(t-T)\,dT \equiv x * y \qquad (4.19)$$

where the integration variable in y runs backward in time! We are not yet ready to digest this strange expression that goes under the even stranger name of *convolution*, but it will turn out to be of the utmost importance later on.

A particular case of equation (4.18) is the DC ($\omega = 0$) term

$$\int_{-\infty}^{\infty} x(t)y(t)\,dt = \frac{1}{2\pi}\int_{\Omega=-\infty}^{\infty} X(\Omega)Y(-\Omega)\,d\Omega$$

and by taking $x(t) = s(t)$, $y(t) = s^*(t)$ and changing the name of the integration variable, we get Parseval's relation for the FT.

$$\int_{-\infty}^{\infty} |s(t)|^2\,dt = \frac{1}{2\pi}\int_{\omega=-\infty}^{\infty} |S(\omega)|^2\,d\omega = \int_{f=-\infty}^{\infty} |S(f)|^2\,df \qquad (4.20)$$

Parseval's relation tells us that the signal's energy is the same whether we look at it in the time domain or the frequency domain. This is an important physical consistency check.

To demonstrate the usefulness of some of these properties, we will now use the integration rule to derive a result regarding signals with discontinuous derivatives. We know that the FT of the impulse is constant, and that its integral is the unit step $u(t)$. Thus we would expect from (4.17) for the FT of the step to be simply $\frac{1}{i\omega}$, which is *not* what we previously found! The reason is that (4.17) breaks down at $\omega = 0$, and so we always have to allow for the possible inclusion of a delta function. Integrating once more we get $f(t) = t\,u(t)$, which is continuous but has a discontinuous first derivative. The integration rule tells us that the FT of this f is $-\omega^{-2}$ (except at $\omega = 0$). Integrating yet another time gives us a signal with continuous first derivative but discontinuous second derivative and $i\omega^{-3}$ behavior. Continuing this way we see that if all derivatives up to order k are continuous but the $(k+1)^{\text{th}}$ is not, then the (nonzero frequency) transform is proportional

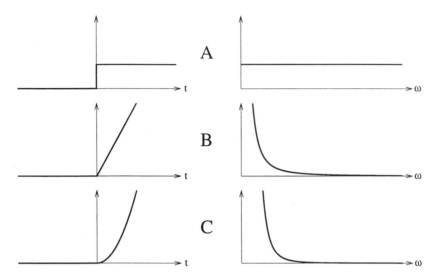

Figure 4.1: The effect of derivative discontinuity on the FT. In (A) the signal itself (the zeroth derivative) is discontinuous and the spectrum is constant. In (B) the first derivative is discontinuous and the spectrum decays as ω^{-2}. In (C) the second derivative jumps and the spectrum decays as ω^{-4}.

to ω^{-k}, and the power spectrum is inversely proportional to ω^{2k}. In other words a discontinuous first derivative contributes a term which decays 6 dB per octave; a second derivative 12 dB per octave, etc. These results, depicted in Figure 4.1, will be useful in Section 13.4.

EXERCISES

4.3.1 Explain why $\int_{-\infty}^{\infty} e^{i\omega t} dt = 2\pi\delta(\omega)$ using a graphical argument.

4.3.2 Show that time reversal causes frequency reversal FT $\left(s(-t)\right) = S(-\omega)$.

4.3.3 Show how differentiation and integration of the spectrum are reflected back to the time domain.

4.3.4 The derivative of $\cos(\omega t)$ is $-\omega \sin(\omega t)$. State this fact from the frequency domain point of view.

4.3.5 Show that we can interchange X and Y in the convolution integral.

4.3.6 Redraw the right-hand side of Figure 4.1 using dB. How does the slope relate to the order of the discontinuity?

4.3.7 Generalize the relationship between spectral slope and discontinuity order to signals with arbitrary size discontinuities not necessarily at the origin. What if there are many discontinuities?

4.4   The Uncertainty Theorem

Another signal with discontinuities is the rectangular window

$$s(t) = \begin{cases} 1 & |t| \leq T \\ 0 & \text{else} \end{cases} \tag{4.21}$$

which is like a single cycle of the rectangular wave. The term 'window' is meant to evoke the picture of the opening a window for a short time. Its FT

$$
\begin{aligned}
\text{FT}\left(s(t)\right) &= \int_{-T}^{T} e^{i\omega t}\, dt \\
&= \frac{e^{+i\omega T} - e^{-i\omega T}}{i\omega} \\
&= 2\frac{\sin(\omega T)}{\omega} \quad = \quad 2T\,\text{sinc}(\omega T)
\end{aligned}
$$

turns out to be a sinc. Now the interesting thing about this sinc is that its bandwidth is inversely proportional to T, as can be seen in Figure 4.2.

The wider the signal is in the time domain, the narrower it is in frequency, and vice versa. In fact if we define the bandwidth to be precisely between the first zeros of the sinc, $\Delta\omega = \frac{2\pi}{T}$, and relate this to the time duration $\Delta t = 2T$, we find that the *uncertainty product*

$$\Delta\omega\,\Delta t \;=\; 4\pi$$

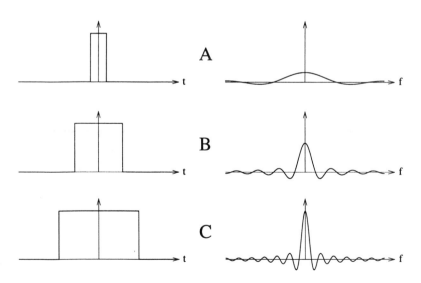

Figure 4.2: Rectangular windows of various widths with their Fourier transforms. Note that the signal energy is not normalized.

although different definitions of bandwidth would change the precise value on the right-hand side.

This is a special case of a more general rule relating time durations to bandwidth. A single sinusoid is defined for all time and has a completely precise line as its spectrum. Signals of finite duration cannot have discrete line spectra since in order build the signal where it is nonzero but cancel it out at $t = \pm\infty$ we need to sum many nearby frequencies. The shorter the time duration the more frequencies we need and so the wider the bandwidth.

It is useful to think of this in a slightly different way. Only if we can observe a sinusoid for an infinite amount of time can we precisely determine its frequency. If we are allowed to see it for a limited time duration we can only determine the frequency to within some tolerance; for all sinusoids with similar frequencies look about the same over this limited time. The less time we are allowed to view the sinusoid, the greater our uncertainty regarding its true frequency. You can convince yourself of this fact by carefully studying Figure 4.3.

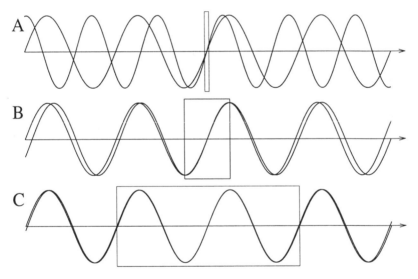

Figure 4.3: The effect of observation window duration on frequency uncertainty. In (A) we only observe the sinusoid for an extremely short time, and hence we can not accurately gauge its frequency. In (B) we observe about half a cycle and can now estimate the frequency, but with relatively large uncertainty. In (C) two full cycles are observed and consequently the uncertainty is much reduced.

As our next example, consider the Gaussian

$$s(t) = Ae^{-\beta t^2} \tag{4.22}$$

whose Fourier transform is

$$S(\omega) = \int_{-\infty}^{\infty} Ae^{-\beta t^2} e^{-i\omega t} dt = \int_{-\infty}^{\infty} Ae^{-\beta t^2 - i\omega t} dt \tag{4.23}$$

which doesn't look hopeful. The mathematical trick to use here is 'completing the square'. The exponent is $-(\beta t^2 + i\omega t)$. We can add and subtract $\frac{\omega^2}{4\beta}$ so that

$$S(\omega) = \int_{-\infty}^{\infty} Ae^{-(\sqrt{\beta}t + \frac{i\omega}{2\sqrt{\beta}})^2} e^{-\frac{i\omega}{4\beta}} dt = Ae^{-\frac{i\omega}{4\beta}} \int_{-\infty}^{\infty} e^{-(\sqrt{\beta}t + \frac{i\omega}{2\sqrt{\beta}})^2} dt \tag{4.24}$$

and a change of variable $u = \sqrt{\beta}t + \frac{i\omega}{2\sqrt{\beta}}$ gives

$$S(\omega) = Ae^{-\frac{i\omega}{4\beta}} \int_{-\infty}^{\infty} e^{-u^2} \frac{du}{\sqrt{\beta}} = A\sqrt{\frac{\pi}{\beta}} e^{-\frac{\omega^2}{4\beta}} \tag{4.25}$$

so the FT of a Gaussian is another Gaussian.

Now let's look at the uncertainty product for this case. The best way of defining Δt here is as the variance of the square of the signal. Why the square? Well, if the signal took on negative values it would be more obvious, but even for the Gaussian the energy is the integral of the square of the signal; the 'center of gravity' is the expected value of the integral of t times the signal squared, etc. Comparing the square of the signal $A^2 e^{-2\beta t^2}$ with equation (A.19) we see that the standard deviation in the time domain is $\Delta t = \frac{1}{2\sqrt{\beta}}$, while the same considerations for equation (4.25) lead us to realize that $\Delta \omega = \sqrt{\beta}$. The uncertainty product follows.

$$\Delta t \, \Delta \omega = \tfrac{1}{2}$$

Now it turns out that no signal has a smaller uncertainty product than this. This theorem is called the *uncertainty theorem*, and it is of importance both in DSP and in quantum physics (where it was first enunciated by Heisenberg). Quantum physics teaches us that the momentum of a particle is the Fourier transform of its position, and hence the uncertainty theorem limits how accurately one can simultaneously measure its position and velocity. Energy and time are similarly related and hence extremely accurate energy measurements necessarily take a long time.

The Uncertainty Theorem

Given any signal $s(t)$ with energy

$$E = \int_{-\infty}^{\infty} s^2(t)dt$$

time center-of-gravity

$$\langle t \rangle \equiv \frac{\int_{-\infty}^{\infty} ts^2(t)dt}{E}$$

squared time uncertainty

$$(\Delta t)^2 \equiv \frac{\int_{-\infty}^{\infty} (t - \langle t \rangle)^2 s^2(t)dt}{E}$$

frequency center-of-gravity

$$\langle \omega \rangle \equiv \frac{\int_{-\infty}^{\infty} \omega S^2(\omega)d\omega}{E}$$

and squared frequency uncertainty

$$(\Delta\omega)^2 \equiv \frac{\int_{-\infty}^{\infty} (\omega - \langle \omega \rangle)^2 S^2(\omega)d\omega}{E}$$

then the uncertainty product

$$\Delta t\, \Delta\omega \geq \tfrac{1}{2}$$

is always greater than one half.   ∎

Although this theorem tells us that mathematics places fundamental limitations on how accurately we are allowed to measure things, there is nothing particularly mystifying about it. It simply says that the longer you are allowed to observe a signal the better you can estimate its frequencies.

Next let's consider the train of Dirac delta functions

$$s(t) = \sum_{n=-\infty}^{\infty} \delta(t - n\tau) \tag{4.26}$$

depicted in Figure 4.4. This signal is truly fundamental to all of DSP, since it is the link between analog signals and their digital representations. We can think of sampling as multiplication of the analog signal by just such a train of impulses,

$$S(\omega) = \int_{t=-\infty}^{\infty} \sum_{n=-\infty}^{\infty} \delta(t - nT)e^{-i\omega t}dt$$

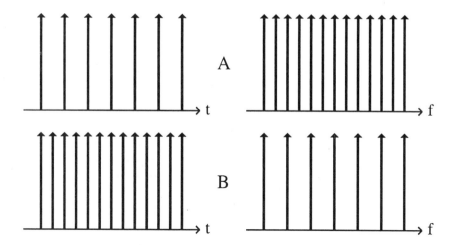

Figure 4.4: Trains of Dirac delta functions in time and frequency domains. Note that the spacing in the time domain is the inverse of that in the frequency domain.

Interchanging the order of summation and integration (we ask permission of the more mathematically sophisticated reader before doing this), we find a sum over the FT of equation (4.8) with $\tau = nT$

$$S(\omega) = \sum_{n=-\infty}^{\infty} \int_{t=-\infty}^{\infty} \delta(t - nT)e^{-i\omega t}dt = \sum_{n=-\infty}^{\infty} e^{-i\omega nT}$$

and once again we are stuck. Looking carefully at the sum we become convinced that for most ω the infinite sum should contain just as many negative contributions as positive ones. These then cancel out leaving zero. At $\omega = 0$, however, we have an infinite sum of ones, which is infinite. Does this mean that the FT of a train of deltas is a single Dirac delta? No, because the same thing happens for all ω of the form $\frac{2\pi}{T}$ as well! So similarly to the Gaussian, a train of impulses has an FT of the same form as itself, a train of impulses in the frequency domain; and when the deltas are close together in the time domain, they are far apart in the frequency domain, and vice versa. The product of the spacings obeys

$$\Delta t\, \Delta \omega \;=\; 2\pi$$

once again a kind of uncertainty relation.

EXERCISES

4.4.1 Prove the *Schwartz inequality* for signals.

$$\left(\int_{-\infty}^{\infty} x^2(t)dt \right) \left(\int_{-\infty}^{\infty} y^2(t)dt \right) \geq \left| \int_{-\infty}^{\infty} x(t)y(t)dt \right|^2$$

4.4.2 Using Parseval's relation and the FT of a derivative prove the following relation involving the uncertainties and the energy E.

$$(\Delta t \, \Delta \omega)^2 = \frac{\int (t - \langle t \rangle)^2 s^2(t)dt \int \left(\frac{ds}{dt} \right)^2 dt}{E^4}$$

4.4.3 Using the Schwartz inequality, the above relation, and integration by parts, prove the uncertainty theorem.

4.5 Power Spectrum

The energy E of a signal $s(t)$ is defined as the integral over all times of the squared values in the time domain. Due to this additive form, we can interpret the integral over some interval of time as the signal's energy during that time. Making the interval smaller and smaller we obtain the power $E(t)$; the signal's energy during a time interval of infinitesimal duration dt centered on time t is $E(t)dt$ where $E(t) = |s(t)|^2$. You can think of the power as the *energy time density*, using the term 'density' as explained at the end of Appendix A.9.

Integrating the power over any finite time interval brings us back to the signal's energy during that time; integrating over all time retrieves the total energy.

$$E = \int_{-\infty}^{\infty} E(t) \, dt = \int_{-\infty}^{\infty} |s(t)|^2 \, dt$$

From Parseval's relation we know that the energy is also computable as the integral of squared values in the frequency domain (except possibly for a normalization factor depending on the FT definition chosen). Hence repeating the above arguments we can define the *energy spectral density* $E(f) = |S(f)|^2$, that specifies how the signal's energy is distributed over frequency. The meaning of $E(f)$ is similar to that of the power; the energy contained in the signal components in an interval of bandwidth df centered on frequency f is $E(f) \, df$.

PSD(f)

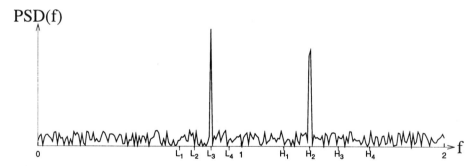

Figure 4.5: Power spectral density for the DTMF digit '8'. The horizontal axis is the frequency in KHz and the vertical axis is a linear measure of the energy density. The eight possible frequencies are marked for convenience.

In the next section we will see that many signals have spectral distributions that vary as time progresses. For such signals we wish to know how much energy is in the frequency range around f at times around t. Since the energy density in the time domain is the power, the desired quantity is called the **Power Spectral Density** (PSD). PSDs that change in time are so common that we almost always use the term *power spectrum* instead of energy spectrum.

Writing the full FT as a magnitude times an angle $S(f) = A(f)e^{i\phi(f)}$, we see that the PSD contains only the magnitude information, all the angle information having been discarded. At this stage of our studies it may not yet be entirely clear why we need the full frequency domain representation, but it is easy to grasp why we would want to know how a signal's energy is divided among the component frequencies. For example, push-button dialing of a phone uses DTMF signals where two tones are transmitted at a time (see Figure 4.5). The lower tone of the two is selected from four candidate frequencies L_1, L_2, L_3, L_4, and the higher is one of H_1, H_2, H_3, H_4. In order to know that an eight was pressed we need only ascertain that there is energy in the vicinities of L_3 and H_2. The phases are completely irrelevant.

As a more complex application, consider a phone line on which several signals coexist. In order for these signals not to interfere with each other they are restricted by 'masks', i.e., specifications of the maximal amount of power they may contain at any given frequency. The masks in Figure 4.6 are specified in dBm/Hz, where dBm is the power in dB relative to a 1 milliwatt signal (see equation (A.16)). The horizontal scale has also been drawn logarithmically in order to accommodate the large range of frequencies from 100 Hz to over 10 MHz. Although the higher frequency signals seem to be

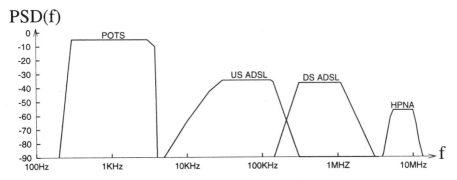

Figure 4.6: PSD masks for several signals on a phone line. The horizontal axis is the frequency in KHz on a logarithmic scale and the vertical axis is the maximum allowed PSD in dBm per Hz. The leftmost signal is the POTS (Plain Old Telephone System) mask, including voice and voicegrade modems. The middle mask is for ADSL, with the lower portion for the 512 Kb/s upstream signal and the upper for the 6 Mb/s downstream signal. At the far right is the mask for the 1 Mb/s Home Phone Network signal.

lower in power, this is only an illusion; the PSD is lower but the bandwidths are much greater.

The mask containing the lowest frequencies is for regular telephone conversations, affectionately called **P**lain **O**ld **T**elephone **S**ervice (POTS). This mask, extending from 200 Hz to about 3.8 KHz, holds for voice signals, signals from fax machines, and voicegrade modems up to 33.6 Kb/s.

The need for high-speed digital communications has led to innovative uses of standard phone lines. The **A**symmetric **D**igital **S**ubscriber **L**ine (ADSL) modem is one such invention. It can deliver a high-speed downstream (from the service provider to the customer) connection of up to 8 Mb/s, and a medium-speed upstream (from the customer to the provider) connection of 640 Kb/s. ADSL was designed in order not to interfere with the POTS signal, so that the standard use of the telephone could continued unaffected. By placing the ADSL signal at higher frequencies, and restricting the amount of power emitted at POTS frequencies, interference is avoided. This restriction may be verified using the power spectrum; the signal phases are irrelevant.

In the same way, after the definition of ADSL the need arose for networking computers and peripherals inside a residence. Of course this can be done by running cables for this purpose, but this may be avoided by using the internal phone wiring but requiring the new 'home phone network' signal to lie strictly above the POTS *and* ADSL signals.

We see that based on the power spectrum alone we may deduce whether signals may coexist without mutual interference. The principle behind this

PSD(f)

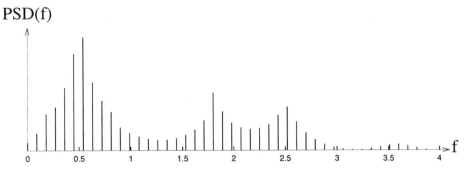

Figure 4.7: Power spectral density for speech, more specifically the sound **eh** pronounced by a male speaker. The horizontal axis is the frequency in KHz and the vertical axis is the energy density in dBm per Hz. The spectrum is obviously made up of discrete lines, and we note that three main resonances at 500, 1820, and 2510 Hz and a weak fourth at a higher frequency.

is that if the frequencies do not overlap the signals may be separated by appropriate filters. Isolation in the frequency domain is a sufficient (but not a necessary) condition for signals to be separable.

A third example is given by the speech signal. Most of the information in speech is encoded in the PSD; in fact our hearing system is almost insensitive to phase, although we use the phase difference between our ears to ascertain direction. In Figure 4.7 we see the spectrum of a (rather drawn out) **eh** sound. The vertical axis is drawn logarithmically, since our hearing system responds approximately logarithmically (see Section 11.2). We can't help noticing three phenomena. First, the spectrum is composed entirely of discrete lines the spacing between which changes with pitch. Second, there is more energy at low frequencies than at high ones; in fact when we average speech over a long time we discover a drop of between 6 and 12 dB per octave. Finally, there seem to be four maxima (called *formants*), three overlapping and one much smaller one at high frequency; for different sounds we find that these formants change in size and location. With appropriate training one can 'read' what is being said by tracking the formants.

In Section 9.3 we will learn that the PSD at a given frequency is itself the FT of a function called the autocorrelation.

$$|S(\omega)|^2 = \int_{-\infty}^{\infty} \left[\int_{-\infty}^{\infty} s(t)s(t-\tau)\, dt \right] e^{-i\omega\tau}\, d\tau \qquad (4.27)$$

The autocorrelation is a generalization of the idea of squaring the signal, and hence this relation tells us that the squaring operation can be performed either before or after the Fourier integral.

EXERCISES

4.5.1 Write a program that finds the PSD by numerical integration (equation (4.1)) and squaring. Use this program to find the PSD of a rectangular window (equation (4.21)) for several different widths. Repeat the exercise for a sinc-shaped pulse for several different pulse widths.

4.5.2 Build 1024 samples of sine waves of 1, 2, 3, 4, 5, 6, 7, and 8 KHz sampled at 8 KHz. Observe the sines in the time domain; can you see the aliasing for $f > 4$ KHz? Extract the PSD (if you didn't write your own program in the first exercise many programs are readily available for this purpose). Can you read off the frequency? What do you see now for $f > 4$ KHz?

4.5.3 Build 1024 sample points of sine waves with frequencies 1.1, 2.2, and 3.3 KHz sampled at 8 KHz. What happened to the spectral line? Try multiplying the signal by a triangular window function that linearly increases from zero at $n = 0$ to one at the center of the interval, and then linearly decreases back to zero).

4.5.4 In exercise 2.6.4 we introduced the V.34 probe signal. Extract its power spectrum. Can you read off the component frequencies? What do you think the probe signal is for?

4.5.5 Find the PSD of the sum of two sinusoids separated by 500 Hz (use 2 KHz $\pm$ 500 Hz) sampled at 8 KHz. Can you distinguish the two peaks? Now reduce the separation to 200 Hz. When do the two peaks merge? Does the triangular window function help?

4.5.6 In the text it was stated that isolation in the frequency domain is a *sufficient* but not a *necessary* condition for signals to be separable. Explain how can signals can be separated when their PSDs overlap.

4.6   Short Time Fourier Transform (STFT)

The Fourier transform is a potent mathematical tool, but not directly relevant for practical analog signal processing, because the integration must be performed from the beginning of time to well after the observer ceases caring about the answer. This certainly seems to limit the number of FTs you will calculate in your lifetime. Of course, one *can* compute the FT for finite time signals, since they were strictly zero yesterday and will be strictly zero tomorrow, and so you only have to observe them today. But that is only the case for signals that are *strictly* zero when you aren't observing them—small isn't good enough when we are integrating over an infinite amount of time!

In Section 4.2 we found the FT for various infinite time signals. Could we have approximated these mathematical results by numerically integrating over a finite amount of time? Other than the restrictions placed by the uncertainty theorem it would seem that this is possible. One needn't observe a simple sinusoid for years and years to be able to guess its spectrum. Of course the longer we observe it the narrower the line becomes, but we will probably catch on after a while. The problem is that we can't be completely sure that the signal doesn't radically change the moment after we give up observing it. Hence we can only give our opinion about what the signal's FT looked like over the time we observed it. Unfortunately, the FT isn't defined that way, so we have to define a new entity—the **Short Time Fourier Transform (STFT)**.

Consider the signal

$$s_1(t) = \begin{cases} \sin(2\pi f_1 t) & t < 0 \\ \sin(2\pi f_2 t) & t \geq 0 \end{cases}$$

which is a pure sine of frequency f_1 from the beginning of time until at time $t = 0$ when, for whatever reason, its frequency abruptly changes to f_2. What is the FT of this signal?

As we have seen, the FT is basically a tool for describing a signal simultaneously at all times. Each frequency component is the sum total of all contributions to this frequency from time $t = -\infty$ to $t = +\infty$. Consequently we expect the power spectrum calculated from the FT to have two equal components, one corresponding to f_1 and the other to f_2.

Now consider the signal

$$s_2(t) = \begin{cases} \sin(2\pi f_2 t) & t < 0 \\ \sin(2\pi f_1 t) & t \geq 0 \end{cases}$$

It is clear that the power spectrum will continue to be composed of two equal components as before since time reversal does not change the frequency composition. Assume now that f_1 and f_2 correspond to a whole number of cycles per second. Then the signal $s_3(t)$

$$s_3(t) = \begin{cases} \sin(2\pi f_1 t) & \lfloor t \rfloor \text{ even} \\ \sin(2\pi f_2 t) & \lfloor t \rfloor \text{ odd} \end{cases}$$

which consists of interleaved intervals of $\sin(2\pi f_1 t)$ and $\sin(2\pi f_2 t)$, must also have the same power spectrum!

The STFT enables us to differentiate between these intuitively different signals, by allowing different spectral compositions at different times. The

FT basically considers all signals to be unvarying, never changing in spectrum, while the STFT is an adaptation of the mathematical idea of the FT to the realities of the real world, where nothing stays unchanged for very long.

The STFT, or more accurately the short time PSD, goes under several different aliases in different fields. A 'musical score' is basically a STFT with a horizontal time axis, a vertical frequency axis and a special notation for durations. The STFT has long been a popular tool in speech analysis and processing, where it goes under the name of *sonogram*. The sonogram is conventionally depicted with a vertical frequency axis, with DC at the bottom, and a horizontal time axis, with time advancing from left to right. Each separate STFT is depicted by a single vertical line, traditionally drawn in a gray-scale. If there is no component at a given frequency at the time being analyzed the appropriate point is left white, while darker shades of gray represent higher energy levels. With the advent of DSP and computer graphics, analog sonographs with their rolls of paper have been replaced with scrolling graphics screens. The modern versions often use color rather than gray-scale, and allow interactive measurement as well.

Figure 4.8 is a sonogram of the author saying 'digital signal processing', with the sounds being uttered registered underneath. With some training one can learn to 'read' sonograms, and forensic scientists use the same sonograms for speaker identification. In the figure the basic frequency (pitch) of about

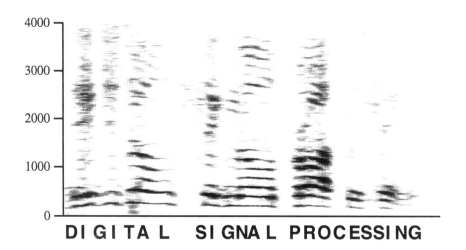

Figure 4.8: Sonogram of author saying 'digital signal processing'. The vertical axis is the frequency from 0 to 4000 Hz, while the horizontal axis is time (approximately 2 seconds). The sounds being uttered at each time are indicated by the writing below.

200 Hz is clearly visible at the bottom, and the difference between vowels and consonants is readily identifiable. You can probably also discern which syllables are accented, and may be able to see similarities among the various i sounds. The two s sounds in the last word seem to be invisible; this is due to their indeed having low energy, and most of that energy being spread out and at high frequencies, above the bandwidth displayed here. The **ing** is also very weak, due to being unaccented.

Rotating the sonogram by 90° we obtain the *falling raster spectrogram* popular in radar signal processing. Here the horizontal axis represents frequencies in the region of interest, time advances from top to bottom, and gray-scale intensity once again represents the square amplitude of the STFT component. Once the desired range of frequencies is selected, falling raster spectral displays provide intuitive real-time pictures; the display scrolling upwards as text does on a computer terminal.

The transition from FT to STFT requires forcing arbitrary signals to become finite time signals. To accomplish this we multiply the signal by a *window* function, that is, a function $w(t)$ that is strictly zero outside the time of interest. The window function itself should not introduce any artifacts to the spectrum of this product, and will be discussed in more detail in Section 13.4. For now you can think of the simplest window, the rectangular window of equation (4.21). Also commonly used are window functions that rise smoothly and continuously from zero to unity and then symmetrically drop back down to zero.

Of course, the uncertainty theorem puts a fundamental limitation on the precision of the STFT. The longer the time during which we observe a signal, the more precise will be our frequency distribution predictions; but the longer the window duration the more we blur the frequency changes that may be taking place in the signal. The uncertainty inequality does not allow us to simultaneously measure to arbitrary accuracy both the spectral composition and the times at which this composition changes.

The sonogram and similar graphic displays are tools to view the signal simultaneously in the time and frequency domains, yet they do not treat time and frequency on equal footing. What we may really want is to find a function $f(t, \omega)$ such that $f(t, \omega) \, dt \, d\omega$ is the energy in the 'time-frequency cell'. This brings us to define joint time-frequency distributions.

These are derived by considering time and frequency to be two characteristics of signals, just as height and weight are two characteristics of humans. In the latter case we can define a joint probability density $p(h, w)$ such that $p(h, w) \, dh \, dw$ is the percentage of people with both height between h and $h + dh$ *and* weight between w and $w + dw$ (see Appendix A.13). For such joint probability distributions we require the so-called 'marginals',

$$p(h) = \int p(h, w)\, dw \qquad\qquad p(w) = \int p(h, w)\, dh$$

where the integrations are over the entire range of possible heights and weights, $p(h)dh$ is the percentage of people with height between h and $h+dh$ regardless of weight, and $p(w)dw$ is the percentage of people with weight between w and $w + dw$ regardless of height.

Similarly, a joint time-frequency distribution is a function of both time and frequency $p(t, w)$. We require that the following marginals hold

$$s(t) = \int_{-\infty}^{\infty} p(t, w)\, dw \qquad\qquad S(w) = \int_{-\infty}^{\infty} p(t, w)\, dt$$

and the integration over both time and frequency must give the total energy, which we normalize to $E = 1$. We may then expect $p(t, w)\, dt\, dw$ to represent the amount of energy the signal has in the range between w and $w + dw$ during the times between t and $t + dt$.

Gabor was the first to express the STFT as a time-frequency distribution

$$p(t, w) = \frac{1}{\sqrt{2\pi}} \left| \int_{-\infty}^{\infty} s(\tau) w(\tau - t) e^{-i w \tau} d\tau \right|^2$$

but he suggested using Gaussian-shaped windows, rather than rectangular ones, since Gaussians have the minimal uncertainty product. Perhaps even simpler than the short-time PSD is the double-square distribution

$$p(t, w) = |s(t)|^2\, |S(w)|^2$$

while more complex is the Wigner-Ville distribution.

$$p(t, w) = \frac{1}{2\pi} \int s^* \left(t - \frac{\tau}{2} \right) e^{-i w \tau} s \left(t + \frac{\tau}{2} \right) d\tau$$

The double square requires computing $|S(w)|^2$ by the FT's integral over all time, and then simply multiplies this by the signal in the time domain. It is obviously zero for times or frequencies for which the signal is zero, but doesn't attempt any more refined time-frequency localization. The Wigner-Ville formula looks similar to equation (4.27) for finding the power spectrum via the autocorrelation, and is only one of an entire family of such *bilinear* distributions.

In addition to these, many other distributions have been proposed; indeed Cohen introduced a general family from which an infinite number of different time-frequency distributions can be derived,

$$p(t, \omega) = \frac{1}{4\pi^2} \int \int \int e^{-i\theta t - i\tau\omega + i\theta u} s^*(u - \frac{\tau}{2}) \varphi(\theta, \tau) s(u + \frac{\tau}{2}) \, du \, d\tau \, d\theta$$

but none are perfect. Although they all satisfy the marginals, unexpected behaviors turn up. For example, when two frequencies exist simultaneously, some distributions display a third in between. When one frequency component ceases and another commences a short time later, some distributions exhibit nonzero components in the gap. These strange phenomena derive from the bilinear nature of the Cohen distributions. Even more bizarre is the fact that while the short-time PSD and the double-square are always positive, most of the others can take on nonintuitive negative values.

EXERCISES

4.6.1 There is another case for which we can compute the FT after only a finite observation time, namely when someone guarantees the signal to be periodic. Do we need the STFT for periodic signals?

4.6.2 In the text, examples were presented of signals with identical power spectra. Doesn't this contradict the very nature of a *transform* as a reversible transformation to another domain? Resolve this paradox by demonstrating explicitly the difference between the three cases.

4.6.3 Compute the FT by numerical integration and plot the empirical PSD of a sinusoid of time duration T. How does the line width change with T?

4.6.4 A FSK signal at any given time is either one of two sinusoids, one of frequency ω_1, and the other of frequency ω_2. Generate a FSK signal that alternates between ω_1 and ω_2 every T seconds, but whose phase is continuous. Using a sampling frequency of 8000 Hz, frequencies 1000 and 2000 Hz, and an alternation rate of 100 per second, numerically compute the power spectrum for various window durations. You may overlap the windows if you so desire. Plot the result as a falling raster spectrogram. What do you get when a transition occurs inside a window? Does the overall picture match what you expect? Can you accurately measure both the frequencies and the times that the frequency changed?

4.6.5 Repeat the previous exercise with the double-square distribution.

4.6.6 Show that the uncertainty theorem does not put any restrictions on joint time-frequency distributions, by proving that any distribution that satisfies the marginals satisfies the uncertainty theorem.

4.7 The Discrete Fourier Transform (DFT)

We have often discussed the fact that signals are functions of time that have pertinent frequency domain interpretation. The importance of being able to transform between time and frequency domains is accordingly evident. For analog signals we have seen that the vehicle for performing the transformation is the Fourier transform (FT), while in DSP it is the **Discrete Fourier Transform (DFT)**.

The DFT can be derived from the FT

$$S(\omega) = \int_{-\infty}^{\infty} s(t)\, e^{-i\omega t}\, dt$$

by discretization of the time variable. To accomplish this we must first determine the entire interval of time $[t_a \ldots t_z]$ wherein $s(t)$ is significantly different from zero. We will call the duration of this interval $T \equiv t_z - t_a$. If this time interval is very large, or even the entire t axis, then we can partition it up in some manner, and calculate the FT separately for each part. Next divide the interval into N equal-sized bins by choosing N equally spaced times $\{t_n\}_{n=0}^{N-1}$ in the following fashion $t_n = t_a + n\Delta t$ where $\Delta t \equiv \frac{T}{N}$. (Note that $t_0 = t_a$ but $t_{N-1} = t_z - \Delta t$; however, $t_{N-1} \approx t_z$ when $N \gg 1$ or equivalently $\Delta t \ll T$.) If we allow negative n, we can always take $t_a = 0$ without limiting generality. In this case we have $t_n = n\Delta t$. For sampled signals we recognize Δt as the basic sample interval (the inverse of the sampling frequency) $t_s = \frac{1}{f_s}$.

Now we also want to discretize the frequency variable. In a similar way we will define $\omega_k = k\Delta\omega$ with $\Delta\omega \equiv \frac{\Omega}{N}$. It is obvious that short time intervals correspond to high frequencies, and vice versa. Hence, if we choose to use a small Δt we will need a high upper frequency limit Ω. The exact correspondence is given by

$$N\Delta\omega = \Omega = \frac{2\pi}{\Delta t} \qquad or \qquad \Delta\omega\Delta t = \frac{2\pi}{N} \qquad (4.28)$$

where we recognize an uncertainty product.

We can now evaluate the FT integral (4.2) as a Riemann sum, substituting t_n and ω_k for the time and frequency variables,

$$S(\omega) = \int_{-\infty}^{\infty} s(t)\, e^{-i\omega t}\, dt \longrightarrow S_k = \sum_{n=0}^{N-1} s_n e^{-i(k\Delta\omega)(n\Delta t)}$$

which upon substitution gives

$$S_k = \sum_{n=0}^{N-1} s_n e^{-i\frac{2\pi nk}{N}} \qquad (4.29)$$

which is the DFT. The power spectrum for the digital case is $|S_k|^2$ and each k represents the energy that the signal has in the corresponding 'frequency bin'.

For a given N, it is useful and customary to define the Nth root of unity W_N. This is a number, in general complex, that yields unity when raised to the Nth power. For example, one square root of unity is -1 since $(-1)^2 = 1$; but $1^2 = 1$ so 1 is a square root of itself as well. Also i is a fourth root of unity since $i^2 = (-10)^2 = 1$, but so are $-i$, -1, and 1. There is a unique *best* choice for W_N, namely the trigonometric constant

$$W_N \equiv e^{-i\frac{2\pi}{N}} = \cos\left(\frac{2\pi}{N}\right) - i\sin\left(\frac{2\pi}{N}\right) \tag{4.30}$$

which for $N = 2$ is as follows.

$$W_2 = e^{-i\frac{\pi}{N}} = -1 \tag{4.31}$$

This is the best choice since its powers W_N^k for $k = 0 \ldots N - 1$ embrace all the N roots. Thinking of the complex numbers as points in the plane, W_N is clearly on the unit circle (since its absolute value is one) and its phase angle is $\frac{1}{N}$ of the way around the circle. Each successive power moves a further $\frac{1}{N}$ around the circle until for $N = 1$ we return to $W_N^0 = 1$. This is illustrated in Figure 4.9 for $N = 8$.

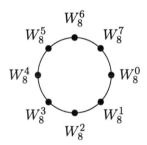

Figure 4.9: The N complex roots of unity displayed graphically. (Here $N = 8$.)

In terms of W_N the DFT can be expressed

$$S_k = \sum_{n=0}^{N-1} s_n W_N^{nk} \tag{4.32}$$

(just note that $(W_N)^{nk} = (e^{-i\frac{2\pi}{N}})^{nk} = e^{-i\frac{2\pi nk}{N}}$). The powers $(W_N)^{nk}$ are also on the unit circle, but at integer multiples of the basic angle. Consequently the set of all the powers of W_N divides the unit circle into N equal pieces.

It is illuminating to investigate the simplest DFT, the two-point transform. Substituting $N = 2$ into equation (4.31) we readily find

$$S_0 = \sum_{n=0}^{1} s_n W_2^0 = s_0 + s_1$$

$$S_1 = \sum_{n=0}^{1} s_n W_2^n = s_0 - s_1 \tag{4.33}$$

which has a simple interpretation. The zeroth (DC) coefficient is simply the sum (i.e., twice the average of s_0 and s_1). The other (high-frequency) coefficient is the difference (the derivative).

How do we return to the time domain given the discrete frequency components S_k?

$$s_n = \frac{1}{N} \sum_{k=0}^{N-1} S_k W_N^{-nk} \tag{4.34}$$

This is easy to show by direct substitution of (4.32).

Equations (4.32) and (4.34) are the main results of this section. We see that the s_n and the S_k can be calculated one from the other, and so contain precisely the same information. They form what is known as the discrete Fourier transform pair. With the equations we have derived one can go back and forth between the time and frequency domains, with absolutely no loss of information.

The DFT as we have derived it looks only at s_n over a finite interval of time. What happens if we take the DFT S_k and try to find s_n for times not in the interval from 0 to $N - 1$? The DC term is obviously the same outside the interval as inside, while all the others are periodic in N. Hence the DFT predicts $s_{N+n} = s_n$, not $s_{N+n} = 0$ as we perhaps expected! There is no way of getting around this paradox; as discussed in Section 2.8 the very act of sampling an analog signal to convert it into a digital one forces the spectrum to become periodic (aliased).

The only way to handle a nonperiodic infinite duration digital signal is to let the DFT's duration N increase without limit. Since the Nyquist frequency range is divided into N intervals by the DFT, the frequency resolution increases until the frequency bins become infinitesimal in size. At this point we have a denumerably infinite number of time samples but a continuous frequency variable $S(\omega)$ (defined only over the Nyquist interval). There is no consensus in the literature as to the name of this Fourier transform. We will sometimes call it the **Long Time DFT** (LTDFT) but only when we absolutely need to differentiate between it and the usual DFT.

The (short time) DFT takes in a finite number of digital values and returns a finite number of digital values. We thus have a true transform designed for digital computation. However, this transform is still a mathematical concept, not a practical tool. In Chapter 14 we will see that the DFT is eminently practical due to the existence of an efficient algorithm for its computation.

EXERCISES

4.7.1 Derive the LTDFT directly from the FT.

4.7.2 Express W_N^{-k} and $W_N^{(N-1)-k}$ in terms of W_N^k. Express $W_N^k + W_N^{-k}$ and $W_N^k - W_N^{-k}$ in terms of sine and cosine. How much is $W_N^{(n+m)k}$? Derive the trigonometric sum formulas (A.23) using these relations.

4.7.3 What is the graphical interpretation of raising a complex number to a positive integer power? What is special about numbers on the unit circle? Give a graphical interpretation of the fact that all powers of W_N are N roots of unity. Write a program that draws the unit circle and all the W_N^k. Connect consecutive powers of each root with straight lines. Describe the pictures you obtain for odd and even N.

4.7.4 What are the equations for 4-point DFT, and what is their interpretation?

4.7.5 Write a straightforward routine for the computation of the DFT, and find the digital estimate of the PSD of various sinusoids. Under what conditions is the estimate good?

4.8 DFT Properties

Some of the DFT's properties parallel those of the FT for continuous signals discussed in Section 4.3, but some are specific to signals with discrete time index. For most of the properties we will assume that the frequency index is discrete as well, but the obvious extensions to the LTDFT will hold.

First, let's review properties that we have already mentioned. We clearly need for the inverse operation defined in equation 4.34 to be a true inverse operation, (i.e., we need a sort of 'Fourier sum theorem').

$$\text{DFT}^{-1}\,\text{DFT}\ s = s \qquad\qquad \text{DFT}\,\text{DFT}^{-1}\,S = S \qquad (4.35)$$

It is equally important for the DFT to be linear.

$$\begin{aligned} \mathrm{DFT}(x_n + y_n) &= X_k + Y_k \\ \mathrm{DFT}(a s_n) &= a S_k \end{aligned} \qquad (4.36)$$

Also important, but not corresponding to any characteristic of the FT, are the facts that the DFT and its inverse are *periodic* with period N. For example, when given a signal $s_0, s_1, \dots s_{N-1}$ we usually compute the DFT for the N frequencies centered around DC. If we want the DFT at some frequency outside this range, then we exploit periodicity.

$$S_{k+mN} = S_k \qquad \text{for all integer } m \qquad (4.37)$$

This leads us to our first implementational issue; how should we put the DFT values into a vector? Let's assume that our signal has $N = 8$ samples, the most commonly used indexation being 0 to $N - 1$ (i.e., $s_0, s_1, \dots s_7$). Since there are only 8 data points we can get no more than 8 independent frequency components, about half of which are negative frequency components.

$$S_{-4}, S_{-3}, S_{-2}, S_{-1}, S_0, S_1, S_2, S_3$$

Why is there an extra negative frequency component? Consider the signals

$$e^{\mathrm{i} 2 \pi f n} = \cos(2\pi f n) + \mathrm{i} \sin(2\pi f n) \qquad \text{where } f = \frac{k}{N}$$

for integer k, which are precisely the signals with only one nonzero DFT component. For all integer k in the range $1 \le k \le \frac{N}{2}$ the signal with frequency $f = +\frac{k}{N}$ and the corresponding signal with negative frequency $f = -\frac{k}{N}$ are different. The real part of the complex exponential is a cosine and so is unchanged by sign reversal, but the imaginary term is a sine and so changes sign. Hence the two signals with the same $|f|$ are complex conjugates. When $k = -\frac{N}{2}$ the frequency is $f = -\frac{1}{2}$ and the imaginary part is identically zero. Since this signal is real, the corresponding $f = +\frac{1}{2}$ signal is indistinguishable. Were we (despite the redundancy) to include both $f = \pm\frac{1}{2}$ signals in a 'basis', the corresponding expansion coefficients of an arbitrary signal would be identical; exactly that which is needed for periodicity to hold.

$$\dots S_{-4}, S_{-3}, S_{-2}, S_{-1}, S_0, S_1, S_2, S_3, S_{-4}, S_{-3}, S_{-2}, S_{-1}, S_0, S_1, S_2, S_3, \dots$$

In fact, any N consecutive Fourier coefficients contain all the information necessary to reconstruct the signal, and the usual convention is for DFT routines to return them in the order

$$S_0, \ S_1, \ S_2, \ S_3, \ S_4 = S_{-4}, \ S_5 = S_{-3}, \ S_6 = S_{-2}, \ S_7 = S_{-1}$$

obtained by swapping the first half $(S_{-4}, S_{-3}, S_{-2}, S_{-1})$ with the second (S_0, S_1, S_2, S_3).

Let's observe a digital signal s_n from time $n = 0$ until time $n = N - 1$ and convert it to the frequency domain S_k. Now using the iDFT we can compute the signal in the time domain for all times n, and as we saw in the previous section the resulting s_n will be periodic. No finite observation duration can completely capture the behavior of nonperiodic signals, and assuming periodicity is as good a guess as any. It is convenient to visualize digital signals as circular buffers, with the periodicity automatically imposed by the buffer mechanics.

Now for some new properties. The DFT of a real signal is Hermitian even,

$$S_{-k} = S_k^* \qquad \text{for real } s_n \tag{4.38}$$

and that of an imaginary signal is Hermitian odd. Evenness (or oddness) for finite duration discrete time signals or spectra is to be interpreted according to the indexation scheme of the previous paragraph. For example, the spectrum $S_0, S_1, S_2, S_3, S_{-4}, S_{-3}, S_{-2}, S_{-1} =$

$$7, \ -1+(\sqrt{2}+1)i, \ -1+i, \ -1+\frac{\sqrt{2}-1}{4}i, \ -1, \ -1-\frac{\sqrt{2}-1}{4}i, \ -1-i, \ -1-(\sqrt{2}+1)i$$

is Hermitian even and hence corresponds to a real signal. This property allows us to save computation time by allowing us to compute only half of the spectrum when the input signal is real.

Conversely, real spectra come from Hermitian even signals ($s_{-n} = s_n^*$) and pure imaginary spectra from Hermitian odd signals. For example, the DFT of the signal $s_0, s_1, s_2, s_3, s_4, s_5, s_6, s_7 =$

$$7, \ -1-(\sqrt{2}+1)i, \ -1-i, \ -1-2(\sqrt{2}-1)i, \ -1, \ -1+2(\sqrt{2}-1)i, \ -1+i, \ -1+(\sqrt{2}+1)i$$

will be real.

The properties that deal with transforming the discrete time and frequency axes are the time shifting property

$$\text{DFT } s_{n-m} = e^{-imk} S_k \tag{4.39}$$

the time reversal property

$$\text{DFT } s_{-n} = S_{-k} \tag{4.40}$$

and the frequency shifting (mixing) property.

$$\text{DFT } \left(s_n e^{in\kappa}\right) = S_{k-\kappa} \tag{4.41}$$

Of course, for finite-duration DFTs, time shifts can move us to times where we haven't observed the signal, and frequency shifts to frequencies where we haven't computed the DFT. When this happens simply use the periodicity properties. When we use the word 'shift' for digital signals we always mean 'circular shift' (i.e., shift in a circular buffer).

Parseval's relation for the DFT is easy to guess

$$N \sum_{n=0}^{N-1} |s_n|^2 = \sum_{k=0}^{N-1} |S_k|^2 \tag{4.42}$$

and for infinite duration signals the sum on the left is over a denumerably infinite number of terms and the right-hand side becomes an integral.

$$\sum_{n=0}^{\infty} |s_n|^2 = \int_{-\infty}^{\infty} |S_k|^2 \, dk \tag{4.43}$$

The simplest application of Parseval's relation for the DFT involves a signal of length two. The DFT is

$$S_0 = s_0 + s_1 \qquad\qquad S_1 = s_0 - s_1$$

and it is easy to see that Parseval's relation holds.

$$S_0^2 + S_1^2 = (s_0 + s_1)^2 + (s_0 - s_1)^2 = 2(s_0^2 + s_1^2)$$

Products of discrete signals or spectra correspond to convolution *sums* rather than convolution integrals.

$$\text{LTDFT}\left(x_n y_n\right) = \sum_{\kappa=-\infty}^{\infty} X_k Y_{k-\kappa} \equiv X * Y \tag{4.44}$$

$$\text{LTDFT}^{-1}\left(X(\omega)Y(\omega)\right) = \sum_{m=-\infty}^{\infty} x_n y_{n-m} \equiv x * y \tag{4.45}$$

When the signals are of finite time duration the periodicity forces us to define a new kind of convolution sum, known as circular (or cyclic) convolution.

$$\text{DFT}\left(x_n y_n\right) = \frac{1}{N} \sum_{\kappa=0}^{N-1} X_k Y_{(k-\kappa) \bmod N} = X \circledast Y \tag{4.46}$$

$$\text{DFT}^{-1}\left(X_k Y_k\right) = \sum_{m=0}^{N-1} x_n y_{(n-m) \bmod N} = x \circledast y \tag{4.47}$$

where the indices $k - \kappa$ and $n - m$ wrap around according to the periodicity. In other words, while the linear (noncircular) convolution of x_0, x_1, x_2, x_3 with y_0, y_1, y_2, y_3 gives

$$
\begin{aligned}
x * y \;=\; & x_0 y_0, \\
& x_0 y_1 + x_1 y_0, \\
& x_0 y_2 + x_1 y_1 + x_2 y_0, \\
& x_0 y_3 + x_1 y_2 + x_2 y_1 + x_3 y_0, \\
& x_1 y_3 + x_2 y_2 + x_3 y_1, \\
& x_2 y_3 + x_3 y_2 \\
& x_3 y_3
\end{aligned}
$$

the circular convolution gives the following periodic signal.

$$
\begin{aligned}
x \circledast y \;=\; & \ldots \\
& x_0 y_0 + x_1 y_3 + x_2 y_2 + x_3 y_1, \\
& x_0 y_1 + x_1 y_0 + x_2 y_3 + x_3 y_2, \\
& x_0 y_2 + x_1 y_1 + x_2 y_0 + x_3 y_3, \\
& x_0 y_3 + x_1 y_2 + x_2 y_1 + x_3 y_0, \\
& x_0 y_0 + x_1 y_3 + x_2 y_2 + x_3 y_1, \\
& x_0 y_1 + x_1 y_0 + x_2 y_3 + x_3 y_2, \\
& x_0 y_2 + x_1 y_1 + x_2 y_0 + x_3 y_3, \\
& \ldots
\end{aligned}
$$

We will return to the circular convolution in Section 15.2.

To demonstrate the use of some of the properties of the FT and DFT we will now prove the sampling theorem. Sampling can be considered to be implemented by multiplying the bandlimited analog signal $s(t)$ by a train of impulses spaced t_s apart. This multiplication in the time domain is equivalent to a convolution in the frequency domain, and since the FT if an impulse train in time is an impulse train in frequency, the convolution leads to a periodic FT. Stated in another way, the multiplication is a sampled signal s_n, and thus we should talk in terms of the DFT, which is periodic. We know that the impulse train in the frequency domain has repetition frequency $f_s = \frac{1}{t_s}$, and so the convolution forces the frequency domain representation to be periodic with this period. The situation is clarified in Figure 4.10 for the case of bandwidth less than half f_s. If the analog signal $s(t)$ has bandwidth wider than $\frac{1}{2} f_s$ the spectra will overlap, resulting in an irreversible loss of information.

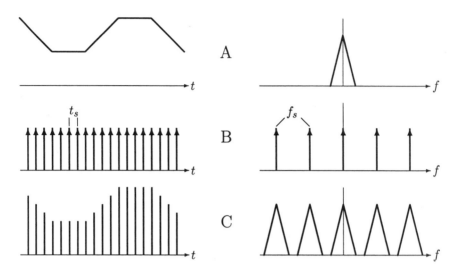

Figure 4.10: The sampling theorem. On the left we see the signals of interest in the time domain, and on the right in the frequency domain. The graphs in (A) depict the original analog signal, those in (B) the sampling impulses, and in (C) the sampled signal.

EXERCISES

4.8.1 Prove all of the DFT's properties stated above.

4.8.2 DFT routines usually return the same number of outputs as inputs, but sometimes we need higher frequency resolution. Assuming that we only have access to N samples, how can we generate $2N$ DFT components? Conversely, assume we have N DFT components and require $2N$ signal values. How can we retrieve them? These tricks seem to create new information that didn't previously exist. How can this be?

4.8.3 Prove that an even time signal has an even DFT, and an odd time signal has an odd DFT. What can you say about real even signals?

4.8.4 Explain why we didn't give the counterparts of several of the properties discussed for the FT (e.g., time scaling and differentiation).

4.8.5 Why does the circular convolution depend on N? (Some people even use the notation $x \otimes y$ to emphasize this fact.)

4.8.6 In Section 2.8 we mentioned the band-pass sampling theorem that holds for a signal with components from frequency $f_0 > 0$ to $f_z > f_0$. Using a figure similar to Figure 4.10 find the precise minimal sampling rate.

4.8.7 What can be said about the FT of a signal that is zero outside the time interval $-T < t < +T$? (Hint: This is the converse of the sampling theorem.)

4.9 Further Insights into the DFT

In this section we wish to gain further insight into the algebraic and computational structure of the DFT. This insight will come from two new ways of understanding the DFT; the first as the product of the W matrix with the signal, and the second as a polynomial in W.

The DFT is a linear transformation of a finite length vector of length N to a finite length vector of the same length. Basic linear algebra tells us that all linear transformations can be represented as matrices. This representation is also quite evident from equation (4.32)! Rather than discussing a function that transforms N signal values s_0 through s_{N-1} into frequency bins S_0 through S_{N-1}, we can talk about the product of an N by N matrix $\underline{\underline{W}}$ with an N-vector $(s_0, \ldots s_{N-1})$ yielding an N-vector $(S_0, \ldots S_{N-1})$.

$$\underline{S} = \underline{\underline{W}}\,\underline{s} \tag{4.48}$$

For example, the simple two-point DFT of equation (4.33) can be written more compactly as

$$\begin{pmatrix} S_0 \\ S_1 \end{pmatrix} = \begin{pmatrix} 1 & 1 \\ 1 & -1 \end{pmatrix} \begin{pmatrix} s_0 \\ s_1 \end{pmatrix}$$

as can be easily seen. More generally, the W_N matrix is

$$\underline{\underline{W}} = \begin{pmatrix} W_N^0 & W_N^0 & W_N^0 & \cdots & W_N^0 \\ W_N^0 & W_N^1 & W_N^2 & \cdots & W_N^{N-1} \\ W_N^0 & W_N^2 & W_N^4 & \cdots & W_N^{2(N-1)} \\ W_N^0 & W_N^3 & W_N^6 & \cdots & W_N^{3(N-1)} \\ \vdots & \vdots & \vdots & & \vdots \\ W_N^0 & W_N^{N-1} & W_N^{2(N-1)} & \cdots & W_N^{(N-1)(N-1)} \end{pmatrix} \tag{4.49}$$

$$= \begin{pmatrix} 1 & 1 & 1 & \cdots & 1 \\ 1 & W_N & W_N^2 & \cdots & W_N^{N-1} \\ 1 & W_N^2 & W_N^4 & \cdots & W_N^{2(N-1)} \\ 1 & W_N^3 & W_N^6 & \cdots & W_N^{3(N-1)} \\ \vdots & \vdots & \vdots & & \vdots \\ 1 & W_N^{N-1} & W_N^{2(N-1)} & \cdots & W_N^{(N-1)(N-1)} \end{pmatrix}$$

and since W_N is the Nth root of unity, the exponents can be reduced modulo N. Thus

$$\underline{\underline{W_2}} = \begin{pmatrix} W_2^0 & W_2^0 \\ W_2^0 & W_2^1 \end{pmatrix} = \begin{pmatrix} 1 & 1 \\ 1 & -1 \end{pmatrix} \tag{4.50}$$

$$
\underline{\underline{W_4}} = \begin{pmatrix} W_4^0 & W_4^0 & W_4^0 & W_4^0 \\ W_4^0 & W_4^1 & W_4^2 & W_4^3 \\ W_4^0 & W_4^2 & W_4^4 & W_4^6 \\ W_4^0 & W_4^3 & W_4^6 & W_4^9 \end{pmatrix} = \begin{pmatrix} 1 & 1 & 1 & 1 \\ 1 & W_4 & W_4^2 & W_4^3 \\ 1 & W_4^2 & W_4^0 & W_4^2 \\ 1 & W_4^3 & W_4^2 & W_4^1 \end{pmatrix} \quad (4.51)
$$

$$
= \begin{pmatrix} 1 & 1 & 1 & 1 \\ 1 & -i & -1 & i \\ 1 & -1 & 1 & -1 \\ 1 & i & -1 & -i \end{pmatrix}
$$

and

$$
\underline{\underline{W_8}} = \begin{pmatrix} W_8^0 & W_8^0 & W_8^0 & W_8^0 & W_8^0 & W_8^0 & W_8^0 & W_8^0 \\ W_8^0 & W_8^1 & W_8^2 & W_8^3 & W_8^4 & W_8^5 & W_8^6 & W_8^7 \\ W_8^0 & W_8^2 & W_8^4 & W_8^6 & W_8^0 & W_8^2 & W_8^4 & W_8^6 \\ W_8^0 & W_8^3 & W_8^6 & W_8^1 & W_8^4 & W_8^7 & W_8^2 & W_8^5 \\ W_8^0 & W_8^4 & W_8^0 & W_8^4 & W_8^0 & W_8^4 & W_8^0 & W_8^4 \\ W_8^0 & W_8^5 & W_8^2 & W_8^7 & W_8^4 & W_8^1 & W_8^6 & W_8^3 \\ W_8^0 & W_8^6 & W_8^4 & W_8^2 & W_8^0 & W_8^6 & W_8^4 & W_8^2 \\ W_8^0 & W_8^7 & W_8^6 & W_8^5 & W_8^4 & W_8^3 & W_8^2 & W_8^1 \end{pmatrix} \quad (4.52)
$$

which can be made explicit using $W_8 = e^{-i\frac{\pi}{4}} = \frac{\sqrt{2}}{2}(1-i)$.

The W matrix is symmetric, as is obvious from the above examples, but there are further, less obvious, symmetries as well. For instance, any two rows of the matrix are orthogonal, and the squared length (sum of squares of the elements) of any row is precisely N. Furthermore, there are relations between the elements of W_N and those of W_M when M divides N. It is these relations that make the FFT possible, as will be explained in Section 14.5.

The matrix representation gives us a simple interpretation for the inverse DFT as well. The IDFT's matrix must be the inverse of the DFT's matrix

$$
\underline{s} = \underline{\underline{W}}^{-1} \underline{S} \quad (4.53)
$$

and

$$
\underline{\underline{W}}^{-1} = \frac{1}{N} \underline{\underline{W}}^* \quad (4.54)
$$

where the Hermitian conjugate of the W_N matrix has elements

$$
(W^*)_N^{nk} = e^{+i\frac{2\pi nk}{N}} = W_N^{-nk}
$$

as can easily be shown.

There is yet another way of writing the basic formula for the DFT (4.32) that provides us with additional insight. For given N and k let us drop the indices and write $W \equiv W_N^k$. Then the DFT takes the form of a polynomial in W with coefficients s_n

$$S_k = \sum_{n=0}^{N-1} s_n W^n \qquad (4.55)$$

which is a viewpoint that is useful for two reasons. First, the connection with polynomials will allow use of efficient algorithms for computation of polynomials to be used here as well. The FFT, although first introduced in signal processing, can be considered to be an algorithm for efficient multiplication of polynomials. Also, use of Horner's rule leads to an efficient recursive computation for the DFT known as Goertzel's algorithm. Second, a more modern approach considers the DFT as the polynomial approximation to the *real* spectrum. When the real spectrum has sharp peaks such a polynomial approximation may not be sufficient and rational function approximation can be more effective.

EXERCISES

4.9.1 Write explicitly the matrices for DFT of sizes 3, 5, 6, 7, and 8.

4.9.2 Invert the DFT matrices for sizes 2, 3, and 4. Can you write the iDFT matrix in terms of the DFT matrix?

4.9.3 Prove that any two rows of the DFT matrix are orthogonal and that the squared length of any row is N. Show that $\frac{1}{\sqrt{N}} \underline{\underline{W_N}}$ is a unitary matrix.

4.10   The z Transform

So far this chapter has dealt exclusively with variations on a theme by Fourier. We extended the FS for periodic analog signals to the FT of arbitrary analog signals, adapted it to the DFT of arbitrary digital signals, and modified it to the STFT of changing signals. In all the acronyms the ubiquitous **F** for Fourier appeared; and for good reason. The concept of spectrum *a la Fourier* is rooted in the basic physics of all signals. From colors of light through the pitch of voices and modes of mechanical vibration to frequencies of radio stations, Fourier's concept of frequency spectrum is so patently useful that it is hard to imagine using anything else.

In the special world of DSP there *is*, however, an alternative. This alternative is entirely meaningless in the analog world, in some ways less meaningful than the Fourier spectrum even in the digital world, and on occasion seems to be a mere artificial, purely mathematical device. It *does* sometimes enhance our understanding of signals, often greatly simplifies calculations, and always includes Fourier's spectrum as a special case.

This alternative is called the *z transform*, which we shall denote zT. This nomenclature is admittedly bizarre since the use of the letter *z* is completely arbitrary (there was no section in the previous chapter named 'Z Discovers Spectrum'), and it is not really a transform at all. Recall that the FS, which maps periodic analog signals to discrete spectra, is not called a transform. The FT, which maps analog signals to continuous spectra, and the DFT, which makes digital signals into discrete spectra, are. The zT takes an arbitrary digital signal and returns a continuous function. This change of form from sequence to function should disqualify it from being called a *transform*, but for some reason doesn't. Even more curious is the fact that outside the DSP sphere of influence the term 'z transform' is entirely unknown; but a closely related entity is universally called the *generating function*.

As we have done in the past, we shall abide by DSP tradition. After all, every field has its own terminology that has developed side by side with its advances and applications, even if these terms seem ridiculous to outsiders. Computer hardware engineers use *flip-flops* without falling. Programmers use *operating systems* without upsetting surgeons. Mathematicians use *irrational* numbers and *nonanalytic* functions, and no one expects either to act illogically. High-energy physicists hypothesize subatomic particles called *quarks* that have *strangeness*, *flavor*, and even *charm*. When lawyers *garnish* they leave people without appetite, while according to their definitions the victim of *battery* can be left quite powerless. So saying *DC* when there is no electric current, *spectral* when we are not scared, and *z transform* pales in comparison with the accepted terminologies of other fields!

The basic idea behind the classic generating function is easy to explain; it is a trick to turn an infinite sequence into a function. Classic mathematics simply knows a lot more about functions than it does about infinite sequences. Sometimes sequences can be bounded from above or below and in this way proven to converge or not. A few sequences even have known limits. However, so much more can be accomplished when we know how to change arbitrary sequences into functions; specifically, recursions involving sequence elements become algebraic equations when using generating functions.

Given a sequence $s_0, s_1, s_2, \ldots$, its generating function is defined to be

$$s(x) \equiv \sum_{n=0}^{\infty} s_n x^n \qquad (4.56)$$

basically an infinite polynomial in x. The variable x itself is entirely artificial, being introduced solely for the purpose of giving the generating function a domain. It is easily seen that the correspondence between a sequence and its generating function is one-to-one; different sequences correspond to different generating functions, and different generating functions generate different sequences. In a way, generating sequences are the opposite of Taylor expansions. A Taylor expansion takes a function $s(x)$ and creates a sequence of coefficients s_n of exactly the form of equation (4.56), while the generating function does just the opposite. The Taylor coefficients give us intuition as to the behavior of the function, while the generating function gives us insight as to the behavior of the sequence.

We can demonstrate the strength of the generating function technique with a simple example, that of the Fibonacci sequence f_n. This famous sequence, invented by Leonardo of Pisa (nicknamed Fibonacci) in 1202, models the number of female rabbits in successive years. We assume that each mature female rabbit produces a female offspring each year and that no rabbit ever dies. We start with a single female rabbit ($f_0 = 1$); there is still only that rabbit after one year ($f_1 = 1$), since it takes a year for the rabbit to reach maturity. In the second year a new baby rabbit is born ($f_2 = 2$), and another in the third ($f_3 = 3$). Thereafter in each year we have the number of rabbits alive in the previous year *plus* those born to rabbits who were alive two years ago. We can deduce the recursive definition

$$f_0 = 1 \qquad f_1 = 1 \qquad f_n = f_{n-1} + f_{n-2} \qquad \text{for } n \geq 2 \qquad (4.57)$$

that produces the values $1, 1, 2, 3, 5, 8, 13, 21, \ldots$. However, were we to need f_{137} we would have no recourse other than to recurse 137 times. Is there an explicit (nonrecursive) formula for f_n? At this point we don't see any way to find one, but this is where the generating function can help. Generating functions convert complex recursions into simple algebraic equations that can often be solved.

The generating function for the Fibonacci sequence is

$$f(x) = \sum_{n=0}^{\infty} f_n x^n = 1 + x + 2x^2 + 3x^3 + 5x^4 + 8x^5 + \ldots$$

and this is what we wish to evaluate. To proceed, take the recursion that defines the Fibonacci sequence, multiply both sides by x^n and sum from $n = 2$ to infinity.

$$\sum_{n=2}^{\infty} f_n x^n = \sum_{n=2}^{\infty} f_{n-1} x^n + \sum_{n=2}^{\infty} f_{n-2} x^n$$

$$= x \sum_{n=2}^{\infty} f_{n-1} x^{n-1} + x^2 \sum_{n=2}^{\infty} f_{n-2} x^{n-2}$$

$$= x \sum_{n=1}^{\infty} f_n x^n + x^2 \sum_{n=0}^{\infty} f_n x^n$$

$$f(x) - f_0 x^0 - f_1 x^1 = x \left(f(x) - f_0 x^0 \right) + x^2 f(x)$$

$$f(x) - 1 - x = f(x)x - x + f(x)x^2$$

Solving the algebraic equation we easily find an explicit expression for the generating function

$$f(x) = \frac{1}{1 - x - x^2}$$

which is plotted in Figure 4.11.

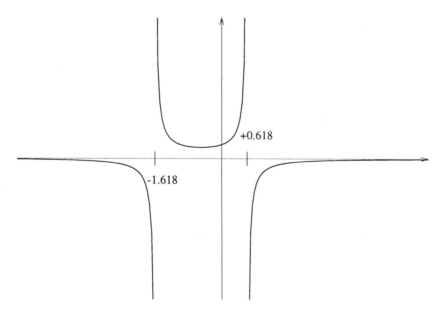

Figure 4.11: The generating function for the Fibonacci sequence. Note the divergences at $-\gamma \approx -1.618$ and $-\gamma' \approx 0.618$.

The zeros of the quadratic in the denominator are easily found to be $-\gamma$ and $-\gamma'$ where $\gamma \equiv \frac{1+\sqrt{5}}{2} = \cos^{-1}(\frac{\pi}{5})$ is the famous 'golden ratio' and $\gamma' \equiv \frac{1-\sqrt{5}}{2} = -\frac{1}{\gamma}$.

We can now return to our original problem. In order to find an explicit formula for the nth Fibonacci element, we need only to rewrite the generating function as an infinite polynomial and pick out the coefficients. To do this we use a 'partial fraction expansion'

$$f(x) = \frac{1}{(x+\gamma)(x+\gamma')} = \frac{1}{a-b}\left(\frac{a}{1-ax} - \frac{b}{1-bx}\right)$$

where $a + b = -ab = 1$. Utilizing the formula for the sum of a geometric progression $\frac{1}{1-ax} = \sum_{n=0}^{\infty}(ax)^n$ and comparing term by term, we find

$$f_n = \frac{1}{\sqrt{5}}\left(\gamma^{n+1} - (\gamma')^{n+1}\right) \tag{4.58}$$

the desired explicit formula for the nth Fibonacci element.

Most people when seeing this formula for the first time are amazed that this combination of irrational numbers yields an integer at all. When that impression wears off, a feeling of being tricked sets in. The two irrational numbers in the numerator contain exactly a factor of $\sqrt{5}$, which is exactly what is being eliminated by the denominator; but if it is all a trick why can't a formula without a $\sqrt{5}$ be devised? So we are now surprised by our prior lack of surprise! Equation (4.58) is *so* astounding that you are strongly encouraged to run to a computer and try it out. Please remember to round the result to the nearest integer in order to compensate for finite precision calculations.

Now that we have become convinced of the great utility of generating functions, we will slightly adapt them for use in DSP. The z-transform is conventionally defined as

$$S(z) = \text{zT}(s_n) = \sum_{n=-\infty}^{\infty} s_n z^{-n} \tag{4.59}$$

and you surely discern two modifications but there is also a third. First, we needed to make the sum run from minus infinity rather than from zero; second, the DSP convention is to use z^{-1} rather than x; and third, we will allow z to be a complex variable rather than merely a real one. The second change is not really significant because of the first; using z instead of z^{-1} is equivalent to interchanging s_n with s_{-n}. The really consequential

change is that of using a complex variable. Unlike the generating function
we saw above, $S(z)$ is defined over the complex plane, called the z-plane.
Sinusoids correspond to z on the unit circle, decaying exponentials to z
inside the unit circle, growing exponentials to z outside the unit circle. The
definition of z in the complex plane makes available even more powerful
analytic techniques. The study of functions of complex variables is one of
the most highly developed disciplines that mathematics has to offer, and
DSP harnesses its strength via the z transform.

Any complex variable z can be written in polar form

$$z = re^{i\omega}$$

where r is the magnitude, and ω the angle. In particular, if z is on the unit
circle $r = 1$, and $z = e^{i\omega}$. If we evaluate the zT on the unit circle in the
z-plane, considering it to be a function of angle, we find

$$s(\omega) = S(z)\Big|_{z=e^{i\omega}} = \sum_{n=-\infty}^{\infty} s_n z^{-n} = \sum_{n=-\infty}^{\infty} s_n e^{-i\omega n} \qquad (4.60)$$

which is precisely the DFT. The zT reduces to the DFT if evaluated on the
unit circle.

For other nonunity magnitudes we can always write $r = e^{\lambda}$ so that
$z = e^{\lambda+i\omega}$ and

$$S(z) = \sum_{n=-\infty}^{\infty} s_n z^{-n} = \sum_{n=-\infty}^{\infty} s_n e^{-(\lambda+i\omega)n} \qquad (4.61)$$

which is a digital version of the **Laplace Transform** (LT). The Laplace transform, which will not be discussed in detail here, expands functions in terms
of exponentially increasing or damped sinusoids, of the type described in
equation (2.11). Its expression is

$$f(s) = \int_{-\infty}^{\infty} f(t)e^{-st}dt \qquad (4.62)$$

where s is understood to be complex (defining the s-plane). Sinusoids correspond to purely imaginary s, decaying exponentials to positive real s, growing exponentials to negative real s. The LT generalizes the FT, since the
FT is simply the LT along the imaginary s axis. This is analogous to the zT
generalizing the DFT, where the DFT is the zT on the unit circle. Although
a large class of analog signals can be expanded using the FT, the LT may
be more convenient, especially for signals that actually increase or decay

with time. This is analogous to the DFT being a sufficient representation for most digital signals but the zT often being more useful.

We have been ignoring a question that always must be raised for infinite series. Does expression (4.59) for the zT *converge?* When there are only a finite number of terms in a series there is no problem with performing the summation, but with an infinite number of terms the terms must decay fast enough with n for the sum not to explode. For complex numbers with large magnitudes the terms will get larger and larger with n, and the whole sum becomes meaningless.

By now you may have become so accustomed to infinities that you may not realize the severity of this problem. The problem with divergent infinite series is that the very idea of adding terms may be called into question. We can see that unconvergent sums can be meaningless by studying the following enigma that purports to prove that $\infty = -1$! Define

$$S = 1 + 2 + 4 + 8 + \ldots$$

so that S is obviously infinite. By pulling out a factor of 2 we get

$$S = 1 + 2(1 + 2 + 4 + 8 + \ldots)$$

and we see that the expression in the parentheses is exactly S. This implies that $S = 1 + 2S$, which can be solved to give $S = -1$. The problem here is that the infinite sum in the parentheses is meaningless, and in particular one cannot rely on normal arithmetical laws (such as $2(a + b) = 2a + 2b$) to be meaningful for it. It's not just that I is infinite; I is truly meaningless and by various regroupings, factorings, and the like, it can seem to be equal to anything you want.

The only truly well-defined infinite series are those that are *absolutely convergent.* The series

$$S = \sum_{n=0}^{\infty} a_n$$

is absolutely convergent when

$$A = \sum_{n=0}^{\infty} |a_n|$$

converges to a finite value. If a series S seems to converge to a finite value but A does not, then by rearranging, regrouping, and the like you can make S equal to just about anything.

Since the zT terms are $a_n = s_n z^n$, our first guess might be that $|z|$ must be very small for the sum to converge absolutely. Note, however, that the sum in the zT is from negative infinity to positive infinity; for absolute convergence we require

$$A = \sum_{n=-\infty}^{\infty} |s_n||z|^n = \sum_{n=-\infty}^{-1} |s_n||z|^n + \sum_{n=0}^{\infty} |s_n||z|^n = \sum_{n=1}^{\infty} |s_{-n}||\zeta|^n + \sum_{n=0}^{\infty} |s_n||z|^n$$

where we defined $\zeta \equiv z^{-1}$. If $|z|$ is small then $|\zeta|$ is large, and consequently small values of $|z|$ can be equally dangerous. In general, the **Region Of Convergence** (ROC) of the z transform will be a ring in the z-plane with the origin at its center (see Figure 4.12). This ring may have $r = 0$ as its lower radius (and so be disk-shaped), or have $r = \infty$ as its upper limit, or even be the entire z-plane. When the signal decays to zero for both $n \to -\infty$ and $n \to \infty$ the ring will include the unit circle.

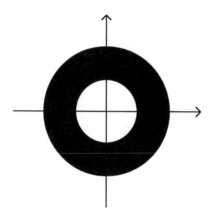

Figure 4.12: In general, the region of convergence (ROC) of the z transform is a ring in the z-plane with the origin at its center.

The z-plane where the zT lives, with its ROCs, poles, and zeros, is a more complex environment than the frequency axis of the FT. We will learn a lot more about it in the coming chapters.

EXERCISES

4.10.1 The zT is an expansion in basis functions $z^n = re^{i\omega n}$. Show that this basis is orthogonal.

4.10.2 Derive the generating function for a Fibonacci sequence with initial conditions $f_0 = 1$, $f_1 = 2$. What is the explicit formula for f_n?

4.10.3 The integer recursions for the two families of sequences g^+ and g^-

$$g_{n+1}^{\pm} = \begin{cases} 3g_n^{\pm} \pm 1 & g_n^{\pm} \text{ odd} \\ \frac{1}{2}g_n^{\pm} & g_n^{\pm} \text{ even} \end{cases}$$

may eventually lead to $g = 1$, or may oscillate wildly. For example, for the g^- case, $g_0 = 5$ leads to a cycle $5, 14, 7, 20, 10, 5$; no cycle has ever been found for the g^+ case (the Collatz problem). Compute numerically the generating functions $g^{\pm}(x)$ for $0 \le x < 1$ and starting values $g_0 = 2 \dots 10$. Can you tell which initial values cycle from the generating function?

4.10.4 Consider the infinite series $S = 1 - 1 + 1 - 1 + \dots$. Writing this $S = (1 - 1) + (1 - 1) + \dots = 0 + 0 + \dots$ it would seem to converge to zero. Regroup to make S equal something other than zero. Is S absolutely convergent?

4.10.5 Show that if the zT of a signal is a rational function of z then the locations of poles and zeros completely specifies that signal to within a gain.

4.10.6 Show that the LT of $s(t)$ is the FT of $s(t)e^{-\lambda t}$.

4.10.7 Find the Laplace transforms of the unit impulse and unit step.

4.10.8 Derive the zT from the LT similarly to our derivation of DFT from FT.

4.10.9 According to the *ratio test* an infinite sum $\sum_{n=0}^{\infty} a_n$ converges absolutely if the ratio $|\frac{a_{n+1}}{a_n}|$ converges to a value less than unity. How does this relate to the zT?

4.11 More on the z Transform

Once again the time has come to roll up our sleeves and calculate a few examples. The first signal to try is the unit impulse $s_n = \delta_{n,0}$, for which

$$S(z) = \text{zT}(\delta_{n,0}) = \sum_{n=-\infty}^{\infty} s_n z^{-n} = 1 \cdot z^0 = 1$$

which is analogous to the FT result. The series converges for all z in the z-plane. Were the impulse to appear at time $m \ne 0$, it is easy to see that we would get $S(z) = z^{-m}$, which has a zero at the origin for negative times and a pole there for positive ones. The ROC is the entire plane for $m \le 0$, and the entire plane except the origin for $m > 0$.

What is the zT of $s_n = \alpha^n u_n$? This signal increases exponentially with time for $\alpha > 1$, decreases exponentially for $0 < \alpha < 1$, and does the same but with oscillating sign for $\alpha < 0$.

$$S(z) = zT(\alpha^n u_n) = \sum_{n=-\infty}^{\infty} \alpha^n u_n z^{-n} = \sum_{n=0}^{\infty} (\alpha z^{-1})^n$$

Using the equation (A.47) for the sum of an infinite geometric series, we find

$$S(z) = \frac{1}{1 - \alpha z^{-1}} = \frac{z}{z - \alpha} \tag{4.63}$$

which has a pole at $z = \alpha$. The ROC is thus $|z| > |\alpha|$, the exterior of disk of radius α. When does the FT exist? As a general rule, poles in the z-plane outside the unit circle indicate explosive signal growth. If $|\alpha| < 1$ the ROC includes the unit circle, and so the FT converges. For the special case of the unit step $s_n = u_n$, we have $\alpha = 1$, so the zT is $\frac{z}{z-1}$ with ROC $|z| > 1$; the FT does not exist.

We can shift the signal step to occur at time m here as well. In this case

$$S(z) = \sum_{n=-\infty}^{\infty} \alpha^{n-m} u_{n-m} z^{-n} = \sum_{n=m}^{\infty} \alpha^{n-m} z^{-m} z^{-(n-m)}$$

which after a change in variable from n to $n - m$ gives

$$S(z) = z^{-m} \sum_{n=0}^{\infty} \alpha^n z^{-n} = z^{-m} \frac{1}{1 - \alpha z^{-1}} = \frac{z^{1-m}}{z - \alpha}$$

with poles at $z = \alpha$ and $z = 0$, and ROC unchanged.

What about $s_n = \alpha^{-n} u_n$? This is a trick question! This is the same as before if we write $s_n = (\frac{1}{\alpha})^n u_n$ so $S(z) = \frac{1}{1-\alpha^{-1}z}$ with ROC $|z| > |\alpha^{-1}|$. Since the sum we performed is true in general, the α used above can be anything, even imaginary or complex. Hence we know, for instance, that the zT of $e^{i\omega n}$ is $\frac{1}{1-e^{i\omega}z^{-1}}$ with ROC $|z| > |e^{i\omega}| = 1$.

We can perform a calculation similar to the above for $s_n = \alpha^n u_{-n}$.

$$S(z) = \sum_{n=-\infty}^{\infty} \alpha^n u_{-n} z^{-n} = \sum_{n=-\infty}^{0} (\alpha z^{-1})^n$$

$$= \sum_{n=0}^{\infty} (\alpha^{-1} z)^n = \frac{1}{1 - \alpha^{-1} z}$$

The ROC is now $|z| < |\alpha|$, the interior of the disk. Shifting the ending time to $n = m$ we get

$$S(z) = \sum_{n=-\infty}^{\infty} \alpha^{n-m} u_{-(n-m)} z^{-n} = \sum_{n=-\infty}^{m} \alpha^{n-m} z^{-m} z^{-(n-m)}$$

$$= \sum_{n=0}^{\infty} (\alpha^{-1}z)^n = z^m \frac{1}{1 - \alpha^{-1}z}$$

with an extra pole if $m < 0$. It will often be more useful to know the zT of $s_n = \alpha^n u_{-n-1}$. This will allow covering the entire range of n with no overlap. It is convenient to remember that the zT of $s_n = -\alpha^n u_{-n-1}$ is exactly that of $s_n = \alpha^n u_n$ but with ROC $|z| < |\alpha|$. The desired transform is obtained by noting that multiplication of s_n by anything, including -1, simply causes the zT to be multiplied by this same amount.

Rather than calculating more special cases directly, let's look at some of the z transform's properties. As usual the most critical is *linearity*, i.e., the zT of $ax_n + by_n$ is $ax(z) + by(z)$. The ROC will always be at least the intersection of the ROCs of the terms taken separately. This result allows us to calculate more transforms, most importantly that of $\cos(\omega n)$. We know that $\cos(\omega n) = \frac{1}{2}(e^{i\omega n} + e^{-i\omega n})$, so the desired result is obtained by exploiting linearity.

$$S(z) = \frac{1}{2} \left(\frac{1}{1 - e^{i\omega}z^{-1}} + \frac{1}{1 - e^{-i\omega}z^{-1}} \right) = \frac{1 - \cos(\omega)z^{-1}}{1 - 2\cos(\omega)z^{-1} + z^{-2}}$$

The next most important property of the zT is the effect of a time shift. For the FS and FT, shifting on the time axis led to phase shifts, here there is something new to be learned. In the cases we saw above, the effect of shifting the time by m digital units was to multiply the zT by z^{-m}. In particular the entire effect of delaying the digital signal by *one* digital unit of time was to multiply the zT by a factor of z^{-1}. This is a general result, as can be easily derived.

$$zT(x_{n-1}) = \sum_{n=-\infty}^{\infty} x_{n-1} z^{-n} = \sum_{n=-\infty}^{\infty} x_n z^{-(n+1)}$$

$$= \sum_{n=-\infty}^{\infty} x_n z^{-1} z^{-n} = z^{-1} \sum_{n=-\infty}^{\infty} x_n z^{-n} = z^{-1} zT(x_n)$$

Accordingly the factor of z^{-1} can be thought of as a unit delay *operator*, as indeed we defined it back in equation (2.21). The origin of the symbol that was arbitrary then is now understood; delaying the signal by one digital unit of time can be accomplished by multiplying it by z^{-1} in the z domain. This interpretation is the basis for much of the use of the zT in DSP.

For example, consider a radioactive material with half-life τ years. At the beginning of an experiment $n = 0$ we have 1 unit of mass $m = 1$ of this material; after one half-life $n = 1$ the mass has dropped to $m = \frac{1}{2}$ units, $\frac{1}{2}$ having been lost. At digital time $n = 2$ its mass has further dropped to $m = \frac{1}{4}$ after losing a further $\frac{1}{4}$, etc. After an infinite wait

$$\tfrac{1}{2} + \tfrac{1}{4} + \tfrac{1}{8} + \tfrac{1}{16} + \ldots = 1$$

all of the material has been lost (actually converted into another material). The mass left as a function of time measured in half-lives is

$$m_n = \tfrac{1}{2}^n$$

an exponentially decreasing signal. Now a scientist measures the amount of mass at some unknown time n and wishes to predict (or is it postdict?) what the mass was one half-life back in time. All that need be done is to double the amount of mass measured, which is to use the operator z^{-1} with z being identified as $\frac{1}{2}$. This example might seem a bit contrived, but we shall see later that many systems when left alone tend to decrease exponentially in just this manner.

What about time reversal? For the FT this caused negation of the frequency; here it is straightforward to show that the zT of s_{-n} has its z variable inverted, $zT(s_{-n}) = S(z^{-1})$. If the original signal had a ROC $R_l < |z| < R_h$, then the time-reversed signal will have a ROC of $R_l^{-1} > |z| > R_h^{-1}$. The meaning of this result is not difficult to comprehend; the inversion of $z = re^{i\omega}$ both negates the ω and inverts r. Thus decaying exponentials are converted to exploding ones and vice versa.

You must be wondering why we haven't yet mentioned the inverse zT (izT). The reason is that it is somewhat more mathematically challenging than the other inverse operations we have seen so far. Remember that the zT's range is a ring in the complex z-plane, not just a one-dimensional line. To regain s_n from $S(z)$ we must perform a *contour integral*

$$s_n = \frac{1}{2\pi i} \oint S(z) z^{n-1} dz \tag{4.64}$$

over any closed counterclockwise contour within the ROC. This type of integral is often calculated using the residue theorem, but we will not need to use this complex mechanism in this book.

Many more special zTs and properties can be derived but this is enough for now. We will return to the zT when we study signal processing systems. Systems are often defined by complex recursions, and the zT will enable us to convert these into simple algebraic equations.

EXERCISES

4.11.1 Write a graphical program that allows one to designate a point in the z-plane and then draws the corresponding signal.

4.11.2 Plot the z transform of $\delta_{n,m}$ for various m.

4.11.3 Prove the linearity of the zT.

4.11.4 Express $zT(\alpha^n x_n)$ in terms of $x(z) = zT(x_n)$.

4.11.5 What is the z transform of the following digital signals? What is the ROC?
1. $\delta_{n,2}$
2. u_{n+2}
3. $a^n u(n)$
4. $a^n u(-n-1)$
5. $\frac{1}{2}^n u_n + \frac{3}{2}^n u_{-n}$

4.11.6 What digital signals have the following z transforms?
1. z^{-2}
2. z^{+2}
3. $\frac{1}{1-2z^{-1}}$ ROC $|z| > |2|$

4.11.7 Prove the following properties of the zT:
1. linearity
2. time shift $zTs_{n-k} = z^{-k}S(z)$
3. time reversal $zTs_{-n} = S(\frac{1}{z})$
4. conjugation $zTs_n^* = S^*(z^*)$
5. rescaling $zT(\alpha^n s_n) = S(\frac{z}{\alpha})$
6. z differentiation $zT(ns_n) = -z\frac{d}{dz}S(z)$

4.12   The Other Meaning of Frequency

We have discussed two quite different representations of functions, the Taylor expansion and the Fourier (or z) transform. There is a third, perhaps less widely known representation that we shall often require in our signal processing work. Like the Fourier transform, this representation is based on frequency, but it uses a fundamentally different way of thinking about the concept of frequency. The two usages coincide for simple sinusoids with a single constant frequency, but differ for more complex signals.

Let us recall the examples with which we introduced the STFT in Section 4.6. There we presented a pure sinusoid of frequency f_1, which abruptly changed frequency at $t = 0$ to become a pure sine of frequency f_2. Intuition tells us that we should have been able to recover an *instantaneous frequency*, defined at every point in time, that would take the value f_1 for negative times, and f_2 for positive times. It was only with difficulty that we managed to convince you that the Fourier transform cannot supply such a frequency value, and that the uncertainty theorem leads us to deny the existence of the very concept of instantaneous frequency. Now we are going to produce just such a concept.

The basic idea is to express the signal in the following way:

$$s(t) = A(t) \cos \Big(\Phi(t) \Big) \tag{4.65}$$

for some $A(t)$ and $\Phi(t)$. This is related to what is known as the *analytic representation* of a signal, but we will call it simply the *instantaneous representation*. The function $A(t)$ is known as the *instantaneous amplitude* of the signal, and the $\Phi(t)$ is the *instantaneous angle*. Often we separate the angle into a linear part and the deviation from linearity

$$s(t) = A(t) \cos \Big(\omega t + \phi(t) \Big) \tag{4.66}$$

where the frequency ω is called the *carrier frequency*, and the residual $\phi(t)$ the *instantaneous phase*.

The *instantaneous frequency* is the derivative of the instantaneous angle

$$2\pi f(t) = \frac{d\Phi(t)}{dt} = \omega + \frac{d\phi(t)}{dt} \tag{4.67}$$

which for a pure sinusoid is exactly the frequency. *This* frequency, unlike the frequencies in the spectrum, is a single function of time, in other words, a signal. This suggests a new world view regarding frequency; rather than understanding signals in a time interval as being made up of many frequencies, we claim that signals are fundamentally sinusoids with well-defined instantaneous amplitude and frequency. One would expect the distribution of different frequencies in the spectrum to be obtained by integration over the time interval of the instantaneous frequency. This is sometimes the case.

Consider, for example, a signal that consists of a sinusoid of frequency f_1 for one second, and then a sinusoid of nearby frequency f_2 for the next second. The instantaneous frequency will be f_1 and then jump to f_2; while the spectrum, calculated over two seconds, will contain two spectral lines at f_1 and f_2. Similarly a sinusoid of slowly increasing instantaneous frequency will have a spectrum that is flat between the initial and final frequencies.

This new definition of frequency seems quite useful for signals that we usually consider to have a single frequency at a time; however, the instantaneous representation of equation (4.65) turns out to very general. A constant DC signal can be represented (using $\omega = 0$), but it is easy to see that a constant plus a sinusoid can't. It turns out (as usual, we will not dwell upon the mathematical details) that all DC-less signals can be represented. This leads to an apparent conflict with the Fourier picture. Consider a signal composed of the *sum* of the two sinusoids with close frequencies f_1 and f_2; what does the instantaneous representation do, jump back and forth between them? No, this is exactly a *beat* signal (discussed in exercise 2.3.3) with instantaneous frequency a constant $\frac{1}{2}(f_1 + f_2)$, and sinusoidally varying amplitude is with frequency $\frac{1}{2}|f_1 - f_2|$. Such a signal is depicted in Figure 4.13. The main frequency that we see in this figure (or hear when listening to such a combined tone) is the instantaneous frequency, and after that the effect of $A(t)$, *not* the Fourier components.

We will see in Chapter 18 that the instantaneous representation is particularly useful for the description of communications signals, where it is the basis of *modulation*. Communications signals commonly carry informa-

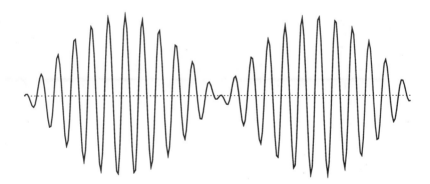

Figure 4.13: The beat signal depicted here is the sum of two sinusoids of relatively close frequencies. The frequencies we see (and hear) are the average and half-difference frequencies, *not* the Fourier components.

tion by varying (modulating) the instantaneous amplitude, phase, and/or frequency of a sinusoidal 'carrier'. The carrier frequency is the frequency one 'tunes in' with the receiver frequency adjustment, while the terms AM (**A**mplitude **M**odulation) and FM (**F**requency **M**odulation) are familiar to all radio listeners.

Let us assume for the moment that the instantaneous representation exists; that is, for any *reasonable* signal $s(t)$ without a DC component, we assume that one can find carrier frequency, amplitude, and phase signals, such that equation (4.65) holds. The question that remains is *how* to find them. The answering of this question is made possible through the use of a mathematical operator known as the Hilbert transform.

The Hilbert transform of a real signal $x(t)$ is a real signal $y(t) = \mathcal{H}x(t)$ obtained by shifting the phases of all the frequency components in the spectrum of $x(t)$ by 90°. Let's understand why such an operator is so remarkable. Assume $x(t)$ to be a simple sinusoid.

$$x(t) = A\cos(\omega t)$$

Obtaining the 90° shifted version

$$y(t) = \mathcal{H}x(t) = A\cos\left(\omega t - \frac{\pi}{2}\right) = A\sin(\omega t)$$

is actually a simple matter, once one notices that

$$y(t) = A\cos\left(\omega\left(t - \frac{\pi}{2\omega}\right)\right) = x\left(t - \frac{\pi}{2\omega}\right)$$

which corresponds to a time delay. So to perform the Hilbert transform of a pure sine one must merely delay the signal for a time corresponding to one quarter of a period. For digital sinusoids of period L samples, we need to use the operator $z^{-\frac{L}{4}}$, which can be implemented using a FIFO of length $L/4$.

However, this delaying tactic will not work for a signal made up of more that one frequency component, e.g., when

$$x(t) = A_1\cos(\omega_1 t) + A_2\cos(\omega_2 t)$$

we have

$$y(t) = \mathcal{H}x(t) = A_1\sin(\omega t) + A_2\sin(\omega t)$$

which does *not* equal $x(t - \tau)$ for any time delay τ.

Hence the Hilbert transform, which shifts all frequency components by a quarter period, independent of frequency, is a nontrivial operator. One way of implementing it is by performing a Fourier transform of the signal, individually shifting all the phases, and then performing an inverse Fourier transform. We will see an alternative implementation (as a *filter*) in Section 7.3.

Now let us return to the instantaneous representation

$$x(t) = A(t) \cos \left(\omega t + \phi(t) \right) \tag{4.68}$$

of a signal, which we now call $x(t)$. Since the Hilbert transform instantaneously shifts all $A \cos(\omega t)$ to $A \sin(\omega t)$, we can explicitly express $y(t)$.

$$y(t) = \mathcal{H} x(t) = A(t) \sin \left(\omega t + \phi(t) \right) \tag{4.69}$$

We can now find the instantaneous amplitude by using

$$A(t) = \sqrt{x^2(t) + y^2(t)} \tag{4.70}$$

the instantaneous phase via the (four-quadrant) arctangent

$$\phi(t) = \tan^{-1} \frac{y(t)}{x(t)} - \omega t \tag{4.71}$$

and the instantaneous frequency by differentiating the latter.

$$\omega(t) = \frac{d\phi(t)}{dt} \tag{4.72}$$

The recovery of amplitude, phase, or frequency components from the original signal is called *demodulation* in communications signal processing.

We have discovered a method of constructing the instantaneous representation of any signal $x(t)$. This method can be carried out in practice for digital signals, assuming that we have a numeric method for calculating the Hilbert transform of an arbitrary signal. The instantaneous frequency similarly requires a numeric method for differentiating an arbitrary signal. Like the Hilbert transform we will see later that differentiation can be implemented as a filter. This type of application of numerical algorithms is what DSP is all about.

EXERCISES

4.12.1 We applied the Hilbert transform to $x(t) = \cos(\omega t + \phi(t))$ and claimed that one obtains $y(t) = \sin(\omega t + \phi(t))$. Using trigonometric identities prove that this is true for a signal with two frequency components.

4.12.2 Even a slowly varying phase may exceed 2π or drop below zero causing nonphysical singularities in its derivative. What should be done to phases derived from equation (4.71) in such a case?

4.12.3 What is the connection between the instantaneous frequency and the spectrum of the signal? Compare the short time power spectrum calculated over a time interval to the histogram of the instantaneous frequency taken over this interval.

4.12.4 Show that given a signal $s(t)$ and any amplitude signal $A(t)$ an appropriate phase $\Phi(t)$ can be found so that equation (4.65) holds. Similarly, show that given any phase an amplitude signal may be found. The amplitude and phase are not unique; the $x(t)$ and $y(t)$ that are related by the Hilbert Transform are the *canonical* (simplest) representation.

4.12.5 Find an explicit direct formula for the instantaneous frequency as a function of $x(t)$ and $y(t)$. What are the advantages and disadvantages of these two methods of finding the instantaneous frequency?

4.12.6 We can rewrite the analytic form of equation (4.68) in quadrature form.

$$x(t) = a(t)\cos(\omega t) + b(t)\sin(\omega t)$$

What is the connection between $a(t)$, $b(t)$ and $A(t)$, $\phi(t)$? We can also write it in sideband form.

$$x(t) = \big(u(t) + l(t)\big)\cos(\omega t) + \big(u(t) - l(t)\big)\sin(\omega t)$$

What are the relationships now?

Bibliographical Notes

The DFT and zT are covered well in many introductory texts, e.g., [187, 252, 167], while the Hilbert transform and analytic representation are confined to the more advanced ones [186, 200]. An early book devoted entirely to the zT is [125], while tables were published even earlier [97].

The uncertainty theorem was introduced in quantum mechanics by Heisenberg [99]. Another physicist, Wigner [282], derived the first example of what we would call a time-frequency distribution in 1932, but this mathematical achievement had to be translated into signal processing terms. The article by Leon Cohen [39] is the best introduction.

Noise

Much of signal processing involves extracting signals of interest from noise. Without noise to combat, a radar receiver could detect an echo by simple energy thresholding. In a noiseless world an infinite amount of information could be transmitted through a communications channel every second. Were it not for noise, signal classification would be reduced to dictionary lookup. Yet signals in the real world are always noisy. Radar echoes are buried under noise, making their detection impossible without sophisticated processing. Modem signals rely on complex modulation and error correction schemes to approach the maximum rate attainable through noisy telephone lines. Due to noise, signal classification is still more an art than a science. Extracting a signal from noise can rely on knowledge of the clean signal and/or knowledge of the noise. Up to now we have learned to characterize clean signals; in this chapter we will study the characteristics of noise.

As discussed in Section 2.3, a stochastic signal cannot be precisely predicted, being bound only by its statistics. What do we mean by 'statistics'? It is jokingly said that *probability* is the science of turning random numbers into mathematical laws, while *statistics* is the art of turning mathematical laws into random numbers. The point of the joke is that most people take 'statistics' to mean a technique for analyzing empirical data that enables one to prove just about anything. In this book 'statistics' refers to something far more tangible, namely the parameters of probabilistic laws that govern a signal. Familiar statistics are the average or *mean value* and the *variance*.

In this chapter we will learn how noisy signals can be characterized and simulated. We will study a naive approach that considers noise to be merely a pathological example of signals not unlike those we have previously met. In particular, we will take the opportunity to examine the fascinating world of chaotic deterministic signals, which for all practical purposes are indistinguishable from stochastic signals but can be approached via periodic signals. Finally, we will briefly discuss the mathematical theory of truly stochastic signals.

5.1 Unpredictable Signals

'Pure noise' is the name we give to a quintessential stochastic signal, one that has only probabilistic elements and no deterministic ones. Put even more simply, pure noise is completely *random*; it obeys only probabilistic laws and can never be perfectly predicted. 'Plain' noise has a softer definition in that we allow signals with some deterministic characteristics, e.g. the sum of a pure noise and a deterministic signal. The ratio of the energy of the deterministic signal to that of the pure noise component is called the **S**ignal to **N**oise **R**atio (SNR), usually specified in dB. A signal with finite SNR is unpredictable to some degree. Our guesses regarding such noisy signals may be better than random, but we can quite never pin them down. An SNR of 0dB (SNR=1) means the signal and noise have equal energies.

There are four distinguishable ways for a signal to appear unpredictable: it may be pseudorandom, incompletely known, chaotic, or genuinely stochastic. The exact boundaries between these four may not always be clear, but there is progressively more known about the signal as we advance from the first to the third. Only the fourth option leads to true noise, but in practice it may be impossible to differentiate even between it and the other three.

A pseudorandom signal is completely deterministic, being generated by some completely defined algorithm. However, this algorithm is assumed to be unknown to us, and is conceivably quite complex. Being ignorant of the algorithm, the signal's behavior seems to us quite arbitrary, jumping capriciously between different values without rhyme or reason; but to the initiated the signal's behavior is entirely reasonable and predictable. If we may assume that there is no correlation between the unknown generating algorithm and systems with which the signal may interact, then for all intents and purposes a pseudorandom signal *is* noise. Pseudorandom signals will be treated in more detail in Section 5.4.

An incompletely known signal is also completely deterministic, being generated by a known algorithm that may depend on several parameters. The details of this algorithm and some, but not all, of these parameters are known to us, the others being *hidden variables*. Were we to know all these parameters the signal would be completely predictable, but our state of knowledge does not allow us to do so. In practice knowing the form and some of the parameters may not help us in the least, and the signal seems to us completely erratic and noise-like. In theory the signal itself is not erratic at all; it's simply a matter of our own ignorance!

A chaotic signal is also completely deterministic, being generated by a completely specified algorithm that may even be completely known to us.

However, a chaotic signal seems noisy because of numeric sensitivity of this algorithm that causes us to rapidly lose information about the signal with the passage of time. Were all initial conditions to be specified to infinite precision, and all calculations to be performed with infinite accuracy, the signal would indeed be perfectly predictable; but *any* imprecision of knowledge or inaccuracy of computation will inevitably lead to complete loss of predictability after enough time has passed. Such *chaotic* signals will be treated in detail in Section 5.5.

A truly stochastic signal is one that is *not* generated by any deterministic algorithm at all. The time between successive clicks of a Geiger counter or the thermal noise measured across a resistor are typical examples. At a fundamental level, quantum mechanics tells us that nature abounds with such genuinely random signals. The philosophical and scientific consequences of this idea are profound [53]. The implications for DSP are also far-reaching, and will be discussed briefly in Section 5.6. However, a formal treatment of stochastic signals is beyond the scope of this book.

EXERCISES

5.1.1 The game of `guessit` is played by two or more people. First the players agree upon a lengthy list of functions of one variable t, each of which is also dependent on one or two parameters. The *inventor* picks function from the list and supplies parameters. Each *analyst* in turn can request a single value of the function and attempt to guess which function has been selected. What strategy should the inventor use to make the analysts' task more difficult? What tactics can the analysts use? Try playing `guessit` with some friends.

5.1.2 Generate a signal x with values in the interval $[0 \ldots 1]$ by starting at an arbitrary value in the interval and iterating $x_{n+1} = \lambda x_n (1 - x_n)$ for $0 \leq \lambda \leq 4$. For what values of λ does this signal look random?

5.1.3 To which of the four types of unpredictable signal does each of the following most closely belong?
    1.   Static noise on shortwave radio
    2.   Sequence of heads (s=1) and tails (s=0) obtained by throwing a coin
    3.   World population as a function of time
    4.   Value of stock portfolio as a function of time
    5.   Sequence produced by your compiler's random number generator
    6.   Distance from earth to a given comet
    7.   Position of a certain drop of water going down a waterfall
    8.   Maximum daily temperature at your location
    9.   The sequence of successive digits of π

5.2   A Naive View of Noise

No matter what its source, a noise-like signal is very different from the signals with which we have dealt so far. Although we can observe it as a function of time, its graph resembles *modern art* as compared to the classical lines of deterministic signals; and every time we observe and plot it we get a completely different graph. In Figures 5.1, 5.2, and 5.3 we plot distinct noise signals in the time domain. All the plots in each figure represent the same noise signal, and are called *realizations* of the underlying noise. No two realizations are precisely the same, yet there are noticeable similarities between realizations of the same noise, and different noise signals may be easily distinguishable by eye.

Were you to be presented with a new, previously unseen realization of one of the noise signals of the figures, and asked to which it belonged, you would probably have little difficulty in classifying it. How do you do it? How can we best characterize noise signals? It will not surprise you to learn that noise signals, like deterministic signals, have characteristics in the time domain and in the frequency domain.

In the time domain we are interested in the statistical attributes of individual signal values v_n, such as the mean (average) $\langle v \rangle$, the variance or standard deviation, and the moments of higher orders. The set of *all* parameters that determine the probabilistic laws is called *sufficient statistics*. Sufficient statistics are not sufficient to enable us to precisely predict the signal's value at any point in time, but they constitute the most complete description of a stochastic signal that there is. Noise signals are called *stationary* when these statistics are not time-dependent. This implies that the probabilistic properties of the noise do not change with time; so if we measure the mean and variance now, or half an hour from now, we will get the same result.

We will almost always assume stationary noise signals to have zero mean, $\langle v \rangle = 0$. This is because noise $v(t)$ of nonzero average can always be written

$$v(t) = \langle v \rangle + \nu(t) \qquad \mathbf{A} \ \Big| \ \mathbf{D} \qquad v_n = \langle v \rangle + \nu_n$$

where the constant $\langle v \rangle$ is of course a (deterministic) DC signal, and ν is noise with zero mean. There is no reason to apply complex techniques for stochastic signals to the completely deterministic DC portion, which can be handled by methods of the previous chapters.

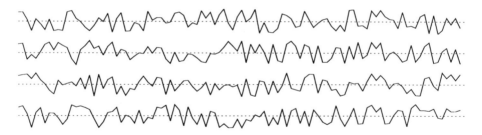

Figure 5.1: A few realizations of a noise signal. The set of all such realizations is called the *ensemble*. Note that each realization is erratic, but although the different realizations are quite varied in detail, there *is* something similar about them.

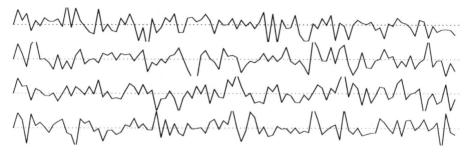

Figure 5.2: A few realizations of another noise signal. Note the differences between this noise signal and the previous one. Although both have zero average and roughly the same standard deviation, the first is uniformly distributed while this signal is Gaussian distributed. A few values are off-scale and thus do not appear.

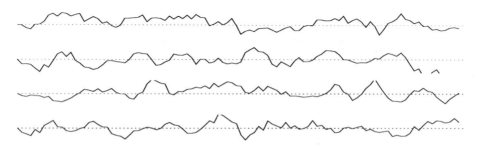

Figure 5.3: A few realizations of a third noise signal. Note the differences between this noise signal and the previous two. Although the signal is also zero average and of the same standard deviation, the first two signals were *white* while this signal has been low-pass filtered and contains less high-frequency energy.

The most detailed information concerning the statistics of individual signal values is given by the complete probability distribution these values. Probability distributions are functions $p(x)$ that tell us the probability of the signals taking on the value x. Digital signals can only take on a finite number of values, and thus (at least in principle) we can record the complete probability distribution as a table. To demonstrate this consider a noise signal that can take on only the values $-1, 0, 1$ and whose probability distribution is the following.

$$p(-1) = \tfrac{1}{4} \qquad p(0) = \tfrac{1}{2} \qquad p(+1) = \tfrac{1}{4} \tag{5.1}$$

Note that the probabilities sum to one since each signal value *must* be either -1, 0, or $+1$. One signal with such a distribution may be

$$\ldots 0, -1, +1, -1, -1, 0, +1, 0, 0, 0, +1, +1, 0, 0, -1, 0, \ldots$$

while another could be

$$\ldots 0, +1, 0, -1, +1, 0, -1, -1, 0, +1, 0, -1, 0, 0, +1, 0, \ldots$$

as the reader may verify.

Given a long enough sample of a digital signal with unknown distribution, we can estimate its probability distribution by simply counting the number of times each value appears and at the end dividing by the number of signal values observed. For example, the noise signal

$$\ldots - 1, 0, +1, +1, 0, -1, 0, -1, 0, +1, +1, -1, +1, 0, -1, \ldots$$

has a probability distribution close to $\tfrac{1}{3}, \tfrac{1}{3}, \tfrac{1}{3}$. The probability distribution of any digital signal must sum to unity (i.e., must be *normalized*)

$$\sum p(x_i) = 1 \tag{5.2}$$

where the sum is over all possible signal values.

We said before that the probability distribution contains the most detailed information available as to individual signal values. This implies that all single signal value statistics can be derived from it. For a digital signal we can express the mean as a sum over time,

$$\mu = \langle s_n \rangle = \frac{1}{N} \sum_{n=1}^{N} s_n \tag{5.3}$$

or we can sort the terms such that smaller s_n appear before larger ones.

This is in turn equivalent to summing each observed signal value s times the relative number of times it was observed $p(s)$,

$$\mu = \sum_{s=-\infty}^{\infty} p(s)s \tag{5.4}$$

which is seen to be a simple sum of the probability distribution. The variance is defined to be the mean-squared deviation from the mean

$$\sigma^2 = \left\langle (s_n - \mu)^2 \right\rangle = \frac{1}{N} \sum_{n=1}^{N} (s_n - \mu)^2 \tag{5.5}$$

which can also be written in terms of the probability distribution.

$$\sigma^2 = \sum_{x=-\infty}^{\infty} p(x)(x - \mu)^2 \tag{5.6}$$

Analog signals have a nondenumerably infinite number of possible signal values, and so a table of probabilities cannot be constructed. In such cases we may resort to using histograms, which is similar to digitizing the analog signal. We quantize the real axis into *bins* of width δx, and similar to the digital case we count the number of times signal values fall into each bin. If the histogram is too rough we can choose a smaller bin-width δx. In the limit of infinitesimal bin-width we obtain the continuous probability distribution $p(x)$, from which all finite width histograms can be recovered by integration. Since the probability distribution does not change appreciably for close values, doubling small enough bin-widths should almost precisely double the number of values falling into each of the respective bins. Put another way, the probability of the signal value x falling into the histogram bin of width δx centered on x_0 is $p(x_0)\delta x$, assuming δx is small enough. For larger bin-widths integration is required, the probability of the signal value being between x_1 and x_2 being $\int_{x_1}^{x_2} p(x)dx$. Since every signal value must be *some* real number, the entire distribution must be normalized.

$$\int_{-\infty}^{\infty} p(x)\, dx = 1 \tag{5.7}$$

In analogy with the digital case, the mean and variance are given by the following.

$$\mu = \int_{-\infty}^{\infty} p(x)\, x\, dx \qquad \sigma^2 = \int_{-\infty}^{\infty} p(x)(x - \mu)^2 dx \tag{5.8}$$

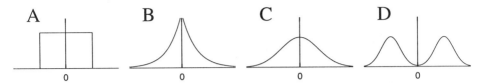

Figure 5.4: Four different probability distributions. (A) represents the uniform distribution. (B) depicts an exponential distribution. (C) is the bell-shaped Gaussian (or normal) distribution. (D) is a representative bimodal distribution, actually the mixture of two Gaussians with different means.

From its very definition, the probability distribution of a random signal must be nonnegative and have an integral of one. There are a large number of such functions! For example, signal values may be uniformly distributed over some range, or exponentially distributed, or have a Gaussian (normal) distribution with some mean and variance, or be multimodal. Uniformly distributed signals only take on values in a certain range, and all of these values are equally probable, even those close to the edges. In Figure 5.4.A we depict graphically the uniform distribution. Gaussian distribution means that all signal values are possible, but that there is a most probable value (called the mean μ) and that the probability decreases as we deviate from the mean forming a bell-shaped curve with some characteristic width (the standard deviation σ). Mathematically,

$$p(x) = \frac{1}{\sqrt{2\pi}\sigma} \, e^{\frac{1}{2}\frac{(x-\mu)^2}{\sigma^2}} \tag{5.9}$$

is the famous Gaussian function, depicted in Figure 5.4.C. It is well known that when many students take an exam, their grades tend to be distributed in just this way. The rather strange constant before the exponent ensures that the Gaussian is normalized.

The frequency domain characteristics of random signals are completely distinct from the single-time signal value characteristics we have discussed so far. This may seem remarkable at first, since in DSP we become accustomed to time and frequency being two ways of looking at one reality. However, the dissimilarity is quite simple to comprehend. Consider a digital signal

$$s_1, s_2, s_3, \ldots s_N$$

with some signal value distribution and a new signal obtained by arbitrarily replicating each signal value

$$s_1, s_1, s_2, s_2, s_3, s_3, \ldots s_N, s_N$$

so that each value appears twice in a row. The new signal obviously has the same single-sample statistics as the original one, but its frequencies have been halved! Alternatively, consider permuting the order of signal values; this once again obviously results in an identical probability distribution, but quite different frequency characteristics! A signal's frequency statistics are determined by the relationship between signal values at various relative positions, and thus contains information different from the signal value statistics.

We will often talk of *white noise*. White noise is similar to white light in that its spectrum is flat (constant, independent of frequency). Having all possible frequencies allows the signal to change very rapidly, indeed even knowing the entire past history of a white noise signal does not contribute anything to prediction of its future. We thus call a discrete time signal s_n white noise if observation of $\{s_n\}_{n=-\infty}^{k-1}$ does not allow us to say anything useful about the value of s_k other than what the single-signal value statistics tell us.

Of course not all noise is white; when the noise signal's spectrum is concentrated in part of the frequency axis we call it colored noise. Colored noise can be made by passing white noise through a band-pass filter, a device that selectively enhances Fourier components in a certain range and rejects others. As we decrease the bandwidth of the filter, the signal more and more resembles a sine wave at the filter's central frequency, and thus becomes more and more predictable.

Since they are independent, time and frequency domain characteristics can be combined in arbitrary ways. For example, white noise may happen to be normally distributed, in which case we speak of Gaussian white noise. However, white noise may be distributed in many other ways, for instance, uniformly, or even limited to a finite number of values. This is possible because the time domain characteristics emanate from the individual signal values, while the frequency domain attributes take into account the relation between values at specific times.

Our naive description of noise is now complete. Noise is just like any other signal—it has well defined time domain and frequency domain properties. Although we have not previously seen a flat spectrum like that of white noise, nothing prevents a deterministic signal from having that spectrum; and colored noise has narrower spectra, more similar to those with which we are familiar. The time domain characterization of noise *is* different from that of regular signals—rather than specifying how to create the signal, we must content ourselves with giving the signal's statistics. From our naive point of view we can think of all noise signals as being pseudorandom or incompletely known; we suppose that if we had more information we *could* describe the 'noise signal' in the time domain just as we describe other signals.

The reader probably realizes from our use of the word *naive* in describing this characterization of noise, that this isn't the entire story. It turns out that stochastic signals don't even *have* a spectrum in the usual sense of the word, and that more sophisticated probabilistic apparatus is required for the description of the time domain properties as well. We will take up these topics in Section 5.6. However, our naive theory is powerful enough to allow us to solve many practical problems. The next section deals with one of the first successful applications of noise removal, the processing of radar returns.

EXERCISES

5.2.1 Write a program to generate digital noise signals with probability distribution (5.1). Estimate the probability distribution using 10, 100, 1000, and 10,000 samples. What is the error of the estimation?

5.2.2 Equation (5.6) for the variance require two passes through the signal values; the first for computation of μ and the second for σ^2. Find a single-pass algorithm.

5.2.3 Using the random number generator supplied with your compiler write a zero-mean and unity variance noise generator. Make a histogram of the values it produces. Is it uniform? Calculate the empirical mean and standard deviation. How close to the desired values are they?

5.2.4 Using the noise generator of the previous exercise, generate pairs of random numbers and plot them as x, y points in the plane. Do you see any patterns? Try skipping L values between the x and y.

5.2.5 The noise generator you built above depends mainly on the most significant bits of the standard random number generator. Write a noise generator that depends on the least significant bits. Is this better or worse?

5.2.6 You are required to build the sample value histogram of a signal that only takes on values in a limited range, based on N samples. If you use too few bins you might miss relevant features, while too many bins will lead to a noisy histogram. What is the 'right' number of bins, assuming the probability distribution is approximately flat? What is the error for 10,000 samples in 100 bins?

5.2.7 What are the average, variance, and standard deviation of a Gaussian signal? What are the sufficient statistics? In what way is a Gaussian noise signal the simplest type of noise?

5.3 Noise Reduction by Averaging

Radar is an acronym for **ra**dio **d**etection **a**nd **r**anging. The basic principle of range finding using radar was first patented in 1935 by Robert Watson-Watt, but practical implementations were perfected by American and British scientists during World War II. Although modern radars are complex signal processing systems, the principles of the basic pulse radar are simple to explain. The radar transmitter periodically sends out a powerful electromagnetic pulse of short time duration; the time between pulses is called the **P**ulse **R**epetition **I**nterval (PRI). The pulse leaves the transmitter at the speed of light c and impinges upon various objects, whereupon minute fractions of the original signal energy are reflected back to the radar receiver. The round-trip time between the transmission of the pulse and the reception of the returned echo can thus be used to determine the distance from the radar to the object

$$r = \tfrac{1}{2}cT \qquad (5.10)$$

where the speed of light c is conveniently expressed as about 300 meters per microsecond.

The radar receiver is responsible for detecting the presence of an echo and measuring its **T**ime **O**f **A**rrival (TOA). The time between the TOA and the previous pulse transmission is called the *lag* which, assuming no ambiguity is possible, should equal the aforementioned round-trip time T. In order to avoid ambiguity the lag should be less than the PRI. Radar receivers must be extremely sensitive in order to detect the minute amounts of energy reflected by the objects to be detected. To avoid damaging its circuitry, the radar receiver is blanked during pulse transmission; and in order to keep the blanking time (and thus distance to the closest detectable target) minimal we try to transmit narrow pulse widths. This limits the amount of energy that may be transmitted, further decreasing the strength of the echo. Unfortunately, large amounts of natural and man-made noise are picked up as well, and the desired reflections may be partially or completely masked. In order to enhance the echo detection various methods have been developed to distinguish between the desired reflection signal and the noise. In general such a method may exploit characteristics of the signal, characteristics of the noise, or both. In this section we show how to utilize the knowledge we have acquired regarding the attributes of noise; the known PRI being the only signal-related information exploited. In Section 9.6 we will see how to improve on our results, notably by embedding easily detectable patterns into the pulses.

We can view the received signal as being the sum of a deterministic periodic signal x_n and an additive zero-mean noise signal ν_n

$$y_n = x_n + \nu_n$$

and our task is to recover x_n to the best of our abilities. The periodicity (with period equal to the PRI) of the desired signal derives from the supposition that the target is stationary or moving sufficiently slowly, and it enables us to observe the same echo signal many times. For sufficiently strong echoes we can simply isolate the echoes and measure the TOA for each pulse transmitted. Then we need only subtract successive TOAs to find the lag. However, this approach is not optimal, and doesn't work at all when the echoes are hidden deep in the noise. We are thus led to seek a stronger technique, one that exploits more knowledge regarding the noise.

The only quantitative statement made about the additive noise ν_n was that it had zero mean. From one PRI to the next the desired signal x_n remains unchanged, but the received signal y_n is seems completely different from x_n, as depicted in Figure 5.5. Sometimes y_n is greater than x_n, but (due to the noise having zero mean) just as frequently it will be less. Mathematically, using the linearity of the expectation operator, we can derive $\langle y_n \rangle = \langle x_n + \nu_n \rangle = \langle x_n \rangle + \langle \nu_n \rangle = x_n$.

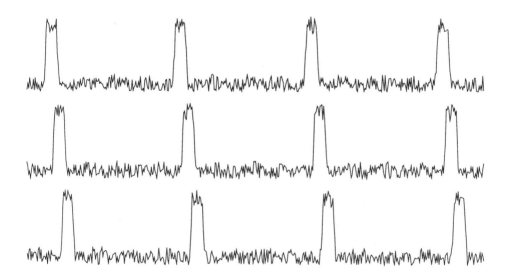

Figure 5.5: A pulsed radar signal contaminated by additive zero-mean noise. Note that from pulse to pulse the noise is different, but the pulse shape stays the same. Thus the uncontaminated signal can be reconstructed by pulse-to-pulse integration.

Hence, although in general the observed y_n is not the desired x_n, its average *is*. We can thus average the observed signals and obtain a much cleaner estimate of x_n. Such averaging over successive pulses is called radar return *integration*. With each new pulse transmitted, the true echo signal becomes stronger and stronger, while the noise cancels out and grows weaker and weaker. Even if the echo was initially completely buried in the noise, after sufficient averaging it will stand out clearly. Once detected, the lag measurement can be made directly on the average signal.

A similar operation can be performed for all periodic phenomena. When the desired underlying signal is periodic, each period observed supplies independent observations, and averaging increases the SNR. Another special case is slowly varying signals. Assuming the additive noise to be white, or at least containing significant spectral components at frequencies above those of x_n, we can average over adjacent values. The time domain interpretation of this operation is clear—since x_n varies more slowly than the noise, adjacent values are close together and tend to reinforce, while the higher-frequency noise tends to average out. The frequency domain interpretation is based on recognizing the averaging as being equivalent to a low-pass filter, which attenuates the high-frequency noise energy, while only minimally distorting the low-frequency signal. So once again just the zero mean assumption is sufficient to enable us to increase the SNR.

These averaging techniques can be understood using our naive theory, but take on deeper meaning in the more sophisticated treatment of noise. For example, we assumed that we could perform the averaging either in time or over separate experiments. This seemingly innocent assumption is known as the *ergodic hypothesis* and turns out to be completely nontrivial. We will return to these issues in Section 5.6.

EXERCISES

5.3.1 Generate M random ± 1 values and sum them up. The average answer will obviously be zero, but what is the standard deviation? Repeat for several different M and find the dependence on M.

5.3.2 In this exercise we will try to recover a constant signal corrupted by strong additive noise. Choose a number x between -1 and $+1$. Generate M random numbers uniformly distributed between -1 and $+1$ and add them to the chosen number, $s_n = x + \nu_n$. Now try to recover the chosen number by averaging over M values $\hat{x} = \sum_{n=1}^{M} s_n$ and observe the error of this procedure $x - \hat{x}$. Perform this many times to determine the *average error*. How does the average error depend on M?

5.3.3 Generate M sets of 1024 points of a sinusoidal signal corrupted by additive zero-mean noise,
$$s_n = \sin(\Omega t) + g\nu_n$$
where ν_n is uniform in the range $[-1 \ldots + 1]$. Average s_n over the M sets to reduce the noise. Use $\Omega = 0.01$, $g = 0.1, 1, 10$ and $M = 10, 100, 1000$. How does the residual noise decrease as a function of M?

5.3.4 Using the same signal as in the previous exercise, replace each s_n value by the average

$$s_{n-L} + s_{n-L+1} + \ldots + s_{n-1} + s_n + s_{n+1} + \ldots + s_{n+L-1} + s_{n+L}$$

How well does this work compared to the previous exercise? Try $\Omega = 0.001$ and $\Omega = 0.1$. What can you say about time averaging?

5.4   Pseudorandom Signals

Although noise is often a nuisance we wish weren't there, we frequently need to generate some of our own. One prevalent motive for this is the building of simulators. After designing a new signal processing algorithm we must check its performance in the presence of noise before deploying it in the real world. The normal procedure (see Section 17.7) requires the building of a simulator that inexpensively provides an unlimited supply of input signals over which we exercise complete control. We can create completely clean signals, or ones with some noise, or a great deal of noise. We can then observe the degradation of our algorithm, and specify ranges of SNR over which it should work well.

We may also desire to generate noise in the actual signal processing algorithm. Some algorithms actually require noise to work! Some produce output with annoying features, which may be masked by adding a small amount of noise. Some are simply more interesting with probabilistic elements than without.

In this section we will discuss methods for generating random numbers using deterministic algorithms. These algorithms will enable us to use our familiar computer environment, rather than having to input truly probabilistic values from some special hardware. You undoubtably already have a function that returns random values in your system library; but it's often best to know how to do this yourself. Perhaps you checked your random number generator in the exercises of the previous section and found that it is not as good as you need. Or perhaps you are designing an embedded application that runs without the benefit of support libraries, and need an efficient noise generator of your own. Or maybe you are given the job of writing just such a library for some new DSP processor.

Before embarking on our exposition of random number generators there is a myth we must dispel. There is no such thing as a random number! If there is no such thing, then why are we trying to generate them? We aren't. What we are trying to generate are *random sequences* or, in DSP terminology, *random digital signals*. Each particular signal value, once generated, is perfectly well known. It's just that the connection *between* the signal values at different times is nontrivial and best described in probabilistic terms. Ideally one should not be able to guess the next value that the generator will produce based on the previous values (unless one knows the algorithm). Unfortunately, the term 'random number generator' has become so entrenched in popular computer science jargon that it would be futile to try to call it something else. You can safely use this term if you remember that these generators are not to be used to generate a single 'random' value; their proper use is always through generating large numbers of values.

There are several relatively good algorithms for generating random sequences of numbers, the most popular of which is the *linear recursion method*, originally suggested by D.H. Lehmer in 1951. This algorithm employs the integer recursion

$$x_{n+1} = (ax_n + b) \bmod m \qquad (5.11)$$

starting from some quite nonrandom initial integer x_0. The integer parameters a, b, and m must be properly chosen for the scheme to work, for instance, by taking large m, and requiring b and m to be relatively prime, and a to be a large 'unusual' number. Real-valued random signals may be obtained by dividing all the integer values by some constant. Thus to create random real-valued signals in the range $[0 \ldots 1)$ one would probably simply use $\nu_n = \frac{x_n}{m}$, yielding quantized values with spacing $\frac{1}{m}$. Subtracting $\frac{1}{2}$ from this yields noise approximately symmetric around the zero.

The signals generated by equation (5.11) are necessarily periodic. This is because the present signal value completely determines the entire future, and since there are only a finite number of integer values, eventually some value must reoccur. Since apparent periodicity is certainly a bad feature for supposedly random signals, we wish the signal's period to be as long (and thus as unnoticeable) as possible. The longest period possible for the linear recursion method is thus the largest integer we can represent on our computer (often called MAXINT).

Long period is not enough. Taking $a = 1$, $b = 1$, and $m =$ MAXINT gives us the sequence $1, 2, 3 \ldots$ MAXINT, which indeed only repeats after MAXINT values, but hardly seems random. This is the reason we suggested that a be relatively large; this allows successive values to be widely separated. Keeping b and m relatively prime makes successive values as unrelated as possible. There is a lot more to say about optimal selection of these parameters, but instead of saying it we refer the reader to the extensive literature.

The implementation of equation (5.11) is quite problematic due to the possibility of overflow. Normally we desire m to be close to MAXINT, but then x may be quite large as well and $ax + b$ would surely overflow. Choosing m to be small enough to prohibit overflow would be overly restrictive, severely limiting period length. In assembly language programming this may sometimes be circumvented by temporarily allocating a larger register, but this option is not available to the writer of a portable or high-level language routine. The constraints can be overcome by restructuring the computation at the expense of slightly increased complexity (in the following / represents integer division without remainder).

```
Given integers m, a, b, x
Precompute:
     q ← m / a
     r ← m − a ∗ q
     l ← m − b
Loop:
     k ← x / q
     x ← a ∗ (x − q ∗ k) − r ∗ k − l
     if x < 0 then x ← x + m
```

By the way, if what you want is random *bits* then it's not a good idea to generate random integers and extract the LSB. This is because a sequence of integers can appear quite random, even when its LSB is considerably less so. Luckily there are good methods for directly generating random bits. The most popular is the **L**inear **F**eedback **S**hift **R**egister (LFSR), which is somewhat similar to linear recursion. A shift register is a collection of bits that can be shifted one bit to the right, thus outputting and discarding the LSB and making room for a new MSB. Linear feedback means that the new bit to be input is built by xoring together some of the bits in the shift register. Starting off with some bits in the shift register, we generate a sequence of bits by shifting to the right one bit at a time. Since the state of the shift register uniquely determines the future of the sequence, the sequence eventually become periodic. If the shift register ever has all zeros it becomes stuck in this state, and so this must be avoided at all costs.

One of the first random number generators was suggested by John von Neumann back in 1946. His method starts with some D digit integer. Squaring this integer produces an integer with $2D$ digits from which the next integer in the sequence is obtained by extracting the middle D digits. This recursion produces a periodic sequence of D digit integers, but this sequence will be considerably less random than one generated by a properly selected linear recursion generator.

Another random number generator does not require a multiplication, but does need more memory

$$x_{n+1} = (x_{n-j} + x_{n-k}) \bmod m$$

where j, k, and m need to be carefully chosen. Of course we need a buffer of length $\max(j, k)$, and must somehow initialize it.

Even if our random number generator turns out to be of inferior performance, there are ways to repair it. The most popular method is to use several different suboptimal generators and to combine their outputs in some way. For example, given three generators with different periods that output b bit integers, we can add the outputs or xor together their respective bits (an operation that is usually fast) and obtain a much better sequence. Given only two generators we can 'whiten' one by placing its values into a FIFO buffer and output a value from the buffer chosen by the second generator. This can even be accomplished by using a single suboptimal generator for both purposes. For example, assume that each call to 'random' returns a new pseudorandom real number between 0 and 1; then

```
Allocate buffer of length N
for i ← 1 to n
      buffer_i ← random
Loop:
      k ← floor(n random) + 1
      output buffer_k
      buffer_k ← random
```

is more random, since it whitens short time correlations.

The algorithms we have discussed so far return uniformly distributed pseudorandom numbers. In practice we frequently require pseudorandom numbers with other distributions, most frequently Gaussian. There are two popular ways of generating Gaussian noise given a source of uniformly distributed noise. The first relies on the 'law of large numbers' (see Appendix A.13) that states that the sum of a large number of independent random numbers, whatever their original distribution, will tend to be Gaussianly distributed. To exploit this law requires generating and adding N (even 12 is often considered large enough) uniform random numbers. Of course the maximum value that can be obtained is N times the maximum value of the uniform generator, so in reality the Gaussian is somewhat truncated, but the true distribution is extremely small there anyway. Often of more concern is the computational burden of computing N uniform random numbers per Gaussian random required.

The second method commonly used to generate Gaussianly distributed numbers, sometimes called the Box-Muller algorithm after its inventors, is best understood in steps. First pick at random a point inside the unit circle, $x + iy = re^{i\theta}$. If we selected the point such that x and y are independent (other than the constraint that the point be inside the circle) then r and θ will be as well. Now θ is uniformly distributed between 0 and 2π; how is r distributed? It is obvious that larger radii are more probable since the circumference increases with radius; in fact it is quite obvious that the probability of having a radius between zero and r increases as r^2. We now create a new point in the plane $u + iv$, whose angle is θ but with radius ρ that obeys $r^2 = e^{-\rho^2/2}$. The probability of such a point having radius less than R is the same as the probability that the original squared radius r^2 is greater than $e^{-R^2/2}$. From this it follows that u and v are Gaussianly distributed.

How do we select a point inside a circle with all points being equally probable? The easiest way is to randomly pick a point inside the square that circumscribes the unit circle, and to discard points outside the circle. Picking a point inside a square involves independently generating two uniformly distributed random numbers x and y. Since u and v are also independent, for every two uniform random numbers that correspond to a point inside the circle we can compute two Gaussianly distributed ones.

Thus we arrive at the following efficient algorithm:

```
generate two uniform random numbers between −1 and +1, x and y
r² ← x² + y²
if r² > 1 return to the beginning
ρ² ← −2 ln r²,  c ← ρ/r
u ← cx and v ← cy
```

EXERCISES

5.4.1 Not only isn't there such a thing as a random number, there really is no such thing as a random sequence of finite length. For example, all sequences of ten digits are equally probable, namely one chance in 10^{10}. Yet we feel viscerally that sequences such as $\{1,1,1,1,1,1,1,1,1,1\}$ or $\{1,2,3,4,5,6,7,8,9\}$ are less random than say $\{1,9,3,6,3,4,5,8,2\}$. Can you explain this feeling?

5.4.2 You can test a **random** function using the following graphical test. Generate successive values $r_1, r_2, \ldots$ and make a scatter plot consisting of points (r_k, r_{k-1}). If the resulting picture has structure (e.g., noticeable lines) the random sequence has short-term correlations. If the plot looks reasonably homogeneous repeat the procedure but plot (r_k, r_{k-m}) instead. Test the integer recursions (equation (5.11)) defined by a=10, b=5, m=50; a=15625, b=0, m=65536; and the generator supplied with your programming environment.

5.4.3 Take inferior random generators from the previous exercise and whiten them using the algorithm given in the text. Perform the graphical test once again.

5.4.4 Code a Gaussian noise generator based on the law of large numbers and check its distribution.

5.4.5 Some people use this algorithm to generate Gaussianly distributed numbers: generate two uniform random numbers, x and y, between 0 and $+1$
$a = \sqrt{-2 \ln x}, \ \phi = 2\pi y$
$u \leftarrow a \sin(\phi)$ and $v \leftarrow a \cos(\phi)$
Is this algorithm correct? What are the advantages and disadvantages relative to the algorithm given in the text?

5.4.6 Other people use the following algorithm:
generate two uniform random numbers, x and y, between 0 and $+1$
$u = \sqrt{-2 \ln x} \sin(2\pi y)$
$v = \sqrt{-2 \ln x} \cos(2\pi y)$
Show that this method is mathematically equivalent to the method given in the text. In addition to requiring calls to sine and cosine functions, this method is numerically inferior to the one given in the text. Why?

5.4.7 Complete the proof of the second algorithm for generating Gaussianly distributed random numbers.

5.4.8 How can we generate random numbers with an arbitrary distribution given a uniform generator?

5.4.9 Show that after an initial transient LFSR sequences are always periodic. What is the maximal period of the sequence from a shift register of length K? Find a maximal length LFSR sequence of length 15.

5.5   Chaotic Signals

Completely specified deterministic signals, that is, signals generated by completely specified deterministic algorithms, can still appear to be entirely random and chaotic. The word 'chaos' comes from the Greek $\xi\alpha o\sigma$, the most ancient of the gods, and refers to the confused primordial state before the creation. The study of chaotic signals is quite the reverse; what can be fruitfully examined is the route taken from orderly (often periodic) behavior to the chaotic. Most of this section will be devoted to the study of the transition from periodic to chaotic behavior in the simplest possible setting.

How can deterministic signals exhibit chaotic behavior? Turbulence of rapidly flowing liquids is one of the prototypes of chaos; although the equations of fluid dynamics are well known, we cannot predict the exact behavior of twisting currents and whirlpools. When the flow is slow the behavior *is* understandable, so we can start with a slowly flowing liquid and gradually increase the flow until chaos sets in. Similarly, the future value of investments may become unpredictable when interest rates are high and the market volatile, but such prediction is straightforward under more subdued conditions. One can forecast the weather for the next day or two when conditions are relatively stable, but prediction becomes impossible over longer periods of time.

There is a simple mathematical explanation for the appearance of chaos in a deterministic setting. Linear equations (whether algebraic, differential, or difference) have the characteristic that small changes in the input lead to bounded changes in output. Nonlinear equations do not necessarily have this attribute. In fact it is known that for nonlinear equations with three or more free parameters there always are values of these parameters for which infinitesimally small changes in initial conditions lead to drastic changes of behavior. Even one or two parameter nonlinear equations *may* become oversensitive. Such equations are said to exhibit chaotic behavior since our knowledge of the initial conditions is never sufficient to allow us to predict the output far enough from the starting point. For example, we may be able to predict tomorrow's weather based on today's, but the fundamental equations are so sensitive to changes in the temperature and air pressure distributions that we have no chance of accurately predicting the weather next week.

Perhaps the simplest example of knowledge loss is the *shift and truncate* recursion

$$x_{n+1} = \text{Trunc}\,(10x_n) \tag{5.12}$$

which shifts the signal value's decimal point to the right, and then removes the integer part. The first few values starting with $x_0 = \pi - 3$ are

$$0.1415926535\ldots, 0.4159265358\ldots, 0.1592653589\ldots,$$
$$0.5926535897\ldots, 0.9265358979\ldots, 0.2653589793\ldots$$

which seem to oscillate wildly over the unit interval. Had we chosen x_0 slightly different from $\pi - 3$, the deviation of the resulting x_n from the above values would exponentially increase; for example, with a difference of 10^{-5} all similarity is lost after only five iterations.

The weather prediction example is similar. It turns out that the equations relating air pressure, temperature, wind velocity, etc. are highly nonlinear, even for rather simplistic models of atmospheric conditions. Weather prediction relies on running such models, with appropriate initial weather conditions, on large computers and observing the resulting weather conditions. The initial specification is rather coarsely defined, since only gross features such as average air temperature and pressure are known. This specification leads to specific predictions of the weather as a function of time. However, slight changes in the specification of the initial weather conditions lead to rather different predictions, the differences becoming more and more significant as time goes on. This is the reason that the weather can be predicted well for the short term, but not weeks in advance. Lorenz, who discovered the instability of weather prediction models in the early 1960s, called this the 'butterfly effect'; a butterfly flapping its wings in Peking will affect the weather in New York a month later!

How can we hope to study such nonlinear equations? Isn't chaos by definition incomprehensible and thus unresearchable? The trick is to study *routes to chaos*; we start at values of parameters for which the nonlinear equations are *not* chaotic, and then to vary the parameters in order to approach the chaotic region. Before entering the chaotic region, the output signal, although increasingly bizarre, *can* be profitably investigated. In this section we will study Feigenbaum's route to chaos. This route is easy to study since it occurs in a simple one-parameter setting, arguably the simplest nonlinear equation possible. It also seems to model well many interesting physical situations, including some of the examples mentioned above.

We'll introduce Feigenbaum's route with a simple example, that of fish in a closed pond. Let us denote by x the present fish population divided by the maximum possible population (thus $0 \leq x \leq 1$). We observe the population every day at the same hour, thus obtaining a digital signal x_n. How does x_n vary with time? Assuming a constant food supply and a small initial number

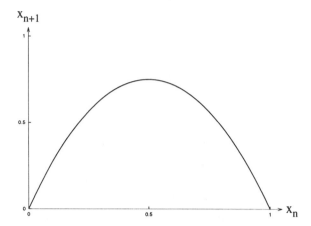

Figure 5.6: The logistics recursion relates the new signal value x_{n+1} to the old one x_n by a inverted parabola. As such it is the simplest nonlinear recursion relation. It can also be used to approximate any recursion with a single smooth maximum.

of fish, we expect an initial exponential increase in population,

$$x_{n+1} = r\, x_n$$

but once the number of fish becomes large, we anticipate an opposite tendency due to overpopulation causing insufficient food and space, and possibly spread of disease. It makes sense to model this latter tendency by a $1 - x_n$ term, since this leads to pressure for population decrease that is negligible for small populations, and increasingly significant as population increases. Thus we predict

$$x_{n+1} = r\, x_n \left(1 - x_n\right) \tag{5.13}$$

which is often called the *logistics equation*. This equation is quadratic (see Figure 5.6) and thus nonlinear. It has a single free parameter r (which is related to the amount we feed the fish daily), which obey $0 \le r \le 4$ in order for the signal x to remain in the required range $0 \le x \le 1$. Although a nonlinear equation with one free parameter is not guaranteed to be chaotic, we will see that there *are* values of r for which small changes in x_0 will lead to dramatic changes in x_n for large n. This means that when we overfeed there will be large unpredictable fluctuations in fish population from day to day.

You may object to studying in depth an equation derived from such a fishy example. In that case consider a socialistic economy wherein the state wishes to close the socioeconomic gap between the poor and the wealthy. It is decided to accomplish this by requiring everyone to deposit their money in a state-controlled bank that pays lower interest rates to the wealthy. Let y_n be the amount invested as a function of time, y_{max} the maximum wealth allowed by law, and i the applicable interest. The usual financial formulas tell us $y_{n+1} = (1+i)y_n$, but here i must be a decreasing function of y, which we take to be $i = i_0(1 - \frac{y_n}{y_{max}})$. Substitution leads to

$$y_{n+1} = \left(1 + i_0\left(1 - \frac{y_n}{y_{max}}\right)\right) y_n$$

which by a simple change of variables becomes the logistics equation (5.13).

We could continue to give examples that lead to the same equation. It is so ubiquitous simply because it is the simplest nonlinear recursion relation for a single bounded signal that contains a single free parameter. Any time we obtain a quadratic relationship we can transform it into the logistics equation by ensuring that the variable is constrained to the unit interval; indeed any time we have any nonlinear recursion with a single smooth maximum we can approximate it by the logistics equation in the vicinity of the maximum.

Now that we are convinced that such a study is worthwhile, let us embark upon it. We expect the signal x_n to eventually approach some limiting value, i.e., that the number of fish or the amount of money would approach a constant for long enough times. This is indeed the case for small enough r values. To find this value as a function of r we need to find a *fixed point* of the recursion, that is, a signal value that once attained forever remains unchanged. Since $x_{n+1} = f(x_n)$ must equal x_n, we conclude that a fixed point x_1 must obey the following equation.

$$x_1 = f(x_1) = r\,x_1\,(1 - x_1) \tag{5.14}$$

Zero is obviously a fixed point of the logistics equation since $x_n = 0$ implies $x_{n+1} = rx_n(1 - x_n) = 0$ as well. When you have no fish at all, none are born, and an empty bank account doesn't grow. Are there any nontrivial fixed points? Solving equation (5.14) we find the nonzero fixed points are given by $x_1 = p_r \equiv 1 - \frac{1}{r}$.

For this simplest of recursions we could algebraically find the fixed points with little trouble. For more complex cases we may fall back to a graphical method for finding them. In the graphical method you first plot the recursion function $x_{n+1} = f(x_n)$ (with x_n on the x axis and x_{n+1} on the y axis). Then you overlay the identity line $x_{n+1} = x_n$. Fixed points must correspond to intersections of the recursion plot with the identity line. In our case the recursion is an inverted parabola, and we look for its intersections with the

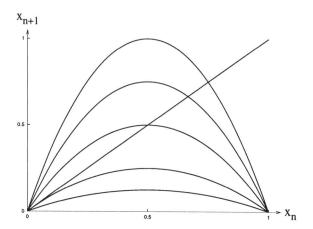

Figure 5.7: Graphical method of finding fixed points of the logistics equation. From bottom to top the inverted parabolas correspond to $r = 0, 1, 2, 3, 4$. We see that for $r < 1$ the parabola intersects the identity line only at $x = 0$, while for larger r there is an additional point of intersection.

$45°$ line (Figure 5.7). It is easy to see that for the parameter region $0 \leq r \leq 1$ the only possible fixed point is $x_0 = 0$, but for $r > 1$ the new fixed point p_r appears. For $r \gtrsim 1$ the new fixed point p_r is close to the old one (zero), gradually moving away with increasing r.

So we have found that the steady state behavior of the recursion is really very simple. For $r < 1$ we are underfeeding our fish, or the interest is negative, and so our fish or money disappear. An example of this behavior is displayed in Figure 5.8.A. For $1 < r < 3$ the number or fish or amount of money approaches a constant value as can be seen in Figure 5.8.B. However, we are in for quite a shock when we plot the behavior of our fish or money for $r > 3$ (Figures 5.8.C and 5.8.D)! In the first case the signal oscillates and in the second it seems to fluctuate chaotically, with no possibility of prediction. In the chaotic case starting at a slightly different initial point produces a completely different signal after enough time has elapsed! We don't yet understand these phenomena since we progressed along the route to chaos too quickly, so let's backtrack and increase r more slowly.

The most important feature of the behavior of the signal for small r is the existence of the fixed point. Fixed points, although perhaps interesting, are not truly significant unless they are *attractive*. An attractive fixed point is one that not only replicates itself under the recursion, but *draws in* neighboring values as well. For $r \leq 1$ the zero fixed point is attractive—no matter where we start we rapidly approach $x = 0$; but for $1 < r < 3$ the new fixed

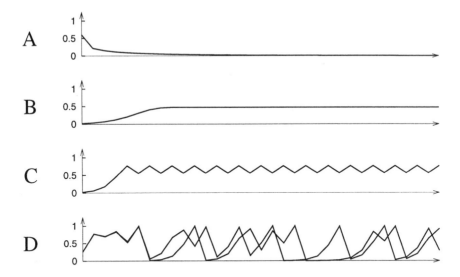

Figure 5.8: Signal produced by recursion of the logistics equation for different values of r. In (A) we have $r = 0.9$ and the signal decays to zero. In (B) we have $r = 1.9$ and the signal approaches a constant value. (C) depicts $r = 3.1$ and the signal oscillates between two values. In (D) we have $r = 4$ with two slightly different initial states; the signal is noise-like and irreproducible.

point 'draws in' all signals that do not begin with $x_0 = 0$ or $x_0 = 1$. This is hinted at in Figure 5.8, but it is both simple and instructive for the reader to experiment with various r and x_0 and become convinced.

The idea of attraction can be made clear by using the 'return map', which is a graphical representation of the dynamics. First, as before, we plot the recursion $x_{n+1} = f(x_n)$ and the 45° line $x_n = x_{n-1}$. We start with a point on the line (x_0, x_0). Now imagine a vertical line that intersects this point; it crosses the recursion curve at some point x_1. We draw the vertical line from (x_0, x_0) to this new point (x_0, x_1). Next we imagine a horizontal line that intersects this new point. It crosses the 45° line at (x_1, x_1), and we proceed to draw a horizontal line to there. The net result of the previous two operations is to draw two lines connecting (x_0, x_0) to (x_1, x_1), corresponding to one iteration from x_0 to x_1.

We can now continue to iterate (as in Figure 5.9) until an attractor is found. In part (A) of that figure we see that when $r < 1$ (no matter where we begin) we converge to the zero fixed point. Part (B) demonstrates that when $1 < r < 3$, we almost always converge on the new fixed point (the exceptions being $x_0 = 0$ and $x_0 = 1$, which remain at the old zero fixed

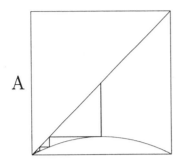 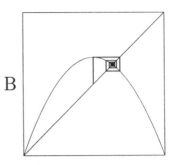

Figure 5.9: Use of return maps to depict the dynamics of a simple recursion. Each iteration starts on the 45° line, proceeds vertically until intersecting the recursion curve, and then returns to the diagonal line. Here we see that after enough iterations we converge on a fixed point, which is the intersection of the recursion curve with the diagonal line. In (A) we have $r = 0.5$ and the only fixed point is zero, while in (B) we see convergence to the nonzero fixed point p_r.

point). We will see shortly that for $r > 3$ even this fixed point ceases being an attractor; if one starts exactly at it, one stays there, but if one strays even slightly the recursion drives the signal away.

How can we mathematically determine if a fixed point p is an attractor? The condition is that the absolute value of the derivative of the recursive relation f must be less than unity at the fixed point.

$$\left| \frac{df(x)}{dx} \right|_{x=p} < 1 \tag{5.15}$$

This ensures that the distance from close points to the fixed point *decreases* with each successive recursion. It is now easy to show that for $r > r_2 = 3$ the fixed point p_r becomes unattractive; but what happens then? No new fixed point can appear this time, since the reasoning that led to the discovery of p_r as the sole nonzero fixed point remains valid for all r! To see what happens we return to the return map. In Figure 5.10.A we see that starting from some initial point we approach a 'square', which translates to alternation between two points. Once the signal reaches its steady state it simply oscillates back and forth between these two values, as can be seen in Figures 5.10.B and 5.11. This dual-valued signal is the new attractor; unless we start with $x_0 = 0, 1, 1 - \frac{1}{r}$ or $f^{-1}(1 - \frac{1}{r})$ we eventually oscillate back and forth between two values. As r increases the distance between the two values that make up this attractor also increases.

So attractors can be more complex than simple fixed points. What happens when we increase r still further? You may have already guessed that this two-valued attractor also eventually becomes unattractive (although if

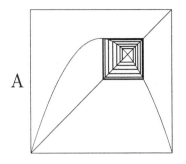 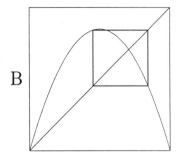

Figure 5.10: Return map representation of the logistics equation for $r > 3$. In (A) $r = 3.2$ and we see that from an arbitrary initial state x_0 we converge on a 'non fixed point' attractor close to p_r. The attractor contains two points, one on either side of p_r. In (B) $r = 3.4$ and we display only the long time behavior (steady state behavior after the transient has died down).

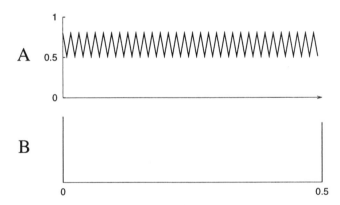

Figure 5.11: The signal resulting from recursion of the logistics equation for $r = 3.2$. In (A) we see the steady state signal in the time domain. It oscillates between the two values that make up the attractor, which means that $x_{n+1} = f(x_n)$ and $x_{n+2} = f(x_{n+1}) = x_n$. In (B) we see the same signal in the frequency domain. The DC component represents the nonzero average of the two points. Since the signal oscillates at the maximum possible frequency, we have a spectral line at digital frequency $\frac{1}{2}$.

one starts at *exactly* one of its points one stays trapped in it) and a new more complex attractor is born. In this case, this happens at $r_3 = 1 + \sqrt{6}$ and the new attractor is composed of a cycle between four signal values, as depicted in Figure 5.12. If we call these points a_1, a_2, a_3, and a_4, the requirement is $a_2 = f(a_1)$, $a_3 = f(a_2)$, $a_4 = f(a_3)$, and $a_1 = f(a_4)$. Note that the 2-cycle's a_1 split up into our present a_1 and a_3, while its a_2 became our new a_2 and a_4. So the attractor's components obey $a_1 < a_3 < a_2 < a_4$, which

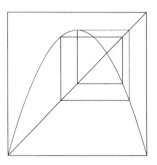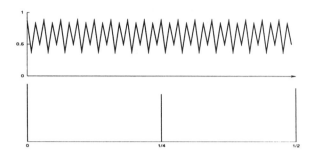

Figure 5.12: The return map, signal, and spectrum for the steady state behavior when $r = 3.5$. The attractor is a 4-cycle.

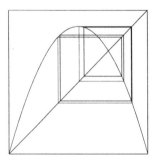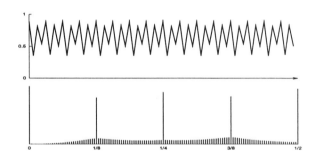

Figure 5.13: The return map, signal, and spectrum for the steady state behavior when $r = 3.55$. The attractor is a 8-cycle.

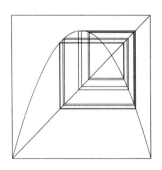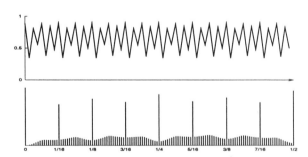

Figure 5.14: The return map, signal, and spectrum for the steady state behavior when $r = 3.5675$. The attractor is a 16-cycle.

means that the closest together in time are the farthest in space and vice versa. This induces a spectrum wherein an additional spectral line appears at twice the period, or half the frequency of the previous line.

We saw that when r increased above r_3 each component of the 2-cycle splits into two, just as the fixed point had earlier split. The same thing happens in turn for the 4-cycle when r goes above r_4 and an 8-cycle is born. The critical feature is that at each stage all components of the present attractor become unattractive simultaneously, a phenomenon known as *pitchfork bifurcation*. Due to the bifurcation, with increasing r we find 16-cycles, 32-cycles, and all possible 2^n-cycles. Examples of such cycles are depicted in Figures 5.12 through 5.14. The rule of 'closest in time are farthest in space' continues to be obeyed, so that new spectral lines continue to appear at harmonics of half the previous basic frequency. Eventually the lines are so close together that the spectrum becomes white, and we have chaotic noise.

The transition from periodicity to chaos can best be envisioned by plotting the attractors as a function of r, as in Figure 5.15. The transition points r_n as a function of n approach a limit

$$r_n \overset{n \to \infty}{\longrightarrow} r_\infty \approx 3.57$$

so that the regions where these cycles exist become smaller and smaller. By the time we reach r_∞ we have finished all the 2^n-cycles.

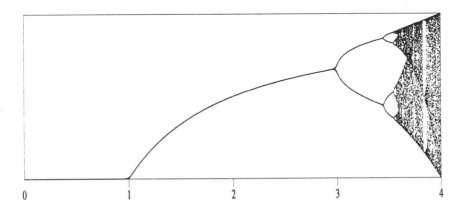

Figure 5.15: The attractors of the recursion as a function of r. Observe the zero attractor for $0 < r < 1$, the fixed point p_r for $1 < r < 3$, the 2-cycle for $3 < r < 1 + \sqrt{6}$, and the 2^n-cycles for $3 < r < r_\infty$. Certain odd cycle regions can also be clearly seen for $r > r_\infty$. At $r = 4$ chaos reigns.

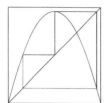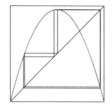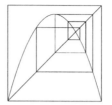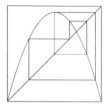

Figure 5.16: Non 2^n-cycle attractors for $r > r_\infty$. We present return maps for a 3-cycle ($r = 3,83$), a 6-cycle that results from bifurcation of that 3-cycle ($r = 3.847$), a quite different 6-cycle ($r = 3.63$), and a 5-cycle ($r = 3.74$).

What happens between here and $r = 4$? It turns out that every length attractor is possible. For example, in Figure 5.16 we see 3-cycles, 5-cycles and 6-cycles. There is a theorem due to Šarkovskii that states that the order of first appearance of any given length is

$$1, 2, 4, 8, \cdots 2^k, \cdots 2^k \cdot 9, 2^k \cdot 7, 2^k \cdot 5, 2^k \cdot 3, \cdots 4 \cdot 9, 4 \cdot 7, 4 \cdot 5, 4 \cdot 3, \cdots 9, 7, 5, 3$$

so that once a 3-cycle has been found we can be certain that all cycle lengths have already appeared.

For $r_\infty < r \leq 4$ there are other types of behavior as well. Let us start at $r = 4$ where all possible x values seem to appear chaotically and decrease r this time. At first x seems to occupy the entire region between $\frac{r}{4}$ and $r\left(1 - \frac{r}{4}\right)$, but below a certain r_1' this band divides into two broad subbands. The signal always oscillates back and forth between the two subbands, but where it falls in each subband is unpredictable. Decreasing r further leads us past r_2' where each subband simultaneously splits into two somewhat narrower subbands. The order of jumping between these four subbands is 'closest in time are farthest in space', but the exact location inside each subband is chaotic. Decreasing further leads us to a cascade of r_n' in between which there are 2^n chaotic subbands, a phenomenon known as 'reverse bifurcation'. Interspersed between the reverse bifurcations are regions of truly periodic behavior (such as the 3-, 5-, and 6-cycles we saw before). The r_n' converge precisely onto r_∞ where the reverse bifurcations meet the previous bifurcations.

We have seen that the simplest possible nonlinear recursion generates an impressive variety of periodic and chaotic signals; but although complex, these signals are still deterministic. In the next section we will see what a truly random signal is.

EXERCISES

5.5.1 What is the change of variables that converts the socialistic economy equation into the fish pond one?

5.5.2 Write a simulator that graphically depicts the behavior of the signal generated by the logistics equation (5.13). Vary r and by trying various starting points identify the attractors in the different regions.

5.5.3 Write a program to plot the attractors as a function of r. For each r go through the possible x_0 systematically and identify when periodic behavior has been reached and plot all points in this attractor. Can you identify the various regions discussed in the text?

5.5.4 Extend the simulator written above to display the spectrum as well. Reproduce the results given in the text.

5.5.5 Prove that equation (5.15) is indeed the criterion for attractiveness. Prove that for $r \leq 1$ zero is indeed an attractor. Prove that for $1 < p \leq 3$ the fixed point p_r is an attractor.

5.5.6 At $r = 4$ a change of variable

$$x = \tfrac{1}{2}(1 - \cos 2\pi\theta)$$

brings us to a variable θ, which is homogeneously distributed. Show that x is distributed according to $\dfrac{1}{\sqrt{x(1-x)}}$.

5.5.7 Plot the signal and spectrum of the 3-, 5-, and 6-cycles.

5.5.8 Henon invented a number of area preserving two-dimensional chaotic signals, for example,

$$
\begin{aligned}
x_{n+1} &= x_n \cos\alpha - (y_n - x_n^2)\sin\alpha \\
y_{n+1} &= x_n \sin\alpha + (y_n - x_n^2)\cos\alpha
\end{aligned}
$$

which is dependent on a single parameter α, which must obey $0 \leq \alpha \leq \pi$. Show that the origin is a fixed point, that large x diverge to infinity, and that there is a symmetry axis at angle $\alpha/2$. Are there any other fixed points?

5.5.9 Write a program to plot in the plane the behavior of the Henon map for various α. For each plot start from a large number of initial states (you can choose these along the 45° line starting at the origin and increasing at constant steps until some maximal value) and recurse a large number of times. What happens at $\alpha = 0$ and $\alpha = \pi$? Increase α from zero and observe the behavior. For what α are 'islands' first formed? Zoom in on these islands. When are islands formed around the islands? Observe the sequence of *multifurcations*. When is chaos achieved?

5.6   Stochastic Signals

In this section we will briefly introduce the formal theory of stochastic signals. This topic is a more advanced mathematically, and we assume the reader has a working knowledge of basic probability theory (see Appendix A.13). Since a full discussion of the theory would require a whole book we will have to content ourselves with presenting only the basic terminology.

Recall that a signal is *deterministic* if we can precisely predict its value at any time; otherwise it is *stochastic*. What do we mean by a random signal? By our original definition a signal must be precisely defined for all times, how can it be random? When we speak of a random signal in the formal sense, we are actually not referring to a single signal at all, but to an infinite number of signals, known as an *ensemble*. Only individual realizations of the ensemble can be actually observed, (recall Figures 5.1, 5.2, and 5.3) and so determining a signal value requires specification of the realization in addition to the time. For this reason many authors, when referring to the entire ensemble, do not use the term 'signal' at all, prefering to speak of a *stochastic process*.

Often a stochastic signal is the sum of a deterministic signal and noise.

$$s^r(t) = x(t) + \nu^r(t) \qquad \mathbf{A} \; \Big| \; \mathbf{D} \qquad s_n^r = x_n + \nu_n^r \qquad (5.16)$$

Here the superscript r specifies the specific realization; for different r the deterministic component is identical, but the noise realization is different. Each specific realization is a signal in the normal sense of the word; although it might not be possible to find an explicit equation that describes such a signal, it nonetheless satisfies the requirements of signalhood, with the exception of those we usually ignore in any case. While taking a specific r results in a signal, taking a specific time t or n furnishes a *random variable*, which is a function of r over the real numbers that can be described via its probability distribution function. Only when both r and t or n are given do we get a numeric value; thus a stochastic signal can be better described as a function of two variables.

When describing a stochastic signal we usually specify its *ensemble statistics*. For example, we can average over the ensemble for each time, thus obtaining an average as a function of time. More generally, for each time we can find the **Probability Density Function** (PDF) of the ensemble.

$$\begin{array}{c|c} \begin{array}{l} f(t,s)ds = \\ \text{Prob}(s \leq s(t) \leq s+ds) \end{array} & \textbf{A} \quad \textbf{D} \quad \begin{array}{l} f_n(s)ds = \\ \text{Prob}(s \leq s_n \leq s+ds) \end{array} \end{array} \qquad (5.17)$$

Here, even for the digital case we assumed that s was unquantized, $f_n(s)ds$ representing the probability that the signal value at time n will be between s and $s+ds$. Note that unlike the statistics of a regular random variable, statistics of stochastic signals are functions of time rather than simple numbers. Only for the special case of *stationary signals* are these statistics constants rather than time-dependent.

As for regular random variables, in addition to the density function $f(s)$ we can define the **Cumulative Distribution Function** (CDF)

$$F(t,s) = \text{Prob}(s(t) \leq s) \qquad \textbf{A} \quad \textbf{D} \qquad F_n(s) = \text{Prob}(s_n \leq s) \qquad (5.18)$$

and it is obvious that for every time instant the density is the derivative of the cumulative distribution. These distribution functions are in practice cumbersome to use and we usually prefer to use statistics such as the mean and variance. These can be derived from the density or cumulative distribution. For example, for analog signals the mean for all times is calculated from the density by integration over all x values

$$m_s(t) = \left\langle s(t) \right\rangle = \int_{-\infty}^{\infty} s(t)\, f(t,s)\, ds \qquad (5.19)$$

and the variance is as one expects.

$$\sigma_s^2(t) = \int_{-\infty}^{\infty} \left(s(t) - m_s(t) \right)^2 f(t,s)\, ds \qquad (5.20)$$

There are also other statistics for stochastic signals, which have no counterpart for simple random variables. The simplest is the correlation between the signal at different times

$$C_s(t_1, t_2) = \left\langle s(t_1)s(t_2) \right\rangle \qquad \textbf{A} \quad \textbf{D} \qquad C_s(n_1, n_2) = \left\langle s_{n_1} s_{n_2} \right\rangle \qquad (5.21)$$

which for stationary signals is a function only of the time difference $\tau = t_1 - t_2$ or $m = n_1 - n_2$.

$$C_s(\tau) = \left\langle s(t)s(t-\tau) \right\rangle \qquad \textbf{A} \quad \textbf{D} \qquad C_s(m) = \left\langle s_n s_{n-m} \right\rangle \qquad (5.22)$$

This correlation is usually called the *autocorrelation*, since there is also a crosscorrelation between two distinct signals $C_{xy}(t_1, t_2) = \langle x(t_1)y(t_2) \rangle$. The autocorrelation tells us how much the value of the signal at time t_1 influences its value at time t_2, and is so important that we will devote all of Chapter 9 to its use. The single time variance is simply the autocorrelation when the two time variables coincide $\sigma_s^2(t) = C_s(t, t)$, and so for stationary signals $\sigma_s^2 = C_s(0)$. We also sometimes use the *autocovariance* $V_s(t_1, t_2) = \langle (s(t_1) - m_s(t_1))(s(t_2) - m_s(t_2)) \rangle$ and it is easy to show that $V_s(t_1, t_2) = C_s(t_1, t_2) - m_s(t_1) m_s(t_2)$.

More generally we can have statistics that depend on three or more time instants. Unlike single time statistics, which can be calculated separately for each time, these multitime statistics require that we simultaneously see the entire stochastic signal (i.e., the entire ensemble for all times). We often use only the mean as a function of time and the correlation as a function of two times. These are adequate when the probability density function for all times is Gaussian distributed, since Gaussians are completely defined by their mean and variance. For more general cases *higher-order signal processing* must be invoked (see Section 9.12), and we define an infinite number of *moment functions*

$$M_s(t_1, t_2, \ldots, t_k) \equiv \langle s(t_1)s(t_2) \cdots s(t_k) \rangle \qquad (5.23)$$

that should not be confused with 'statistical moments'

$$m_s = \langle t^s s(t) \rangle$$

which are simply numbers. A stochastic signal is said to be 'stationary to order k' if its moments up to order k obey

$$M_s(t_1, t_2, \ldots, t_k) = M_s(t_1 + \tau, t_2 + \tau, \ldots, t_k + \tau)$$

and stationarity implies stationarity to order k for all finite k.

A few concrete examples will be helpful at this point. An analog *Markov* signal is a stationary, zero mean stochastic signal for which the autocorrelation dies down exponentially.

$$C(t_1, t_2) = \frac{e^{-\frac{|t_2 - t_1|}{\tau}}}{\tau}$$

Thus there is essentially only correlation between signal values separated by about τ in time; for much larger time differences the signal values are essentially uncorrelated. When τ approaches zero we obtain white noise, which thus has a delta function autocorrelation.

$$C(t_1, t_2) = \sigma^2 \delta(t_1 - t_2)$$

This means that for any two distinct times, no matter how close these times are, there is no correlation at all between signal values of white noise.

For discrete time signals we define Markov signals of different orders. A *first-order Markov* signal is one for which s_n depends on s_{n-1} but not directly on any previous value. A *second-order Markov* signal has the signal value depending on the two previous values.

There is an important connection between white noise and Markov signals; A Markov signal s_n can be generated by filtering white noise ν_n. We cannot fully explain this result as our study of filters will only begin in the next chapter, but the main idea can be easily understood. Signal values of white noise at different times can be independent because of the high-frequency components in the noise spectrum. Filtering out these high frequencies thus implies forcing the signal value at time n to depend on those at previous instants. A particular type of low-pass filtering produces precisely Markovian behavior.

$$s_n = \alpha s_{n-1} + \nu_n \tag{5.24}$$

Low-pass filtering of white noise returns us to a Markov signal; band-pass filtering results in what is often called 'colored noise'. These signals have nonflat power spectra and nondelta autocorrelations.

Note that although we often use Gaussian white noise, these two characteristics are quite independent. Noise can be white without being Gaussian and vice versa. If for any two times the signal is uncorrelated, and all moments above the second-order ones are identically zero, we have Gaussian white noise. However, when the signal values at any two distinct times are statistically independent, but the distributions although identical at all times are not necessarily Gaussian, we can only say that we have an Independent Identically Distributed (IID) signal. Conversely, when there are correlations between the signal values at various times, but the joint probability function of n signal values is n-dimensional Gaussian, then the signal is Gaussian noise that is not white.

Stochastic signals are truly complex, but it is reassuring to know the most general stationary stochastic signal can be built from the elements we have already discussed. In the 1930s Wold proved the following theorem.

Theorem: Wold's Decomposition

Every stationary stochastic signal s can be written

$$s_n = x_n + \sum_{m=0}^{\infty} h_m w_{n-m}$$

as the sum of a deterministic signal x and filtered white noise.   ∎

In addition to the ensemble statistics we have been discussing, there is another type of statistics that can be computed for stochastic signals, namely *time statistics*. For these statistics we consider a single realization and average over the time variable, rather than hold the time constant and averaging over the ensemble. Thus the time average of a signal s at time zero is

$$\langle s \rangle = \frac{1}{T} \int_{\tau-\frac{T}{2}}^{\tau+\frac{T}{2}} s(t)\, dt \qquad \mathbf{A} \,\Big|\, \mathbf{D} \qquad \langle s \rangle = \frac{1}{N} \sum_{n=\nu-\frac{N}{2}}^{\nu+\frac{N}{2}} s_n \qquad (5.25)$$

where T or N are called the 'integration windows'. This type of averaging is often simpler to carry out than ensemble averaging since for $s(t)$ and s_n we can use any realization of the signal s that is available to us, and we needn't expend the effort of collecting multiple realizations. When we previously suggested combating noise for a narrow-band signal by averaging over time, we were actually exploiting time statistics rather than ensemble statistics.

What is the connection between ensemble statistics and time statistics? In general, there needn't be any relation between them; however, we often assume a very simple association. We say that a signal is *ergodic* if the time and ensemble statistics coincide. The name 'ergodic' has only historical significance, deriving from the 'ergodic hypothesis' in statistical physics that (wrongly) posited that the two types of statistics must always coincide. To see that in general this will not be the case, consider the ensemble of all different DC signals. The ensemble average will be zero, since for every signal in the ensemble there is another signal that has precisely the opposite value. The time average over any one signal is simply its constant value, and *not* zero! A less trivial example is given by the digital sinusoid

$$s_n = A \sin(\omega n)$$

with A chosen *in the ensemble* with equal probability to be either 1 or -1. Here both the ensemble and time averages are zero; but were we to have chosen A to be either 0 or 1 with equal probability, then the time average would remain zero, while the ensemble average becomes the time-dependent $\frac{1}{2}\sin(\omega n)$.

What does ergodicity really mean? Simply that rather than acquiring an ensemble of N signal generators we can use only a single generator but restart our experiment N times. If the signal with all the possible different initial times reproduces the entire ensemble of the stochastic signal, then the signal is ergodic. Not only must all possible realizations be reproduced, they must be reproduced the same number of times. When we thinking about it this way, ergodicity is rather too strong a statement; no signal can really be so random that a single realization completely samples all the possibilities of the ensemble! The number of realizations generated by restarting the experiment at all possible times equals the number of points on the real line, while there are many more different functions of time! However, ergodicity makes life so simple that we most often assume it anyway.

For ergodic signals we can redefine the correlations in terms of time averages. For example, the autocorrelation becomes

$$ C_s(\tau) = \int s(t)s(t-\tau)dt \quad \mathbf{A} \,\Big|\, \mathbf{D} \quad C_s(m) = \sum_n s_n s_{n-m} \qquad (5.26) $$

and it is these forms that we shall use in Chapter 9.

EXERCISES

5.6.1 Consider a stationary signal that can only take the values 0 and 1. What is the probability that the signal is nonzero at two times τ apart? What is the meaning of moments for this type of signal?

5.6.2 Derive the relation between autocovariance and autocorrelation $V_s(t_1, t_2) = C_s(t_1, t_2) - m_s(t_1)m_s(t_2)$.

5.6.3 Show that for white signals (for which all times are independent) the autocovariance is zero except for when $t_1 = t_2$.

5.6.4 In the text we discussed the filtering of white noise, although white noise is not a signal and thus we have never properly defined what it means to filter it. Can you give a plausible meaning to the filtering of a stochastic signal?

5.6.5 Prove that a first-order Markov signal can be obtained by low-pass filtering white noise. Assuming that s_n is created from a noise signal v_n by equation 5.24 with $|\alpha| < 1$, what is the probability distribution of s_n given that we already observed s_{n-1}?

5.6.6 Show that the signal s_n generated by the recursion $s_n = \alpha_1 s_{n-1} + \alpha_2 s_{n-2} + v_n$ (where v_n is white) is a second-order Markov signal.

5.6.7 Given the Markov signals of equation (5.24) and the previous exercise, can you recover the white noise signal ν_n? What can you learn from the expression for ν_n?

5.6.8 What is the power $\langle s_n^2 \rangle$ of the Markov signal of equation (5.24)? Why did we require $|\alpha| < 1$? The special case $\alpha = 1$ is called the *random walk* or *Wiener signal*. What happens here?

5.6.9 Pink noise is a term often used for a noise whose power spectrum decreases 3 dB per octave (doubling of frequency). What is the spectral density's dependence on frequency? How does the power per octave depend on frequency?

5.6.10 Blue noise is the opposite of pink, with power spectrum increasing 3 dB per octave; red noise has a 6 dB drop per octave. How do these spectral densities depend on frequency?

5.7   Spectrum of Random Signals

We know what the spectrum of a signal is, and thus we know what the spectrum of a single realization of a stochastic signal is; but can we give meaning to the spectrum of the entire stochastic signal? The importance of the frequency domain in signal processing requires us to find some consistent definition for the spectrum of noisy signals. Without such an interpretation the concept of filtering would break down, and the usefulness of DSP to real signals (all of which are noisy to some degree) would be cast in doubt. Fortunately, although a much more formidable task than it would seem, it *is* possible to define (and compute) the spectrum of a stochastic signal. Unfortunately, there are several different ways to do so.

If we consider the entire ensemble and take the FT of each realization individually we obtain an ensemble of transforms. Well, almost all realizations of a *stationary* stochastic signal will have infinite energy and therefore the FT won't converge, but we already know (see Section 4.6) to use the STFT for this case. Similarly, for nonstationary signals whose statistics vary slowly enough we can use the STFT over short enough times that the signal is approximately stationary. Thus from here on we shall concentrate on the STFT of stationary random signals.

We could consider the entire ensemble of spectra as *the spectrum*. Such a 'spectrum' is itself stochastic, that is, for every frequency we have a complex random variable representing the magnitude and angle. To see why these are truly random variables consider a realization of a white noise signal. Many more realizations of the same stochastic signal can be created by shifting this one in time by any arbitrary interval. Thus the phases of the spectrum of such a signal should be uniformly distributed random variables. There is no way to resolve this problem other than to avoid it. Thus we concentrate on the short time power spectrum of stationary stochastic signals. Returning to our ensemble of transforms we square the values and discard the phases and obtain an ensemble of power spectra.

For well-behaved stationary stochastic signals (the type we are interested in) a unique (nonrandom) power spectrum can be defined. In practice we do not have access to the entire ensemble of signals but can observe one particular realization of the stationary signal for some amount of time. Assuming ergodicity, this can be just as good. Thus, if we compute the short time power spectrum of the realization we happen to have, we expect to obtain a good estimate of the aforementioned true power spectrum.

What do we mean by a 'good' estimate? An estimator is considered good if it is unbiased and has a small variance. For example, consider the mean value of a stationary signal s. Were we to have access to the entire ensemble we could take any single moment of time, and calculate the mean of the signal values in all realizations of the ensemble at that time. This calculation provides the true mean. Since the signal is assumed stationary, we could repeat this at any other time and would obtain precisely the same result. Alternately, assuming ergodicity, we could perform the average over time in a single realization. For a digital signal this entails adding all signal values from $n = -\infty$ to $n = \infty$, which would take quite a long time to carry out. Instead we could estimate the mean by

$$m = \frac{1}{N} \sum_{n=1}^{N} s_n$$

summing over N consecutive signal values. Such an estimator is unbiased; it will be too large just as many times as it will be too small. More precisely, if we carry out the estimation process many times, the mean of the results will be the true mean. Also this estimator has a variance that decreases with increasing N as $\frac{1}{N}$. That is, if we double the number of times we estimate the mean, the average variance will drop by half; the variance vanishes in the limit $N \to \infty$.

Returning to power spectra, we expect that our estimation of the power spectrum based on the STFT of a single realization to be unbiased and have variance that vanishes asymptotically. Unfortunately, neither of these expectations is warranted. If we calculate the power spectrum based on a single realization, the estimated power spectrum thus obtained will be biased and will have a standard deviation of about the same size as the value being estimated. Increasing the size of the window of the STFT does reduce the bias but doesn't reduce the variance at all!

It is informative to understand the reasons for these enigmas. The bias problem is the less severe of the two and the easier one to understand. Simply stated, the bias comes from comparing two different entities. When we use the STFT to estimate the energy at a given frequency, we are actually dividing the frequency axis into bins, each of width determined by the number of signal points in the transform. The STFT estimated spectrum averages together the true spectrum's values for all frequencies in the bin. Thus the STFT power spectrum's value at some frequency f should not be expected to precisely replicate the true spectrum's value there. However, as the number of points in the STFT becomes larger, the bins become smaller and the difference between the two decreases. Another way of looking at this is to think of the STFT as the FT of the original signal multiplied by the data window. This will of course equal the desired FT convolved with the FT of the window function. For any given window duration use of good window functions can help (see Section 13.4), but the fundamental uncertainty remains. As the duration of the window increases the FT of the window function approaches a delta function and the bias disappears.

The true problem is the variance of our estimator. The spectral variance, unlike the variance of the mean, does not decrease with increasing the number of data points used. At first this seems puzzling but the reason (as first realized by Tukey in the late 1940s) is quite simple. When we double the size of the STFT we automatically double the number of frequency bins. All the information in the new data goes toward providing more frequency resolution and not toward improving the accuracy of the existing estimates. In order to decrease the variance we must find a way to exploit more of the signal without increasing the frequency resolution. Two such methods come to mind.

Assume that we increase the number of input signal values by a factor of M. Bartlett proposed performing M separate power spectra and averaging the results rather than performing a single (M times larger) STFT. This averaging is similar to the mean estimator discussed above, and reduces the estimator's variance by a factor of M. Welch further improved this method by overlapping the data (with 50% overlap being about ideal). Of course performing multiple transforms rather than a single large transform is somewhat less efficient if the FFT is being used, but this is a small price to pay for the variance reduction. The second way to reduce the variance does perform a single STFT but then sums adjacent bins to reduce the resolution. This effectively smooths the estimated power spectrum resulting in a similar variance reduction. We will delve further into these techniques in Section 13.3.

Earlier we stated that a unique (nonrandom) power spectrum can be defined. This was first done by Wiener and Khintchine based on the following theorem.

Theorem: Wiener-Khintchine

The autocorrelation and the power spectral density are an FT pair.   ∎

In Chapter 9 we will prove this theorem for the deterministic case (after properly defining the autocorrelation for deterministic signals). Here we take this theorem as the definition for the stationary stochastic case. The basic idea behind the theorem is clear. If we are only interested in the square of the spectrum then we should only have to look at second-order entities in the time domain; and the autocorrelation is the most basic of these.

Basing ourselves on Wiener-Khintchine we can now compute power spectra of noisy signals in a new way, due to Blackman and Tukey. Rather than directly computing the signal's FT and squaring, we calculate the autocorrelation and then take the FT. All that we have seen above about bias and variance still holds, but averaging the computed spectra still helps. Since we can use the FFT here as well, the Blackman-Tukey technique is similar in computational complexity to the more direct Bartlett and Welch methods.

EXERCISES

5.7.1 Generate a finite-duration digital signal consisting of a small number of sinusoids and create K realizations by adding zero-mean Gaussian noise of variance σ^2. Compute the power spectrum in the following three ways. Compute the FT of each of the realizations, average, and then square. Compute the FT, square, and then average. Compute the autocorrelation from the realizations and find the power spectrum from Wiener-Khintchine. Compare your results and explain.

5.7.2 Generate a single long (a power of two is best) realization of a signal as above. Compare power spectrum estimates using windows without overlap, overlapping windows, and smoothing of a single long FFT.

5.8   Stochastic Approximation Methods

Sometimes we *are* allowed access to the ensemble of signals, in which case rather different techniques can be employed. As a concrete example we will briefly consider the Robbins-Monro algorithm for finding a zero of a function corrupted by additive noise. The zero of a function $f(t)$ is a z such that $f(z) = 0$. Finding the zero of a purely deterministic function is relatively straightforward. The standard way is to search for intervals $[t_1 \ldots t_2]$ where the sign of $f(t)$ changes, i.e., $f(t_1) < 0$ and $f(t_2) > 0$ or $f(t_1) > 0$ and $f(t_2) < 0$. Then we look at some t in the interval $t_1 < t < t_2$, and check if $f(t) = 0$ to within the desired accuracy. If not, we replace either t_1 or t_2 with t, depending on the sign of $f(t)$. The various algorithms differ only in the method of choosing t.

In the Robbins-Monro scenario we can only observe the noisy signal $g(t) = f(t) + \nu(t)$, where the noise is assumed to be zero-mean $\langle \nu(t) \rangle = 0$ and of finite variance $\langle \nu^2(t) \rangle < \infty$. However, we are allowed to make as many measurements of $g(t)$ as we desire, at any t we wish. One way to proceed would be to imitate the standard procedure, but averaging out the noise by sampling $g(t)$ a sufficient number of times. However, the smaller the absolute value of $g(t)$, the more susceptible is its sign to noise. This causes the number of samples required to diverge.

The Robbins-Munro algorithm recursively updates the present estimate z_k for the zero instead.

$$z_{k+1} = z - \frac{g(z_k)}{k} \tag{5.27}$$

It can be shown that this procedure both converges to the desired root in the mean square, i.e.,

$$\lim_{k \to \infty} \left\langle (z_k - z)^2 \right\rangle = 0$$

and converges with probability 1, i.e.,

$$\text{Prob}(\lim_{k \to \infty} z_k = z) = 1$$

although the convergence may, in practice, be very slow.

EXERCISES

5.8.1 Is the division by k required for the deterministic case? Code the algorithm and check for a few polynomials and a sinusoid.

5.8.2 Add noise to the signals you used in the previous exercise and run the full algorithm. How does the error in the zero location depend on the noise level?

5.9 Probabilistic Algorithms

The Robbins-Monro algorithm is a way to *combat* noise, but we have mentioned that there are probabilistic algorithms that actually *exploit* noise. The usual definition of 'algorithm' is a precisely defined (i.e., deterministic) prescription of the solution of a problem; why would we want to make an algorithm probabilistic? The reason has to do with practicalities; sometimes the standard deterministic algorithm takes too long to compute its answer, while a probabilistic algorithm may be able to come up with an usable estimate much faster.

Numerical integration is a good example. The deterministic approach requires dividing the x axis into small intervals and summing the value of the function in these intervals. The function needs to be approximately constant over each interval so for rapidly varying functions many functional values must be evaluated and summed. Multidimensional integration is much more demanding; here all of the axes corresponding to independent variables must be divided into sufficiently small intervals, so that the computational complexity increases exponentially with the dimensionality.

As a concrete example consider finding the area of a circle, which can be expressed as a two-dimensional integral. The standard numeric approach requires dividing two-dimensional space into a large number of small squares, and the integration is carried out by counting the number of squares inside the circle. Of course there will always be the problem of those squares that straddle the circumference of the circle; only by using small enough squares can we ensure that these questionable cases do not overly effect the answer.

How can a probabilistic algorithm find the area? Circumscribe the circle by a square and choose at random any point inside this square. The probability that this point is inside the circle is exactly the ratio of the area of the circle to that of the square. So by generating a large number of random points (using any of the random number generators of section 5.4) and counting up how many fall inside the circle we can get an estimate of the area. Note that there isn't a well-defined end to this computation; each new random point simply improves the previous estimate. So there is a natural trade-off between accuracy and computational complexity.

This lucky integration technique is often called Monte-Carlo integration (for obvious reasons), and you can bet that it can be generalized to any integration problem in any number of dimensions.

EXERCISES

5.9.1 Compute π by Monte-Carlo determination of the area of the unit-radius circle. Monitor the error as a function of the number of points generated. How does the computation required to obtain a given accuracy compare with that of direct numerical integration?

5.9.2 Find the volume of a unit-radius sphere and the hypervolume of a unit-radius hypersphere in four dimensions. Make the same computational complexity comparisons as in the previous exercise.

5.9.3 In certain cases deterministic and probabilistic approaches to integration can be combined to obtain a faster and more accurate method. Explain the idea and apply to the previous exercise. (Hint: Inscribe the circle with a second square.)

Bibliographical Notes

Our treatment of noise has been very different, and a good deal less pedantic, than that found in engineering textbooks. For those who miss the formalistic treatment there are several good books on stochastic processes. The classic text is that of Papoulis [190]; only slightly less classic but much less friendly is van Trees [264]; but any text that has the words 'stochastic' and 'process' in its title will probably do. There are also texts with a major emphasis on stochastic processes that mix in a certain amount of straight signal processing, e.g., [250], and others with the opposite stress, such as [188].

Those interested in more information regarding radar systems can try anything by Skolnik [245, 243, 244] or the book by Levanon [145].

The generation of pseudorandom signals is discussed at length in the second volume of Knuth [136]. The transformation from uniform to Gaussian distributed random numbers (also found in Knuth) was discovered by Box and Muller [22]. The standard text on shift register sequences is by Golomb [81].

Deterministic chaos is quite a popular subject, with many books, each with its own approach. A suitable text for physicists is [234], while there are other books suitable for engineers or for mathematicians. The popular account by Gleick [75] is accessible and interesting.

Perhaps the earliest mathematical account of noise is [219, 220], which presented a complete theory including power spectra, statistical properties, and the effect of nonlinear systems on noise. Many would claim that the most important book on stochastic processes is that of Papoulis mentioned above [190].

An accessible source from which one can gain insight regarding the true meaning of noise is to be found in much of what is called modern music.

Part II

Signal Processing Systems

Systems

The study of signals, their properties in time and frequency domains, their fundamental mathematical and physical limitations, the design of signals for specific purposes, and how to uncover a signal's capabilities through observation belong to *signal analysis*. We now turn to *signal processing*, which requires adding a new concept, that of the *signal processing system*.

A signal processing system is a device that processes input signals and/or produces output signals. Signal processing systems were once purely analog devices. Older household radio receivers input analog radio frequency signals from an antenna, amplify, filter, and extract the desired audio from them using analog circuits, and then output analog audio to speakers. The original telephone system consisted of analog telephone sets connected via copper wire lines, with just the switching (dialing and connecting to the desired party) discrete. Even complex radar and electronic warfare systems were once purely analog in nature.

Recent advances in microelectronics have made DSP an attractive alternative to analog signal processing. Digital signal processing systems are employed in a large variety of applications where analog processing once reigned, and of course newer purely digital applications such as modems, speech synthesis and recognition, and biomedical electronics abound. There still remain applications where analog signal processing systems prevail, mainly applications for which present-day DSP processors are not yet fast enough; yet the number of such applications is diminishing rapidly.

In this chapter we introduce systems analogously to our introduction of signals in Chapter 2. First we define analog and digital signal processing systems. Then we introduce the simplest possible systems, and important classes of systems. This will lead us to the definition of a *filter* that will become a central theme in our studies. Once the concept of filter is understood we can learn about MA, AR, and combined ARMA filters. Finally we consider the problem of *system identification* which leads us to the concepts of *frequency response, impulse response,* and *transfer function.*

6.1 *System* Defined

The first question we must ask when approaching the concept of signal processing is 'What exactly do we mean by a signal processing *system*?'

Definition: signal processing system
A signal processing system is any device that takes in zero or more signals as input, and returns zero or more signals as outputs.   ■

According to this definition systems deal only with signals. Of course images may be considered two-dimensional signals and thus image processing *is* automatically included. Nevertheless, we will often extend the definition to include systems that may also input other entities, such as numeric or logical values. A system may output such other entities as well. An important output entity is a multiclass classification identifier, by which we mean that various signals may be input to the system as a function of time, and the system classifies them as they arrive as belonging to a particular class. The only practical requirement is that there should be at least one output, either signal, numeric, logical, or classification. Were one to build a system with no outputs, after possibly sophisticated processing of the input, the *system* would know the result (but you wouldn't).

What kind of system has no input signals? An example would be an *oscillator* or tone generator, which outputs a sinusoidal signal of constant frequency, irrespective of whatever may be happening around it. A simple modification would be to add a numeric input to control the amplitude of the sine, or a logical input to reset the phase. Such an oscillator is a basic building block in communications transmitters, radars, signaling systems, and music synthesizers.

What kind of system has no signal output? An example would be a *detector* that outputs a logical *false* until a signal of specified parameters is detected. A simple modification would be to output a numeric value that relates the time of detection to a reference time, while a more challenging extension would continually output the degree to which the present input matches the desired signal (with 0 standing for no match, 1 for perfect match). Such a system is the basis for modem demodulators, radar receivers, telephone switch signaling detectors, and pattern analyzers. Systems that output only multiclass classifications are the subject of a discipline known as *pattern recognition*.

EXERCISES

6.1.1 Which of the following are signal processing systems (we shall use x for inputs and y for outputs)? Explain.
  1. The identity $y = x$
  2. The constant $y = k$ irrespective of x
  3. $y = \pm\sqrt{x}$
  4. A device that inputs a pizza and outputs a list of its ingredients
  5. $y = \sin(\frac{1}{t})$
  6. $y(t) = \int_{-\infty}^{t} x(t)$
  7. The Fourier transform
  8. A television
  9. A D/A converter

6.1.2 Given any two signals $x(t)$ and $y(t)$, is there always a system that inputs x and outputs y? Given a system that inputs $x(t)$ and outputs $y(t)$, is there always a system that inputs y and outputs x?

6.2 The Simplest Systems

Let us now present a few systems that will be useful throughout our studies. The simplest system with both an input signal x and an output signal y is the *constant*, $y(t) = k$ in analog time or $y_n = k$ in digital time. This type of system may model a power supply that strives to output a constant voltage independent of its input voltage. We can not learn much from this trivial system, which completely ignores its input. The next simplest system is the *identity*, whose output exactly replicates its input, $y(t) = x(t)$ or $y_n = x_n$.

The first truly nontrivial system is the *amplifier*, which in the analog world is $y(t) = Ax(t)$ and in the digital world $y_n = Ax_n$. A is called the *gain*. When $A > 1$ we say the system *amplifies* the input, since the output as a function of time looks like the input, only larger. For the same reason, when $A < 1$ we say the system *attenuates*. Analog amplifiers are vital for broadcast transmitters, music electronics (the reader probably has a stereo amplifier at home), public address systems, and measurement apparatus.

The ideal amplifier is a *linear system*, that is, the amplification of the sum of two signals is the sum of the amplifications, and the amplification of a constant times a signal is the constant times the amplification of the signal.

$$A\Big(x_1(t) + x_2(t)\Big) = Ax_1(t) + Ax_2(t) \quad \text{and} \quad A\Big(cx(t)\Big) = cAx(t)$$

Such perfect linear amplification can only be approximated in analog circuits; analog amplifiers saturate at high amplitudes, lose amplification at high frequencies, and do not respond linearly for very high amplitudes. Digitally amplification is simply multiplication by a constant, a calculation that may be performed reliably for all inputs, unless overflow occurs.

We can generalize the concept of the amplifier/attenuator by allowing deviations from linearity. For example, real analog amplifiers cannot output voltages higher than their power supply voltage, thus inducing *clipping*. This type of nonlinearity

$$y(t) = \text{Clip}_\theta \left(Ax(t) \right) \qquad \mathbf{A} \,\big|\, \mathbf{D} \qquad y_n = \text{Clip}_\theta \left(Ax_n \right) \qquad (6.1)$$

where

$$\text{Clip}_\theta(x) \equiv \begin{cases} \theta & x \geq \theta \\ x & -\theta < x < \theta \\ -\theta & -\theta \leq x \end{cases}$$

is depicted in Figure 6.1. On the left side of the figure we see the output versus input for an ideal linear amplifier, and on the right side the output when

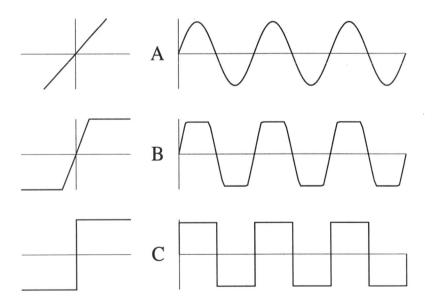

Figure 6.1: The effect of clipping amplifiers with different gains on sinusoidal signals. In (A) there is no clipping, in (B) intermediate gain and clipping, while (C) represents the infinite gain case (hard limiter).

a sinusoid is input. Figure 6.1.B represents an amplifier of somewhat higher gain, with a limitation on maximal output. The region where the output no longer increases with increasing input is called the region of *saturation*. Once the amplifier starts to saturate, we get 'flat-topping' of the output, as is seen on the ride side. The flat-topping gets worse as the gain is increased, until in 6.1.C the gain has become infinite and thus the system is always saturated (except for exactly zero input). This system is known as a 'hard limiter', and it essentially computes the sign of its input.

$$y(t) = \text{sgn}\left(x(t)\right) \qquad \mathbf{A} \,\Big|\, \mathbf{D} \qquad y_n = \text{sgn}\left(x_n\right) \qquad (6.2)$$

Hard limiting changes sinusoids into square waves, and is frequently employed to obtain precisely this effect.

These clipping amplifiers deal symmetrically with positive and negative signal values; another form of nonlinearity arises when the sign explicitly affects the output. For example, the gain of an amplifier can depend on whether the signal is above or below zero. Extreme cases are the *half-wave rectifier*, whose output is nonzero only for positive signal values,

$$y(t) = \theta\left(x(t)\right) x(t) \qquad \mathbf{A} \,\Big|\, \mathbf{D} \qquad y_n = \theta\left(x_n\right) x_n \qquad (6.3)$$

and the *full-wave rectifier*, whose output is always positive

$$y(t) = |x(t)| \qquad \mathbf{A} \,\Big|\, \mathbf{D} \qquad y_n = |x_n| \qquad (6.4)$$

as depicted in Figure 6.2.

Yet another deviation from linearity is termed *power-law distortion*; for example, quadratic power distortion is

$$y(t) = x(t) + \epsilon x^2(t) \qquad \mathbf{A} \,\Big|\, \mathbf{D} \qquad y_n = x_n + \epsilon x_n^2 \qquad (6.5)$$

for small $\epsilon > 0$. More generally higher powers may contribute as well. Real amplifiers always deviate from linearity to some degree, and power law distortion is a prevalent approximation to their behavior. Another name for power law distortion is *harmonic generation*; for example, quadratic power

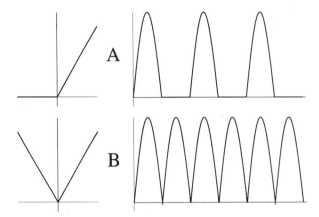

Figure 6.2: Half-wave and full-wave rectifiers. (A) depicts the output as a function of input of a half-wave rectifier, as well as its effect on a sinusoid. (B) depicts the same for a full-wave rectifier.

distortion is called second-harmonic generation. The reasoning behind this name will become clear in Section 8.1.

Let us summarize some of the systems we have seen so far:

constant	$y(t) = k$	$y_n = k$
identity	$y(t) = x(t)$	$y_n = x_n$
amplification	$y(t) = Ax(t)$	$y_n = Ax_n$
clipping	$y(t) = \text{Clip}_\theta\left(Ax(t)\right)$	$y_n = \text{Clip}_\theta\left(Ax_n\right)$
hard limiter	$y(t) = \text{sgn}\left(x(t)\right)$	$y_n = \text{sgn}\left(x_n\right)$
half-wave rectification	$y(t) = \theta\left(x(t)\right)x(t)$	$y_n = \theta\left(x_n\right)x_n$
full-wave rectification	$y(t) = |x(t)|$	$y_n = |x_n|$
quadratic distortion	$y(t) = x(t) + \epsilon x^2(t)$	$y_n = x_n + \epsilon x_n^2$

This is quite an impressive collection. The maximal extension of this type of system is the *general point transformation* $y(t) = f\left(x(t)\right)$ or $y_n = f\left(x_n\right)$. Here f is a completely general function, and the uninitiated to DSP might be led to believe that we have exhausted all possible signal processing systems. Notwithstanding, such a system is still extremely simple in at least two senses. First, the output at any time depends only on the input at that same time and nothing else. Such a system is *memoryless* (i.e., does not retain memory of previous inputs). Second, this type of system is *time-*

invariant (i.e., the behavior of the system does not change with time). Classical mathematical analysis and most non-DSP numerical computation deal almost exclusively with memoryless systems, while DSP almost universally requires combining values of the input signal at many different times. Time invariance, the norm outside DSP, is common in many DSP systems as well. However, certain important DSP systems *do* change as time goes on, and may even change in response to the input. We will see such systems mainly in Chapter 10.

EXERCISES

6.2.1 The digital amplifier is a linear system as long as no *overflow* or *underflow* occur. What is the effect of each of these computational problems? Which is worse? Can anything be done to prevent these problems?

6.2.2 Logarithmic companding laws are often used on speech signals to be quantized in order to reduce the required dynamic range. In North America the standard is called μ-law and is given by

$$y = \text{sgn}(x)\frac{\log(1 + \mu|x|)}{\log(1 + \mu)} \tag{6.6}$$

where x is assumed to be between -1 and $+1$ and $\mu = 255$. In Europe A-law is used

$$y = \begin{cases} \text{sgn}(x)\frac{\log(A|x|)}{\log(1+\log(A))} & 0 < |x| < \frac{1}{A} \\ \text{sgn}(x)\frac{1+\log(A|x|)}{1+\log(A)} & \frac{1}{A} < |x| < 1 \end{cases} \tag{6.7}$$

where $A = 87.6$. Why are logarithmic curves used? How much difference is there between the two curves?

6.2.3 Time-independent point transformations can nontrivially modify a signal's spectrum. What does squaring signal values do to the spectrum of a pure sinusoid? To the sum of two sinusoids? If point operations can modify a signal's spectrum why do you think systems with memory are needed?

6.3 The Simplest Systems with Memory

There are two slightly different ways of thinking about systems with memory. The one we will usually adopt is to consider the present output to be a function of the present input, previous inputs, and previous outputs.

$$y_n = f\left(x_n, x_{n-1}, x_{n-2}, \ldots y_{n-1}, y_{n-2} \ldots\right) \tag{6.8}$$

The other line of thought, called the *state-space description*, considers the output to be calculated based on the present input and the present *internal state* of the system.

$$y_n = f\left(x_n, \underline{S}\right) \tag{6.9}$$

In the state-space description the effect of the input on the system is twofold, it causes an output to be generated and it changes the state of the system. These two ways of thinking are clearly compatible, since we could always define the internal state to contain precisely the previous inputs and outputs. This is even the best way of defining the system's state for systems that explicitly remember these values. However, many systems do not actually remember this history; rather this history influences their behavior.

The simplest system with memory is the *simple delay*

$$y(t) = x(t - \tau) \qquad \mathbf{A} \;\bigg|\; \mathbf{D} \qquad y_n = x_{n-m} \tag{6.10}$$

where the time τ or m is called the *lag*. From the signal processing point of view the simple delay is only slightly less trivial than the identity. The delay's output still depends on the input at only *one* time, that time just happens not to be the present time, rather the present time minus the lag.

We have said that the use of delays is one of the criteria for contrasting simple numeric processing from signal processing. Recall from Chapter 2 that what makes signal processing special is the schizophrenic jumping back and forth between the *time domain* and the *frequency domain*. It is thus natural to inquire what the simple delay does to the frequency domain representation of signals upon which it operates. One way to specify what any signal processing system does in the frequency domain is to input simple sinusoids of all frequencies of interest and observe the system's output for each. For the simple delay, when a sinusoid of amplitude A and frequency ω is input, a sinusoid of identical amplitude and frequency is output. We will see later on that a system that does not change the frequency of sinusoids and does not create new frequencies is called a *filter*. A filter that does not change the amplitude of arbitrary sinusoids, that is, one that passes all frequencies without attenuation or amplification, is called an *all-pass filter*. Thus the simple delay is an all-pass filter. Although an all-pass filter leaves the *power spectrum* unchanged, this does *not* imply that the spectrum remains unchanged. For the case of the delay it is obvious that the phase of the output sinusoid will usually be different from that of the input. Only if the lag is precisely a whole number of periods will the phase shift be zero; otherwise the phase may be shifted either positively or negatively.

After a little consideration we can deduce that the phase is shifted by the frequency times the delay lag. When the phase shift is proportional to the frequency, and thus is a straight line when plotted as a function of frequency, we say that the system is *linear-phase*. The identity system $y = x$ is also linear-phase, albeit with a trivial constant zero phase shift. Any time delay (even if unintentional or unavoidable such as a processing time delay) introduces a linear phase shift relation. Indeed any time-invariant linear-phase system is equivalent to a zero phase shift system plus a simple delay. Since simple delays are considered trivial in signal processing, linear-phase systems are to be considered 'good' or 'simple' in some sense. In contrast when the phase shift is not linear in frequency, some frequencies are delayed more than others, causing phase distortion. To appreciate the havoc this can cause imagine a nonlinear-phase concert hall. In any large concert hall a person in the balcony hears the music a short time after someone seated up front. When the room acoustics are approximately linear-phase this delay is not particularly important, and is more than compensated for by the reduction in ticket price. When nonlinear phase effects become important the situation is quite different. Although the music may be harmonious near the stage, the listener in the balcony hears different frequencies arriving after *different* time delays. Since the components don't arrive together they sum up to quite a different piece of music, generally less pleasant to the ear. Such a concert hall would probably have to pay people to sit in the balcony, and the noises of indignation made by these people would affect the musical experience of the people up front as well.

How can the simple delay system be implemented? The laws of relativity physics limit signals, like all information-carrying phenomena, from traveling at velocities exceeding that of light. Thus small analog delays can be implemented by *delay lines*, which are essentially appropriately chosen lengths of cable (see Figure 6.3.A). A voltage signal that exits such a delay line cable is delayed with respect to that input by the amount of time it took for the electric signal to travel the length of the cable. Since electric signals tend to travel quickly, in practice only very short delays can be implemented using analog techniques. Such short delays are only an appreciable fraction of a period for very high-frequency signals. The delay, which is a critical processing element for all signal processing, is difficult to implement for low-frequency analog signals.

Digital delays of integer multiples of the sampling rate can be simply implemented using a FIFO buffer (see Figure 6.3.B). The content of this FIFO buffer is precisely the system's internal state from the state-space point of view. The effect of the arrival of an input is to cause the oldest value

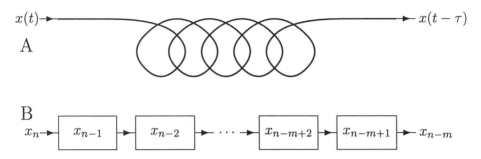

Figure 6.3: Implementation of the simple delay. In (A) we see how an analog delay of lag τ can be obtained by allowing a voltage of current signal to travel at finite velocity through a sufficiently long delay line. In (B) a digital delay of lag m is implemented using a FIFO buffer.

stored in the FIFO to be output and promptly discarded, for all the other values to 'move over', and for the present input to be placed in the buffer. Of course long delays will require large amounts of memory, but memory tends to drop in price with time, making DSP more and more attractive vis-a-vis analog processing. DSP does tend to break down at high frequencies, which is exactly where analog delay lines become practical.

Leaving the simple delay, we now introduce a somewhat more complex system. Think back to the last time you were in a large empty room (or a tunnel or cave) where there were strong echoes. Whenever you called out you heard your voice again after a delay (that we will call τ), which was basically the time it took for your voice to reach the wall from which it was reflected and return. If you tried singing or whistling a steady tone you would notice that some tones 'resonate' and seem very strong, while others seem to be absorbed. We are going to model such a room by a system whose output depends on the input at two different times, the present time and some previous time $t - \tau$. Our simple 'echo system' adds the signal values at the two times

$$y(t) = x(t) + x(t - \tau) \qquad \mathbf{A} \,\big|\, \mathbf{D} \qquad y_n = x_n + x_{n-m} \qquad (6.11)$$

and is easily implemented digitally by a FIFO buffer and an adder.

In the frequency domain this system is *not* all-pass; the frequency dependence arising from the time lag τ (or m) corresponding to different phase differences at different frequencies. When we input a sinusoidal signal with angular frequency ω such that τ corresponds to precisely one period (i.e., $\omega\tau = 2\pi$), the net effect of this system is to simply double the signal's amplitude. If, however, the input signal is such that τ corresponds to a half period ($\omega\tau = \pi$), then the output of the system will be zero. This is the reason some frequencies resonate while others seem to be absorbed.

More generally, we can find the *frequency response*, by which we mean the response of the system to any sinusoid as a function of its frequency. To find the frequency response we apply an input of the form $\sin(\omega t)$. The output, which is the sum of the input and its delayed copy, will be

$$\sin(\omega t) + \sin\left(\omega(t - \tau)\right) = 2\cos\left(\frac{\omega\tau}{2}\right)\sin\left(\omega(t - \frac{\tau}{2})\right)$$

$$= A(\omega\tau)\sin\left(\omega(t - \frac{\tau}{2})\right)$$

which is easily seen to be a sinusoid of the same frequency as the input. It is, however, delayed by half the time lag (linear-phase!), and has an amplitude that depends on the product $\omega\tau$. This amplitude is maximal whenever $\omega\tau = 2k\pi$ and zero when it is an odd multiple of π. We have thus completely specified the frequency response; every input sine causes a sinusoidal output of the same frequency, but with a linear phase delay and a periodic amplification. A frequency that is canceled out by a system (i.e., for which the amplification of the frequency response is zero) is called a *zero* of the system. For this system all odd multiples of π are zeros, and all even multiples are maxima of the frequency response.

Our next system is only slightly more complex than the previous one. The 'echo system' we just studied assumed that the echo's amplitude was exactly equal to that of the original signal. Now we wish to add an echo or delayed version of the signal to itself, only this time we allow a multiplicative coefficient (a *gain* term).

$$y(t) = x(t) + hx(t - \tau) \qquad \mathbf{A} \;\Big|\; \mathbf{D} \qquad y_n = x_n + hx_{n-m} \qquad (6.12)$$

When $h = 1$ we return to the previous case, while $h < 1$ corresponds to an attenuated echo, while $h > 1$ would be an amplified echo. We can also consider the case of negative h, corresponding to an echo that returns with phase reversal.

$$y(t) = x(t) - |h|x(t - \tau) \qquad \mathbf{A} \;\Big|\; \mathbf{D} \qquad y_n = x_n - |h|x_{n-m}$$

We leave the full mathematical derivation of the frequency response of our generalized echo system as an exercise. Still we can say a lot based on a little experimentation (using pen and paper or a computer graphing program). The first thing we notice is that a sinusoidal input will produce a sinusoidal output of the same frequency, but with amplitude between $1 - |h|$ and $1 + |h|$. Thus when $|h| \neq 1$ we can never perfectly cancel out a sinusoidal input signal, no matter what frequency we try, and thus the frequency response will have no zeros. Of course when $|h| < 1$ we can't double the amplitude either; the best we can do is to amplify the signal by $1 + |h|$. Yet this should be considered a mere quantitative difference, while the ability or inability to exactly zero out a signal is qualitative. The minima of the frequency response still occur when the echo is exactly out of phase with the input, and so for positive h occur whenever $\omega\tau$ is an odd multiple of π, while for negative h even multiples are needed. We present the graphs of amplification as a function of frequency for various positive h in Figure 6.4.

We can generalize our system even further by allowing the addition of multiple echoes. Such a system combines the input signal (possibly multiplied by a coefficient) with delayed copies, each multiplied by its own coefficient. Concentrating on digital signals, we can even consider having an echo from every possible time lag up to a certain maximum delay.

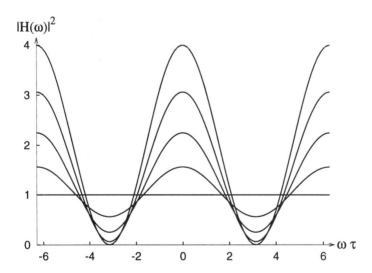

Figure 6.4: Amplitude of the frequency response for the echo system with positive coefficients. The amplitude is plotted as a function of $\omega\tau$. The coefficients are $h = 0$ (the straight line), $0.25, 0.5, 0.75$, and 1.0 (the plot with zeros).

$$y_n = h_0 x_n + h_1 x_{n-1} + h_2 x_{n-2} + \cdots + h_L x_{n-L} = \sum_{l=0}^{L} h_l x_{n-l} \qquad (6.13)$$

This system goes under many different names, including **M**oving **A**verage (MA) filter, *FIR* filter, and *all-zero* filter, the reasoning behind all of which will be elucidated in due course. The mathematical operation of summing over products of indexed terms with one index advancing and one retreating is called *convolution.*

Now this system may seem awesome at first, but it's really quite simple. It is of course linear (this you can check by multiplying x by a constant, and by adding $x_1 + x_2$). If the input signal is a pure sine then the output is a pure sine of the same frequency! Using linearity we conclude that if the input signal is the sum of sinusoids of certain frequencies, the output contains only these same frequencies. Although certain frequencies may be zeros of the frequency response, no new frequencies are ever created. In this way this system is simpler than the nonlinear point transformations we saw in the previous section. Although limited, the FIR filter will turn out to be one of the most useful tools in DSP.

What should be our next step in our quest for ever-more-complex digital signal processing systems? Consider what happens if echoes from the distant past are still heard—we end up with a nonterminating convolution!

$$y_n = \sum_{l=-\infty}^{\infty} h_l x_{n-l} \qquad (6.14)$$

In a real concert hall or cave the gain coefficients h_l get smaller and smaller for large enough l, so that the signal becomes imperceptible after a while. When an amplifier is involved the echoes can remain finite, and if they are timed just right they can all add up and the signal can become extremely strong. This is what happens when a microphone connected to an amplifier is pointed in the direction of the loudspeaker. The 'squeal' frequency depends on the time it takes for the sound to travel from the speaker to the microphone (through the air) and back (through the wires).

The FIR filter owes its strength to the idea of iteration, looping on all input signal values from the present time back to some previous time.

```
y_n = 0
for  i = 0 to  L
     y_n ← y_n + h_l x_{n-l}
```

More general than *iteration* is *recursion*,

$$y_n = f(x_n, x_{n-1}, x_{n-2}, \ldots, y_{n-1}, y_{n-2}, \ldots)$$

and the *IIR filter* exploits this by allowing y_n to be a weighted sum of all previous *outputs* as well as inputs.

$$\begin{aligned} y_n &= a_0 x_n + a_1 x_{n-1} + \cdots + a_L x_{n-L} + b_1 y_{n-1} + b_2 y_{n-2} + \cdots + b_M y_{n-M} \\ &= \sum_{l=0}^{L} a_l x_{n-l} + \sum_{m=1}^{M} b_m y_{n-m} \end{aligned} \tag{6.15}$$

We see here two convolution sums, one on the inputs and one on the (previous) outputs. Although even this system cannot create sinusoids of frequencies that do not exist in the input at all, it *can* magnify out of all proportion components that barely exist (see exercises). Of course even IIR filters are simple in a sense since the coefficients a_l and b_m do not vary with time. More complex systems may have coefficients that depend on time, on other signals, and even on the input signal itself. We will see examples of such systems when we discuss adaptive filters.

Are these time- and signal-dependent systems the most complex systems DSP has to offer? All I can say is 'I hope not.'

EXERCISES

6.3.1 Prove the following characteristics of the convolution.

existence of identity	$s_1 * \delta$	$= s_1$
commutative law	$s_1 * s_2$	$= s_2 * s_1$
associative law	$s_1 * (s_2 * s_3)$	$= (s_1 * s_2) * s_3$
distributive law	$s_1 * (s_2 + s_3)$	$= (s_1 * s_2) * (s_1 * s_3)$

6.3.2 We saw that a generalized echo system $y(t) = x(t) + hx(t - \tau)$ has no zeros in its frequency response for $|h| < 1$; i.e., there are no sinusoids that are exactly canceled out. Are there signals that are canceled out by this system?

6.3.3 Find the frequency response (both amplitude and phase) for the generalized echo system. Use the trigonometric identity for the sine of a sum, and then convert $a \sin(\omega t) + b \cos(\omega t)$ to $A \sin(\omega t + \phi)$. Check that you regain the known result for $h = 1$. Show that the amplitude is indeed between $1 - |h|$ and $1 + |h|$.

6.3.4 Plot the amplitude found in the previous exercise for positive coefficients and check that Figure 6.4 is reproduced. Now plot for negative h. Explain. Plot the phase found in the previous exercise. Is the system always linear-phase?

6.3.5 The digital generalized echo system $y_n = x_n + hx_{n-m}$ can only implement an echo whose delay is an integer number of sample intervals t_s. How can a fractional sample delay echo be accommodated?

6.3.6 Show that an IIR filter can 'blow up', that is, increase without limit even with constant input.

6.3.7 Show that the IIR filter

$$y_n = x_n - a_1 y_{n-1} - y_{n-2} \qquad\qquad y_n = 0 \text{ for } n < 0$$

when triggered with a unit impulse $x_n = \delta_{n,0}$ can sustain a sinusoid. What is its frequency?

6.3.8 The sound made by a plucked guitar string is almost periodic, but starts loud and dies out with time. This is similar to what we would get at the output of an IIR system with a delayed and attenuated echo of the output $y_n = x_n + g y_{n-m}$ with $0 < g < 1$. What is the frequency response of this system? (Hint: It is easier to use $x_n = e^{i\omega n}$ for $n > 0$ and zero for $n < 0$, rather than a real sinusoid.)

6.3.9 All the systems with memory we have seen have been *causal*, that is, the output at time T depends on the input at previous times $t \leq T$. What can you say about the output of a causal system when the input is a unit impulse at time zero? Why are causal systems *sensible*? One of the advantages of DSP over analog signal processing is the possibility of implementing noncausal systems. How (and when) can this be done?

6.4 Characteristics of Systems

Now that we have seen a variety of signal processing systems, both with memory and without, it is worthwhile to note some general characteristics a system might have. We will often use operator notation for systems with a single input and a single output signal.

$$y(t) = Hx(t) \qquad \mathbf{A} \,\big|\, \mathbf{D} \qquad y_n = Hx_n \qquad\qquad (6.16)$$

Here H is a *system* that converts one signal into another, not merely a function that changes numbers into numbers.

A memoryless system is called *invertible* if distinct input values lead to distinct output values. The system $y_n = 2x_n$ is thus invertible since every finite value of x_n leads to a unique y_n. Such systems are called invertible since one can produce an *inverse system* H^{-1} such that $x_n = H^{-1}y_n$. For the system just mentioned it is obvious that $x_n = \frac{1}{2}y_n$. Since

$$x_n = H^{-1} y_n = H^{-1} H x_n \qquad (6.17)$$

we can formally write

$$H^{-1} H = 1 \qquad (6.18)$$

where 1 is the identity system. The system $y_n = x_n^2$ is noninvertible since both $x_n = -1$ and $x_n = +1$ lead to $y_n = +1$. Thus there is no system H^{-1} that maps y_n back to x_n.

The notion of invertibility is relevant for systems *with* memory as well. For example, the simple FIR filter

$$y_n = x_n - x_{n-1}$$

has an inverse system

$$x_n = y_n + x_{n-1}$$

which is an IIR filter. Unraveling this further we can write

$$
\begin{aligned}
x_n &= y_n + (y_{n-1} + x_{n-2}) \\
&= y_n + y_{n-1} + (y_{n-2} + x_{n-3}) \\
&= y_n + y_{n-1} + y_{n-2} + y_{n-3} + \cdots
\end{aligned}
$$

and assuming that the input signal was zero for $n = 0$ we get an infinite sum.

$$x_n = \sum_{i=0}^{\infty} y_i \qquad (6.19)$$

Inverse systems are often needed when signals are distorted by a system and we are called upon to counteract this distortion. Such an inverse system is called an *equalizer*. An equalizer with which you may be familiar is the adjustable or preset equalizer for high-fidelity music systems. In order to reproduce the original music as accurately as possible, we need to cancel out distortions introduced by the recording process as well as resonances introduced by room acoustics. This is accomplished by dividing the audio spectrum into a small number of bands, the amplification of which can be individually adjusted. Another equalizer you may use a great deal, but without realizing it, is the equalizer in a modem. Phone lines terribly distort data signals and without equalization data transmission speeds would be around

2400 bits per second. By employing sophisticated adaptive equalization techniques to counteract the distortion, transmission speeds more than ten times faster can be attained.

In a Section 6.2 we mentioned *linearity*, although in the restricted context of memoryless systems. The definition remains the same in the general case, namely

$$H(x + y) = Hx + Hy \qquad \text{and} \qquad H(cx) = cHx \qquad (6.20)$$

that is, H is a *linear* system if its output, when the input is a sum of two signals, is precisely the sum of the two signals that would have been the outputs had each signal been inputed to H separately. The second part states that when the input is a constant times a signal the output must be the constant times the output that would have been obtained were the unamplified signal input instead. We have already seen quite a few nonlinear systems, such as the squaring operation and the hard limiter. Nonlinear systems require special care since they can behave chaotically. We use the term chaos here in a technical sense—small changes to the input may cause major output changes.

This last remark leads us to the subject of *stability*. A system is said to be stable if bounded input signals induce bounded output signals. For example, the system

$$y_n = \tan\left(x_n - \frac{\pi}{2}\right)$$

is unstable near $x_n = 0$ since the output explodes there while the input is zero. However, even linear systems can be unstable according to the above definition. For instance, the system

$$y_n = \sum_{l=0}^{L} x_l$$

is linear, but when presented with a constant input signal the output grows (linearly) without limit.

We generally wish to avoid instability as much as possible, although the above definition is somewhat constraining. Systems with sudden singularities or exponentially increasing outputs should be avoided at all costs; but milder divergences are not as damaging. In any case true analog systems are always stable (since real power supplies can only generate voltages up to a certain level), and digital systems can not support signal values larger than the maximum representable number. The problem with this compelled stability is that it comes at the expense of nonlinearity.

The next characteristic of importance is *time-invariance*. A system H is said to be time-invariant if its operation is not time-dependent. This means that applying time delay or time advance operators to the input of a system is equivalent to applying them to the output.

$$y(t) = Hx(t) \longrightarrow y(t+T) = Hx(t+T) \qquad (6.21)$$

A time-variant system has some internal clock that influences its behavior. For example,

$$y_n = \left(1 + \sin(\Omega t)\right) x_n$$

is time-variant, as is any system that is turned on at some time (i.e., that has zero output before this time no matter what the input, but output dependent on the input after this time).

The combination of linearity and time invariance is important enough to receive a name of its own. Some DSP engineers call a linear and time-invariant systems *LTI* systems, but most use the simpler name *filter*.

Definition: filter
A filter is a system H with a single input and single output signal that is both linear (obeys (6.20)) and time-invariant (obeys equation (6.21)). ∎

As usual we often deviate from the precise definition and speak of *nonlinear filters*, *time-variant filters*, and *multidimensional filters*, but when used without such qualifications the term 'filter' will be taken to be equivalent to LTI.

We already know about systems with memory and without. The output value of a system without memory depends only on the input value at the same time. Two weaker characteristics that restrict the time dependence of the output are *causality* and *streamability*. A system is termed causal if the output signal value at time T is only dependent on the input signal values for that time and previous times $t \leq T$. It is obvious that a memoryless system is always causal, and it is easy to show that a filter is causal if and only if a unit impulse input produces zero output for all negative times. Noncausal systems seem somewhat unreasonable, or at least necessitate time travel, since they require the system to correctly guess what the input signal is going to do at some future time. The philosophical aspects of this dubious behavior are explored in an exercise below. Streamable systems are either causal or can be made causal by adding an overall delay. For example, neither $y_n = x_{-n}$ nor $y_n = x_{n+1}$ are causal, but the latter is streamable while the former is not.

When working off-line, for instance with an input signal that is available as a file or known as an explicit function, one can easily implement noncausal systems. One need only *peek ahead* or precompute the needed input values, and then place the output value in the proper memory or file location. Analog systems can realize only causal systems since they must output values immediately without peeking forward in time, or going back in time to correct the output values. Since analog systems are also required to be stable, stable causal systems are called *realizable*, meaning simply that they may be built in analog electronics. Real-time digital systems can realize only stable streamable systems; the amount of delay allowed is application dependent, but the real-time constraint requires the required delay to be constant.

EXERCISES

6.4.1 Find the inverse system for the following systems. If this is in IIR form find the FIR form as well (take $x_n = 0$ for $n \leq 0$).

1. $y_n = x_n + x_{n-1}$
2. $y_n = x_n - \frac{1}{2}x_{n-1}$
3. $y_n = x_n - x_{n-1} - x_{n-2}$
4. $y_n = \sum_{i=-\infty}^{n} x_n$
5. $y_n = x_n + y_{n-1}$
6. $y_n = x_n - x_{n-1} + y_{n-1}$

6.4.2 What can you say about the FIR and IIR characteristics of inverse systems?

6.4.3 Which of the following systems are filters? Explain. 1. $y_n = x_{n-1} + k$

2. $y_n = x_{n+1}x_{n-1}$
3. $y_n = x_{ln}$
4. $y_n = 0$
5. $y_n = y_{n-1}$

6.4.4 Show that a filter is causal if and only if its output, when the input is a unit impulse centered on time zero, is nonzero only for positive times.

6.4.5 Show that if two signals are identical up to time t, then the output of a causal system to which these are input will be the same up to time t.

6.4.6 Show that the smoothing operation $y_n = \frac{1}{2}(x_n + x_{n-1})$ is causal while the similar $y_n = \frac{1}{2}(x_{n+1} + x_n)$ is not. The system $y_n = \frac{1}{2}(x_{n+1} - x_{n-1})$ is an approximation to the derivative. Show that it is not causal but is streamable. Find a causal approximation for the derivative.

6.4.7 Consider the philosophical repercussions of noncausal systems by reflecting on the following case. The system in question outputs -1 for two seconds if its input will be positive one second from now, but $+1$ for two seconds if its input will be negative. Now feed the output of the system back to its input.

6.4.8 Explain why streamable systems can be realized in DSP but not in analog electronics. What does the delay do to the phase response?

6.4.9 The systems $y_n = x_n + a$ (which adds a DC term) and $y_n = x_n^2$ (which squares its input) do not commute. Show that any two *filters* do commute.

6.4.10 Systems do not have to be deterministic. The **M**odulated **N**oise **R**eference Unit (MNRU) system, defined by $y_n = \left(1 + 10^{-\frac{Q}{20}} \nu_n\right) x_n$ (where ν is wideband noise) models audio quality degradation under logarithmic companding (exercise 6.2.2). Which of the characteristics defined in this section does the MNRU have? Can you explain how the MNRU works?

6.5 Filters

In the previous section we mentioned that the combination of linearity and time invariance is important enough to deserve a distinctive name, but did not explain why this is so. The explanation is singularly DSP, linking characteristics in the time domain with a simple frequency domain interpretation. We shall show shortly that the spectrum of a filter's output signal is the input signal's spectrum multiplied by a frequency-dependent weighting function. This means that some frequencies may be amplified, while others may be attenuated or even removed; the amplification as a function of frequency being determined by the particular filter being used. For example, an ideal low-pass filter takes the input signal spectrum, multiplies all frequency components below a cutoff frequency by unity, but multiplies all frequency components over that frequency by zero. It thus passes low frequencies while removing all high-frequency components. A band-pass filter may zero out all frequency components of the input signal except those in a range of frequencies that are passed unchanged.

Only filters (LTI systems) can be given such simple frequency domain interpretations. Systems that are not linear and time-invariant can create new frequency components where none existed in the input signal. For example, we mentioned at the end of Section 6.2 and saw in exercise 6.2.3 that the squaring operation generated harmonics when a sinusoidal signal was input, and generated combination frequencies when presented with the sum of two sinusoids. This is a general feature of non-LTI systems; the spectrum of the output will have frequency components that arise from complex combinations of input frequency components. Just as the light emerging from a optical filter does not contain colors lacking in the light impinging upon it, just as when pouring water into a coffee filter brandy never emerges, just so you can be sure that the output of a signal processing filter does not contain frequencies absent in the input.

Let's prove this important characteristic of filters. First, we expand the input in the SUI basis (as the sum of unit impulses weighted by the signal value at that time).

$$x_n = \sum_{m=-\infty}^{\infty} x_m \delta_{n,m}$$

Next, using the linearity of the filter H, we can show that

$$y_n = H x_n = H \left(\sum_{m=-\infty}^{\infty} x_m \delta_{n,m} \right) = \sum_{m=-\infty}^{\infty} H \left(x_m \delta_{n,m} \right)$$

but since the x_m are simply constants multiplying the SUIs, linearity also implies that we can move them outside the system operator.

$$y_n = \sum_{m=-\infty}^{\infty} x_m H \delta_{n,m}$$

Now the time has come to exploit the time invariance. The operation of the system on the SUI $H\delta_{n,m}$ is precisely the same as its operation on the unit impulse at time zero, only shifted m time units. The *impulse response* h_n is defined to be the response of a system at time n to the unit impulse.

$$h_n = H \delta_{n,0} \tag{6.22}$$

For causal systems $h_n = 0$ for $n < 0$, and for practical systems h_n must become small for large enough n. So time invariance means $H\delta_{n,m} = h_{n-m}$ and we have found the following expression for a filter's output.

$$y_n = \sum_{m=-\infty}^{\infty} x_m h_{n-m} \tag{6.23}$$

For causal filters future inputs cannot affect the output.

$$y_n = \sum_{m=-\infty}^{0} x_m h_{n-m} \tag{6.24}$$

We have seen this type of sum before! We called it a *convolution* sum and saw in Section 4.8 that its LTDFT was particularly simple. Taking the LTDFT of both sides of equation (6.23) and using equation (4.44) we find

$$Y_k = H_k X_k \tag{6.25}$$

which states that the output at frequency k is the input at that frequency multiplied by a frequency-dependent factor H_k. This factor is the digital version of what we previously called the frequency response $H(\omega)$.

Using terminology borrowed from linear algebra, what we have proved is that the sinusoids are *eigenfunctions* or, using more fitting terminology, *eigensignals* of filters. If we allow complex signals we can prove the same for the complex sinusoids $x_n = e^{i\omega n}$.

EXERCISES

6.5.1 In exercise 6.3.7 we saw that the system $y_n = x_n - a_1 y_{n-1} - y_{n-2}$ could sustain a sinusoidal oscillation even with no input. Yet this is a filter, and thus should not be able to create frequencies not in the input! Explain.

6.5.2 Show that the system $y_n = x_n + x_{-n}$ is not a filter. Show that it indeed doesn't act as a filter by considering the inputs $x_n = \sin(\omega n + \phi)$ and $x_n = \cos(\omega n + \phi)$.

6.5.3 Prove the general result that z^n are eigenfunctions of filters.

6.5.4 Prove the filter property for analog signals and filters.

6.6   Moving Averages in the Time Domain

We originally encountered the FIR filter as a natural way of modeling a sequence of echoes, each attenuated or strengthened and delayed in time. We now return to the FIR filter and ask why it is so popular in DSP applications. As usual in DSP there are two answers to this question, one related to the time domain and the other to the frequency domain. In this section we delve into the former and ask why it is natural for a system's output to depend on the input at more than one time. We will motivate this dependency in steps.

Consider the following problem. There is a signal z_n that is known to be constant $z_n = z$, and we are interested in determining this constant. We are not allowed to directly observe the signal z_n, only the signal

$$x_n = z_n + \nu_n$$

where ν_n is some noise signal. We know nothing about the noise save that its average is zero, and that its variance is finite.

Since the noise averages to zero and the observed signal is the sum of the desired constant signal and this noise, the observed signal's average value must be z. Our path is clear; we need to average the observed signal

$$\frac{1}{L}\sum_{l=0}^{L-1} x_l \approx z \qquad (6.26)$$

with the sum approaching z more and more closely as we increase L. For finite L our estimate of z will be not be exact, but for large enough L (the required size depending on the noise variance) we will be close enough.

Now let us assume that z_n is not a constant, but a slowly varying signal. By *slowly varying* we mean that z_n is essentially the same for a great many consecutive samples. Once again we can only observe the noisy x_n, and are interested in recovering z_n. We still need to average somehow, but we can no longer average as much as we please, since we will start 'blurring' the desired nonconstant signal. We thus must be content with averaging over several x_n values,

$$y_n = \frac{1}{L}\sum_{l=0}^{L-1} x_{n+l} \approx z_n \qquad (6.27)$$

and repeating this operation every j samples in order to track z_n.

$$\overbrace{x_0, x_1, \ldots x_{L-1}}^{y_0}, x_L, x_{L+1}, \ldots \overbrace{x_j, x_{j+1}, \ldots x_{j+L-1}}^{y_j}, x_{j+L}, x_{j+L+1}, \ldots$$

We must take j small enough to track variations in z_n, while $L \leq j$ must be chosen large enough to efficiently average out the noise. Actually, unless there is some good reason not to, we usually take $L = j$ precisely.

$$\overbrace{x_0, x_1, \ldots x_{L-1}}^{y_0}, \overbrace{x_L, x_{L+1}, \ldots x_{2L-1}}^{y_L}, \overbrace{x_{2L}, x_{2L+1}, \ldots x_{3L-1}}^{y_{2L}}, \ldots$$

Now we assume that z_n varies a bit faster. We must reduce j in order to track z_n sufficiently well, but we cannot afford to reduce L this much unless the noise is very small. So why can't the averaging intervals overlap? Why can't we even calculate a new average every sample?

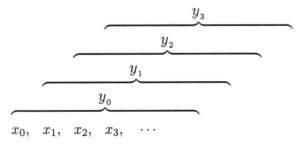

Well, we can; this type of averaging is called a *moving average*, which is often abbreviated MA. The moving average operation produces a new signal y_n which is an approximation to the original z_n. Upon closer inspection we discover that we have introduced a delay of $\frac{L}{2}$ in our estimates of z_n. We could avoid this by using

$$y_n = \frac{1}{2L+1} \sum_{l=-L}^{L} x_{n+l} \approx z_n \qquad (6.28)$$

but this requires breaking of causality.

Our final step is to assume that z_n may vary very fast. Using the moving average as defined above will indeed remove the noise, but it will also intolerably average out significant variations in the desired signal itself. In general it may be impossible to significantly attenuate the noise without harming the signal, but we must strive to minimize this harm. One remedy is to notice that the above averaging applies equal weight to all L points in its sum. We may be able to minimize the blurring that this causes by weighting the center of the interval more than the edges. Consider the difference between the following noncausal moving averages.

$$y_n = \tfrac{1}{3}x_{n-1} + \tfrac{1}{3}x_n + \tfrac{1}{3}x_{n+1} \qquad \text{and} \qquad y_n = \tfrac{1}{4}x_{n-1} + \tfrac{1}{2}x_n + \tfrac{1}{4}x_{n+1}$$

The latter more strongly emphasizes the center term, de-emphasizing the influence of inputs from different times. Similarly we can define longer moving averages with coefficients becoming smaller as we move away from the middle (zero) terms.

The most general moving average (MA) filter is

$$y_n = \sum_{l=-L}^{L} h_l x_{n+l} \qquad (6.29)$$

where the coefficients h_l need to be chosen to maximize the noise suppression while minimizing the signal distortion. If we are required to be realizable we use

$$y_n = \sum_{l=-L}^{0} h_l x_{n+l} = \sum_{l=0}^{L} h_{l-L} x_{n+l-L} \qquad (6.30)$$

although L here will need to be about twice as large, and the output y_n will be delayed with respect to z_n.

EXERCISES

6.6.1 Experiment with the ideas presented in this section as practical techniques for removing noise from a signal. Start with a signal that is constant 1 to which a small amount of Gaussian white noise has been added $x_n = 1 + \epsilon \nu_n$. Try to estimate the constant by adding N consecutive signal values and dividing by N. How does the estimation error depend on ϵ and N?

6.6.2 Perform the same experiment again only this time take the clean signal to be a sinusoid rather than a constant. Attempt to reconstruct the original signal from the noisy copy by using a noncausal moving average with all coefficients equal. What happens when the MA filter is too short or too long?

6.6.3 Now use an MA filter with different coefficients. Take the center coefficient (that which multiplies the present signal value) to be maximal and the others to decrease linearly. Thus for length-three use $(\frac{1}{4}, \frac{1}{2}, \frac{1}{4})$, for length-five use $\frac{1}{9}(1, 2, 3, 2, 1)$, etc. Does this perform better?

6.6.4 Find a noncausal MA differentiator filter, that is, one that approximates the signal's derivative rather than its value. How are this filter's coefficients different from those of the others we have discussed?

6.6.5 A parabola in digital time is defined by $p(n) = an^2 + bn + c$. Given any three signal values x_{-1}, x_0, x_{+1} there is a unique parabola that goes through these points. Given five values $x_{-2}, x_{-1}, x_0, x_{+1}, x_{+2}$ we can find coefficients a, b and c of the *best fitting* parabola $p(n)$, that parabola for which the squared error $\epsilon^2 = (p(-2) - x_{-2})^2 + (p(-1) - x_{-1})^2 + (p(0) - x_0)^2 + (p(+1) - x_{+1})^2 + (p(+2) - x_{+2})^2$ is minimized. We can use this best fitting parabola as a MA *smoothing filter*; for each n we find the best fitting parabola for the 5 signal values $x_{n-2}, x_{n-1}, x_n, x_{n+1}, x_{n+2}$ and output the center value of this parabola. Show that the five-point parabola smoothing filter is an MA filter. What are its coefficients?

6.6.6 After finding the best fitting parabola we can output the value of its derivative at the center. Find the coefficients of this five-point differentiator filter.

6.7 Moving Averages in the Frequency Domain

The operation of an MA filter in the time domain is simple to understand. The filter's input is a signal in the time domain, its output is once again a time domain signal, and the filter coefficients contain all the information needed to transform the former into the latter. What do we mean by the frequency domain description of a filter? Recall that the operation of a filter on a signal has a simple frequency domain interpretation. The spectrum of a filter's output signal is the input signal's spectrum multiplied by a frequency-dependent weighting function. This weighting function is what we defined in Section 6.3 as the filter's *frequency response*. In Section 6.12 we will justify this identification of the frequency response as the fundamental frequency domain description. For now we shall just assume that the frequency response is the proper attribute to explore.

We originally defined the frequency response as the output of a filter given a real sinusoid of arbitrary frequency as input. In this section we extend our original definition by substituting *complex exponential* for *sinusoid*. As usual the main reason for this modification is mathematical simplicity; it is just easier to manipulate exponents than trigonometric functions. We know that at the end we can always extract the real part and the result will be mathematically identical to that we would have found using sinusoids.

Let's start with one of the simplest MA filters, the noncausal, equally weighted, three-point average.

$$y_n = \tfrac{1}{3}(x_{n-1} + x_n + x_{n+1}) \tag{6.31}$$

In order to find its frequency response $H(\omega)$ we need to substitute

$$x_n = e^{i\omega n}$$

and since the moving average is a filter, we know that the output will be a complex exponential of the same frequency.

$$y_n = H(\omega)e^{i\omega n}$$

Substituting

$$y_n = \tfrac{1}{3}\left(e^{i\omega(n-1)} + e^{i\omega n} + e^{i\omega(n+1)}\right) = \tfrac{1}{3}\left(e^{-i\omega} + 1 + e^{i\omega}\right)e^{i\omega n}$$

we immediately identify

$$H(\omega) = \tfrac{1}{3}\left(1 + e^{-i\omega} + e^{i\omega}\right) = \tfrac{1}{3}\left(1 + 2\cos(\omega)\right) \tag{6.32}$$

as the desired frequency response. If we are interested in the energy at the various frequencies, we need the square of this, as depicted in Figure 6.5. We see that this system is somewhat low-pass in character (i.e., lower frequencies are passed while higher frequencies are attenuated). However, the attenuation does not increase monotonically with frequency, and in fact the highest possible frequency $\tfrac{1}{2}f_s$ is not well attenuated at all!

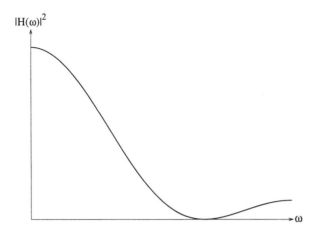

Figure 6.5: The (squared) frequency response of the simple three-point average filter. The response is clearly that of a low-pass filter, but not an ideal one.

At the end of the previous section we mentioned another three-point moving average.

$$y_n = \tfrac{1}{4}x_{n-1} + \tfrac{1}{2}x_n + \tfrac{1}{4}x_{n+1} \qquad (6.33)$$

Proceeding as before we find

$$y_n = \tfrac{1}{4}e^{i\omega(n-1)} + \tfrac{1}{2}e^{i\omega n} + \tfrac{1}{4}e^{i\omega(n+1)} = \left(\tfrac{1}{4}e^{-i\omega} + \tfrac{1}{2} + \tfrac{1}{4}e^{i\omega}\right)e^{i\omega n}$$

and can identify

$$H(\omega) = \left(\tfrac{1}{4}e^{-i\omega} + \tfrac{1}{2} + \tfrac{1}{4}e^{i\omega}\right) = \tfrac{1}{2}\left(1 + \cos(\omega)\right) \qquad (6.34)$$

a form known as a 'raised cosine'.

This frequency response, contrasted with the previous one in Figure 6.6 is also low-pass in character, and is more satisfying since it *does* go to zero at $\tfrac{1}{2}f_s$. However it is far from being an ideal low-pass filter that drops to zero response above some frequency; in fact it is wider than the frequency response of the simple average.

What happens to the frequency response when we average over more signal values? It is straightforward to show that for the simplest case

$$y_n = \frac{1}{2L+1}\sum_{l=-L}^{L} x_{n+l} \qquad (6.35)$$

the frequency response is

$$\frac{\sin(\frac{Lx}{2})}{L\sin(\frac{x}{2})} \qquad (6.36)$$

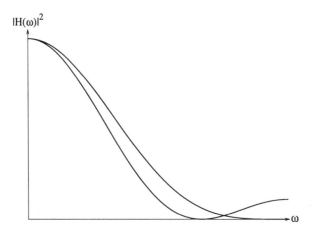

Figure 6.6: The (squared) frequency responses of two simple three-point average filters. Both responses are clearly low-pass but not ideal. The average with coefficients goes to zero at $\frac{1}{2}f_s$, but is 'wider' than the simple average.

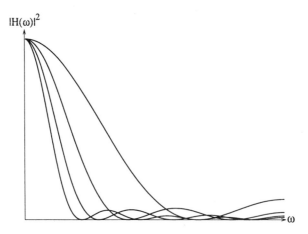

Figure 6.7: The (squared) frequency responses of simple averaging filters for $L = 3, 5, 7$ and 9. We see that as L increases the pass-band becomes narrower, but oscillations continue.

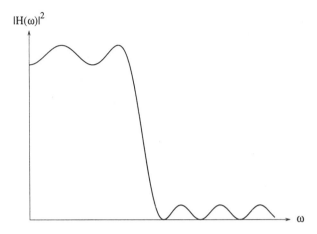

Figure 6.8: The (squared) frequency responses of a 16-coefficient low-pass filter. With these coefficients the lower frequency components are passed essentially unattenuated, while the higher components are strongly attenuated.

as is depicted in Figure 6.7 for $L = 3, 5, 7, 9$. We see that as L increases the filter becomes more and more narrow, so that for large L only very low frequencies are passed. However, this is only part of the story, since even for large L the oscillatory behavior persists. Filters with higher L have a narrower main lobe but more sidelobes.

By using different coefficients we can get different frequency responses. For example, suppose that we need to pass frequencies below half the Nyquist frequency essentially unattenuated, but need to block those above this frequency as much as possible. We could use a 16-point moving average with the following magically determined coefficients

$$
\begin{array}{rrrr}
0.003936, & -0.080864, & 0.100790, & 0.012206, \\
-0.090287, & -0.057807, & 0.175444, & 0.421732, \\
0.421732, & 0.175444, & -0.057807, & -0.090287, \\
0.012206, & 0.100790, & -0.080864, & 0.003936
\end{array}
$$

the frequency response of which is depicted in Figure 6.8. While some oscillation exists in both the pass-band and the stop-band, these coefficients perform the desired task relatively well.

Similarly we could find coefficients that attenuate low frequencies but pass high ones, or pass only in a certain range, etc. For example, another simple MA filter can be built up from the finite difference.

$$y_n = \Delta x_n = x_n - x_{n-1} \tag{6.37}$$

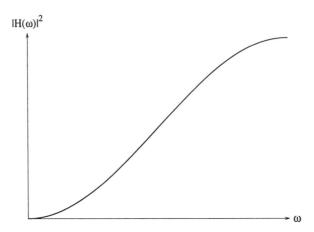

Figure 6.9: The (squared) frequency response of a finite difference filter. With these coefficients the lower frequency components are passed essentially unattenuated, while the higher components are strongly attenuated.

It is easy to show that its frequency response (see Figure 6.9) attenuates low and amplifies high frequencies.

EXERCISES

6.7.1 Calculate the frequency response for the simple causal moving average.

$$y_n = \frac{1}{L} \sum_{k=0}^{L-1} x_{n-k}$$

Express your result as the product of an amplitude response and a phase response. Compare the amplitude response to the one derived in the text for equally weighted samples? Explain the phase response.

6.7.2 Repeat the previous exercise for the noncausal case with an even number of signal values. What is the meaning of the phase response now?

6.7.3 Verify numerically that the 16-point MA filter given in the text has the frequency response depicted in Figure 6.8 by injecting sinusoids of various frequencies.

6.7.4 Find the squared frequency response of equation (6.37).

6.7.5 Find an MA filter that passes intermediate frequencies but attenuates highs and lows.

6.7.6 Find nontrivial MA filters that pass all frequencies unattenuated.

6.7.7 The second finite difference Δ^2 is the finite difference of the finite difference, i.e., $\Delta^2 x_n = \Delta(x_n - x_{n-1}) = x_n - 2x_{n-1} + x_{n-2}$. Give explicit formulas for the third and fourth finite differences. Generalize your results to the kth order finite difference. Prove that $y_n = \Delta^k x_n$ is an MA filter with $k + 1$ coefficients.

6.8 Why Convolve?

The first time one meets the convolution sum

$$x * y = \sum_i x_i \, y_{k-i}$$

one thinks of the algorithm

```
Given x, y, k
Initialize: conv ← 0;  i, j
Loop:
        increment conv by x_i y_j
        increment i
        decrement j
```

and can't conceive of any good reason to have the two indices moving in opposite directions. Surely we can always redefine y_j and rephrase this as our original MA filter *moving average*

```
Given x, y
Initialize: conv ← 0;  i, j
Loop:
        increment conv by x_i y_i
        increment i
```

saving a lot of confusion. There must be some really compelling reason for people to prefer this strange (or should I say *convoluted*) way of doing things. We will only fully understand why the convolution way of indexing is more prevalent in DSP in Section 6.12, but for now we can somewhat demystify the idea.

We'll start by considering two polynomials of the second degree.

$$\begin{aligned} A(x) &= a_0 + a_1\, x + a_2\, x^2 \\ B(x) &= b_0 + b_1\, x + b_2\, x^2 \end{aligned}$$

Their product is easily found,

$$
\begin{aligned}
A(x)B(x) \;=\;& a_0 b_0 + (a_0 b_1 + a_1 b_0)x + (a_0 b_2 + a_1 b_1 + a_2 b_0)x^2 \\
& + (a_1 b_2 + a_2 b_1)x^3 + a_2 b_2\, x^4
\end{aligned}
$$

and the connection between the indices seems somewhat familiar. More generally, for any two polynomials

$$
A(x) \;=\; \sum_{i=0}^{N} a_i\, x^i
$$

$$
B(x) \;=\; \sum_{j=0}^{M} b_j\, x^j
$$

we have

$$
A(x)B(x) = \sum_{k=0}^{N+M} \left(\sum_{\substack{i,j \\ i+j=k}} a_i b_j \right) x^k = \sum_{k=0}^{N+M} (a * b)_k\, x^k
$$

and the fundamental reason for these indices to run in opposite directions is obvious—the two exponents must sum to a constant!

Put this way the idea of indices running in opposite directions isn't so new after all. In fact you probably first came across it in grade school. Remember that an integer is represented in base-10 as a polynomial in 10, $A = \sum_{i=0}^{d_A} a_i 10^i$ (where d_A is the number of digits). Thus multiplication of two *integers* A and B is also really a convolution.

$$
AB = \sum_{k=0}^{d_A + d_B} (a * b)_k 10^k
$$

The algorithm we all learned as *long multiplication* is simply a tabular device for mechanizing the calculation of the convolution.

		A_N	A_{N-1}	$\cdots$	A_1	A_0
	$*$	B_N	B_{N-1}	$\cdots$	B_1	B_0
		$B_0 A_N$	$B_0 A_{N-1}$	$\cdots$	$B_0 A_1$	$B_0 A_0$
	$B_1 A_N$	$B_1 A_{N-1}$		$\cdots$	$B_1 A_0$	
				$\vdots$		
$B_N A_N$	$\cdots$	$B_N A_1$	$B_N A_0$			
C_{2N}		C_{N+1}	C_N	C_{N-1}	$\cdots$	C_1	C_0

We now understand why the symbol '$*$' is used for convolution; there is a simple connection between convolution and multiplication!

Anyone who is comfortable with multiplication of integers and polynomials is automatically at home with convolutions. There is even a formalism for turning *every* convolution into a polynomial multiplication, namely the z *transform* (see section 4.10). The basic idea is to convert every digital signal x_n into an equivalent polynomial

$$x_n \longleftarrow X(d) = \sum_n d^n$$

although it is conventional in DSP to use z^{-1} instead of d. Then the convolution of two digital signals can be performed by multiplying their respective z transforms. You can think of z transforms as being similar to logarithms. Just as logarithms transform multiplications into additions, z transforms transform convolutions into multiplications.

We have seen an isomorphism between convolution and *polynomial* products, justifying our statement that convolution is analogous to multiplication. There is also an isomorphism with yet another kind of multiplication that comes in handy. The idea is to view the signal x_n as a vector in N-dimensional space, and the process of convolving it with some vector h_n as an operator that takes x_n and produces some new vector y_n. Now since linear operators can always be represented as matrices, convolution is also related to *matrix* multiplication. To see this explicitly let's take the simple case of a signal x_n that is nonzero only between times $n = 0$ and $n = 4$ so that it is analogous to the vector $(x_0, x_1, x_2, x_3, x_4)$. Let the filter h_n have three nonzero coefficients h_{-1}, h_0, h_1 so that it becomes the vector (h_{-1}, h_0, h_1). The convolution $y = h * x$ can only be nonzero between times $n = -1$ and $n = 5$, but we will restrict our attention to times that correspond to nonzero x_n. These are given by

$$
\begin{aligned}
y_0 &= h_0 x_0 + h_{-1} x_1 \\
y_1 &= h_{-1} x_0 + h_0 x_1 + h_{-1} x_2 \\
y_2 &= h_{-1} x_1 + h_0 x_2 + h_{-1} x_3 \\
y_3 &= h_{-1} x_2 + h_0 x_3 + h_{-1} x_4 \\
y_4 &= h_{-1} x_3 + h_0 x_4
\end{aligned}
$$

and more compactly written in matrix form.

$$
\begin{pmatrix} y_0 \\ y_1 \\ y_2 \\ y_3 \\ y_4 \end{pmatrix} = \begin{pmatrix} h_0 & h_1 & 0 & 0 & 0 \\ h_{-1} & h_0 & h_1 & 0 & 0 \\ 0 & h_{-1} & h_0 & h_1 & 0 \\ 0 & 0 & h_{-1} & h_0 & h_1 \\ 0 & 0 & 0 & h_{-1} & h_0 \end{pmatrix} \begin{pmatrix} x_0 \\ x_1 \\ x_2 \\ x_3 \\ x_4 \end{pmatrix}
$$

The matrix has quite a distinctive form, all elements on each diagonal being equal. Such a matrix is said to be *Toeplitz* in structure, and Toeplitz matrices tend to appear quite a lot in DSP.

EXERCISES

6.8.1 N.G. Kneer, the chief DSP hardware engineer at NeverWorks Incorporated, purchases pre-owned DSP processors on an as-is basis from two suppliers, Alpha Numerics and Beta Million. From Alpha Numerics one gets a perfect lot 30% of the time, but 10% of the time all five chips are bad. From Beta one never gets a completely bad lot, but only gets a perfect lot 10% of the time. In fact, N.G. has come up with the following data regarding the probability of k defective chips out of a lot of five.

k	A_k	B_k
0	0.3	0.1
1	0.2	0.2
2	0.2	0.4
3	0.1	0.2
4	0.1	0.1
5	0.1	0.0

In order to reduce his risk, N.G. buys from both suppliers. Assuming he buys ten chips, five chips from each, what should he expect to be the distribution of defective chips? What is the connection between this and convolution?

6.8.2 Dee Espy has to purchase 100 DSPs, and decides that she should strive to minimize the number of defective chips she purchases. How many should she buy from each of the above suppliers?

6.8.3 Convolve the signal $x_n = \ldots, 0, 0, 1, 1, 1, 1, 1, 1, 1, 1, 0, 0 \ldots$ with the filter $h = (1, 1, 1)$ and plot x_n, h_n and the output y_n. Convolve y_n with h_n resulting in $y^{[2]}1_n$ and again this new signal to get $y_n^{[3]}$, etc. Plot $x_n = y_n^{[0]}$, $y_n = y_n^{[1]}$, $y^{[2]}1_n$, $y^{[3]}1_n$, one under the other. What can be said about the effect of this consecutive filtering?

6.9 Purely Recursive Systems

In this section we will deal with purely recursive systems, that is systems for which y_n depends on previous y values and the present x_n, but not on previous x values. You should know that DSP engineers call such systems **AutoRegressive** (AR) systems, a name coined by G. Udny Yule in 1927. The word regression here refers to *regression analysis*, a well-known statistical method for finding the relationship of a variable to other variables. Yule was studying the number of sunspots observed as a function of time and decided to attempt to relate the present sunspot activity to previous values of the *same* quantity using regression analysis. He thus called this technique *autoregression* analysis. We prefer the name 'purely recursive' to autoregressive, but will nonetheless adopt the prevalent abbreviation 'AR'.

For AR systems the output y_n is obtained from the input x_n by

$$y_n = x_n + \sum_{m=1}^{M} b_m y_{n-m} \tag{6.38}$$

and to start up the recursion we have to make some assumption as to earlier outputs (e.g., take them to be zero). If the input signal was zero before time $n = 0$ then any causal system will have $y_n = 0$ for all negative n. However, if we choose to start the recursion at $n = 0$ but the input actually preexisted, the zeroing of the previous outputs is contrived.

Let's return to the problem introduced in Section 6.6 of finding the true value of a constant signal obscured by additive noise. Our first attempt was to simply average up some large number L of signal values.

$$y_{L-1} = \frac{1}{L} \sum_{l=0}^{L-1} x_l$$

Were we to determine that L values were not sufficient and we wished to try $L+1$, we do not have to add up all these values once again. It is easy to see that we need only multiply y_{L-1} by L to regain the sum, add the next input x_L, and divide by the new number of signal values $L + 1$.

$$y_L = \frac{1}{L+1} \left(L\, y_{L-1} + x_L \right)$$

This manipulation has converted the original iteration into a recursion

$$y_L = \alpha\, x_L + \beta\, y_{L-1} \qquad \text{where } \alpha = \frac{1}{L}, \ \beta = \frac{L}{L+1}$$

and we have the relation $\alpha + \beta = 1$. Changing the index to our more usual n we can now write

$$y_n = (1 - \beta)x_n + \beta y_{n-1} \qquad\qquad 0 \le \beta < 1 \qquad (6.39)$$

which is of the AR form with $M = 1$.

The nice thing about equation (6.38) is that it is already suitable for rapidly varying signals. We needn't go through all the stages that lead to the MA filter with coefficients; by changing β this AR filter can be set to track rapidly varying signals or to do a better job of removing noise from slowly varying ones. When $\beta = 0$ (corresponding to $L = 0$) the AR filter output y_n is simply equal to the input, no noise is averaged out but no bandwidth lost either. As β increases the past values assume more importance, and the averaging kicks in at the expense of not losing the ability to track the input as rapidly. When $\beta \to 1$ (corresponding to infinite L) the filter paradoxically doesn't look at the current input at all!

Equation (6.39) is similar to the causal version of the moving average filter of equation (6.30) in that it moves along the signal immediately outputting the filtered signal. However, unlike the moving average filter, equation (6.39) never explicitly removes a signal value that it has seen from its consideration. Instead, past values are slowly 'forgotten' (at least for $\beta < 1$). For large β signal values from relatively long ago are still relatively important, while for small β past values lose their influence rapidly. You can think of this AR filter as being similar to an MA filter operating on L previous values, the times before $n - L$ having been forgotten. To see this, unravel the recursion in equation (6.39).

$$y_n = (1-\beta)x_n + \beta(1-\beta)x_{n-1} + \beta^2(1-\beta)x_{n-2} + \beta^3(1-\beta)x_{n-3} + \ldots \quad (6.40)$$

We see that the coefficient corresponding to x_{n-l} is smaller than that of x_n by a factor of β^l, and so for all practical purposes we can neglect the contributions for times before some l. For example, if $\beta = 0.99$ and we neglect terms that are attenuated by e^{-1}, we need to retain about 100 terms; however for $\beta = 0.95$ only about 20 terms are needed, for $\beta = 0.9$ we are down to ten terms, and for $\beta = 0.8$ to 5 terms. It is not uncommon to use $\beta = 0.5$ where only the x_n and x_{n-1} terms are truly relevant, the x_{n-2} term being divided by 4.

We should now explore the frequency response $H(\omega)$ of our AR filter. Using the technique of Section 6.7 we assume

$$x_n = e^{i\omega n}$$

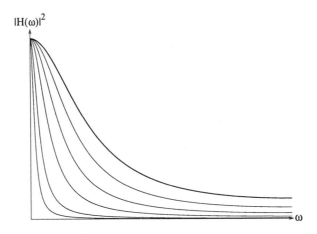

Figure 6.10: The (squared) frequency response of the simple AR low-pass filter for several different values of β. From top to bottom $\beta = 0.5, 0.6, 0.7, 0.8, 0.9, 0.95$.

and since the AR filter is a filter, we know that the output will be a complex exponential of the same frequency.

$$y_n = H(\omega)e^{i\omega n}$$

Substituting from equation (6.40) and using equation (A.47)

$$
\begin{aligned}
y_n &= (1 - \beta)e^{i\omega n} + \beta(1 - \beta)e^{i\omega(n-1)} + \beta^2(1 - \beta)e^{i\omega(n-2)} + \dots \\
&= (1 - \beta)\sum_{k=0}^{\infty}(\beta e^{-i\omega})^k e^{i\omega n} \\
&= \frac{(1 - \beta)}{(1 - \beta e^{-i\omega})}e^{i\omega n}
\end{aligned}
$$

and we immediately identify

$$H(\omega) = \frac{(1 - \beta)}{(1 - \beta e^{-i\omega})} \tag{6.41}$$

as the desired frequency response, and

$$|H(\omega)|^2 = \frac{1 - 2\beta + \beta^2}{1 - 2\beta \cos(\omega) + \beta^2} \tag{6.42}$$

as its square. We plot this squared frequency response, for several values of β, in Figure 6.10.

Another useful AR filter is

$$y_n = x_n + y_{n-1} \tag{6.43}$$

which unravels to the following infinite sum:

$$y_n = x_n + x_{n-1} + x_{n-2} + \ldots = \sum_{m=0}^{\infty} x_{n-m}$$

We can write this in terms of the time delay operator

$$y = (1 + z^{-1} + z^{-2} + \ldots)x = \Upsilon x$$

where we have defined the infinite accumulator operator

$$\Upsilon \equiv \sum_{m=0}^{\infty} z^{-m} x_n \tag{6.44}$$

which roughly corresponds to the integration operator for continuous signals. The finite difference $\Delta \equiv (1 - z^{-1})$ and the infinite accumulator are related through $\Delta\Upsilon = 1$ and $\Upsilon\Delta = 1$, where 1 is the identity operator.

What happens when the infinite accumulator operates on a constant signal? Since we are summing the same constant over and over again the sum obviously gets larger and larger in absolute value. This is what we called *instability* in Section 6.4, since the output of the filter grows without limit although the input stays small. Such unstable behavior could never happen with an MA filter; and it is almost always an unwelcome occurrence, since all practical computational devices will eventually fail when signal values grow without limit.

EXERCISES

6.9.1 What is the exact relation between β in equation (6.39) and the amount of past time τ that is still influential? Define 'influential' until the decrease is by a factor of e^{-1}. Graph the result.

6.9.2 Quantify the bandwidth BW of the AR filter as a function of β and compare it with the influence time τ.

6.9.3 Contrast the squared frequency response of the AR filter (as depicted in Figure 6.10) with that of the simple averaging MA filter (Figure 6.7). What can you say about the amount of computation required for a given bandwidth?

6.9.4 Find an AR filter that passes high frequencies but attenuates low ones. Find an AR filter that passes intermediate frequencies but attenuates highs and lows.

6.9.5 Calculate the effect of the infinite accumulator operator Υ on the following signals, and then check by generating the first 10 values.
1. $x_n = u_n$ where u_n is the unit step of equation (2.4)
2. $x_n = (-1)^n u_n = 1, -1, 1, -1, \ldots$
3. $x_n = n u_n = 0, 1, 2, 3, \ldots$
4. $x_n = \alpha^n u_n$ where $\alpha \neq 1$

6.9.6 Apply the finite difference operator to the results obtained in the previous exercise, and show that $\Delta \Upsilon = 1$.

6.9.7 Prove that $\Upsilon \Delta = 1$. (Hint: Prove that the 'telescoping' series $x_1 - x_0 + x_2 - x_1 \ldots a_n - a_{n-1} = x_0 + x_n$.)

6.9.8 What is the condition for $y_n = \alpha x_n + \beta y_{n-1}$ to be stable? (Hint: Take a constant input $x_n = 1$ and compute y_n when $n \to \infty$.)

6.10 Difference Equations

We have seen that there are MA filters, with output dependent on the present and previous inputs, and AR filters, with output dependent on the present input and previous outputs. More general still are combined ARMA filters, with output dependent on the present input, L previous inputs, and M previous outputs.

$$y_n = \sum_{l=0}^{L} a_l x_{n-l} + \sum_{m=1}^{M} b_m y_{n-m} \tag{6.45}$$

When $M = 0$ (i.e., all b_m are zero), we have an MA filter $y_n = \sum_{l=0}^{L} a_l x_{n-l}$, while $L = 0$ (i.e., all $a_l = 0$ except a_0), corresponds to the AR filter $y_n = x_n + \sum_{m=1}^{M} b_m y_{n-m}$.

To convince yourself that ARMA relationships are natural consider the amount of money y_n in a bank account at the end of month n, during which x_n is the total amount deposited (if $x_n < 0$ more was withdrawn than deposited) and interest from the previous month is credited according to a rate of i.

$$y_n = y_{n-1} + x_n + i y_{n-2}$$

A slightly more complex example is that of a store-room that at the end of day n contains y_n DSP chips, after x_n chips have been withdrawn from stock that day. The requisitions clerk orders new chips based on the average usage over the past two days $\frac{1}{2}(x_n + x_{n-1})$, but these are only delivered the next day. The number of chips in stock is thus given by an ARMA system.

$$y_n = y_{n-1} - x_n + \tfrac{1}{2}(x_{n-1} + x_{n-2})$$

Equation (6.45) can also be written in a more symmetric form

$$\sum_{m=0}^{M} \beta_m y_{n-m} = \sum_{l=0}^{L} \alpha_l x_{n-l} \tag{6.46}$$

where

$$\alpha_l = a_l \qquad \beta_0 \equiv 1 \qquad \beta_m = -b_m \quad \text{for} \quad m = 1 \ldots M$$

although this way of expressing the relationship between y and x hides the fact that y_n can be simply derived from previous x and y values. This form seems to be saying that the x and y signals are both equally independent, but happen to obey a complex relationship involving present and past values of both x and y. In fact the symmetric form equally well describes the inverse system.

$$x_n = y_n - \sum_{m=1}^{M} b_m y_{n-m} - \sum_{l=1}^{L} a_l x_{n-l}$$

We can formally express equation (6.46) using the time delay operator

$$\sum_{m=0}^{M} \beta_m z^{-m} y_n = \sum_{l=0}^{L} \alpha_l z^{-l} x_n \tag{6.47}$$

and (as you will demonstrate in the exercises) in terms of finite differences.

$$\sum_{m=0}^{M} B_m \Delta^m y_n = \sum_{l=0}^{L} A_l \Delta^l x_n \tag{6.48}$$

Recalling from Section 2.4 that the finite difference operator bears some resemblance to the derivative, this form bears some resemblance to a linear differential equation

$$\sum_{m=0}^{M} \beta_m y^{[n-m]}(t) = \sum_{l=0}^{L} \alpha_l x^{[n-l]}(t) \tag{6.49}$$

where $x^{[k]}$ is the kth derivative of $x(t)$ with respect to t. For this reason ARMA systems are often called *difference equations*.

Derivatives are defined using a limiting process over differences. In DSP the time differences cannot be made smaller than the sampling interval T, and thus finite differences take the place of differentials, and difference equations replace differential equations.

There are many similarities between differential equations and difference equations. The recursion (6.45) is a prescription for generating y_n given initial conditions (e.g., all x_n and y_n are zero for $n < 0$); similarly solutions to differential equations need initial conditions to be specified. The most general solution of a difference equation can be written as a solution to the equation with zero input $x_n = 0$ and any particular solution with the actual input; readers familiar with differential equations know that general solutions to linear differential equations are obtained from the homogeneous solution plus a particular solution. Linear differential equations with constant coefficients can be solved by assuming solutions of the form $y(t) = Ae^{\lambda t}$; solutions to linear difference equations can be similarly found by assuming $y_n = z^n$, which is why we have been finding frequency responses by assuming $y_n = e^{i\omega t}$ all along.

Differential equations arise naturally in the analysis and processing of analog signals, because derivatives describe changes in signals over short time periods. For example, analog signals that have a limited number of frequency components have short time predictability, implying that only a small number of derivatives are required for their description. More complex signals involve more derivatives and extremely noisy analog signals require many derivatives to describe. Similarly digital signals that contain only a few sinusoids can be described by difference equations of low order while more complex difference equations are required for high-bandwidth signals.

EXERCISES

6.10.1 Difference equations are not the only tool for describing ARMA systems; the state-space description explicitly uses the system's memory (internal state). Denoting the input x_n, the output y_n, and the vector of internal state variables at time n by s_n, the state equation description relates the output to the present input and system state and furthermore describes how to update the state given the input.

$$
\begin{aligned}
y_n &= \underline{f} \cdot \underline{s}_{n-1} + g\,x_n \\
\underline{s}_{n+1} &= \underline{\underline{A}}\,\underline{s}_n + x_n\,\underline{c}
\end{aligned}
$$

Relate the state equation parameters $\underline{f}$, g, $\underline{\underline{A}}$, and $\underline{c}$ to those of the ARMA description, $\underline{a}$ and $\underline{b}$.

6.10.2 Devise a circumstance that leads to an ARMA system with $L = 2$ and $M = 2$. $L = 1$ and $M = 3$.

6.10.3 For any digital signal s, we recursively define $s^{[m]}$, the *finite difference of order m*.

$$s_n^{[0]} = s_n$$
$$s_n^{[n]} = s_n^{[n-1]} - s_{n-1}^{[n-1]}$$

For example, given the sequence $a_n = 2n + 1$ we find the following.

$$
\begin{array}{llllllll}
a_0 & = & 1 & 3 & 5 & 7 & 9 & \cdots \\
a_1 & = & & 2 & 2 & 2 & 2 & \cdots \\
a_2 & = & & & 0 & 0 & 0 & \cdots \\
a_3 & = & & & & 0 & 0 & \cdots \\
a_4 & = & & & & & 0 & \cdots \\
\end{array}
$$

We see here that the second and higher finite differences are all zero. In general, when the sequence is of the form $s_n = c_m n^m + c_{m-1} n^{m-1} + \cdots + c_1 n + c_0$, the $m + 1$ order finite differences are zero. This fact can be used to identify sequences. Find the finite differences for the following sequences, and then identify the sequence.

$$
\begin{array}{llllllll}
3 & 6 & 9 & 12 & 15 & 18 & \cdots \\
3 & 6 & 11 & 18 & 27 & 38 & \cdots \\
3 & 6 & 13 & 24 & 39 & 58 & \cdots \\
3 & 6 & 17 & 42 & 87 & 158 & \cdots \\
3 & 6 & 27 & 84 & 195 & 378 & \cdots \\
\end{array}
$$

6.10.4 Find the first, second, and third finite difference sequences for the following sequences.
  1. $s_n = an$
  2. $s_n = bn^2$
  3. $s_n = cn^3$
  4. $s_n = bn^2 + an$
  5. $s_n = cn^3 + bn^2 + an$

6.10.5 Show that if $x_n = \sum_k^L a_k n^k$ then the $(L+1)^{\text{th}}$ finite difference is zero.

6.10.6 Plot the first, second and third differences of $s_n = \sin(2\pi f n)$ for frequencies $f = 0.1, 0.2, 0.3, 0.4$.

6.10.7 $a_0 x_n + a_1 x_{n-1}$ can be written $A_0 x_n + A_1 \Delta x_n$ where $a_1 = -A_1$ and $a_0 = A_0 + A_1$. What is the connection between the coefficients of $a_0 x_n + a_1 x_{n-1} + a_2 x_{n-2}$ and $A_0 x_n + A_1 \Delta x_n + A_2 \Delta^2 x_n$? What about $a_0 x_n + a_1 x_{n-1} + a_2 x_{n-2} + a_3 x_{n-3}$ and $A_0 x_n + A_1 \Delta x_n + A_2 \Delta^2 x_n + A_3 \Delta^3 x_n$? Generalize and prove that all ARMA equations can be expressed as difference equations.

6.11   The Sinusoid's Equation

The usefulness of MA and AR filters can be clarified via a simple example. The analog sinusoid $s(t) = A\sin(\Omega t + \phi)$ not only has a simple spectral interpretation, but also obeys the second-order *differential* equation

$$\ddot{s}(t) + \Omega^2 s(t) = 0 \qquad (6.50)$$

commonly called the equation of simple harmonic motion. Indeed this equation can be considered to be the defining equation for the family of analog sinusoidal signals, and its simplicity can be used as an alternate explanation of the importance of these signals.

In the digital domain we would expect digital sinusoids to obey a second-order *difference* equation. That this is indeed the case can be shown using the trigonometric identities (A.23)

$$
\begin{aligned}
\sin\Big(\Omega(t - 2T)\Big) &= \sin\Omega t \cos 2\Omega T - \cos\Omega t \sin 2\Omega T \\
&= \sin\Omega t \left(2\cos^2\Omega T - 1\right) + \cos\Omega t \left(2\sin\Omega T \cos\Omega T\right) \\
&= -\sin\Omega t + 2\cos\Omega T \left(2\sin\Omega T \cos\Omega T\right) \\
&= -\sin\Omega t + 2\cos\Omega T \sin\Omega(t - T)
\end{aligned}
$$

which can easily be shown to be

$$s(t - 2T) - 2\cos(\Omega T)s(t - T) + s(t) = 0 \qquad (6.51)$$

or in digital form

$$s_n + c_1 s_{n-1} + c_2 s_{n-2} = 0 \qquad (6.52)$$

where $c_1 = -2\cos\Omega T$ and $c_2 = 1$.

This difference equation, obeyed by all sinusoids, can be exploited in several different ways. In the most direct implementation it can be used as a *digital oscillator* or tone generator, i.e., an algorithm to generate sinusoidal signals. Given the desired digital oscillation frequency ω_d, amplitude A, and initial phase ϕ, we precompute the coefficient

$$c_1 = -2\cos\Omega T = -2\cos\frac{\Omega}{f_s} = -2\cos\Omega_d$$

and the first two signal values

$$
\begin{aligned}
s_0 &= A\sin(\phi) \\
s_1 &= A\sin(\Omega T + \phi) = A\sin(\Omega_d + \phi)
\end{aligned}
$$

where Ω_d is the digital frequency (we will omit the subscript from here on). The difference equations now recursively supply all signal values:

$$s_2 = -(c_1 s_1 + s_0)$$
$$s_3 = -(c_1 s_2 + s_1)$$
$$s_4 = -(c_1 s_3 + s_2)$$

and so on. This digital oscillator requires only one multiplication, one addition, and one sign reversal per sample point! This is remarkably efficient when compared with alternative oscillator implementations such as approximation of the sine function by a polynomial, table lookup and interpolation, or direct application of trigonometric addition formulas. The main problems with this implementation, like those of all purely recursive algorithms, are those of accuracy and stability. Since each result depends on the previous two, numerical errors tend to add up, and eventually swamp the actual calculation. This disadvantage can be rectified in practice by occasionally resetting with precise values.

Another application of the difference equation (6.51) is the removal of an interfering sinusoid. Given an input signal x_n contaminated with an interfering tone at known frequency Ω; we can subtract the sinusoidal component at this frequency by the following MA filter

$$y_n = x_n + c_1 x_{n-1} + x_{n-2}$$

where c_1 is found from Ω. The frequency response is found by substituting an arbitrary complex exponential $e^{i\omega n}$

$$
\begin{aligned}
y_n &= e^{i\omega n} - 2\cos\Omega e^{i\omega(n-1)} + e^{i\omega(n-2)} \\
&= \left(1 - 2\cos\Omega e^{-i\omega} + e^{-2i\omega}\right) e^{i\omega n} \\
&= e^{-i\omega}\left(e^{i\omega} - 2\cos\Omega + e^{-i\omega}\right) e^{i\omega n} \\
&= e^{-i\omega}\left(2\cos\omega - 2\cos\Omega\right) e^{i\omega n}
\end{aligned}
$$

which can be written in the form $y_n = H(\omega)x_n$. The square of $H(\omega)$ is depicted in Figure 6.11 for digital frequency $\frac{1}{2}$. Note that no energy remains at the interfering frequency; the system is a *notch filter*.

Finally the difference equation (6.52) can be used to estimate the frequency of a sine buried in noise. The idea is to reverse the equation, and using observed signal values to estimate the value of c_1. From this the frequency ω can be derived. Were no noise to be present we could guarantee

$$c_1 = -\frac{x_n + x_{n-2}}{x_{n-1}}$$

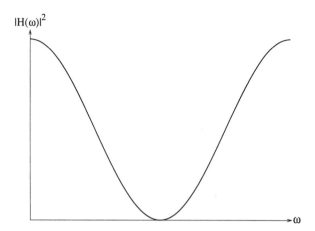

Figure 6.11: The (squared) frequency response of the MA notch filter set to one-half the Nyquist frequency. Note that no energy remains at the notched frequency.

but with noise this only holds on average

$$c_1 = - \left\langle \frac{x_n + x_{n-2}}{x_{n-1}} \right\rangle = - \left\langle \frac{(x_n + x_{n-2})x_{n-1}}{|x_{n-1}|^2} \right\rangle$$

the second form being that most commonly used.

EXERCISES

6.11.1 Show that the signal $s_n = e^{qn}$ obeys the equation $s_n = as_{n-1}$ where $a = e^q$.

6.11.2 Show that the signal $s_n = \sin(\Omega n)$ obeys the equation $s_n = a_1 s_{n-1} + a_2 s_{n-2}$ with coefficients a_i determined by the equation $1 - a_1 z^{-1} - a_2 z^{-2} = 0$ having solutions $z = e^{\pm i\Omega}$.

6.11.3 Show that if a signal is the sum of p exponentials

$$s_n = \sum_{i=1}^{p} A_i e^{q_i n}$$

then the equation $1 - \sum_k a_k z^{-k} = 0$ has roots $z = e^{q_i}$.

6.11.4 Generalize the previous exercises and demonstrate that the sum of p sines obeys a recursion involving $2p$ previous values. What is the equation and how are its coefficients determined?

6.12 System Identification—The Easy Case

Assume that someone brings you a signal processing system enclosed in a black box. The box has two connectors, one marked *input* and the other *output*. Other than these labels there are no identifying marks or documentation, and nothing else is known about what is hidden inside. What can you learn about such a system? Is there some set of measurements and calculations that will enable you to accurately predict the system's output when an arbitrary input is applied? This task is known as *system identification*.

You can consider system identification as a kind of game between yourself and an opponent. The game is played in the following manner. Your opponent brings you the black box (which may have been specifically fabricated for the purpose of the game). You are given a specified finite amount of time to experiment with the system. Next your opponent specifies a test input and asks you for your prediction—were this signal to be applied what output would result? The test input is now applied and your prediction put to the test.

Since your opponent is an antagonist you can expect the test input to be totally unlike any input you have previously tried (after all, you don't have time to try *every possible* input). Your opponent may be trying to trick you in many ways. Is it possible to win this game?

This game has two levels of play. In this section we will learn how to play the easy version; in the next section we will make a first attempt at a strategy for the more difficult level. The easy case is when you are given complete control over the black box. You are allowed to apply controlled inputs and observe the resulting output. The difficult case is when you are not allowed to control the box at all. The box is already hooked up and operating. You are only allowed to observe the input and output.

The latter case is not only more difficult, it may not even be possible to pass the prediction test. For instance, you may be unlucky and during the entire time you observe the system the input may be zero. Or the input may contain only a single sinusoid and you are asked to predict the output when the input is a sinusoid of a different frequency. In such cases it is quite unreasonable to expect to be able to completely identify the hidden system. Indeed, this case is so much harder than the first that the term *system identification* is often reserved for it.

However, even the easy case is far from trivial in general. To see this consider a system that is not time-invariant. Your opponent knows that precisely at noon the system will shut down and its output will be zero thereafter. You are given until 11:59 to observe the system and give your prediction a few seconds before noon. Of course when the system is tested after noon your prediction turns out to be completely wrong! I think you will agree that the game is only fair if we limit ourselves to the identification of time-invariant systems.

Your opponent may still have a trick or two left! The system may have been built to be sensitive to a very specific trigger. For example, for almost every input signal the box may pass the signal unchanged; but for the trigger signal the output will be quite different! A signal that is different from the trigger signal in any way, even only having a slightly different amplitude or having an infinitesimal amount of additive noise, does not trigger the mechanism and is passed unchanged. You toil away trying a large variety of signals and your best prediction is that the system is simply an identity system. Then your opponent supplies the trigger as the test input and the system's output quite astounds you.

The only sensible way to avoid this kind of pitfall is to limit ourselves to linear systems. Linear systems may still be sensitive to specific signals. For example, think of a box that contains the identity system and in parallel a narrow band-pass filter with a strong amplifier. For most signals the output equals the input, but for signals in the band-pass filter's range the output is strongly amplified. However, for linear systems it is not possible to hide the trigger signal. Changing the amplitude or adding some noise will still allow triggering to occur, and once the effect is observed you may home in on it.

So the system identification game is really only fair for linear time-invariant systems, that is, for filters. It doesn't matter to us whether the filters are MA, AR, ARMA, or even without memory; that can be determined from your measurements. Of course since the black box is a real system, it is of necessity realizable as well, and in particular causal. Therefore from now on we will assume that the black box contains an unknown causal filter. If anyone offers to play the game without promising that the box contains a causal filter, don't accept the challenge!

Our task in this section is to develop a winning strategy for the easy case. Let's assume you are given one hour to examine the box in any way you wish (short of prying off the top). At the end of precisely one hour your opponent will reappear, present you with an input signal and ask you what you believe the box's response will be. The most straightforward way of proceeding would be to quickly apply as many different input signals as you can and to record the corresponding outputs. Then you win the game if your opponent's input signal turns out to be essentially one of the inputs you have checked. Unfortunately, there are very many possible inputs, and an hour is to short a time to test even a small fraction of them. To economize we can exploit the fact that the box contains a linear time-invariant system. If we have already tried input x_n there is no point in trying ax_n or x_{n-m}, but this still leaves a tremendous number of signals to check.

Our job can be made more manageable in two different ways, one of which relies on the time domain description of the input signal, and the other on its frequency domain representation. The frequency domain approach is based on Fourier's theorem that every signal can be written as the weighted sum (or integral) of basic sinusoids. Assume that you apply to the unknown system not every possible signal, but only every possible sinusoid. You store the system's response to each of these and wait for your opponent to appear. When presented with the test input you can simply break it down to its Fourier components, and exploit the filter's linearity to add the stored system responses with the appropriate Fourier coefficients.

Now this task of recording the system outputs is not as hard as it appears, since sinusoids are eigensignals of filters. When a sinusoid is input to a filter the output is a single sinusoid of the same frequency, only the amplitude and phase may be different. So you need only record these amplitudes and phases and use them to predict the system output for the test signal. For example, suppose the test signal turns out to be the sum of three sinusoids

$$x_n = X_1 \sin(\omega_1 n) + X_2 \sin(\omega_2 n) + X_3 \sin(\omega_3 n)$$

the responses of which had been measured to be

$$H_1 \sin(\omega_1 n + \phi_1), \qquad H_2 \sin(\omega_2 n + \phi_2), \qquad \text{and} \qquad H_3 \sin(\omega_3 n + \phi_3)$$

respectively. Then, since the filter is linear, the output is the sum of the three responses, with the Fourier coefficients.

$$y_n = H_1 X_1 \sin(\omega_1 n + \phi_1) + H_2 X_2 \sin(\omega_2 n + \phi_2) + H_3 X_3 \sin(\omega_3 n + \phi_3)$$

More generally, any finite duration or periodic test digital signal can be broken down by the DFT into the sum of a denumerable number of complex exponentials

$$x_n = \frac{1}{N} \sum_{k=0}^{N-1} X_k e^{i \frac{2\pi k}{N} n}$$

and the response of the system to each complex exponential is the same complex exponential multiplied by a number H_k.

$$H_k e^{i \frac{2\pi k}{N} n}$$

Using these H_k we can predict the response to the test signal.

$$y_n = \frac{1}{N} \sum_{k=0}^{N-1} X_k H_k e^{i\frac{2\pi k}{N} n}$$

The H_k are in general complex (representing the gains and phase shifts) and are precisely the elements of the frequency response. A similar decomposition solves the problem for nonperiodic analog signals, only now we have to test a nondenumerable set of sinusoids.

The above discussion proves that the frequency response provides a complete description of a filter. Given the entire frequency response (i.e., the response of the system to all sinusoids), we can always win the game of predicting the response for an arbitrary input.

The frequency response is obviously a frequency domain quantity; the duality of time and frequency domains leads us to believe that there should be a complete description in the time domain as well. There is, and we previously called it the *impulse response*. To measure it we excite the system with a unit impulse (a Dirac delta function $\delta(t)$ for analog systems or a unit impulse signal $\delta_{n,0}$ for digital systems) and measure the output as a function of time (see equation 6.22). For systems without memory there will only be output for time $t = 0$, but in general the output will be nonzero over an entire time interval. A causal system will have its impulse response zero for times $t < 0$ but nonzero for $t \geq 0$. A system that is time-variant (and hence not a filter) requires measuring the response to all the SUIs, a quantity known as the Green's function.

Like the frequency response, the impulse response may be used to predict the output of a filter when an arbitrary input is applied. The strategy is similar to that we developed above, only this time we break down the test signal in the basis of SUIs (equation (2.26)) rather than using the Fourier expansion. We need only record the system's response to each SUI, expand the input signal in SUIs, and exploit the linearity of the system (as we have already done in Section 6.5). Unfortunately, the SUIs are not generally eigensignals of filters, and so the system's outputs will not be SUIs, and we need to record the entire output. However, unlike the frequency response where we needed to observe the system's output for an infinite number of basis functions, here we can capitalize on the fact that all SUIs are related by time shifts. Exploiting the time-invariance property of filters we realize that after measuring the response of an unknown system to a single SUI (e.g., the unit impulse at time zero), we may immediately deduce its response to all SUIs! Hence we need only apply a single input and record a single response in order to be able to predict the output of a filter when an arbitrary input is applied! The set of signals we must test in order to be able to predict the output of the system to an arbitrary input has been reduced to a single signal! This is the strength of the impulse response.

The impulse response may be nonzero only over a finite interval of time but exactly zero for all times outside this interval. In this case we say the system has a *finite impulse response*, or more commonly we simply call it an FIR filter. The MA systems studied in Sections 6.6 and 6.7 are FIR filters. To see this consider the noncausal three-point averaging system of equation (6.33).

$$y_n = \tfrac{1}{4}x_{n-1} + \tfrac{1}{2}x_n + \tfrac{1}{4}x_{n+1}$$

As time advances so does this window of time, always staying centered on the present. What happens when the input is an impulse? At time $n = \pm 1$ we find a $\tfrac{1}{4}$ multiplying the nonzero signal value at the origin, returning $\tfrac{1}{4}$; of course, the $n = 0$ has maximum output $\tfrac{1}{2}$. At any other time the output will be zero simply because the window does not overlap any nonzero input signal values. The same is the case for any finite combination of input signal values. Thus all the systems that have the form of equation (6.13), which we previously called FIR filters, are indeed FIR.

Let's explicitly calculate the impulse response for the most general causal moving average filter. Starting from equation (6.30) (but momentarily renaming the coefficients) and using the unit impulse as input yields

$$
\begin{aligned}
y_n &= \sum_{l=0}^{L} g_l \delta_{n-L+l,0} \\
&= g_0 \delta_{n-L,0} + g_1 \delta_{n-L+1,0} + g_2 \delta_{n-L+2,0} + \ldots + g_{L-1}\delta_{n-1,0} + g_L \delta_{n,0}
\end{aligned}
$$

which is nonzero only when $n = 0$ or $n = 1$ or ... or $n = L$. Furthermore, when $n = 0$ the output is precisely $h_0 = g_L$, when $n = 1$ the output is precisely $h_1 = g_{L-1}$, etc., until $h_L = g_0$. Thus the impulse response of a general MA filter consists exactly of the coefficients that appear in the moving average sum, but in reverse order!

The impulse response is such an important attribute of a filter that it is conventional to reverse the definition of the moving average, and define the FIR filter via the *convolution* in which the indices run in opposite directions, as we did in equation (6.13).

It is evident that were we to calculate the impulse response of the nonterminating convolution of equation (6.14) it would consist of the coefficients as well; but in this case the impulse response would never quite become zero.

If we apply a unit impulse to a system and its output never dies down to zero, we say that the system is **Infinite Impulse Response** (IIR). Systems of the form (6.15), which we previously called IIR filters, can indeed sustain an impulse response that is nonzero for an infinite amount of time. To see this consider the simple case

$$y_n = x_n + \tfrac{1}{2}y_{n-1}$$

which is of the type of equation (6.15). For negative times n the output is zero, $y_n = 0$, but at time zero $y_0 = 1$, at time one $y_1 = \tfrac{1}{2}$ and thereafter y_n is halved every time. It is obvious that the output at time n is precisely $y_n = 2^{-n}$, which for large n is extremely small, but never zero.

Suppose we have been handed a black box and measure its impulse response. Although there may be many systems with this response to the unit impulse, there will be only one filter that matches, and the coefficients of equation (6.14) are precisely the impulse response in reverse order. This means that if we know that the box contains a filter, then measuring the impulse response is sufficient to uniquely define the system. In particular, we needn't measure the frequency response since it is mathematically derivable from the impulse response.

It is instructive to find this connection between the impulse response (the time domain description) and the frequency response (the frequency domain description) of a filter. The frequency response of the nonterminating convolution system

$$y_n = \sum_{i=-\infty}^{\infty} h_i x_{n-i}$$

is found by substituting a sinusoidal input for x_n, and for mathematical convenience we will use a complex sinusoid $x_n = e^{i\omega n}$. We thus obtain

$$
\begin{aligned}
H(\omega)\,x_n = \quad y_n \;&=\; \sum_{k=-\infty}^{\infty} h_k\, e^{i\omega(n-k)} \\
&=\; \sum_{k=-\infty}^{\infty} h_k\, e^{-i\omega k}\, e^{i\omega n} \\
&=\qquad H_k \qquad\quad x_n
\end{aligned}
\tag{6.53}
$$

where we identified the Fourier transform of the impulse response h_k and the input signal. We have once again shown that when the convolution system has a sinusoidal input its output is the same sinusoid multiplied by a (frequency-dependent) gain. This gain is the frequency response, but here we have found the FT of the impulse response; hence the frequency response and the impulse response are an FT pair. Just as the time and frequency domain representations of signals are connected by the Fourier transform, the simplest representations of filters in the time and frequency domains are related by the FT.

EXERCISES

6.12.1 Find the impulse response for the following systems.
1. $y_n = x_n$
2. $y_n = x_n + x_{n-2} + x_{n-4}$
3. $y_n = x_n + 2x_{n-1} + 3x_{n-2}$
4. $y_n = \sum_i a_i x_{n-i}$
5. $y_n = x_n + y_{n-1}$
6. $y_n = x_n + \frac{1}{2}(y_{n-1} + y_{n-2})$

6.12.2 An ideal low-pass filter (i.e., one that passes without change signals under some frequency but entirely blocks those above it) is unrealizable. Prove this by arguing that the Fourier transform of a step function is nonzero over the entire axis and then invoking the connection between frequency response and impulse response.

6.12.3 When determining the frequency response we needn't apply each sinusoidal input separately; sinusoid orthogonality and filter linearity allow us to apply multiple sinusoids at the same time. This is what is done in probe signals (cf. exercise 2.6.4). Can we apply all possible sinusoids at the same time and reduce the number of input signals to one?

6.12.4 Since white noise contains all frequencies with the same amplitude, applying white noise to the system is somehow equivalent to applying all possible sinusoids. The *white noise response* is the response of a system to white noise. Prove that for linear systems the spectral amplitude of the white noise response is the amplitude of the frequency response. What about the phase delay portion of the frequency response?

6.12.5 The fact that the impulse and frequency responses are an FT pair derives from the general rule that the FT relates convolution and multiplication $FT(x * y) = FT(x)FT(y)$. Prove this general statement and relate it to the Wiener-Khintchine theorem.

6.12.6 Donald S. Perfectionist tries to measure the frequency response of a system by measuring the output power while injecting a slowly sweeping tone of constant amplitude. Unbeknownst to him the system contains a filter that passes most frequencies unattenuated, and amplifies a small band of frequencies. However, following the filter is a fast **A**utomatic **G**ain **C**ontrol (AGC) that causes all Donald's test outputs to have the same amplitude, thus completely masking the filter. What's wrong?

6.13   System Identification—The Hard Case

Returning to our system identification game, assume that your opponent presents you with a black box that is already connected to an input. We will assume first that the system is known to be an FIR filter of known length $L+1$. If the system is FIR of unknown length we need simply assume some extremely large $L+1$, find the coefficients, and discard all the zero coefficients above the true length.

The above assumption implies that the system's output at time n is

$$y_n = a_0 x_n + a_1 x_{n-1} + a_2 x_{n-2} + \cdots + a_L x_{n-L}$$

and your job is to determine these coefficients a_l by simultaneously observing the system's input and output. It is clear that this game is riskier than the previous one. You may be very unlucky and during the entire time we observe it the system's input may be identically zero; or you may be very lucky and the input may be a unit impulse and we readily derive the impulse response.

Let's assume that the input signal was zero for some long time (and the output is consequently zero as well) and then suddenly it is turned on. We'll reset our clock to call the time of the first nonzero input time zero (i.e., x_n is identically zero for $n < 0$, but nonzero at $n = 0$). According to the defining equation the first output must be

$$y_0 = a_0 x_0$$

and since we observe both x_0 and y_0 we can easily find

$$a_0 = \frac{y_0}{x_0}$$

which is well defined since by definition $x_0 \neq 0$. Next, observing the input and output at time $n = 1$, we have

$$y_1 = a_0 x_1 + a_1 x_0$$

which can be solved

$$a_1 = \frac{y_1 - a_0 x_1}{x_0}$$

since everything needed is known, and once again $x_0 \neq 0$.

Continuing in this fashion we can express the coefficient a_n at time n in terms of $x_0 \ldots x_n$, $y_0 \ldots y_n$, and $a_0 \ldots a_{n-1}$, all of which are known. To see this explicitly write the equations

$$
\begin{aligned}
y_0 &= a_0 x_0 \\
y_1 &= a_0 x_1 + a_1 x_0 \\
y_2 &= a_0 x_2 + a_1 x_1 + a_2 x_0 \\
y_3 &= a_0 x_3 + a_1 x_2 + a_2 x_1 + a_3 x_0 \\
y_4 &= a_0 x_4 + a_1 x_3 + a_2 x_2 + a_3 x_1 + a_4 x_0
\end{aligned}
\tag{6.54}
$$

and so on, and note that these can be recursively solved

$$
\begin{aligned}
a_0 &= \frac{y_0}{x_0} \\
a_1 &= \frac{y_1 - a_0 x_1}{x_0} \\
a_2 &= \frac{y_2 - a_0 x_2 - a_1 x_1}{x_0} \\
a_3 &= \frac{y_3 - a_0 x_3 - a_1 x_2 - a_2 x_1}{x_0} \\
a_4 &= \frac{y_4 - a_0 x_4 - a_1 x_3 - a_2 x_2 - a_3 x_1}{x_0}
\end{aligned}
\tag{6.55}
$$

one coefficient at a time.

In order to simplify the arithmetic it is worthwhile to use linear algebra notation. We can write equation (6.54) in matrix form, with the desired coefficients on the right-hand side

$$
\begin{pmatrix} y_0 \\ y_1 \\ y_2 \\ \vdots \end{pmatrix} = \begin{pmatrix} x_0 & 0 & 0 & 0 & \cdots \\ x_1 & x_0 & 0 & 0 & \cdots \\ x_2 & x_1 & x_0 & 0 & \cdots \\ \vdots & \vdots & \vdots & \vdots & \vdots \end{pmatrix} \begin{pmatrix} a_0 \\ a_1 \\ a_2 \\ \vdots \end{pmatrix}
\tag{6.56}
$$

and identify the matrix containing the input values as being lower triangular and Toeplitz. The solution of (6.55) is simple due to the matrix being lower triangular. Finding the lth coefficient requires l multiplications and subtractions and one division, so that finding all $L + 1$ coefficients involves $\frac{1}{2}L(L+1)$ multiplications and subtractions and $L + 1$ divisions.

The above solution to the 'hard' system identification problem was based on the assumption that the input signal was exactly zero for $n < 0$. What can we do in the common case when we start observing the signals at an arbitrary time before which the input was not zero? For notational simplicity let's assume that the system is known to be FIR with $L = 2$. Since we need to find three coefficients we will need three equations, so we observe three outputs, y_n, y_{n+1} and y_{n+2}. Now these outputs depend on five inputs, $x_{n-2}, x_{n-1}, x_n, x_{n+1}$, and x_{n+2} in the following way

$$
\begin{aligned}
y_n &= a_0 x_n + a_1 x_{n-1} + a_2 x_{n-2} \\
y_{n+1} &= a_0 x_{n+1} + a_1 x_n + a_2 x_{n-1} \\
y_{n+2} &= a_0 x_{n+2} + a_1 x_{n+1} + a_2 x_n
\end{aligned}
\tag{6.57}
$$

which in matrix notation can be written

$$
\begin{pmatrix} y_n \\ y_{n+1} \\ y_{n+2} \end{pmatrix} =
\begin{pmatrix} x_n & x_{n-1} & x_{n-2} \\ x_{n+1} & x_n & x_{n-1} \\ x_{n+2} & x_{n+1} & x_n \end{pmatrix}
\begin{pmatrix} a_0 \\ a_1 \\ a_2 \end{pmatrix}
\tag{6.58}
$$

or in other words $\underline{y} = \underline{\underline{X}}\,\underline{a}$, where $\underline{\underline{X}}$ is a nonsymmetric Toeplitz matrix. The solution is obviously $\underline{a} = \underline{\underline{X}}^{-1}\underline{y}$ but the three-by-three matrix is not lower triangular, and so its inversion is no longer trivial. For larger number of coefficients L we have to invert an $N = L+1$ square matrix; although most direct N-by-N matrix inversion algorithms have computational complexity $O(N^3)$, it is possible to invert a general matrix in $O(N^{\log_2 7}) \sim O(N^{2.807})$ time. Exploiting the special characteristics of Toeplitz matrices reduces the computational load to $O(N^2)$.

What about AR filters?

$$
y_n = x_n + \sum_{m=1}^{M} b_m y_{n-m}
$$

Can we similarly find their coefficients in the hard system identification case? Once again, for notational simplicity we'll take $M = 3$. We have three unknown b coefficients, so we write down three equations,

$$
\begin{aligned}
y_n &= x_n + b_1 y_{n-1} + b_2 y_{n-2} + b_3 y_{n-3} \\
y_{n+1} &= x_{n+1} + b_1 y_n + b_2 y_{n-1} + b_3 y_{n-2} \\
y_{n+2} &= x_{n+2} + b_1 y_{n+1} + b_2 y_n + b_3 y_{n-1}
\end{aligned}
\tag{6.59}
$$

or in matrix notation

$$
\begin{pmatrix} y_n \\ y_{n+1} \\ y_{n+2} \end{pmatrix} =
\begin{pmatrix} x_n \\ x_{n+1} \\ x_{n+2} \end{pmatrix} +
\begin{pmatrix} y_{n-1} & y_{n-2} & y_{n-3} \\ y_n & y_{n-1} & y_{n-2} \\ y_{n+1} & y_n & y_{n-1} \end{pmatrix}
\begin{pmatrix} b_1 \\ b_2 \\ b_3 \end{pmatrix}
\tag{6.60}
$$

or simply $\underline{y} = \underline{x} + \underline{\underline{Y}}\,\underline{b}$. The answer this time is $\underline{b} = \underline{\underline{Y}}^{-1}(\underline{y} - \underline{x})$, which once again necessitates inverting a nonsymmetric Toeplitz matrix.

Finally, the full ARMA with $L = 2$ and $M = 3$

$$y_n = \sum_{l=0}^{L} a_l x_{n-l} + \sum_{m=1}^{M} b_m y_{n-m}$$

has six unknowns, and so we need to take six equations.

$$
\begin{aligned}
y_n &= a_0 x_n + a_1 x_{n-1} + a_2 x_{n-2} + b_1 y_{n-1} + b_2 y_{n-2} + b_3 y_{n-3} \\
y_{n+1} &= a_0 x_{n+1} + a_1 x_n + a_2 x_{n-1} + b_1 y_n + b_2 y_{n-1} + b_3 y_{n-2} \\
y_{n+2} &= a_0 x_{n+2} + a_1 x_{n+1} + a_2 x_n + b_1 y_{n+1} + b_2 y_n + b_3 y_{n-1} \\
y_{n+3} &= a_0 x_{n+3} + a_1 x_{n+2} + a_2 x_{n+1} + b_1 y_{n+2} + b_2 y_{n+1} + b_3 y_n \\
y_{n+4} &= a_0 x_{n+4} + a_1 x_{n+3} + a_2 x_{n+2} + b_1 y_{n+3} + b_2 y_{n+2} + b_3 y_{n+1} \\
y_{n+5} &= a_0 x_{n+5} + a_1 x_{n+4} + a_2 x_{n+3} + b_1 y_{n+4} + b_2 y_{n+3} + b_3 y_{n+2}
\end{aligned}
$$

This can be written compactly

$$
\begin{pmatrix} y_n \\ y_{n+1} \\ y_{n+2} \\ y_{n+3} \\ y_{n+4} \\ y_{n+5} \end{pmatrix}
=
\begin{pmatrix}
x_n & x_{n-1} & x_{n-2} & y_{n-1} & y_{n-2} & y_{n-3} \\
x_{n+1} & x_n & x_{n-1} & y_n & y_{n-1} & y_{n-2} \\
x_{n+2} & x_{n+1} & x_n & y_{n+1} & y_n & y_{n-1} \\
x_{n+3} & x_{n+2} & x_{n+1} & y_{n+2} & y_{n+1} & y_n \\
x_{n+4} & x_{n+3} & x_{n+2} & y_{n+3} & y_{n+2} & y_{n+1} \\
x_{n+5} & x_{n+4} & x_{n+3} & y_{n+4} & y_{n+3} & y_{n+2}
\end{pmatrix}
\begin{pmatrix} a_0 \\ a_1 \\ a_2 \\ b_1 \\ b_2 \\ b_3 \end{pmatrix}
\quad (6.61)
$$

and the solution requires inverting a six-by-six nonsymmetric non-Toeplitz matrix. The ARMA case is thus more computationally demanding than the pure MA or AR cases.

Up to now we have assumed that we observe x_n and y_n with no noise whatsoever. In all practical cases there will be at least some quantization noise, and most of the time there will be many other sources of additive noise. Due to this noise we will not get precisely the same answers when solving equations (6.58), (6.60), or (6.61) for two different times. One rather obvious tactic is to solve the equations many times and average the resulting coefficients. However, the matrix inversion would have to be performed a very large number of times and the equations (especially (6.60) and (6.61)) often turn out to be rather sensitive to noise. A much more successful tactic is to average *before* solving the equations, which has the advantages of providing more stable equations and requiring only a single matrix inversion.

Let's demonstrate how this is carried out for the MA case.

$$y_n = \sum_{k=0}^{L} a_k x_{n-k} \tag{6.62}$$

In order to average we multiply both sides by x_{n-q} and sum over as many n as we can get our hands on.

$$\sum_n y_n x_{n-q} = \sum_{k=0}^{L} a_k \sum_n x_{n-k} x_{n-q}$$

We define the x autocorrelation and the x-y crosscorrelation (see Chapter 9)

$$C_x(k) = \sum_n x_n x_{n-k} \qquad\qquad C_{yx}(k) = \sum_n y_n x_{n-k}$$

and note the following obvious symmetry.

$$C_x(-k) = C_x(k)$$

The deconvolution equations can now be written simply as

$$C_{yx}(q) = \sum_k a_k C_x(q-k) \tag{6.63}$$

and are called the Wiener-Hopf equations. For $L = 2$ the Wiener-Hopf equations look like this:

$$\begin{pmatrix} C_{yx}(0) \\ C_{yx}(1) \\ C_{yx}(2) \end{pmatrix} = \begin{pmatrix} C_x(0) & C_x(-1) & C_x(-2) \\ C_x(1) & C_x(0) & C_x(-1) \\ C_x(2) & C_x(1) & C_x(0) \end{pmatrix} \begin{pmatrix} a_0 \\ a_1 \\ a_2 \end{pmatrix}$$

and from the aforementioned symmetry we immediately recognize the matrix as *symmetric* Toeplitz, a fact that makes them more stable and even faster to solve.

For a black box containing an AR filter, there is a special case where the input signal dies out (or perhaps the input happens to be an impulse). Once the input is zero

$$y_n = \sum_{m=1}^{M} b_m y_{n-m}$$

multiplying by y_{n-q} and summing over n we find

$$\sum_n y_n y_{n-q} = \sum_{m=1}^{M} b_m \sum_n y_{n-m} y_{n-q}$$

in which we identify y autocorrelations.

$$\sum_{m=1}^{M} C_y(|m-q|)b_m = C_y(q) \tag{6.64}$$

For $M = 3$ these equations look like this.

$$\begin{pmatrix} C_y(0) & C_y(1) & C_y(2) \\ C_y(1) & C_y(0) & C_y(1) \\ C_y(2) & C_y(1) & C_y(0) \end{pmatrix} \begin{pmatrix} b_1 \\ b_2 \\ b_3 \end{pmatrix} = \begin{pmatrix} C_y(1) \\ C_y(2) \\ C_y(3) \end{pmatrix}$$

These are the celebrated Yule-Walker equations, which will turn up again in Sections 9.8 and 9.9.

EXERCISES

6.13.1 Write a program that numerically solves equation (6.55) for the coefficients of a causal MA filter given arbitrary inputs and outputs. Pick such a filter and generate outputs for a pseudorandom input. Run your program for several different input sequences and compare the predicted coefficients with the true ones (e.g., calculate the squared difference). What happens if you try predicting with too long a filter? Too short a filter? If the input is a sinusoid instead of pseudorandom?

6.13.2 Repeat the previous exercise for AR filters (i.e., solve equation (6.60)). If the filter seems to be seriously wrong, try exciting it with a new pseudorandom input and comparing its output with the output of the intended system.

6.13.3 In the text we assumed that we knew the order L and M. How can we find the order of the system being identified?

6.13.4 Assume that y_n is related to x_n by a noncausal MA filter with coefficients $a_{-M} \ldots a_M$. Derive equations for the coefficients in terms of the appropriate number of inputs and outputs.

6.13.5 In deriving the Wiener-Hopf equations we could have multiplied by y_{n-q} to get the equations

$$C_y(q) = \sum_k h_k C_{xy}(q-k)$$

rather than multiplying by x_{n-q}. Why didn't we?

6.13.6 In the derivation of the Wiener-Hopf equations we assumed that C_x and C_{yx} depend on k but not n. What assumption were we making about the noisy signals?

6.13.7 In an even harder system identification problem not only don't you have control over the system's input, you can't even observe it. Can the system be identified based on observation of the output only?

6.13.8 Assume that the observed input sequence x_n is white, i.e., that all autocorrelations are zero except for $C_x(0)$. What is the relationship between the crosscorrelation $C_{yx}(n)$ and the system's impulse response? How does this simplify the hard system identification task?

6.14 System Identification in the z Domain

In the previous section we solved the hard system identification problem in the time domain. The solution involved solving sets of linear equations, although for many cases of interest these equations turn out to be relatively simple. Is there a method of solving the hard system identification problem without the need for solving equations? For the easy problem we could inject an impulse as input, and simply measure the impulse response. For the hard problem we extended this technique by considering the input to be the sum of SUIs $x_n = \sum x_m \delta n - m$ and exploiting linearity. Realizing that each output value consists of intertwined contributions from many SUI inputs, we are forced to solve linear equations to isolate these individual contributions. There is no other way to disentangle the various contributions since although the SUI basis functions from which the input can be considered to be composed are orthogonal and thus easily separable by projection without solving equations, the time-shifted impulse responses are not.

This gives us an idea; we know that sinusoids *are* eigenfunctions of filters, and that they are mutually orthogonal. Hence the input at ω can be derived from the output at that same frequency, with no other frequencies interfering. We can thus recover the frequency response by converting the input and output signals into the frequency domain and merely dividing the output at every frequency by the input at that same frequency.

$$Y(\omega) = H(\omega)X(\omega) \qquad \longrightarrow \qquad H(\omega) = \frac{Y(\omega)}{X(\omega)}$$

What could be easier? If we wish we can even recover the impulse response from the frequency response, by using equation (6.53). So it would seem best to solve the easy system identification problem in the time domain and the hard problem in the frequency domain.

One must be careful when using this frequency domain approach to the hard system identification problem. To see why, think of an MA all-pass system that simply delays the input by L samples $y_n = x_{n-L}$. Had we observed the system when the input was of the form $x_n = \sin(2\pi \frac{kn}{L} + \phi)$ for any k we would conclude that we were observing the identity system $y_n = x_n$! This mistake is obviously due to the input being periodic and the system looking back in time by precisely an integer number of periods; equivalently in the frequency domain this input consists of a single line, and thus we can only learn about $H(\omega)$ at this single frequency. The lesson to be learned is more general than this simple example. In order to uniquely specify a system the input must excite it at all frequencies. There are often frequencies for which the system produces no output at all, and based on these we certainly would not be able to identify the system. The unit impulse is a single excitation that squeezes all possible information out of the system; due to orthogonality and the eigenfunction property a single sinusoid contributes only an infinitesimal amount of information about the system.

The frequency response is a great tool for FIR systems, but not as good for IIR systems since they may become unstable. When an IIR system's output increases without limit for a frequency ω, this is a sign that its frequency response is infinite there. For example, a typical frequency response is depicted in Figure 6.12. As usual, the horizontal axis is from DC to half the sampling rate. We see that the frequency response goes to zero for a digital frequency of 0.2. This means that when the input is a sinusoid of

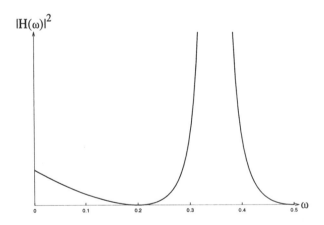

Figure 6.12: The frequency response of an ARMA filter with a zero at frequency $0.2f_s$ and a pole at $0.35f_s$.

this frequency there will be no output. We call such frequencies *zeros* of the frequency response. Around digital frequency 0.35 something different occurs. Input signals in this region are strongly amplified, and at 0.35 itself the output grows without limit. We recall from Section (4.11) that the Fourier transform is *not* the proper tool to describe this type of behavior; the z transform *is*.

So let's see how the zT can be used in system identification. We know that the effect of any filter can be expressed as convolution by the impulse response $y = h * x$, although for non-FIR systems this convolution is in principle infinite. In view of the connection between convolution and multiplication (see Section 6.8) we would like somehow to write $h = y/x$, a process known as *deconvolution*. The zT is the tool that transforms convolutions into algebraic multiplications, so we apply it now.

Similarly to the result for FT and DFT, convolution in the time domain becomes multiplication in the z domain

$$Y(z) = H(z)\, X(z) \tag{6.65}$$

and this is simple to prove.

$$
\begin{aligned}
Y(z) &= \sum_{n=-\infty}^{\infty} y_n z^{-n} \\
&= \sum_{n=-\infty}^{\infty} \left\{ \sum_{k=-\infty}^{\infty} h_k x_{n-k} \right\} z^{-n} \\
&= \sum_{k=-\infty}^{\infty} h_k \sum_{n=-\infty}^{\infty} x_{n-k} z^{-n} \\
&= \sum_{k=-\infty}^{\infty} h_k z^{-k} \sum_{m=-\infty}^{\infty} x_m z^{-m} \\
&= \quad H(z) \qquad\quad X(z)
\end{aligned}
$$

The z transform of the impulse response h is called the *transfer function*, and this name implies that we can think of the operation of the system as transferring $X(z)$ into $Y(z)$ by a simple multiplication. Of course, thinking of $z = re^{i\omega}$, the transfer function is seen to be a generalization of the frequency response. Evaluated on the unit circle $r = 1$ (i.e., $z = e^{i\omega}$) the transfer function is precisely the frequency response, while for other radii we obtain the response of the system to decaying $(r < 1)$ or growing $(r > 1)$ sinusoids as in equation (2.12).

So the complete solution of the hard system identification problem is easy. The transfer function of the system in question is

$$H(z) = \frac{Y(z)}{X(z)}$$

with frequency response obtainable by evaluating $H(z)$ on the unit circle $z = e^{i\omega}$, and impulse response derivable from the frequency response.

Let's use this method to identify a nontrivial system. Assume that the unknown system is equation (6.39) with $\beta = \frac{1}{2}$

$$y_n = \frac{1}{2}(y_{n-1} + x_n) \tag{6.66}$$

and that the input is observed to turn on at $n = 0$ and is DC from then on $x_n = u_n$. Observing the input and output you compose the following table:

n	0	1	2	$\cdots$	n	$\cdots$
x_n	1	1	1	$\cdots$	1	$\cdots$
y_n	$\frac{1}{2}$	$\frac{3}{4}$	$\frac{7}{8}$	$\cdots$	$\frac{2^{n+1}-1}{2^{n+1}}$	$\cdots$

Using the time domain method of equation (6.55) after some work you could deduce that the coefficients of the nonterminating convolution are $h_n = \frac{1}{2}^{-(n+1)}$, but some inspiration would still be required before the ARMA form could be discovered. So let's try the zT line of attack. The zTs of the input and output are easily found by using equation (4.63).

$$X(z) = \quad zT\, u_n \quad = \frac{1}{1 - z^{-1}} \quad \text{ROC } |z| > 1$$

$$Y(z) = \quad zT\, \frac{2^{n+1}-1}{2^{n+1}} u_n \quad = \frac{1}{1 - z^{-1}} - \frac{1}{2}\frac{1}{1 - \frac{1}{2}z^{-1}}$$

$$= \frac{\frac{1}{2}}{(1 - z^{-1})(1 - \frac{1}{2}z^{-1})} \quad \text{ROC } |z| > \frac{1}{2}$$

Now the transfer function is the ratio

$$H(z) = \frac{Y(z)}{X(z)} = \frac{\frac{1}{2}}{1 - \frac{1}{2}z^{-1}}$$

so that the difference equation $Y(z) = H(z)X(z)$ is

$$\left(1 - \frac{1}{2}z^{-1}\right) Y(z) = \frac{1}{2}X(z)$$

which from the meaning of z^{-1} in the time domain is simply

$$y_n - \tfrac{1}{2}y_{n-1} = \tfrac{1}{2}x_n$$

and equation (6.66) has magically appeared!

EXERCISES

6.14.1 As we shall learn in Chapter 18, modem signals are distorted by the telephone lines through which they travel. This distortion can be modeled as a filter, and removed by appropriate inverse filtering (equalizing). Can you explain why modems transmit known pseudonoise signals during their start-up procedures?

6.14.2 You observe a system when its input consists of the sum of two different sinusoids. Find two systems that cannot be distinguished based on this input. Do the same for an input composed of M sinusoids.

6.14.3 What is the transfer function of two systems connected in series (cascaded so that $y = H_2 w, w = H_1 x$)? Of two systems connected in parallel (i.e., so that $y = H_1 x + H_2 y$)?

6.14.4 Prove that ARMA systems commute.

6.14.5 Deconvolution is equivalent to finding the inverse system for a filter. Explain how to carry out deconvolution using the transfer function.

6.14.6 Prove (as in exercise 6.4.1) that the inverse system of an MA filter is AR and vice versa.

6.14.7 Many communication channels both distort the information carrying signal by an unknown filter and add nonwhite noise to it. The frequency characteristics of both the channel filter and the noise can be directly measured by inputting a signal consisting of a comb of equidistant sinusoids each with known amplitude and phase, and measuring the output at these frequencies. The input signal is conveniently generated and the output recovered using the DFT. In order to combat noise the procedure should be repeated N times and the output averaged. We denote the input in bin k by X_k and the measured output at repetition m by $Y_k^{[m]}$. Explain how to measure the SNR of bin k.

6.14.8 Continuing the previous exercise, a **Frequency EQualizer** (FEQ) tries to remove the frequency distortion introduced by the channel filter by directly multiplying each output Y_k by complex number e_k in order to recover the input $X_k = e_k Y_k$. Explain how to find the FEQ coefficients e_k.

6.14.9 Continuing the previous exercises, the frequency magnitude response $|H_k|^2$ (the ratio of the output to input energies) as measured at repetition m is $|H_k^{[m]}|^2 = \frac{(Y_k^{[m]})^2}{X_k^2}$. Express $|H_k|^2$ in terms of e_k and SNR_k.

Bibliographical Notes

Signal processing systems are treated in all the standard signal processing texts, [186, 185, 187, 200, 189, 252, 159, 167], as well as books specifically on system design [126].

The word convolution was used as early as 1935 by mathematicians, but seems to have been picked up by the signal processing community rather later. Norbert Wiener, in his classic 1933 text [277], uses the German *Faltung* noting the lack of an appropriate English-language word. In his later book of the 1940s [278] there is a conspicuous absence of the word. Rice, in an influential 1944–45 article on noise [220] gives the FT of a product in an appendix, calling the convolution simply 'the integral on the right'. In 1958 Blackman and Tukey [19] use the word convolution freely, although they mention several other possible names as well.

The impulse response, known in other fields as the Green's function, was first published by Green in 1828 [86].

Amazingly, the Wiener-Hopf equations were originally derived in the early 1930s to solve a problem involving radiation equilibrium in stars [281]. While working on defense-related problems during World War II, Wiener discovered that these same equations were useful for prediction and filtering. Several years before Eberhard Hopf had returned to Nazi Germany in order to accept a professorship at Leipzig that had been vacated by a cousin of Wiener's who had fled Germany after the rise of Hitler ([280]). Despite this turn of events Wiener always referred to the 'Hopf-Wiener' equations.

The great Cambridge statistician George Udny Yule formulated the Yule-Walker equations for signals containing one or two sinusoidal components in the late 1920s, in an attempt to explain the periodicity of sunspot numbers [289]. A few years later Sir Gilbert Walker expanded on this work [267], discovering that the autocorrelations were much smoother than the noisy signal itself, and applying this technique to a meteorological problem.

Otto Toeplitz was one of the founders of operator theory, as well as a great teacher and historian of math [21, 14]. In operator theory he was one of Hilbert's principle students, emphasizing matrix methods and considering Banach's methods too abstract. In teaching he was a disciple of Felix Klein (who considered group theory too abstract). In Bonn he would lecture to packed audiences of over 200 students, and was said to recognize each student's handwriting and writing style. He indirectly influenced the development of Quantum Mechanics by teaching his friend Max Born matrix methods; Born later recognized that these were the mathematical basis of Heisenberg's theory. Toeplitz was dismissed from his university position on racial grounds after the Nürnberg laws of 1935, but stayed on in Germany until 1939 representing the Jewish community and helping minority students emigrate. He finally left Germany in 1939, traveling to Jerusalem where he assumed the post of scientific advisor to the Hebrew University, a position he held less than a year until his death in 1940.

Filters

In everyday parlance a 'filter' is a device that removes some component from whatever is passed through it. A drinking-water filter removes salts and bacteria; a coffee filter removes coffee grinds; an air filter removes pollutants and dust. In electronics the word 'filter' evokes thoughts of a system that removes components of the input signal based on frequency. A notch filter may be employed to remove a narrow-band tone from a received transmission; a noise filter may remove high-frequency hiss or low-frequency hum from recordings; antialiasing filters are needed to remove frequencies above Nyquist before A/D conversion.

Less prevalent in everyday usage is the concept of a filter that emphasizes components rather than removing them. Colored light is created by placing a filter over a white light source; one filters flour retaining the finely ground meal; entrance exams filter to find the best applicants. The electronic equivalent is more common. Radar filters capture the desired echo signals; deblurring filters are used to bring out unrecognizable details in images; narrow-band audio filters lift Morse code signals above the interference.

In signal processing usage a filter is any system whose output spectrum is derived from the input's spectrum via multiplication by a time-invariant weighting function. This function may be zero in some range of frequencies and as a result remove these frequencies; or it may be large in certain spectral regions, consequently emphasizing these components. Or it may half the energy of some components while doubling others, or perform any other arbitrary characteristic.

However, just as a chemical filter cannot create gold from lead, a signal processing filter cannot create frequency components that did not exist in the input signal. Although definitely a limitation, this should not lead one to conclude that filters are uninteresting and their output trivial manipulation of the input. To do so would be tantamount to concluding that sculptors are not creative because the sculpture preexisted in the stone and they only removed extraneous material.

In this chapter we will learn how filters are specified in both frequency and time domains. We will learn about fundamental limitations that make the job of designing a filter to meet specifications difficult, but will not cover the theory and implementation of filter design in great detail. Whole books are devoted to this subject and excellent software is readily available that automates the filter design task. We will only attempt to provide insight into the basic principles of the theory so that the reader may easily use any of the available programs.

7.1 Filter Specification

Given an input signal, different filters will produce different output signals. Although there are an infinite number of different filters, not every output signal can be produced from a given input signal by a filter. The restrictions arise from the definition of a filter as a linear time-invariant operator. Filters never produce frequency components that did not exist in the input signal, they merely attenuate or accentuate the frequency components that exist in the input signal.

Low-pass filters are filters that pass DC and low frequencies, but block or strongly attenuate high frequencies. *High-pass* filters pass high frequencies but block or strongly attenuate low frequencies and DC. *Band-pass* filters block both low and high frequencies, passing only frequencies in some 'pass-band' range. *Band-stop* filters do the opposite, passing everything not in a defined 'stop-band'. *Notch* filters are extreme examples of band-stop filters, they pass all frequencies with the exception of one well defined frequency (and its immediate vicinity). *All-pass* filters have the same gain magnitude for all frequencies but need not be the identity system since phases may still be altered.

The above definitions as stated are valid for analog filters. In order to adapt them for DSP we need to specify that only frequencies between zero and half the sampling rate are to be considered. Thus a digital system that blocks low frequencies and passes frequencies from quarter to half the sampling frequency is a high-pass filter.

An ideal filter is one for which every frequency is either in its pass-band or stop-band, and has unity gain in its pass-band and zero gain in its stop-band. Unfortunately, ideal filters are unrealizable; we can't buy one or even write a DSP routine that implements one. The problem is caused by the sharp jump discontinuities at transitions in the frequency domain that cannot be precisely implemented without peeking infinitely into the

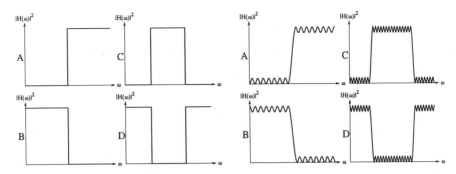

Figure 7.1: Frequency response of ideal and nonideal filters. In (A) we see the low-pass filters, in (B) the high-pass filters, in (C) the band-pass filters and in (D) the band-stop (notch) filters.

future. On the left side of Figure 7.1 we show the frequency response of ideal filters, while the right side depicts more realistic approximations to the ideal response. Realistic filters will always have a finite *transition region* between pass-bands and stop-bands, and often exhibit *ripple* in some or all of these areas. When designing a filter for a particular application one has to specify what amount of ripple and how much transition width can be tolerated. There are many techniques for building both analog and digital filters to specification, but all depend on the same basic principles.

Not all filters are low-pass, high-pass, band-pass, or band-stop, any frequency dependent gain is admissible. The gain of a pre-emphasis filter increases monotonically with frequency, while that of a de-emphasis filter decreases monotonically. Such filters are often needed to compensate for or eliminate the effects of various other signal processing systems.

Filtering in the analog world depends on the existence of components whose impedance is dependent on frequency, usually capacitors and inductors. A capacitor looks like an open circuit to DC but its impedance decreases with increasing frequency. Thus a series-connected capacitor effectively blocks DC current but passes high frequencies, and is thus a low-pass filter. A parallel-connected capacitor short circuits high frequencies but not DC or low frequencies and is thus a high-pass filter. The converse can be said about series- and parallel-connected inductors.

Filtering in DSP depends on mathematical operations that remove or emphasize different frequencies. Averaging adjacent signal values passes DC and low frequencies while canceling out high frequencies. Thus averaging behaves as a low-pass filter. Adding differences of adjacent values cancels out DC and low frequencies but will pass signals with rapidly changing signs. Thus such operations are essentially high-pass filters.

One obvious way to filter a digital signal is to 'window' it in the frequency domain. This requires transforming the input signal to the frequency domain, multiplying it there by the desired frequency response (called a 'window function'), and then transforming back to the time domain. In practice the transformations can be carried out by the FFT algorithm in $O(N \log N)$ time (N being the number of signal points), while the multiplication only requires $O(N)$ operations; hence this method is $O(N \log N)$ in complexity. This method as stated is only suitable when the entire signal is available in a single, sufficiently short vector. When there are too many points for a single DFT computation, or when we need to begin processing the signal before it has completely arrived, we may perform this process on successive blocks of the signal. How the individually filtered blocks are recombined into a single signal will be discussed in Section 15.2.

The frequency domain windowing method is indeed a straightforward and efficient method of digital filtering, but not a panacea. The most significant drawback is that it is not well suited to real-time processing, where we are given a single input sample, and are expected to return an output sample. Not that it is impossible to use frequency domain windowing for real-time filtering. It may be possible to keep up with real-time constraints, but a processing delay *must* be introduced. This delay consists of the time it takes to fill the buffer (the buffer delay) plus the time it takes to perform the FFT, multiplication, and iFFT (the computation delay). When this delay cannot be tolerated there is no alternative to time domain filtering.

EXERCISES

7.1.1 Classify the following filters as low-pass, high-pass, band-pass, or notch.
   1. Human visual system, which has a persistence of $\frac{1}{20}$ of a second
   2. Human hearing, which cannot hear under 30 Hz or above 25KHz
   3. Line noise filter used to remove 50 or 60 Hz AC hum
   4. Soda bottle amplifying a specific frequency when air is blown above it
   5. Telephone line, which rejects below 200 Hz and above 3800 Hz

7.1.2 Design an MA filter, with an even number of coefficients N, that passes a DC signal $(a, a, a, \ldots)$ unchanged but completely kills a maximal frequency signal $(a, -a, a, -a, \ldots)$. For example, for $N = 2$ you must find two numbers g_1 and g_2 such that $g_1 a + g_2 a = a$ but $g_a a + g_2(-a) = 0$. Write equations that the g_i must obey for arbitrary N. Can you find a solution for odd N?

7.1.3 Design a moving average digital filter, with an even number of coefficients N, that passes a maximal frequency signal unchanged but completely kills DC. What equations must the g_i obey now? What about odd N?

7.1.4 The squared frequency response of the ideal low-pass filter is unity below the cutoff frequency and zero above.

$$|H(\omega)|^2 = \begin{cases} 1 & \omega < \omega_c \\ 0 & \text{else} \end{cases}$$

What is the full frequency response assuming a delay of N samples?

7.1.5 Show that the ideal low-pass filter is not realizable. To do this start with the frequency response of the previous exercise and find the impulse response using the result from Section 6.12 that the impulse response is the FT of the frequency response. Show that the impulse response exists for negative times (i.e., before the impulse is applied), and that no amount of delay will make the system causal.

7.1.6 Show that results similar to that of the previous exercise hold for other ideal filter types. (Hint: Find a connection between the impulse response of ideal band-pass or band-stop filters and that of ideal low-pass filters.)

7.1.7 The Paley-Wiener theorem states that if the impulse response h_n of a filter has a finite square sum then the filter is causal if and only if $\int |\ln|H(\omega)||\, d\omega$ is finite. Use this theorem to prove that ideal low-pass filters are not realizable.

7.1.8 Prove the converse to the above, namely that any signal that is nonzero over some time can't be band-limited.

7.2   Phase and Group Delay

The previous section concentrated on the specification of the magnitude of the frequency response, completely neglecting its angle. For many applications power spectrum specification is sufficient, but sometimes the spectral phase can be important, or even critical. A signal's phase can be used for carrying information, and passing such a phase-modulated signal through a filter that distorts phase may cause this information to be lost. There are even many uses for all-pass filters, filters that have unity gain for all frequencies but varying spectral phase!

Let's return to fundamentals. The frequency response $H(\omega)$ is defined by the relation

$$Y(\omega) = H(\omega)X(\omega)$$

which means that

$$|Y(\omega)| = |H(\omega)||X(\omega)| \qquad \text{and} \qquad \angle Y(\omega) = \angle X(\omega) + \angle H(\omega)$$

or in words, the input spectral magnitude at each frequency is multiplied by the frequency response gain there, while the spectral phase is delayed by the angle of the frequency response at each frequency. If the spectral phase is unchanged by the filter, we say that the filter introduces no phase distortion; but this is a needlessly harsh requirement.

For example, consider the simple delay $y_n = x_{n-m}$. This FIR filter is all-pass (i.e., the absolute value of its frequency response is a constant unity), but delaying sinusoids effectively changes their phases. By how much is the phase delayed? The sinusoid $x_n A \sin(\omega n)$ becomes

$$ y_n = x_{n-m} = A \sin \left(\omega(n - m) \right) = A \sin(\omega n - \omega m) $$

so the phase delay is ωm, which is frequency-dependent. When the signal being delayed is composed of many sinusoids, each has a phase delay proportional to its frequency, so the simple delay causes a spectral phase shift proportional to frequency, a characteristic known as *linear phase*.

Some time delay is often unavoidable; the noncausal FIR filter $y = h * x$ with coefficients

$$ h_{-L}, h_{-L+1}, \ldots h_{-1}, h_0, h_1, \ldots h_{L-1}, h_L $$

introduces no time delay since the output y_n corresponds to the present input x_n. If we require this same filter to be causal, we cannot output y_n until the input x_L is observed, and so a time delay of L, half the filter length, is introduced.

$$ g_0 = h_{-L}, \quad g_1 = h_{-L+1}, \quad g_L = h_0, \quad \ldots \quad g_{2L} = h_L $$

This type of delay is called *buffer delay* since it results from buffering the inputs.

It is not difficult to show that if the impulse response is symmetric (or antisymmetric) then the linear phase shift resulting from buffer delay is the only phase distortion. Applying the symmetric noncausal FIR filter with an odd number of coefficients

$$ h_L, h_{L-1}, \ldots h_1, h_0, h_1, \ldots h_{L-1}, h_L $$

to a complex exponential $e^{i\omega n}$ we get

$$ y_n = \sum_{m=-L}^{+L} h_{|m|} e^{i\omega(n-m)} = h_0 e^{i\omega n} + 2e^{i\omega n} \sum_{m=1}^{L} h_{|m|} \cos(m\omega) $$

so that the frequency response is real and thus has zero phase delay.

$$H(\omega) = h_0 + 2 \sum_{m=1}^{L} h_{|m|} \cos(m\omega)$$

We can force this filter to be causal by shifting it by L

$$g_0 = h_L, \quad g_1 = h_{L-1}, \quad \cdots \quad g_L = h_0, \quad \cdots \quad g_{2L} = h_L$$

and the symmetry is now somewhat hidden.

$$g_0 = g_{2L}, \quad g_1 = g_{2L-1}, \quad \cdots \quad g_m = g_{2L-m}$$

Once again applying the filter to a complex exponential leads to

$$y_n = \sum_{m=0}^{2L} g_m e^{i\omega(n-m)} = g_L e^{i\omega(n-L)} + 2e^{i\omega n} e^{-i\omega L} \sum_{m=0}^{L-1} g_m \cos(m\omega)$$

so that the frequency response is

$$H(\omega) = \left(g_L + 2 \sum_{m=0}^{L-1} g_m \cos(m\omega) \right) e^{-i\omega L} = |H(\omega)| e^{-i\omega L}$$

(the important step is isolating the imaginary portion) and the filter is seen to be linear-phase, with phase shift corresponding to a time delay of L.

The converse is true as well, namely all linear-phase filters have impulse responses that are either symmetric or antisymmetric. We can immediately conclude that causal IIR filters cannot be linear-phase, since if the impulse response continues to the end of time, and must be symmetric, then it must have started at the beginning of time. This rules out the filter being causal.

From now on we will not consider a linear phase delay (constant time delay) to be phase 'distortion'. True phase distortion corresponds to nonlinearities in the phase as a function of frequency. To test for deviation from linearity it is useful to look at the first derivative, since linear phase response will have a constant derivative, and deviations from linearity will show up as deviations from a constant value. It is customary to define the *group delay*

$$\tau(\omega) = -\frac{d}{d\omega} \angle H(\omega) \tag{7.1}$$

where the phase must be unwrapped (i.e., the artificial discontinuities of 2π removed) before differentiation. What is the difference between phase delay and group delay?

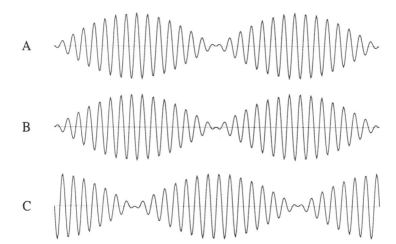

Figure 7.2: The difference between phase delay and group delay. In (A) we see the input signal, consisting of the sum of two sinusoids of nearly the same frequency. (B) depicts the output of a filter with unity gain, phase delay of π, and zero group delay, while the graph in (C) is the output of a filter with unity gain, no phase delay, but nonzero group delay. Note that the local phase in (C) is the same as that of the input, but the position of the beat amplitude peak has shifted.

In Figure 7.2 we see the effect of passing a signal consisting of the sum of two sinusoids of nearly the same frequency through two filters. Both filters have unity gain in the spectral area of interest, but the first has maximal phase delay and zero derivative (group delay) there. The second filter has zero phase delay but a group delay of one-half the beat period. Both filters distort phase, but the phase distortions are different at the frequency of the input signal.

EXERCISES

7.2.1 Show that an antisymmetric FIR filter ($h_n = -h_{-n}$) has zero phase and when made causal has linear phase.

7.2.2 Prove that all linear-phase filters have impulse responses that are either symmetric or antisymmetric.

7.2.3 Assume that two filters have phase delay as a function of frequency $\Phi_1(\omega)$ and $\Phi_2(\omega)$. What is the phase delay of the two filters in series? What about the group delay?

7.2.4 In a non-real-time application a nonlinear-phase filter is run from the end of the signal buffer toward the beginning. What phase delay is introduced?

7.2.5 Stable IIR filters cannot be truly linear-phase. How can the result of the previous exercise be used to create a filter with linear phase based on IIR filtering? How can this technique be used for real-time linear-phase IIR filtering with delay? (Hint: Run the filter first from the beginning of the buffer to the end, and then back from the end toward the beginning.)

7.2.6 What is the phase delay of the IIR filter of equation (6.39)? What is the group delay?

7.2.7 Can you think of a use for all-pass filters?

7.3   Special Filters

From the previous section you may have received the mistaken impression that all filters are used to emphasize some frequencies and attenuate others. In DSP we use filters to implement almost every conceivable mathematical operation. Sometimes we filter in order to alter the time domain characteristics of a signal; for example, the simple delay is an FIR filter, although its specification is most natural in the time domain. The DSP method of detecting a narrow pulse-like signal that may be overlooked is to build a filter that emphasizes the pulse's particular shape. Conversely, a signal may decay too slowly and be in danger of overlapping other signals, in which case we can narrow it by filtering. In this section we will learn how to implement several mathematical operations, such as differentiation and integration, as filters.

A simple task often required is smoothing, that is, removing extraneous noise in order to recover the essential signal values. In the numerical analysis approach smoothing is normally carried out by approximating the data by some appropriate function (usually a polynomial) and returning the value of this function at the point of interest. This strategy works well when the chosen function is smooth and the number of free parameters limited so that the approximation is not able to follow all the fluctuations of the observed data. Polynomials are natural in most numeric analysis contexts since they are related to the Taylor expansion of the function in the region of interest. Polynomials are not as relevant to DSP work since they have no simple frequency domain explanation. The pertinent functional form is of course the sum of sinusoids in the Fourier expansion, and limiting the possible oscillation of the function is equivalent to requiring these sinusoids to be of low frequency. Hence the task of smoothing is carried out in DSP by low-pass filtering. The new interpretation of smoothing is that of blocking the high-frequency noise while passing the signal's energy.

The numerical analysis and DSP approaches are not truly incompatible. For the usual case of evenly sampled data, polynomial smoothing can be implemented as a filter, as was shown for the special case of a five-point parabola in exercise 6.6.5. For that case the smoothed value at time n was found to be the linear combination of the five surrounding input values,

$$y_n = a_2 x_{n-2} + a_1 x_{n-1} + a_0 x_n + a_1 x_{n+1} + a_2 x_{n+2}$$

which is precisely a symmetric MA filter. Let's consider the more general case of optimally approximating $2L + 1$ input points x_n for $n = -L \ldots + L$ by a parabola in discrete time.

$$y_n = a_2 n^2 + a_1 n + a_0$$

For notational simplicity we will only consider retrieving the smoothed value for $n = 0$, all other times simply requiring shifting the time axis.

The essence of the numerical analysis approach is to find the coefficients a_2, a_1, and a_0 that make y_n as close as possible to the $2L + 1$ given x_n ($n = -L \ldots + L$). This is done by requiring the squared error

$$\epsilon = \sum_{n=-L}^{+L} (y_n - x_n)^2 = \sum_{n=-L}^{+L} (a_2 n^2 + a_1 n + a_0 - x_n)^2$$

to be minimal. Differentiating with respect to a, b, and c and setting equal to zero brings us to three equations, known as the *normal equations*

$$\begin{aligned}
B_{00}a_0 + B_{01}a_1 + B_{02}a_2 &= C_0 \\
B_{10}a_0 + B_{11}a_1 + B_{12}a_2 &= C_1 \\
B_{20}a_0 + B_{21}a_1 + B_{22}a_2 &= C_2
\end{aligned} \qquad (7.2)$$

where we have defined two shorthand notations.

$$B_{ij} = \sum_{n=-L}^{+L} n^{i+j} \qquad \text{and} \qquad C_i = \sum_{n=-L}^{+L} n^i x_n$$

The B coefficients are universal, i.e., do not depend on the input x_n, and can be precalculated given L. It is obvious that if the data are evenly distributed around zero ($n = -L, -L+1, \ldots -1, 0, +1, \ldots L-1, L$) then

$B_{ij} = 0$ when $i + j$ is odd, and the other required values can be looked up in a good mathematical handbook.

$$B_{00} = \sum_{n=-L}^{+L} 1 = 2L + 1 \equiv \mathcal{B}_0$$

$$B_{02} = B_{20} = B_{11} = \sum_{n=-L}^{+L} n^2 = \frac{L(L+1)(2L+1)}{3} \equiv \mathcal{B}_2$$

$$B_{22} = \sum_{n=-L}^{+L} n^4 = \frac{L(L+1)(2L+1)(3L^2 + 3L - 1)}{15} \equiv \mathcal{B}_4$$

The three C values are simple to compute given the inputs.

$$C_0 = \sum_{n=-L}^{+L} x_n$$

$$C_1 = \sum_{n=-L}^{+L} n x_n$$

$$C_2 = \sum_{n=-L}^{+L} n^2 x_n$$

In matrix notation the normal equations are now

$$\begin{pmatrix} \mathcal{B}_0 & 0 & \mathcal{B}_2 \\ 0 & \mathcal{B}_2 & 0 \\ \mathcal{B}_2 & 0 & \mathcal{B}_4 \end{pmatrix} \begin{pmatrix} a_0 \\ a_1 \\ a_2 \end{pmatrix} = \begin{pmatrix} C_0 \\ C_1 \\ C_2 \end{pmatrix} \tag{7.3}$$

and can be readily solved by inverting the matrix

$$\begin{pmatrix} a_0 \\ a_1 \\ a_2 \end{pmatrix} = \begin{pmatrix} \mathcal{D}_0 & 0 & \mathcal{D}_2 \\ 0 & \mathcal{D}_1 & 0 \\ \mathcal{D}_2 & 0 & \mathcal{D}_4 \end{pmatrix} \begin{pmatrix} C_0 \\ C_1 \\ C_2 \end{pmatrix} \tag{7.4}$$

and the precise expressions for the $\mathcal{D}$ elements are also universal and can be found by straightforward algebra.

$$D = \mathcal{B}_0 \mathcal{B}_2 \mathcal{B}_4 - \mathcal{B}_2^3$$

$$\mathcal{D}_0 = \frac{\mathcal{B}_2 \mathcal{B}_4}{D}$$

$$\mathcal{D}_1 = \frac{1}{\mathcal{B}_2^2}$$

$$\mathcal{D}_2 = -\frac{\mathcal{B}_2^2}{D}$$

$$\mathcal{D}_3 = \frac{\mathcal{B}_0 \mathcal{B}_2}{D}$$

Now that we have found the coefficients a_0, a_1, and a_2, we can finally find the desired smoothed value at $n = 0$

$$y_0 = a_2 = \mathcal{D}_0 C_0 + \mathcal{D}_2 C_2 = \sum_{n=-L}^{+L} (\mathcal{D}_0 + \mathcal{D}_2 n^2) x_n$$

which is seen to be a symmetric MA filter. So the numerical analysis approach of smoothing by parabolic approximation is equivalent to a particular symmetric MA filter, which has only a single adjustable parameter, L.

Another common task is the differentiation of a signal,

$$y(t) = \frac{d}{d\tau} x(\tau) \tag{7.5}$$

a common use being the computation of the instantaneous frequency from the phase using equation (4.67). The first approximation to the derivative is the finite difference,

$$y_n = x_n - x_{n-1}$$

but for signals sampled at the Nyquist rate or only slightly above the sample times are much too far apart for this approximation to be satisfactory. The standard numerical analysis approach to differentiation is derived from that for smoothing; first one approximates the input by some function, and then one returns the value of the derivative of that function. Using the formalism developed above we can find that in the parabolic approximation, the derivative at $n = 0$ is given by

$$y_0 = a_1 = \mathcal{D}_1 C_1 = \sum_{n=-L}^{+L} (\mathcal{D}_1 n) x_n$$

which is an antisymmetric MA filter, with coefficients proportional to $|n|$! The antisymmetry is understandable as a generalization of the finite difference, but the idea of the remote coefficients being more important than the adjacent ones is somewhat hard to embrace. In fact the whole idea of assuming that values of the derivative to be accurate just because we required the polynomial to approximate the signal values is completely ridiculous. If we do not require the derivative values to be close there is no good reason to believe that they will be; quite the contrary, requiring the polynomial approximation to be good at sampling instants will cause the polynomial to oscillate wildly in between these times, resulting in meaningless derivative estimates.

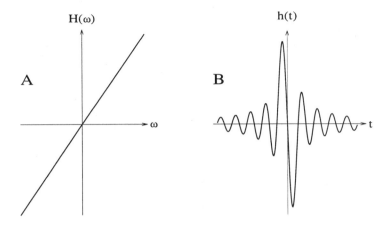

Figure 7.3: Frequency and impulse responses of the ideal differentiation filter.

Differentiation is obviously a linear and time-invariant operation and hence it is not surprising that it can be performed by a filter. To understand this filter in the frequency domain note that the derivative of $s(t) = e^{i\omega t}$ is $i\omega s(t)$, so that the derivative's frequency response increases linearly with frequency (see Figure 7.3.A) and its phase rotation is a constant 90°.

$$H(\omega) = i\omega \tag{7.6}$$

This phase rotation is quite expected considering that the derivative of sine is cosine, which is precisely such a 90° rotation. The impulse response, given by the iFT of the frequency response,

$$
\begin{aligned}
h(t) &= \frac{1}{2\pi} \int_{-\pi}^{\pi} i\omega e^{i\omega t} d\omega \\
&= \frac{i}{2\pi} \left(\frac{e^{i\pi t}}{it} (\pi - \frac{1}{in}) - \frac{e^{-i\pi t}}{it} (-\pi - \frac{1}{in}) \right) \\
&= \frac{\cos(\pi t)}{t} - \frac{\sin(\pi t)}{\pi t^2}
\end{aligned}
$$

is plotted in Figure 7.3.B.

We are more interested in digital differentiators than in the analog one just derived. When trying to convert the frequency response to the digital domain we run into several small snags. First, from the impulse response we see that the ideal differentiator is unrealizable. Second, since the frequency response is now required to be periodic, it can no longer be strictly linear, but instead must be sawtooth with discontinuities. Finally, if the filter has an

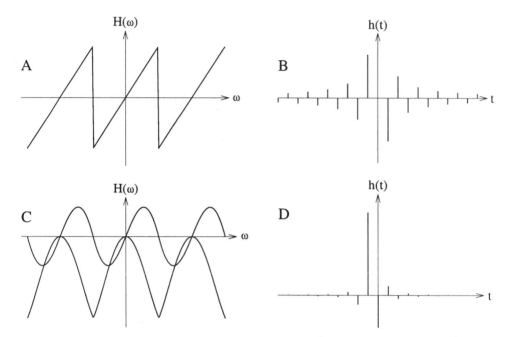

Figure 7.4: Frequency and impulse responses of digital differentiation filters with even and odd numbers of coefficients. In (A) we see the frequency response of an odd length differentiator; note the linearity and discontinuities. (B) is the impulse response for this case. In (C) we see the real and imaginary parts of the frequency response of an even length differentiator. (D) is its impulse response; note that fewer coefficients are required.

even number of coefficients it can never reproduce the derivative at precisely time $t = 0$, but only one-half sample before or after. The frequency response for a time delay of $-\frac{1}{2}$ is

$$H(\omega) = i\omega e^{i\omega(t+\frac{1}{2})} \qquad (7.7)$$

which has both real and imaginary parts but is no longer discontinuous. We now need to recalculate the impulse response.

$$
\begin{aligned}
h(t) &= \frac{1}{2\pi} \int_{-\pi}^{\pi} i\omega e^{i\omega(t+\frac{1}{2})} d\omega \\
&= \frac{-\cos(\pi t)}{\pi(t + \frac{1}{2})^2}
\end{aligned}
$$

The frequency and impulse responses for the odd and even cases are depicted in Figure 7.4. We see that FIR differentiators with an even number of coefficients have no discontinuities in their frequency response, and hence their coefficients vanish quickly. In practical applications we must truncate after a finite number of coefficients. For a given amount of computation an even-order differentiator has smaller error than an odd-order one.

After studying the problem of differentiation it will come as no surprise that the converse problem of integration

$$y(t) = \int_{-\infty}^{t} x(\tau)\, d\tau \qquad (7.8)$$

can be implemented by filtering as well. Integration is needed for the recovery of running phase from instantaneous frequency, and for discovering the cumulative effects of slowly varying signals. Integration is also a popular function in analog signal processing where capacitors are natural integrators; DSP integration is therefore useful for simulating analog circuits.

The signal processing approach to integration starts by noting that the integral of $s(t) = e^{i\omega t}$ is $\frac{1}{i\omega} s(t)$, so that the required frequency response is inversely proportional to the frequency and has a phase shift of 90°.

$$H(\omega) = \frac{1}{i\omega} \qquad (7.9)$$

The standard Riemann sum approximation to the integral

$$\int_{0}^{nT} x(t)\, dt \approx T(x_0 + x_1 + \ldots x_{n-1})$$

is easily seen to be an IIR filter

$$y_n = y_{n-1} + T x_n \qquad (7.10)$$

and we'll take $T = 1$ from here on. What is the frequency response of this filter? If the input is $x_n = e^{i\omega n}$ the output must be $y_n = H(\omega)e^{i\omega n}$ where $H(\omega)$ is a complex number that contains the gain and phase shift. Substituting into the previous equation

$$H(\omega)e^{i\omega n} = y_n = y_{n-1} + x_n = H(\omega)e^{i\omega(n-1)} + e^{i\omega n}$$

we find that

$$
\begin{aligned}
H(\omega) &= \frac{1}{1 - e^{-i\omega}} \\
|H(\omega)|^2 &= \frac{1}{2(1 - \cos(\omega))} \\
\angle H(\omega) &= \tfrac{1}{2}(\pi + \omega)
\end{aligned}
$$

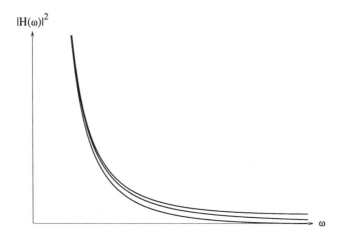

Figure 7.5: The (squared) frequency response of integrators. The middle curve is that of the ideal integrator, the Riemann sum approximation is above it, and the trapezoidal approximation below.

which isn't quite what we wanted. The phase is only the desired $\frac{\pi}{2}$ at DC and deviates linearly with w. For small w, where $\cos(\omega) \sim 1 - \frac{1}{2}w^2$, the gain is very close to the desired $\frac{1}{w}$, but it too diverges at higher frequencies (see Figure 7.5). What this means is that this simple numeric integration is relatively good when the signal is extremely oversampled, but as we approach Nyquist both gain and phase response strongly deviate.

A slightly more complex numeric integration technique is the trapezoidal rule, which takes the average signal value $(x_{n-1} + x_n)$ for the Riemann rectangle, rather than the initial or final value. It too can be written as an IIR filter.

$$y_n = y_{n-1} + \tfrac{1}{2}(x_{n-1} + x_n) \tag{7.11}$$

Using the same technique we find

$$H(\omega)e^{i\omega n} = y_n = y_{n-1} + \tfrac{1}{2}(x_{n-1} + x_n) = H(\omega)e^{i\omega(n-1)} + \tfrac{1}{2}(e^{i\omega(n-1)} + e^{i\omega n})$$

which means that

$$H(\omega) = \frac{i}{2}\frac{1}{\tan(\frac{\omega}{2})}$$

$$|H(\omega)|^2 = \frac{1}{4\tan^2(\frac{\omega}{2})}$$

$$\angle H(\omega) = \frac{\pi}{2}$$

so that the phase is correct, and the gain (also depicted in Figure 7.5) is about the same as before. This is not surprising since previous signal values contribute just as in the Riemann sum, only the first and last values having half weight.

Integrators are always approximated by IIR filters. FIR filters cannot be used for true integration from the beginning of all time, since they forget everything that happened before their first coefficient. Integration over a finite period of time is usually performed by a 'leaky integrator' that gradually forgets, which is most easily implemented by an IIR filter like that of equation (6.39). While integration has a singular frequency response at DC, the frequency response of leaky integration is finite.

Our final special filter is the Hilbert transform, which we introduced in Section 4.12. There are two slightly different ways of presenting the Hilbert transform as a filter. We can consider a real filter that operates on $x(t)$ creating $y(t)$ such that $z(t) = x(t) + iy(t)$ is the analytic representation, or as a complex filter that directly creates $z(t)$ from $x(t)$. The first form has an antisymmetric frequency response

$$H(\omega) = -i\,\mathrm{sgn}(\omega) = \begin{cases} -i & \omega > 0 \\ 0 & \omega = 0 \\ i & \omega < 0 \end{cases} \tag{7.12}$$

which means $|H(\omega)|^2 = 1$ and its phase is $\pm\frac{\pi}{2}$. The impulse response for delay τ is not hard to derive

$$h(t) = \frac{2}{\pi}\frac{\sin^2\left(\frac{\pi}{2}(t - \tau)\right)}{t - \tau} \tag{7.13}$$

except for at $t = 0$ where it is zero. Of course the ideal Hilbert filter is unrealizable. The frequency response of the second form is obtained by summing $X(\omega)$ with i times the above.

$$H(\omega) = \begin{cases} 2 & \omega > 0 \\ 0 & \omega \leq 0 \end{cases} \tag{7.14}$$

The factor of two derives from our desire to retain the original energy after removing half of the spectral components.

The Hilbert transform can be implemented as a filter in a variety of ways. We can implement it as a noncausal FIR filter with an odd number of coefficients arranged to be antisymmetric around zero. Its impulse response

$$h(t) = \frac{2}{\pi}\frac{\sin^2(\frac{\pi}{2}t)}{t}$$

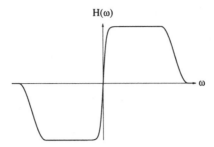

Figure 7.6: Imaginary portion of the frequency response of a realizable digital Hilbert filter with zero delay. The ideal filter would have discontinuities at both DC and half the sampling frequency.

decays slowly due to the frequency response discontinuities at $\omega = 0$ and $\omega = \pi$. With an even number of coefficients and a delay of $\tau = -\frac{1}{2}$ the frequency response

$$H(\omega) = -\mathrm{i}\,\mathrm{sgn}(\omega)e^{-\mathrm{i}\frac{\omega}{2}}$$

leads to a simpler-looking expression;

$$h(t) = \frac{1}{\pi(t + \frac{1}{2})}$$

but simplicity can be deceptive, and for the same amount of computation odd order Hilbert filters have less error than even ones.

The trick in designing a Hilbert filter is bandwidth reduction, that is, requiring that it perform the 90° phase shift only for the frequencies absolutely required. Then the frequency response plotted in Figure 7.6 can be used as the design goal, rather than the discontinuous one of equation (7.12).

EXERCISES

7.3.1 Generate a signal composed of a small number of sinusoids and approximate it in a small interval by a polynomial. Compare the true derivative to the polynomial's derivative.

7.3.2 What are the frequency responses of the polynomial smoother and differentiator? How does the filter length affect the frequency response?

7.3.3 What is the ratio between the Riemann sum integration gain and the gain of an ideal integrator? Can you explain this result?

7.3.4 Show that the odd order Hilbert filter when discretized to integer times has all even coefficients zero.

7.4   Feedback

While FIR filters can be implemented in a *feedforward* manner, with the input signal flowing through the system in the forward direction, IIR filters employ *feedback*. Feedforward systems are simple in principle. An FIR with N coefficients is simply a function from its N inputs to a single output; but feedback systems are not static functions; they have dynamics that make them hard to predict and even unstable. However, we needn't despair as there are properties of feedback systems that can be easily understood.

In order to better understand the effect of feedback we will consider the simplest case, that of a simple amplifier with instantaneous feedback. It is helpful to use a graphical representation of DSP systems that will be studied in detail in Chapter 12; for now you need only know that in Figure 7.7 an arrow with a symbol above it represents a gain, and a circle with a plus sign depicts an adder.

Were it not for the feedback path (i.e., were $a = 0$) the system would be a simple amplifier $y = Gx$; but with the feedback we have

$$y = Gw \tag{7.15}$$

where the intermediate signal is the sum of the input and the feedback.

$$w = x + ay \tag{7.16}$$

Substituting

$$y = G(x + ay) = Gx + aGy$$

and solving for the output

$$y = \frac{G}{1 - aG} x \tag{7.17}$$

we see that the overall system is an amplifier like before, only the gain has been enhanced by a denominator. This gain obtained by closing the feedback

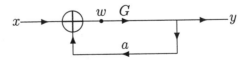

Figure 7.7: The DSP diagram of an amplifier with instantaneous feedback. As will be explained in detail in Chapter 12, an arrow with a symbol above it represents a gain, a symbol above a filled circle names a signal, and a circle with a plus sign depicts an adder. The feedforward amplifier's gain is G while the feedback path has gain (or attenuation) a.

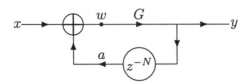

Figure 7.8: An amplifier with delayed feedback. As will be explained in detail in Chapter 12, a circle with z^{-N} stands for a delay of N time units. Here the feedforward amplifier's gain is G while the feedback path has delay of N time units and gain (or attenuation) a.

loop is called the *closed loop gain*. When a is increased above zero the closed loop gain increases.

What if a takes precisely the value $a = \frac{1}{G}$? Then the closed loop gain explodes! We see that even this simplest of examples produces an instability or 'pole'. Physically this means that the system can maintain a finite output even with zero input. This behavior is quite unlike a normal amplifier; actually our system has become an *oscillator* rather than an amplifier. What if we subtract the feedback from the input rather than adding it? Then for $a = \frac{1}{G}$ the output is exactly zero.

The next step in understanding feedback is to add some delay to the feedback path, as depicted in Figure 7.8. Now

$$y_n = Gw_n$$

with

$$w_n = x_n + ay_{n-N}$$

where N is the delay time. Combining

$$y_n = G(x_n + ay_{n-N}) = Gx_n + aGy_{n-N} \tag{7.18}$$

and we see that for constant signals nothing has changed. What happens to time-varying signals? A periodic signal x_n that goes through a whole cycle, or any integer number of whole cycles, during the delay time will cause the feedback to precisely track the input. In this case the amplification will be exactly like that of a constant signal. However, consider a sinusoid that goes through a half cycle (or any odd multiple of half cycles) during the delay time. Then y_{n-N} will be of opposite sign to y_n and the feedback will destructively combine with the input; for $aG = 1$ the output will be zero! The same is true for a periodic signal that goes through a full cycle (or any multiple) during the delay time, with negative feedback (i.e., when the feedback term is *subtracted from* rather than *added to* the input).

$$w_n = x_n - ay_{n-N} \tag{7.19}$$

So feedback with delay causes some signals to be emphasized and others to be attenuated, in other words, feedback can *filter*. When the feedback produces a pole, that pole corresponds to some frequency, and only that frequency will build up without limit. When a 'zero' is evoked, no matter how much energy we input at the particular frequency that is blocked, no output will result. Of course nearby frequencies are also affected. Near a pole sinusoids experience very large but finite gains, while sinusoids close to a zero are attenuated but not eliminated.

With unity gain negative feedback it is possible to completely block a sinusoid; can this be done with $aG \neq 1$? For definiteness let's take $G = 1, a = \frac{1}{2}$. Starting at the peak of the sinusoid $x_0 = 1$ the feedback term to be subtracted a cycle later is only $ay_{n-N} = \frac{1}{2}$. Subtracting this leads to $w = \frac{1}{2}$, which a cycle later leads to the subtraction of only $ay_{n-N} = \frac{1}{4}$. In the steady state the gain settles down to $\frac{2}{3}$, the prediction of equation (7.17) with a taken to be negative. So nonunity gain in the negative feedback path causes the sinusoid to be attenuated, but not notched out. You may easily convince yourself that the gain can only be zero if $aG = 1$. Similarly nonunity gain in a positive feedback path causes the sinusoid to be amplified, but not by an infinite amount.

So a sinusoid cannot be completed blocked by a system with a delayed negative feedback path and nonunity feedback gain, but is there a nonsinusoidal signal that *is* completely notched out? The only way to compensate for nonunity gain in the feedback term to be subtracted is by having the signal vary in the same way. Hence for $aG > 1$ we need a signal that increases by a factor of aG after the delay time N, i.e.,

$$s_n = e^{+(\ln aG)\frac{n}{N}} \sin\left(2\pi \frac{n}{N}\right)$$

while for $aG < 1$ the signal needs to decrease in the same fashion. This is a general result; when the feedback gain is not unity the signals that are optimally amplified or notched are exponentially growing or damped sinusoids.

Continuing our argument it is easy to predict that if there are several delayed feedback paths in parallel then there will be several frequency regions that are amplified or attenuated. We may even put filters in the feedback path, allowing feedback at certain frequencies and blocking it at others. Indeed this is the way filters are designed in analog signal processing; feedback paths of various gains and phases are combined until the desired effect is approximated.

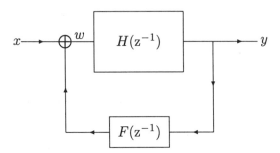

Figure 7.9: The general feedback amplifier. The boxes represent general filters, with transfer functions as marked. The amplifier's transfer function is H while that of the feedback path is F.

In the most general setting, consider a digital system with transfer function $H(z^{-1})$ to which we add a feedback loop with transfer function $F(z^{-1})$, as depicted in Figure 7.9. The closed loop transfer function is given by

$$H'(z^{-1}) = \frac{H(z^{-1})}{1 - F(z^{-1})H(z^{-1})} \qquad (7.20)$$

which has a pole whenever the denominator becomes zero (i.e., for those z for which $F(z^{-1})H(z^{-1}) = 1$). The value of z determines the frequency of the oscillation.

EXERCISES

7.4.1 When the microphone of an amplification system is pointed toward the speaker a squealing noise results. What determines the frequency of the squeal? Test your answer. What waveform would you expect?

7.4.2 A feedback pole causes an oscillation with frequency determined by the delay time. This oscillation is sustained even without any input. The system is linear and time-invariant, and so is a filter; as a filter it cannot create energy at a frequency where there was no energy in the input. Resolve this paradox.

7.4.3 What is the effect of a delayed feedback path with unity gain on a sinusoid of frequency close, but not equal, to the instability? Plot the gain as a function of frequency (the frequency response).

7.4.4 Find a signal that destabilizes a system with a delayed positive feedback path and nonunity feedback gain.

7.4.5 Show that for $G = \frac{1}{2}, a = 1$ a sinusoid of frequency corresponding to the delay is amplified by the gain predicted by equation (7.17).

7.4.6 What is the effect of a delayed feedback path with nonunity gain G on a sinusoid of frequency corresponding to the delay? Plot the effective gain as a function of G.

7.4.7 Simulate a system that has a causal MA filter in the feedback path. Start with a low-pass filter, then a high-pass, and finally a band-pass. Plot the frequency response.

7.5 The ARMA Transfer Function

In Section 6.14 we defined the transfer function of a filter. The transfer function obeys

$$Y(z) = H(z)X(z)$$

where $X(z)$ is the zT of the input to the filter and $Y(z)$ is the zT of the output. Let's find the transfer function of an ARMA filter. The easiest way to accomplish this is to take the z transform of both sides of the general ARMA filter in the symmetric form (equation (6.46))

$$\sum_{m=0}^{M} \beta_m y_{n-m} = \sum_{l=0}^{L} \alpha_l x_{n-l}$$

the zT of the left side being

$$\sum_{n=-\infty}^{\infty} \left\{ \sum_{m=0}^{M} \beta_m y_{n-m} \right\} z^{-n} = \sum_{m=0}^{M} \beta_m \sum_{n=-\infty}^{\infty} y_{n-m} z^{-n} =$$

$$\sum_{m=0}^{M} \beta_m z^{-m} \sum_{\nu=-\infty}^{\infty} y_\nu z^{-\nu} = \left\{ \sum_{m=0}^{M} \beta_m z^{-m} \right\} Y(z)$$

and similarly that of the right side.

$$\sum_{n=-\infty}^{\infty} \left\{ \sum_{l=0}^{L} \alpha_l x_{n-l} \right\} z^{-n} = \sum_{l=0}^{L} \alpha_l \sum_{n=-\infty}^{\infty} x_{n-l} z^{-n} =$$

$$\sum_{l=0}^{L} \alpha_l z^{-l} \sum_{\nu=-\infty}^{\infty} x_\nu z^{-\nu} = \left\{ \sum_{l=0}^{L} \alpha_l z^{-l} \right\} X(z)$$

Putting these together

$$\left\{ \sum_{m=0}^{M} \beta_m z^{-m} \right\} Y(z) = \left\{ \sum_{l=0}^{L} \alpha_l z^{-l} \right\} X(z)$$

and comparing with equation (6.65) we find that the transfer function is the ratio of two polynomials in z^{-1}.

$$H(z) = \frac{\sum_{l=0}^{L} \alpha_l z^{-l}}{\sum_{m=0}^{M} \beta_m z^{-m}} \tag{7.21}$$

This can be also expressed in terms of the coefficients in equation (6.45)

$$H(z) = \frac{\sum_{l=0}^{L} a_l z^{-l}}{1 - \sum_{m=1}^{M} b_m z^{-m}} \tag{7.22}$$

a form that enables one to build the transfer function 'by inspection' from the usual type of difference equation.

For an AR filter $L = 0$ and neglecting an uninteresting gain (i.e., taking $a_0 = 1$)

$$H(z) = \frac{1}{1 - \sum_{m=1}^{M} b_m z^{-m}} \tag{7.23}$$

while for an MA filter all the b_m are zero and the transfer function is a polynomial.

$$H(z) = \sum_{l=0}^{L} a_l z^{-l} \tag{7.24}$$

It is often burdensome to have to deal with polynomials in z^{-1}, so we express the transfer function in terms of z instead.

$$H(z) = z^{M-L} \frac{\sum_{l=0}^{L} \alpha_l z^{L-l}}{\sum_{m=0}^{M} \beta_m z^{M-m}} \tag{7.25}$$

We see that $H(z)$ is a rational function of z.

The fact that the transfer function $H(z)$ of the general ARMA filter is a rational function, has interesting and important ramifications. The fundamental theorem of algebra tells us that any polynomial of degree M can be completely factored over the complex numbers

$$\sum_{i=0}^{D} c_i x^i = G \prod_{i=1}^{D} (x - \zeta_i)$$

where the D roots ζ_i are in general complex numbers. When the coefficients c_i are real, the sum itself is always real, and so the roots must either be real, or appear in complex conjugate pairs. Thus we can rewrite the transfer function of the general ARMA filter as

$$H(z) = G \frac{\prod_{l=1}^{L} (z - \zeta_l)}{\prod_{m=1}^{M} (z - \pi_m)} \tag{7.26}$$

where the roots of the numerator ζ_l are called 'zeros' of the transfer function, and the roots of the denominator π_m its 'poles'. So other than an simple overall gain G, we need only specify the zeros and poles to completely determine the transfer function; no further information is needed.

From equations 7.23 and 7.24 we see that the transfer function of the MA filter has zeros but no poles while that of the AR filter has poles but no zeros. Hence the MA filter is also called an all-zero filter and the AR filter is called an all-pole filter.

What is the meaning of these zeros and poles? A zero of the transfer function is a complex number $\zeta = re^{i\omega}$ that represents a complex (possibly decaying or increasing) exponential signal that is attenuated by the ARMA filter. Poles π represent complex exponential signals that are amplified by the filter. If a zero or pole is on the unit circle, it represents a sinusoid that is either completely blocked by the filter or destabilizes it.

Since the positions in the complex plane of the zeros and poles provide a complete description of the transfer function of the general ARMA system, it is conventional to graphically depict them using a *pole-zero plot*. In such plots the position of a zero is shown by a small filled circle and a pole is marked with an X. Poles or zeros at $z = 0$ or $z = \infty$ that derive from the z^{M-L} factor in equation (7.25) are not depicted, but multiple poles and/or zeros at the same position are. This single diagram captures everything one needs to know about a filter, except for the overall gain.

A few examples are in order. First consider the causal equally weighted $L+1$-point average MA filter (since we intend to discard the gain we needn't normalize the sum).

$$y_n = \sum_{l=0}^{L} x_{n-l}$$

By inspection the transfer function is

$$H(z) = \sum_{l=0}^{L} z^{-l} = \frac{1 - z^{-1-L-1}}{1 - z^{-1}} = \frac{1}{z^L}\frac{z^{L+1} - 1}{z - 1}$$

and we seem to see L poles at the origin, the $L+1$ zeros of $z^{L+1} - 1$ and a pole at $z = 1$. The zeros are the $L + 1$ roots of unity, $z = e^{i2\pi \frac{k}{L+1}}$, one of which is $z = 1$ itself; hence that zero cancels the putative pole at $z = 1$. The L poles at the origin are meaningless and may be ignored. We are therefore left with L zeros equally spaced around the unit circle (not including $z = 1$), as displayed in Figure 7.10.A. It is not difficult to verify that the corresponding sinusoids are indeed blocked by the averaging MA filter.

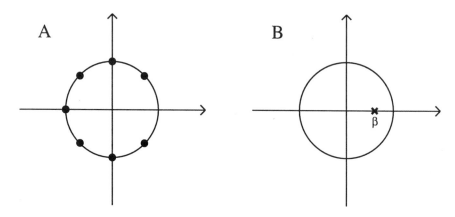

Figure 7.10: The pole-zero plots of two simple systems. In (A) we see the pole-zero plot for the MA filter that averages with equal weights eight consecutive input values. In (B) is the simple AR low-pass filter $y_n = (1 - \beta)x_n + \beta y_{n-1}$.

Our second example is our favorite AR filter of equation (6.39).

$$y_n = (1 - \beta)x_n + \beta y_{n-1} \qquad 0 \le \beta < 1$$

By inspection we can write

$$H(z) = \frac{(1 - \beta)}{1 - \beta z^{-1}} = z\frac{1 - \beta}{z - \beta}$$

which has a trivial zero at the origin and a single pole at β, as depicted in Figure 7.10.B.

As our last example we choose a general first-order section, that is, an ARMA system with a single zero and a single pole.

$$y_n = a_0 x_n + a_1 x_{n-1} + b_1 y_{n-1}$$

This is a useful system since by factorization of the polynomials in both the numerator and denominator of the transfer function we can break down any ARMA filter into a sequence of first-order sections in cascade. By inspection the transfer function

$$H(z) = \frac{a_0 + a_1 z^{-1}}{1 - b_1 z^{-1}} = a_0 \frac{z + \frac{a_1}{a_0}}{z - b_1}$$

has its zero at $z = -\frac{a_1}{a_0}$ and its pole at $z = b_1$. To find the frequency response we substitute $z = e^{i\omega}$

$$H(\omega) = \frac{a_0 + a_1 e^{-i\omega}}{1 - b_1 e^{-i\omega}}$$

which at DC is $\frac{a_0+a_1}{1-b_1}$ and at Nyquist $\omega = \pi$ is $\frac{a_0-a_1}{1+b_1}$. To find the impulse response we need the inverse zT, which generally is difficult to calculate. Here it can be carried out using a trick

$$H(z) = a_0 \frac{(z - b_1) + (\frac{a_1}{a_0} + b_1)}{z - b_1}$$

$$= a_0 \left(1 + \frac{(\frac{a_1}{a_0} + b_1)z^{-1}}{1 - b_1 z^{-1}} \right)$$

$$= a_0 \left(1 + \left(\frac{a_1}{a_0} + b_1\right) z^{-1} + \left(\frac{a_1}{a_0} + b_1\right) b_1 z^{-2} + \left(\frac{a_1}{a_0} + b_1\right) b_1^2 z^{-3} + \dots \right)$$

and the desired result is obtained.

$$h_n = \begin{cases} a_0 & n = 0 \\ (a_1 + a_0 b_1)b_1^{n-1} & n \neq 0 \end{cases}$$

EXERCISES

7.5.1 Sometimes it is useful to write difference equations as $y_n = Gx_n + \sum a_l x_{n-l} + \sum b_m y_{n-m}$ where G is called the *gain*. Write the transfer function in rational-function- and factored-form for this case.

7.5.2 Derive equation (7.21) more simply than in the text by using the time shift relation for the zT.

7.5.3 Consider the system with a single real pole or zero. What signal is maximally amplified or attenuated? Repeat for a complex pole or zero.

7.5.4 Calculate the transfer function $H(z)$ for the noncausal MA system of equation (6.35). Relate this to the transfer function of the causal version and to the frequency response (equation (6.36)) previously calculated.

7.5.5 Show that stable ARMA filters have all their poles inside the unit circle.

7.5.6 Prove that real all-pass filters have poles and zeros in conjugate reciprocal locations.

7.5.7 Show that the first-order section is stable when $|b_1| < 1$ both by considering the pole and by checking the impulse response.

7.5.8 Plot the absolute value of the frequency response of the first-order section for frequencies between DC and Nyquist. When is the filter low-pass (passes low frequencies better than highs)?

7.5.9 If the maximum input absolute value is 1, what is the maximal output absolute value for the first-order section? If the input is white noise of variance 1, what is the variance of the output of the first-order section?

7.5.10 In general, when breaking down ARMA systems into first-order sections the zeros and poles may be complex. In such cases we most often use real-valued second-order sections instead.

$$H(z) = \frac{a_0 + a_1 z^{-1} + a_2 z^{-2}}{1 - b_1^{-1} - b_2 z^{-2}}$$

What is the frequency response for the second-order section with complex conjugate poles?

7.6   Pole-Zero Plots

The main lesson from the previous section was that the positions of the zeros and poles of the transfer function determine an ARMA filter to within a multiplicative gain. The graphical depiction of these positions such as in Figure 7.10 is called a pole-zero plot. Representing filters by pole-zero plots is analogous to depicting signals by the z-plane plots introduced in Section 4.10. Indeed there is a unique correspondence between the two since z-plane plots contain a complete frequency domain description of signals, and filters are specified by their effect in the frequency domain.

The pole-zero plot completely specifies an ARMA filter except for the overall gain. At first sight it may seem strange that the positions of the zeros and poles are enough to completely specify the transfer function of an ARMA filter. Why can't there be two transfer functions that have the same zeros and poles but are different somewhere far from these points? The fact is that were we to allow arbitrary systems then there could indeed be two different systems that share zeros and poles; but the transfer function of an ARMA filter is constrained to be a rational function and the family of rational functions does not have that much freedom. For instance, suppose we are given the position of the zeros of an MA filter, $\zeta_1, \zeta_2 \ldots \zeta_L$. Since the transfer function is a polynomial, is must be

$$H(z) = G \prod_{l=1}^{L} (z - \zeta_l)$$

since any other polynomial will have different zeros.

In addition to being mathematically sufficient, pole-zero plots are graphically descriptive. The pole-zero plot provides the initiated at a glance everything there is to know about the filter. You might say that the pole-zero plot picture is worth a thousand equations. It is therefore worthwhile to become proficient in 'reading' pole-zero plots.

We can place restrictions on the poles and zeros before we even start. Since we wish real inputs to produce real outputs, we require all the coefficients of the ARMA filter to be real. Now real-valued rational functions will have poles and zeros that are either real valued, or that come in complex conjugates. For example, the three zeros 1, $1 + i$ and $1 - i$ form the real polynomial $(z - 1)(z - i)(z + i) = z^3 - z^2 + z - 1$, while were the two complex zeros *not* complex conjugates the resulting polynomial would be complex! So the pole-zero plots of ARMA systems with real-valued coefficients are always mirror-symmetric around the real axis.

What is the connection between the pole-zero plots of a system and its inverse? Recall from equation (6.17) that when the output of a system is input to its inverse system the original signal is recovered. In exercise 6.14.3 we saw that the transfer function of the concatenation of two systems is the product of their respective transfer functions. So the product of the transfer functions of a system and its inverse must be unity, and hence the transfer functions reciprocals of each other. Hence the pole-zero plot of the inverse system is obtained by replacing all poles with zeros and all zeros with poles. In particular it is easy now to see that the inverse of an all-zero system is all-pole and vice versa.

In Section 7.4 we saw what it means when a pole or a zero is on the unit circle. A zero means that the frequency in question is swallowed up by the system, and nearby frequencies are attenuated. A pole means that the system is capable of steady state output without input at this frequency, and nearby frequencies are strongly amplified. For this reason poles on the unit circle are almost always to be avoided at all costs.

What if a pole or zero is inside the unit circle? Once again Section 7.4 supplied the answer. The signal that is optimally amplified or blocked is a damped sinusoid, exactly the basic signal represented by the pole or zero's position in the z-plane. If the pole or zero is outside the unit circle the signal most affected is the growing sinusoid represented by that point. Although we don't want poles on the unit circle, we want them even less outside it. A pole corresponding to an exponentially growing sinusoid would mean that we have an unstable system that could explode without notice. Thus IIR system designers must always ensure that all poles are inside the unit circle.

The pole-zero plot directly depicts the transfer function, but the frequency response is also easily inferred. Think of the unit circle as a circular railroad track with its height above sea level representing the gain at the corresponding frequency. In this analogy poles are steep mountains and zeros are craters. As the train travels around the track its height increases and decreases because of proximity to a mountain or crater. Of course at any position there may be several poles and/or craters nearby, and the overall height is influenced by each of them according to its distance from the train. Now let's justify this analogy. Substituting $z = e^{i\omega}$ into equation (7.26) we find that the frequency response of an ARMA systems is

$$H(\omega) = G \frac{\prod_{l=1}^{L}(e^{i\omega} - \zeta_l)}{\prod_{m=1}^{M}(e^{i\omega} - \pi_m)} \tag{7.27}$$

with magnitude and angle given by the following.

$$|H(\omega)| = G \frac{\prod_{l=1}^{L}|e^{i\omega} - \zeta_l|}{\prod_{m=1}^{M}|e^{i\omega} - \pi_m|}$$

$$\angle H(\omega) = \sum_{l=1}^{L} \angle(e^{i\omega} - \zeta_l) - \sum_{m=1}^{M} \angle(e^{i\omega} - \pi_m)$$

The lth factor in the numerator of the magnitude is the distance between the point on the unit circle and the lth zero, and the mth factor in the denominator is the distance to the mth pole. The magnitude is seen to be the product of the distances to all the zeros divided by the product to all the poles. If one of the zeros or poles is very close it tends to dominate, but in general the train's height is influenced by all the mountains and craters according to their distances from it. The lth term in the numerator of the angle is the direction of the vector between the point on the unit circle and the lth zero, and the mth term in the denominator is the angle to the mth pole. Therefore the phase of the frequency response is seen to be the sum of the angles to all the zeros minus the sum of the angles to the poles. If one of the zeros or poles is very close its angle changes rapidly as the train progresses, causing it to dominate the group delay.

Suppose we design a filter by some technique and find that a pole is outside the unit circle. Is there some way to stabilize the system by moving it back inside the unit circle, without changing the frequency response? The answer is affirmative. Let the pole in question be $\pi_0 = Pe^{i\theta}$. You can convince yourself that the distance from any point on the unit circle to π_0 is exactly P^2 times the distance to $\pi_0' = \frac{1}{P}e^{i\theta}$, the point along the same radius but with reciprocal magnitude. Thus to within a gain term (that we have been neglecting here) we can replace any pole outside the unit circle with its 'reciprocal conjugate' π_0'. This operation is known as 'reflecting a pole'. We can also reflect a zero from outside the unit circle inward, or from the inside out if we so desire. For real filters we must of course reflect both the pole and its complex conjugate.

Let's see how the concept of a pole-zero plot enables us to design some useful filters. Assume we want a DC blocker, that is, a filter that blocks DC but passes AC frequencies. A first attempt might be to simply place a zero at DC

$$H(z) = z - 1 = z(1 - z^{-1}) \qquad \Rightarrow \qquad y_n = x_n - x_{n-1}$$

discarding the term representing a zero at $z = 0$; but this filter is simply the finite difference, with frequency response

$$|H(\omega)|^2 = |1 - e^{-i\omega}|^2 = 2(1 - \cos\omega)$$

not corresponding to a sharp notch. We can sharpen the response by placing a pole on the real axis close to, but inside, the unit circle. The reasoning behind this tactic is simple. The zero causes the DC frequency response to be zero, but as we move away from $\omega = 0$ on the unit circle we immediately start feeling the effects of the pole.

$$H(z) = \frac{z - 1}{z - \beta} = \frac{1 - z^{-1}}{1 - \beta z^{-1}} \qquad \Longrightarrow \qquad y_n = \beta y_{n-1} + (x_n - x_{n-1})$$

Here $\beta < 1$ but the closer β is to unity the sharper the notch will be. There is a minor problem regarding the gain of this filter. We would like the gain to be unity far away from DC, but of course pole-zero methods cannot control the gain. At $z = -1$ our DC blocker has a gain of

$$\frac{z - 1}{z - \beta} = \frac{-1 - 1}{-1 - \beta} = \frac{1}{1 - \frac{\alpha}{2}}$$

where we defined the small positive number α via $\beta = 1 - \alpha$. We can compensate for this gain by multiplying the x terms by a factor $g = 1 - \frac{\alpha}{2}$.

$$y_n = (1 - \alpha)\, y_{n-1} + (1 - \tfrac{1}{2}\alpha)\,(x_n - x_{n-1})$$

In addition to a DC blocker we can use the same technique to make a notch at any frequency Ω. We need only put a conjugate pair of zeros on the unit circle at angles corresponding to $\pm\Omega$ and a pair of poles at the same angles but slightly reduced radius. We can also make a sharp band-pass filter by reversing the roles of the zeros and poles. Wider band-pass or band-stop filters can be approximated by placing several poles and/or zeros along the desired band. Every type of frequency-selective filter you can imagine can be designed by careful placement of poles and zeros.

EXERCISES

7.6.1 Although practically every filter you meet in practice is ARMA, they are *not* the most general LTI system. Give an example of a linear time-invariant system that is not ARMA.

7.6.2 Make a pole-zero plot for the system

$$H(z) = \frac{(z - \alpha)(z - \frac{1}{\alpha})}{(z - r\alpha)(z - \frac{r}{\alpha})}$$

where $\alpha = e^{i\Omega}$ and $r \lesssim 1$. Sketch the frequency response. What kind of filter is this?

7.6.3 Why did we call π_0' the reciprocal conjugate? Prove that the distance from any point on the unit circle to π_0 is exactly P^2 times the distance to the reciprocal conjugate π_0'.

7.6.4 A stable system whose inverse is stable as well is said to be *minimum phase*. What can you say about the pole-zero plot of a minimum phase system?

7.6.5 Prove that reflecting poles (or zeros) does not change the frequency response.

7.6.6 What can be said about the poles and zeros of an all-pass filter? What is the connection between this question and the previous one?

7.6.7 A notch filter can be designed by adding the outputs of two all-pass filters that have the same phase everywhere except in the vicinity of the frequency to be blocked, where they differ by 180°. Design a notch filter of the form $H(z) = \frac{1}{2}\left(1 + A(z)\right)$ where $A(z)$ is the transfer function of an all-pass filter. How can you control the position and width of the notch?

7.6.8 Consider the DC blocking IIR filter $y_k = 0.9992(x_k - x_{k-1}) + 0.9985y_{k-1}$. Draw its frequency response by inputting pure sinusoids and measuring the amplitude of the output. What is its pole-zero plot?

7.7   Classical Filter Design

Classical filter design means analog filter design. Why are we devoting a section in a book on DSP to analog filter design? There are two reasons. First, filtering is one of the few select subjects in analog signal processing about which every DSP expert should know something. Not only are there always analog antialiasing filters and reconstruction filters, but it is often worthwhile to perform other filtering in the analog domain. Good digital filters are notoriously computationally intensive, and in high-bandwidth systems there may be no alternative to performing at least some of the filtering using analog components. Second, the discipline of analog filter design was already well-developed when the more complex field of digital filter design was first developing. It strongly influenced much of the terminology and algorithms, although its stranglehold was eventually broken.

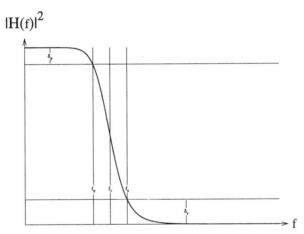

Figure 7.11: Desired frequency response of the analog low-pass filter to be designed. The pass-band is from $f = 0$ to the pass-band edge f_p, the transition region from f_p to f_s, and the stop-band from the top-band edge f_s to infinity. The frequency response is halfway between that of the pass-band and that of the stop-band at the cutoff frequency f_c. The maximal ripple in the pass-band is δ_p and in the stop-band δ_s.

We will first focus on the simplest case, that of an analog low-pass filter. Our ideal will be the ideal low-pass filter, but that being unobtainable we strive toward its best approximation. The most important specification is the cutoff frequency f_c, below which we wish the signal to be passed, above which we wish the signal to be blocked. The pass-band and stop-band are separated by a transition region where we do not place stringent requirements on the frequency response. The end of the pass-band is called f_p and the beginning of the stop-band f_s. Other specifications for a practical implementation are the transition width $\Delta = f_s - f_p$, the maximal deviation from unity gain in the pass-band δ_p, and the maximal amplitude in the stop-band δ_s. In a typical analog filter design problem f_c (or f_p or f_s) and the maximal allowed values for Δ, δ_p, and δ_s are given. Figure 7.11 depicts the ideal and approximate analog low-pass filters with these parameters.

Designing an analog filter essentially amounts to specifying the function $H(f)$ whose square is depicted in the figure. From the figure and our previous analysis we see that

$$
\begin{aligned}
|H(0)|^2 &= 1 \\
|H(f)|^2 &\approx 1 \qquad \text{for} \quad f < f_c \\
|H(f)^2 &\approx 0 \qquad \text{for} \quad f > f_c \\
|H(f)|^2 &\to 0 \qquad \text{for} \quad f \to \infty
\end{aligned}
$$

are the requirements for an analog low-pass filter. The first functional forms that come to mind are based on arctangents and hyperbolic tangents, but these are natural when the constraints are at plus and minus infinity, rather than zero and infinity. Classical filter design relies on the form

$$
|H(f)|^2 = \frac{1}{1 + p(f)} \tag{7.28}
$$

where $p(f)$ is a polynomial that must obey

$$
\begin{aligned}
p(0) &= 0 \\
p(f) &\xrightarrow{f \to \infty} \infty
\end{aligned}
$$

and be well behaved. The classical design problem is therefore reduced to the finding of this polynomial.

In Figure 7.11 the deviation of the amplitude response from the ideal response is due entirely to its smoothly decreasing from unity at $f = 0$ in order to approach zero at high frequencies. One polynomial that obeys the constraints and has no extraneous extrema is the simple quadratic

$$
p(f) = \left(\frac{f}{f_c}\right)^2
$$

which when substituted back into equation (7.28) gives the 'slowest' filter depicted in Figure 7.12. The other filters there are derived from

$$
p(f) = \left(\frac{f}{f_c}\right)^{2N}
$$

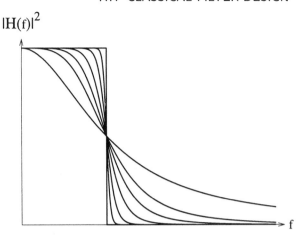

$|H(f)|^2$

Figure 7.12: Frequency response of analog Butterworth low-pass filters. From bottom to top at low frequencies we have order $N = 1, 2, 3, 5, 10, 25, \infty$.

and are called the Butterworth low-pass filters of order n. It is obvious from the figure that the higher n is the narrower the transition.

Butterworth filters have advantages and disadvantages. The attenuation monotonically increases from DC to infinite frequency; in fact the first $2N-1$ derivatives of $|H(f)|^2$ are identically zero at these two points, a property known as 'maximal flatness'. An analog Butterworth filter has only poles and is straightforward to design. However, returning to the design specifications, for the transition region Δ to be small enough the order N usually has to be quite high; and there is no way of independently specifying the rest of the parameters.

In order to obtain faster rolloff in the filter skirt we have to give something up, and that something is the monotonicity of $|H(f)|^2$. A Butterworth filter 'wastes' a lot of effort in being maximally flat, effort that could be put to good use in reducing the size of the transition region. A filter that is allowed to oscillate up and down a little in either the pass-band, the stop-band or both can have appreciably smaller Δ. Of course we want the deviation from our specification to be minimal in some sense. We could require a minimal squared error between the specification and the implemented filter

$$\epsilon^2 = \int |H_{spec}(\omega) - H_{impl}(\omega)|^2 \, d\omega$$

but this would still allow large deviation from specification at some frequencies, at the expense of overexactness at others. It makes more sense to require *minimax error*, i.e., to require that the maximal deviation from specification

$$\max_{\omega} |H_{spec}(\omega) - H_{impl}(\omega)|$$

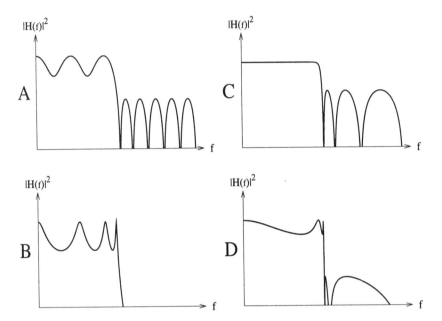

Figure 7.13: Frequency response of low-pass equiripple designs. In (A) we see an FIR filter designed using the Remez algorithm for comparison purposes. In (B) we the IIR Chebyshev design, in (C) the inverse Chebyshev and in (D) the elliptical design.

be minimal. Achieving true minimax approximation is notoriously difficult in general, but approximation using Chebyshev polynomials (see Appendix A.10) is almost the same and straightforward to realize. This approximation naturally leads to equiripple behavior, where the error oscillates around the desired level with equal error amplitude, as shown in Figure 7.13.

The Chebyshev (also known as Chebyshev I) filter is equiripple in the pass-band, but maximally flat in the stop-band. It corresponds to choosing the polynomial

$$p(f) = \delta^2 \, T_N^2 \left(\frac{f}{f_p} \right)$$

and like the Butterworth approximation, the analog Chebyshev filter is all-pole. The inverse Chebyshev (or Chebyshev II) filter is equiripple in the stop-band but maximally flat in the pass-band. Its polynomial is

$$p(f) = \delta^2 \, \frac{T_N^2 \left(\frac{f_s}{f_p} \right)}{T_N^2 \left(\frac{f_s}{f} \right)}$$

The Chebyshev filter minimax approximates the desired response in the pass-band but not in the stop-band, while the inverse Chebyshev does just the opposite. For both types of Chebyshev filters the parameter δ sets the ripple in the equiripple band. For the inverse Chebyshev, where the equiripple property holds in the stop-band, the attenuation is determined by the ripple; lower ripple means higher stop-band rejection.

Finally, the *elliptical* filter is equiripple in both pass-band and stop-band, and so approximates the desired response in the minimax sense for all frequencies. Its 'polynomial' is not a polynomial at all, but rather a rational function $U_N(\frac{f}{f_p})$. These functions are defined using the elliptical functions (see Appendices A.8 and A.10). Taking the idea from equation (A.59), we define the function

$$U_{r;k,q}(u) \equiv \mathrm{sn}_k\left(r\,\mathrm{sn}_q^{-1}(u)\right) \qquad (7.29)$$

and when r and the complete elliptical integrals K_k and K_q obey certain relations that we will not go into here, this function becomes a rational function.

$$U_N(u) = a^2 \begin{cases} \dfrac{(u_1^2-u^2)(u_3^2-u^2)\cdots(u_{2N-1}^2-u^2)}{(1-u_1^2u^2)(1-u_3^2u^2)\cdots(1-u_{2N-1}^2u^2)} & N \text{ even} \\[3mm] \dfrac{u(u_2^2-u^2)(u_4^2-u^2)\cdots(u_{2N}^2-u^2)}{(1-u_2^2u^2)(1-u_4^2u^2)\cdots(1-u_{2N}^2u^2)} & N \text{ odd} \end{cases} \qquad (7.30)$$

This rational function has several related interesting characteristics. For $u < 1$ the function lies between -1 and $+1$. Next,

$$U_N\left(\frac{1}{u}\right) = \frac{1}{U_N(u)}$$

and its zeros and poles are reciprocals of each other. Choosing all the N zeros in the range $0 < \zeta < 1$ forces all N poles to fall in the range $1 < \pi < \infty$. Although the zeros and poles are not equally spaced, the behavior of

$$|H(f)|^2 = \frac{1}{1+U_N(\frac{f}{f_p})}$$

is equiripple in both the pass-band and the stop-band.

It is useful to compare the four types of analog filter—Butterworth, Chebyshev, inverse Chebyshev, and elliptical. A very strong statement can be made (but will not be proven here) regarding the elliptical filter; given any three of the four parameters of interest (pass-band ripple, stop-band ripple, transition width, and filter order) the elliptical filter minimizes the remaining parameter. In particular, for given order N and ripple tolerances the elliptical filter can provide the steepest pass-band to stop-band transition. The Butterworth filter is the weakest in this regard, and the two Chebyshev types are intermediate. The Butterworth filter, however, is the best approximation to the Taylor expansion of the ideal response at both DC and infinite frequency. The Chebyshev design minimizes the maximum pass-band ripple, while the inverse Chebyshev maximizes the minimum stop-band rejection.

The design criteria as we stated them do not address the issue of phase response, and none of these filters is linear-phase. The elliptical has the worst phase response, oscillating wildly in the pass-band and transition region (phase response in the stop-band is usually unimportant). The Butterworth is the smoothest in this regard, followed by the Chebyshev and inverse Chebyshev.

Although this entire section focused on analog low-pass filter, the principles are more general. All analog filters with a single pass-band and/or stop-band can be derived from the low-pass designs discussed above. For example, we can convert analog low-pass filter designs into high-pass filters by the simple transformation $f \to \frac{1}{f}$. Digital filters are a somewhat more complex issue, to be discussed in the next section. For now it is sufficient to say that IIR filters are often derived from analog Butterworth, Chebyshev, inverse Chebyshev, or elliptical designs. The reasoning is not that such designs are optimal; rather that the theory of the present section predated DSP and early practitioners prefered to exploit well-developed theory whenever possible.

EXERCISES

7.7.1 Show that a Butterworth filter of order N is maximally flat.

7.7.2 All Butterworth filters have their half gain (3 dB down) point at f_c. Higher order N makes the filter gain decrease faster, and the speed of decrease is called the 'rolloff'. Show that for high frequencies the rolloff of the Butterworth filter is 6 dB per octave (i.e., the gain decreases 6 dB for every doubling in frequency) or 20 dB per decade. How should N be set to meet a specification involving a pass-band end frequency f_p, a stop-band start frequency f_s, and a maximum error tolerance δ?

7.7.3 Show that the $2N$ poles of $|H(f)|^2$ for the analog Butterworth filter all lie on a circle of radius f_c in the s-plane, are equally spaced, and are symmetric with respect to the imaginary axis. Show that the poles of the Chebyshev I filter lie on an ellipse in the s-plane.

7.7.4 The HPNA 1.0 specification calls for a pulse consisting of 4 cycles of a 7.5 MHz square wave filtered by a five-pole Butterworth filter that extends from 5.5 MHz to 9.5 MHz. Plot this pulse in the time domain.

7.7.5 The frequency response of a certain filter is given by

$$|H(f)|^2 = \frac{f_c^\alpha}{f^\alpha + f_c^\alpha}$$

where α and f_c are parameters. What type of filter is this and what is the meaning of the parameters?

7.7.6 Repeat the previous exercise for these filters.

$$|H(f)|^2 = \frac{f^\alpha}{f^\alpha + f_c^\alpha}$$
$$|H(f)|^2 = \frac{f^\alpha + f_d^\alpha}{f^\alpha + f_c^\alpha}$$

7.7.7 Show that in the pass-band the Chebyshev filter gain is always between $\frac{1}{\sqrt{1-\delta^2}}$ and $\frac{1}{\sqrt{1+\delta^2}}$ so that the ripple is about $4\delta^2$ dB. Show that the gain falls monotonically in the stop-band with rolloff $20N$ dB per decade but always higher than the Butterworth filter of the same order.

7.7.8 We stated that an analog low-pass filter can be converted into a high-pass filter by a simple transformation of the frequency variable. How can band-pass and band-stop filters be similarly designed by transformation?

7.8 Digital Filter Design

We will devote only a single section to the subject of digital filter design, although many DSP texts devote several chapters to this subject. Although the theory of digital filter design is highly developed, it tends to be highly uninspiring, mainly consisting of techniques for constrained minimization of approximation error. In addition, the availability of excellent digital filter design software, both full graphic applications and user-callable packages, makes it highly unlikely that you will ever need to design on your own. The aim of this section is the clarification of the principles behind such programs, in order for the reader to be able to use them to full advantage.

Your first reaction to the challenge of filter design may be to feel that it is a trivial pursuit. It is true that finding the frequency response of a given filter is a simple task, yet like so many other inverse problems, finding a filter that conforms to a frequency specification is a more difficult problem. From a frequency domain specification we can indeed directly derive the impulse response by the FT, and the numeric values of the impulse response are undeniably FIR filter coefficients; but such an approach is only helpful when the impulse response quickly dies down to zero. Also numeric transformation of N frequency values will lead to a filter that obeys the specification at the exact frequencies we specified, but at in-between frequencies the response may be far from what is desired. The main trick behind filter design is how to constrain the frequency response of the filter so that it does not significantly deviate from the specification at *any* frequency.

We should note that this malady is not specific to time domain filtering. Frequency domain filtering uses the FT to transfer the signal to the frequency domain, performs there any needed filtering operation, and then uses the iFT to return to the time domain. We can only numerically perform a DFT for a finite number of signal values, and thus only get a finite frequency resolution. Multiplying the signal in the frequency domain enforces the desired filter specification at these frequencies only, but at intermediate frequencies anything can happen. Of course we can decide to double the number of signal times used thus doubling the frequency resolution, but there would still remain intermediate frequencies where we have no control. Only in the limit of the LTDFT can we completely enforce the filter specification, but that requires knowing the signal values at all times and so is an unrealizable process.

At its very outset the theory and practice of digital filter design splits into two distinct domains, one devoted to general IIR filters, and the other restricted to linear-phase FIR filters. In theory the general IIR problem is the harder one, and we do not even know how to select the minimum number of coefficients that meet a given specification, let alone find the optimal coefficients. Yet in practice the FIR problem is considered the more challenging one, since slightly suboptimal solutions based on the methods of the previous section can be exploited for the IIR problem, but not for the FIR one.

Let's start with IIR filter design. As we mentioned before we will not attempt to directly optimize filter size and coefficients; rather we start with a classical analog filter design and bring it into the digital domain. In order to convert a classical analog filter design to a digital one, we would like to somehow digitize. The problem is that the z-plane is not like the analog (Laplace) s-plane. From Section 4.10 we know that the sinusoids live on the imaginary axis in the s-plane, while the periodicity of digital spectra force them to be on the unit circle in the z-plane. So although the filter was originally specified in the frequency domain we are forced to digitize it in the time domain.

The simplest time domain property of a filter is its impulse response, and we can create a digital filter by evenly sampling the impulse response of any of the classical designs. The new digital filter's transfer function can then be recovered by z transforming this sampled impulse response. It is not hard to show that a transfer function thus found will be a rational function, and thus the digital filter will be ARMA. Furthermore the number of poles is preserved, and stable analog filters generate stable digital filters. Unfortunately, the frequency response of the digital filter will not be identical to that of the original analog filter, because of aliasing. In particular, the classical designs do not become identically zero at high frequencies, and so aliasing cannot be avoided. Therefore the optimal frequency domain properties of the analog designs are not preserved by impulse response sampling.

An alternative method of transforming analog filters into digital ones is the bilinear mapping method. The basic idea is to find a mapping from the s-plane to the z-plane and to convert the analog poles and zeros into the appropriate digital ones. For such a mapping to be valid it must map the imaginary axis $s = i\omega$ onto the unit circle $z = e^{i\omega}$, and the left half plane into the interior of the unit circle. The mapping (called 'bilinear' since the numerator and denominator are both linear in s)

$$z = \frac{1 + s}{1 - s} \tag{7.31}$$

does just that. Unfortunately, being nonlinear it doesn't preserve frequency, but it is not hard to find that the analog frequency can be mapped to the digital frequency by

$$\omega_{\text{analog}} = \tan(\tfrac{1}{2}\omega_{\text{digital}}) \tag{7.32}$$

thus compressing the analog frequency axis from $-\infty$ to ∞ onto the digital frequency axis from $-\pi$ to $+\pi$ in a one-to-one manner. So the bilinear mapping method of IIR filter design goes something like this. First 'prewarp' the frequencies of interest (e.g., f_p, f_c, f_s) using equation (7.32). Then design an analog filter using a Butterworth, Chebyshev, inverse Chebyshev, or elliptical design. Finally, transform the analog transfer function into a digital one by using the bilinear mapping of equation (7.31) on all the poles and zeros.

FIR filters do not directly correspond to any of the classical designs, and hence we have no recourse but to return to first principles. We know that given the required frequency response of a filter we can derive its impulse response by taking the iLTDFT

$$h_n = \frac{1}{2\pi} \int_{-\pi}^{\pi} H(e^{i\omega}) e^{i\omega n} \, d\omega \tag{7.33}$$

and that these h_n are the coefficients of the convolution in the time domain. Therefore, the theoretical frequency responses of the ideal low-pass, high-pass, band-pass, and band-stop filters already imply the coefficients of the ideal digital implementation. Assuming a noncausal filter with an odd number of coefficients, it is straightforward to find the following.

$$
\begin{aligned}
\text{low-pass:} \quad h_n &= \begin{cases} \frac{\omega_c}{\pi} & n = 0 \\ \frac{\omega_c}{\pi}\operatorname{sinc}(n\omega_c) & n \neq 0 \end{cases} \\[2mm]
\text{high-pass:} \quad h_n &= \begin{cases} 1 - \frac{\omega_c}{\pi} & n = 0 \\ -\frac{\omega_c}{\pi}\operatorname{sinc}(n\omega_c) & n \neq 0 \end{cases} \\[2mm]
\text{band-pass:} \quad h_n &= \begin{cases} \frac{\omega_2-\omega_1}{\pi} & n = 0 \\ \frac{\omega_2}{\pi}\operatorname{sinc}(n\omega_2) - \frac{\omega_1}{\pi}\operatorname{sinc}(n\omega_1) & n \neq 0 \end{cases} \\[2mm]
\text{band-stop:} \quad h_n &= \begin{cases} 1 + \frac{\omega_1-\omega_2}{\pi} & n = 0 \\ \frac{\omega_1}{\pi}\operatorname{sinc}(n\omega_1) - \frac{\omega_2}{\pi}\operatorname{sinc}(n\omega_2) & n \neq 0 \end{cases}
\end{aligned}
\tag{7.34}
$$

Unfortunately these h_n do not vanish as $|n|$ increases, so in order to implement a *finite* impulse response filter we have to truncate them after some $|n|$.

Truncating the FIR coefficients in the time domain means multiplying the time samples by a rectangular function and hence is equivalent to a convolution in the frequency domain by a sinc. Such a frequency domain convolution causes blurring of the original frequency specification as well as the addition of sidelobes. Recalling the Gibbs effect of Section 3.5 and the results of Section 4.2 regarding the transforms of signals with discontinuities, we can guess that multiplying the input signal by a smooth window

$$
h'_n = w_n\, h_n \tag{7.35}
$$

rather than by a sharply discontinuous rectangle should reduce (but not eliminate) the ill effects.

What type of window should be used? In Section 13.4 we will compare different window functions in the context of power spectrum estimation. Everything to be said there holds here as well, namely that the window function should smoothly increase from zero to unity and thence decrease smoothly back to zero. Making the window smooth reduces the sidelobes of the window's FT, but at the expense of widening its main lobe, and thus widening the transition band of the filter. From the computational complexity standpoint, we would like the window to be nonzero over only a short time duration; yet even nonrectangular windows distort the frequency response by convolving with the window's FT, and thus we would like this FT to be as narrow as possible. These two wishes must be traded off because the uncertainty theorem limits how confined the window can simultaneously be in the time and frequency domains. In order to facilitate this trade-off there are window families (e.g., kaiser and Dolph-Chebyshev) with continuously variable parameters.

So the windowing method of FIR filter design goes something like this.

```
Decide on the frequency response specification
Compute the infinite extent impulse response
Choose a window function:
      trade off transition width against stop-band rejection
      trade off complexity against distortion
Multiply the infinite extent impulse response by the window
```

The window design technique is useful when simple programming or quick results are required. However, FIR filters designed in this way are not optimal. In general it is possible to find other filters with higher stop-band rejection and/or lower pass-band ripple for the same number of coefficients. The reason for the suboptimality is not hard to find, as can be readily observed in Figure 7.14. The ripple, especially that of the stop-band, decreases as we move away from the transition. The stop-band attenuation specification that must be met constrains only the first sidelobe, and the stronger rejection provided by all the others is basically wasted. Were we able to find

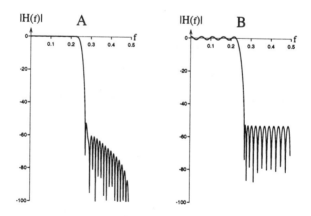

Figure 7.14: FIR design by window method vs. by Remez algorithm. (A) is the frequency response of a 71-coefficient low-pass filter designed by the window method. (B) is a 41-coefficient filter designed by the Remez algorithm using the same specification. Note the equiripple characteristic.

an *equiripple* approximation we could either reduce the maximum error or alternatively reduce the required number of coefficients.

As in classical filter design the equiripple property ensures that the maximal deviation from our amplitude specification be minimal. Techniques for solving the minimax polynomial approximation problem are reviewed in Appendix A.12. In the early seventies McClellan, Parks, and Rabiner published a paper and computer program that used the Remez exchange algorithm for FIR design. This program has become the most widely used tool in FIR design, and it is suggested that the reader obtain a copy (or a full up-to-date program with user interface and graphics based on the original program) and become proficient in its use.

Before concluding this chapter we should answer the question that must have occurred to you. When should FIR filters be used and when IIR? As a general rule integrators are IIR, while differentiators are FIR. Hilbert transforms are usually FIR although IIR designs are sometimes used. As to frequency-selective filters, the answer to this question is often (but not always) easy. First recall from Section 7.2 that FIR filters can be linear-phase, while IIR filters can only approach this behavior. Hence, if phase response is critical, as in many communications systems (see Chapter 18), you may be forced to use FIR filters (although the trick of exercise 7.2.5 may be of use). If phase response is not of major importance, we can generally meet a specification using either FIR or IIR filters. From the computational complexity point of view, IIR filters almost always end up being significantly more efficient, with elliptical filters having the lowest computational requirements. The narrower the transitions the more pronounced this effect becomes. However, these elliptical filters also have the worst phase response, erratically varying in the vicinity of transitions.

EXERCISES

7.8.1 Some digital filter design programs assume a sampling frequency (e.g., 8000 Hz). Can these programs be used to design filters for systems with different sampling frequencies?

7.8.2 Obtain a good filter design program and design an IIR low-pass filter using the four classical types from Section 7.7. What happens as you force the transition region to shrink in size? What is the effect of f_p for a given transition region width? Plot the phase response and group delay. How can you make the phase response more linear?

7.8.3 Design a low-pass FIR using the same criteria as in the previous exercise. Compare the amount of computation required for similar gain characteristics.

7.8.4 Repeat the previous two questions for a narrow band-pass filter.

7.8.5 An extremely narrow FIR low-pass filter requires a large number of coefficients, and hence a large amount of computation. How can this be reduced?

7.8.6 The impulse responses in equations (7.34) were for odd N. For even N the ideal frequency responses must be shifted by a half-integer delay $e^{i\frac{\omega}{2}}$ before applying equation (7.33). Find the ideal impulse responses for even N.

7.8.7 What are the coefficients of the ideal differentiator and Hilbert filters for even and odd N?

7.9 Spatial Filtering

Up to now we have dealt with filters that are frequency selective—filters that pass or block, amplify or attenuate signals based on frequency. In some applications there are other signal characteristics that help differentiate between signals, and these can be used along with frequency domain filtering, or by even by themselves when we need to separate signals of the same frequency. One such characteristic is the geographical position of the signal's source; if we could distinguish between signals on that basis we could emphasize a specific signal while eliminating interference from others not colocated with it.

A *wave* is a signal that travels in space as well as varying in time, and consequently is a function of the three-dimensional spatial coördinates s as well as being a function of time t. At any particular spatial coördinate the wave is a signal, and at any particular time we see a three-dimensional spatially varying function. A wave that travels at a constant velocity v without distortion is a function of the combination $s - vt$; traveling at exactly the right speed you 'move with the wave'. The distance a periodic wave travels during a single period is called the wavelength λ. Light and radio waves travel at the speed of light (approximately $3 \cdot 10^8$ meters per second), so that a wavelength of one meter corresponds to a frequency of 300 MHz.

Directional antennas, such as the TV antennas that clutter rooftops, are spatially selective devices for the reception and/or transmission of radio waves. Using carefully spaced conducting elements of precise lengths, transmitted radiation can be focused in the desired direction, and received signals arriving from a certain direction can be amplified with respect to those from other angles. The problem with such directional antennas is that changing the preferred direction involves physically rotating the antenna to point the desired way. *Beamforming* is a technique, mainly utilized in transmission and reception of sonar and radar signals, for focusing transmitted energy or amplifying received energy without having to physically rotate antennas. This feat is performed by combining a number of omnidirectional sensors (antennas, microphones, hydrophones, or loudspeakers depending on the type of wave).

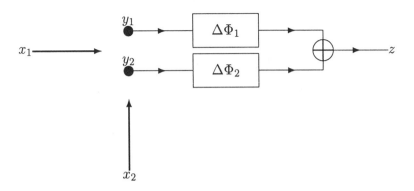

Figure 7.15: Beamforming to separate two sinusoidal signals of the same frequency. The sensor array consists of two antennas separated by the distance traveled by the wave during half a period. Each sensor is connected to a phase shifter and the phase shifted signals are summed.

In the simplest example of the principle involved we need to discriminate between two sinusoidal waves of precisely the same frequency and amplitude but with two orthogonal directions of arrival (DOAs) as depicted in Figure 7.15. Wave x_1 impinges upon the two sensors at the same time, and therefore induces identical signals y_1 and y_2. Wave x_2 arrives at the lower sensor before the upper, and accordingly y_1 is delayed with respect to y_2 by a half period. Were the reception of wave x_1 to be preferred we would set both phase shifters to zero shift; y_1 and y_2 would sum when x_1 is received, but would cancel out when x_2 arrives. Were we to be interested in wave x_2 we could set $\Delta\Phi_2$ to delay y_2 by one half period, while $\Delta\Phi_1$ would remain zero; in this fashion x_1 would cause y_1 and y_2 to cancel out, while x_2 would cause them to constructively interact. For waves with DOA separations other than $90°$ the same idea applies, but different phase shifts need to be employed.

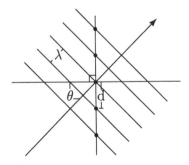

Figure 7.16: A wave impinging upon an array of $M = 5$ sensors spaced d apart. The parallel lines represent the peaks of the sinusoids and hence there is one wavelength λ between each pair. The wave arrives at angle θ from the normal to the line of the sensors. It is obvious from the geometry that when $\lambda = d \sin \theta$ the wave takes the same value at all of the sensors.

The device just described is a rudimentary example of a *phased array*, and it has the advantage of eliminating mechanical motors and control mechanisms. Switching between different directions can be accomplished essentially instantaneously, and we may also simultaneously recover signals with multiple DOAs with the same array, by utilizing several different phase shifters. We can enhance directivity and gain of a phased array by using more than two sensors in the array. With an array with M sensors, as in Figure 7.16, at every time n we receive M signals y_{mn} that can be considered a vector signal $\underline{y}_n$. To enhance a signal of frequency ω impinging at angle θ we need a phase delay of $\kappa = 2\pi \frac{d}{\lambda} \sin \theta$ between each two consecutive sensors. We could do this by successive time delays (resulting in a *timed array*) but in a phased array we multiply the mth component of the vector signal by a phase delay $e^{-\mathrm{i}\kappa m}$ before the components are combined together into the output z_n.

$$z_n = \sum_{m=0}^{M-1} y_{mn} e^{-\mathrm{i}\kappa m} \qquad (7.36)$$

Forgetting the time dependence for the moment, and considering this as a function of the DOA variable κ, this is seen to be a spatial DFT! The sensor number m takes the place of the time variable, and the DOA κ stands in for the frequency. We see here the beginnings of the strong formal resemblance between spatial filtering and frequency filtering.

Now what happens when a sinusoidal wave of frequency ω and DOA ϕ

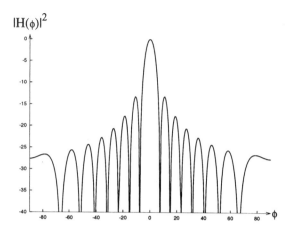

Figure 7.17: Angle response of a phased array. We depict the square of the response in dB referenced to the zero degree response for a phased array with $M = 32$ and $\pi\frac{d}{\lambda} = \frac{3}{4}$. The phased array is pointed to $\theta = 0$ and the horizontal axis is the angle ϕ in degrees.

is received? Sensor m sees at time n

$$
\begin{aligned}
y_{mn} &= A\, e^{+\mathrm{i}\varphi+\mathrm{i}\omega n+\mathrm{i}m\frac{2\pi d}{\lambda}\sin\phi} \\
&= A\, e^{\mathrm{i}\varphi} e^{\mathrm{i}\omega n} e^{\mathrm{i}km}
\end{aligned}
$$

where φ is the phase at the first sensor, and k is the DOA variable corresponding to angle ϕ. Substituting this into equation (7.36)

$$
\begin{aligned}
z_n &= \sum_{m=0}^{M-1} A\, e^{\mathrm{i}\varphi} e^{\mathrm{i}\omega n} e^{\mathrm{i}km} e^{-\mathrm{i}\kappa m} \\
&= A\, e^{\mathrm{i}\varphi} e^{\mathrm{i}\omega n} \sum_{m=0}^{M-1} e^{\mathrm{i}(k-\kappa)m} \\
&= A\, e^{\mathrm{i}\varphi} e^{\mathrm{i}\omega n}\, \frac{1 - e^{\mathrm{i}M(k-\kappa)}}{1 - e^{\mathrm{i}(k-\kappa)}} \\
&= A\, e^{\mathrm{i}\varphi} e^{\mathrm{i}\omega n} e^{\mathrm{i}\frac{1}{2}M(k-\kappa)} e^{-\mathrm{i}\frac{1}{2}(k-\kappa)}\, \frac{\sin\frac{1}{2}M(k-\kappa)}{\sin\frac{1}{2}(k-\kappa)}
\end{aligned}
$$

where we have performed the sum using (A.48), symmetrized, and substituted (A.8). The phased array *angle response* is the square of this expression

$$
|z|^2 = \left(\frac{\sin\frac{1}{2}M(k-\kappa)}{\sin\frac{1}{2}(k-\kappa)}\right)^2 = \left(\frac{\sin M\pi\frac{d}{\lambda}(\sin\phi-\sin\theta)}{\sin\pi\frac{d}{\lambda}(\sin\phi-\sin\theta)}\right)^2 \tag{7.37}
$$

and is plotted in Figure 7.17.

So the phased array acts as a spatial filter that is really quite similar to a regular frequency domain filter. The angle response of equation (7.37) is analogous to the frequency response of a frequency filter, and the high side-lobes in Figure 7.17 can be attenuated using techniques from filter design, such as windowing.

Our discussion has focused on simple sinusoidal waves; what if we need to pull in a complex wave? If the wave consists of only two frequency components, we can build two separate phased arrays based on the same sensors and add their results, or equivalently a single phased array with two delays per sensor. A little thought should be sufficient to convince you that arbitrary waves can be accommodated by replacing the simple phase delay with full FIR filters. In this way we can combine spatial and frequency filtering. Such a combined filter can select or reject a signal based on both its spectral and spatial characteristics.

EXERCISES

7.9.1 Direction fixing can also be performed using time of arrival (TOA) techniques, where the time a signal arrives at multiple sensors is compared. We use both phase differences and TOA to locate sound sources with our two ears, depending on the frequency (wavelength) of the sound. When is each used? How is elevation determined? (Hint: The external ear is not symmetric.) Can similar principles be exploited for SONAR echolocation systems?

7.9.2 Bats use biological sonar as their primary tool of perception, and are able to hunt insects at night (making the expression *blind as a bat* somewhat frivolous). At first, while searching for insects, they emit signals with basic frequency sweeping from 28 KHz down to 22 KHz and duration of about 10 milliseconds. Once a target is detected the sounds become shorter (about 3 milliseconds) in duration but scan from 50 KHz down to 25 KHz. While attempting to capture the prey, yet a third mode appears, of lower bandwidth and duration of below 1 millisecond. What is the purpose of these different cries? Can similar principles be used for fighter aircraft radar?

7.9.3 Figure 7.17 is not the conventional way of displaying antenna directivity patterns. Plot the same data in polar coördinates. Plot the frequency response of the above phased array at zero angle.

7.9.4 Show how to enhance a wave consisting of several weighted frequency components. Generalize this result to arbitrary waves.

Bibliographical Notes

Filters and filter design are covered in all standard DSP texts [186, 185, 200, 167], and chapters 4 and 5 of [241], as well as books devoted specifically to the subject [191]. Many original papers are reprinted in [209, 40, 41].

The original Parks-McClellan FIR design program is described and (FORTRAN) source code provided in [192, 165]. Extensions and portings of this code to various languages are widely available. After the original article appeared, much follow-on work appeared that treated the practical points of designing filters, including differentiators, Hilbert transforms, etc. [208, 213, 212, 78, 207].

Exercise 7.2.5 is based on [197].

8

Nonfilters

Filters have a lot going for them. In the previous chapter we have seen that they are simple to design, describe and implement. So why bother devoting an entire chapter to the subject of systems that are *not* filters?

There are two good reasons to study nonfilters—systems that are either nonlinear, or not time-invariant, or both. First, no system in the real world is ever perfectly linear; all 'linear' analog systems are nonlinear if you look carefully enough, and digital signals become nonlinear due to round-off error and overflow. Even relatively small analog nonlinearities can lead to observable results and unexpected major nonlinearities can lead to disastrous results. A signal processing professional needs to know how to identify these nonlinearities and how to correct them. Second, linear systems are limited in their capabilities, and one often requires processing functions that simply cannot be produced using purely linear systems. Also, linear systems are predictable; a small change in the input signal will always lead to a bounded change in the output signal. Nonlinear systems, however, may behave chaotically, that is, very small changes in the input leading to completely different behavior!

We start the chapter with a discussion of the effects of small nonlinearities on otherwise linear systems. Next we discuss several 'nonlinear filters', a term that is definitely an oxymoron. We *defined* a 'filter' as a linear and time-invariant system, so how can there be a 'nonlinear filter'? Well, once again, we are not the kind of people to be held back by our own definitions. Just as we say delta 'function', or talk about infinite energy 'signals', we allow ourselves to call systems that are obviously not filters, just that.

The mixer and the phase locked loop are two systems that are not filters due to not being time-invariant. These systems turn out to very important in signal processing for telecommunications. Our final topic, time warping, is an even more blatant example of the breakdown of time invariance.

8.1 Nonlinearities

Let's see what makes nonlinear systems interesting. We start by considering the simplest possible nonlinearity, a small additive quadratic term, which for analog signals reads

$$y(t) = x(t) + \epsilon x^2(t) \tag{8.1}$$

(assume $\epsilon \ll 1$). The spectral consequences can be made clear by considering an arbitrary sinusoidal input

$$x(t) = A\cos(\omega t) \tag{8.2}$$

for which the system will output

$$y(t) = A\cos(\omega t) + \epsilon A^2 \cos^2(\omega t) \tag{8.3}$$

which can be simplified by substituting from equation (A.25).

$$y(t) = A\sin(\omega t) + \frac{\epsilon A^2}{2} + \tfrac{1}{2}cos(2\omega t) \tag{8.4}$$

We see here three terms; the first being simply the original unscathed signal, the other two going to zero as $\epsilon \to 0$. The second term is a small DC component that we should have expected, since $\cos^2$ is always positive and thus has a nonzero mean. The final term is an attenuated replica of the original signal, but at twice the original frequency! This component is known as the *second harmonic* of the signal, and the phenomenon of creating new frequencies which are integer multiples of the original is called *harmonic generation*. Harmonic generation will always take place when a nonlinearity is present, the energy of the harmonic depending directly on the strength of the nonlinearity. In some cases the harmonic is unwanted (as when a nonlinearity causes a transmitter to interfere with a receiver at a different frequency), while in other cases nonlinearities are introduced precisely to obtain the harmonic.

We see here a fundamental difference between linear and nonlinear systems. Time-invariant linear systems are limited to filtering the spectrum of the incoming signal, while nonlinear systems can generate new frequencies.

What would have happened had the nonlinearity been cubic rather than quadratic?

$$y(t) = x(t) + \epsilon x^3(t)$$

You can easily find that there is *third harmonic generation* (i.e., a signal with thrice the original frequency appears from nowhere). A fourth order nonlinearity

$$y(t) = x(t) + \epsilon x^4(t)$$

will generate both second and fourth harmonics (see equation (A.33)); and n^{th} order nonlinearities generate harmonics up to order n. Of course a general nonlinearity that can be expanded in a Taylor expansion

$$y(t) = x(t) + \epsilon_2 x^2(t) + \epsilon_3 x^3(t) + \epsilon_4 x^4(t) + \cdots \tag{8.5}$$

will produce many different harmonics.

We can learn more about nonlinear systems by observing the effect of simple nonlinearities on signals composed of two different sinusoids.

$$x(t) = A_1 \cos(\omega_1 t) + A_2 \cos(\omega_2 t) \tag{8.6}$$

Inputing this signal into a system with a small quadratic nonlinearity

$$
\begin{aligned}
y(t) &= A_1 \cos(\omega_1 t) + A_2 \cos(\omega_2 t) \\
&\quad + A_1^2 \cos^2(\omega_1 t) + A_2^2 \cos^2(\omega_2 t) \\
&\quad + 2 A_1 A_2 \cos(\omega_1 t) \cos(\omega_2 t) \\
&= A_1 \cos(\omega_1 t) + A_2 \cos(\omega_2 t) \\
&\quad + A_1^2 \cos^2(\omega_1 t) + A_2^2 \cos^2(\omega_2 t) \\
&\quad + A_1 A_2 \cos\left((\omega_1 + \omega_2)t\right) \\
&\quad + A_1 A_2 \cos\left(|\omega_1 - \omega_2|t\right)
\end{aligned}
$$

we see harmonic generation for both frequencies, but there is also a new nonlinear term, called the *intermodulation product*, that is responsible for the generation of sum and difference frequencies. Once again we see that nonlinearities cause energy to migrate to frequencies where there was none before.

More general nonlinearities generate higher harmonics plus more complex intermodulation frequencies such as

$$
\begin{aligned}
&\omega_1 + \omega_2, \quad |\omega_1 - \omega_2|, \\
&\omega_1 + 2\omega_2, \quad 2\omega_1 + \omega_2, \\
&|2\omega_1 - \omega_2|, \quad |2\omega_2 - \omega_1|, \\
&\omega_1 + 3\omega_2, \quad 3\omega_1 + \omega_2, \\
&2\omega_1 + 3\omega_2, \quad 2\omega_1 + 3\omega_2,
\end{aligned}
$$

etc.

This phenomenon of intermodulation can be both useful and trouble-some. We will see a use in Section 8.5; the negative side is that it can cause hard-to-locate **R**adio **F**requency **I**nterference (RFI). For example, a tele-vision set may have never experienced any interference even though it is situated not far from a high-power radio transmitter. Then one day a taxi cab passes by a rusty fence that can act as a nonlinear device, and the com-bination of the cab's transmission and the radio station can cause a signal that interferes with TV reception.

EXERCISES

8.1.1 Show exactly which harmonics and intermodulation products are generated by a power law nonlinearity $y(t) = x(t) + \epsilon x^n(t)$.

8.1.2 Assume that the nonlinearity is exponential $y(t) = x(t) + \epsilon e^{x(t)}$ rather than a power law. What harmonics and intermodulation frequencies appear now?

8.2   Clippers and Slicers

One of the first systems we learned about was the clipping amplifier, or peak clipper, defined in equation (6.1). The peak clipper is obviously strongly nonlinear and hence generates harmonics, intermodulation products, etc. What is less obvious is that sometimes we use a clipper to *prevent* nonlinear effects. For example, if a signal to be transmitted has a strong peak value that will cause problems when input to a nonlinear medium, we may elect to artificially clip it to the maximal value that can be safely sent.

The opposite of this type of clipper is the *center clipper*, which zeros out signal values *smaller* than some threshold.

$$y = C_\theta(x) = \begin{cases} 0 & |x| < \theta \\ x & \text{else} \end{cases} \tag{8.7}$$

The center clipper is also obviously nonlinear, and although at first sight its purpose is hard to imagine, it has several uses in speech processing. The first relates to the removal of unwanted zero crossings. As we will see in Section 13.1 there are algorithms that exploit the number of times a signal crosses the time axis, and/or the time between two such successive zero crossings. These algorithms work very well on clean signals, but fail in the presence of noise that introduces extraneous zero crossings. The problem is not severe for strong signals but when the signal amplitude is low the noise may dominate and we find many extraneous zero crossings. Center clipping can remove unwanted zero crossings, restoring the proper number of zero crossings, at the price of introducing uncertainty in the precise time between them. In fact center clipping has become so popular in this scenario that it is used even when more complex algorithms, not based on zero crossings, are employed.

A related application is motivated by something we will learn in Chapter 11, namely that our hearing system responds approximately logarithmically to signal amplitude. Thus small amounts of noise that are not noticeable when the desired signal is strong become annoying when the signal is weak or nonexistent. A case of particular interest is echo over long distance telephone connections; linear echo cancellers do a good job at removing most of the echo, but when the other party is silent we can still hear our own voice returning after the round-trip delay, even if it has been substantially suppressed. This small but noticeable residual echo can be removed by a center clipper, which in this application goes under the uninformative name of NonLinear Processor (NLP). Unfortunately this leaves the line sounding too quiet, leading one to believe that the connection has been lost; this defect can be overcome by injecting artificial 'comfort noise' of the appropriate level.

The peak clipper and center clipper are just two special cases of a more general nonfilter called a *slicer*. Consider a signal known to be restricted to integer values that is received corrupted by noise. The obvious recourse is to clip each real signal value to the closest integer. This in effect slices up the space of possible received values into slices of unity width, the slice between $n - \frac{1}{2}$ and $n + \frac{1}{2}$ being mapped to n. The nonlinear system that performs this function is called a *slicer*.

Up to now we have discussed slicers that operate on a signal's amplitude, but more general slicers are in common use as well. For example, we may know that a signal transmitted to us is a sinusoid of given frequency but with phase of either $+\pi$ or $-\pi$. When measuring this phase we will in general find some other value, and must decide on the proper phase by slicing to the closest allowed value. Even more complex slicers must make decisions based on both phase and amplitude values. Such slicers are basic building blocks of modern high-speed modems and will be discussed in Section 18.18. You may wish to peek at Figure 18.26 to see the complexity of some slicers.

EXERCISES

8.2.1 Apply a center clipper with a small threshold to clean sampled speech. Do you hear any effect? What about noisy speech? What happens as you increase the threshold? At what point does the speech start to sound distorted?

8.2.2 Determine experimentally the type of harmonic generation performed by the clipper and the center clipper.

8.2.3 There is a variant of the center clipper with continuous output as a function of input, but discontinuous derivative. Plot the response of this system. What are its advantages and disadvantages?

8.2.4 When a slicer operates on sampled values a question arises regarding values exactly equidistant between two integer values. Discuss possible tactics.

8.2.5 A 'resetting filter' is a nonlinear system governed by the following equations.

$$y_n = x_n + w_n$$

$$r_n = \begin{cases} -\Theta & y_n < -\Theta \\ 0 & |y_n| < \Theta \\ \Theta & y_n > \Theta \end{cases}$$

$$w_n = y_{n-1} - r_{n-1}$$

Explain what the resetting filter does and how it can be used.

8.3   Median Filters

Filters are optimal at recovery of signals masked by additive Gaussian noise, but less adept at removing other types of unwanted interference. One case of interest is that of unreliable data. Here we believe that the signal samples are generally received without additive noise, but now and then may be completely corrupted. For example, consider what happens when we send a digital signal as bits through a unreliable communications channel. Every now and then a bit is received incorrectly, corrupting some signal value. If this bit happens to correspond to the least significant bit of the signal value, this corruption may not even be detected. If, however, it corresponds to the most significant bit there is a isolated major disruption of the signal. Such isolated incorrect signal values are sometimes called *outliers*.

An instructive example of the destructive effect of outliers is depicted in Figure 8.1. The original signal was a square wave, but four isolated signal values were strongly corrupted. Using a low-pass filter indeed brings the

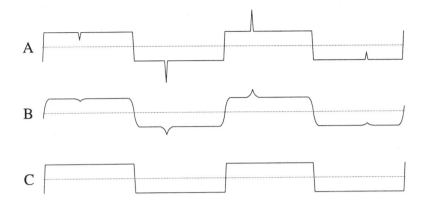

Figure 8.1: Comparison of a linear filter with a median filter. In (A) we have the original corrupted square wave signal; in (B) the signal has been filtered using a symmetric noncausal FIR low-pass filter; and in (C) we see the effect of a median filter.

corrupted signal values closer to their correct levels, but also changes signal values that were not corrupted at all. In particular, low-pass filtering smooths sharp transitions (making the square wave edges less pronounced) and disturbs the signal in the vicinity of the outlier. The closer we wish the outlier to approach its proper level, the stronger this undesirable smoothing effect will be.

An alternative to the low-pass filter is the median filter, whose effect is seen in Figure 8.1.C. At every time instant the median filter observes signal values in a region around that time, similar to a noncausal FIR filter. However, instead of multiplying the signal values in this region by coefficients, the median filter sorts the signal values (in ascending order) and selects 'median', i.e., the value precisely in the center of the sorted buffer. For example, if a median filter of length five overlaps the values $1, 5, 4, 3, 2$, it sorts them into $1, 2, 3, 4, 5$ and returns 3. In a more typical case the median filter overlaps something like $2, 2, 2, 15, 2$, sorts this to $2, 2, 2, 2, 15$ and returns 2; and at the next time instant the filter sees $2, 2, 15, 2, 2$ and returns 2 again. Any isolated outlier in a constant or slowly varying signal is completely removed.

Why doesn't a median filter smooth a sharp transition between two constant plateaus? As long as more than half the signal values belong to one side or the other, the median filter returns the correct value. Using an odd-order noncausal filter ensures that the changeover happens at precisely the right time.

What happens when the original signal is not constant? Were the linearly increasing signal $\ldots 1, 2, 3, 4, 5, 6, 7, 8, 9, 10 \ldots$ to become corrupted to

...$1, 2, 3, 4, 99, 6, 7, 8, 9, 10, \ldots$, a median filter of length 5 would be able to correct this to ...$1, 2, 3, 4, 6, 7, 8, 8, 9, 10, \ldots$ by effectively skipping the corrupted value and replicating a later value in order to resynchronize. Similarly, were the corrupted signal to be ...$1, 2, 3, 4, -99, 6, 7, 8, 9, 10, \ldots$, the median filter would return the sequence ...$1, 2, 2, 3, 4, 6, 7, 8, 9, 10, \ldots$ replicating a previous value and skipping to catch up. Although the corrupted value never explicitly appears, it leaves its mark as a phase shift that lasts for a short time interval.

What if there is additive noise in addition to outliers? The simplest thing to do is to use a median filter *and* a linear low-pass filter. If we apply these as two separate operations we should probably first median filter in order to correct the gross errors and only then low-pass to take care of the noise. However, since median filters and FIR filters are applied to the input signal in similar ways, we can combine them to achieve higher computational efficiency and perhaps more interesting effects. One such combination is the *outlier-trimmed FIR filter*. This system sorts the signal in the observation window just like a median filter, but then removes the m highest and lowest values. It then adds together the remaining values and divides by their number returning an MA-smoothed result. More generally, an *order statistic filter* first sorts the buffer and then combines the sorted values as a weighted linear sum as in an FIR filter. Usually such filters have their maximal coefficient at the center of the buffer and decrease monotonically toward the buffer ends.

The novelty of the median filter lies in the sorting operation, and we can exploit this same idea for processing other than noise removal. A *dilation filter* outputs the maximal value in the moving buffer, while an *erosion filter* returns the minimal value. These are useful for emphasizing constant positive-valued signals that appear for short time durations, over a background of zero. Dilation expands the region of the signal at the expense of the background while erosion eats away at the signal. Dilation and erosion are often applied to signals that can take on only the values 0 or 1. Dilation is used to fill in holes in long runs of 1s while erosion clips a single spike in the midst of silence. For very noisy signals with large holes or spikes dilation or erosion can be performed multiple times. We can also define two new operations. An *opening* filter is an erosion followed by a dilation while a *closing* filter is a dilation followed by an erosion. The names are meaningful for holes in 0, 1-valued signals. These four operations are most commonly used in image processing, where they are collectively called *morphological processing*.

EXERCISES

8.3.1 Prove that the median filter is not linear.

8.3.2 Median filtering is very popular in image processing. What properties of common images make the median filter more appropriate than linear filtering?

8.3.3 The *conditional median filter* is similar to the median filter, but only replaces the input value with the median if the difference between the two is above a threshold, otherwise it returns the input value. Explain the motivation behind this variant.

8.3.4 Graphically explain the names dilation, erosion, opening, and closing by considering 0, 1-valued signals.

8.3.5 Explain how morphological operations are implemented for image processing of binary images (such as fax documents). Consider 'kernels' of different shapes, such as a 3*3 square and a 5-pixel cross. Program the four operations and show their effect on simple images.

8.4 Multilayer Nonlinear Systems

Complex filters are often built up from simpler ones placed in series, a process known as *cascading*. For example, if we have a notch filter with 10 dB attenuation at the unwanted frequencies, but require 40 dB attenuation, the specification can be met by cascading four identical filters. Assume that each of N cascaded subfilters is a causal FIR filter of length L, then the combined filter's output at time n depends on its input at time $n - NL$. For example, assume that a finite duration signal x_n is input to a filter h producing y_n that is input into a second filter g resulting in z_n. Then

$$
\begin{aligned}
y_n &= h_0 x_n + h_1 x_{n-1} + h_2 x_{n-2} + \ldots + h_{L-1} x_{L-1} \\
z_n &= g_0 y_n + g_1 y_{n-1} + g_2 y_{n-2} + \ldots + g_{L-1} x_{L-1} \\
&= g_0 \left(h_0 x_n + h_1 x_{n-1} + h_2 x_{n-2} + \ldots + h_{L-1} x_{L-1} \right) \\
&\quad + g_1 \left(h_0 x_{n-1} + h_1 x_{n-2} + h_2 x_{n-3} + \ldots + h_{L-2} x_{L-1} \right) \\
&= g_0 h_0 x_n + \left(g_0 h_1 + g_1 h_0 \right) x_{n-1} + \left(g_0 h2 + g_1 h_1 + g_2 h_0 \right) x_{n-2} + \ldots
\end{aligned}
$$

which is equivalent to a single FIR filter with coefficients equal to the convolution $g * h$.

In order for a cascaded system to be essentially different from its constituents we must introduce nonlinearity. Augmenting the FIR filter with a hard limiter we obtain a 'linear threshold unit' known more commonly as the *binary perceptron*

$$y_n = \text{sgn}\left(\sum_n w_n x_n\right) \qquad (8.8)$$

while using a less drastic smooth nonlinearity we obtain the *sigmoid perceptron.*

$$y_n = \tanh\left(\beta \sum_n w_n x_n\right) \qquad (8.9)$$

As β increases the sigmoid perceptron approaches the threshold one. In some applications $0, 1$ variables are preferable to ± 1 ones, and so we use the step function

$$y_n = \Theta\left(\sum_n w_n x_n\right) \qquad (8.10)$$

or the smooth version

$$y_n = \sigma\left(\sum_n w_n x_n\right) \qquad (8.11)$$

where we defined the 'logistic sigmoid'.

$$\sigma(x) \equiv \frac{e^x}{1 + e^x} = 1 + \tfrac{1}{2}\tanh x \qquad (8.12)$$

Cascading these nonlinear systems results in truly new systems; a single perceptron can only approximate a small fraction of all possible systems, while it can be shown that arbitrary systems can be realized as cascaded sigmoid perceptrons.

In Figure 8.2 we depict a **MultiLayer Perceptron** (MLP). This particular MLP has two 'layers'; the first computes L weighted sum of the N input values and then hard or soft limits these to compute the values of L 'hidden units', while the second immediately thereafter computes a single weighted sum over the L hidden units, creating the desired output. To create a three-layer perceptron one need only produce many second-layer sigmoid weighted sums rather than only one, and afterward combine these together using one final perceptron. A theorem due to Kolmogorov states that three layers are sufficient to realize arbitrary systems.

The perceptron was originally proposed as a classifier, that is, a system with a single signal as input and a logical output or outputs that identify the signal as either belonging to a certain class. Consider classifying spoken digits as belonging to one of the classes named $0, 1, 2 \ldots 9$. Our MLP could look at all the nonzero speech signal samples, compute several layers of

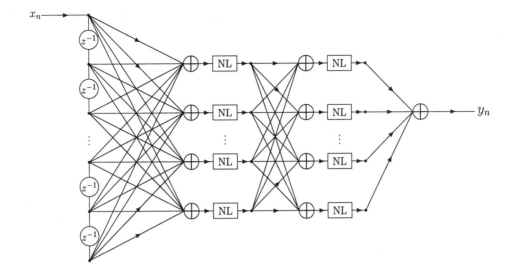

Figure 8.2: A general nonlinear two-layer feedforward system. Although not explicitly shown, each connection arc represents a weight. NL stands for the nonlinearity, for example, the sgn or tanh function.

hidden values, and finally activate one of 10 output units, thereby expressing its opinion as to the digit that was uttered. Since humans can perform this task we are confident that there is *some* system that can implement the desired function from input samples to output logical values. Since the aforementioned theorem states that (assuming a sufficient number of hidden units) three-layer MLPs can implement arbitrary systems, there must be a three-layer MLP that imitates human behavior and properly classifies the spoken digits.

How are MLP systems designed? The discussion of this topic would lead us too far astray. Suffice it to say that there are *training algorithms* that when presented with a sufficient amount of data can accomplish the required system identification. The most popular of these algorithms is 'backpropagation', ('backprop') which iteratively presents an input, computes the present output, corrects the internal weights in order to decrease the output error, and then proceeds to the next input-output pair.

How many hidden units are needed to implement a given system? There are few practical rules here. The aforementioned theorem only says that there is some number of hidden units that allows a given system to be emulated; it does not inform us as to the minimum number needed for all specific cases, or whether one, two, or three layers are needed. In practice these architectural parameters are often determined by trial and error.

EXERCISES

8.4.1 Using linear threshold units we can design systems that implement various logic operations, where signal value 0 represents 'false' and 1 'true'. Find parameters w_1, w_2, and φ such that $y = \Theta\left(w_1 x_1 + w_2 x_2 - \varphi\right)$ implements the logical AND and logical OR operations. Can we implement these logical operations with linear systems?

8.4.2 Of the 16 logical operations between two logical variables, which can and which can't be implemented?

8.4.3 Find a multilayer system can implements XOR.

8.4.4 What is the form of curves of equal output for the perceptron of equation (8.9)? What is the form of areas of the same value of equation (8.8)? What is the form of these areas for multilayer perceptrons formed by AND or OR of different simple perceptrons? What types of sets cannot be implemented? How can this limitation be lifted?

8.4.5 What are the derivatives of the sigmoid functions (equations (8.11) and (8.9))? Show that $\sigma'(x) = \sigma(x)\left(1 - \sigma(x)\right)$. Can you say something similar regarding the tanh sigmoid?

8.4.6 Another nonlinear system element is $y(x) = e^{\beta \sum_n (x_n - \mu_n)^2}$, known as the Gaussian radial unit. What is the form of curves of equal output for this unit? What can be said about implementing arbitrary decision functions by radial units?

8.5   Mixers

A *mixer* is a system that takes a band-pass signal centered around some frequency f_0, and moves it along the frequency axis (without otherwise changing it) until it is centered around some other frequency f_1. Some mixers may also invert the spectrum of the mixed signal. In Figures 8.3 and 8.4 we depict the situation in stylized fashion, where the triangular spectrum has become prevalent in such diagrams, mainly because spectral inversions are obvious. In older analog signal processing textbooks mixing is sometimes called 'heterodyning'. In many audio applications the term 'mixing' is used when simple weighted addition of signals is intended; thus when speaking to audio professionals always say 'frequency mixing' when you refer to the subject of this section.

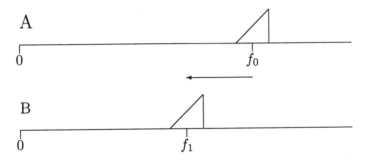

Figure 8.3: The effect of mixing a narrow-band analog signal without spectral inversion. In (A) we see the spectrum of the original signal centered at frequency f_0, and in (B) that of the mixed signal at frequency f_1.

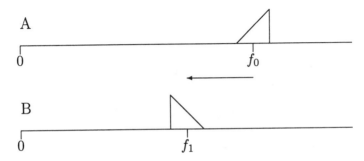

Figure 8.4: The effect of mixing a narrow-band analog signal with spectral inversion. In (A) we see the spectrum of the original signal centered at frequency f_0, and in (B) the mixed and inverted signal at frequency f_1. Note how the triangular spectral shape assists in visualizing the inversion.

It is obvious that a mixer *cannot* be a filter, since it can create frequencies where none existed before. In Section 8.1 we saw that harmonics could be generated by introducing nonlinearity. Here there is no obvious nonlinearity; indeed we expect that shifting the frequency of a sum signal will result in the sum of the shifted components. Thus we must conclude that a mixer must be a linear but not a time-invariant system.

Mixers have so many practical applications that we can only mention a few of them here. Mixers are crucial elements in telecommunications systems which transmit signals of the form given in equation (4.66)

$$s(t) = A(t) \sin \left(2\pi f_c t + \phi(t) \right)$$

where the frequency f_c is called the *carrier frequency*. The information to be sent is contained either in the amplitude component $A(t)$, the phase component $\phi(t)$, or both; the purpose of a receiver is to recover this information.

Many receivers start by mixing the received signal down by f_c to obtain the simpler form of equation (4.65)

$$s(t) = A(t) \sin \big(\phi(t)\big)$$

from which the amplitude and phase can be recovered using the techniques of Section 4.12.

The phase is intimately connected with the carrier frequency so that the mixing stage is obviously required for proper phase recovery. Even were there to be a mixer but its frequency to be off by some small amount Δf the phase would be misinterpreted as $2\pi\Delta ft + \phi(t)$ along with the unavoidable jumps of 2π. The amplitude signal is apparently independent of the carrier frequency; can we conclude that no mixer is required for the recovery of amplitude-modulated signals? No, although mistuning is much less destructive. The reason a mixer is required is that the receiver sees many possible transmitted signals, each with its own carrier frequency f_c. Isolation of the desired signal is accomplished by downmixing it and injecting it into a narrow low-pass filter. The output of this filter now contains only the signal of interest and demodulation can continue without interference. When you tune an AM or FM radio in order to hear your favorite station you are actually adjusting a mixer. Older and simpler receivers allow this downmix frequency to be controlled by a continuously rotatable (i.e., analog) knob, while more modern and complex receivers use digital frequency control.

Telephone-quality speech requires less than 4 KHz of bandwidth, while telephone cables can carry a great deal more bandwidth than this. In the interest of economy the telephone network compels a single cable to simultaneously carry many speech signals, a process known as *multiplexing*. It is obvious that we cannot simply add together all the signals corresponding to the different conversations, since there would be no way to separate them at the other end of the cable. One solution, known as **Frequency Domain Multiplexing (FDM)**, consists of upmixing each speech signal by a different offset frequency before adding all the signals together. This results in each signal being confined to its own frequency band, and thus simple band-pass filtering and mixing back down (or mixing first and then low-pass filtering) allows complete recovery of each signal. The operation of building the FDM signal from its components involves upmixing and addition, while the extraction of a single signal requires downmixing and filtering.

Sometimes we need a mixer to compensate for the imperfections of other mixers. For example, a modem signal transmitted via telephone may be upmixed to place it in a FDM transmission, and then downmixed before delivery to the customer. There will inevitably be a slight difference between the frequency shifts of the mixers at the two ends, resulting in a small residual frequency shift. This tiny shift would never be noticed for speech, but modem signals use frequency and phase information to carry information and even slight shifts cannot be tolerated. For this reason the demodulator part of the modem must first detect this frequency shift and then employ a mixer to correct for it.

Mixers may even appear without our explicitly building them. We saw in Section 8.1 that transmitted signals that pass through nonlinearities may give rise to intermodulation frequencies; we now realize that this is due to unintentional mixing.

A first attempt at numerically implementing a mixer might be to Fourier analyze the signal (e.g., with the FFT), translate the signal in the frequency domain to its new place, and then return to the time domain with the iFT. Such a strategy may indeed work, but has many disadvantages. The digital implementation would be quite computationally intensive, require block processing and so not be real-time-oriented, and only admits mixing by relatively large jumps of the order $\frac{f_s}{N}$. What we require is a real-time-oriented time-domain algorithm that allows arbitrary frequency shifts.

As in many such cases, inspiration comes from traditional hardware implementations. Mixers are traditionally implemented by injecting the output of an oscillator (often called the local oscillator) and the signal to be mixed into a nonlinearity. This nonlinearity generates a product signal that has frequency components that are sums and differences of the frequencies of the signal to be mixed and the local oscillator. The mixer is completed by filtering out all components other than the desired one. The essential part of the technique is the forming of a product signal and then filtering.

Consider an analog complex exponential of frequency ω.

$$s(t) = Ae^{i\omega t}$$

In order to transform it into an exponential of frequency ω'

$$s'(t) = Ae^{i\omega' t}$$

we need only multiply it by $e^{i(\omega'-\omega)t}$.

$$s(t)e^{i(\omega'-\omega)t} = Ae^{i\omega t}e^{i(\omega'-\omega)t} = Ae^{i\omega' t} = s'(t)$$

Note that the multiplying signal is sinusoidal at the frequency shift frequency and thus the system is not time-invariant.

Similarly, a signal that is composed of many frequency components

$$s(t) = \sum_k A_k e^{i\omega_k t}$$

will be rigidly translated in frequency when multiplied by a complex exponential.

$$s(t) e^{-i\Delta\omega t} = \sum_k A_k e^{i(\omega_k - \Delta\omega)t}$$

When a signal is mixed down in frequency until it occupies the range from DC up to its bandwidth, it is said to have been 'downmixed to low-pass'. When we go even further and set the signal's center frequency to zero, we have 'downmixed to zero'.

So it seems that mixing is actually quite simple. The problems arise when we try to mix real-valued signals rather than analytic ones, or digital signals rather than analog ones. To illustrate the problems that arise, consider first the mixing of real signals. Since real signals have symmetric spectra, we have to look at both positive and negative frequencies to understand the whole story.

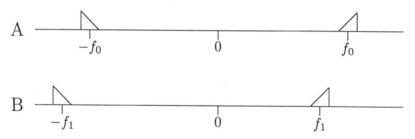

Figure 8.5: A real signal at frequency f_0, whose spectrum is depicted in (A), is moved to frequency f_1 by complex mixing. When a signal is multiplied by a complex exponential all frequency components are shifted in the same direction, as seen in (B).

In Figure 8.5 we see the effect of mixing a real-valued signal using a complex exponential local oscillator. The mixer's effect is precisely as before, but the resulting signal is no longer real! What we really want to do is to mix a real signal using a real oscillator, which is depicted in Figure 8.6. Here the mixer no longer rigidly moves the whole spectrum; rather it compresses or expands it around the DC. In particular we must be careful with downmixing signals past the DC to where the two sides overlap, as in Figure 8.7. Once different parts of the spectrum overlap information is irrevocably lost, and we can no longer reverse the operation by upmixing.

A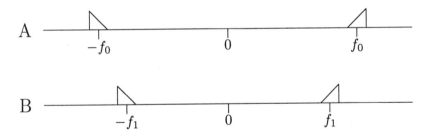

Figure 8.6: A real signal at frequency f_0, whose spectrum is depicted in (A), is moved to frequency f_1 by real mixing. When a real signal is multiplied by a real sinusoid its positive and negative frequency approach each other, as seen in (B).

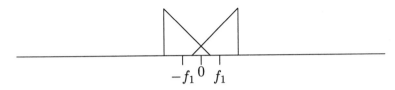

Figure 8.7: A real signal after destructive downmixing. Once the spectrum overlaps itself information is lost.

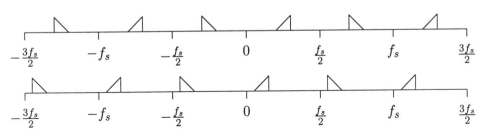

Figure 8.8: A digital mixer. Here a real-valued digital signal is mixed by a complex digital exponential.

What about digital signals? The spectrum of a digital signal is periodic and mixing moves all of the replicas, as depicted in Figure 8.8 for a real digital signal being mixed downward in frequency by a complex exponential. Note that there is a new phenomenon that may occur. Even when mixing with a complex oscillator downmixing to zero causes other spectral components to enter the Nyquist spectral region. This is a kind of aliasing but is both reversible and correctable by appropriate filtering.

EXERCISES

8.5.1 Diagram all the cases of mixing real or complex digital signals by real or complex oscillators.

8.5.2 We originally claimed that a mixer generates new frequencies due to its being time-invariant but *linear*. Afterward when discussing its analog implementation we noted that the product is generated by a time-invariant *nonlinearity*. Reconcile these two statements.

8.5.3 There are two techniques to mix a real signal down to zero. The signal can be converted to the analytic representation and then multiplied by a complex exponential, or multiplied by the same complex exponential and then low-pass filtered. Demonstrate the equivalence of these two methods. What are the practical advantages and disadvantages of each approach?

8.6   Phase-Locked Loops

Another common system that fulfills a function similar to that of a filter, but is not itself a filter, is the **Phase-Locked Loop** (PLL). This is a system that can 'lock on' to a sinusoidal signal whose frequency is approximately known, even when this signal is only a small component of the total input. Although the basic idea is to filter out noise and retain the sinusoidal signal of interest, such 'locking on' is definitely a nonlinear and time-variant phenomenon and as such cannot be performed by a filter.

Why do we need such a system? One common use is clock recovery in digital communications systems. As a simple example consider someone sending you digital information at a constant rate of 1 bit every T seconds (presumably T would be some small number so that a large number of bits may be sent per second). Now the transmitter has a clock that causes a bit to be sent every T seconds. The receiver, knowing the sender's intentions, expects a bit every T seconds. However, the receiver's clock, being an independent electronic device, will in general run at a slightly different rate than that of the transmitter. So in effect the receiver looks for a bit every T' seconds instead of every T seconds. This problem may not be evident at first, but after enough time has passed the receiver is in effect looking for bits at the wrong times, and will either miss bits or report extraneous ones. For high bit rates it doesn't take long for this to start happening!

In order to avoid this problem the sender can transmit a second signal, for example, a sinusoid of frequency f generated by the transmitter's internal clock. The receiver need only set its clock precisely according to this sinusoid and the discrepancy problem vanishes. This operation of matching clocks is called *synchronization*, often shortened to 'synching' (pronounced *sinking*) or 'synching up'. Synchronization of the receiver's clock to the transmitter's has to be maintained continuously; even if properly initially matched, non-synched clocks will drift apart with time, introducing bit slips and insertions.

The accurate synching up of the receiver's clock depends critically on obtaining a clean signal from the transmitter. A naive DSP approach would be to use a very narrow-band band-pass filter centered on f to recover the clock signal and reject as much noise as possible. Such an attempt is doomed to failure since we don't know f (were we to know f there wouldn't be anything to do). Setting an extremely sharp band-pass filter centered on the receiver's estimate f' may leave the true f outside the filter bandwidth. Of course we could use a wider filter bandwidth, but that would increase the noise. What we really need is to find and track the received signal's center frequency. That is what the PLL does.

In order to build a PLL we first need some basic building blocks. The first is traditionally called a **V**oltage-**C**ontrolled **O**scillator (VCO). Like an ordinary oscillator the VCO outputs a real sinusoid, but unlike the oscillators we have seen before the VCO has an input as well. With zero input the VCO oscillates at its 'natural frequency' ω_0, but with nonzero input $x(t)$ the VCO output's instantaneous frequency changes to $\omega_0 + \nu(t)$. It is now straightforward to express the VCO output $y(t)$ in terms of its input $x(t)$.

$$y(t) = A\sin\left(\omega_0 t + \varphi(t)\right) \qquad \text{where} \qquad x(t) = \frac{d\varphi(t)}{dt} \qquad (8.13)$$

The analog VCO is thus controlled by the voltage at its input, and hence its name; the digital version should properly be called a **N**umerically-**C**ontrolled **O**scillator (NCO), but the name VCO is often used even when no voltages are evident.

The next basic subsystem has two inputs where it expects two pure sinusoids; its output is proportional to the difference in frequency between the two. There are many ways to implement this block, e.g., one could use two frequency demodulators (Section 4.12) and an adder with one input negated. A more devious implementation uses a mixer, a special notch filter and an amplitude demodulator. The VCO output is used to downmix the input to zero; the mixer output is input to a filter with gain $|\omega|$ so that when the input frequency matches the VCO there is no output, while as the deviation increases so does the amplitude; finally the amplitude demodulator outputs the desired frequency difference.

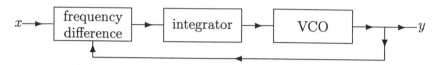

Figure 8.9: The frequency-locked loop (FLL). The output is a sinusoid that tracks the frequency of the input signal.

Using the two special blocks we have defined so far we can already make a first attempt at a system that tracks sinusoidal components (see Figure 8.9). We will call this system a **Frequency-Locked Loop** (FLL), as its feedback loop causes it to lock onto the frequency of the input signal. Consider what happens when a sinusoid with frequency $\omega > \omega_0$ is applied to the input (previously zero). At first the frequency difference block outputs $\omega - \omega_0$, and were this to be input to the VCO it would change its frequency from ω_0 to $\omega_0 + \omega - \omega_0 = \omega$. Unfortunately, this correct response is just an instantaneous spike since the difference would then become zero and the VCO would immediately return to its natural frequency. The only escape from this predicament is to integrate the difference signal before passing it to the VCO. The integral maintains a constant value when the difference becomes zero, forcing the VCO to remain at ω.

The FLL can be useful in some applications but it has a major drawback. Even if the input is a pure sinusoid the FLL output will not in general precisely duplicate it. The reason being that there is no direct relationship between the input and output *phases*. Thus in our bit rate recovery example the FLL would accurately report the rate at which the bits are arriving, but could not tell us precisely when to expect them. In order to track the input signal in both frequency and phase, we need the more sensitive phase-locked loop. Looking carefully at our FLL we see that the frequency difference is integrated, returning a phase difference; the PLL replaces the frequency difference block of the FLL with an explicit phase difference one.

The phase difference subsystem expects two sinusoidal inputs of approximately the same frequency and outputs the phase difference between them. One could be built similarly to the frequency difference block by using two phase demodulators and an adder with one input negated; however, there are approximations that are much easier to build for analog signals and cheaper to compute for digital ones. The most common approximate difference block shifts the phase of one input by 90° and multiplies the two signals.

$$
\begin{aligned}
s_1(t) &= \sin(\omega_0 t + \phi_1) \\
s_2(t) &= \sin(\omega_0 t + \phi_2) \\
\tilde{s}_2(t) &= \cos(\omega_0 t + \phi_2) \\
s_1(t)\tilde{s}_2(t) &= \tfrac{1}{2}\Big(\sin(\phi_1 - \phi_2) + \sin(2\omega_0 + \phi_1 + \phi_2)\Big)
\end{aligned}
$$

It the low-pass filters the output to remove the double frequency component. The filtered product is proportional to

$$
\sin(\phi_1 - \phi_2) \sim \phi_1 - \phi_2
$$

where the approximation is good for small phase differences.

You may question the wisdom of limiting the range of the phase difference approximation to small values, but recall that even the ideal phase difference is limited to $\pm\pi$! So the ideal phase difference block has a sawtooth characteristic while the approximation has a sinusoidal one. If you really prefer piecewise linear characteristics the *xor phase comparator* is implemented by hard limiting s_1 and $\tilde{s}_2$ before multiplying them and then averaging over a single cycle. When s_1 and s_2 are precisely in phase, s_1 and $\tilde{s}_2$ are 90° out of phase and thus their product is positive just as much as it is negative, and so averages to zero. When they move out of phase in either direction the duty cycle of the product becomes nonzero. The characteristics of the three phase difference blocks are contrasted in Figure 8.10.

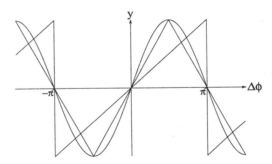

Figure 8.10: Characteristics of three phase difference blocks. The ideal phase difference subsystem has its output vary like a sawtooth as a function of the phase difference $\Delta\phi = \phi_1 - \phi_2$. The simple product subsystem has sinusoidal characteristic, while the xor comparator has a triangular one. The important feature of all these blocks is that for small phase differences the characteristic is linear.

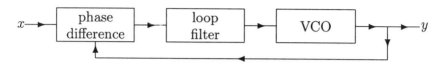

Figure 8.11: The phase-locked loop, or PLL. The output is a sinusoid that tracks the phase of the input signal.

No matter how we build the phase detector, the proper way to use it is depicted in Figure 8.11. If the input is truly a sinusoid of the VCO's natural frequency, the phase difference output causes the VCO frequency to momentarily increase in order to catch up with the input or decrease to let the input catch up. The PLL is even more useful when the input is noisy. In this case the phase difference varies erratically but the low-pass filter smooths the jumps so that the VCO only tracks the average input phase. Quite noisy signals can be applied provided the low-pass filter is sufficiently narrow.

What if the input frequency doesn't equal the VCO natural frequency? Small constant frequency differences can be thought of as constantly changing phase differences, and the phase corrections will cause the VCO to oscillate at the average input frequency. If the frequency difference is larger than the low-pass filter bandwidth the VCO will receive zero input and remain at its natural frequency, completely oblivious to the input. For input frequencies in the *capture range* the VCO does get some input and starts moving toward the input frequency. The difference then further decreases, allowing more energy through the filter, and the PLL 'snaps' into lock. Once locked the phase difference is DC and completely passed by the filter, thus maintaining lock. If the input frequency varies the VCO automatically tracks it as long as it remains in the *tracking range*.

The low-pass filter used in the PLL is usually of the IIR type. When the phase detector is of the product type a single low-pass filter can be used both for the filtering needed for the PLL's noise rejection and for rejecting the double frequency component. When the double frequency rejection is not required we may be able to skip the filter altogether. In this case there is still a feedback path provided by the PLL architecture, and so the PLL is said to be of *first order*. If the IIR filter has a single pole the additional pole-like behavior leads us to say that the PLL is of *second order*. Higher orders are seldom used because of stability problems.

EXERCISES

8.6.1 Simulate the FLL's frequency behavior by assuming a VCO natural frequency, inputting some other frequency, and using simple addition to integrate. Simulate a slowly varying input frequency. How far can the input frequency be from the natural frequency?

8.6.2 Adding a clipping amplifier between the frequency difference and the integrator of the FLL makes the FLL have two operating regions, acquisition and tracking. Analyze the behavior of the system in these two regions.

8.6.3 Compare the PLL and FLL from the aspects of frequency acquisition range, steady state frequency, and steady state phase error.

8.6.4 Explain how to use the PLL to build a frequency synthesizer, that is, an oscillator with selectable accurate frequency.

8.6.5 What effect does decreasing a PLL's low-pass filter bandwidth have on the capture range, the acquisition time, and robustness to noise?

8.7 Time Warping

Say 'pneumonoultramicroscopicsilicovolcanoconiosis'. I bet you can't say it again! I mean pronounce precisely the same thing again. It might sound the same to you, but that is only because your brain corrects for the phenomenon to which I am referring; but were you to record both audio signals and compare them you would find that your pacing was different. In the first recording you may have dwelled on the second syllable slightly longer while in the second recording the fourth syllable may have more stress. This relative stretching and compressing of time is called 'time warping', and it is one of the main reasons that automatic speech recognition is so difficult a problem.

For sinusoidal signals making time speed up and then slow down is exactly equivalent to changing the instantaneous frequency, but for more complex signals the effect is somewhat harder to describe using the tools we have developed so far. A system that dynamically warps time is obviously not time-invariant, and hence not a filter; but we are not usually interested in building such a system anyway. The truly important problem is how to compare two signals that would be similar were it not for their undergoing somewhat different time warping.

One approach to solving this problem is called **D**ynamic **T**ime **W**arping (DTW). DTW is a specific application of the more general theory of *dynamic* *programming* and essentially equivalent to the *Viterbi algorithm* that we will discuss in Section 18.11. In order to facilitate understanding of the basic concepts of dynamic programming we will first consider the problem of spelling checking. Spelling checkers have become commonplace in word processors as a means of detecting errant words and offering the best alternatives. We will assume that the checker has a precompiled word list (dictionary) and is presented with a string of characters. If the string is a dictionary word then it is returned, otherwise an error has occurred and the *closest* word on the list should be returned.

Three types of errors may occur. First, there may be a *deletion*, that is, a character of the dictionary word may have been left out. Next there may be an *insertion*, where an extra character is added to the text. Finally there may be a *substitution* error, where an incorrect character is substituted for that in the dictionary word. As an example, the word `digital` with a single deletion (of the `a`) becomes `digitl`, and with an additional substitution of `j` for `g` becomes `dijitl`. Were there only substitution errors the number of letters would be preserved, but deletions and insertions cause the matching problem to be similar to DTW.

The *Levenshtein distance* between two character strings is defined to be the minimal number of such errors that must have occurred for one of the strings to become the other. In other words, the Levenshtein distance is the least number of deletions, insertions and substitutions that must be performed on one string to make it become the other. As we saw above `dijitl` is distance *two* from `digital`; of course we could have arrived at `dijitl` by two deletions and an insertion, but this would not have been the *minimal* number of operations. The Levenshtein distance is thus an ideal candidate for the idea of 'closeness' needed for our spelling checker. When the given string is not in the dictionary we return the dictionary word separated from the input string by minimal Levenshtein distance.

In order to be able to use this distance in practice, we must now produce an algorithm that efficiently computes it. To see that this is not a trivial task let's try to find the distance between `prossesing` and `processing`. Simple counting shows that it is better to substitute a `c` for the first `s`, delete the second and then add another `s` (3 operations) rather than deleting the `es` and adding `ce` (4 operations). But how did we come up with this set of operations and how can we prove that this is the best that can be done? The problem is that the Levenshtein distance is a cost function for changing an entire string into another, and thus a global optimization seems to be required. Dynamic programming is an algorithm that reduces this global optimization to a sequence of local calculations and decisions.

Dynamic programming is best understood graphically. Write the dictionary word from left to right at the bottom of a piece of graph paper, and write the input string from bottom to top at the left of the word. For our previous example you should get something like this.

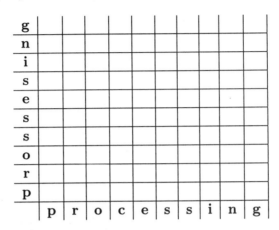

Now we fill in each of the blank squares with the minimal cost to get to that square. The bottom-left square is initialized to zero since we start there, and all the rest of the squares will get values that can be computed recursively. We finally arrive at the top right square, and the value there will be the total cost, namely the Levenshtein distance.

The recursive step involves comparing three components. One can enter a square from its left, corresponding to a deletion from the dictionary word, by taking the value to its left and adding one. One can enter a square from underneath, corresponding to an insertion into the dictionary word, by taking the value underneath it and incrementing. Finally, one can enter a square from the square diagonally to the left and down; if the letter in the dictionary word at the bottom of the column is the same as the letter in the string at the beginning of the row, then there is no additional cost and we simply copy the value from the diagonal square. If the letters differ, a substitution is needed and so we increment the value diagonally beneath. In this fashion each square gets three possible values, and we always choose the minimum of these three.

Let's try this out on our example. We start with the table from above, initialize the bottom left square, and trivially fill in the lowest row and leftmost column.

g	9									
n	8									
i	7									
s	6									
e	5									
s	4									
s	3									
o	2									
r	1									
p	0	1	2	3	4	5	6	7	8	9
0	p	r	o	c	e	s	s	i	n	g

Now we can continue filling in the entire table, and find (as we previously discovered in a rather undisciplined fashion) that the Levenshtein distance is indeed 3.

g	9	8	7	7	6	5	5	5	5	3
n	8	7	6	6	5	4	4	4	3	5
i	7	6	5	5	4	3	3	3	5	5
s	6	5	4	4	3	2	3	4	4	5
e	5	4	3	3	2	3	3	3	4	5
s	4	3	2	2	2	2	2	3	4	5
s	3	2	1	1	2	2	3	4	5	6
o	2	1	0	1	2	3	4	5	6	7
r	1	0	1	2	3	4	5	6	7	8
p	0	1	2	3	4	5	6	7	8	9
0	p	r	o	c	e	s	s	i	n	g

From the table we can discover more than simply the total distance, we can actually reconstruct the optimal sequence of operations. Indeed the optimal set of deletions, insertions, and substitutions pops out to the eye as the path of minimal cost through the table. At first there seem to be many optimal paths, but quite a few of these correspond to making a deletion and insertion instead of some substitution. The true path segments are the ones that contributed the minimal cost transitions. Thus to find the true path you start at the end point and retrace your steps backward through the table; we can save redundant computation by storing in each square not only its cost but the previous square visited. The only ambiguities that remain correspond to squares where more than one transition produced the same minimal cost; in our example changing the dictionary **c** to the incorrect ss could be accomplished by changing the **c** to **s** and then inserting an **s**, or by first inserting an **s** and then changing the **c** to **s**.

Up to now we have assumed that all errors have the same cost, but that is not always the case. Some mistaken keypresses are more prevalent then others, and there is really very little reason to assume a deletion is as likely as an insertion. However, it is not difficult to generalize the Levenshtein distance to take this into account; one need only add specific penalties rather than simply incrementing by one.

This algorithm for finding the generalized Levenshtein distance is exactly the DTW algorithm for comparing two spoken words. The word from the dictionary is placed horizontally from left to right at the bottom of a table, and the word to be compared is stretched vertically from bottom to top. We then compare short segments of the two words using some cost function (e.g., correlation, difference in spectral description, etc.) that is small for similar sounding segments. When noise contaminates a segment we may make a substitution error, while time warping causes deletions and insertions of segments. In order to identify a word we compare it to all words in the dictionary and return the word with the lowest Levenshtein distance.

EXERCISES

8.7.1 The game of *doublets* was invented in 1879 by Lewis Carroll (the mathematician Charles Lutwidge Dodgson 1832-1898). The aim of the game is to convert a word into a related word in the minimal number of *substitution* steps; However, each step must leave an actual word. For example, we can change **hate** *into* **love**, in three steps in the following way: **hate have lave love**. Show how to make a **cat** *into a* **dog** in three steps, how an **ape** *can evolve into a* **man** in five steps, and how to *raise* **four** *to* **five** by a seven step procedure. **four foul fool foot fort fore fire five**. How many steps does it take to *drive the* **pig** *into the* **sty**?

8.7.2 In more complex implementations of spelling checkers further types of errors may be added (e.g., reversal of the order of two letters). Can the dynamical programming algorithm still be used to determine the Levenshtein distance?

8.7.3 An alternative method for comparing time-warped signals is the Markov model approach. Here we assume that the signal is generated by a Markov model with states $O_1, O_2 \ldots O_M$. When the model is in state O_m it has probability $a_{m,m}$ of staying in the same state, probability $a_{m,m+1}$ of transitioning to state O_{m+1}, and probability $a_{m,m+2}$ of skipping over state O_{m+1} directly to state O_{m+2}. When the model is in state O_m it outputs a characteristic signal segment s_m. Write a program that simulates a Markov model and run it several times. Do you see how the time warping arises?

8.7.4 An extension to the above model is the **Hidden Markov Model** (HMM). The HMM states are hidden since they do not uniquely correspond to an output signal segment; rather when the model is in a state O_m it has probability b_{ml} of outputting signal s_l. Extend the program of the previous exercise to generate HMM signals. Why is the HMM more realistic for speech?

Bibliographical Notes

Although there are a lot of books that deal with things that are not filters, there are very few such that happen to treat signal processing.

Median and morphological filters are mostly discussed in books on image processing, but see [68, 258, 157].

Multilayer perceptrons were introduced in [225], and popularized in the books by the same authors [168, 169], although the basic idea had beed previously discovered by several researchers. A popular short introduction is [150].

Phase-locked loops are usually discussed in books on digital communications, e.g., [242, 199].

Time warping and HMM are discussed in texts on speech recognition, e.g., [204, 176].

Correlation

Our study of signal processing systems has been dominated by the concept of 'convolution', and we have somewhat neglected its close relative the 'correlation'. While formally similar (in fact convolution by a symmetric FIR filter can be considered a correlation as well), the way one should think about the two is different. Convolution is usually between a signal and a filter; we think of it as a system with a single input and stored coefficients. Crosscorrelation is usually between two signals; we think of a system with two inputs and no stored coefficients. The difference may be only in our minds, but nonetheless this mind-set influences the way the two are most often used.

Although somewhat neglected we weren't able to get this far without mentioning correlations at all. We have already learned that crosscorrelation is a measure of similarity between two signals, while autocorrelation is a measure of how similar a signal is to itself. In Section 5.6 we met the autocorrelation for stochastic signals (which are often quite unlike themselves), and in Section 6.13 we used the crosscorrelation between input and output signals to help identify an unknown system.

Correlations are the main theme that links together the present chapter. We first motivate the concept of correlation by considering how to compare an input signal to a reference signal. We find that the best signal detector is the correlator. After formally defining both crosscorrelation and autocorrelation and calculating some examples, we prove the important Wiener-Khintchine theorem, which relates the autocorrelation to the power spectral density (PSD).

Next we compare correlation with convolution and discover that the optimal signal detector can be implemented as a matched filter. The matched filter was invented for radar and a digression into this important application is worthwhile. The matched filter is good for signal detection, but for cleaning up a partially unknown signal we need the Wiener filter, which is also based on correlations.

There is also a close connection between correlation and prediction. Linear predictive coding is crucial in speech processing, and we present it here in preparation for our later studies.

The Wiener-Khintchine theorem states that correlations are second-order entities. Although these are sufficient for a wide variety of tasks, we end this chapter with a short introduction to the more general higher-order signal processing.

9.1   Signal Comparison and Detection

A *signal detector* is a device that alerts us when a desired signal appears. Radar and sonar operate by transmitting a signal and detecting its return after having being reflected by a distant target. The return signal is often extremely weak in amplitude, while interference and noise are strong. In order to be able to reliably detect the presence of the return signal we employ a signal detector whose output is maximized when a true reflection appears. Similar signal detectors are employed in telephony call progress processing, medical alert devices, and in numerous other applications.

Envision a system with a single input that must sound an alarm when this input consists of some specified signal. It is important not to miss any events even when the signal is weak compared to the noise, but at the same time we don't want to encourage false alarms (reporting detection when the desired signal was not really there). In addition, we may need to know as accurately as possible precisely *when* the expected signal arrived.

The signal to be detected may be as simple as a sinusoid of given frequency, but is more often a rather complex, but known signal. It is evident that signal detection is closely related to *signal comparison*, the determination of how closely a signal resembles a reference signal. Signal comparison is also a critically important element in its own right, for example, in digital communications systems. In the simplest of such systems one of several basic signals is transmitted every T seconds and the receiver must determine which. This can be accomplished by building signal detectors for each of the basic signals and choosing the signal whose respective detector's output is the highest. A more complex example is speech recognition, where we may build detectors for a multitude of different basic sounds and convert the input audio into a string of best matches. Generalization of this technique to images produces a multitude of further applications, including optical character recognition.

From these examples we see that comparison and detection are essentially the same. The simplest detector is implemented by comparing the output of a comparator to a threshold. Complex detectors may employ more sophisticated decision elements, but still require the basic comparison mechanism to function.

Signal detection and comparison are nontrivial problems due to the presence of noise. We know how to build filters that selectively enhance defined frequency components as compared to noise; but how do we build a system that selectively responds to a known but arbitrary reference signal? Our first inclination would be to subtract the input signal s_n from the desired reference r_n, thus forming an error signal $\epsilon_n = r_n - s_n$. Were the error signal to be identically zero, this would imply that the input precisely matches the reference, thus triggering the signal detector or maximizing the output of the signal comparator. However, for an input signal contaminated by noise $s_n = r_n + \nu_n$, we can not expect the instantaneous error to be identically zero, but the lower the energy of the error signal the better the implied match. So a system that computes the energy of the difference signal is a natural comparator.

This idea of using a simple difference is a step in the right direction, but only the first step. The problem is that we have assumed that the input signal is simply the reference signal plus additive noise; and this is too strong an assumption. The most obvious reason for this discrepancy is that the amplitude of the input signal is usually arbitrary. The strength of a radar return signal depends on the cross-sectional area of the target, the distance from the transmitter to the target and the target to the receiver, the type and size of the radar antenna, etc. Communications signals are received after path loss, and in the receiver probably go through several stages of analog amplification, including automatic gain control. A more reasonable representation of the input signal is

$$s_n = A\,r_n + \nu_n$$

where A is some unknown gain parameter.

In order to compare the received signal s_n with the reference signal r_n it is no longer sufficient to simply form the difference; instead we now have to find a gain parameter g such that $r_n - gs_n$ is minimized. We can then use the energy of the resulting error signal

$$\epsilon_n = \min_g (r_n - gs_n)$$

as the final match criterion. How can we find this g? Assuming for the moment that there is no noise, then for every time n we require

$$g = \frac{r_n}{s_n} = \frac{1}{A} \tag{9.1}$$

in addition to the weaker constraint that the error energy be zero.

$$\sum_n (r_n - gs_n)^2 = 0 \tag{9.2}$$

By opening the square the latter becomes

$$\sum_n r_n^2 - 2g \sum_n r_n s_n + g^2 \sum_n s_n^2 = 0$$

which can be rewritten in the following way.

$$E_r - 2gC_{rs} + g^2 E_s = 0 \tag{9.3}$$

Here E_r is the energy of the reference signal, E_s is the energy of the input signal, and $C_{rs} = \sum_n r_n s_n$ is the *crosscorrelation* between the reference and the input. Among all input signals of given energy the correlation is maximal exactly when the energy of the difference signal is minimal.

Now, from equation (9.1) we can deduce that

$$g^2 = \frac{\sum_n r_n^2}{\sum_n s_n^2} = \frac{E_r}{E_s}$$

which when substituted into (9.3) brings us to the conclusion that

$$C_{rs} = \sqrt{E_r E_s} \tag{9.4}$$

in the absence of noise. When the input signal does not precisely match the reference, due to distortion or noise, we have $|C_{rs}| < +\sqrt{E_r E_s}$. The crosscorrelation C_{rs} is thus an easily computed quantity that compares the input signal to the reference, even when the amplitudes are not equal. A comparator can thus be realized by simply computing the correlation, and a signal detector can be implemented by comparing it to $\sqrt{E_r E_s}$ (e.g., requiring $|C_{rs}| > \sqrt{E_r E_s} - e$ or $|C_{rs}| > \gamma \sqrt{E_r E_s}$).

Unfortunately we have not yet considered all that happens to the reference signal before it becomes an input signal. In addition to the additive noise and unknown gain, there will also usually be an unknown time shift. For communications signals we receive a stream of signals to compare, each offset by an unknown time delay. For the radar signal the time delay derives from the round-trip time of the signal from the transmitter to the target and back, and is precisely the quantity we wish to measure. When there is a time shift, a reasonable representation of the input signal is

$$s_n = Ar_{n+m} + \nu_n \qquad \forall n$$

where A is the gain and $m < 0$ the time shift parameter.

In order to compare the received signal s_n with the reference signal r_n we can no longer simply compute a single crosscorrelation; instead we now have to find the time shift parameter m such that

$$C_{rs}(m) = \sum_n r_{n+m}s_n = \sum_n r_n s_{n-m}$$

is maximal. How do we find m? The only way is to compute the crosscorrelation $C_{rs}(m)$ for all relevant time shifts (also called time 'lags') m and choose the maximal one. It is this

$$C_{rs} = \max_m C_{rs}(m)$$

that must be compared with $\sqrt{E_r E_s}$ in order to decide whether a signal has been detected.

EXERCISES

9.1.1 Formulate the concept of correlation in the frequency domain starting from spectral difference and taking into account an arbitrary gain of the spectral distribution. What happens if we need to allow an arbitrary spectral shift?

9.1.2 Give a complete algorithm for the optimal detection of a radar return s_n given that the transmitted signal r_n was sent at time T_1, returns are expected to be received before time T_2, and the correlation is required to be at least γ. Note that you can precompute E_r and compute E_s and $C_{rs}(m)$ in one loop.

9.1.3 Design an optimal detector for the V.34 probe signal introduced in exercise 2.6.4. The basic idea is to perform a DFT and implement a correlator in the frequency domain by multiplying the spectrum by a comb with 21 pass-bands (of suitable bandwidth). However, note that this is not independent of signal strength. You might try correcting this defect by requiring the correlation to be over 80% of the total signal energy, but this wouldn't work properly since, e.g., *answer tone* (a pure 2100 Hz tone) would trigger it, being one of the frequencies of the probe signal. What is wrong? How can this problem be solved?

9.2   Crosscorrelation and Autocorrelation

The time has come to formally define correlation.

Definition: crosscorrelation
The crosscorrelation between two real signals x and y is given by

$$C_{xy}(\tau) \equiv \int_{-\infty}^{\infty} x(t)y(t-\tau)dt \qquad \mathbf{A} \ \bigg|\ \mathbf{D} \qquad C_{xy}(m) \equiv \sum_{n=-\infty}^{\infty} x_n y_{n-m} \quad (9.5)$$

where the time shift τ or m is called the *lag*.   ■

There is an important special case, called *autocorrelation*, when y is taken to be x. It might seem strange to compare a signal with itself, but the lag in equation (9.5) means that we are actually comparing the signal at different times. Thus autocorrelation can assist in detecting periodicities.

Definition: autocorrelation
The autocorrelation of a real signal s is given by

$$C_s(\tau) \equiv \int_{-\infty}^{\infty} s(t)s(t-\tau)dt \qquad \mathbf{A} \ \bigg|\ \mathbf{D} \qquad C_s(m) \equiv \sum_{n=-\infty}^{\infty} s_n s_{n-m} \quad (9.6)$$

and the normalized autocorrelation is defined to be

$$c_s(\tau) \equiv \frac{C_s(\tau)}{C_s(0)} \qquad \mathbf{A} \ \bigg|\ \mathbf{D} \qquad c_s(m) \equiv \frac{C_s(m)}{C_s(0)} \qquad (9.7)$$

where τ or m is called the *lag*.   ■

These definitions are consistent with those of Section 5.6 for the case of stationary ergodic signals. In practice we often approximate the autocorrelation of equation (5.22) by using equation (9.6) but with the sum only over a finite amount of time. The resulting quantity is called the *empirical autocorrelation*. The correlation is also somewhat related to the *covariance matrix* of vector random variables, and strongly related to the convolution, as will be discussed in the next section.

Before discussing properties of the correlations, let's try calculating a few. The analog rectangular window

$$s(t) = \begin{cases} 1 & |t| < 1 \\ 0 & \text{else} \end{cases}$$

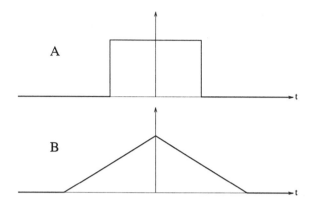

Figure 9.1: The autocorrelation of an analog rectangularly shaped signal. In (A) the signal is depicted while the autocorrelation is in (B). Note that the autocorrelation is symmetric and has its maximal value at the origin.

is depicted in Figure 9.1.A. Its autocorrelation is given by the triangular

$$C_s(\tau) = \int_{-\infty}^{\infty} s(t)s(t-\tau)dt = \int_{\max(-1,\tau-1)}^{\min(1,\tau+1)} dt = (2 - |\tau|) \qquad (9.8)$$

depicted in Figure 9.1.B. In that figure we see several features that are readily shown to be more general. The autocorrelation is symmetric around time lag zero, and it takes on its maximum value at lag zero, where it is simply the energy E_s. The autocorrelation is also wider than the original signal, but attacks and decays more slowly.

Had we used an inverted rectangle (which differs from the original signal by a phase shift)

$$s(t) = \begin{cases} -1 & |t| < 1 \\ 0 & \text{else} \end{cases}$$

we would have found the same autocorrelation. Indeed the generalization of autocorrelation to complex signals,

$$C_s(\tau) \equiv \int_{-\infty}^{\infty} s^*(t)s(t-\tau)dt \qquad \mathbf{A} \;\Big|\; \mathbf{D} \qquad C_s(m) \equiv \sum_{n=-\infty}^{\infty} s_n^* s_{n-m} \qquad (9.9)$$

can be shown to be *phase blind* (unchanged by multiplying s by a common phase factor).

What is the autocorrelation of the periodic square wave $\Box(t)$? Generalizing our previous result we can show that the autocorrelation is a periodic triangular wave of the same period. This too is quite general—the autocorrelation of a periodic signal is periodic with the same period; and since the lag-zero autocorrelation is a global maximum, all lags that are multiples of the period have globally maximal autocorrelations. This fact is precisely the secret behind using autocorrelation for determining the period of a periodic phenomenon. One looks for the first nonzero peak in the autocorrelation as an indication of the period. The same idea can be used for finding Fourier components as well; each component contributes a local peak to the autocorrelation.

As our final example, let's try a digital autocorrelation. The signal b_n is assumed to be zero except for $n = 1 \ldots 13$ where it takes on the values ± 1.

$$\ldots 0, 0, +1, +1, +1, +1, +1, -1, -1, +1, +1, -1, +1, -1, +1, 0, 0, \ldots \quad (9.10)$$

Its autocorrelation is easily computed to be $C(0) = 13$, $C(m) = 0$ for odd m in the range $-13 < m < 13$, $C(m) = 1$ for even nonzero m in this range, and all other autocorrelations are zero. We see that the autocorrelation is indeed maximal at $m = 0$ and symmetric, and in addition the highest nonzero-lag correlations are only 1. Signals consisting of ± 1 values with this last property (i.e., with maximal nontrivial autocorrelation of $\frac{C(0)}{N}$ or less) are called *Barker codes*, and are useful for timing and synchronization. There is no known way of generating Barker codes and none longer than this one are known.

The definitions for autocorrelation or crosscorrelation given above involve integrating or summing over all times, and hence are not amenable to computation in practice. In any case we would like to allow signals to change behavior with time, and thus would like to allow correlations that are defined for finite time durations. The situation is analogous to the problem that led to the definition of the STFT, and we follow the same tactic here. Assuming a rectangular window of length N, there are N terms in the expression for the zero lag, but only $N - 1$ terms contribute to the lag 1 correlation $s_1 s_0 + s_2 s_1 + \ldots + s_{N-1} s_{N-2}$, and only $N - m$ terms in the lag m sum. So we define the short-time autocorrelation

$$C_s(m) = \frac{1}{N - m} \sum_{n=1}^{N-m} s_n s_{n-m} \quad (9.11)$$

where now the zero lag is the power rather than the energy. This quantity is often called the unbiased empirical autocorrelation when it is looked upon as a numerical estimate of the full autocorrelation.

EXERCISES

9.2.1 What is the connection between autocorrelation defined here for deterministic signals and the autocorrelation we earlier defined for stochastic signals (equation (5.22))?

9.2.2 What is the crosscorrelation between a signal $s(t)$ and the impulse $\delta(t)$?

9.2.3 Compute and draw the crosscorrelation between two analog rectangular signals of different widths.

9.2.4 Compute and draw the crosscorrelation between two analog triangular signals.

9.2.5 Show that $C_{yx}(m) = C_{xy}(-m)$.

9.2.6 Prove that the autocorrelation is symmetric and takes its maximum value at the origin, where it is the energy. Show that $|c_{xy}(m)| \leq 1$.

9.2.7 Can you find Barker codes of length 5, 7, and 11? What are their autocorrelations?

9.2.8 What is the proper generalization of crosscorrelation and autocorrelation to complex signals? (Hint: The autocorrelation should be phase independent.)

9.2.9 Prove that the autocorrelation of a periodic signal is periodic with the same period.

9.2.10 Prove that zero mean symmetric signals have zero odd lag autocorrelations.

9.2.11 Assume $y_n = x_{n-l}$. What are the connections between $C_{xy}(m)$, $C_y(m)$ and $C_x(m)$?

9.2.12 Derive the first few autocorrelation values for $s_n = A\sin(\omega n + \phi)$.

9.2.13 Generalize the previous exercise and derive the following expression for the general autocorrelation of the sinusoid.

$$C_s(m) = \langle s_n s_{n+m} \rangle = \frac{A^2}{2}\cos(\omega m)$$

9.3 The Wiener-Khintchine Theorem

The applications of correlation that we have seen so far derive from its connection with the difference between two signals. Another class of applications originate in the relationship between autocorrelation and power spectrum (see Section 4.5), a relationship known as the Wiener-Khintchine Theorem.

The PSD of a signal is the absolute square of its FT, but it is also can be considered to be the FT of some function. Parseval's relation tells us that integrating the PSD over all frequencies is the same as integrating the square of the signal over all times, so it seems reasonable that the iFT of the PSD is somehow related to the square of the signal.

Could it be that the PSD is simply the FT of the signal squared? The DC term works because of Parseval, but what about the rest? We don't have to actually integrate or sum to find out since we can use the connection between convolution and FT of a product $FT(xy) = X * Y$ (equation (4.18) or (4.46)). Using the signal s for both x and y we see that the FT of $s^2(t)$ is $S * S = \int S(\omega - \Omega)S(\Omega)d\Omega$, which is not quite the PSD $|S|^2 = S^*S = S(-\omega)S(\omega)$ (for real signals), but has an additional integration. We want to move this integration to the time side of the equation, so let's try $s * s$. From equation (4.19) or (4.47) we see that the FT of $s * s$ is $S^2(\omega)$ which is even closer, but has both frequency variables positive, instead of one positive and one negative. So we need something very much like $s * s$ but with some kind of time variable inversion; that sounds like the autocorrelation!

So let's find the FT of the autocorrelation.

$$
\begin{aligned}
\mathrm{FT}\left(C_s(t)\right) &= \mathrm{FT}\left(\int_{-\infty}^{\infty} s(\tau)s(\tau - t)d\tau\right) \\
&= \int_{-\infty}^{\infty}\left(\int_{-\infty}^{\infty} s(\tau)s(\tau - t)d\tau\right) e^{-i\omega t}dt \\
&= \int_{-\infty}^{\infty}\int_{-\infty}^{\infty} s(\tau)s(\tau - t)e^{-i\omega t}dt d\tau \\
&= \int_{-\infty}^{\infty}\int_{-\infty}^{\infty} s(\tau)s(T)e^{-i\omega(\tau - T)}dT d\tau \\
&= \int_{-\infty}^{\infty} s(T)e^{i\omega T}dT \int_{-\infty}^{\infty} s(\tau)e^{-i\omega \tau}d\tau \\
&= \quad S^*(\omega) \qquad\qquad S(\omega) \qquad = |S(\omega)|^2
\end{aligned}
$$

The PSD at last!

We have thus proven the following celebrated theorem.

The Wiener-Khintchine Theorem
The autocorrelation $C_s(t)$ and the power spectrum $S(\omega)$ are an FT pair. ∎

Although we proved the theorem for deterministic analog signals, it is more general. In fact, in Section 5.7 we used the Wiener-Khintchine theorem as the definition of spectrum for random signals.

As a corollary to the theorem we can again prove that the autocorrelation is 'phase blind', that is, independent of the spectral phase. Two signals with the same power spectral density but different spectral phase will have the same autocorrelation function, and hence an infinite number of signals have the same autocorrelation. Methods of signal analysis that are based on autocorrelation can not differentiate between such signals, no matter how different they may look in the time domain. If we need to differentiate between such signals we need to use the higher-order statistics of Section 9.12.

EXERCISES

9.3.1 The period of a pure sinusoid is evident as a peak in the autocorrelation and hence its frequency is manifested as a peak in the power spectrum. This is the true basis for the connection between autocorrelation and PSD. What can you say about the autocorrelation of a general periodic signal? What is the autocorrelation of the sum of two sinusoidal components? Can you see the PSD connection?

9.3.2 Express and prove the Wiener-Khintchine theorem for digital signals.

9.3.3 Generalize the Wiener-Khintchine theorem by finding the FT of the cross-correlation of two signals $x(t)$ and $y(t)$.

9.4 The Frequency Domain Signal Detector

Simply observing the input signal in the time domain is not a very sensitive method of detecting low-SNR signals, a fact made obvious by looking back at Figure 2.9. Since correlation is a method for detecting weak signals, and correlation is related to spectrum by the Wiener-Khintchine theorem, there should be a way of exploiting the frequency domain for signal detection.

In Section 5.3 we saw how to reduce noise by averaging it out. This would seem to be a purely time domain activity, but there is a frequency domain connection. To see this, consider the simplest case, that of a pure sinusoid in noise. For averaging to optimally reinforce the signal we must first ensure that all the times intervals commence at precisely the same phase in a period, an operation called 'time registration'. Without registration the signal cancels out just like the noise; with inaccurate registration the signal is only partially reinforced. If we wish to take successive time intervals, accurate registration requires the intervals to be precise multiples of the sinusoid's basic period. Thus signal emphasis by averaging requires precise knowledge of the signal's frequency.

Now let's see how we can emphasize signals working directly in the frequency domain. In a digital implementation of the above averaging each time interval corresponds to a buffer of samples. Assume that the period is L samples and let's use a buffer with exactly k periods. We start filling up the buffer with the input signal consisting of signal plus noise. Once the buffer is filled we return to its beginning, adding the next signal sample to that already there. Performing this addition M times increases the sinusoidal component by M but the noise component only by $\sqrt{M}$ (see exercise 5.3.1). Hence the SNR, defined as the ratio of the signal to noise *energies*, is improved by M. This SNR increase is called the *processing gain*.

How many input samples did we use in the above process? We filled the buffer of length kL exactly M times; thus $N = kLM$ input samples were needed. We can use a buffer with length corresponding to any integer number of periods k, but the N input signal samples are used most efficiently when the buffer contains a single cycle $k = 1$. This is because the processing gain $M = \frac{N}{kL}$ will be maximal for a given N when $k = 1$. However, it is possible to do even better! It is possible to effectively reduce the 'buffer' to a single sample such that $M = N$, and obtain the maximal processing gain of N.

All we have to do is to downmix the signal to DC, by multiplying by a complex exponential and low-pass filtering. The noise will remain zero mean while the sinusoid becomes a complex constant, so that averaging as in Section 6.6 cancels out the noise but reinforces the constant signal. Now, as explained in Section 13.2, this complex downmixing can be performed using the DFT. So by performing a DFT the energy in the bin corresponding to the desired signal frequency increases much faster than all the other bins. In the frequency domain interpretation the processing gain is realized due to the signal being concentrated in this single bin, while the white noise is spread out over N bins. Thus were the signal and noise energies initially equal, the ratio of the energy in the bin corresponding to the signal frequency to that of the other bins would be N, the same processing gain deduced from time domain arguments.

So we see that our presumption based on the Wiener-Khintchine theorem was correct; the frequency domain interpretation is indeed useful in signal detection. Although we discussed only the simple case of a single pure sinusoid, it is relatively easy to extend the ideas of this section to more general signals by defining distinctive spectral signatures. Instead of doing this we will return to the time domain and see how to build there a signal detection system for arbitrary signals.

EXERCISES

9.4.1 Express the processing gain in decibels when the DFT is performed using a 2^m point FFT.

9.4.2 In the text we tacitly assumed the signal frequency to be precisely at a bin center. If this is not the case a window function w_n (see Section 13.4) must be employed. Show that with a window the signal energy is enhanced by $(\sum_n w_n)^2$ while the noise energy is increased by $\sum_n w_n^2$ thus resulting in a processing gain of the ratio of these two expressions.

9.4.3 Build a detector for a signal that consists of the equally weighted sum of two sinusoids. Is it worthwhile taking the phases into account? What if the signal is the weighted sum of the two sinusoids?

9.4.4 Extend the technique of the previous exercise and build a DFT-based detector for a completely general signal.

9.5 Correlation and Convolution

Although we have not mentioned it until now, you have no doubt noticed the similarity between the expression for digital crosscorrelation in equation (9.5) and that for convolution in equation (6.13). The only difference between them is that in correlation both indices run in the same direction, while in convolution they run in opposite directions. Realizing this, we can now realize our signal comparator as a filter. The filter's coefficients will be the reference signal reversed in time, as in equation (2.16). Such a filter is called a *matched filter*, or a correlator. The name matched filter refers to the fact that the filter coefficients are matched to the signal values, although in reverse order.

What is the frequency response of the matched filter? Reversing a signal in time results in frequency components $\mathrm{FT}\left(s(-t)\right) = S(-\omega)$, and if the signal is real this equals $S^*(\omega)$, so the magnitude of the FT remains unchanged but the phase is reversed.

From the arguments of Section 9.1 the correlator, and hence the theoretically identical matched filter, is the optimum solution to the problem of detecting the appearance of a known signal s_n contaminated by additive white noise $x_n = s_n + \nu_n$.

Can we extend this idea to optimally detect a signal in colored noise? To answer this question recall the joke about the mathematician who wanted a cup of tea. Usually he would take the kettle from the cupboard, fill it with water, put it on the fire, and when the water boiled, pour it into a cup and drop in a tea bag. One day he found that someone had already boiled the water. He stared perplexed at the kettle and then smiled. He went to the sink, poured out the boiling water, returned the kettle to the cupboard and declared triumphantly: 'The problem has been reduced to one we know how to solve.'

How can we reduce the problem of a signal in colored noise to the one for which the matched filter is the optimal answer? All we have to do is filter the contaminated signal x_n by a filter whose frequency response is the inverse of this noise spectrum. Such a filter is called a *whitening filter*, because it flattens the noise spectrum. The filtered signal $x'_n = s'_n + \nu'_n$ now contains an additive white noise component ν'_n, and the conditions required for the matched filter to be optimal are satisfied. Of course the reference signal s'_n is no longer our original signal s_n; but finding the matched filter for s'_n is straightforward.

EXERCISES

9.5.1 Create a sinusoid and add Gaussian white noise of equal energy. Recover the sinusoid by averaging. Experiment with inaccurate registration. Now recover the sinusoid by a DFT. What advantages and disadvantages are there to this method? What happens if the frequency is inaccurately known?

9.5.2 Build a matched filter to detect the HPNA 1.0 pulse (see exercise 7.7.4). Try it out by synthesizing pulses at random times and adding Gaussian noise. HPNA 1.0 uses PPM where the information is in the pulse position. How precisely can you detect the pulse's time of arrival?

9.5.3 Compare the time domain matched filter with a frequency domain detector based on the FFT algorithm. Consider computational complexity, processing delay, and programming difficulty.

9.6   Application to Radar

Matched filters were invented in order to improve the detection of radar returns. We learned the basic principles of radar in Section 5.3 but were limited to explaining relatively primitive radar processing techniques. With our newly acquired knowledge of matched filters we can now present improved radar signals and receivers.

Radar pulses need to have as much energy as possible in order to increase the probability of being detected, and thus should be long in duration. In order to increase a radar's range resolution we prefer narrow pulses since it's hard to tell when exactly a wide pulse arrives. How can we resolve this conflict of interests? The basic idea is to use a wide pulse but to *modulate* it (i.e., to change its characteristics with time). The output of a filter matched to this modulation can be made to be very short in duration, but containing all the energy of the original pulse.

To this end some radars vary their instantaneous frequency linearly with time over the duration of the pulse, a technique known as *FM chirp*. We demonstrate in Figure 9.2 the improvement chirp can bring in range resolution. The pulse in Figure 9.2.A is unmodulated and hence the matched filter can do no better than to lock onto the basic frequency. The output of such a matched filter is the autocorrelation of this pulse, and is displayed in Figure 9.2.B. Although theoretically there is a maximum corresponding to the perfect match when the entire pulse is overlapped by the matched filter, in practice the false maxima at shifts corresponding to the basic period

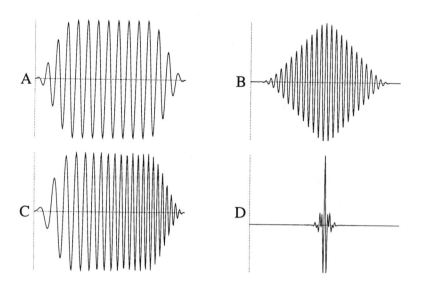

Figure 9.2: The autocorrelation of pulses with and without chirp. In (A) a pulse with constant instantaneous frequency is depicted, and its wide autocorrelation is displayed in (B). In (C) we present a pulse with frequency chirp; its much narrower autocorrelation is displayed in (D).

make it difficult to determine the precise TOA. In contrast the chirped pulse of Figure 9.2.C does not match itself well at any nontrivial shifts, and so its autocorrelation (Figure 9.2.D) is much narrower. Hence a matched filter built for a chirped radar pulse will have a much more precise response.

Chirped frequency is not the only way to sharpen a radar pulse's autocorrelation. Barker codes are often used because of their optimal autocorrelation properties, and the best way to embed a Barker code into a pulse is by changing its instantaneous phase. **Binary Phase Shift Keying** (BPSK), to be discussed in Section 18.13, is generated by changing a sinusoidal signal's phase by $180°$, or equivalently multiplying the sinusoid by -1. To use the 13-bit Barker code we divide the pulse width into 13 equal time intervals, and assign a value ±1 to each. When the Barker code element is $+1$ we transmit $+\sin(\omega t)$, while when it is -1 we send $-\sin(\omega t)$. This Barker BPSK sharpens the pulse's autocorrelation by a factor of 13.

Not all radars utilize pulses; a **Continuous Wave** (CW) radar transmits continuously with constant amplitude. How can range be determined if echo arrives continuously? Once again by modulating the signal, and if we want constant amplitude we can only modulate the frequency or phase (e.g., by chirp or BPSK). Both chirp and BPSK modulation are popular for CW radars, with the modulation sequence repeating over and over again without stopping. CW radars use LFSR sequences rather than Barker codes for a very simple reason. Barker codes have optimal *linear* autocorrelation properties, while maximal-length LFSR sequences can be shown to have optimal *circular* autocorrelation characteristics. Circular correlation is analogous to circular convolution; instead of overlapping zero when one signal extends past the other, we wrap the other signal around periodically. A matched filter that runs over a periodically repeated BPSK sequence essentially reproduces the circular autocorrelation.

EXERCISES

9.6.1 Plot, analogously to Figure 9.2, the autocorrelation of a pulse with a 13-bit Barker code BPSK.

9.6.2 What is the circular autocorrelation of the LFSR15 sequence?

9.6.3 What is the difference between coherent and incoherent pulse radars? In what way are coherent radars better?

9.7   The Wiener Filter

The matched filter provides the optimum solution to the problem of detecting the arrival of a known signal contaminated by noise; but correlation-based filters are useful for other problems as well, for example, removing noise from an unknown signal.

If the signal is known in the matched filter problem, then why do we need to clean it up? The reason is that the signal may be only partially known, and we must remove noise to learn the unknown portion. In one common situation we expect a signal from a family of signals and are required to discover which specific signal was received. Or we might know that the signal is a pure sinusoid, but be required to measure its precise frequency; this is the case for Doppler radars which determine a target's velocity from the Doppler frequency shift.

Let's see how to build a filter to optimally remove noise and recover a signal. Our strategy is straightforward. It is simple to recover a sufficiently strong signal in the presence of sufficiently weak noise (i.e., when the SNR is sufficiently high). When the SNR is low we will design a filter to enhance it; such a filter's design must take into account everything known about the signal and the noise spectra.

Before starting we need some notation. For simplicity we observe the spectrum from DC to some frequency F. We will denote the original analog signal in time as $s(t)$ and in frequency as $S(f)$. We will call its total energy E_s. We denote the same quantities for the additive noise, $\nu(t)$, $\mathcal{V}(f)$, and E_ν, respectively. These quantities are obviously related by

$$E_s = \int_{-\infty}^{\infty} |s(t)|^2 \, dt = \int_{-\infty}^{\infty} |S(f)|^2 \, df$$

$$E_\nu = \int_{-\infty}^{\infty} |\nu(t)|^2 \, dt = \int_{-\infty}^{\infty} |\mathcal{V}(f)|^2 \, df$$

and if the noise is white then we further define its constant power spectral density to be $\mathcal{V}_0 = \frac{E_\nu}{F}$ watt per Hz. The overall signal-to-noise ratio is the ratio of the energies

$$SNR = \frac{E_s}{E_\nu} \tag{9.12}$$

but we can define time- and frequency-dependent SNRs as well.

$$SNR(t) = \frac{|s(t)|^2}{|\nu(t)|^2} \qquad\qquad SNR(f) = \frac{|S(f)|^2}{|\mathcal{V}(f)|^2} \tag{9.13}$$

Finally, the observed signal is the sum of the signal plus the noise.

$$x(t) = s(t) + \nu(t) \qquad\qquad X(f) = S(f) + \mathcal{V}(f) \qquad (9.14)$$

We'll start with the simple case of a relatively pure sinusoid of frequency f_0 in white noise. The signal PSD consists of a single narrow line (and its negative frequency conjugate), while the noise PSD is a constant $\mathcal{V}_0$; accordingly the SNR is $\frac{E_s}{\mathcal{V}_0}$. What filter will optimally detect this signal given this noise? Looking at the frequency-dependent SNR we see that the signal stands out above the noise at f_0; so it makes sense to use a narrow band-pass filter centered on the sinusoid's frequency f_0. The narrower the filter bandwidth BW, the less noise energy is picked up, so we want BW to be as small as possible. The situation is depicted in Figure 9.3.A where we see the signal PSD represented as a single vertical line, the noise as a horizontal line, and the optimum filter as the smooth curve peaked around the signal. The signal-to-noise ratio at the output of the filter

$$
\begin{aligned}
SNR_{out} &= \frac{\int_{-\infty}^{\infty} |H(f)S(f)|^2\, df}{\int_{-\infty}^{\infty} |H(f)\mathcal{V}(f)|^2\, df} \qquad\qquad (9.15)\\[2mm]
&= \frac{\int_{f_0-\frac{BW}{2}}^{f_0+\frac{BW}{2}} |H(f)S(f)|^2\, df}{\int_{f_0-\frac{BW}{2}}^{f_0+\frac{BW}{2}} |H(f)\mathcal{V}(f)|^2\, df} = \frac{E_s}{\mathcal{V}_0 BW}
\end{aligned}
$$

is greater than that at the input by a factor of $\frac{F}{BW}$. For small BW this is a great improvement in SNR and allows us to detect the reference signal even when buried in very high noise levels.

Now let's complicate matters a bit by considering a signal with two equal spectral components, as in Figure 9.3.B. Should we use a filter that captures both spectral lines or be content with observing only one of them? The two-component filter will pass twice the signal energy but twice the noise energy as well. However, a filter that matches the signal spectrum may enhance the time-dependent SNR; the two signal components will add constructively at some time, and by choosing the relative phases of the filter components we can make this peak occur whenever we want. Also, for finite times the noise spectrum will have local fluctuations that may cause a false alarm in a single filter, but the probability of that happening simultaneously in both filters is much smaller. Finally, the two-component filter can differentiate better between the desired signal and a single frequency sinusoid masquerading as the desired signal.

Were one of the frequency components to be more prominent than the other, we would have to compensate by having the filter response $H(f)$ as

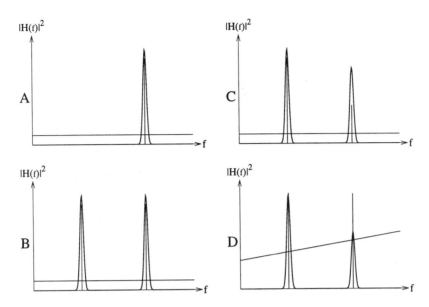

Figure 9.3: Expected behavior of an optimum filter in the frequency domain. In all the figures we see the PSD of the reference signal and noise, as well as the Wiener filter. The various cases are discussed in the text.

depicted in Figure 9.3.C. This seems like the right thing to do, since such a filter emphasizes frequencies with high SNR. Likewise Figure 9.3.D depicts what we expect the optimal filter to look like for the case of two equal signal components, but non-white noise.

How do we actually construct this optimum filter? It's easier than it looks. From equation (9.14) the spectrum at the filter input is $S(f) + \mathcal{V}(f)$, so the filter's frequency response must be

$$H(f) = \frac{S(f)}{S(f) + \mathcal{V}(f)} \tag{9.16}$$

in order for the desired spectrum $S(f)$ to appear at its output. This frequency response was depicted in Figure 9.3. Note that we can think of this filter as being built of two parts: the denominator corresponds to a whitening filter, while the numerator is matched to the signal's spectrum. Unlike the whitening filter that we met in the matched filter detector, here the entire signal plus noise must be whitened, not just the noise.

This filter is a special case of the *Wiener filter* derived by Norbert Wiener during World War II for optimal detection of radar signals. It is a special case because we have been implicitly assuming that the noise and signal are uncorrelated. When the noise can be correlated to the signal we have to be more careful.

This is not the first time we have attempted to find an unknown FIR filter. In Section 6.13 we found that the hard system identification problem for FIR filters was solved by the Wiener-Hopf equations (6.63). At first it seems that the two problems have nothing in common, since in the Wiener filter problem only the input is available, the output being completely unknown (otherwise we wouldn't need the filter), while in the system identification case both the input and output were available for measurement! However, neither of these statements is quite true. Were the output of the Wiener filter completely unspecified the trivial filter that passes the input straight through would be a legitimate solution. We do know certain characteristics of the desired output, namely its spectral density or correlations. In the hard system identification problem we indeed posited that we intimately knew the input and output signals, but the solution does not exploit this much detail. Recall that only the correlations were required to find the unknown system.

So let's capitalize on our previous results. In our present notation the input is $x_n = s_n + \nu_n$ and the desired output s_n. We can immediately state the Wiener-Hopf equations in the time domain

$$C_{sx}(q) = \sum_k h_k C_x(q - k)$$

so that given C_{sx} and C_x we can solve for h, the Wiener filter in the time domain. To compare this filter with our previous results we need to transfer the equations to the frequency domain, using equation (4.47) for the FT of a convolution.

$$P_{sx}(\omega) = H(\omega)P_x(\omega)$$

Here $P_{sx}(\omega)$ is the FT of the crosscorrelation between $s(t)$ and $x(t)$, and $P_x(\omega)$ is the PSD of $x(t)$ (i.e., FT of its autocorrelation). Dividing we find the full Wiener filter.

$$H(\omega) = \frac{P_{sx}(\omega)}{P_x(\omega)} \qquad (9.17)$$

For uncorrelated noise $P_{sx}(\omega) = P_s(\omega)$ and $P_x(\omega) = P_s(\omega) + P_n(\omega)$ and so the full Wiener filter reduces to equation (9.16).

The Wiener filter only functions when the signals being treated are stationary (i.e., P_{sx} and P_s are not functions of time). This restriction too can be lifted, resulting in the *Kalman filter*, but any attempt at explaining its principles would lead us too far astray.

EXERCISES

9.7.1 Assume that the signal $s(t)$ has constant PSD in some range but the noise $\nu(t)$ is narrow-band. Explain why we expect a Wiener filter to have a notch at the disturbing frequency.

9.7.2 An alternative to the SNR is the 'signal-plus-noise-to-noise-ratio' S+NNR. Why is this ratio of importance? What is the relationship between the overall S+NNR and SNR? What is the relationship between the Wiener filter and the frequency-dependent S+NNR and SNR?

9.8 Correlation and Prediction

A common problem in DSP is to predict the next signal value s_n based on the values we have observed so far. If s_n represents the closing value of a particular stock on day n the importance of accurate prediction is obvious. Less obvious is the importance of predicting the next value of a speech signal. It's not that I impolitely do not wish to wait for you to finish whatever you have to say; rather the ability to predict the next sample enables the *compression* of digitized speech, as will be discussed at length in Chapter 19. Any ability to predict the future implies that less information needs to be transferred or stored in order to completely specify the signal.

If the signal s is white noise then there is no correlation between its value s_n and its previous history (i.e., $C_s(m) = 0 \quad \forall m \neq 0$), and hence no prediction can improve on a guess based on single sample statistics. However, when the autocorrelation is nontrivial we can use past values to improve our predictions. So there is a direct connection between correlation and prediction; we can exploit the autocorrelation to predict what the signal will must probably do.

The connection between correlation and prediction is not limited to autocorrelation. If two signals x and y have a nontrivial crosscorrelation this can be exploited to help predict y_n given x_n. More generally, the causal prediction of y_n could depend on previous y values, x_n, and previous x values. An obvious example is when the crosscorrelation has a noticeable peak at lag m, and much information about y_n can be gleaned from x_{n-m}.

We can further clarify the connection between autocorrelation and signal prediction with a simple example. Assume that the present signal value s_n depends strongly on the previous value s_{n-1} but only weakly on older values. We further assume that this dependence is *linear*, $s_n \approx b\,s_{n-1}$ (were we to take $s_n \approx b\, s_{n-1} + c$ we would be forced to conclude $c = 0$ since otherwise the signal would diverge after enough time). Now we are left with the problem of finding b given an observed signal. Even if our assumptions are not very good, that is, even if s_n *does* depend on still earlier values, and/or the dependence is not really linear, and even if s_n depends on other signals as well, we are still interested in finding that b that gives the best linear prediction given only the previous value.

$$\tilde{s}_n = b\, s_{n-1} \tag{9.18}$$

What do we mean by *best* prediction? The best definition of *best* is for the Mean Squared Error (MSE)

$$d_n^2 = (s_n - \tilde{s}_n)^2 = (s_n - b s_{n-1})^2 = s_n^2 - 2 b\, s_n\, s_{n-1} + b^2 s_{n-1}^2$$

to be as small as possible, on the average. We are now in familiar territory. Assuming the signal to be time-invariant we average over all time

$$\left\langle d_n^2 \right\rangle = \left\langle s_n^2 \right\rangle - 2b \left\langle s_n s_{n-1} \right\rangle + b^2 \left\langle s_{n-1}^2 \right\rangle = (1 + b^2)\, C_s(0) - 2b\, C_s(1)$$

and then differentiate and set equal to zero. We find that the optimal linear prediction is

$$b = \frac{C_s(1)}{C_s(0)} = c_s(1) \tag{9.19}$$

the normalized autocorrelation coefficient for lag 1. Substituting this back into the expression for the average square error, we find

$$\left\langle d_n^2 \right\rangle = \frac{C_s^2(0) - C_s^2(1)}{C_s(0)} \tag{9.20}$$

so that the error vanishes when the lag 1 correlation equals the energy.

EXERCISES

9.8.1 Wiener named his book *The Extrapolation, Interpolation and Smoothing of Stationary Time Series with Engineering Applications*. Wiener's 'extrapolation' is what we have called 'prediction'. What did he mean by 'interpolation' and 'smoothing'?

9.8.2 Find the optimal linear prediction coefficients when two lags are taken into account.

$$\tilde{s}_n = b_1 s_{n-1} + b_2 s_{n-2}$$

9.9   Linear Predictive Coding

Signal coding, that is, compression of the amount of information needed to represent a signal, is an important application of DSP. To see why, consider the important application of digital speech. A bandwidth of 4 KHz is required so we must sample at 8000 samples per second; with 16 bit samples this requires 128 Kb/s, or just under 1 MB of data every minute. This data rate cannot be transferred over a telephone connection using a modem (the fastest telephone-grade modems reach 56 Kb/s) and would even be a tremendous strain on storage facilities. Yet modern speech compression techniques (see Chapter 19) can reduce the required rate to 8 Kb/s or less with only barely noticeable quality degradation.

Let's call the signal to be compressed s_n. If s is not white noise then it is at least partially linearly predictable based on its M previous values.

$$s_n = Ge_n + \sum_{m=1}^{M} b_m s_{n-m} \qquad (9.21)$$

Here e_n is the portion of the signal not predictable based on the signal's own history, G is an arbitrarily introduced gain, and b_m are called the Linear Predictive Coding (LPC) coefficients. Note that most people use a for these coefficients, but we reserve a for FIR coefficients; some people use a minus sign before the sum (i.e., use what we call β coefficients).

Equation (9.21) has a simple interpretation; the signal s_n is obtained by filtering the unpredictable signal e_n by a all-pole filter with gain G and coefficients b_m. The e_n is called the 'excitation' signal since it 'excites' the filter into operation. Since the filter is all-pole it enhances certain excited frequencies; these amplified frequencies are responsible for the non-flat spectrum and nontrivial autocorrelation of predictable signals. For speech signals (see Section 11.3) the excitation e_n is the 'glottal excitation'; for voiced speech (e.g., vowels) this is a periodic set of pulses created by the vocal chords, while for unvoiced speech (e.g., **h**) it is a noise-like signal created by constricting the passage of air. For both cases the mouth and nasal cavities act as a filter, enhancing frequencies according to their geometry.

In order to compress the signal we need an algorithm for finding the $M + 1$ parameters G and b_m given a buffer of N samples of the signal $\{s_n\}_{n=0}^{N-1}$. Looking carefully at equation (9.21) we note a problem. There are too many unknowns. In order to uniquely determine the coefficients b_m we need to know both the observed speech signal s_n and the excitation e_n. Unfortunately, the latter signal is usually inaccessible; for speech signals obtaining it would require swallowing a microphone so that it would be close to the vocal chords and before the vocal tract. We thus venture forth under the assumption that the excitation is identically zero. This is true most of the time for a pulse train excitation, only erring for those time instants when the impulse appears. It is obviously not a good approximation for many other cases.

Under the assumption of zero excitation we get the homogeneous recursion

$$s_n = \sum_{m=1}^{M} b_m s_{n-m} \tag{9.22}$$

for which $s = 0$ (the zero signal) is a solution. It is the only solution if the excitation was truly always zero; but due to the IIR nature of the filter, other possibilities exist if the excitation was once nonzero, even if zero during the duration of the present buffer. For speech the excitation is not truly zero, so even when we find the coefficients b_m we can only approximately predict the next signal value.

$$\tilde{s}_n = \sum_{m=1}^{M} b_m s_{n-m} \tag{9.23}$$

The error of this approximation is called the *residual signal*

$$r_n = s_n - \tilde{s}_n = s_n - \sum_{m=1}^{M} b_m s_{n-m} = \sum_{m=0}^{M} \beta_m s_{n-m} \tag{9.24}$$

(where $\beta_0 \equiv 1$ and $\beta_m = -b_m$), and the correct LPC coefficients minimize this residual. Note that the residual is obtained by FIR filtering the input signal, with the filter coefficients being precisely β_m. This all-zero filter is usually called the 'LPC analysis filter' and it is the inverse filter of the 'LPC synthesis filter' that synthesizes the speech from the excitation (see Figure 9.4). The analysis filter is also called the 'LPC whitening filter', the residual being much whiter than the original speech signal, since the linear predictability has been removed.

There is another way of looking at the residual signal. Rather than taking no excitation and treating the residual as an error signal, we can pretend that there *is* excitation but take the error to be precisely zero. What must the excitation be for $\tilde{s}_n$ to be the correct signal value? Comparing equations (9.24) and (9.21) we see that $r_n = Ge_n$, the residual is simply the excitation amplified by the gain. Thus when analyzing voiced speech we see that the residual is usually small but displays peaks corresponding to the vocal chord pulses.

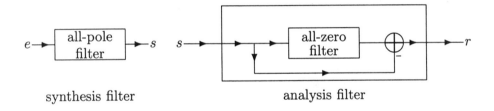

synthesis filter   analysis filter

Figure 9.4: LPC synthesis and analysis filters. The synthesis filter synthesizes the signal s_n from the excitation e_n, while the analysis filter analyzes incoming signal s_n and outputs the residual error signal r_n. The synthesis and analysis filters are inverse systems to within a gain.

One final remark regarding the residual. In speech compression terminology the residual we defined is called the *open-loop* residual. It can be calculated only if the original speech samples s_n are available. When decompressing previously compressed speech these samples are no longer available, and we can only attempt to predict the present signal value based on past *predicted values*. It is then better to define the *closed-loop* residual

$$r_n^c = s_n + \sum_{m=1}^{M} b_m \tilde{s}_{n-m}$$

and minimize *it* instead.

Returning to our mission, we wish to find coefficients b_m that minimize the residual of equation (9.24). In order to simultaneously minimize the residual r_n for all times of interest n, we calculate the MSE

$$E = \sum_n r_n^2 = \sum_n (s_n - \sum_{m=1}^{M} b_m s_{n-m})^2 \tag{9.25}$$

and minimize it with respect to the b_m $(m = 1 \dots M)$. This minimization is carried out by setting all M partial derivatives equal to zero

$$\frac{\partial E}{\partial b_m} = 0$$

which leads us to the following set of M equations.

$$\sum_n \left(\sum_{l=1}^{M} b_l s_{n-m} s_{n-l} - s_n s_{n-m} \right) = 0 \tag{9.26}$$

Moving the sum on n inside we can rewrite these

$$\sum_{l=1}^{M} b_l \sum_n s_{n-m}s_{n-l} = \sum_n s_n s_{n-m} \tag{9.27}$$

in which the signal enters only via autocorrelations C_s.

$$\sum_{l=1}^{M} C_s(|m-l|)\, b_l = C_s(m) \tag{9.28}$$

These are, of course, the Yule-Walker equations for the LPC coefficients.

The sum in the autocorrelations *should* run over all times n. This is problematic for two reasons. First, we are usually only given an input signal buffer of length N, and even if we are willing to look at speech samples outside this buffer, we cannot wait forever. Second, many signals including speech are stationary only for short time durations, and it is only sensible to compute autocorrelations over such durations. Thus we must somehow limit the range of times taken into account in the autocorrelation sums. This can be done in two ways. The brute-force way is to artificially take all signal values outside the buffer to be zero for the purposes of the sums. A somewhat more gentle variant of the same approach uses a window function (see Section 13.4) that smoothly reduces the signal to zero. The second way is to retain the required values from the previous buffer. The first way is called the autocorrelation method and is by far the most popular; the second is called the covariance method and is less popular due to potential numerical stability problems.

The autocorrelation method allows the sum in the MSE to be over all times, but takes all signal values outside the buffer $s_0 \ldots s_{N-1}$ to be zero. Since the error e_n in equation (9.24) depends on $M+1$ signal values, it can only be nonzero for $n = 0 \ldots N+M-1$. Accordingly, the MSE is

$$E = \sum_{n=0}^{N+M-1} r_n^2$$

and the correlations appearing in it have these limits.

$$C_s(m,l) \equiv \sum_{n=0}^{N+M-1} s_{n-m}s_{n-l} = C_s(|m-l|)$$

Writing the Yule-Walker equations in matrix notation

$$
\begin{pmatrix}
C_s(0) & C_s(1) & \cdots & C_s(M-1) \\
C_s(1) & C_s(0) & \cdots & C_s(M-2) \\
C_s(2) & C_s(1) & \cdots & C_s(M-3) \\
\vdots & \vdots & \vdots & \vdots \\
C_s(M-1) & C_s(M-2) & \cdots & C_s(0)
\end{pmatrix}
\begin{pmatrix}
b_1 \\ b_2 \\ b_3 \\ \vdots \\ b_M
\end{pmatrix}
=
\begin{pmatrix}
C_s(1) \\ C_s(2) \\ C_s(3) \\ \vdots \\ C_s(M)
\end{pmatrix}
\tag{9.29}
$$

we see that the matrix is symmetric and Toeplitz. In the next section we will study a fast method for solving such equations.

The MSE in the covariance method is taken to be

$$
E = \sum_{n=0}^{N-1} r_n^2
$$

and here we don't assume that the signal was zero for $n < 0$. We must thus access $N + M$ signal values, including M values from the previous buffer. Equations (9.27) are still correct, but now the sums over n no longer lead to genuine autocorrelations due to the limits of the sums being constrained differently.

$$
\mathcal{C}_s(m,l) \equiv \sum_{n=0}^{N-1} s_{n-m}s_{n-l} = \mathcal{C}_s(l,m)
$$

In particular $\mathcal{C}_s$ although symmetric is no longer a function of $|l - m|$, but rather a function of l and m separately. Writing these equations in matrix form we get a matrix that is symmetric but not Toeplitz.

$$
\begin{pmatrix}
\mathcal{C}_s(1,1) & \mathcal{C}_s(1,2) & \cdots & \mathcal{C}_s(1,M) \\
\mathcal{C}_s(1,2) & \mathcal{C}_s(2,2) & \cdots & \mathcal{C}_s(2,M) \\
\mathcal{C}_s(1,3) & \mathcal{C}_s(2,3) & \cdots & \mathcal{C}_s(3,M) \\
\vdots & \vdots & \vdots & \vdots \\
\mathcal{C}_s(1,M) & \mathcal{C}_s(2,M) & \cdots & \mathcal{C}_s(M,M)
\end{pmatrix}
\begin{pmatrix}
b_1 \\ b_2 \\ b_3 \\ \vdots \\ b_M
\end{pmatrix}
=
\begin{pmatrix}
\mathcal{C}_s(0,1) \\ \mathcal{C}_s(0,2) \\ \mathcal{C}_s(0,3) \\ \vdots \\ \mathcal{C}_s(0,M)
\end{pmatrix}
\tag{9.30}
$$

The fast methods of solving Toeplitz equations are no longer available, and the Cholesky decomposition (equation (A.94)) is usually employed.

Since general covariance matrices are of this form this method is called the covariance method, although no covariances are obviously present. For $N \gg M$ the difference between using N samples and using $N + M$ samples becomes insignificant, and the two methods converge to the same solution. For small buffers the LPC equations can be highly sensitive to the boundary conditions and the two methods may produce quite different results.

EXERCISES

9.9.1 What is the approximation error for the covariance method?

9.9.2 Equation (9.22) predicts s_n based on M previous values $s_{n-1}, s_{n-2}, \ldots s_{n-M}$ and is called the *forward predictor*. We can also 'predict' (postdict?) s_{n-M} based on the next M values $s_{n-M+1}, \ldots, s_{n-1}, s_n$. This surprising twist on LPC is called backward linear prediction. Modify equation (9.22) for this case (call the coefficients c_m). What is the residual?

9.9.3 Show that the MSE error can be written $E = \sum_n s_n^2 + \sum_{m=1}^M b_m \sum_n s_n s_{n-m}$ and thus for the autocorrelation method $E = C_s(0) + \sum_{m=1}^M b_m C_s(m)$.

9.9.4 Show that assuming the input to be an impulse $G\delta_{n,0}$ the gain is given by the error as given in the previous exercise.

9.9.5 Use the LPC method to predict the next term in the sequence $1, \alpha, \alpha^2, \alpha^3, \ldots$ for various $0 < \alpha < 1$. Repeat for $\alpha > 1$. Does the LPC method always correctly predict the next signal value?

9.10   The Levinson-Durbin Recursion

Take an empty glass soft-drink bottle and blow over its mouth. Now put a little water in the bottle and blow again. The frequency produced is higher since the wavelength that resonates in the cavity is shorter (recall our discussion of wavelength in Section 7.9). By tuning a collection of bottles you can create a musical instrument and play recognizable tunes.

The bottle in this experiment acts as a filter that is excited by breath noise. Modeling the bottle as a simple cylinder, the frequency it enhances is uniquely determined by its height. What if we want to create a signal containing two different frequencies? One way would be to blow over two different bottles separately (i.e., to place the filters in parallel). From our studies of filters we suspect that there may be a way of putting the filters in series (cascade) as well, but putting two cylinders one after the other only makes a single long cylinder. In order to get multiple frequencies we can use cylinders of different cross-sectional areas, the resonant frequencies being determined by the widths rather than the heights.

If we send a sound wave down a pipe that consists of a sequence of cylinders of different cross-sectional areas A_i, at each interface a certain amount of acoustic energy continues to travel down the pipe while some is reflected back toward its beginning. Let's send a sinusoidal acoustic wave down a pipe consisting of two cylinders. Recalling from Section 7.9 that traveling waves are functions of $s - vt$, we can express the incoming wave for this one directional case as follows.

$$\psi_{inc}^{[1]}(x, t) = A \sin(x - vt) \tag{9.31}$$

The reflected wave in the first cylinder will be sinusoid of the same frequency but traveling in the opposite direction and reduced in amplitude

$$\psi_{ref}^{[1]}(x, t) = kA \sin(x + vt) \tag{9.32}$$

where the reflection coefficient k is the fraction of the wave that is reflected. Since the energy is proportional to the signal squared, the fraction of the wave's energy that is reflected is k^2, while the wave energy that continues on to the second cylinder is whatever remains.

$$E_2 = (1 - k^2)E_1 \tag{9.33}$$

Now for a little physics. The ψ for sound waves can represent many different physical quantities (e.g., the average air particle displacement, the air particle velocity, the pressure). We'll assume here that it represents the velocity. Physically this velocity must be continuous across the interface between the two sections, so at the interface the following must hold.

$$\psi_{inc}^{[1]}(x, t) - \psi_{ref}^{[1]}(x, t) = \psi^{[2]}(x, t)$$

The derivative of the velocity is the acceleration, which is proportional to the force exerted on the air particles. The pressure, defined as the force per unit area, must be continuous at the interface, implying that the following must hold there.

$$\frac{\dot\psi_{inc}^{[1]}(x, t) - \dot\psi_{ref}^{[1]}(x, t)}{A_1} = \frac{\dot\psi^{[2]}(x, t)}{A_2}$$

Combining these two equations results in

$$\frac{1 + k}{A_1} = \frac{1 - k}{A_2}$$

and rearranging we find an expression for the reflection coefficient in terms of the cross-sectional areas.

$$k = \frac{A_1 - A_2}{A_1 + A_2}$$

Let's check to see that this result is reasonable. If the second cylinder shrinks to zero area (closing off the pipe) then $k = 1$ and the wave is entirely reflected, as it should be. If there really is no interface at all (i.e., $A_1 = A_2$) then $k = 0$ and no energy is reflected. If $A_2 \gg A_1$ then $k = -1$, which seems unreasonable at first; but an open-ended pipe has zero pressure at its end, and so the wave reflects but with a phase reversal.

It isn't hard to generalize our last result to a pipe with many sections. The reflection coefficient at the interface between section i and section $i+1$ is

$$k_i = \frac{A_i - A_{i+1}}{A_i + A_{i+1}} \tag{9.34}$$

What does all this have to do with solving the Yule-Walker equations for the LPC coefficients in the autocorrelation method? The LPC coefficients b_m are not the only way of describing an all-pole system; the area ratios, the reflection coefficients, and many others (including an interesting set to be discussed in the next section) can be used instead. Since all of these parameter sets contain exactly the same information, it follows that we can derive any set from any other set. Many of the parameter sets are related by linear transformations, and hence the conversion is equivalent to multiplying by a matrix. We will now show that the connection between the reflection and LPC coefficients can be expressed as a recursion that is the most efficient way of deriving both.

How can equation (9.29) be solved recursively? For simplicity we'll drop the subscript identifying the signal, but we have to add superscripts identifying the recursion depth. The first case is simple (for further simplicity we have dropped the subscript)

$$C(0)\, b_1^{[1]} = C(1) \qquad \longrightarrow \qquad b_1^{[1]} = \frac{C(1)}{C(0)}$$

and its MSE is

$$E_1 = C(0) - b_1^{[1]} C(1) = C(0)(1 - k_1^2)$$

where we have defined $k_1 \equiv b_1^{[1]}$. Let's assume we have already solved the mth case

$$\begin{pmatrix} C(0) & C(1) & \cdots & C(m-1) \\ C(1) & C(0) & \cdots & C(m-2) \\ \vdots & \vdots & \vdots & \vdots \\ C(m-1) & C(m-2) & \cdots & C(0) \end{pmatrix} \begin{pmatrix} b_1^{[m]} \\ b_2^{[m]} \\ \vdots \\ b_m^{[m]} \end{pmatrix} = \begin{pmatrix} C(1) \\ C(2) \\ \vdots \\ C(m) \end{pmatrix}$$

and let's write this $\underline{\underline{C}}^{[m]}\underline{b}^{[m]} = \underline{c}^{[m]}$. We are now interested in the $(m+1)^{\text{th}}$ case

$$
\begin{pmatrix}
C(0) & C(1) & \cdots & C(m-1) & C(m) \\
C(1) & C(0) & \cdots & C(m-2) & C(m-1) \\
\vdots & \vdots & \vdots & \vdots & \vdots \\
C(m-1) & C(m-2) & \cdots & C(0) & C(1) \\
\hline
C(m) & C(m-1) & \cdots & C(1) & C(0)
\end{pmatrix}
\begin{pmatrix}
b_1^{[m+1]} \\
b_2^{[m+1]} \\
\vdots \\
b_m^{[m+1]} \\
b_{m+1}^{[m+1]}
\end{pmatrix}
=
\begin{pmatrix}
C(1) \\
C(2) \\
\vdots \\
C(m) \\
C(m+1)
\end{pmatrix}
$$

where we have drawn in delimiters that divide the equations into two parts:

$$
\begin{pmatrix}
C(0) & C(1) & \cdots & C(m-1) \\
C(1) & C(0) & \cdots & C(m-2) \\
\vdots & \vdots & \vdots & \vdots \\
C(m-1) & C(m-2) & \cdots & C(0)
\end{pmatrix}
\begin{pmatrix}
b_1^{[m+1]} \\
b_2^{[m+1]} \\
\vdots \\
b_m^{[m+1]}
\end{pmatrix}
+ b_{m+1}^{[m+1]}
\begin{pmatrix}
C(m) \\
C(m-1) \\
\vdots \\
C(1)
\end{pmatrix}
$$

$$
=
\begin{pmatrix}
C(1) \\
C(2) \\
\vdots \\
C(m)
\end{pmatrix}
\tag{9.35}
$$

and

$$
\begin{pmatrix} C(m) & C(m-1) & \cdots & C(1) & C(0) \end{pmatrix}
\begin{pmatrix}
b_1^{[m+1]} \\
b_2^{[m+1]} \\
\vdots \\
b_m^{[m+1]} \\
b_{m+1}^{[m+1]}
\end{pmatrix}
= C(m+1)
\tag{9.36}
$$

Now multiply equation (9.35) by the inverse of the autocorrelation matrix of the mth iteration $(\underline{\underline{C}}^{[m]})^{-1}$ and use the results of that iteration.

$$
\begin{pmatrix}
b_1^{[m+1]} \\
b_2^{[m+1]} \\
\vdots \\
b_m^{[m+1]}
\end{pmatrix}
+ b_{m+1}^{[m+1]}(\underline{\underline{C}}^{[m]})^{-1}
\begin{pmatrix}
C(m) \\
C(m-1) \\
\vdots \\
C(1)
\end{pmatrix}
= (\underline{\underline{C}}^{[m]})^{-1}
\begin{pmatrix}
C(1) \\
C(2) \\
\vdots \\
C(m)
\end{pmatrix}
$$

Defining $\underset{=}{J}$ as the matrix that reverses row order

$$\underset{=}{J} = \begin{pmatrix} 0 & 0 & \cdots & 0 & 1 \\ 0 & 0 & \cdots & 1 & 0 \\ \vdots & \vdots & \vdots & \vdots & \vdots \\ 0 & 1 & \cdots & 0 & 0 \\ 1 & 0 & \cdots & 0 & 0 \end{pmatrix}$$

and noting that it commutes with Toeplitz matrices, we can finally write the following recursion for the LPC coefficients

$$\begin{pmatrix} b_1^{[m+1]} \\ b_2^{[m+1]} \\ \vdots \\ b_m^{[m+1]} \end{pmatrix} = \left(\underset{=}{I} - k_{m+1} \underset{=}{J} \right) \begin{pmatrix} b_1^{[m]} \\ b_2^{[m]} \\ \vdots \\ b_m^{[m]} \end{pmatrix} \tag{9.37}$$

where $k_m \equiv b_m^{[m]}$.

In the statistics literature the k variables are called 'partial correlation' or PARCOR coefficients, since they can be shown to measure the correlation between the forward and backward prediction errors (see exercise 9.9.2). Later we will show that they are exactly the reflection coefficients.

Were we to know k_{m+1} this recursion would produce all the other new $b^{[m+1]}$ given the old $b^{[m]}$. So we have reduced the problem of finding the LPC coefficients to the problem of finding the PARCOR coefficients. Yet it is obvious from equation (9.36) that the converse is also true, $k_{[m+1]}$ can be derived from the lower $b^{[m+1]}$ coefficients. So let's derive a recursion for the ks and try to eliminate the bs.

First we rewrite equation (9.36) as

$$\begin{pmatrix} C(1) & C(2) & \cdots & C(m) \end{pmatrix} \underset{=}{J} \begin{pmatrix} b_1^{[m+1]} \\ b_2^{[m+1]} \\ \vdots \\ b_m^{[m+1]} \end{pmatrix} + k_{m+1} C(0) = C(m+1)$$

which can be written (with obvious notation) as follows.

$$\underset{-}{c} \cdot \underset{==}{J} b^{[m+1]} + k_{m+1} C(0) = C(m+1)$$

Now we substitute the $b^{[m+1]}$ vector from equation (9.37)

$$\underline{c} \cdot \underline{\underline{J}}(\underline{\underline{I}} - k_{m+1}\underline{\underline{J}})\underline{b}^{[m]} + k_{m+1}C(0) = C(m+1)$$

and finally solve for k_{m+1} (noting $\underline{\underline{J}}^2 = \underline{\underline{I}}$)

$$k_{m+1} = \frac{C(m+1) - \underline{c} \cdot \underline{\underline{J}}\underline{b}^{[m]}}{C(0) - \underline{c} \cdot \underline{b}^{[m]}}$$

$$= \frac{C(m+1) - \underline{c} \cdot \underline{\underline{J}}\underline{b}^{[m]}}{E_m}$$

identifying the MSE in the denominator. After following all the above the reader will have no problem proving that the MSE obeys the simplest recursion of all.

$$E_{m+1} = (1 - k_{m+1}^2)E_m \tag{9.38}$$

Let's now group together all the recursive equations into one algorithm that computes the k and b coefficients for successively higher orders until we reach the desired order M.

Given the signal autocorrelations $C(0)$ through $C(M)$
Start with $E_0 = C(0)$
for $m \leftarrow 1$ to M
$\qquad b^{[m]} \leftarrow k_m = \dfrac{C(m) - \sum_{\mu=1}^{m-1} C(m-\mu)b_\mu^{[m-1]}}{E_{m-1}}$
$\qquad$ for $\mu = m - 1$ down to 1
$\qquad\qquad b_\mu^{[m]} \leftarrow b_\mu^{[m-1]} - k_m b_{m-\mu}^{[m-1]}$
$\qquad E_m \leftarrow (1 - k_m^2)E_{m-1}$
for $\mu \leftarrow 1$ to M
$\qquad b_\mu \leftarrow b_\mu^{[M]}$

To see how the algorithm works let's run through it for the case of two coefficients.

$$\begin{pmatrix} C(0) & C(1) \\ C(1) & C(0) \end{pmatrix} \begin{pmatrix} b_1 \\ b_2 \end{pmatrix} = \begin{pmatrix} C(1) \\ C(2) \end{pmatrix}$$

The first iteration is easy.

$$E_0 = C(0)$$
$$b_1^{[1]} = k_1 = \frac{C(1)}{C(0)}$$

Now we can perform the second iteration

$$E_1 = \frac{C^2(0) - C^2(1)}{C(0)}$$

$$b_2^{[2]} = k_2 = \frac{C(2)C(0) - C^2(1)}{C^2(0) - C^2(1)}$$

$$b_1^{[2]} = b_1^{[1]} - k_2 = \frac{C(1)C(0) - C(1)C(2)}{C^2(0) - C^2(1)}$$

and we have found the desired coefficients b_1 and b_2.

We finish the section by fulfilling our promise to show that the k are the reflection coefficients. If we implement the LPC analysis filter (the FIR filter that converts the signal into the residual as a multistage lattice filter) then equation (9.38) tells us how the energy of the residual decreases. Comparing this with equation (9.33) completes the identification.

EXERCISES

9.10.1 Prove equation (9.34) for a pipe with multiple sections taking into account the reflected wave from the next interface.

9.10.2 Transmission lines have both voltage and current traveling waves, the ratio between the voltage and current being the impedance Z. At a splice where the impedance changes a reflected wave is generated. Express the reflection coefficient in terms of the impedances. Explain the limiting cases of shorted and open circuited cables.

9.10.3 Prove equation (9.38) for the MSE.

9.10.4 Solve the three-coefficient problem on paper using the Levinson-Durbin recursion.

9.10.5 Show that the complexity of the Levinson-Durbin algorithm is $O(M^2)$ rather than $O(M^3)$ as for non-Toeplitz systems.

9.10.6 Levinson originally solved the more general problem of solving the equations $\underline{\underline{T}}\,\underline{x} = \underline{y}$ where $\underline{\underline{T}}$ is Toeplitz but unrelated to $\underline{y}$. Generalize the recursion to solve this problem. (Hint: You will need another set of recursions.) How much more computationally complex is the solution?

9.11 Line Spectral Pairs

Another set of parameters that contain exactly the same information as the LPC coefficients are the **Line Spectral Pair** (LSP) frequencies. To introduce them we need to learn a mathematical trick that can be performed on the polynomial in the denominator of the LPC system function.

A polynomial of degree M

$$a(x) = \sum_{m=0}^{M} p_m x^m = a_0 + a_1 x + a_2 x^2 + \ldots a_{M-2} x^{M-2} + a_{M-1} x^{M-1} + a_M x^M$$

is called 'palindromic' if $a_m = a_{M-m}$, i.e.,

$$a_0 = a_M \qquad a_1 = a_{M-1} \qquad a_2 = a_{M-2} \qquad \text{etc.}$$

and 'antipalindromic' if $a_m = -a_{M-m}$, i.e.,

$$a_0 = -a_M \qquad a_1 = -a_{M-1} \qquad a_2 = -a_{M-2} \qquad \text{etc.}$$

so $1 + 2x + x^2$ is palindromic, while $x + x^2 - x^3$ is antipalindromic. It is not hard to show that the product of two palindromic or two antipalindromic polynomials is palindromic, while the product of an antipalindromic polynomial with a palindromic one is antipalindromic.

We will now prove that every real polynomial that has all of its zeros on the unit circle is either palindromic or antipalindromic. The simplest cases are $x+1$ and $x-1$, which are obviously palindromic and antipalindromic, respectively. Next consider a second degree polynomial with a pair of complex conjugate zeros on the unit circle.

$$\begin{aligned} a(x) &= \left(x - e^{i\phi}\right)\left(x - e^{-i\phi}\right) \\ &= x^2 - e^{-i\phi}x - e^{i\phi}x + e^{i\phi}e^{-i\phi} \\ &= x^2 - 2\cos(\phi) + 1 \end{aligned}$$

This is obviously palindromic.

Any real polynomial that has k pairs of complex conjugate zeros will be the product of k palindromic polynomials, and thus palindromic. If a polynomial has k pairs of complex conjugate zeros and the root $+1$ it will also be palindromic, while if it has -1 as a root it will be antipalindromic. This completes the proof.

The converse of this statement is not necessarily true; not every palindromic polynomial has all its zeros on the unit circle. The idea behind the

LSPs is to define palindromic and antipalindromic polynomials that *do* obey the converse rule. Let's see how this is done.

Any arbitrary polynomial $a(x)$ can be written as the sum of a palindromic polynomial $p(x)$ and an antipalindromic polynomial $q(x)$

$$a_m = \tfrac{1}{2}(p_m + q_m) \qquad \text{where} \qquad \begin{aligned} p_m &= a_m + a_{M-m} \\ q_m &= a_m - a_{M-m} \end{aligned} \qquad (9.39)$$

(if M is even the middle coefficient appears in p_m only). When we are dealing with polynomials that have their constant term equal to unity, we would like the polynomials p_m and q_m to share this property. To accomplish this we need only pretend for a moment that a_m is a polynomial of order $M+1$ and use the above equation with $a_{M+1} = 0$.

$$a_m = \tfrac{1}{2}(p_m + q_m) \qquad \text{where} \qquad \begin{aligned} p_m &= a_m + a_{M+1-m} \\ q_m &= a_m - a_{M+1-m} \end{aligned} \qquad (9.40)$$

Now $a_0 = p_0 = q_0 = 1$ but p_m and q_m are polynomials of degree $M+1$.

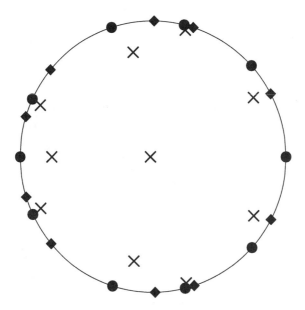

Figure 9.5: The zeros of a polynomial and of its palindromic and antipalindromic components. The **X**s are the zeros of a randomly chosen tenth order polynomial (constrained to have its zeros inside the unit circle). The circles and diamonds are the zeros of the $p(x)$ and $q(x)$. Note that they are all on the unit circle and are intertwined.

Formally we can write the relationships between the polynomials

$$a(x) = \tfrac{1}{2}\left(p(x) + q(x)\right) \qquad \text{where} \qquad \begin{pmatrix} p(x) \\ q(x) \end{pmatrix} = a(x) \pm x^{M+1}a(x^{-1})$$

and it is not hard to show that if all the zeros of $a(x)$ are inside the unit circle, then all the zeros of $p(x)$ and of $q(x)$ are *on* the unit circle. Furthermore, the zeros of $p(x)$ and $q(x)$ are intertwined, i.e., between every two zeros of $p(x)$ there is a zero of $q(x)$ and vice versa. Since these zeros are on the unit circle they are uniquely specified by their angles. For the polynomial in the denominator of the LPC frequency response these angles represent frequencies, and are called the LSP frequencies.

Why are the LSP frequencies a useful representation of the all-pole filter? The LPC coefficients are not a very homogeneous set, the higher-order b_m being more sensitive than the lower-order ones. LPC coefficients do not quantize well; small quantization error may lead to large spectral distortion. Also the LPC coefficients do not interpolate well; we can't compute them at two distinct times and expect to accurately predict them in between. The zeros of the LPC polynomial are a better choice, since they all have the same physical interpretation. However, finding these zeros numerically entails a complex two-dimensional search, while the zeros of $p(x)$ and $q(x)$ can be found by simple one-dimensional search techniques. In speech applications it has been found empirically that the LSP frequencies quantize well and interpolate better than all other parameters that have been tried.

EXERCISES

9.11.1 Let's create a random polynomial of degree M by generating $M + 1$ random numbers and using them as coefficients. We can now find the zeros of this polynomial and plot them in the complex plane. Verify empirically the hard-to-believe fact that for large M most of the zeros are close to the unit circle (except for large negative real zeros). Change the distribution of the random number generator. Did anything change? Can you explain why?

9.11.2 Prove that if all the zeros of $a(x)$ are inside the unit circle, then all the zeros of $p(x)$ and of $q(x)$ are *on* the unit circle. (Hint: One way is write the p and q polynomials as $a(x)\left(1 \pm h(x)\right)$ where $h(x)$ is an all-pass filter.) Prove that the zeros of $p(x)$ and $q(x)$ are intertwined. (Hint: Show that the phase of all-pass filter is monotonic, and alternately becomes π (zero of p) and 0 (zero of q).)

9.11.3 A pipe consisting of $M + 1$ cylinders that is completely open or completely closed at the end has its last reflection coefficient $k_{M+1} = \pm 1$. How does this relate to the LSP representation?

9.11.4 Generate random polynomials and find their zeros. Now build $p(x)$ and $q(x)$ and find their zeros. Verify that if the polynomial zeros are inside the unit circle, then those of p and q are on the unit circle. Is there a connection between the angles of the polynomial zeros and those of the LSPs?

9.11.5 The Greek mathematician Apollonius of Perga discovered that given two points in the plane z_1 and z_2, the locus of points with distances to z_1 and z_2 in a fixed ratio is circle (except when the ratio is fixed at one when it is a straight line). Prove this theorem. What is the connection to LSPs?

9.12   Higher-Order Signal Processing

The main consequence of the Wiener-Khintchine theorem is that most of the signal processing that we have learned is actually only 'power spectrum' processing. For example, when we use frequency selective filters to enhance signals we cannot discriminate between signals with the same power spectrum but different spectral phase characteristics. When we use correlations to solve system identification problems, we are really only recovering the square of the frequency response. We have yet to see methods for dealing with signals with non-Gaussian distributions or non-minimum-phase attributes of systems.

   In this section we will take a brief look at a theory of signal processing that *does* extend beyond the power spectrum. We will assume that our signals are stochastic and stationary and accordingly use the probabilistic interpretation of correlations, first introduced in Section 5.6. There we defined the moment functions, definitions we repeat here in slightly modified form.

$$M_s^{[k]}(m_1, m_2, \ldots, m_k) \equiv \langle s_n s_{n+m_1} \cdots s_{n+m_k} \rangle \qquad (9.41)$$

The kth moment function of the digital stationary stochastic signal s is the average of the product of $k + 1$ signal values, at time lags defined by the moment function's parameters.

   The first-order moment function is simply

$$M_s^{[1]} = \langle s_n \rangle$$

the signal's average (DC) value. The second-order moment function is

$$M_s^{[2]}(m) = \langle s_n s_{n+m} \rangle = C_s(-m)$$

the autocorrelation (recall the probabilistic interpretation of autocorrelation of equation (5.22)). The third-order moment function is a new entity.

$$M_s^{[3]}(m_1, m_2) = \langle s_n s_{n+m_1} s_{n+m_2} \rangle$$

Normal signal processing exploits only the first and second moments; higher-order signal processing utilizes the third and higher moments as well.

The moment functions have an especially simple interpretation for the special case of a stochastic signal that can only take on values 0 and 1. The first moment, the signal's average value, can be interpreted as the probability that the signal takes on the value 1; if the average is $\frac{1}{2}$ this means that the 0 and 1 values are equally probable. The second moment, the autocorrelation, relates the signal's values at two different times separated by m. Its interpretation is the probability that the signal takes on value 1 at any two times separated by m. If the signal is white (i.e., the 0 or 1 value is chosen independently at each time instant), then the autocorrelation will be $\frac{1}{2}$ (fifty-fifty) for all nonzero time lags. A higher correlation at time lag m means that the signal's being 1 at time n encourages the probability that it will be 1 at time $n + m$ as well, while a lower correlation indicates that a 1 at one time inhibits a second 1. A periodic signal will have its second moment function equal to unity for a lag equaling the period of any multiple thereof, since the probability of matching values is a certainty. As correlations of nonperiodic function normally die down for large enough lags, the two events become independent for large m.

$$M_s^{[2]}(m) \overset{m \to \infty}{\Longrightarrow} (M_s^{[1]})^2 \tag{9.42}$$

The interpretation of the third moment function is now clear. It is the probability that the 0-1 stochastic signal takes on the value 1 at all three times n, $n + m_1$, and $n + m_2$. If both m_1 and m_2 are very large we expect the third moment to equal the mean cubed, while if m_1 is small enough for there to be nontrivial correlations, but m_2 still large, then we expect a slightly more complex expression.

$$M_s^{[3]}(m_1, m_2) \overset{m_2 \to \infty}{\Longrightarrow} M_s^{[2]}(m_1) M_s^{[1]} \tag{9.43}$$

However, the third moment can be significantly different from this as well. For instance, a signal that is generated by

$$s_n = \Theta\left(\nu_n + \alpha_1 s_{n-m_1} + \alpha_2 s(n - m_2) + \ldots\right)$$

(where ν_n is some driving noise signal) will have a nontrivial third moment function with just these lags.

Similarly the fourth and higher moment functions give the probability that the signal takes on 1 values at four or more times. In practice, interpretation of numeric moment function data is complex because of the contributions from lower-order moments, as in equations (9.42) and (9.43). For example, if 0 and 1 are equally probable, we expect to observe 1 at two different times with a probability of one-quarter; only deviations from this value signify that there is something special about the lag between the two times. Likewise, to really understand how connected four different times are, we must subtract from the fourth moment function all the contributions from the third-order moments, but these in turn contain portions of second-order moments and so on. The way to escape this maize of twisty little passages is to define *cumulants*.

The exact definition of the cumulant is a bit tricky since we have to keep track of all possible groupings of the time instants that appear in the moment function. For this purpose we use the mathematical concept of a 'partition', which is a collection of nonempty sets whose union is a given set. For example, in the third moment there are three time instances $n_0 = n$, $n_1 = n + m_1$, and $n_2 = n + m_2$, and these can be grouped into five different partitions. $P_1 = \{(n_1, n_2, n_3)\}$, $P_2 = \{(n_1), (n_2, n_3)\}$, $P_3 = \{(n_2), (n_1, n_3)\}$, $P_4 = \{(n_2), (n_1, n_2)\}$, and $P_5 = \{(n_1), (n_2), (n_3)\}$. We'll use the symbol S_{ij} for the jth set of partition P_i (e.g., $S_{21} = (n_1)$ and $S_{42} = (n_1, n_2)$), N_i the number of such sets ($N_1 = 1$, $N_2 = N_3 = N_4 = 2$ and $N_5 = 3$), and N_{ij} for the number of elements in a set (e.g., $N_{11} = 3$, $N_{51} = 1$). We can now define the cumulant

$$C_s^{[k]} = \sum_i (-1)^{N_i - 1}(N_i - 1)! \prod_{j=1}^{i} M_s^{[N_{ij}]}(S_{ij}) \qquad (9.44)$$

where the sum is over all possible partitions of the k time instants.

It will be convenient to have a special notation for the signal with its DC component removed, $\hat{s} \equiv s - \langle s \rangle$. The first few cumulants can now be expressed as follows:

$$C_s^{[1]} = M_s^{[1]} = \langle s_n \rangle$$

as expected,

$$C_s^{[2]}(m) = M_s^{[2]}(m) - (M_s^{[1]})^2 = \langle \hat{s}_n \hat{s}_{n+m} \rangle$$

which is the autocovariance rather than the autocorrelation,

$$C_s^{[3]}(m_1, m_2) =$$
$$M_s^{[3]}(m_1, m_2) - M_s^{[1]} \left(M_s^{[2]}(n) + M_s^{[2]}(n+m_1) + M_s^{[2]}(n+m_2) \right) + 2(M_s^{[1]})^3$$
$$= \langle \hat{s}_n \hat{s}_{n+m_1} \hat{s}_{n+m_2} \rangle$$

a surprisingly simple result,

$$C_s^{[4]}(m_1, m_2, m_3) = M_s^{[4]}(m_1, m_2, m_3) - \ldots - 6(M_s^{[1]})^4 =$$
$$\langle \hat{s}_n \hat{s}_{n+m_1} \hat{s}_{n+m_2} \hat{s}_{n+m_3} \rangle$$
$$- C_{\hat{s}}^{[2]}(m_1) C_{\hat{s}}^{[2]}(m_2 - m_3) - C_{\hat{s}}^{[2]}(m_2) C_{\hat{s}}^{[2]}(m_3 - m_1) - C_{\hat{s}}^{[2]}(m_3) C_{\hat{s}}^{[2]}(m_1 - m_2)$$

which is somewhat more complex. For the special case of a zero mean signal and $m_1 = m_2 = m_3$, $C^{[2]}$ is the variance, $C^{[3]}$ the 'skew', and $C^{[4]}$ the 'kurtosis'.

Other than their interpretability, the cumulants are advantageous due to their convenient characteristics. The most important of these, and the reason they are called 'cumulants', is their additivity.

$$C_{x+y}^{[k]}(m_1, m_2, \ldots m_{k-1}) = C_x^{[k]}(m_1, m_2, \ldots m_{k-1}) + C_y^{[k]}(m_1, m_2, \ldots m_{k-1})$$

It is easy to see that this characteristic is not shared by the moment functions. Another nice feature is their blindness to DC components

$$C_{x+\alpha}^{[k]}(m, m_2, \ldots m_{k-1}) = C_x^{[k]}(m_1, m_2, \ldots m_{k-1})$$

where α is any constant. Like the moments, cumulants are permutation blind

$$C_s^{[k]}(m_{\sigma_1}, m_{\sigma_2}, \ldots m_{\sigma_{k-1}}) = C_s^{[k]}(m_1, m_2, \ldots m_{k-1})$$

where σ_i is any permutation of $1 \ldots k-1$; and scale according to their order.

$$C_{gs}^{[k]}(m_1, m_2, \ldots m_{k-1}) = g^k C_s^{[k]}(m_1, m_2, \ldots m_{k-1})$$

If the signal is symmetrically distributed then all odd-order cumulants vanish. If a signal is Gaussianly distributed all cumulants above the second-order vanish.

Higher-order spectra are defined in analogy with the Wiener-Khintchine theorem. Just as the spectrum is the FT of $C_s^{[2]}(m)$, the *bispectrum* is defined to be the two-dimensional FT of $C_s^{[3]}(m_1, m_2)$, and the *trispectrum* the three-dimensional FT of $C_s^{[4]}(m_1, m_2, m_3)$. It can be shown that for signals with finite energy, the general *polyspectrum* is given by a product of FTs.

$$S^{[k]}(\omega_1, \omega_2 \ldots \omega_{k-1}) = S(\omega_1) S(\omega_2) \ldots S(\omega_{k-1}) S^*(\omega_1 + \omega_2 + \ldots + \omega_{k-1})$$

Now that we have defined them, we can show that cumulants are truly useful. Assume that we have a non-Gaussian signal distorted by Gaussian noise. Standard signal processing does not take advantage of the higher-order statistics of the signal, and can only attempt to separate the signal from the noise in the power spectral domain. However, cumulants of the third and higher orders of the noise will vanish exactly, while those of the signal will not, thus providing a more powerful tool for recovery of such a signal. For example, higher-order matched filters can be used as sensitive detectors of the arrival of non-Gaussian signals in Gaussian noise.

We know from Section 8.1 that intermodulation products are produced when two sinusoids enter a nonlinearity. Assume we observe several frequency components in the output of a possibly nonlinear system; is there any way to tell if they are intermodulation frequencies rather than independent signals that happen to be there? The fingerprint of the phenomenon is that intermodulation products are necessarily phase coupled to the inputs; but such subtle phase relations are lost in classical correlation-based analysis. By using higher-order cumulants intermodulation frequencies can be identified and the precise nature of system nonlinearities classified.

In Sections 6.12 and 6.13 we saw how to perform correlation-based system identification when we had access to a system's input and output. Sometimes we may desire to identify a system, but can only observe its output. Amazingly, this problem may be tractable if the input signal is non-Gaussian. For example, if the unknown system is an N tap FIR filter,

$$y_n = \sum_{m=0}^{N-1} h_m x_{n-m} + \nu_n$$

the input x is zero mean but with nonzero third-order cumulant, and the output y is observed contaminated by additive Gaussian (but not necessarily white) noise ν, then the system's impulse response can be derived solely from the output's third-order cumulants.

$$h_m = \frac{C_y^{[3]}(N-1, m)}{C_y^{[3]}(N-1, 0)} \tag{9.45}$$

This amazing result is due to the input's third-order cumulant (assumed nonzero) appearing in the numerator and denominator and hence cancelling out, and can be generalized to higher-order cumulants if needed. A related result is that cumulant techniques can be used for *blind equalization*, that is, constructing the inverse of an unknown distorting system, without access to the undistorted input.

EXERCISES

9.12.1 Find all the partitions of four time instants and express $C_s^{[4]}(m_1, m_2, m_3)$ in terms of moments.

9.12.2 Consider the three systems $(0 < a, b < 1)$

$$
\begin{aligned}
y_n &= x_n - (a + b)x_{n-1} + abx_{n-2} \\
y_n &= x_n - (a + b)x_{n+1} + abx_{n+2} \\
y_n &= -ax_{n+1} + (1 + ab)x_n - bx_{n-1}
\end{aligned}
$$

What are the system functions for these systems? Which system is minimum phase, which maximum phase, and which mixed phase? Take x_n to be a zero mean stationary white noise signal, with $\langle x_n x_{n+m} \rangle = \delta_{m0}$ and $\langle x_n x_{n+m_1} x_{n+m_2} \rangle = \delta_{m_1 m_2}$. Show that the output signals from all three systems have the same autocorrelations. Prove that for all three systems the same frequency response is measured. Why is this result expected? Show that the third-order moments *are* different.

9.12.3 Prove equation (9.45).

9.12.4 There is another way of defining cumulants. Given the k signal values

$$s_n, s_{n+m_1}, \ldots, s_{n+m_{k-1}}$$

we posit k dummy variables $w_0 \ldots w_{k-1}$ and define the following function, known as the characteristic function.

$$\Phi(w_0 \ldots w_{k-1}) = \left\langle e^{\mathrm{i}(w_0 s_n + w_1 s_{n+m_1} + \cdots + w_{k-1} s_{n+m_{k-1}})} \right\rangle$$

The cumulants are the coefficients of the Taylor expansion of the logarithm of this function. Derive the first few cumulants according to this definition and show that they agree with those in the text. Derive the additivity property from this new definition.

9.12.5 In the text we mentioned the application of higher-order signal processing to the identification of intermodulation products. Let φ_1, φ_2 and φ_3 be independent uniformly distributed random variables and define two stochastic signals

$$
\begin{aligned}
s_n^{[a]} &= \cos\left(\omega_1 n + \varphi_1\right) + \cos\left(\omega_2 n + \varphi_2\right) \cos\left((\omega_1 + \omega_2)n + (\varphi_1 + \varphi_2)\right) \\
s_n^{[b]} &= \cos\left(\omega_1 n + \varphi_1\right) + \cos\left(\omega_2 n + \varphi_2\right) \cos\left((\omega_1 + \omega_2)n + \varphi_3\right)
\end{aligned}
$$

each of which has three spectral lines, the highest frequency being the sum of the lower two. The highest component of $s^{[a]}$ could be an intermodulation product since it is phase-locked with the other two, while that of $s^{[b]}$ is an unrelated signal. Show that both signals have the same autocorrelation and power spectrum, but differ in their third-order cumulants.

Bibliographical Notes

Matched filters are covered in most books on communications theory, e.g. [242, 95].

Wiener's first expositions of the Wiener-Khintchine theorem were in mathematical journals [276] but he later wrote an entire book on his discoveries [277]. The co-discoverer of the theorem was Aleksandr Khintchine (or Khinchin), whose *Mathematical Foundations of Information Theory* was translated into English from the original Russian in 1957.

The second volume of Norbert Wiener's autobiography [280] has fascinating background information on Wiener's work at MIT during the World War II years. His 1942 report, entitled *Extrapolation, Interpolation and Smoothing of Stationary Time Series*, was suppressed because of possible military applications, and finally released only in 1949 [278]. Even though written to be more understandable than the former paper, its mathematics, more familiar to physicists than engineers, was so difficult to the latter audience that it was commonly called the 'yellow peril'. Levinson both explained Wiener's results to a wider audience [146] and translated the formalism to the digital domain. While accomplishing this second task he invented his recursion [147], although digital hardware capable of computing it did not exist at the time.

The invention of LPC is due to Bishnu Atal of Bell Labs [10], who was mainly interested in its use for compression of speech [9]. The LSP frequencies are due to Itakura of NTT Research Labs [109] (but don't bother checking the original reference, it's only an abstract).

Higher-order signal processing is the subject of a book [181] and numerous review articles [173, 182]. [33] discusses partitions in a simple way, and includes source code for computing the number of partitions of n objects. Cumulants were introduced in statistics by Fisher in the 1930s and in use in physics at about the same time. The idea of higher-order spectra as the FT of cumulants dates back to Kolmogorov, but the nomenclature 'polyspectra' is due to Tukey. The use of cumulants for output-only system identification is due to Georgios Giannakis [72]. A few references to the extensive literature on applications of cumulants include noise cancellation [49]; system identification [73, 65]; blind equalization [235, 236]; and signal separation [286, 287, 108].

Adaptation

We have already learned about many different types of systems. We started with frequency selective filters and filters designed for their time-domain properties. Then we saw nonfilters that had capabilities that filters lack, such as PLLs that can lock onto desired frequency components. Next we saw how to match a filter to a prespecified signal in order to best detect that signal. We have even glimpsed higher-order signal processing systems that can differentiate between signals with identical power spectra. Yet all these systems are simple in the sense that their design characteristics are known ahead of time. Nothing we have studied so far can treat problems where we are constantly changing our minds as to what the system should do.

In this chapter we briefly discuss adaptive filters, that is, filters that vary in time, adapting their coefficients according to some reference. Of course the term 'adaptive *filter*' is a misnomer since by definition filters must be time-invariant and thus cannot vary at all! However, we allow this shameful usage when the filter coefficients vary much more slowly than the input signal.

You may think that these adaptive filters would be only needed on rare occasions but in practice they are extremely commonplace. In order to understand how and why they turn up we disregard our usual custom and present three applications before tackling the more general theory. These applications, noise cancellation, echo cancellation, and equalization turn out to have a lot in common.

After this motivation we can introduce the more general problem, stressing the connection with the Wiener-Hopf equations. Direct solution of these equations is usually impossible, and so we will learn how to iteratively approximate a solution using the Widrow-Hoff equations and the LMS algorithm. We then briefly present several of the variants to vanilla LMS, and the alternative RLS algorithm.

10.1   Adaptive Noise Cancellation

A lecture is to be recorded using a microphone placed at some distance from the lecturer. It is a hot summer day and the lecture hall is packed; a large air-conditioning unit is running noisily, and the fluorescent fixtures are emitting a low buzzing noise. As the lecturer begins to speak the crowd hushes and a tape-recorder starts to record. What exactly is being recorded?

Were we to listen to the recording we would certainly hear the lecturer, but we would soon notice other sounds as well. Fluorescent lamp noise is spectrally localized at harmonics of the AC supply frequency and if truly annoying could be filtered out using techniques we have discussed previously. The air-conditioner sounds and the background talking from the audience are not as easy to remove. They are neither spectrally localized nor stationary in character. Humans are extremely good at 'tuning out' such noises, but our brains use filtering based on content, a difficult feat to duplicate. Is there a practical way to remove these interferences from the recording?

Let's focus on the air-conditioner noise, although the audience's babble could be similarly treated. We propose using a second microphone placed near the air-conditioner so that it picks up mainly its noise and not the speaker's voice. Now since the first microphone is picking up the sum of two signals (the desired speech and the air-conditioner noise) we need to *subtract* the air-conditioner noise signal as picked up by the second microphone from the first signal. If done correctly the speech signal alone will remain.

Simplifying for the sake of presentation, we will assume that the second microphone hears the air-conditioner noise q_n alone. The lecturer's microphone signal y_n contains both the desired speech signal x_n and the air-conditioner noise. However, y_n will not be simply the sum $x_n + q_n$ for at least two reasons. First, the amplitude of the air-conditioner noise at the lecturer's microphone will most probably be weaker than that of the microphone directly in front of the unit. Second, the speed of sound is finite, and thus the air-conditioner noise as detected at the lecturer's microphone is delayed as compared to the close microphone. This delay is far from negligible; for example, assume the lecturer's microphone is 15 meters from that of the air-conditioner, take the speed of sound to be 300 meters per second, and let's sample at 48 kilosamples per second. Using these numbers it takes 50 milliseconds for the sound to travel from the air-conditioner microphone to the lecturer's, a delay that corresponds to 2,400 samples! Thus, at least as a rough approximation we believe that

$$y_n = x_n + h q_{n-k} \qquad (10.1)$$

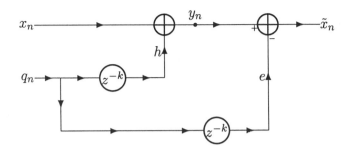

Figure 10.1: Cancelling delayed and attenuated noise by subtracting.

with $k \approx 2400$ and $h < 1$. Of course the delay need not be an integer number of samples, and indeed in a closed room we will get multiple noise echoes due to the sound waves bouncing off the walls and other surfaces. Each such echo will arrive at a different time and with a different amplitude, and the total effect is obtained by adding up all these contributions. We will return to the effect of multiple echoes later.

Let's try to regain the desired clean lecturer's voice signal from the noisy received signal y_n and the reference signal q_n. Let's assume at first that we know the delay k having measured the distance between the microphones, but have no information regarding the gain h. We can try to subtract out the interference

$$\tilde{x}_n = y_n - e q_{n-k} \tag{10.2}$$

with $\tilde{x}_n$ representing our attempt at recovering x_n. This attempt is depicted in Figure 10.1, using a self-explanatory graphical technique to be presented more fully in Chapter 12. We know that this *could* work; were we to know h we could set $e = h$ and

$$\tilde{x}_n = y_n - e q_{n-k} = (x_n + h q_{n-k}) - h q_{n-k} = x_n$$

as required; but since we *don't* know h we have to *find* e. When e is improperly chosen we get the desired signal plus a residual interference,

$$\tilde{x}_n = y_n - e q_{n-k} = x_n + (h - e) q_{n-k} = x_n + r_{n-k} \tag{10.3}$$

with the amplitude of the residual r_n depending on the value of e.

In order to find e we will make the assumption that the speech signal x_n and the interference signal q_n (delayed by any amount) are not correlated. By uncorrelated we mean that the correlation between x_n and q_{n-l}, as measured over a certain time interval,

$$C_{xq}(l) = \sum_n x_n q_{n-l}$$

is zero for every lag l. This is a reasonable assumption since correlation would imply some connection between the signals that links their values. We believe that the air-conditioner doesn't care what the lecturer is saying, and indeed would be making essentially the same noise were the lecturer not to have started speaking. Now it is true that when the compressor kicks in and the air-conditioner becomes suddenly louder the lecturer might start speaking more loudly, causing some correlation between the speech and the noise, but this is a very slow and weak effect. So we shall assume for now that x_n and q_n are uncorrelated.

How does this assumption help us? The lack of correlation is significant because when we sum uncorrelated signals their energies add. Think of taking two flashlights and shining them on the same spot on a wall. It is clear from the conservation of energy that the energy of the spot is the sum of each flashlight's energy. You may recall seeing experiments where two light beams combine and destructively interfere leaving darkness, but for this to happen the beams must be *correlated*. When the light beams are *uncorrelated* their energies add, not their amplitudes, and the same is true for sounds. In large rooms there may be places where echoes constructively or destructively interfere, making localized spots where sounds can be heard from afar or mysteriously disappear; but this is because different echoes of the *same* sound *are* correlated.

Returning to $\tilde{x}_n = x_n + r_{n-k}$, since r_n is q_n to within a multiplicative constant, x_n and r_n are also uncorrelated. Thus the energy of our recovered $\tilde{x}_n$ signal is the sum of the energy of the original x_n and that of the residual r_n. However, the energy of the residual is dependent on our estimate for the coefficient e; the residual has large energy when this estimate is poor, but when we are close to the proper value the residual's energy is close to zero. Of course the energy of x_n is not affected by our choice of e. Thus we can minimize the energy of the sum signal $\tilde{x}_n$ by correctly choosing the coefficient e!

To see this mathematically, we write the energy of $\tilde{x}_n$

$$E_{\tilde{x}} = \sum_n \tilde{x}_n^2 = \sum_n (x_n + r_{n-k})^2 = \sum_n x_n^2 + 2\sum_n x_n r_{n-k} + \sum_n r_{n-k}^2$$

but the cross term is precisely lag k of the correlation between x_n and r_n that was assumed to be zero.

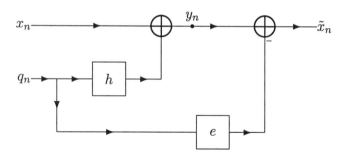

Figure 10.2: Cancellation of filtered noise by subtraction. The filter e_k is adapted to equal the distorting filter h_k. When successfully adapted the output of e_k equals that of h_k so that the interference is subtracted from the desired signal.

Continuing

$$E_{\tilde{x}} = \sum_n x_n^2 + \sum_n r_{n-k}^2 = \sum_n x_n^2 + (h-e)^2 \sum_n q_{n-k}^2 = E_x + E_q(h-e)^2$$

which as a function of e, is a parabola, with its minimum corresponding to E_x, the energy of the speech signal.

So to find the proper coefficient e all we need to do is to vary it until we find the minimal energy of the reconstructed signal. Since the energy is a parabola there is a single global minimum that is guaranteed to correspond to the original lecturer's voice.

Now, what can we do if the delay k is unknown? And what if the delay is not a integer number of samples? We might as well consider the more general problem of many different paths from the air-conditioner to the lecturer's microphone that all combine with different k and h. In such a case we have

$$y_n = x_n + \sum_k h_k q_{n-k} \tag{10.4}$$

which we recognize as corresponding to the adding of a filtered version of the air-conditioner noise q_n to the desired signal. We try to recover x_n by looking for the unknown filter

$$\tilde{x}_n = y_n - \sum_k e_k q_{n-k} \tag{10.5}$$

as depicted in Figure 10.2. Once again we are assured that this *can* be successful, since selecting $e_k = h_k$ will guarantee $\tilde{x}_n = x_n$. Viewed in this light, the problem of noise removal is equivalent to the finding of an unknown filter, with the filter coefficients possibly varying in time.

Following the same path as before we find that due to the assumption of lack of correlation between x_n and q_n, the energy of the attempted reconstruction is the sum of two parts.

$$E_{\tilde{x}} = \sum_n x_n^2 + \sum_n \left(\sum_k (h_k - e_k) q_{n-k} \right)^2$$

The first is the energy of the desired signal x_n and the second is the energy of the residual interference. As a function of the vector of coefficients, the energy $E(e_1, e_2, \ldots e_N)$ is a hyperparaboloid with a single global minimum to be found. Once again this minimum corresponds to the desired signal.

How does one find this minimum in practice? When there was only a single coefficient e to be found, this was a relatively easy job. For example, we could start with any arbitrary e and then try moving along the e axis by some positive or negative amount. If the energy decreases then we keep moving in the same direction; otherwise we move in the opposite direction. If after several steps that decrease the energy, it starts to rise again, then we have gone too far; so we reduce the step size and 'home in' on the minimum.

The more general case can also be solved by arbitrarily moving around and checking the energy, but such a strategy would take a long time. With one variable there were just two directions in which to move, while with N coefficients there are an infinite number of directions. However, since we know that the energy surface in e_k space is a hyperparaboloid, we can (with only a little extra work) make a good guess regarding the best direction. The extra work is the calculation of the gradient of the energy in e_k space, $\nabla E(e_1, e_2, \ldots e_N)$. Recall that the gradient of a surface is the multidimensional extension of the derivative. The gradient of a function is a vector that points in the direction the function increases most rapidly, and whose length is proportional to the steepness of the function. At a maximum or minimum (like the base of the energy paraboloid) the gradient is the zero vector. Were we to be interested in finding a maximum of the energy, the best strategy would be to move in the direction of the gradient. Any other direction would not be moving to higher energy values as quickly. In order to find the energy's *minimum* we have to reverse this strategy and move in the direction opposite the gradient. This technique of finding a minimum of a function in N-dimensional space is called *steepest descent* or *gradient descent*, and will be more fully explained in Section 10.5.

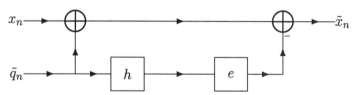

Figure 10.3: Cancelling filtered noise by an inverse filter (equalizer). This time the filter e_k is adapted to equal the inverse of the distorting filter h_k. When successfully adapted the output of filter e equals the input of h so that the interference is subtracted from the desired signal.

Before concluding this section we wish to note an alternative solution to the noise cancellation problem. We could have considered the basic noise signal to be that which is added at the lecturer's microphone, and the noise picked up by the reference microphone to be the filtered noise. According to this interpretation the problem is solved when the constructed filter approximates the *inverse filter*, as depicted in Figure 10.3. The desired signal is recovered due to the noise going through a filter and its inverse in series and then being subtracted. Both direct and inverse interpretations are useful, the best one to adopt depending on the application.

EXERCISES

10.1.1 Unlike the air-conditioner, the audience is not located at one well-defined location. Can the audience noise be removed in a manner similar to the air-conditioner noise?

10.1.2 Build a random signal and measure its energy. Add to it a sinusoid and measure the resulting energy. Did the energies add? Subtract from the combined signal the same sinusoid with varying amplitudes (but correct phase). Graph the energy as a function of amplitude. What curve did you get? Keep the correct amplitude but vary the phase. Is the behavior the same?

10.1.3 Electrocardiographs are required to record weak low-frequency signals and are often plagued by AC line frequency pickup (50 or 60 Hz). Were there are no desired signal components near this frequency a sharp notch filter would suffice, however generally an adaptive technique should be employed. Since we can directly measure the AC line sinusoid, the problem is reduced to finding the optimum gain and phase delay. Explain how to solve this problem. Simulate your solution using a stored waveform as the desired signal and a slowly amplitude- and phase-varying sinusoid as interference.

10.1.4 A 'frequency agile notch filter' can remove periodic interference (of unknown frequency) from a nonperiodic desired signal without a separate reference signal. Explain how this can be done.

10.2   Adaptive Echo Cancellation

Communications systems can be classified as one-way (simplex) or two-way (full-duplex); radio broadcasts and fax machines are of the former type, while telephones and modems are of the latter. Half-duplex systems, with each side transmitting in turn, lie in between; radio transceivers with push-to-talk microphones are good examples of this mode. True two-way communications systems are often plagued by echo, caused by some of the signal sent in one direction leaking back and being received by the side that transmitted it. This echo signal is always delayed, usually attenuated, and possibly filtered.

For telephones it is useful to differentiate between two types of echo. *Acoustic echo* is caused by acoustic waves from a loudspeaker being reflected from surfaces such as walls and being picked up by the microphone; this type of echo is particularly annoying for hands-free mobile phones. A device that attempts to mitigate this type of echo is called an *acoustic echo canceller*. *Line echo* is caused by reflection of electric signals traveling along the telephone line, and is caused by imperfect impedance matching. The most prevalent source of line echo is the *hybrid*, the device that connects the subscriber's single two-wire full-duplex telephone line to the four-wire (two simplex) channels used by the telephone company, as depicted in Figure 10.4. We will concentrate on line echo in this section.

Actually, telephones purposely leave some echo to sound natural, i.e., a small amount of the talker's voice as picked up at the handset's microphone is intentionally fed back to the earpiece. This feedback is called 'sidetone' and if not present the line sounds 'dead' and the subscriber may hang up. If there is too little sidetone in his telephone, John will believe that Joan barely hears his voice and compensates by speaking more loudly. When this happens Joan instinctively speaks more softly reinforcing John's impression that he is speaking too softly, resulting in his speaking even more loudly. If there is too much sidetone in Joan's telephone, she will speak more softly causing John to raise his voice, etc.

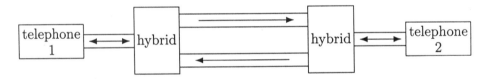

Figure 10.4: The telephone hybrid. At both ends of the telephone connection are two wire channels, but in between the conversation is carried over four-wire circuits.

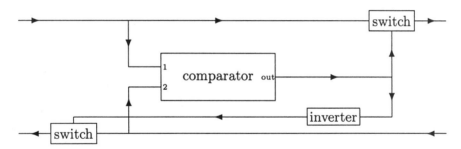

Figure 10.5: The telephone echo suppressor. The top line represents one direction of the four wire telephone channel, and the bottom line the other. When the upper signal is greater than the lower one the comparator gives positive output, thus keeping the upper path open but suppressing the lower signal. When the lower signal is greater the comparator output is negative and the switches open the lower path but shut the upper.

When the delay of line echo is short, it simply combines with the sidetone and is not noticeable. However, when the delay becomes appreciable line echo becomes quite annoying. Most people find it disconcerting to hear their own voice echoing back in their ear if the delay is over 30 milliseconds. An *echo suppressor* is a simple device that combats line echo by disconnecting one side of the conversation while the other side is talking. The functioning of an echo suppressor is clarified in Figure 10.5. Echo suppressors often cause conversations to be carried out as if the telephone infrastructure were half-duplex rather than full-duplex. Such conversations are unnatural, with each side lecturing the other without interruption, rather than engaging in true dialog. In addition, echo suppressors totally disrupt the operation of data communications devices such as faxes and modems, and must be disabled before these devices can be used. A **L**ine **E**cho **C**anceller (LEC) is a more complex device than an echo suppressor; it enables full-duplex conversations by employing adaptive DSP algorithms.

How does an echo canceller work? Like the adaptive noise canceller, the basic idea is that of subtraction; since we know the original signal that has been fed back, we need only subtract it out again. However, we need to know the delay, attenuation, and, more generally, the filter coefficients before such subtraction can be carried out.

Full-duplex modems that fill all of the available bandwidth and use a single pair of wires for both directions always experience echo. Indeed the echo from the nearby modulator may be as strong as the received signal, and demodulation would be completely impossible were it not to be removed effectively. Hence a modem must remove its own transmitted signal from the received signal before attempting demodulation.

Modems typically determine the echo canceller parameters during a short initialization phase before data is transferred. Consider the following common technique to measure the delay. The modem on one side sends an agreed-upon event (e.g., a phase jump of 180° in an otherwise unmodulated sinusoid) while the other side waits for this event to occur. As soon as the event is detected the second modem sends an event of its own (e.g., a phase reversal in its sinusoid), while the first waits. The time the first modem measures between its original event and detecting the other modem's event is precisely the round-trip delay. Similarly, the finding of the filter coefficients can be reduced to a system identification problem, each side transmitting known signals and receiving the filtered echo. While the system identification approach is indeed useful, its results are accurate only at the beginning of the session; in order to remain accurate the echo canceller must continuously adapt to changing line conditions. For this reason modem echo cancellers are initialized using system identification but thereafter become adaptive.

Returning to telephone conversations, it is impractical to require humans to start their conversations with agreed-upon events (although starting with 'hello' may be almost universal), but on the other hand the requirements are not as severe. You will probably not notice hearing an echo of your own voice when the delay is less than 20 milliseconds, and international standards recommend controlling echo when the round-trip delay exceeds 50 milliseconds. This 50 milliseconds corresponds to the round-trip propagation delay of a New York to Los Angeles call, but modern digital networks introduce processing delay as well, and satellite links introduce very annoying half-second round-trip delays. Even when absolutely required voice echo cancellers needn't remove echo as completely as their modem counterparts and are allowed to be even less successful for a short amount of time at the beginning of the conversation.

In the late 1970s the phone companies introduced phone network LECs, an implementation of which is depicted in Figure 10.6. Its philosophy is exactly opposite that of the modem's internal echo canceller discussed above. It filters the signal arriving over the phone network from the far-end (the reference) and subtracts it from the near-end signal to be sent out to the network, aspiring to send only clean echo-free near-end speech. Echo is completely controlled by placing LECs at both ends of the four-wire network.

Figure 10.6 is not hard to understand. After the hybrid in the local telephone company office, the signal to be sent is digitized in order to send it to its destination over the phone system's digital infrastructure. Before the signal is sent out it undergoes two processes, namely subtraction of the echo estimate and NonLinear Processing (NLP).

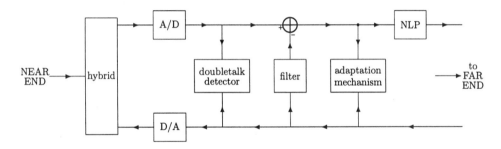

Figure 10.6: A digital telephone network line echo canceller (LEC). In this diagram only signal flow lines are indicated; invisible are the logic indications sent from the double-talk detector to the adaptation mechanism and NLP, and the fact that the adaptation mechanism sets the filter coefficients and the NLP threshold.

The filter processor places digital samples from the far-end into a static buffer (called the 'X register' in LEC terminology), convolves them with the filter (called the H register), and outputs the echo estimate to be subtracted from the near-end samples.

The adaptation mechanism is responsible for adapting the filter coefficients in order to reproduce the echo as accurately as possible. Assume that the far-end subscriber is talking and the near-end silent. In this case the entire signal at the input to the subtracter is unwanted echo generated by the nearby hybrid and the near-end telephone. Consequently, the adaptation mechanism varies the filter coefficients in order to minimize the energy at the output of the subtracter (the place where the energy is measured is marked in the figure). If the far-end is quiet the adaptation algorithm automatically abstains from updating the coefficients.

When the double-talk detector detects that both the near-end and far-end subscribers are talking at the same time, it informs the adaptation mechanism to freeze the coefficients. The Geigel algorithm compares the absolute value of the near-end speech plus echo to half the maximum absolute value in the filter's static buffer. Whenever the near-end exceeds the far-end according to this test, we can assume that only the near-end is speaking.

The nonlinear processor (NLP) is a center clipper (see equation (8.7)), that enables the LEC to remove the last tiny bit of perceived echo. For optimal functioning the center clipping threshold should also be adapted.

Although the LEC just described is somewhat complex, the basic filter is essentially the same as that of the adaptive noise canceller. In both cases a filtered reference signal is subtracted from the signal we wish to clean up, and in both cases the criterion for setting the coefficients is energy minimization. These two characteristics are quite general features of adaptive filters.

EXERCISES

10.2.1 Why is an acoustic echo canceller usually more complex than an LEC?

10.2.2 Why is the phone network LEC designed to cancel echo from the transmitted signal, rather than from the received signal?

10.2.3 Describe the following performance criteria for echo cancellers: convergence speed, ERLE (echo return loss enhancement), and stability (when presented with narrow-band signals). The minimum performance of acoustic echo cancellers is detailed in ITU-T standard G.167, and that of LECs in G.165 and G.168. Research, compare, and contrast these standards.

10.2.4 Assume that each tap of the echo cancelling FIR filter takes a single instruction cycle to calculate, that each coefficient update takes a single cycle as well, and that all the other elements are negligible. Estimate the maximum and typical computational complexities (in MIPs) required to echo cancel a standard voice channel (8000 samples per second) assuming a 16-millisecond 'tail' in which echoes can occur.

10.2.5 Explain the Geigel algorithm for double-talk detection. Why isn't it sufficient to compare the present near-end to a single far-end value? Why compare to *half* the maximum far-end? How does it differ from the comparator in the echo suppressor? How can it be improved?

10.3   Adaptive Equalization

As a third and final example of adaptive signal processing we will consider adaptive equalization of digital communications signals. We previously defined an *equalizer* as a filter that counteracts the unwanted effects of another filter. For communications signals (to be treated in Chapter 18) this invariably means trying to overcome destructive effects of the communications channel; this channel being universally modeled as a filter followed by addition of noise, as depicted in Figure 10.7.

In general the equalizer cannot overcome noise, and so the optimal equalizer is the inverse filter of the channel. Recall from the previous section how modems calculate their echo cancellers; in similar fashion they use system identification techniques during an initialization phase in order to learn the channel and hence the optimum equalizer. Adaptive equalization is needed thereafter to track changes in the channel characteristics.

Is channel equalization really needed? Let's consider the simplest possible digital communications signal, one that takes on one value for each 0 bit

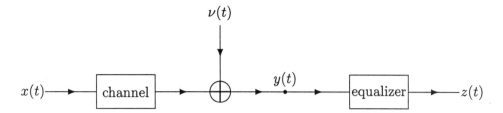

Figure 10.7: An equalizer for digital communications signals. The original signal $x(t)$ transmitted through the communications channel, and subject to additive noise $\nu(t)$, is received as signal $y(t)$. The purpose of the equalizer is to construct a signal $z(t)$ that is as close to $x(t)$ as possible.

to be transmitted, and another for each 1 bit. These transmitted values are referred to as 'symbols', and each such symbol is transmitted during a symbol interval. Ideally the signal would be constant at the proper symbol value during each symbol interval, and jump instantaneously from symbol to symbol; in reality it is sufficient for the signal value at the center of the symbol interval to be closer to the correct symbol than to the alternative. When this is the case the receiver, by focusing on times far from transitions, can make correct decisions as to the symbols that were transmitted.

When the modem signal traverses a channel it becomes distorted and the ability of the receiver to properly retrieve the original information is impaired. This effect is conventionally tested using the *eye pattern* (see Figure 10.8). The eye pattern is constructed by collecting multiple traces of the signal at the output of the equalizer. When the 'eye is open' information retrieval is possible, but when the 'eye is closed' it is not. In terms of the eye pattern, the purpose of an equalizer is to open the eye.

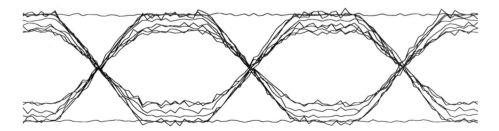

Figure 10.8: The eye pattern display graphically portrays the effect of ISI, noise and possibly other impairments on the receiver's capability to properly decode the symbol. In the present diagram the eye is 'open' and proper decoding is possible.

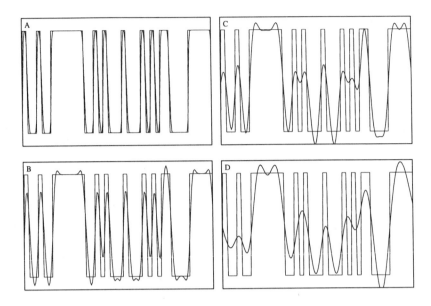

Figure 10.9: The effect of increasing intersymbol interference. The filtered channel output is superposed over the original signal. In (A) (the mildest channel) the received signal is close to the ideal signal. In (B) the bandwidth has been reduced and symbol recovery has become harder. In (C) proper symbol recovery is not always likely. In (D) (the harshest channel) symbol recovery has become impossible.

Why do channels cause the eyes to close? Channels limit the bandwidth of signals that pass through them, and so ideal symbols will never be observed at the channel output. Mild channels merely smooth the symbol-to-symbol jumps, without impairing our ability to observe the proper symbol value far from transitions, but channels with long impulse responses smear each symbol over many symbol intervals, as seen in Figure 10.9. As a result the channel output at any given time is composed not only of the desired symbol, but of contributions of many previous symbols as well, a phenomenon known as **InterSymbol Interference** (ISI). When the ISI is strong the original information cannot be recovered without equalization.

At first glance the adaptation of an equalizer would seem to be completely different from the applications we discussed in previous sections. In the previous cases there was an interfering signal that contaminated the signal of interest; here the source of contamination is the signal itself! In the previous cases there was a reference signal highly correlated to the contaminating signal; here we observe only a single signal! Notwithstanding these apparent differences, we can exploit the same underlying principles. The trick is to devise a new signal (based on our knowledge of the original signal) to play the role of the reference signal.

Assuming that the equalizer was initially acquired using system identification techniques, we can presume that the receiver can make proper decisions regarding the symbols that were transmitted, even after some drift in channel characteristics. If proper decisions can be made we can reconstruct a model of the originally transmitted signal and use this artificially reconstructed signal as the reference. This trick is known as **Decision Directed Equalization** (DDE). Using DDE makes adaptive equalization similar to adaptive noise cancellation and adaptive echo cancellation.

EXERCISES

10.3.1 An alternative to equalization at the receiver as illustrated in Figure 10.7 is 'Tomlinson equalization', where the inverse filter is placed at the transmitter. What are the advantages and disadvantages of this approach? (Hints: What happens if the channel's frequency response has zeros? How can the equalizer be adapted?)

10.3.2 DDE is not the only way to adapt an equalizer. Blind equalization uses general characteristics of the signal, without making explicit decisions. Assume the symbol for a 0 bit is -1 and that for a 1 bit is $+1$. How can the fact that the square of both symbols is unity be used for blind equalization? Describe a blind equalizer for a constant amplitude signal that encodes information in its phase.

10.3.3 Signal separation is a generalization of both equalization and echo cancellation. The task is to separate the signal mixtures and recover the original signals. Let x_i be the original signals we wish to recover, and y_i the observed combination signals. The most general linear two-signal case is

$$
\begin{aligned}
y_1 &= h_{11} * x_1 + h_{12} * x_2 \\
y_2 &= h_{21} * x_1 + h_{22} * x_2
\end{aligned}
\tag{10.6}
$$

where h_{ii} are the self-filters (which need to be equalized) and the $h_{i \neq j}$ the cross-filters (which need to be echo-cancelled). Generalize this to N combinations of N signals. What conditions must hold for such problems to be solvable?

10.4   Weight Space

After seeing several applications where adaptive filters are commonly used, the time has come to develop the conceptual formalism of adaptive signal processing. In Section 10.1 we saw how to adapt a noise cancellation filter by minimization of energy; in this section we will see that a large family of problems can be solved by finding the minimum of a *cost function*. A cost function, or loss function, is simply a function that we wish to minimize. If you have to buy a new computer in order to accomplish various tasks, and the computer comes in many configurations and with many different peripherals, you would probably try to purchase the package of minimum cost that satisfies all your needs. Some people, apparently with a more positive mind-set, like to speak of maximizing gain functions rather than minimizing loss functions, but the two approaches are equivalent.

We start by reformulating the difficult FIR system identification problem of Section 6.13. Your opponent has an FIR filter v that produces a desired output signal $d_m = \sum_{n=1}^{N} v_n x_{m-n}$. We can rewrite this using a new notation that stresses the fact that the output is the weighted combination of its inputs.

$$d^{[m]} = \sum_{n=1}^{N} v_n x_n^{[m]} = \underline{v} \cdot \underline{x}^{[m]} \tag{10.7}$$

We have introduced this rather unusual vector notation in order to keep our discussion as general as possible. Using the dot product we can consider d to be the output of an FIR filter, in which case x are N consecutive values of a signal; the output of a phased array (see Section 7.9), in which case x are values of N different signals received simultaneously by N sensors; or a two-class linearly separable pattern recognition discrimination function. In this last application there are objects, each of which has N measurable numerical features, $x_1 \ldots x_N$. Each object belongs to one of two classes, and pattern recognition involves identifying an object's class. Two classes are called linearly separable when there is a linear function $d(x^{[m]}) = \sum_{n=1}^{N} v_n x_n^{[m]}$ that is positive for all objects belonging to one class and negative for all those belonging to the other.

When using this new notation the N coefficients are called 'weights', and v a 'weight vector'. In all three cases, the adaptive filter, the adaptive beamformer, and the two-class discriminator, our task is to find this weight vector given example inputs $x^{[m]}$ and outputs $d^{[m]}$. Since this is still the system identification problem, you know that the optimum solution will be given by the Wiener-Hopf equations (6.63). However, we beg your indulgence as our aim is to rederive these equations in a way that will be more suitable for adaptive filters.

Assume that after seeing $m-1$ inputs and respective desired outputs we manage to come up with some weight vector $\underline{w}$. Then upon observing the mth example, we predict the output to be

$$y^{[m]} = \sum_{n=1}^{N} w_n x_n^{[m]} = \underline{w} \cdot \underline{x}^{[m]} \tag{10.8}$$

and if the desired output is really $d^{[m]}$ our output y_m is in error by $\delta^{[m]} = d^{[m]} - y^{[m]}$.

Consider now the abstract N-dimensional vector space of all possible weight vectors $\underline{w}$. Before our opponent allows us to observe the system all weight vectors are possible, and all points in weight space are equally plausible. After we have observed a single input-output example only a small subset of weight space remains as plausible weight vectors, since most weight vectors would produce outputs differing significantly from the observed one. We can pick any point in this subset of plausible weight vectors as our guess $\underline{w}$. Each successive input-output example we observe reduces the size of the subset of plausible weight vectors; indeed, were there no noise, after seeing N different examples the subset would have been reduced to a single point.

This picture is encouraging, but doesn't provide us a practical heuristic with which to find good weight vectors. To do so we now define the cost (or loss) function $L(\underline{w})$. This cost function is defined for every weight vector in weight space, and is simply a measure of how plausible a weight vector $\underline{w}$ really is. A highly plausible weight vector should have a low cost, while one that noticeably violates the desired examples would be assigned a high cost. An obvious candidate for the cost function is the **Mean Squared Error (MSE)**

$$L(\underline{w}) = \left\langle (\delta^{[m]})^2 \right\rangle \tag{10.9}$$

the averaging being done over all the observed examples. From its definition the MSE is always nonnegative, and in the absence of noise there is a single weight vector for which the MSE is precisely zero. This weight vector is precisely the weight vector your opponent used, and by finding it you win the game. In the presence of noise there will generally not be any weight vectors with precisely zero MSE, but your best guess will be the weight vector with the **Minimum Mean Squared Error (MMSE)**.

Now you have a strategy with which to proceed. For each example m take the input $\underline{x}^{[m]}$, calculate the corresponding output $y^{[m]}$ for every weight vector in weight space according to equation (10.8), and furthermore compute the square of the error $(\delta^{[m]})^2 = (d^{[m]} - y^{[m]})^2$. Repeat this procedure for all the examples and compute the average error for each weight vector. In so doing you have assigned a nonnegative number to every point in weight space. You then need only look for the weight vector with the minimum cost, and you're done.

Of course it would be quite time consuming to compute this MSE cost function for *all* points in weight space, so let's use a little mathematical analysis to zoom in on the MMSE. The MSE cost function is

$$
\begin{aligned}
L(\underline{w}) &\equiv \left\langle (\delta^{[m]})^2 \right\rangle = \left\langle (d^{[m]} - y^{[m]})^2 \right\rangle \\
&= \left\langle (d^{[m]})^2 - 2d^{[m]}y^{[m]} + (y^{[m]})^2 \right\rangle \qquad (10.10) \\
&= \left\langle (d^{[m]})^2 \right\rangle - 2 \left\langle d^{[m]}y^{[m]} \right\rangle + \left\langle (y^{[m]})^2 \right\rangle
\end{aligned}
$$

where the expectation $\left\langle (d^{[m]})^2 \right\rangle$ simply means adding up all the errors and dividing by the number of examples. Substituting the basic relation (10.8) we find

$$
\begin{aligned}
L(\underline{w}) &= \left\langle (d^{[m]})^2 \right\rangle - 2 \left\langle d^{[m]} \sum_n w_n x_n \right\rangle + \left\langle \sum_n \sum_l w_n w_l x_n x_l \right\rangle \\
&= \left\langle (d^{[m]})^2 \right\rangle - 2 \sum_n w_n \left\langle d^{[m]} x_n \right\rangle + \sum_n \sum_l w_n w_l \left\langle x_n x_l \right\rangle
\end{aligned}
$$

where the sums are all from 1 to N and the expectation on m.

The expressions in the last line have simple interpretations. The first term is the average of the square of the desired outputs; we'll call it D^2. The second term contains N crosscorrelations between each of the input components x_n and the desired output $d^{[m]}$, $C_{dx}(n) \equiv \left\langle d^{[m]} x_n \right\rangle$. The third term contains all the input autocorrelations $C_x(n, l) \equiv \left\langle x_n x_l \right\rangle$. Considering the crosscorrelation to be a vector (with index n) and the autocorrelation to be a matrix (with indices n and l), we can write the following matrix equation for the cost function as a function of the weight vectors.

$$
\begin{aligned}
L(\underline{w}) &= D^2 - 2 \sum_n w_n (C_{dx})_n + \sum_n \sum_l w_n w_l (C_x)_{nl} \qquad (10.11) \\
&= D^2 - 2\underline{w} \cdot \underline{C_{dx}} + \underline{w}\, \underline{\underline{C_x}}\, \underline{w}
\end{aligned}
$$

To find the minimum of the cost function we need to use the gradient operator

$$\nabla \equiv \left(\frac{\partial}{\partial w_1}, \frac{\partial}{\partial w_2}, \ldots, \frac{\partial}{\partial w_N} \right) \tag{10.12}$$

and set the gradient of the cost equal to zero.

$$0 = \nabla L(\underline{w}) = -2(C_{dx})_n + 2 \sum_l (C_x)_{nl} w_l = -2\underline{C_{dx}} + 2\underline{\underline{C_x}}\,\underline{w}$$

Solving, we find the following set of N equations

$$(C_{dx})_n = \sum_l (C_x)_{nl} w_l \qquad \text{i.e.} \qquad \underline{C_{dx}} = \underline{\underline{C_x}}\,\underline{w} \tag{10.13}$$

which we immediately recognize as the Wiener-Hopf equations (6.63). The solution to these equations is immediate.

$$\underline{w}^* = \underline{\underline{C_x}}^{-1} \underline{C_{dx}} \tag{10.14}$$

To recap, given M input-output examples, we compute N input-output crosscorrelations $(C_{dx})_n$ and N^2 input autocorrelations $(C_x)_{nl}$. We then write down N coupled algebraic equations that can be solved for w_n. For realistically large N these equations are difficult to solve explicitly, and it is usually worthwhile to find the MMSE iteratively.

Finding the minimum of a function in high-dimensional space is a hard problem, but one that has been extensively studied. The major problem with numeric methods for finding a global minima is the fact that they tend to get stuck in local minima; in our case, the cost function in weight space defined in equation (10.11) is a quadratic function that can never become negative; as such it is always a hyperparaboloid with a single global minimum.

One family of minima (or maxima) finding methods is *iterative descent*. These methods start with some initial guess and repeatedly update this guess using

$$\underline{w}' = \underline{w} + \delta \underline{w} \tag{10.15}$$

choosing the correction term such that the cost function decreases.

$$L(\underline{w}') < L(\underline{w}) \tag{10.16}$$

If the cost function indeed decreases at each step, we must eventually arrive at a minimum.

The simplest type of iterative step is *gradient descent*, where the new guess is found by moving in the direction in which the cost function decreases the fastest. To do this we compute the gradient $\nabla L(\underline{w})$, which is

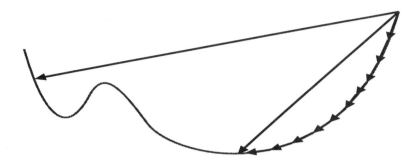

Figure 10.10: The effect of different step sizes on constant step size gradient descent. Small steps waste iterations while large steps may overshoot the minimum.

the direction in which the cost function increases most rapidly, and move in the opposite direction. More sophisticated methods exploit the matrix of second derivatives (the Hessian) as well, but even just calculating and storing the N-by-N matrix can be prohibitive in high dimensions. All of these methods require inverting the Hessian matrix, an operation that is not only computationally costly, but numerically problematic.

In the simplest type of gradient descent we move some arbitrary step size ρ at every step of the algorithm.

$$\underline{w}' = \underline{w} - \rho \frac{\nabla L(w)}{|\nabla L(w)|} \tag{10.17}$$

In general, this is often not a good idea (see Figure 10.10) since where the gradient is steep this step size may be overly small requiring us to take many small steps where one large one would have sufficed, while where the gradient is shallow we may overshoot the minimum and need to reverse direction at the next iteration. Alternatively, we can save computation by moving some fraction of the value of the gradient

$$\underline{w}' = \underline{w} - \lambda \nabla L(w) \tag{10.18}$$

which is a logical thing to do if the gradient gets larger as we go further from the minimum. There are more complex techniques that search along the line to determine how far to move (requiring much more computation), or vary the step size depending on the absolute value of the gradient or the difference between the present gradient direction and that of the previous iteration.

We have seen that the MMSE weight vector can be found explicitly via the Wiener-Hopf equations, or numerically using minimum finding techniques such as gradient descent. Both of these methods assume that the underlying system is time-invariant. When the system can constantly change we require MMSE finding methods that can dynamically adapt to these changes. The rest of the chapter will be devoted to such methods. It is an interesting coincidence of alliteration that the equations that constitute the simplest adaptive adaptation of the Wiener-Hopf equations is called the Widrow-Hoff equation.

EXERCISES

10.4.1 Assume that there is but a single weight w, so that the Wiener-Hopf equation is simply $C_{dx} = w^* C_x$. Show that the cost function as a function of this w is a simple nonnegative parabola with a single minimum. For what weight is the cost precisely zero?

10.4.2 Assume that there are two weights w_1 and w_2. Show that the cost function surface is a paraboloid with a single minimum.

10.4.3 What is the computational complexity of the solution in (10.14)?

10.4.4 Try directly solving the Wiener-Hopf equations for the case of simple averaging (i.e., the unknown coefficients are all $\frac{1}{N}$). Generate some large number of input-output pairs, compute the correlations, and use the matrix inversion technique of the previous exercise to solve. Have an opponent supply some random $\underline{w}$ and try to discover it.

10.4.5 Show that the MMSE weight vector decorrelates the error from the input vector, (i.e., for $\underline{w}^*$ the error $\delta^{[m]}$ and the input $\underline{x}^{[m]}$ obey $\left\langle \delta^{[m]} \underline{x}^{[m]} \right\rangle = 0$). What is the deeper meaning of this statement, sometimes called the *orthogonality principle*? What can be said about the error-output correlation?

10.5 The LMS Algorithm

In the previous section we saw that by using gradient descent we could approximate the solution to the Wiener-Hopf equations without inverting the autocorrelation matrix. However, we still have to set aside memory and compute the autocorrelation matrix and crosscorrelation vector for some large N. We would really like to avoid these as well. Accordingly we make a further approximation; we assume that we can iteratively update the weight vector based on each input-output example *taken in isolation*. In this way each time we observe a new input-output example, we make an independent estimate of the gradient, perform gradient descent, and then discard the example before the next one is presented. Of course in general such a gradient estimate may not be very good, and we will often take 'pseudo-gradient descent' steps in the wrong direction! Unfortunately, there is no way to avoid this, but if we take small enough steps, and observe enough input-output examples, then the majority tendency toward lower cost will eventually dominate although there will be some small steps in the wrong direction.

Now it really isn't so incredible that the gradient can be approximated by quantities that relate solely to a single input-output example. We originally defined the cost function as the average error; assuming that we are given some finite number of samples M, we could equally well have defined it as the sum of the errors, or half that sum.

$$L(\underline{w}) \equiv \tfrac{1}{2} \sum_{m=1}^{M} (\delta^{[m]})^2 = \tfrac{1}{2} \sum_{m=1}^{M} (d^{[m]} - y^{[m]})^2 \qquad (10.19)$$

We can thus expressed the MSE cost function as the sum of M nonnegative single example terms, which can be zero only if all the individual terms are zero. As an inference the gradient of this cost function must also be the sum of single example terms! The problem is that moving $\underline{w}$ in the direction dictated by one example, although decreasing the present contribution, may increase the contributions from other examples! In particular we may move the weight vector in order to optimize for some input-output example, and then move it right back for the next example; but when we get close enough to the global minimum everything should work out fine.

So let's investigate the single example gradient. Its nth coördinate is

$$
\begin{aligned}
\left(\nabla L^{[m]} \right)_n &= \frac{\partial L(\underline{w})}{\partial w_n} \\
&= \frac{\partial L(\underline{w})}{\partial y^{[m]}} \frac{\partial y^{[m]}}{\partial w_n} \\
&= -\delta^{[m]} x_n^{[m]}
\end{aligned}
$$

where we have used the chain rule, equation (10.19) and equation (10.8). Substituting this into equation (10.18) we find

$$\underline{w}^{[m]} = \underline{w}^{[m-1]} + \lambda \delta^{[m]} \underline{x}^{[m]} \qquad (10.20)$$

This is the Widrow-Hoff equation. In neural network terminology it is often called the 'delta rule', referring to the δ that figures in it so prominently. The iterative algorithm for finding best weight vector based on the Widrow-Hoff equation

```
Initialize: w^[0] = 0
Loop until converged:
    get new input x^[m] and desired output d^[m]
    compute new output y^[m] = w^[m] · x^[m]
    calculate error δ^[m] = d^[m] − y^[m]
    correct weight vector w^[m+1] = w^[m] + λδ^[m] x^[m]
```

is called the LMS algorithm. LMS stands for **L**east **M**ean **S**quared, referring to our attempt at finding an MMSE solution.

Unlike our attempts at finding the MMSE in the previous section, the LMS algorithm is an *adaptive* algorithm. If the true weights v vary slowly in time, the LMS algorithm will follow these changes, approximating at each instant the best weight vector for that time. Of course if the underlying system varies too rapidly, even an adaptive algorithm may not be able to keep up.

The LMS algorithm is by far the most popular adaptive algorithm, and the Widrow-Hoff equation appears in many contexts, although sometimes it may be hidden. There is an easy way to recognize Widrow-Hoff in disguise; all the correction terms contain the same output error term, while each weight correction term multiplies it by its own input. Remember that the complete correction term is a constant times the input times the output error.

In order to get a 'feel' for the use of the LMS algorithm, let's try a simple example. We'll take a three-dimensional case with the true weight vector $w^0 = (\frac{1}{3}, \frac{1}{3}, \frac{1}{3})$, and start with an initial guess of $w = (0,0,0)$. Now assuming that the first input is $x = (1,1,1)$, we'll be told that the desired output is

$$d = w^0 \cdot x = (\tfrac{1}{3}, \tfrac{1}{3}, \tfrac{1}{3}) \cdot (1,1,1) = 1$$

while the present system outputs $w \cdot x = (0,0,0) \cdot (1,1,1) = 0$. The output error is thus $\delta = d - y = 1$. If we use $\lambda = \frac{1}{4}$ the corrections will be as follows.

$$w \leftarrow w + \lambda\delta x = (0,0,0) + \tfrac{1}{4} \cdot 1 \cdot (1,1,1) = (\tfrac{1}{4}, \tfrac{1}{4}, \tfrac{1}{4})$$

Let's take the same input and perform another iteration. The output is

$$\underline{w} \cdot \underline{x} = (\tfrac{1}{4}, \tfrac{1}{4}, \tfrac{1}{4}) \cdot (1, 1, 1) = \tfrac{3}{4}$$

so that the error is $\delta = d - y = 1 - \tfrac{3}{4} = \tfrac{1}{4}$. The new weight vector is

$$\underline{w} \leftarrow \underline{w} + \lambda \delta \underline{x} = (\tfrac{1}{4}, \tfrac{1}{4}, \tfrac{1}{4}) + \tfrac{1}{4} \cdot \tfrac{1}{4} \cdot (1, 1, 1) = \left(\frac{5}{16}, \frac{5}{16}, \frac{5}{16} \right)$$

with each component deviating only about 2% from the true value.

In this case two iterations were sufficient to obtain a weight vector quite close to the correct one. Of course all of the components were equal at every iteration, since both the initial guess and inputs had this symmetry. In Figure 10.11 we plot the convergence of the LMS algorithm for a slightly harder problem. The correct weight vector is as before, but we select inputs randomly, and observe the desired output in 10% uniform additive noise. We decided to use a smaller $\lambda = 0.1$ here, in order to better average out the noise and randomness. We see in the figure the three weights are no longer identical, but nevertheless remain close to each other. The convergence takes longer, partially because of the noise but mainly due to the lower λ, but the weights consistently approach their proper values.

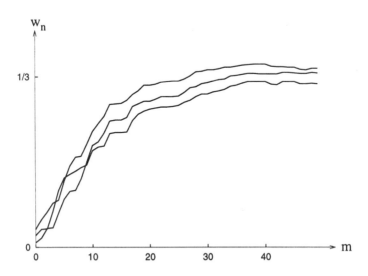

Figure 10.11: The convergence of the Widrow-Hoff algorithm. Here the correct weight vector has three components equal to $\tfrac{1}{3}$, the initial guess is zero, the inputs are random, the desired output is observed with 10% noise, and $\lambda = 0.1$. We see that the weights w_1, w_2, and w_3 converge slowly but are close to the proper values after $m \approx 50$ iterations.

We just used the word 'convergence' without specifying what this means. We may mean that the difference between the derived weight vector and the true one must converge to zero, or that the MSE must converge to zero, but for the LMS algorithm we usually mean that the mean squared error should converge to some small value (nonzero due to the noise). How do we know when the weights have converged? In Figure 10.11 we see that the changes have become much less drastic after about thirty cycles, but can we really be sure that this isn't a temporary phenomenon? Around cycle 15 there was a short stretch where the weights did not change much for a few cycles, but afterward the changes returned. In practice it really can be a tricky decision, and usually the convergence criterion can be concocted only after in-depth study of the particular problem at hand.

Assuming that the LMS algorithm does indeed converge, can we be sure that it will converge to the right answer? Happily we can prove that the expectation of the weight vector approaches the MMSE solution as M increases. To show this we first slightly rewrite the nth component of the Widrow-Hoff equation.

$$
\begin{aligned}
w_n^{[m+1]} &= w_n^{[m]} + \lambda \delta^{[m]} x_n^{[m]} \\
&= w_n^{[m]} + \lambda \left(d^{[m]} - y^{[m]} \right) x_n^{[m]} \\
&= w_n^{[m]} + \lambda \left(d^{[m]} - \sum_l w_l^{[m]} x_l^{[m]} \right) x_n^{[m]} \\
&= \lambda d^{[m]} x_n^{[m]} + \sum_l \left(\delta_{nl} - \lambda x_n^{[m]} x_l^{[m]} \right) w_l^{[m]}
\end{aligned}
$$

In matrix notation we can write this

$$
\underline{w}^{[m+1]} = \lambda d^{[m]} \underline{x}^{[m]} + \left(\underline{\underline{I}} - \lambda \underline{x}^{[m]} \underline{x}^{[m]} \right) \underline{w}^{[m]} \tag{10.21}
$$

where the two input vectors form an outer product. Now we unfold this recursion into an iteration

$$
\underline{w}^{[m+1]} = \lambda \sum_{\mu=0}^{m} \left(\underline{\underline{I}} - \lambda \underline{x}^{[m]} \underline{x}^{[m]} \right)^{m-\mu} d^{[\mu]} \underline{x}^{[\mu]} + \left(\underline{\underline{I}} - \lambda \underline{x}^{[m]} \underline{x}^{[m]} \right)^{m} \underline{w}^{[0]}
$$

and take the expectation of both sides.

$$
\left\langle \underline{w}^{[m+1]} \right\rangle = \lambda \sum_{\mu=0}^{m} \left(\underline{\underline{I}} - \lambda \underline{\underline{C}}_x \right)^{\mu} \underline{\underline{C}}_{dx} + \left(\underline{\underline{I}} - \lambda \underline{\underline{C}}_x \right)^{m} \underline{w}^{[0]}
$$

This last equation, being a matrix equation, must be true in all coördinate systems, in particular in a coördinate system where the input autocorrelation matrix is diagonal. In this system $\underline{\underline{I}} - \lambda \underline{\underline{C_x}}$ is also diagonalized. The diagonal elements $1 - \lambda c_n$ will be less than unity in absolute value if λ obeys

$$0 < \lambda < \frac{2}{c} \tag{10.22}$$

where $c = \max_n c_n$ is the maximum eigenvalue of the input autocorrelation matrix. Assuming that λ obeys this criterion the mth power of this matrix approaches the zero matrix as m increases without limit. Hence, with this proviso on λ, the second term above vanishes as $m \to \infty$. In the same coördinate system and limit we can sum the geometric series in the first term according to (A.47)

$$\lambda \sum_{\mu=0}^{\infty} (1 - \lambda c_n)^{\mu} = \frac{\lambda}{1 - (1 - \lambda c_n)} = \frac{1}{\lambda c_n}$$

which is the inverse of the correlation matrix in these coördinates. Plugging this back in we find

$$\left\langle \underline{w}^{[m+1]} \right\rangle \to \underline{\underline{C_x}}^{-1} \underline{\underline{C_{dx}}} \tag{10.23}$$

which is precisely the solution to the Wiener-Hopf equations, and so the MMSE solution!

This proof is not only reassuring, it also incidentally provides the maximal step size for convergence. We noted above that we may choose a small step size because of noise, but if λ is chosen too low it will take ages for the weight vector to converge. In adaptive applications we may not even find the weight vector before it changes! So we wish to use as large a λ as possible, but no larger than dictated by equation (10.22) since otherwise the LMS algorithm may diverge. Of course only in unusual cases do we know the value of the largest input autocorrelation eigenvalue, but if it is significantly larger than the rest of the eigenvalues, we may take it to be approximately equal to the trace of the autocorrelation, namely

$$v_{max} \approx \text{Tr}\, \underline{\underline{C_x}} = N E_x$$

where E_x is the energy of the input signal. This leads to a useful approximate range for λ.

$$0 < \lambda < \frac{2}{\langle E_x \rangle} \tag{10.24}$$

Up to now our discussion has been completely general; we end this section by restricting the general discussion of input-output pairs to the system identification case. By using the observed recent inputs and output of an FIR system we can combine FIR convolution and LMS adaptation, thus defining the standard adaptive FIR filter. We can make a new input-output example for *every* new time instant

$$
\begin{array}{cccccl}
x_{-N+1}, & x_{-N+2}, & \cdots & x_{-1}, & x_0 & \longrightarrow & y_0 \\
x_{-N+2}, & x_{-N+3}, & \cdots & x_0, & x_1 & \longrightarrow & y_1 \\
x_{-N+3}, & x_{-N+4}, & \cdots & x_1, & x_2 & \longrightarrow & y_2 \\
x_{-N+4}, & x_{-N+5}, & \cdots & x_2, & x_3 & \longrightarrow & y_3 \\
x_{-N+5}, & x_{-N+6}, & \cdots & x_3, & x_4 & \longrightarrow & y_4
\end{array}
$$

etc., or we can use only some of these possibilities. By choosing examples at the maximum rate we get the most information for adaptation and track changes in the signal at the highest time resolution. However, this requires the most computational power as well.

EXERCISES

10.5.1 It is easy to extend the derivation of the delta rule to nonlinear combinations, the most important of which is the sigmoidal nonlinearity of equation (8.11). Show that in this case the delta rule reads $w_n \leftarrow w_n + \lambda \delta y (1 - y) x_n$. (Hint: Use a further chain rule and exercise 8.4.5).

10.5.2 Assume the unknown weights of a three-parameter linear combination are $w^0 = (\frac{2}{3}, \frac{1}{3}, 0)$ and that the inputs are $x = (1, 0, 0), (0, 1, 0), (0, 0, 1)$ over and over again. Simulate this system with $\lambda = \frac{1}{2}$ and no noise. Try other values for λ and add noise. How fast can you make the LMS algorithm converge? What happens if λ is too large? What happens if we multiply all the inputs by 100?

10.5.3 Although LMS finds the best direction, its choice of constant step size seems overly primitive. A more sophisticated approach would be to search for minimal cost along the gradient direction. Compare LMS and this line-search gradient algorithm on a simulated problem. How many cycles are required for convergence? How many output and error evaluations?

10.5.4 Equation (10.20) seems to require two multiplications and one addition for each tap update. Show how this can be reduced. Compare the computational complexity of LMS update to that of running the FIR filter. Why do some DSP processors have an LMS instruction?

10.5.5 What is necessary to make the LMS algorithm work for complex-valued signals? What is the complexity compared to a real signal?

10.5.6 In exercise 10.1.3 we discussed the cancellation of power line noise from weak medical signals. Even when a narrow notch filter could be used LMS filters may require less computation. For example, assume that an ECG is a quasiperiodic with period about one Hz. For the purposes of simulation, model the ECG signal as two sinusoidal cycles one after the other, the first with period one-tenth of the whole period, and the second filling the remaining 0.9, but with amplitude one-tenth of the first. Add some white noise and a nominal power line frequency with total energy about the same as the desired signal. Try to remove the power line signal with a static FIR notch filter; how many coefficients are required? Now use an LMS filter; how many taps are required now?

10.6   Other Adaptive Algorithms

Although vanilla LMS is the most popular adaptive algorithm, it is certainly not the only one. There are both countless variants on the LMS theme, and also a few completely different algorithms. The LMS variants all start off with standard LMS and try to rectify some potential problem.

What problems does LMS potentially have? The need to guess the best step size, the possibly slow speed of convergence, dependence on initial conditions, and numerical instability are related but distinct problems that many variants try to resolve.

One LMS variant that frequently converges faster and that helps in the step size problem is **Normalized LMS** (NLMS). In the spirit of equation (10.17) we normalize the input vectors

$$\underline{w} \leftarrow \underline{w} + \rho\delta\frac{\underline{x}}{E_x} \tag{10.25}$$

where E_x is the input signal's energy. One way of thinking about NLMS is to cast it in standard LMS form with a normalized λ;

$$\lambda \equiv \frac{\rho}{E_x}$$

accordingly NLMS is LMS with the step size tuned individually for each input's energy. In many applications NLMS converges faster than vanilla LMS. More good news about NLMS is that it converges when $0 < \rho < 2$, so we needn't estimate input energy or autocorrelation eigenvalues. In fact, $\rho = 1$ is just about always best. One drawback is that NLMS requires the additional computation of the input signal's energy and a division. Even more worrisome is the fact that for low-energy signals the division is by a small number, causing numeric problems for fixed-point implementations.

Another popular LMS variant, called *block LMS*, strives to speed convergence by smoothing out the weight vector fluctuations while still allowing a relatively large λ. BLMS is less computationally expensive than conventional LMS since it does not perform the actual correction for every input. Instead an averaged estimate of the gradient in weight space is computed

$$\nabla = \sum_m \delta^{[m]} \underline{x}^{[m]}$$

by adding up the error times input for all m in the block; then once the entire block has been seen a single correction

$$\underline{w} \leftarrow \underline{w} + \Lambda \nabla \tag{10.26}$$

is performed.

Block LMS is preferable to vanilla LMS when the input signal fluctuates rapidly, but converges more slowly for relatively stationary signals. To convince yourself of this latter fact think of a block of length M in which the signal is constant. Standard LMS will perform M separate iterations while block LMS essentially performs only the first of these.

A compromise between BLMS and vanilla LMS is LMS with momentum. In this variant we smooth the weight changes by a kind of AR filtering

$$\underline{w}^{[m+1]} = \underline{w}^{[m]} + \lambda \delta^{[m]} \underline{x}^{[m]} + \alpha \left(\underline{w}^{[m]} - \underline{w}^{[m-1]} \right) \tag{10.27}$$

the new term approximating the derivative of the movement in weight space. If $\alpha = 0$ we have vanilla LMS, while for larger α the new term tends to cause the weight vector to continue as in the previous iteration (hence the term 'momentum').

After seeing all these LMS variants the time has come to discuss a completely different algorithm. In deriving the LMS algorithm we wrote the MSE as a sum of single example terms. This allowed us to adapt to time-varying systems, but is not the only way to acquire this adaptability. An alternative policy is to take the MSE as the average over the last M examples seen,

$$\frac{1}{M} \sum_{m=1}^{M} |d^{[m]} - \underline{w} \cdot \underline{x}^{[m]}|^2 \tag{10.28}$$

where M is taken small enough that the underlying system does not vary appreciably during the M time instants. This policy automatically provides a certain amount of averaging, curing one of the aforementioned potential problems. If the underlying system changes rapidly, it is more appropriate to use a recursive filter rather than a plain average. Calling the forgetting factor φ, we have

$$\sum_{m=1}^{M} \varphi^{M-m} |d^{[m]} - \underline{w} \cdot \underline{x}^{[m]}|^2 \tag{10.29}$$

where we have discarded any normalization factors that will not affect the minimization. For this to be exponentially *decaying* weighting, we require $0 < \mu < 1$. We can use the arguments that lead up to equations (10.13) almost without change to show that the MMSE solution here is

$$\underline{w} = \underline{\underline{C}}_x^{-1} \underline{C}_{dx} \tag{10.30}$$

where the correlations appearing here are only slightly different from the usual ones.

$$\underline{\underline{C}}_x = \sum_{m=1}^{M} \varphi^{M-m} \underline{x}^{[m]} \underline{x}^{[m]}$$

$$\underline{C}_{dx} = \sum_{m=1}^{M} \varphi^{M-m} d^{[m]} \underline{x}^{[m]}$$

The useful thing about exponential weighting is that these quantities can be built up recursively.

$$\underline{\underline{C}}_x^{[m]} = \varphi \underline{\underline{C}}_x^{[m-1]} + \underline{x}^{[m]} \underline{x}^{[m]} \tag{10.31}$$

$$\underline{C}_{dx}^{[m]} = \varphi \underline{C}_{dx}^{[m-1]} + d^{[m]} \underline{x}^{[m]}$$

Now if we only knew how to recursively update the inverse autocorrelation matrix $\underline{\underline{P}} \equiv \underline{\underline{C}}_x^{-1}$ we could substitute these recursions into (10.30) to obtain a recursion for the weight vector. Luckily, there *are* ways to recursively update the inverse of a matrix of this sort. Using the matrix inversion lemma (A.100) with $\underline{\underline{A}} = \varphi \underline{\underline{C}}_x^{[m-1]}$, $\underline{\underline{B}} = \underline{\underline{D}} = \underline{x}^{[m]}$ and $\underline{\underline{C}} = 1$ we find

$$\left(\varphi \underline{\underline{C}}_x^{[m-1]} + \underline{x}^{[m]} \underline{x}^{[m]} \right)^{-1} = \varphi^{-1} \underline{\underline{P}}^{[m-1]} - \frac{\varphi^{-2} \underline{\underline{P}}^{[m-1]} \underline{x}^{[m]} \underline{x}^{[m]} \underline{\underline{P}}^{[m-1]}}{\varphi^{-1} \underline{x}^{[m]} \underline{\underline{P}}^{[m-1]} \underline{x}^{[m]} + 1}$$

which looks messy but contains only quantities known at step m of the recursion. In order to clean it up a little we define the *gain vector*

$$\underline{k}^{[m]} \equiv \frac{\underline{\underline{P}}^{[m-1]} \underline{x}^{[m]}}{\underline{x}^{[m]} \underline{\underline{P}}^{[m-1]} \underline{x}^{[m]} + \varphi} \tag{10.32}$$

in terms of which we can rewrite the recursion for the inverse autocorrelation.

$$\underline{\underline{P}}^{[m]} = \varphi^{-1} \left(\underline{\underline{P}}^{[m-1]} - \underline{k}^{[m]} \underline{x}^{[m]} \underline{\underline{P}}^{[m-1]} \right) \tag{10.33}$$

Now the expression for the gain vector looks terrible, but it really has a simple interpretation. Rearranging equation (10.32)

$$\underline{k}^{[m]} = \left(\varphi^{-1} \underline{\underline{P}}^{[m-1]} - \varphi^{-1} \underline{k}^{[m]} \underline{x}^{[m]} \underline{\underline{P}}^{[m-1]} \right) \underline{x}^{[m]}$$

and from equation (10.33) we recognize the factor in the parenthesis to be precisely $\underline{\underline{P}}^{[m]}$. As a result the gain vector

$$\underline{k}^{[m]} = \underline{\underline{P}}^{[m]} \underline{x}^{[m]} = \underline{\underline{C_x}}^{[m]} \underline{x}^{[m]} \tag{10.34}$$

is the input partially decorrelated by its own inverse autocorrelation.

Now we can finally substitute the recursions (10.31) back into equation (10.30).

$$\begin{aligned}
\underline{w}^{[m]} &= (\underline{\underline{C_x}}^{[m]})^{-1} \underline{C_{dx}}^{[m]} \\
&= \underline{\underline{P}}^{[m]} \left(\varphi \underline{C_{dx}}^{[m-1]} + d^{[m]} \underline{x}^{[m]} \right) \\
&= \varphi \underline{\underline{P}}^{[m]} \underline{C_{dx}}^{[m-1]} + d^{[m]} \underline{\underline{P}}^{[m]} \underline{x}^{[m]}
\end{aligned}$$

We now substitute the recursive update of the inverse autocorrelation (10.33) and use equations (10.30) and (10.34).

$$\begin{aligned}
\underline{w}^{[m]} &= \underline{\underline{P}}^{[m-1]} \underline{C_{dx}}^{[m-1]} - \underline{k}^{[m]} \underline{x}^{[m]} \underline{\underline{P}}^{[m-1]} \underline{C_{dx}}^{[m-1]} + d^{[m]} \underline{\underline{P}}^{[m]} \underline{x}^{[m]} \\
&= \underline{w}^{[m-1]} - \underline{k}^{[m]} \underline{x}^{[m]} \underline{w}^{[m-1]} + d^{[m]} \underline{\underline{P}}^{[m]} \underline{x}^{[m]} \\
&= \underline{w}^{[m-1]} - \underline{k}^{[m]} \left(\underline{x}^{[m]} \underline{w}^{[m-1]} - d^{[m]} \right)
\end{aligned}$$

The final step is to recognize the error $\delta^{[m]}$ in the parentheses and we have found the desired recursion.

$$\underline{w}^{[m]} = \underline{w}^{[m-1]} + \delta^{[m]} \underline{k}^{[m]} \tag{10.35}$$

We now understand why we called $\underline{k}$ the 'gain vector'; it is a directed gain that multiplies the error in the weight update recursion.

We can at last give the **R**ecursive **L**east **S**quares (RLS) algorithm.

```
Initialize: w[0] = 0
Loop until converged:
     get new input x[m] and desired output δ[m]
     compute new output y[m] = w[m] · x[m]
     calculate error δ[m] = d[m] − y[m]
     compute gain vector k[m] using equation (10.32)
     correct weight vector using equation (10.35)
     update inverse autocorrelation P using equation (10.33)
```

Comparing this to the LMS algorithm we see some differences but a strong similarity. Recalling equation (10.34) we can write the weight update as

$$\underline{w}^{[m]} = \underline{w}^{[m-1]} + \underline{\underline{P}}^{[m]}\delta^{[m]}\underline{x}^{[m]}$$

Hence RLS can be thought of as LMS with a very intelligent adaptive step size. This step size is a matrix, and hence takes care of cost function surfaces in weight space that are steep in some directions but flat in others. The step size is optimized at every step to ensure rapid convergence.

Each iteration of the RLS algorithm is more complex than an iteration of the LMS algorithm, and indeed RLS is often impractical for real-time applications. However, the RLS algorithm will normally converge faster than the LMS one, at least when the noise is small. When there is strong additive noise a long period of averaging is necessary in order to average out the noise, and so RLS cannot significantly decrease the number of iterations needed.

As with many recursive update formulas, the RLS updates can accumulate numerical error, eventually leading to a noninvertible $\underline{\underline{C}}_x$. This usually isn't a problem when only a few $\underline{w}$ are needed, but becomes intolerable when we must continuously update weight vectors. One solution to this problem is to iteratively update the Cholesky decomposition of $\underline{\underline{C}}_x$ (see equation (A.94)) rather than the matrix itself. Another is the so-called QR-RLS algorithm, which multiplies the equations by an orthogonal matrix in order to keep them triangular; but further discussion of these topics would take us too far astray.

EXERCISES

10.6.1 Compare the LMS, NLMS, and RLS algorithms on benchmarks of your choice. Which is fastest? Which is most robust?

10.6.2 One way of ameliorating the numeric difficulties of NLMS is by using the following regularization.

$$\underline{w}' = \underline{w} + \rho\delta\frac{\underline{x}}{\epsilon + E_x}$$

Experiment with NLMS and regularized NLMS for signals with large dynamic range.

Bibliographical Notes

There are many good textbooks on adaptive signal processing. A classic text is that of Widrow and Stearns [275] and a more recent text is that of Haykin [96].

The invention of adaptive filters was necessitated by the conversion of the phone to digital technologies and the consequent problems of echo cancellation and adaptive differential coding of speech. Adaptive beamforming [272] and equalization were not far behind, and the use of LMS in the related field of pattern recognition helped enrich the field.

In 1960, Bernard Widrow and one of his graduate students, Marcian (Ted) Hoff, presented the Widrow-Hoff approach [274] which is basically the LMS algorithm. Widrow went on to adapt this approach to become one of the fathers of the neural network; Hoff joined Intel and went on to become one of the fathers of the microprocessor.

The applications we presented are covered by review articles of note; adaptive noise cancelling in [273, 59], echo cancellation in [87, 178], and equalizers in [202, 203].

Adaptive equalization for digital communications was originally developed by Lucky at Bell Labs [152, 153] and was the main reason for the increase of speed of telephone modems from 2400 b/s to 9600 b/s.

The technique for equalization at the transmitter discussed in exercise 10.3.1 was developed in the early 1970s by Tomlinson and by Harashima in Japan [260, 92].

Biological Signal Processing

At first it may seem a bit unusual to find a chapter on *biological* signal processing in a book dedicated to *digital* signal processing; yet this is in reality no more peculiar than motivating DSP by starting with the analogous principles of *analog* signal processing. Indeed the biological motivation should be somewhat closer to our hearts (or eyes, ears and brains). In this book we have chosen to introduce analog and digital signal processing together, but have confined our discussion of biological signal processing to this chapter.

In the first two sections we examine how we map external signal parameters into internal (biological/psychological) representations. This question belongs to the realm of psychophysics, the birth of which we describe. Our senses are highly sensitive and yet have a remarkably large dynamic range; we would like to understand and emulate this ability. We will see that a form of universal compression is employed, one that is useful in many DSP contexts.

The majority of the signals we acquire from the outside world and process in our brains are visual, and much interesting signal processing takes place in our visual system. Much has been discovered about the functioning of this system but here we concentrate on audio biological mechanisms since the focus of this book is one-dimensional signals. Hearing is the sense with the second largest bandwidth, and speech is our primary method of communications. We will devote a section each to speech production and perception mechanisms. In a later chapter we will study a DSP model of speech production that is based on this simplified biology.

After studying the signal input and output mechanisms we proceed to the processing apparatus, namely the brain. We discuss the basic processor, the neuron, and compare its architecture with that of processors with which we are more familiar. We introduce a simple model neuron and the concept of a neural network, and conclude with a performance comparison of man vs. machine.

11.1   Weber's Discovery

Ernst Weber was professor of physiology and anatomy at the university of Leipzig in the first half of the nineteenth century. His investigations involved the sensitivity of the senses. His initial studies dealt with the tactile sense, for example, the effect of temperature, pressure and location on the sense of touch. One of his discoveries was that cold objects felt subjectively heavier than hot objects of the same weight.

In his laboratory Weber would study the effect of different stimuli on human subjects. In order to measure subjective sensitivity he invented the idea of the **J**ust **N**oticeable **D**ifference (JND), which is the minimal change in the physical world that produces a noticeable difference to the subject's senses. For example, he studied the minimal separation required between two points of contact with the skin, in order to be noticeable. He found that this varied widely, with large separations required on the back while very small separations could be distinguished on the fingertips. From this he could infer the relative densities of neural coverage.

In order to study the subjective feeling of weight he defined the JND to be the minimal weight that must be added in order for a subject to perceive them as different. In a typical experiment (from about 1830) a subject would be given two bags of coins to hold, one placed on each hand. Let's assume that there were 29 coins on the left hand and 30 coins on the right. If most subjects could reliably report the right-hand bag as heavier than the left, Weber would be able to conclude that the threshold was equal or less than the weight of a single coin.

Weber's most important discovery that the JND varied with total weight. Adding a single coin to 29 coins produced a discernible difference, but 59 coins were indistinguishable from 58. Albeit subjects could reliably and repeatably distinguish between 58 and 60 coins. Likewise, most subjects could not reliably feel the difference between 116 coins in one hand and 118 or 119 in the other, only the addition of 4 coins caused a reliably distinguishable effect. Thus the JND definitely increased with increasing total weight.

Upon closer examination Weber noticed something even more significant. The threshold was a single coin when the total weight was that of 29 coins, two coins for 58, 4 coins for 116. The conclusion was obvious—the ratios (1:29, 2:58, 4:116) were all the same. Weber stated this result as 'the sensitivity of a subject to weight is in direct proportion to the weight itself', which translated into mathematics looks like this.

$$\Delta W = KW \qquad K \approx 0.034$$

This means that in order for a change in weight to be noticeable, one has to add a specific *percentage* of the present weight, *not* an *absolute* weight value.

This radically changed the way Weber understood the JND. He set out to check the dependence of other sensitivity thresholds on total stimulus intensity and found similar relationships.

$$\Delta I = K_I I \qquad\qquad (11.1)$$

In each case the ratio K_I, called Weber's constant, was different, but the linear dependence of the JND on total stimulus was universal.

Although this relationship surprised Weber it really is quite familiar. Have you ever lain awake in the middle of the night and heard the ticking of a clock or the barking of a distant dog? These sounds are not heard during the day when the ambient noise is higher, but seem quite loud at night when the total stimulus is low. Yet they *must* be there during the day even if not discernible. It is simply that the addition of the ticking or distant barking to the other sounds does not increase the total sound by a sufficient percentage.

You get out of bed and open the window. You remember how the stars were so bright when you were a child, yet seem so dim now. During the day you can't see them at all. Yet they *must* be there during the day even if not discernible. It is simply that with no light from the sun the starlight is a more significant fraction of the total stimulus. With the expansion of cities and the resulting 'light pollution' the stars are disappearing, and one has to go further and further out into the countryside in order to see them. You close the window and strike a match in the dark room. The entire room seems to light up, yet had you struck the same match during the day no change in illumination would have been noticed.

Let's now consider the sequence of physical values that are perceivably different. Think of turning on the radio and slowly increasing the volume until you just begin to hear something. You then turn a bit more until you notice that the sound has definitely grown louder. Continuing this way we can mark the points on the volume control where the sound has become noticeably louder. A direct application of Weber's law tells us that these marks will not be evenly spaced.

Assume for the purpose of argument that the particular stimulus we are studying just becomes detectable at one physical unit $I_0 = 1$ and that Weber's constant for this stimulus is a whopping 100%. Then the second distinguishable level will be $I_1 = 2$ because any value of I that adds less than one unit is indistinguishable from I_0. Continuing, we must now add $K_I I = 2$ units to the existing two in order to obtain the third distinguishable level $I_3 = 4$. It is easy to see that $I_l = 2^l$, i.e., that the levels of **J**ust **N**oticeable **D**ifferences (JNDs) form a geometric progression. Similarly, the distinguishable intensity levels for a stimulus that just becomes detectable at I_0 physical units, and for which Weber's constant is K_I, obey

$$I_l = I_0(1 + K_I)^l \tag{11.2}$$

which is an alternative statement of Weber's law.

Weber's law, equation (11.1) or (11.2), has been found to hold, at least approximately, for hundreds of different stimuli. Scientists have measured the required increase in the length of lines, the amount of salt that must be added to soup, and even the extra potency perfume requires. At extremely low and high stimuli there *are* deviations from Weber's law, but over most of the range the linear relationship between threshold and stimulus holds astonishingly well.

EXERCISES

11.1.1 Try Weber's coin experiment. Can you measure Weber's constant?

11.1.2 Write a computer program that presents a random rectangle on one part of the graphics screen, and allows subjects to reproduce it as closely as possible somewhere else on the screen. What is K here?

11.1.3 Allow a subject to listen for a few seconds to a pure sinusoid of constant frequency and then attempt to adjust the frequency of a sinusoid to match it. What is K here? Repeat the experiment with amplitude instead of frequency.

11.1.4 Patterns of dots can be hidden by randomly placing large numbers of dots around them. The original pictures stand out if the dots are of different color or size, are made to slowly move, etc. Devise an experiment to determine different people's thresholds for detecting patterns in random dot pictures.

11.2   The Birth of Psychophysics

Psychophysics is precisely what its name implies, the subject that combines psychology and physics. At first, such a combination sounds ridiculous, how could there possibly be any relationship between *physics*, the queen of the rationalistic empirical sciences, and *psychology*, the most subjective and hard to predict study? On second thought scientists learn everything they know by observing the world with their senses. So even scientists are completely dependent on the subjective in order to arrive at the objective.

The English philosopher Berkeley was fond of saying 'esse est percipi', that is, 'existence is being perceived'. We have all heard the famous conundrum about a tree falling in a forest not making a sound if there is no one around to hear it. A physical signal that is not captured by our senses might as well not exist. This capturing of physical signals and their translation into internal representations is called *perception*.

The connection between physical signals and psychological manifestations is by no means simple. The cover of this book looks the same in direct sunlight, under a fluorescent lamp, and by the light of a candle. Your mother's voice sounds the same outside, in a train car, and over the phone. Your friend seems the same height when he is standing close to you, when he has walked across the street, and even on television. In all these cases the physical signals varied widely but the internal psychological representation remained the same. Our perception of quite different physical phenomena may be the nearly the same.

Is it possible to say anything quantitative about internal psychological representations? Can feelings be measured? Surely our perceptions and thoughts are personal and unobservable to the outside world. How then can we talk about representing them quantitatively? Although consideration of such questions has convinced many sages to completely reject psychophysics, these very same questions can be raised regarding much of modern science. We cannot directly observe quarks, electrons, protons, or even atoms, but we become convinced of their existence by indirectly perceiving their effects. Individual cells cannot be seen, but biologists are convinced of their existence. We cannot hold the Milky way galaxy in our hand, yet astronomers have deduced its existence. Feelings may not be openly witnessed, but their existence may be inferred from psychophysical experiments.

Notwithstanding the importance and wide applicability of Weber's law, it is not a true psychophysical law. Psychophysical laws should relate external physical signals to internal psychological representations. Weber's law relates the intensity threshold ΔI to the total stimulus I, both of which are physical entities. Yet another step is needed to make a true psychophysical law.

The first direct attempt to quantitatively pin down feelings was made by one of Weber's students, Gustav Theodor Fechner. Fechner initially studied medicine, but after graduation was more involved in physics. Weber's discoveries retriggered his interest in psychophysics. Fechner started studying color perception, and later performed a series of experiments on the persistence of color after a bright light has been removed.

One series of experiments involved viewing sunlight filtered through colored lenses. Fechner, who acted as his own subject, was tragically blinded from the prolonged exposure to direct sunlight. Without his eyesight his promising scientific career was finished. Fechner became depressed and took up the study of philosophy, religion, and mysticism. His main interest was in the so-called 'body and mind' problem. Unlike many of his contemporaries, Fechner believed that the external physical world and the world as viewed internally by the mind were two aspects of one entity.

Then, in 1850, his eyesight miraculously returned. Fechner was convinced that this was a sign that he was to complete the solution to the body and mind problem once and for all. His unique background, combining medicine, physics, and philosophy, allowed him to make a mental leap that his contemporaries were not able or willing to achieve. The solution came to him in what is called a 'Eureka experience' while lying in bed on the morning of October 22, 1850. The anniversaries of this day are celebrated the world over as 'Fechner day'.

Fechner's solution was made up of two parts, a physical part and a psychological part. For the physical part Fechner assumed that Weber's law was correct, namely that equation (11.2) regarding the geometric progression of JND levels holds. For the psychological part Fechner made the simple assumption that all just noticeable changes were somehow equivalent. When we feel that the music has become noticeably louder, or that the light has become brighter, or the soup just a little saltier, or the joke just noticeably funnier, these all indicate an internal change of one unit.

Fechner invented three different methods of experimentally determining the connection between physical and psychological variables. We will demonstrate one by considering a scientist sitting on a mountaintop waiting for the sun to rise. The scientist has brought along nothing save a light meter (which measures physical units I) and a pair of eyes (which register psychological units Y). Sometime before the scientist notices anything happening the light meter shows an increase in the illumination. Suddenly the scientist perceives the light and records that $Y = 0$ corresponds to the physical reading I_0. When the light becomes just noticeably brighter the scientist records that $Y = 1$ corresponds to $I_1 = I_0(1 + K_I)$. The next event is recorded as $Y = 2$, which corresponds to $I_2 = I_0(1 + K_I)^2$. In general we see that the scientist's personal feeling of Y corresponds to a physical reading of $I_p = I_0(1 + K_I)^Y$. We are more interested in knowing the converse connection—given the physical event of intensity I, what is the psychological intensity Y? It is easy to show that

$$Y = A \log I + B \qquad (11.3)$$

i.e., that apart from an additive constant that derives from the minimum biological sensitivity, the psychological intensity is proportional to the logarithm of the physical intensity.

We know that the logarithm is an extremely compressive function. A logarithmic psychophysical connection would explain the fantastic ranges that our senses can handle. Under proper conditions we can hear a sound that corresponds to our ear drum moving less than the diameter of a hydrogen atom, and we can actually see single photons. Yet we can also tolerate the sound of a jet engine corresponding to 10^{12} times the minimum intensity and see (for short periods of time as Fechner learned) direct sunlight 15 orders of magnitude stronger. In order to quantitatively compare two signals that may differ by such large amounts we introduce the *Bel* (named after Alexander Graham), defined as the base 10 logarithm of the ratio of the powers of the two signals. In other words, if the power of the second signal is greater than that of the first by a factor of ten, we say that it is one Bel (1 B) stronger. In turns out that the Bel is a bit too large a unit for most purposes, and so we usually use the decibel (dB), which is ten times smaller.

$$d(dB) = 10 \log \frac{P_1}{P_2} \qquad (11.4)$$

Since power is the integral of the square of the signal values, if we know RMS signal values we can directly compute the difference between two signals.

$$d(dB) = 20 \log \frac{S_1}{S_2} \qquad (11.5)$$

The JND for strong sounds is about 0.5 dB, while at the threshold of hearing about 3 dB is needed.

An audio signal's amplitude is not the only characteristic that is perceived approximately logarithmically. Humans can hear from about 20 Hz (lower than that is felt rather than heard) to over 20 KHz (the precise upper limit depending on age). This corresponds to about 10 *octaves*, each octave being a doubling of frequency. Sinusoids separated by whole octaves sound similar to us, this fact being the principle behind the musical scale. Inside each octave the conventional western ('well-tempered') division is into twelve chromatic keys, each having frequency $\sqrt[12]{2}$ higher than the previous one. These keys sound to us approximately equally spaced, pointing once again to a logarithmic perception scale.

The *mel* (from 'melody') frequency scale is designed to correspond to the subjective psychophysical sensation of a tone's pitch. The perceived pitch of a 1 KHz tone at 40 dB above the hearing threshold is defined to be 1000 mels. Equal mel intervals correspond to equal pitch perception differences; under about 1 KHz the mel scale is approximately linear in frequency, but at higher frequencies it is approximately logarithmic.

$$M \approx 1000 \log_2(f_{KHz} + 1)$$

The *Bark* (named after the acoustician H.G. Barkhausen) scale approximates the natural frequency scale of the auditory system. Psychophysically, signals heard simultaneously are perceived as separate sounds when separated by one Bark or more since they excite different basilar membrane regions. A Bark is about 100 Hz for frequencies under 500 Hz, is about 150 Hz at 1 KHz, and a full KHz at about 5 KHz.

$$1\,\text{Bark}_{Hz} \approx 25 + 75(1 + 1.4 f_{KHz}^2)^{0.69}$$

If we divide the entire audio range into nonoverlapping regions of one Bark bandwidth we get 24 'critical bands'. Both the mel and Bark scales are approximately logarithmic in frequency.

EXERCISES

11.2.1 Derive equation (11.3). What is the meaning of A and B? What should be the base of the logarithm?

11.2.2 How long does a tone have to be on for its frequency to be identifiable? Experiment!

11.2.3 The well-tempered scale is a relatively recent invention, having become popular with the invention of keyboard-based instruments such as the piano. Using a computer with a programmable sound generator, test the difference between a linearly divided scale and a well-tempered one. Play a series of notes each higher than the previous one by 50 Hz. Do the differences sound to same? Play a simple tune on the well-tempered scale and on a linearly divided octave scale. Can you hear the difference? Can you describe it?

11.2.4 Since we perceive sound amplitudes logarithmically, we should quantize them on a logarithmic scale as well. Compare the μ-law and A-law quantizations prevalent in the public telephone system (equations (19.3) and (19.4)) with logarithmic response. How are negative values handled? Can you guess why these particular forms are used?

11.2.5 Two approximations to the Bark warping of frequency are

$$
\begin{aligned}
B &\approx 13\tan^{-1}(0.76 f_{KHz}) + 3.5\tan^{-1}\left(\frac{f_{KHz}}{7.5}\right)^2 \\
&\approx 7\sinh^{-1}(f_{KHz}/0.65)
\end{aligned}
$$

while the Mel warping was given in the text. Compare these three empirical formulas with true logarithmic behavior $\alpha \ln(1+x)$ in the range from 50 Hz to 5 KHz.

11.2.6 Recent research has shown that Fechner's law is only correct over a certain range, failing when the stimuli are either very weak or very strong. Stevens proposed a power law $Y = kI^n$ where k and n are parameters dependent on the sense being described. Research Stevens' law. For what cases does Stevens' law fit the empirical data better than Fechner's law?

11.2.7 Toward the end of his life Fechner studied æsthetically pleasing shapes. Write a program that allows the user to vary the ratio of the sides of a rectangle and allow a large number of people to find the 'nicest' rectangle that is not a square. What ratio do people like? (This ratio has been employed in architecture since the Greeks.)

11.3 Speech Production

In this section we introduce the biological generation mechanism for one of the most important signals we process, namely human speech. We give a quick overview of how we use our lungs, throats, and mouths to produce speech signals. The next section will describe speech perception, i.e., how we use our ears, cochlea, and auditory nerves to detect speech.

It is a curious fact that although we can input and process much more visual information than acoustic, the main mode of communications between humans is speech. Wouldn't it have been more efficient for us to communicate via some elaborate sign language or perhaps by creating rapidly changing color patterns on our skin? Apparently the main reason for our preferring acoustic waves is their long wavelengths and thus their diffraction around obstacles. We can broadcast our speech to many people in different places; we can hear someone talking without looking at the mouth and indeed without even being in the same room. These advantages are so great that we are willing to give up bandwidth for them; and speech is so crucial to the human race that we are even willing to risk our lives for it.

To understand this risk we have to compare our mouth and throat regions with those of the other primates. Comparing the profile of a human with that of a chimpanzee reveals that the chimpanzee's muzzle protrudes much further, while the human has a longer pharynx (throat) and a lower larynx (voice box). These changes make it easy for the human to change the resonances of the vocal cavity, but at the expense of causing the respiratory and alimentary tracts to overlap. Thus food can 'go down the wrong way', impeding breathing and possibly even leading to death by choking. However, despite this importance of spoken communication, the speech generation mechanism is still basically an adapted breathing and eating apparatus, and the speech acquisition mechanism is still essentially the acoustic predator/prey detection apparatus.

It is convenient to think of speech as being composed of a sequence of basic units called *phonemes*. A phoneme is supposed to be the smallest unit of speech that has independent meaning, and thus can be operationally defined as the minimal amount of speech that if replaced could change the meaning of what has been said. Thus **b** and **k** are distinct phonemes in English (e.g., 'book' and 'cook' have different meanings), while l and **r** are indistinguishable to speakers of many oriental languages, **b** and **p** are the same in Arabic, and various gutturals and clicks are not recognized by speakers of Latin-based languages. English speakers replace the French or Spanish **r** with their own because the originals do not exist in English and are thus not properly distinguished. Different sources claim that there are between 42 and 64 phonemes in spoken English, with other languages having typically between 25 and 100. Although the concept of a phoneme is an approximation to the whole story, we will posit speech generation and perception to be the production and detection of sequences of phonemes.

Speech generation commences with air being exhaled from the lungs through the 'trachea' (windpipe) to the 'larynx' (voice box). The 'vocal cords' are situated in the larynx. While simply breathing these folds of tissue are held open and air passes through them unimpeded, but when the laryngeal muscles stretch them taut air must pass through the narrow opening between the cords known as the 'glottis'. The air flow is interrupted by the opening and closing of the glottis, producing a periodic series of pulses, the basic pulse rate being between 2.5 and 20 milliseconds. The frequency corresponding to this pulse interval is called the *pitch*. The tighter the cords are stretched, the faster the cycle of opening the cords, releasing the air, and reclosing, and so the higher the pitch. Voice intensities result from the pressure with which the expelled air is forced through the vocal cords. The roughly triangular-shaped pulses of air then pass into the *vocal tract* consisting of the 'pharynx' (throat), mouth cavity, tongue, lips, and nose and are finally expelled. There are two door-like mechanisms that prohibit or allow air to flow. Two passages proceed downward from the pharynx, the 'esophagus' (food pipe) and trachea. The 'epiglottis' separates the two by closing the air passage during swallowing. In addition, air can enter the nasal tract only when the 'velum' is open.

The air pulses exiting the vibrating vocal cords can be thought of as a signal with a basic periodicity of between 50 and 400 Hz (typically 50–250 for men, 150–400 for women) but rich in harmonics. Thus the spectrum of this signal consists of a set of equally spaced lines, typically decreasing in amplitude between 6 and 12 dB per octave. Because of its physical dimensions, the vocal tract resonates at various frequencies called formants, corresponding to the length of the throat (between 200 and 800 Hz), length of the nasal passage (500–1500 Hz), and size of the mouth between throat and teeth (1000–3000 Hz). These resonances enhance applicable frequencies in the glottal signal, in the manner of a set of filters. The result is the complex waveform that carries the speech information. The spectrum thus consists of a set of lines at harmonics of the pitch frequency, with amplitudes dependent on the phoneme being spoken.

The vocal cords do not vibrate for all speech sounds. We call phonemes for which they vibrate *voiced* while the others are *unvoiced*. Vowels (e.g., **a**, **e**, **i**, **o**, **u**) are always voiced unless spoken in a whisper, while some consonants are voiced while others are not. You can tell when a sound is voiced by placing your fingers on your larynx and feeling the vibration. For example, the sound **s** is unvoiced while the sound **z** is voiced. The vocal tract is the same in both cases, and thus the formant frequencies are identical, but **z** has a pitch frequency while **s** doesn't. Similarly the sounds **t** and **d** share vocal tract positions and hence formants, but the former is unvoiced and the latter voiced. When there is no voicing the excitation of the vocal tract is created by restricting the air flow at some point. Such an excitation is noise-like, and hence the spectrum of unvoiced sounds is continuous rather than discrete. The filtering of a noise-like signal by vocal tract resonances results in a continuous spectrum with peaks at the formant frequencies.

The unvoiced fricatives **f**, **s**, and **h** are good examples of this; **f** is generated by constricting the air flow between the teeth and lip, **s** by constricting the air flow between the tongue and back of the teeth, and **h** results from a glottal constriction. The **h** spectrum contains all formants since the excitation is at the beginning of the vocal tract, while other fricatives only excite part of the tract and thus do not exhibit all the formants.

Nasal phonemes, such as **m** and **n**, are generated by closing the mouth and forcing voiced excitation through the nose. They are weaker than the vowels because the nasal tract is smaller in cross sectional area than the mouth. The closed mouth also results in a spectral zero, but this is not well detected by the human speech recognition apparatus. Glides and liquids, such as **w** and l, are also voiced but weaker than vowels, this time because the vocal tract is more closed than for vowels. They also tend to be shorter in duration than vowels. Stops, such as **b** and **t**, may be voiced or unvoiced, and are created by first completely blocking the vocal tract and then suddenly opening it. Thus recognition of stops requires observing the signal in the time domain.

We have seen that all phonemes, and thus all speech, can be created by using a relatively small number of basic building blocks. We need to create an excitation signal, either voiced or unvoiced, and to filter this signal in order to create formants. In 1791, Wolfgang von Kempelen described a mechanical mechanism that could produce speech in this fashion, and Charles Wheatstone built such a device in the early 1800s. A bellows represented the lungs, a vibrating reed simulated the vocal cords, and leather pipes performed as mouth and nasal passages. By placing and removing the reed, varying the cross-sectional area of the pipes, constricting it in various places, blocking it and releasing, etc., Wheatstone was able to create intelligible short sentences. Bell Labs demonstrated an electronic synthesizer at the 1939 World's Fair in New York. Modern speech synthesizers are electronic and computerized, digitally creating the excitation and filtering using methods of DSP. We will return to this subject in Section 19.1.

EXERCISES

11.3.1 What are the main differences between normal speaking on the one hand and whispering, singing, and shouting on the other?

11.3.2 Why do some boys' voices change during adolescence?

11.3.3 Match the following unvoiced consonants with their voiced counterparts: t, s, k, p, f, ch, sh, th (as in think), **wh**.

11.3.4 Simulate the speech production mechanism by creating a triangle pulse train of variable pitch and filtering with a 3–4 pole AR filter. Can you produce signals that sound natural?

11.3.5 Experiment with a more sophisticated software speech synthesizer (source code may be found on the Internet). How difficult is it to produce natural-sounding sentences?

11.4 Speech Perception

The human ear along with the human brain are a most impressive sound receiver. We can actually detect sounds that are so weak that the air pressure density fluctuations are less than one billionth of the average density. These sounds are so weak that the ear drum moves only about the diameter of a single hydrogen atom! But we can also hear very strong sounds, sounds so strong that the ear drum moves a millimeter. The frequency range of the ear is also quite remarkable. Not only can we hear over ten octaves (our visual system is sensitive over only about one octave), most people can distinguish between 998 Hz and 1002 Hz, a difference of a few parts per thousand. Piano tuners tune to within much better than this by using beat frequencies. Even the most tone deaf can easily distinguish a great variety of timbres, which are effects of lack of sinusoidality.

Sound perception commences with sound waves impinging on the outer ear, and being funneled into the 'auditory canal' toward the middle ear. The sound waves are amplified as they progress along the somewhat narrowing canal, and at its end hit the 'tympanic membrane' or eardrum and set it into vibration. The physical dimensions of the outer ear also tend to band-pass the sound waves, enhancing frequencies in the range required for speech. The eardrum separates the outer ear from the middle ear, which is a small air-filled space, with an opening called the 'Eustachian tube' that leads to the nasal tract. The Eustachian tube equalizes the air pressure on both sides of the eardrum, thus allowing it to vibrate unimpeded. A chain of three movable bones called 'ossicles' (and further named the 'hammer', 'anvil' and 'stirrup') traverses the middle ear connecting the eardrum with the inner ear. The vibrations of the eardrum set the hammer ossicle into motion, and that in turn moves the anvil and it the stirrup. The vibrations are eventually transmitted to a second membrane, called the 'oval window', which forms the boundary between the middle and inner ear. Since the base of the stirrup is much smaller than the surface of the eardrum, the overall effect of this chain of relay stations is once again to amplify the sound signal.

From the oval window the vibrations are transmitted into a liquid-filled tube, coiled up like a snail, called the 'cochlea'. Were the cochlear tube to be straightened out it would be about 3 centimeters in length, but coiled up as a $2\frac{1}{2}$- to 3-turn spiral it is only about 0.5 cm. The cochlea is divided in half along its length by the 'basilar membrane', and contains the organ of Corti; both the basilar membrane and the 'organ of Corti' spiral the length of the cochlea. Vibrations of the oval window excite waves in the liquid in the cochlea setting the basilar membrane into mechanical vibration. Were we to straighten the cochlea out we would observe that its width tapers from about a half-centimeter near the oval window to very small at its apex; however, the basilar membrane is stiff near the oval window and more flexible near the apex. Combined, these two characteristics make the basilar membrane frequency selective. High frequencies cause the basilar membrane to vibrate most strongly near the oval window, and as the frequency is lowered the point of strongest vibration moves along the length of the basilar membrane toward the apex of the cochlea.

The organ of Corti transduces the mechanical vibrations into electric signals. It has about 15,000 sensory receptors called 'hair cells' that contact the basilar membrane and stimulate over 30,000 motion sensitive neurons that create electric pulses that are transmitted along the auditory nerve to the brain. There are two types of hair cells, three rows of 'outer' hair cells and one row of 'inner' hair cells. Motion of the basilar membrane moves the hair cells back and forth causing them to release neurotransmitter chemicals that cause auditory neurons to fire. Since different parts of the membrane respond to different frequencies, auditory neurons that are activated by inner hair cells that contact a particular location on the basilar membrane respond mainly to the frequency appropriate to that location. Complex sounds activate the basilar membrane to different degrees along its entire length, thus creating an entire pattern of electric auditory response. Similarly the outer hair cells are intensity selective, different sound intensities stimulate different hair cells and create different neuron activity patterns.

We can roughly describe the operation of the cochlea as a bank of filters spectral decomposition with separate gain measurement. As different sounds arrive at the inner ear the hair cell response changes creating a varying spatial representation. The neural outputs are passed along the auditory nerve toward the cortex without disturbing this representation; the spatial layout of the neurons in the nuclei (groups of nerve cells that work together) closely resembles that of the hair cells in the inner ear. Indeed in all nuclei along this path tonotopic organization is observed; this means that nearby neurons respond to similar frequencies, and as one moves across the nucleus the frequency of optimal response smoothly varies.

The auditory nerve from each ear feeds a cochlear nucleus in the auditory brainstem for that ear. From both cochlear nuclei signals are sent both upward toward the primary auditory cortex and sideways to the superior olivary complex, from which they proceed to the pathway belonging to the opposite ear. This pathway mixing enables binaural hearing as well as mechanisms for location and focus.

What about the auditory cortex itself? We started the previous section by contrasting the vocal tracts of the human with those of other primates, yet the difference in our brain structure between ourselves and the apes is even more remarkable. The human brain is not the most massive of any animal's, but our brain mass divided by body mass is truly extraordinary, and our neocortex is much larger than that of any other animal. There are two cortical regions that deal specifically with speech, Broca's area and Wernicke's area, and these areas are much more highly developed in humans than in other species. Broca's area is connected with motor control of speech production apparatus, while Wernicke's area is somehow involved in speech comprehension.

To summarize, the early stages of the biological auditory system perform a highly overlapped bank of filters spectral analysis, and it is this representation that is passed on to the auditory cortex. This seems to be a rather general-purpose system, and is not necessarily the optimal match to the speech generation mechanism. For example, there is no low-level extraction of pitch or formants, and these features have to be derived based on the spectral representation. While the biology of speech generation has historically had a profound influence on speech synthesis systems, we are only now beginning to explore how to exploit knowledge of the hearing system in speech recognition systems.

EXERCISES

11.4.1 Experiment to find if the ear is sensitive to phase. Generate combinations of evenly spaced sines with different phase differences. Do they sound the same?

11.4.2 Masking in the context of hearing refers to the psychophysical phenomenon whereby weak sounds are covered up by stronger ones at nearby frequencies. Generate a strong tone at 1 KHz and a weaker one with variable frequency. How far removed in frequency does the tone have to be for detection? Attenuate the weaker signal further and repeat the experiment.

11.4.3 Sit in a room with a constant background noise (e.g., an air-conditioner) and perform some simple task (e.g., read this book). How much time elapses until you no longer notice the noise?

11.4.4 Go to a (cocktail or non-drinking) party and listen to people speaking around the room. What effects your ability to separate different voices (e.g. physical separation, pitch, gender, topic discussed)?

11.4.5 Have someone who speaks a language with which you are unfamiliar speak a few sentences. Listen carefully and try to transcribe what is being said as accurately as you can. How well did you do?

11.4.6 Talk to someone about speech recognition and then quickly ask 'Do you know how to wreck a nice peach?'. Ask your partner to repeat what you just said. What does this prove?

11.4.7 Most of the time and energy of the speech signal is spent in the vowels, and hence the speech perception mechanism performs best in them. But do vowels carry most of the information? You can find out by performing the following experiment. Select several sentences at random from this book. From each sentence create two character strings, one in which all consonants are replaced by question marks, and one in which all vowels are. Now present these strings to subjects and ask them to fill in the blanks. What are your findings?

11.4.8 Explain the possible mechanisms for acoustic source location. Take into account the width of the head, the fact that localization is most accurate for high-frequency sounds with sharp attack times, and the idea that the head will absorb some sounds casting an acoustic shadow (about 3 dB at 500 Hz, 20 dB at 6 KHz). How is height determined? Devise a neurobiologically plausible model for a time-of-arrival crosscorrelator.

11.4.9 Simulate the sound perception mechanism by building a bank of overlapping band-pass filters (at least 100 are needed) and graphically displaying the output power with time as horizontal axis and filter number as vertical axis. Test by inputting a sinusoid with slowly varying frequency. Input speech and try to segment the words on the graphic display.

11.5   Brains and Neurons

The human brain is certainly a remarkable computer and signal processor. We have seen above how it can communicate with other brains using audio frequency waves in the air by coercing the mouth (an organ developed for eating and breathing) to broadcast messages and obliging the ear (originally for detecting predators and prey) to capture these messages. It can also communicate by using the hands to write and eyes to read; it can process visual information at high speed, recognizing human faces and familiar objects in real time; it can instruct the hands to manipulate objects, and enable the body to avoid obstacles and navigating in order to get to wherever it wants. The brain can use tools, create new tools, find rules in complex phenomena; it can write music and poetry, do mathematical calculations, learn to do things it didn't know how to do previously. It can even create new thinking machines and signal processing machines that excel in areas where it itself is limited.

This section contains a short introduction to the brain's hardware architecture. Our purpose is not the study the brain's physiology per se; rather we wish to understand its prominent features in order to gain inspiration that may lead to the building of better signal processing and computing machines. In the following section we will introduce *artificial neural networks*, which are models that attempt to capture the essential properties of the brain's computational architecture and are used both to explain the functioning of the biological brain and to solve practical problems.

The brain as an organ had been studied by the ancients, and by the mid-nineteenth century it was known that certain well-defined areas of the brain were responsible for specific functions such as speech. It was Santiago Ramón y Cajal who first convincingly demonstrated, in the latter half of the 1880s, that the brain is not simply a large mass of fibers, but a vast number of richly interconnected brain cells, which we call 'neurons'. Using a cell staining technique earlier developed by Camillo Golgi (with whom he shared the 1906 Nobel prize for Physiology or Medicine) he both observed neurons and mapped their anatomy.

Neurons are a specialized type of cell, of which there are over 10 billion (10^{10}) in the human brain. From a functional point of view we can roughly categorize neurons into three classes, namely 'sensory neurons' (such as those in the retina of the eye that are sensitive to light), 'motor neurons' (e.g., those which activate and control the motion of our fingers), and 'higher processing neurons' (those in the neocortex). We will discuss mainly the last of these categories, but even of these neurobiologists have discovered many different varieties, such as pyramidal cells, Golgi cells, spiny stellate cells, smooth stellate cells, interneurons, etc. Our description will be so simplified that the differences between these various varieties will be unimportant.

The classical processing neuron, depicted schematically in Figure 11.1, is made up of three anatomical structures—the 'soma', the 'axons', and the 'dendritic tree'. The soma is the cell body and is responsible for the processing itself. The dendrites supply inputs to the neuron, while the axon carries the neuron's output. We can think of the neuron as a simple processing element that inputs multiple signals (about 10^4 to 10^5 is typical) and outputs a single signal.

What kind of signals are input and output? The interior of a neuron is usually electrically negative relative to the outside, due to the cell membrane selectively passing ions from inside the cell outward and from outside inward. The membrane's electric potential is not constant however, and its behavior as a function of time can be viewed as a signal. Perhaps the most significant type of behavior is the 'action potential'. This is a very fast event,

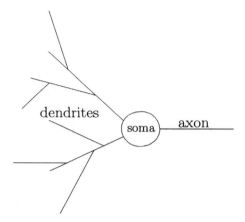

Figure 11.1: A highly schematic diagram of the classic higher processing neuron. The dendrites at the left are the inputs, the axon at the right is the output, and the processing is performed by the soma.

occupying about a millisecond, although afterward slow oscillations can occupy a further 100 milliseconds. The event itself starts with the membrane potential becoming even more negative than usual followed by short sign reversal. This spike can travel along the neuron's axon without decrease in amplitude; this propagation is not like electric current in a conductor, rather it is due to the axon being made of active material with each section exciting the next to spike. Due to the all-or-none nature of the action potential we will usually refer to the neuron as 'firing' if it has developed an action potential, or 'quiescent' if it has not. After a spike there is an 'absolute refractory period' during which the neuron cannot fire again, and a 'relative refractory period' during which the neuron is less susceptible to spiking. Although rates of several hundred spikes per second are possible, more typical frequencies are on the order of 10 Hz.

A 'synapse' is formed where one neuron touches and influences another. While there are other possibilities we will discuss synapses formed by the axon of the presynaptic neuron touching a dendrite of the postsynaptic neuron. At the synapse the two cell membranes touch but cellular material does not indiscriminately flow between the cells, the influence of the presynaptic potential being indirect. For example, in chemical synapses a presynaptic spike causes a transmitter substance to be released that changes the postsynaptic membrane permeability. Synapses may be 'excitatory', meaning that the firing of the presynaptic neuron increases the probability of the postsynaptic neuron firing as well; or 'inhibitory' if that firing decreases the postsynaptic neuron's chance of firing.

No single neuron is really that significant; only the 'network' of myriad neurons is of consequence. Each neuron's dendrites are contacted by axons of other neurons, and its own axon contacts, in turn, many other neurons. Thus the emerging picture is that of a huge number (about 10^{10}) of neurons, each firing or quiescent. The decision whether to fire is made in the soma, based on the input from all the (about 10^5) neurons that make synapses upon its dendrites. Once fired the action potential rapidly propagates from the soma down the axon to influence the firing of yet more (about 10^5) neurons.

Neurons in contact need not be physically close, nor do physically close neurons need to be in direct contact. Thus in our quest to understand brain function we are led to consider large areas of brain matter. The brain is highly organized, with specific locales responsible for specific functions, and various task geometries mapped onto brain geometries. As early as 1861, Paul Broca described a patient who had lost his ability to speak although he did understand spoken language. After the patient's death he tied this to a lesion in a specific position in the brain, now called Broca's area. Hubel and Wiesel earned the 1981 Nobel prize in medicine for their description of the early visual system. They discovered formations they called hypercolumns. The neurons in each hypercolumn respond to lines in certain areas of the visual field, with nearby hypercolumns responding to lines in nearby locations. As one travels along a hypercolumn the angle of the detected line slowly rotates. These facts and more lead us to conclude that the entire brain is a large interconnected network of neurons that can be broken down into task-specific subnetworks that are tightly connected with other subnetworks.

Since the days of Cajal and Golgi, neurobiologists have studied in depth the characteristics of single isolated neurons, and although much progress has been made, this study does not seem to lead to any deep explanation of brain function. Others have studied the larger-scale structure of the brain and discovered the mappings between function and specific areas of the brain, but even this immensely valuable information explains *where* but not *how* or *why*. In order to gain insight into the connection between the brain's anatomy and its function it is necessary to simplify things.

EXERCISES

11.5.1 The central nervous system is composed of the spinal cord, the brain stem, the cerebellum, the midbrain, and the left and right hemispheres of the neocortex. What are the functions of these different components? How do we know? What creatures (reptiles, mammals, primates) have each component?

11.5.2 In addition to neurons the brain also contains glia cells; in fact there are more glia than neurons. Why did we focus on neurons and neglect glia cells in our explanation?

11.5.3 There are many morphologies other than the classical model described in this section. For example, there are neurons with no axon or many axons, synapses may occur between two axons, or between two dendrites, or even on the cell body. Research several such variations. What is the function of these cells?

11.5.4 Research the Nobel prize-winning contribution of Hubel and Wiesel to the understanding of the neurons in the mammalian visual system. What are simple cells, complex cells, and hypercomplex cells? What is a hypercolumn and how are the cells arranged geometrically?

11.6   The Essential Neural Network

The single neuron does not perform any significant amount of computation; computation is performed by large collections of neurons organized into 'neural networks'. The term 'neural network' is actually misleading; the concept is not that of a network that has neural characteristics, but simply a network *of* neurons. Perhaps the term should be 'neuron network', but the original term has become entrenched. By association, other collections of interconnected processors, including ones we can make ourselves, are often called neural networks as well. However, the term is only fitting when the collection of processors is somehow inspired by the brain. A LAN of personal computers is a collection of interconnected processors that would not normally be considered a neural network.

When does a collection of processors become a neural network? Any definition we give will be subjective, and probably the number of different definitions equals the number of people working in the field. However, there are a number of requirements that most researchers would agree upon. My own definition goes something like this.

Definition: neural network
A *neural network* is a large set of simple, richly interconnected processing units that exhibits collective behavior after learning.   ∎

There must be a *large* number of processors, at least in the hundreds, before we leave the more conventional 'parallel processing' and reach the regime where collective behavior is meaningful. The individual neuron must be *simple*, performing one basic operation like calculating the dot product of its input with stored weights, or finding the distance between its input and a stored pattern. We definitely don't want to depend on multifunctional, highly precise, processors here. Some people would require nonlinearity of the neuron's operation, but we will be lax in this regard. To make up for the simplicity of the individual processor, and to exploit the large number of processors, we want them to be *richly interconnected*. Conventional parallel processing techniques prefer to connect processors to nearest neighbors on a grid or with hypercubic geometry. Biological networks may not be fully interconnected, but the connectivity is quite high.

The robustness to failure of conventional computers is infinitesimal. Were one to remove a randomly selected circuit from a personal computer or even simply cut a randomly selected conductor, the probability of total system failure is very high. This should be contrasted with the brain which loses large numbers of neurons daily without serious performance degradation. How is this robustness obtained?

A clue is the fundamentally different methods of storing information in the two competing architectures. The conventional computer uses Location Addressable Memory (LAM) wherein information is stored in a particular location. In order to retrieve this information the location must be known. The brain uses *content addressable memory* (CAM). For example, once an image is stored we can present it and ask whether it is a known image. A generalization of this idea is *associative memory*, by which we mean that we can present an image and ask if there is a stored picture that is similar (the association). In this fashion we can recognize a friend's face even with sunglasses and a different hair cut.

We can now try to piece the puzzle together. The real motive for the high connectivity of neural networks is to obtain *collective behavior*, also called *self-organization* and related to *distributed representations*. Were each memory to be stored, as in an LAM, in a specific neuron or definite small set of neurons, then failure of that neuron would wipe out that memory. Instead it seems plausible that memories are stored as eigenstates of the entire network. The mechanism that brings this about is spontaneous collective behavior, or self-organization.

Learning refers to the method of introducing memories and storing procedures. Conventional computers must be laboriously programmed; each new task requires expensive and time-consuming outside intervention. Brains learn from experience, automatically adapt to changing environments, and tend to be much more forgiving to 'bugs'.

EXERCISES

11.6.1 Which of the following are neural networks according to the definition given in the text?
   1. transistors on the substrate of an integrated circuit
   2. arithmetic registers in a microprocessor
   3. CPUs in a parallel processing environment
   4. cells in the spinal column
   5. neurons in an aplysia

11.6.2 By introspection, make a rough order-of-magnitude estimate of the amount of information (in bits) passed to the brain by the various senses. For vision, for example, estimate the size of the main field of vision, the pixel density, the dynamic range, and the number of pictures transferred per second. Based on the above estimates, how much information must the brain store in a day? A year? A lifetime? The brain contains about 10^{10} neurons. Does the above estimate make sense?

11.7   The Simplest Model Neuron

In this section we will consider a simple model neuron. This model does not do justice to the real biological neuron. Even using a single model, no matter how complex, is a gross simplification. Real neurons have complex time-dependent properties that we will completely ignore in this simple model; and the functioning of our model will be a mere caricature of the real thing.

So why should we attempt to model the neuron? An analogy is useful here. The reader will remember the *ideal gas law $PV = nk_BT$*, which reliably relates the pressure, volume and temperature for a large number of gases. This law is only approximate, and indeed it breaks down at very high pressures or a temperatures close to the condensation temperature of the particular gas. However, it is a good approximation for a very large number of gases over a large regions of P, V, and T, and furthermore corrections can be added to better approximate the actual behavior. The ideal gas law can be derived in statistical physics from the microscopic behavior of the gas molecules under the assumption that they are essentially ping pong balls. By this we mean that the gas molecules are assumed to be small spheres of definite size, which only interact with other molecules by colliding with them. Upon collision the molecules change their velocities as colliding ping pong balls would. Using techniques of statistical physics, which is a mathematical formalism designed to derive macroscopic 'average' laws from the behavior of huge numbers of simple particles, the ideal gas can be derived.

No-one really believes that the gas molecules are ping pong balls. They are definitely not spheres of well-defined radius—they are composed of a nucleus with protons and neutrons surrounded by electron 'clouds'. They definitely do not interacting like ping pong balls—there are electromagnetic fields that act at a distance and the dynamics is inherently quantum mechanical. So why does the ideal gas law work?

The answer is that it doesn't. When the pressure is high or the temperature low, the molecules are close together and the model breaks down miserably. The gas condenses into a liquid, and the temperature at which this happens is different for different gases. But for a large range of parameters the most important thing is that there are a very large number of molecules that interact only weakly with the others, except for short periods of time when they are close. Thus any model that obeys these constraints will give approximately the same behavior, so we might as well pick the easiest model to work with. Since the ping pong ball model is the simplest to handle mathematically, it is the natural starting point.

Let's return to the neuron. There is a large variety of types, and each is an extremely complex entity; but we believe that as a first approximation the most important features are the huge number of neurons, and the fact that these are so richly interconnected. In the spirit of statistical physics we search for the simplest 'ping pong ball' model of a neuron. This is the McCulloch-Pitts model, first proposed in the early 1940s.

The McCulloch-Pitts neuron was originally designed to show that a simple neuron-like device could calculate logical functions such as AND and OR. The neuronal output is calculated by comparing a weighted linear combination (convolution) of the inputs to a threshold. Only if the linear combination is above the threshold will the neuron fire. Such a function is often called a linear threshold function.

In order to state this description mathematically, we must introduce some notation. The output of the neuron under consideration will be called O, while its N inputs will be called I_j with $j = 1 \ldots N$. At this early stage the neural computation community already divides into two rival camps. Both camps represent the neuron firing as $O = +1$, but one uses $O = 0$ for quiescence, while the other prefers $O = -1$. The synaptic efficacy of the connection from input j will be represented by a real number W_j. For excitatory synapses $W_j > 0$, while for inhibitory ones $W_j < 0$. The absolute value of W_j is also important since not all inputs affect the output in the same measure. If $W_j = 0$ then the input does not affect the output at all (there is no synapse). If $|W_j|$ is large then the effect of this input is significant, while small $|W_j|$ means the input only marginally affects the output.

The linear combination of the McCulloch-Pitts neuron means that the cell's potential is assumed to be

$$h = \sum_{j=1}^{N} W_j I_j \tag{11.6}$$

and the neuron will fire when this potential surpasses the threshold potential required to excite an action potential θ. Using the $0, 1$ representation we can write

$$O = \Theta(h - \theta) = \Theta \left(\sum_{j=1}^{N} W_j I_j - \theta \right) \tag{11.7}$$

where Θ is the step function. With the ± 1 representation we write

$$O = \text{sgn}(h - \theta) \doteq \text{sgn} \left(\sum_{j=1}^{N} W_j I_j - \theta \right) \tag{11.8}$$

where the signum function sgn returns the sign of its argument. Note that the meaning of this latter equation is somewhat different from the previous one; here neurons that are *not* firing also contribute to the sum.

The McCulloch-Pitts neuron is the simplest possible neuron model. The inputs are combined in a simple deterministic way. The decision is made based on a simple deterministic comparison. The summation and comparison are instantaneous and time-independent. Yet this completely nonbiological formal neuron is already strong enough to perform nontrivial computations. For example, consider the following image processing problem. We are presented with a black-and-white image, such as a fax, that has been degraded by noise. The classic DSP approach would be to filter this binary-valued image with a two-dimensional low-pass filter, which could be implemented by averaging neighboring pixels with appropriate coefficients. This would leave a gray-scale image that could be converted back to a black-and-white image by thresholding. This combination of the linear combination of input pixels in neighborhoods followed by thresholding can be implemented by a two-dimensional array of McCulloch-Pitts neurons. This same architecture can implement a large variety of other image processing operators.

What is the most general operation a single linear threshold function can implement? For every possible input vector the McCulloch-Pitts neuron outputs 0 or 1 (or ± 1). Such a function is called a 'decision function' or a dichotomy. Thinking of the 2^N possible input configurations as points in N-dimensional space, the linear combination is obviously the equation for a $N - 1$-dimensional hyperplane, and the thresholding separates the inputs that cause positive output (which are all on one side of the hyperplane) from the others. Thus we see that the threshold linear function implements linearly separable dichotomies.

This is only a tiny fraction of all the dichotomizations we might need to use. If this were all neuron models could do, they would find little practical use. The modern reincarnation of neural networks exploits such architectures as the multilayer perceptron (see Section 8.4), which broaden the scope of implementable dichotomies. In fact, feedforward networks of neurons can implement arbitrarily complex functions.

How does the brain learn the weights it needs to function? Hebb proposed a principle that can be interpreted at the neuron level in the following way.

Theorem: Hebb's Principle
The synaptic weight increases when the input to a neuron and its output tend to fire simultaneously. Using the notation W_{ij} for the weight that connects presynaptic neuron s_j with postsynaptic neuron s_i,

$$W_{ij} \leftarrow W_{ij} + \lambda s_i s_j \qquad (11.9)$$

where the s_i are either $0, 1$ or ± 1.   ∎

This form, where weights are updated accordingly to a constant times the product of the input and output, strongly reminds us of the LMS rule, only there the product is of the input and error. However, this difference is only apparent since if the postsynaptic neuron fires $s_i = +1$ when it shouldn't have the error is $1 - (-1) = 2$ while in the opposite case the error is $-1 - (+1) = -2$ and the difference is only a factor of two that can be absorbed into λ. The true difference is that the desired s_i can only take on the discrete values $(0, 1$ or $\pm 1)$; so rather than converging to the true answer like LMS, we expect a neuron-motivated adaptation algorithm to eventually attain precisely the right answer. The first such algorithm, the 'perceptron learning algorithm' was actually discovered before the LMS algorithm. It can be shown to converge to an answer in a finite number of steps, assuming there is an answer to the problem at hand.

The problem with the perceptron learning algorithm is that it does not readily generalize to the more capable architectures, such as the multilayer perceptron. The most popular of the modern algorithms is based on a variant of LMS.

EXERCISES

11.7.1 One can convert $0, 1$ neurons into ± 1 neurons by using the transformation $S \rightarrow 2S - 1$. Show that equations (11.7) and (11.8) are equivalent by thus transforming the I_j and finding transformations for W_j and θ.

11.7.2 Show that **AND** and **OR** gates of in-degree N can be implemented using McCulloch-Pitts neurons. That is, specify W_j and θ such that the neuron will fire only if all inputs fire, or if at least one input fires.

11.7.3 Draw and label the possible input configurations of a three-input linear threshold function as vertices of a cube. Show graphically which dichotomies can be implemented with zero threshold.

11.7.4 Extend the results of the previous problem to general McCulloch-Pitts neurons (nonzero threshold).

11.7.5 The McCulloch-Pitts neuron can be used as a signal detection mechanism. Assume we wish to detect a signal s_n of N samples and report the time that it appears. One builds a neuron with N *continuous* inputs and weights precisely $W_j = s_j$. We then have the input flow past the neuron, so that at any time the neuron sees N consecutive inputs. Consider the ± 1 representation and show that the PSP will be maximal when the required signal is precisely lined up with the neural inputs. How should the threshold be set (take noise and false alarms into account)? To what signal processing technique does this correspond?

11.7.6 Extend the results of the previous exercise to image recognition. Would such an approach be useful for recognition of printed characters on a page? If not why not? What about cursive handwriting?

11.7.7 Discuss the use of McCulloch-Pitts neurons for associative memory.

11.7.8 What is the difference between Hebb's principle for $0, 1$ neurons and ± 1 neurons?

11.8 Man vs. Machine

Now that we have a basic understanding of the brain's computational architecture we can attempt a quantitative comparison between the brain and the conventional computer. The pioneers of the modern computer were aware of the basic facts of the previous section, and were interested in eventually building a brain-like device. However, the prospect of 10^{10} parallel processing elements was quite daunting to these early computer engineers, who thus compromised on a single processing element as a kind of first approximation.

Surprisingly, by making this one processing element faster and more powerful, this computer evolved into a completely different architecture, quite powerful in its own right. Only now, with computer speeds approaching the absolute limits that physics imposes, is parallel processing being once again seriously considered; but even now when computer engineers talk about parallel processing they are referring to small numbers of CPUs, such as two, four, or eight. A comparison of the human brain to a conventional computer, based on the information of the last few sections, is to be found in Table 11.1.

	Brain	Computer
number of processors	$\approx$ 10 billion *neurons* (massively parallel)	1 *CPU* (intrinsically serial)
processor complexity	simple inaccurate	complex accurate
processor speed	slow (millisec)	fast (nanosec)
inter-processor communications	fast (μsec)	slow (millisec)
learning mode	learn from experience	manual programming
failure robustness	many neurons die without drastic effect	single fault often leads to system failure
memory organization	content addressable (CAM)	location addressable (LAM)

Table 11.1: A quantitative and functional comparison of the human brain and a serial processing computer.

The term *architecture* as applied to computers was invented to describe all the aspects of the computer's hardware that software must take into account. Two computers of identical architectures but different speeds can be uniquely compared as to strength—if the clock speed of one is twice that of the other, every program will run on it twice as fast. Two computers of similar, but not identical architectures are not uniquely comparable, and thus different programs will run at slightly different speed-up ratios. The more the architectures differ, the greater will be the divergence of the benchmarking results. This does not mean that we cannot say that a supercomputer is stronger than a desktop computer! There is another way to define the concept of 'stronger'.

Many engineering workstations today come with software emulations of personal computer environments. These emulations can run actual applications designed for the personal computer by emulating that entire platform in software. When a PC program is input to the emulation, all the PC opcodes must be read and interpreted and the required operation precisely simulated by an appropriate workstation command or routine. Were the program to run on the workstation emulation faster than on the target PC, we would be justified in concluding that the workstation is stronger than the PC, even though their architectures are dissimilar. Can we make such a comparison between the conventional computer and the brain?

To answer this question definitively we require an estimate as to the number of computers required to emulate a human brain at the hardware level. From the previous section we know that to emulate the simplest possible neuron, we would have to carry out N multiplications and accumulate operations (see Section 17.1) every 'clock' period. Here the number of synapses $N \approx 10^5$ and the clock period is about 5 milliseconds, and so a single neuron would require at least $2 10^7$ MACs per second, and 10^{10} neurons would require over $2 10^1 7$ MAC/sec. Assuming even an extremely fast CPU that could carry out a MAC in 5 nanoseconds, we would require 10^9 such computers in parallel to simulate a single human brain!

So the brain is equivalent to 1 gigacomputer! This sounds quite impressive, even without taking the small physical size and low power requirements into account. Now let's ask the converse question. How many humans would be required to emulate this same 5 nanosecond computer? Even assuming that a human could carry out the average operation in five seconds (and it is doubtful that many of us can perform an arbitrary 16-bit multiplication in this time, let alone 32-bit divisions) the computer would have carried out 10^9 operations in this same 5 seconds, and so we would need 10^9 humans in parallel to emulate the computer! So the computer is equivalent to a gigahuman as well.

How could these comparisons have turned out so perverse? The reason is that the underlying architectures are so very different. In such a case cross-emulation is extremely inefficient and direct comparison essentially meaningless. Certain benchmarking programs will run much faster on one machine while others will demonstrate the reverse behavior. The concept of 'stronger' must be replaced with the idea of 'best suited'.

Thus when two quite different computational architectures are available and a new problem presents itself, the would-be solver must first ask 'Which architecture is more suited to this problem?'. Although it may indeed be *possible* to solve the problem using either architecture, choosing the wrong one may make the solution extremely inefficient or even unsolvable in practice. For example, were we required to calculate the 137th root of a 50-digit number, I believe that most readers would agree that the conventional number-crunching computer is more suited to the task than the human (or chimpanzee). However, when the problem is the understanding of spoken language, the reading of cursive handwriting, or the recognition of faces, the architecture of the brain has proved the more viable. Indeed for these tasks there is at least an existence proof for a neural network solution, but none has yet been proffered for the serial computer!

Despite the above argument for neural computation, the neural network approach has had only limited success so far. **O**ptical **C**haracter **R**ecognition (OCR) engines based on neural networks have indeed eclipsed other technologies, yet progress on speech recognition has been slow. At least part of the fault lies in the size of network we can presently build, see Figure 11.2. Our largest systems seem to be on the level of a mentally retarded mosquito! We are forced to conclude that our previous 'existence proof' for neural solutions to ASR, OCR, face recognition, and other problems is contrived at best. The only way our present-day artificial neural networks will be able to solve practical problems is by being more efficient than biology by many orders of magnitude.

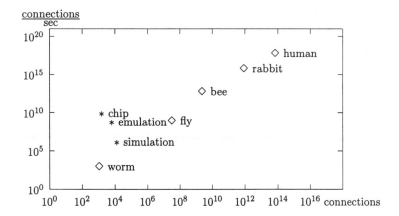

Figure 11.2: The speed and complexity of various neural networks. The horizontal axis is the number of synapses in the network, a number that determines both the information capacity and the complexity of processing attainable. The vertical axis is the number of synaptic calculations that must be performed per second to emulate the network on a serial computer, and is an estimate of the network's speed.

EXERCISES

11.8.1 The 'expert system' approach to artificial intelligence is based on the tenet that anything a human (or other rational agent) can do a standard computer can be programmed to do. Discuss the veracity and relevance of this hypothesis. What types of problems have been solved using expert systems? What AI problems have eluded prolonged attempts at solution?

11.8.2 In the Hopfield model there are N neurons $\{s_i\}_{i=1...N}$ each of which can take on the values ±1, where $s_i = +1$ means that the ith neurons is firing. The synapse from presynaptic neuron j to postsynaptic neuron i is called W_{ij}, and the matrix of synaptic efficacies has zeros on the diagonal $W_{ii} = 0$ and is symmetric $W_{ij} = W_{ji}$. At any given time only one neuron updates its state; the updating of the ith neuron is according to

$$s_i(t+1) = \text{sgn} \sum_{j=1}^{N} W_{ij} s_j(t)$$

after which some other neuron updates. Write a program that generates a random symmetric zero diagonal synaptic matrix, starts at random initial states, and implements this dynamics. Display the state of the network graphically as a rectangular image, with $s_i = \pm1$ represented as different colored pixels. What can you say about the behavior of the network after a long enough time? What happens if you update all the neurons simultaneously based on the previous values of all the neurons? What happens in both of these cases if the synaptic matrix is antisymmetric? General asymmetric?

11.8.3 Given P N-bit *memories* $\{\xi_i^\mu\}_{i=1...N}^{\mu=1...P}$ to be stored, the Hebbian synaptic matrix is defined as

$$W_{ij} = \sum_{\mu=1}^{P} \xi_i^\mu \xi_j^\mu$$

which is the sum of outer products of the memories. Enhance the program written for the previous exercise by adding a routine that inputs desired memory images and computes the Hebbian matrix. Store $P < 0.1N$ memories and run the dynamics starting near one of the memories. What happens? What happens if you start far from any of the memories? Store $P > 0.2N$ memories and run the dynamics again. What happens now?

Bibliographical Notes

Good general introductions to psychophysics can be found in [50, 71].

Alexander Graham Bell's original vocation was speech production and much of the early research on speech generation mechanisms was performed at Bell Labs [54, 55]. The classic formant tables [193] and the ear sensitivity curves [62] also originated there.

Speech production mechanisms are presented in many books on speech processing, such as [211] and in more depth in [253]. Speech perception is covered in [253, 195, 129]. The Bark scale is formally presented in [290, 232, 6] and the mel scale was defined in [254]. Cochlear modeling is reviewed in [5]. The application of the psychophysical principle of masking to speech compression is discussed in [232].

The McCulloch-Pitts neuron was introduced in [171]. In 1957 a team led by Frank Rosenblatt and Charles Wightman built an electronic neural network, which they called the *Mark I Perceptron*. This device was designed to perform character recognition. It was Rosenblatt who discovered and popularized the perceptron learning algorithm [224]. Minsky and Papert's charming book [174] both thoroughly analyzed the algorithm and dampened all interest in neural network research by its gloomy predictions regarding the possibility of algorithms for more capable networks.

For a light introduction to the functioning of the brain, I suggest [32], while a more complete treatment can be found in [180]. [8] is a thorough introduction to neuron modeling from a physicist's point of view. [168] is a seminal work on neural network modeling and [169] is the companion book of computer exercises.

Part III

Architectures and Algorithms

Graphical Techniques

Digital signal processing means *algorithmic* processing, representing signals as streams of numbers that can be manipulated by a programmable computer. Since DSP algorithms are programmed, standard computer languages may be used in principle for their implementation. In particular, block diagrams, that are conventionally used to help one grasp the essential elements of complex conventional computer programs, may be useful as DSP description and specification tools as well.

It is difficult for people to capture and comprehend the structure of large pieces of algorithmic code, with the difficulty increasing rapidly with the length of uninterrupted code, the number of conditionals and branches, and the inherent complexity of the algorithm. In block diagrams, rectangles represent calculations the program may perform, straight lines represent possible paths between the calculations, and there are also special symbols for control structures. The proponents of block diagrams claim that by looking at a skillfully prepared block diagram the program structure becomes clear. Detractors say that these diagrams are useful only for a certain paradigm of programming that went out with the 'goto'; and that they only describe the control structures and not the data structures. Both sides agree that they are essentially a second language (in addition to the language in which the program is coded) to describe the same functionality, and as such the task of keeping them up to date and accurate is arduous.

In computer science the use of block diagrams was once pervasive but has gone out of style. In DSP *flow graphs*, which are similar to block diagrams, are still very popular. This is not because DSP is old-fashioned or less developed than computer science. This is not because DSP lacks other formalisms and tools to describe signals and systems. It is simply because the block diagram is a much more useful tool in DSP than it ever was in programming. DSP flow graphs graphically depict a DSP system's signal structure; rectangles and circles represent systems and directed lines represent signals. We thus capture the dual nature of systems and signals in one graphic portrait.

In addition, many common DSP tasks are highly structured in time and/or frequency; this results in DSP block diagrams that have easily perceived geometric structure. Finally, an algebra of graphic transformations can be developed that allow one to simplify DSP block diagrams. Such transformations often result in reductions in computational complexity that would be hard to derive without the graphical manipulations.

In this chapter we will consider DSP graphical techniques. The word *graphical* is not used here as in 'computer graphics' (although we advocate the use of graphical displays for depicting DSP graphs), rather as in *graph theory*. The term *graph* refers to a collection of points and lines between these points. We start with a historical introduction to graph theory. Thereafter we learn about DSP flow graphs and how to manipulate them. RAX is a visual programming block diagram system. We describe the operation and internals of RAX in order to acquaint the reader with this important class of DSP tools.

12.1   Graph Theory

Graphic representations have doubtless been used in science and technology for as long as humankind has pursued these subjects. The earliest uses were probably simple geometric constructions; it is easy to envision chief engineers in primitive civilizations making rough drawings before embarking on major projects; we can imagine sages in ancient civilizations studying figures and charts and then surprising kings with their predictions. We know that thousands of years ago diagrams were used for engineering and education. What the ancients grasped was that one can capture the essential elements of a complex problem using simple graphical representations.

A diagram obviously does not capture all the features of the original. A map of a city is not of the original size nor does it reveal the wonders of architecture, the smells of the restaurants, the sounds of honking horns, etc. Still the map is extremely useful when navigating around town, even if it omits which streets are one-way and which tend to have traffic jams. Maps of the entire world are even more abstract representations since the world is spherical and the map is flat. Yet maps can be designed to correctly portray distances between cities, or bearing from one spot to another (but not both). As long as one realizes that a diagram can only capture certain elements of the original, and selects a diagrammatic method that captures the elements needed to solve the problem at hand, diagrams can be helpful.

In Euclidean geometry we consider two triangles to be equivalent if one could slide one on top of the other and they would coincide. The color or line width of such triangles is not taken into account, and neither is their orientation or position on the page. The transformations that are considered unimportant include arbitrary translations and rotations. When two triangles are related by such a transformation they are considered to be the same triangle. Much of high school geometry deals with methods to show two triangles are equivalent in this sense. A simple extension would be to allow transformations that include a change of scale. This would make a triangle on a map equivalent to the triangle on the ground. In this type of geometry any two triangles are considered equivalent if all of their angles are the same. In affine geometry even more general transformations are allowed, namely those which scale the x axis and y axis differently. In affine geometry all triangles are the same, but they are different from all the rectangles (which are all equivalent to each other).

Topology is even more general than affine geometry. It allows completely arbitrary transformations as long as they do not rip apart the plane or glue it different points together. You can think of this as drawing the figure on a sheet of rubber and stretching it however you want—as long as it doesn't rip or stick to another part of itself. In topology a triangle is equivalent to a rectangle or a circle, but different from a figure-eight. Graph theory is the study of points and the lines between them in topological space. In graph theory almost all the original geometry is thrown away, and we are left with a single abstraction, the *graph*.

The word *graph* as used in graph theory means a collection of points and lines that connect these points. In the mathematical terminology the points are called *vertices* and the lines *edges*; in computer science the designations *nodes* and *arcs* are more common. We shall require arcs to connect distinct nodes (no arc loops back to the same node) and rule out multiple arcs between identical nodes. The distances between nodes, the lengths or thicknesses of the arcs, and the geometric orientations are meaningless in graph theory. All that counts is which nodes are connected to which.

In Figure 12.1 we see all possible types of graphs with up to four nodes. Two nodes are said to be 'adjacent' if they are connected by an arc. A 'path' is a disjoint collection of arcs that leads from one node to another. For example, in G_4^4 there is a path of length 2 from the top-left node to the bottom-right, but no path to the top-right node. A 'cycle' is a path that leads from a node to itself. In G_6^4 there is a cycle, but not in G_7^4. The number of arcs emanating from a given node is called its *degree*; there are always an even number of nodes of odd degree.

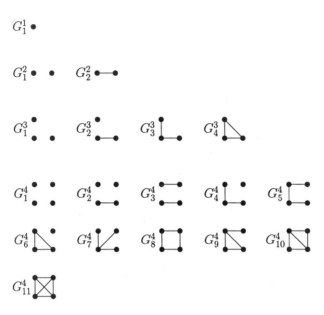

Figure 12.1: All graphs with up to four nodes. We only allow arcs connecting distinct nodes and disallow self connections.

Many of the most interesting problems in graph theory involve the number of graphs of a certain kind. A graph in which there is a path from any node to every other node (e.g., $G_1^1, G_2^2, G_3^3, G_4^3, G_7^4, G_8^4, G_9^4, G_{10}^4, G_{11}^4$) is called 'connected', while one that has all nodes connected to all others (e.g., $G_1^1, G_2^2, G_4^3, G_{11}^4$) is called complete.

The beginnings of graph theory are usually traced back to 1736. In that year the famous mathematician, Leonhard Euler, considered the father of analysis, published his solution to a puzzle that he had been working on. Euler, who was born in Switzerland, was professor of mathematics at the academy of St. Petersburg (founded in 1725 by Catherine, the wife of Peter the Great). The cold weather so adversely affected his eyesight that in 1736 we find him living in the capital of East Prussia, Königsberg (German for 'the Kings city'). This city, founded in 1255 by Teutonic knights, was the seat of the dukes of Prussia from 1525 through 1618. After World War II the city was annexed to the USSR and renamed Kaliningrad (Russian for 'Kalinin's city') after the Soviet leader M.I. Kalinin. Today it is the capital of the Kaliningrad Oblast and is Russia's sole port that does not freeze-over in winter.

Königsberg's topography is even more interesting than its history. The Pregel river (Pregolya in Russian) flows through the city from east to west, on its way to the Frisches Haff (German for 'freshwater bay', called Wislany Zalew in Polish, and Vistula in Lithuanian), a lagoon of the Baltic Sea. Not only does the river divide the city in two, but the river itself splits into northern and southern branches that later reconverge, forming an island in the center of town. The island is connected to the other parts of town by seven bridges. The question that puzzled Euler was this. Is it possible to leave your home for a walk, cross all the bridges exactly once, and return home? In terms of graph theory the question is, 'Is it possible to start at some node, traverse all the arcs exactly once, returning to the initial node?' If possible, such a path is called an *Euler cycle*. Euler recognized the puzzle as being a specific example of a general question—does a given graph have an Euler cycle or not? Today we call such graphs *Eulerian*.

It is obvious that the distance between the nodes of the graph do not affect the answer to this question; this is truly a topological problem. What Euler discovered is that *degree is* important. A connected graph is 'Eulerian' if and only if every node has even degree. Only Eulerian graphs have paths of the desired type. With this insight Euler simultaneously founded the disciplines of graph theory and topology.

We pick up our story once again 120 years later in Great Britain. Sir William Rowan Hamilton, well known for his extensive contributions to physics (the Hamiltonian function, the Hamilton-Jacobi equation) and mathematics (complex numbers, vector algebra, group theory, 'quaternions'), was also studying a puzzle. This puzzle involved the nodes of the regular dodecahedron (a solid with twelve regular pentagonal faces). Unlike Euler's problem where each *arc* of a graph must be traversed exactly once, in Hamilton's puzzle one is required to visit each *node* exactly once. A graph is 'Hamiltonian' if and only if it contains a Hamiltonian cycle, that is, a cycle that contains each node exactly once. What Hamilton discovered is that the dodecahedron is Hamiltonian, that is, there is a path from node to node that returns to the starting point after visiting each node exactly once. Hamilton considered finding this path so challenging that he attempted to market the puzzle. The determination of necessary and sufficient conditions for a graph to be Hamiltonian turned out to be even more challenging; it remains one of the major unsolved problems in graph theory.

'Directed graphs' or 'digraphs' are just like graphs, only the arcs have an associated direction (which we depict with an arrow). In Figure 12.2 we see all possible types of digraphs with one or two nodes. Nodes in digraphs have in-degree and out-degree (also called fan-in and fan-out), and two nodes may

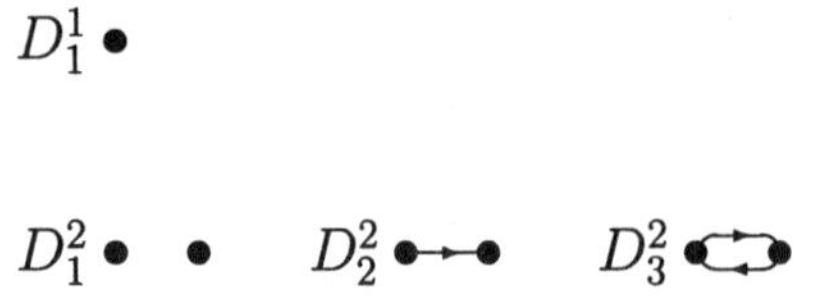

$$D_1^1 \bullet$$

$$D_1^2 \bullet \quad \bullet \qquad D_2^2 \bullet\!\!-\!\!\bullet \qquad D_3^2 \; \text{⊂⊃}$$

Figure 12.2: All digraphs with up to two nodes.

be connected by two directed arcs in opposite directions. Paths and cycles must traverse arcs in the direction of the arrows.

A digraph is a 'forest' if it does not contain a cycle. For forests the direction of the arrow dictates a certain priority, so rather than saying that connecting nodes are adjacent, we speak of parent and child nodes. The pre-arc node is the parent while the post-arc one is the child. All nodes that can be reached from a given node are called descendants and the original node the ancestor. Since forests do not have cycles we never have to worry about a node being its own ancestor. A forest that has a single ancestor node (the 'root') from which all other nodes descend is called a tree. Actually it is easier to think of the tree as being the basic graph and the forest as being a collection of trees.

Digraphs are the basis of a computational model used extensively in DSP called the flow graph (or 'flow diagram', 'dataflow network', 'DSP block diagram', 'graphical flow programming', 'visual programming language', etc.). The directed arcs of the digraph represent signals while the nodes stand for processing subsystems. If the digraph is a forest we say that the system is a *feedforward system*, while digraphs with cycles are called *feedback systems*. The study of flow graphs will be the subject of the next section.

EXERCISES

12.1.1 Special types of graphs are used in electronics (schematic diagrams), physics (Feynman diagrams), computer science (search trees), and many other fields. Research and explain at least three such uses.

12.1.2 Why isn't ⊠ on the list of graphs with four nodes? What about ⋈ ?

12.1.3 How many different kinds of graphs are there for five nodes?

12.1.4 Draw all digraphs for 3 nodes.

12.1.5 Explain Euler's rule intuitively.

12.1.6 A graph is called nonplanar if it cannot be drawn on a piece of paper without arcs crossing each other. Draw a nonplanar graph.

12.1.7 An Euler path is similar to an Euler cycle except that one needn't return to the same node. Similarly we can define a Hamiltonian path. Draw Euler and Hamilton path for points on a two-dimensional grid. Which paths are cycles? Find Euler paths that are not Hamilton paths and vice versa.

12.1.8 A *trellis* is a digraph created by mapping possible *transitions* between N *states* as a function of time. Conventionally the time axis is from left to right and the N states are drawn vertically. There is an arc between each state at time t and a several possible states at time $t + 1$. Assume a trellis with four states (called 0, 1, 2 and 3) with states 0 and 1 at time n being able to transition to even states at time $n + 1$, while 2 and 3 can only transition to odd states. Draw this trellis from time $n = 0$ through 4. How many different trellises of length L are there? How may a trellis be stored in a file? What data structure may be used in a program?

12.2 DSP Flow Graphs

Superficially DSP flow graphs look similar to the block diagrams used to describe algorithms in computer science. In computer science the arcs indicate control paths, and computation is performed or decisions taken at nodes. Depending on decisions taken the processing will continue down different arcs to different computational nodes. Thus by the use of this graphical technique we can capture the control structure of computer programs. Other aspects of the program, such as data types and memory requirements, are not captured in these diagrams, and must be documented in some other way.

The metaphor behind the use of DSP flow graphs is that of signals 'flowing' between processing subsystems. At each of the nodes input signals are processed to produce output signals that are passed along arcs to the following nodes. Thus DSP flow graphs capture both the signal and processing system aspects of a problem. Since the vertices contain processing elements that we must identify, we will have to enrich the graphic notation of the previous section.

Let's see how to make DSP flow graphs. When a signal x is the input to some system we will depict this

and similarly we depict y as the output of a system in the following way.

Thus the identity system, which leaves a signal x unchanged, is depicted

$$x \longrightarrow y$$

which means precisely $y = x$. A *hidden* signal, that is, a signal that is neither a system input nor output, can be named by placing a symbol near the corresponding arc. All the above are standard digraph figures.

As we mentioned before, nodes correspond to processing, which must be identified. We do this by drawing circles (for simple common processes) or squares (for more general processes). For example, $y = f(x)$ is depicted

and $z = g(y) = g\left(f(x)\right)$ is shown

where the hidden signal y has been identified.

A very common operation is to multiply a signal by a real number. The standard digraph would have a multiplication node perform this function, i.e., we would expect $y = Gx$ to be depicted something like

but since the operation is so common, we introduce a short-hand notation.

$$x \xrightarrow{G} y$$

Whenever a symbol appears near an arrow we understand an implicit multiplication node $y = Gx$. Do not confuse this with the symbol representing a hidden signal that is placed close to an arc but not near an arrow. In such cases no multiplication is intended, and eliminating this symbol would not change the system's operation at all.

Many times we wish to have the signal x reach more than one processing system. In regular digraphs each arc connects a single pre-arc node with a unique post-arc node; in DSP we allow the connection of a single pre-arc node to multiple post-arc nodes with a single arc. The point where the signal splits into two is called a *branch point* or a *tee connector*.

This means that the same signal is delivered to both nodes, $y_1 = y_2 = x$.

Of course it is meaningless to connect more than one pre-arc node together; but we *can* add two signals x_1 and x_2 to get $y = x_1 + x_2$. This is depicted in the standard notation using an addition node.

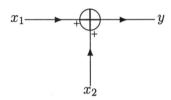

The small + signs mean that addition is to be performed. Subtraction $y = x_1 - x_2$ is depicted

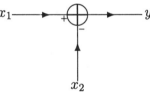

and other combinations of signs are possible.

Value-by-value multiplication $y = x_1 x_2$ is depicted as you would expect.

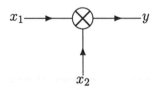

We can combine these basic elements in many ways. For example, the basic $N = 2$ DFT of equation (4.33) is depicted

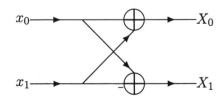

and signifies $X_0 = x_0 + x_1$ and $X_1 = x_0 - x_1$.

One of the most important processing nodes is the unit delay, which is depicted by z^{-1} inscribed in a circle.

This diagram means that the *signal* y is the same as the *signal* x delayed one digital unit of time, that is, $y_n = x_{n-1}$ for all n. Often we loosely think of the signal value at some time n as entering the delay, and its previous value exiting. Since we only represent time-invariant systems with flow graphs this interpretation is acceptable, as long as it is remembered that this same operation is performed for every unit of time. Also note that we shall never see a z in a signal flow diagram. We represent only causal, realizable systems.

Using the unit delay we can easily represent the simple difference approximation to the derivative $y_n = \Delta x_n = x_n - x_{n-1}$

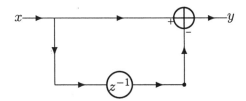

or a general single delay convolution $y_n = a_0 x_n + a_1 x_{n-1}$.

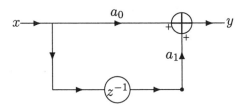

You will notice that we have drawn a small filled circle in each of these diagrams. This circle does not represent a processing system, rather it is a reminder that a memory location must be set aside to store a signal value. In order to continuously calculate the simple difference or single delay convolution, we must store x_n at every time n so that at the next time step it will be available as x_{n-1}. We do not usually explicitly mark these memory locations, only stressing them when the additional emphasis is desired.

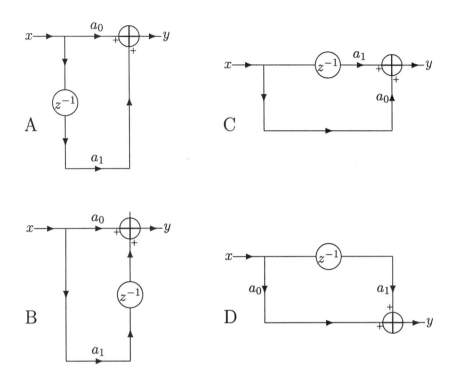

Figure 12.3: Four other ways of drawing a basic MA (FIR) block as a DSP flow graph.

The beauty of using graphs is that we can redraw them any way we consider æsthetic or useful, as long as the topology stays the same. Thus, the basic single delay FIR block also can be drawn in the ways depicted in Figure 12.3 and in many other ways.

All the different DSP graphs that depict the same DSP process are called *implementations* of this process. Note that the implementation of Figure 12.3.A is topologically identical to the previous graph, but has the gains appearing more symmetrically. In Figure 12.3.B we interchanged the order of the gain and the delay. Thus this implementation is not identical from the pure graph-theoretic point of view, but is an identical DSP process since the gain and delay operators commute. Figure 12.3.C looks different but one can easily convince oneself that it too represents the same block. Figure 12.3.D is similar to Figure 12.3.C but with the gains positioned symmetrically.

How can we implement the more general convolution?

$$y_n = \sum_{l=0}^{L} a_l x_{n-l}$$

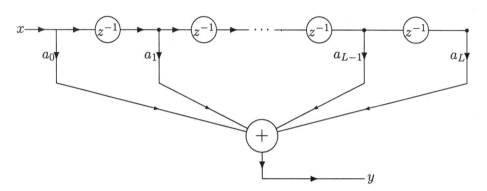

Figure 12.4: A straightforward implementation of the FIR filter using an adder of in-degree $L + 1$.

In Figure 12.4 we see a straightforward implementation, where the large node at the bottom is an adder of in-degree $L + 1$. Such an adder is not always available, and is not really required, since we can also implement the FIR filter using the standard two-input adder, as in Figure 12.5.

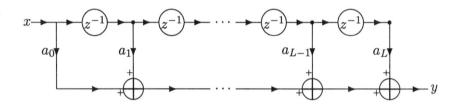

Figure 12.5: A straightforward implementation of the FIR filter using standard two-input adders.

This figure is worth studying. To assure yourself that you understand it completely, mark all the vertical arcs, especially those marked with filled circles. For example, the arc that descends after the first delay and then is summed with $a_0 x_n$ splits should be marked x_{n-1}. By this we mean the signal that for all n is equal to the incoming signal x delayed by one time unit. When you consider a complex DSP graph of this type it is worthwhile interpreting it in two stages. First think of analog signals flowing through the graph. In order to assist in this interpretation assume that every processing node corresponds to a separate hardware component. Ignore any filled circles and treat z^{-1} nodes as time delays that happen to correspond to the sampling interval t_s.

Once you understand the graph at this level you can return to the world of DSP programming. The delays are now single sample delays, the processing nodes are computations that may all be carried out by the same processor, and the filled circles are memory locations. When thinking in this mode we often think of 'typical values', such as x_n and x_{n-1}, rather than entire signals such as x. We implicitly allow the same computation to be carried out over and over by a single processor. The basic computation to be performed repeatedly consists of multiplication of a delayed input value by a filter coefficient and adding; this combination is called a **Multiply-And-Accumulate (MAC)** operation.

Looking closely at Figure 12.5 we see that this FIR implementation is based on the block from Figure 12.3.D. Several of these blocks are concatenated in order to form the entire convolution. This is a widely used technique—after perfecting an implementation we replicate it and use it again and again. You can think of an implementation as being similar in this regard to a subroutine in standard procedural programming languages.

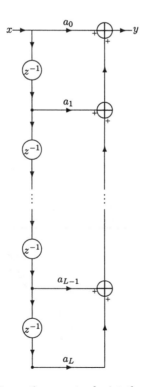

Figure 12.6: An alternative way to depict the all-zero (MA) filter.

Of course this is not the only way to draw an FIR filter. A particularly interesting way to depict the same graph is shown in Figure 12.6. In this implementation we replicate the FIR block of Figure 12.3.A. It is easy to see that this graph is topologically identical to the previous one.

Up to now we have only seen graphs without cycles, graphs that correspond to feedforward systems. The simple feedback system, $y_n = by_{n-1} + x_n$, is depicted as

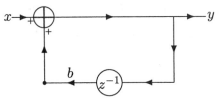

while a full all-pole system

$$y_n = x_n - \sum_{m=1}^{M} b_m y_{n-m}$$

can be depicted as in Figure 12.7.

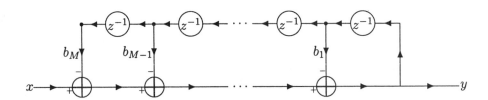

Figure 12.7: A full all-pole filter implemented using MAC operations.

Once again it is worthwhile to carefully mark all the arcs to be sure that you understand how this implementation works. Don't be concerned that signal values are transported *backward* in time and then influence their own values like characters in science fiction time-travel stories. This is precisely the purpose of using feedback (remember Section 7.4).

Of course this is not the only way to draw this AR filter. A particularly interesting implementation is depicted in Figure 12.8.A. We purposely made this implementation a mirror reflection of the FIR implementation of Figure 12.6. Now by concatenating the MA and AR portions we can at last implement

$$y_n = \sum_{l=0}^{L} a_l x_{n-l} - \sum_{m=1}^{M} b_m y_{n-m}$$

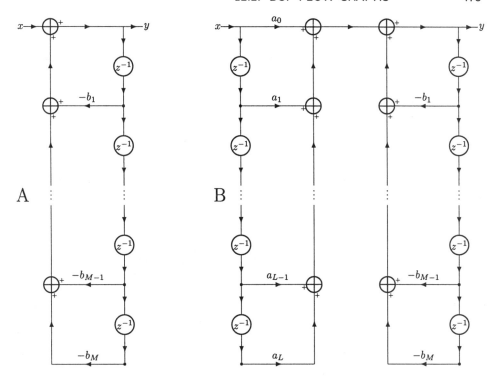

Figure 12.8: In (A) we present an alternative graphical representation of the all-pole (AR) filter. (B) is an implementation of the full ARMA filter.

the full ARMA filter. We do this in Figure 12.8.B.

Take a few moments to appreciate this diagram. First, by its very construction, it graphically demonstrates how ARMA filters can be decomposed into separate MA and AR subsystems. These FIR and all-pole systems are seen to be quite different in character. Of course the order of the AR and MA subsystems are not necessarily equal; they only look the same here since an ellipsis can hide different heights. Second, it's pleasingly symmetric; there *are* computational commonalities between the subsystems. Thus computational hardware or software designed for one may be modified to compute the other as well. Finally, this diagram gives us a novel way of understanding filters. Looking at the analog level we cannot avoid imagining the signal flowing in on the left, traveling down, splitting up, and recombining with differently weighted and delayed versions of itself. It then enters the feedback portion where it loops around endlessly, each time delayed, weighted, and combined with itself, until it finally exits at the right.

EXERCISES

12.2.1 The following examples demonstrate simplification of diagrams with gains. In all cases identify the gain(s) that appear in the right diagram in terms of those that appear in the left diagram.

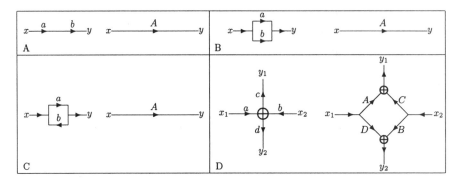

12.2.2 Draw an ARMA filter with MA order 3 and AR order 4. How many memory locations are required? Label all the signals.

12.2.3 A filter implementation with minimal number of memory allocations is called *canonical*. What is the number of memory locations in the canonical ARMA filter with MA order p and AR order q?

12.2.4 The *transposition theorem* states that reversing all the arc directions, changing adders to tee connections and vice-versa, and interchanging the input and output, does not alter the system's transfer function. Prove this theorem for the simple case of $y_n = x_n + by_{n-1}$.

12.3   DSP Graph Manipulation

Let's summarize all that we have learned so far. Flow graphs are used to represent realizable, time-invariant signal processing systems. The most important graphic elements are depicted in Figure 12.9. Combining these basic elements in various ways we can depict many different systems. Every DSP flow graph corresponds to a unique signal processing system, but every system can be implemented by many seemingly different graphs.

Different implementations may have somewhat different characteristics, and may correspond to hardware implementations of differing cost and software implementations of varying complexity. It is thus useful to learn ways of manipulating flow graphs, that is, to change the graph without changing

 x————————————	x is the input to a calculation
 ——————————y	y is the output from a calculation
 x——————————y	$y = x$
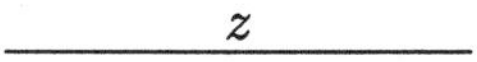 ——————z——————	the hidden signal z
 ————————•————	a memory location
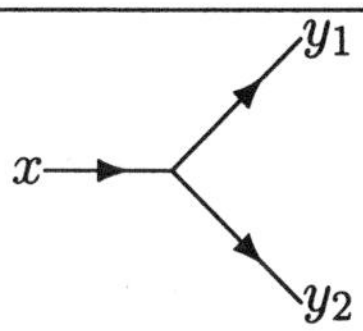 x splitting to y_1 and y_2	splitting a signal $y_1 = x$ and $y_2 = x$
 x———G———y	a gain $y = Gx$
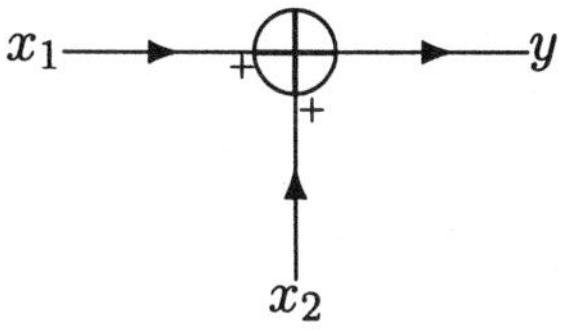 x_1, x_2 summed to y	adding signals $y = x_1 + x_2$
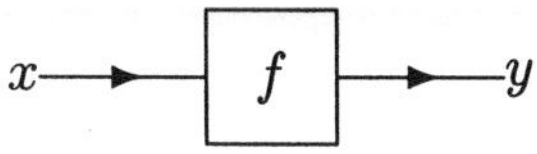 x——$\boxed{f}$——y	an arbitrary system $y = f(x)$

Figure 12.9: The most important DSP graph elements.

the system implemented. Remember that a system is defined only by the outputs generated for all inputs. As long as these remain unchanged the system is unchanged, no matter what the flow diagram looks like. We will often call graph operations that leave the system unchanged 'symmetries'.

The first symmetry, which we have already stressed, is that graphs are to be understood topologically. Geometric quantities such as arc length, angles, and such are irrelevant. We can even perform mirror reflections drawing the whole picture backward as long as all the arrows are reversed (but please print the alphanumeric characters in the conventional orientation).

Since the topology remains unchanged one can always move a gain along an arc to a convenient place. You will doubtless recall that we did this when redrawing the basic FIR block. More generally, you can move a gain along an arc until the first addition or nonlinear system as long as you replicate it at tee connectors. You can also combine consecutive gains into a single gain or split a single gain into two in series.

These operations are special cases of the 'like signal merging' symmetry. Whenever we find identical hidden signals in two different places, we can consolidate the graph, eliminating extraneous arcs and nodes, as long as no input or output signals are affected. For example, consider the graph

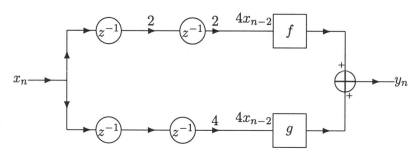

where we have identified the signal $4x_{n-1}$ on two different arcs. We can consolidate everything between the input and these two signals.

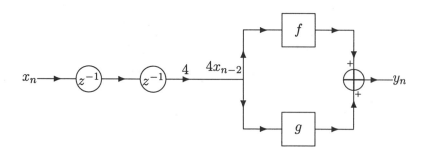

Two signal processing systems f and g are said to 'commute' if one can interchange their order without changing the overall system.

In particular any two linear systems, and thus any two filters commute. We often exploit commutation of filters to simplify DSP flow graphs.

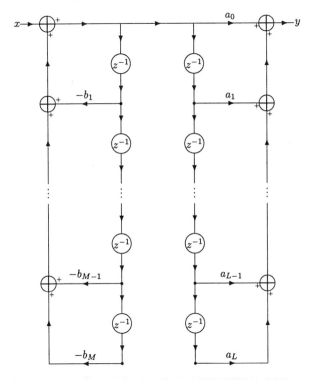

Figure 12.10: An alternative way to depict the ARMA filter. Here we perform the autoregressive (all-pole) filter before the moving average (FIR) filter.

As an example, let's simplify Figure 12.8.B for the full ARMA filter. Note that there we actually performed the MA portion before the AR, which would make this an MAAR filter (were this possible to pronounce). Since the MA and AR subsystems are filters and thus commute we can place the AR portion before the MA part, and obtain Figure 12.10.

This diagram is just as symmetric as the first but seems to portray a different story. The signal first enters the infinite loop, cycling around and around inside the left subsystem, and only the signal that manages to leak

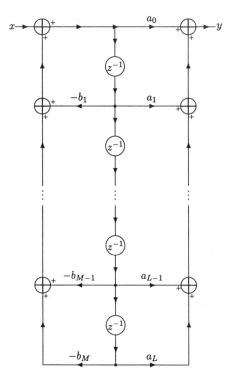

Figure 12.11: Yet another way to depict the ARMA filter. This graph requires the minimal number of internal memory locations.

out proceeds to the more straightforward convolution subsystem. Once again we suggest taking the time to label the arcs, for example, calling the output of the all-pole subsystem w. Comparing arcs we discover a common hidden signal and can consolidate to obtain the more efficient graph of Figure 12.11.

The final symmetry we will discuss is that of grouping and ungrouping in *hierarchical flow graphs*. Up to now we have seen graphs made up of primitive processes such as delays, gains, and additions. Although we spoke of using rectangles to represent general systems, we have not yet discussed how *these* subsystems are to be specified. One way would be to describe them ad hoc as algorithms in pseudocode or some programming language; but a more consistent description would be in terms of flow graphs! For example, once we have presented the flow graph for a general ARMA filter, we can represent it from then on as a single rectangle, or 'black box'. The processes of grouping elements of a flow graph together to form a new subsystem, and of ungrouping a black box to its lower-level components, are symmetries.

The grouping into higher-level subsystems results in a simplification similar to that of using subroutines in programming languages. Of course these new black boxes can be used in turn to recursively build up yet more complex subsystems. Such a system, built up from various levels of subsystems that can be graphically decomposed into simpler subsystems, is called a hierarchical flow graph. Recursively applying the ungrouping symmetries reduces hierarchical graphs to graphs composed solely of primitive processes.

EXERCISES

12.3.1 Draw the basic all-pole block in four different ways.

12.3.2 Recall that the main entity of the state-space description is the system's internal state. One way to encode the internal state involves specifying hidden signals. Which hidden signals are required to be identified? How does the state thus specified evolve with time?

12.3.3 Give an example of two systems that do *not* commute.

12.3.4 Draw high-level hierarchical flow graphs for the following systems, and then decompose the high-level description into primitives.

- a filter composed of several copies of the same FIR filter in series
- a band-pass filter that mixes a signal down to zero frequency and then low-pass filters it
- a circuit that computes the instantaneous frequency

12.4 RAX Externals

In the early eighties the author was working on signal analysis in a sophisticated signal processing lab. This lab, like most at that time, was composed largely of complex analog signal processing equipment mounted vertically in 19-inch racks. Each rack would typically house between five and ten different pieces of equipment, including function generators, amplifiers, filters, precision synthesizers, and oscilloscopes. Each piece of equipment conventionally had buttons and knobs on its front panel, and input and output connections on its back. These back panel connections would be routed to *patch panels* where the users could rapidly connect them up.

The lab had several analysis stations, and a typical station consisted of two or three racks full of complex and expensive equipment. Each individual piece of equipment could cost tens to hundreds of thousands of dollars, would have to be calibrated and serviced regularly, and would usually take about two to three days to master. Just mounting a new box would take several hours, including placing it onto slides, screwing the slides into the racks, routing all the cables from its back panel to the patch panels, testing these cables (which would always seem to fail), and properly labeling the patch panel connectors. While the veteran lab staff could set up quite complex signal processing functions in minutes, someone new to the lab would go through a learning process of several months before feeling confident enough to work alone.

This lab was considered both modern and efficient. Outdated equipment was continually replaced with the most modern and sophisticated available; the lab staff was the most competent that could be found. However, trouble was definitely on the horizon. Maintenance costs were skyrocketing, the training of new lab staff was getting harder and lengthier, and even the most sophisticated equipment was not always sufficient for all the new challenges the lab faced. For these reasons we embarked on the development of an experimental software system.

The system was originally called *RACKS*, supposedly an acronym for **R**eplace **A**nalog **C**omponents with **K**nowledge-based *S*oftware, but actually referring to the *racks* of equipment the system emulated. The name was later shortened to RAX as an acronym for **R**eally **A**wesome bo**X**es. RAX was a visual programming environment that simulated the operation of an analysis station. Using a pointing device (originally a joystick, but you can think of it as a mouse if you prefer), equipment could be instantly taken out of a virtual store room, placed into virtual racks, connected by virtual cables, and operated. The operation of RAX was not always real-time, but it enabled useful analyses to be easily performed.

RAX was never commercially available, and is hardly state-of-the-art, not having been updated since its initial development. It was quite limited, for example, not allowing hierarchical definition of blocks. It was also not very run-time efficient, generally passing single samples between blocks. The host computer and DSP cards used as the platform for RAX are by now museum pieces. However, I have several reasons for expounding on it here. First and foremost is my own familiarity with its internals; many of the issues that we will examine are quite general, and I can discuss them with maximum knowledge regarding RAX. Second, RAX is relatively simple and thus easy to grasp, but at the same time general and easily extensible. Third, I promised my coworkers that one day I would finish the documentation, and better fifteen years late than never.

The RAX model of the world is that of *racks* of equipment, which are vertical rectangular arrays. These vertical arrays are called *racks*; racks are made up of *slots*, and each slot can hold a single piece of *equipment*. The piece of equipment in a specific rack and slot position is called a *box*, while what a type of equipment does is called its *function*. Each box has *input connectors*, *output connectors*, and *buttons*. *Cables* can be connected between a single output connector and any number of input connectors. Buttons are used to set internal parameters of the different boxes (e.g., input filenames and amplification gains).

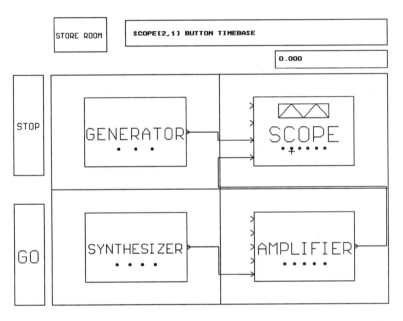

Figure 12.12: The graphics screen of RAX for a simple setup. Note that this figure's resolution and fonts are representative of 1983 computer graphic displays (the original had 16 colors).

An example of a working RAX system is depicted in Figure 12.12. The resolution and fonts are representative of the technology of graphic displays circa 1985; the original screens had up to 16 colors, which are not observable here. In the figure we see a function generator, a synthesizer, an amplifier, and a scope connected in a frivolous way. The function generator is a box with no inputs and a single output. This output can be a square, triangular, or sawtooth wave, with buttons to select the signal type and control the amplitude and frequency. The synthesizer generates (real or complex) sinusoids of given amplitude, frequency, and phase. The amplifier (which would never actually be used in this way since the synthesizer has adjustable amplitude) can take up to five inputs. Its output is the sum of these inputs, each multiplied by its own gain (selected by appropriate buttons). The scope has input channels for one or two signals to be displayed. Its other inputs are for external trigger and clock. The scope has no output, but buttons that set the volts per division, timebase, trigger mode and level, clock rate, number of sweeps. The scope also has a display, called CRT for **C**athode **R**ay **T**ube. The output of the function generator is connected to channel B of the scope, while the synthesizer feeds the amplifier which in turn is connected to channel A of the scope. A sample scope display from this setup is depicted in Figure 12.13.

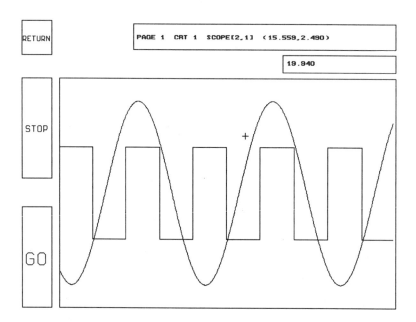

Figure 12.13: The graphics screen of the scope for the simple RAX setup depicted in the previous figure. The small + represents the position of the pointing device, and this position is indicated in the message window.

Although completely general, for reasons of efficiency RAX boxes are usually relatively high-level functions. For example, an FIR filter would be hand coded and called as a box, and not built up from individual multipliers and adders. It is relatively easy to add new functions as required; one need only code them using special conventions and link them into RAX. Computationally demanding functions may actually run on DSP processors, if they have been coded to exploit such processors and the processors are available.

RAX can be integrated into the real world in two ways. For non-real-time use there are 'input file' and 'output file' boxes. These boxes are stream-oriented, reading and writing as required. The input file box has one output and a button with which the user specifies the name of the file to read. Similarly, the output file box has a single input and a button to specify the file to write. For simple processing, or when DSP processors are used, there are also A/D and D/A boxes that stream to and from the true devices.

In RAX the same piece of equipment can be placed into many different rack-slot positions, which is interpreted as different boxes that happen to have the function. Boxes are identified by giving the function, the rack number, and the slot number (e.g., SCOPE[2,1]). Connectors and buttons have their own notations. When the pointing device enters a rack-slot that houses a piece of equipment its identifier is displayed in the message area at the upper right. When the pointer is close enough to a button or connector, its identifier is displayed as well. Pressing the pointer's actuator (similar to clicking a mouse) over a button causes a pop-up menu to appear where the user can edit the corresponding parameter. Pressing the actuator near a connector causes a 'rubber band line' to be drawn from that connector to the pointer, which can then be placed near another connector and pressed again. If the connection is valid the rubber band line disappears and in its place a connection route is drawn. Valid connections connect a single output to any number of inputs. The connection route is drawn to avoid existing routes, and is color coded for optimal distinguishability.

After bringing up the application, the user specifies the number of racks and the number of slots per rack. These numbers can be changed at any time, with the restriction that no mounted equipment should fall onto the floor. Next the user opens the store room and drags pieces of equipment from there, placing them into rack-slots. The user can then connect output connectors to input connectors and set parameters using the buttons. When satisfied the user points and depresses the run button. At that point time starts to run and the signal processing begins. The user may at any time select any display (e.g., from a scope, spectrum analyzer, or voltmeter) and view the graphic results. When such a display is active, the message area continuously displays the pointer's coördinates, for example, volts and time for the scope. To return to the racks display the user can then press the return button, and equipment buttons can be adjusted while time is running. To stop time from running there is a stop button. If the user considers the setup to be useful it can be saved to a *netlist* file, from which it can be loaded at some later date. The netlist for the simple setup of Figure 12.12 is printed out as Figure 12.14.

```
SAMPLING FREQUENCY =   100.00000
RACK  1
  SLOT  1  GENERATOR
                              FUNC=SQUARE
                              FREQ=1
                              VOLT=1
                              OUT>SCOPE[2,1].CHANB
  SLOT  2  SYNTHESIZER
                              FREQ=1
                              VOLT=1
                              PHASE=1
                              COMPL=REAL
                              OUT>AMPLIFIER[2,2].IN
RACK  2
  SLOT  1  SCOPE
                              VOLTDIV=1
                              TIMEBASE=1
                              TRIGMODE=FREE
                              CLOCK=0
                              SWEEPS=1
                              TRIGLEV=0
                              CHANA<AMPLIFIER[2,2].OUT
                              CHANB<GENERATOR[1,1].OUT
  SLOT  2  AMPLIFIER
                              GAIN=1
                              GAIN2=0
                              GAIN3=0
                              GAIN4=0
                              GAIN5=0
                              IN<SYNTHESIZER[1,2].OUT
                              OUT>SCOPE[2,1].CHANA
```

Figure 12.14: The *netlist* of the simple example.

EXERCISES

12.4.1 RAX is a 'clock-driven' system, meaning that some external concept of time causes the scheduler to operate. Alternatives include 'data-driven' and 'control-driven' systems. In the former, external inputs are the trigger for everything to happen; each input is followed through causing box after box to operate in turn. In the latter, whenever a box cannot run due to an input not being ready, the box connected to it is run in order to generate that input (and recursively all boxes before it). Discuss the advantages and disadvantages of these three techniques.

12.4.2 Observe in Figure 12.12 that different cables never overlap, at most they cross at a point. Give a simple algorithm for accomplishing this.

12.4.3 Find some visual programming language to experiment with. How is it similar to and how is it different from RAX?

12.4.4 What are RAX's main functional deficiencies?

12.5   RAX Internals

Now let's start to peek behind the scenes to see how RAX accomplishes its magic. The first thing we must explain is that RAX is an interpreter rather than a compiler. The entire RAX run-time system must be present for anything to happen and the GUI, IO, signal display, task scheduling, and processing are all supplied by RAX itself. Modern systems will usually allow the user to operate in interpreted mode in order to debug the system and then to compile to some standalone language such as C or DSP assembly language. This point understood, let us proceed to the program's structure.

The main program looks like this in pseudocode:

```
initializations
main loop
      handle user events
      update graphics
      if RunMode
            schedule tasks
            increment time
finalizations
```

where initializations and finalizations refer to the opening and closing of files, allocation and deallocation of memory, starting and terminating the graphics environment, and other mundane computer tasks. Handling user events refers to checking for motion of the pointing device and updating the message area accordingly; and checking for pointer device button presses or keyboard entry and the corresponding changing of parameters and program modes. RunMode is true whenever the user has pressed the START button, and stays true until the STOP button is pressed.

In the analog world every box is operating all the time. This has to be emulated in RAX since the main and all the boxes run on a single CPU.

This is the responsibility of the *task scheduler*. The scheduler runs through every piece of equipment in the racks, and decides whether it is ready to run. This decision is based on continuity of time and an assumption of causality. Each box remembers when it last ran, and each input to each box contains a time-stamped value. A box can run only when the present time is strictly after the time it last ran, and only when the time is not before the time of its inputs. Assuming a box can run, the scheduler is responsible for loading the box state, connecting input and output cables, calling the proper task (perhaps running on a DSP), and storing the new state. These duties determine the *context switch time*, the minimum time it takes to switch from running one box to another. RAX was somewhat wasteful in this regard, having originally been designed for simulation purposes and only later being retrofitted with DSP boards for real-time use. An alternative strategy (one that was employed for the DSP code) is for each box to keep its state information internally. This cannot be done using static arrays for host code since one equipment type can be used multiple times in a single setup.

Finally, when all boxes have run for the specified time, the time is incremented according to the present sampling rate. One of the major limitations of RAX is the use of a single sampling rate. Although the sampling rate can be changed, there cannot be simultaneously more than one rate, and all boxes must use the same clock. This is both an efficiency problem (some processes might only need updating very infrequently, but must be scheduled every time) and a real constraint (resampling processes, such as those required for modems, cannot be implemented). This problem could be fixed by simulating real time using a common multiple of all desired sampling rates and dividing as required. Hardware systems commonly implement this same policy by using a high-frequency crystal oscillator and various frequency dividers.

Behind the simple description of the handling of user events and scheduler are a plethora of infrastructure functions. For example, there is a function that given the rack and slot numbers determines whether the rack-slot is occupied, and if so retrieves the type of equipment. Another, given a type of equipment, finds the meaning of the inputs or buttons. Given a cable identifier, the output number, origin rack-slot that feeds it, and all inputs it feeds can be found. Given an entire configuration, the minimum number of racks and slots may be calculated. Of course there are functions to place a piece of equipment in a given position, to remove a piece of equipment, to connect a cable between connectors, etc.

Behind every graphics display (scope, spectrum analyzer, etc.) there is a 'display list' that contains all lines drawn to that display. Whenever the display is to be shown the display list lines are translated into screen coördinates and plotted. Every motion of the pointer device in such a screen requires translation from screen coördinates back to world coördinates.

Now let's discuss how rack-slots are populated and how equipment is described. Every box placed in a rack-slot position is assigned a unique identification number, starting from 1 and reaching the total number of occupied positions. This identifier is used by the scheduler, which loops from 1 to the total number of boxes in its main loop. There is a vector of this length that holds the position and an equipment pointer, this pointer in turn bringing us to the equipment type, state, parameters, cabling, time last run, processor (0 for host, otherwise DSP number), and equipment function.

How is this function defined? When we wish to add a new function we must define a constructor that returns the allocated and initialized state, buttons, inputs, and outputs; and a destructor that undoes all of the above. Then we write the *run* routine. This routine takes the state, buttons, and inputs, and returns the updated state, and outputs (and possibly updates the graphic display). All run routines tend to have the same form. First the buttons are read. Then the inputs are checked for correctness and type (since boxes react differently depending on type, for example, amplification of a complex input returns a complex output, while a real returns a real). One time step is then performed, generating values that are placed into the appropriate outputs. If there is a graphics display, its display list is updated. Finally, the state and time variables are updated, and control is returned to the scheduler.

Functions that run on DSP processors are built slightly differently. These functions store their state locally and are not as tightly controlled as their native counterparts. For them the constructor consists of downloading the object code to the appropriate processor and noting this fact. The run routine on the host simply passes the inputs to the processor through shared memory (SHAM) and collects the outputs. The DSP code is written as an infinite loop that checks for the appearance of new inputs in the SHAM, processes, and copies the outputs to the SHAM.

How are all the disjoint entities coupled to make a single coherent system? This is a general problem in systems with many functional parts, and there are in general four possibilities. These possibilities are usually known as compile-time, link-time, download-time, and run-time bonding. The simplest method is to gather all the program code for all the different kinds of equipment together into a single file, and compile this file together. In such an implementation there can be one global constructor and destructor procedure, which allocates and deallocates all memory required for all boxes. Such compile-time bonding results in very fast run-time code, but is extremely inefficient from the memory utilization point of view. The object code for every sort of equipment is present even if we need only a few boxes, although in modern paging virtual memory systems this may not actually impact performance. A more serious design flaw is the complexity and lack of flexibility of the code.

The next possibility is link-time bonding. Here each function is defined in a separate file that is separately compiled into an object file. The linker is responsible for bonding all the object files together. Simple link-time bonding may still waste memory for unused functions, but with proper operating system support the unused functions may take up executable file size but not actually sit in run-time memory. Also, the reduction of the interdependence of the different functions reduces system complexity by forcing object-oriented techniques.

Download-time bonding involves compiling and linking each function into a separately executable program. When the system is run the control system selects which function programs need to be downloaded and launched, and then either the control system, the operating system, or hardware are responsible for moving data between these programs. For example, the data may be passed between these programs using 'interprocess communications' or 'sockets' or 'pipes', or each program may run on separate DSP processors with hardware communications links between them.

The most complex and most memory-efficient form of bonding is run-time, also known as dynamic allocation. Like download-time bonding, each function is a separate program unit. However, functions may be loaded and launched during the running of the system. This is a particularly useful feature when the functioning of the system depends on the input signal. For example, consider a voicemail system that must decode DTMF tones, compress and store speech, and demodulate, decode, and store facsimile transmissions. Since DTMF tones are used to control the system and may be used at any time, the DTMF decoder must be continuously available. When a session commences the speech function may be loaded by default, but upon detection of fax tones the facsimile function must take its place.

In the original implementation of RAX the host code for each piece of equipment was compiled separately, but the entire program was linked together before execution. This link-time bonding was chosen since the programming system used did not support dynamic allocation for user routines. The DSP code, however, was compiled into individually executable programs, and downloaded upon pressing START. This download-time bonding could be efficiently performed since it only required loading the appropriate DSP code and data, and releasing the processor.

The full algorithm for bonding went something like this. Each function for which there was DSP code was assigned a complexity number between 1 and 10; functions with no DSP code received a zero. For example, the N sample delay had complexity 0, the amplifier was given a 1, while the spectrum analyzer (which had to window, perform FFT, square, average, take logarithms, perform graphics, etc.) rated a 10. In no case did the functions exceed the capabilities of a single DSP processor. Whenever the configuration changed and START was pressed, the functions with nonzero complexity were sorted in descending order of complexity. Assuming there were P available processors, the first P functions would be downloaded to DSPs, while the rest of the functions would run on the host processor.

We have yet to fully explain the cable mechanism. Cables are internally arranged in a array, every cable having one source and any number of sinks. Each element in the cable array consists of three parts, namely a time, a type (boolean, integer, real, or complex), and a value. When a box computes an output value, the time, this value and its type are written into the cable array where they can be read by all those boxes that require it. In RAX only one value can be placed into a given cable at a time; once a value is placed in the cable it remains there until overwritten. As mentioned before, boxes test the time and value on the cable before using the value.

This implementation of cables, which we call *overwrite*, allows multiple boxes to receive a single cable as input, but assumes that boxes always take values they will need, even if they are not yet ready to use them. For example, think of a Fourier analysis box that collects an entire buffer of signal values before calculating an entire spectrum. This box must be activated each time an input value appears, just in order to store away this value into its state. This wastes context switches, potentially slowing the system down. There are several alternative strategies that can be used.

An alternative, called *buffered write*, would be for the cable to collect past values into an ordered list, and for the Fourier analysis box to be called only when the desired number of inputs are all ready. In such an alternative representation we must consider what is to be done about clearing past values that are no longer needed. One possibility is for cables to implement a fixed-length buffer, always holding some number of past values. This is easy to implement, for example, by a circular buffer, but has two problems. First it limits the generality of what a box can do, since no box can use more history than what the cable stores. Second many boxes might only require the present value or a few past ones, and we might be wasting memory by storing the maximum number. Of course we *could* configure each cable differently, choosing the number of values to store according to the maximum needed by boxes to which the cable is connected.

Another possibility is to allow only one box to be fed by a given cable and to make it that box's responsibility to clear unneeded values from the cable. In such an implementation the precable box *writes* to the cable, and the post-cable box *reads* from it deleting the value; thus we call this method 'write-read'. This is similar to the mechanisms of 'pipes' and 'sockets' that are provided by many operating systems, and these mechanisms can be exploited in implementing such cables. In a system with cables that are written and read we could allow several boxes to write to a single cable, but only one box can read a cable since once a value has been read it is no longer available for other boxes to use. This is not really an insurmountable limitation since we could easily create a tee connector equipment type, which takes in a value from one input and makes it available on two or more outputs.

We have still not completely specified what happens in systems with cables of the latter type. One possibility is for the read attempt to fail if the desired values are not ready, or equivalently, to give the reading box the ability to test for readiness before attempting to read from a cable. Such 'write-test-read' systems can seem to act nondeterministically, even without explicit randomness built in. For example, consider a piece of equipment built to merge two inputs into a single output. This equipment checks its inputs until it finds one which is ready and writes its value to the output. It then reads and discards all other inputs that might be simultaneously available. Even if it always sweeps through its inputs in the same order, its output depends on the detailed operation of the scheduler, and thus seems unpredictable. A second possibility is for the reading box to become blocked until the desired input value is ready. Indeed the blocking mechanism can be used as the heart of the system instead of an explicitly encoded scheduler. All boxes in a 'blocked-read' system simply run in parallel, with the unblocked boxes preparing outputs that in time unblock other processes. Finally, the writing process may become blocked until the cable is ready to receive the value to be sent. Although this 'blocked-write' method seems strange at first, it shares with blocked-read the advantage of automatically synchronizing truly parallel processes.

Since RAX was an 'overwrite' system, DSP processors could only be employed for their relative speed as compared with the host processor available at that time. The potential for parallel processing could not be exploited since the scheduler was responsible for sending data to each processor, waiting for it to complete its computation, and then collecting its output. In order to allow the processors to truly run in parallel some method of synchronization, either that inherent in blocked-read and blocked-write or an explicit interprocess communications method, must be employed. One model that has been exploited for the parallelization of DSP tasks is Hoare's communicating sequential processes. In this model a collection of computational processes, each of which separately runs sequentially, run truly in parallel and communicate via unidirectional blocked-write channels.

This completes our description of RAX internals. While some of the details are specific to this system, many of the concepts are applicable to any visual programming system. When using such a system for simple tasks the analogy with analog equipment is enough to get you started, but for more complex problems a basic understanding of the internals may mean the difference between success and frustration.

EXERCISES

12.5.1 Write a package to implement graphics display lists as singly linked lists of commands. At minimum there must be MOVE (x,y) and DRAW (x,y) commands, while more elaborate implementations will have other opcodes such as SETCOLOR c, DRAWRECTANGLE (left, right, bottom, top), and CIRCLE (x,y,r). Remember to include routines to construct a new display list, clear the display list (as, for example, when a scope retriggers), add a command to the display list (take into account that the display may or not be currently showing), show the display on screen, translate between real world coördinates and integer screen coördinates, and free the list.

12.5.2 Write a package to handle netlist files. You will need at least one routine to read a netlist file and translate it into an internal representation of boxes and parameters and one routine to write a netlist file.

12.5.3 Write a RAX-like GUI.

12.5.4 Write a RAX-like scheduler.

12.5.5 Implement a basic RAX system for functions that all run on the host. You should supply at least a sine wave generator, a filter, and a scope.

Bibliographical Notes

A good general reference on graph theory is [91], which is a newer version of a classic text.

The present author has not found any mention of the use of flow graphs for signal processing before the 1953 article of Mason [160, 161].

One of the earliest uses of visual programming in DSP was BLODI (the **BLO**ck **DI**agram compiler), which was developed Bell Labs in late 1960. Although without a true graphic interface, it was said to be easier to learn than FORTRAN, and at times easier to use even for the experienced programmer. BLODI had blocks for IO, signal and noise generation, arithmetic operations between signals, delay and FIR filtering, sampling and quantization, and even a flip-flop. Other documented flow languages for signal processing include SIGNAL [88] and LUSTRE [89]. Probably the most popular visual programming environment for signal processing is what was once called BOSS (**B**lock-**O**riented **S**ystem **S**imulator) but was later renamed SPW (**S**ignal **P**rocessing **W**orkSystem).

Hoare's communicating sequential processes, presented in [103], also motivated several DSP systems. For a good discussion of implementational issues for data-flow-oriented languages consult [2].

13

Spectral Analysis

It is easy enough to measure the frequency of a clean sinusoid, assuming that we have seen enough of the signal for its frequency to be determinable. For more complex signals the whole concept of frequency becomes more complex. We previously saw two distinct meanings, the spectrum and the instantaneous frequency. The concept of spectrum extends the single frequency of the sinusoid to a simultaneous combination of many frequencies for a general signal; as we saw in Section 4.5 the power spectral density (PSD) defines how much each frequency contributes to the overall signal. Instantaneous frequency takes the alternative approach of assuming only one frequency at any one time, but allowing this frequency to vary rapidly. The tools that enable us to numerically determine the instantaneous frequency are the Hilbert transform and the differentiation filter.

There is yet a third definition about which we have not spoken until now. Model based spectral estimation methods assume a particular mathematical expression for the signal and estimate the parameters of this expression. This technique extends the idea of estimating the frequency of a signal assumed to be a perfect sinusoid. The difference here is that the assumed functional form is more complex. One popular model is to assume the signal to be one or more sinusoids in additive noise, while another takes it to be the output of a filter. This approach is truly novel, and the uncertainty theorem does not directly apply to its frequency measurements.

This chapter deals with the practical problem of numerically estimating the frequency domain description of a signal. We begin with simple methods and cover the popular FFT-based methods. We describe various window functions and how these affect the spectral estimation. We then present Pisarenko's Harmonic Decomposition and several related super-resolution methods. We comment on how it is possible to break the uncertainty barrier. We then briefly discuss ARMA (maximum entropy) models and how they are fundamentally different from periodogram methods. We finish off with a brief introduction to wavelets.

13.1   Zero Crossings

Sophisticated methods of spectral estimation are not always necessary. Perhaps the signal to noise ratio is high, or we don't need very high accuracy. Perhaps we know that the signal consists of a single sinusoid, or are only interested in the most important frequency component. Even more frequently we don't have the real-time to spare for computationally intensive algorithms. In such cases we can sometimes get away with very simple methods.

The quintessence of simplicity is the zero crossing detector. The frequency of a clean analog sinusoid can be measured by looking for times when it crosses the t axis (zero signal value). The interval between two successive zero crossings represents a half cycle, and hence the frequency is half the reciprocal of this interval. Alternatively, we can look for zero crossings of the same type (i.e., both 'rising' or both 'falling'). The reciprocal of the time interval between two rising (or falling) zero crossings is precisely the frequency. Zero crossings can be employed to determine the basic frequency even if the signal's amplitude varies.

In practice there are two distinct problems with the simple implementation of zero crossing detection. First, observing the signal at discrete times reduces the precision of the observed zero crossing times; second, any amount of noise makes it hard to accurately pin down the exact moment of zero crossing. Let's deal with the precision problem first. Using only the sign of the signal (we assume any DC has been removed), the best we can do is to say the zero is somewhere between time n and time $n + 1$. However, by exploiting the signal values we can obtain a more precise estimate. The simplest approach assumes that the signal traces a straight line between the two values straddling the zero. Although in general sinusoids do not look very much like straight lines, the approximation is not unreasonable near the zero crossings for a sufficiently high sampling rate (see Figure 13.1). It is easy to derive an expression for the fractional correction under this assumption, and expressions based on polynomial interpolation can be derived as well.

Returning to the noise problem, were the signal more 'observable' the Robins-Munro algorithm would be helpful. For the more usual case we need to rely on stationarity and ergodicity and remove the noise through a suitable averaging process. The simplest approach is to average interpolated time intervals between zero crossings.

The time duration between zero crossings predicts the basic frequency, only assuming this basic frequency is constant. If it does vary, but sufficiently slowly, it makes sense to monitor the so-called 'zero crossing derivative', the sequence of time differences between successive zero crossing intervals.

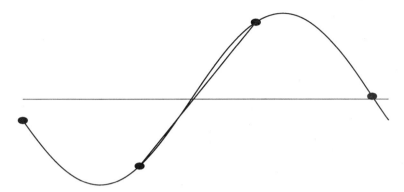

Figure 13.1: Zero crossing detector for clean sinusoid with no DC offset. The sampling rate is about double Nyquist (four samples per period). Note that the linear approximation is reasonable but not perfect.

Given the signal we first compute the sequence of interpolated zero crossing instants $t_0, t_1, t_2, t_3 \ldots$ and then compute the zero crossing intervals by subtraction of successive times (the finite difference sequence) $\Delta_1 = t_1 - t_0$, $\Delta_2 = t_2 - t_1$, $\Delta_3 = t_3 - t_2$ and so on. Next we find the zero crossing derivative as the second finite difference $\Delta_2^{[2]} = \Delta_2 - \Delta_1$, $\Delta_3^{[2]} = \Delta_3 - \Delta_2$, $\Delta_4^{[2]} = \Delta_4 - \Delta_3$ etc. If the underlying frequency is truly constant the Δ sequence averages to the true frequency reciprocal and the $\Delta^{[2]}$ sequence is close to zero. Frequency variations show up in the derivative sequence.

This is about as far as it is worth going in this direction. If the zero crossing derivatives are not sufficient then we probably have to do something completely different. Actually zero crossings and their derivatives are frequently used to derive features for pattern recognition purposes but almost never used as frequency estimators. As feature extractors they are relatively robust, fast to calculate, and contain a lot of information about the signal. As frequency estimators they are not reliable in noise, not particularly computationally efficient, and cannot compete with the optimal methods we will present later on in this chapter.

EXERCISES

13.1.1 What is the condition for two signal values s_n and s_{n+1} to straddle a rising zero crossing? A falling zero crossing? Any zero crossing?

13.1.2 Assume that we have located a rising zero crossing between times n and $n+1$. Derive an expression for δt, the fractional correction to be added to $t = n$, assuming that the signal traces a straight line between s_n and s_{n+1}. Extend to an arbitrary (rising or falling) zero crossing.

13.2   Bank of Filters

The zero crossing approach is based on the premise of well-defined instantaneous frequency, what we once called the 'other meaning' of frequency. Shifting tactics we return to the idea of a well-defined spectrum and seek an algorithm that measures the distribution of energy as a function of frequency. The simplest approach here is the 'bank of filters', inspired by the analog spectrum analyzer of that name. Think of the frequency band of interest, let's say from 0 to F Hz, as being composed of N equal-size nonoverlapping frequency subbands. Employing N band-pass filters we extract the signal components in these subbands, which we denote $\tilde{s}^0$ through $\tilde{s}^{N-1}$. We have thus reduced a single signal of bandwidth F into N signals each of bandwidth $\frac{F}{N}$; see Figure 13.2.

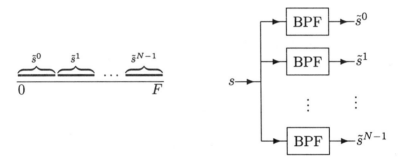

Figure 13.2: A bank of filters dividing the frequency band from 0 to F into N subbands, each containing a band-pass signal. On the left the spectrum is depicted, while the right shows the bank of filters that accomplishes this division.

At this point we could simply add the filter outputs $\tilde{s}^k$ together and reconstruct the original signal s; thus the set of signals $\tilde{s}^k$ contains all the information contained in the original signal. Such an equivalent way of encoding the information in a signal is called a *representation*. The original signal s is the time domain representation, the spectrum is the frequency domain representation, and this new set of signals is the *subband representation*. Subband representations are useful in many contexts, but for now we will only compute the energies of all the subband signals $\tilde{s}^k$, obtaining an estimate of the power spectrum. The precision of this estimate is improved when using a larger number of subbands, but the computational burden goes up as well.

The bank of filters approach to the PSD does not differentiate between a clean sinusoid and narrow-band noise, as long as both are contained in the same subband. Even if the signal is a clean sinusoid this approach cannot provide an estimate of its frequency more precise than the bandwidth of the subband.

We have taken the subbands to be equal in size (i.e., we have divided the total spectral domain into N equal parts), but this need not be the case. For instance, speech spectra are often divided equally on a logarithmic scale, such that lower frequencies are determined more precisely than higher ones. This is no more difficult to do, since it only requires proper design of the filters. In fact it is computationally lighter if we build up the representation recursively. First we divide the entire domain in two using one low-pass and one high-pass filter. The energy at the output of the high-pass filter is measured, while the signal at the output of the low-pass filter is decimated by two and then input to a low-pass and a high-pass filter. This process is repeated until the desired precision of the lowest-frequency bin is attained.

Returning to the case of equal size subbands, we note that although all the signals $\tilde{s}^0$ through $\tilde{s}^{N-1}$ have equal bandwidth, there is nonetheless a striking lack of equality. The lowest subband s^0 is a low-pass signal, existing in the range from 0 to $\frac{F}{N}$. It can be easily sampled and stored using the low-pass sampling theorem. All the other $\tilde{s}^k$ are band-pass signals and hence require special treatment. For example, were we required to store the signal in the subband representation rather than merely compute its power spectrum, it would be worthwhile to downmix all the band-pass signals to the frequency range of 0 to $\frac{F}{N}$. Doing this we obtain a new set of signals we now call simply s^k; s^0 is exactly $\tilde{s}^0$, while all the other s^k are obtained from the respective $\tilde{s}^k$ by mixing down by $\frac{kF}{N}$. This new set of signals also contains all the information of the original signal, and is thus a representation as well. We can call it the *low-pass subband representation* to be contrasted with the previous *band-pass subband representation*. The original signal s is reconstructed by mixing up each subband to its proper position and then summing as before. The power spectrum is computed exactly as before since the operation of mixing does not affect the energy of the subband signals.

The low-pass subband representation of a signal can be found without designing and running N different band-pass filters. Rather than filtering with band-pass filters and then downmixing, one can downmix first and then low-pass filter the resulting signals (see Figure 13.3). In sequential computation this reduces to a single mixer-filter routine called N times on the same input with different downmix frequencies. This is the digital counterpart of the *swept-frequency spectral analyzer* that continuously sweeps in sawtooth fashion the local oscillator of a mixer, plotting the energy at the output of a low-pass filter as a function of this frequency.

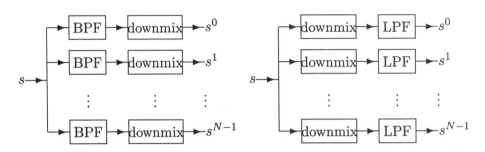

Figure 13.3: Two equivalent implementations of a bank of filters dividing the frequency range into N low-pass signals. In (A) the band-pass signals are band-pass filtered and then downmixed using a real mixer, while in (B) the input signal is downmixed by a complex mixer and then low-pass filtered.

Although the two methods of computing the band-pass representation provide exactly the same signals s^k, there *is* an implementational difference between them. While the former method employed band-pass filters with real-valued multiplications and a real mixer (multiplication by a sine function), the latter requires a complex mixer (multiplication by a complex exponential) and then complex multiplications. The complex mixer is required in order to shift the entire frequency range without spectral aliasing (see Section 8.5). Once such a complex mixer is employed the signal becomes complex-valued, and thus even if the filter coefficients are real two real multiplications are needed.

Since all our computation is complex, we can just as easily input complex-valued signals, as long as we cover the frequency range up to the sampling frequency, rather than half f_s. For N subbands, the analog downmix frequencies for such a complex input are $0, \frac{f_s}{N}, \frac{2f_s}{N}, \ldots \frac{(N-1)f_s}{N}$, and therefore the digital complex downmixed signals are

$$s_n e^{-i\frac{2\pi}{N}kn} = s_n W_N^{nk} \qquad \text{for } k = 0 \ldots N - 1$$

where W_N is the Nth root of unity (see equation (4.30)). These products need to be low-pass filtered in order to build the s^k. If we choose to implement the low-pass filter as a causal FIR filter, what should its length be? From an information theoretic point of view it is most satisfying to choose length N, since then N input samples are used to determine N subband representation values. Thus we find that the kth low-pass signal is given by

$$s_n^k = \sum_{n=0}^{N-1} h_n s_n e^{-i\frac{2\pi}{N}kn} = \sum_{n=0}^{N-1} h_n s_n W_N^{nk} \qquad (13.1)$$

which looks somewhat familiar. In fact we can decide to use as our low-pass filter a simple moving average with all coefficients equal to one (see Section 6.6). Recall that this *is* a low-pass filter; perhaps not a very good one (its frequency response is a sinc), but a low-pass filter all the same. Now we can write

$$s_n^k = \sum_{n=0}^{N-1} s_n e^{-i\frac{2\pi}{N}kn} = \sum_{n=0}^{N-1} s_n W_N^{nk} = S_k \qquad (13.2)$$

which is, of course, precisely the DFT. However, instead of thinking of the DFT as providing the frequency domain representation of a signal, here we consider it as calculating the low-pass subband representation. In this fashion the DFT becomes a tool for efficiently simultaneously downmixing and filtering the signal. The mixers are easily seen in the definition of the DFT; the filtering is implicit in the sum over N input values.

We have to acclimate ourselves to this new interpretation of the DFT. Rather than understanding S_k to be a frequency component, we interpret s_n^k as a time domain sample of a subband signal. For instance, an input signal consisting of a few sinusoids corresponds to a spectrum with a few discrete lines. All subband signals corresponding to empty DFT bins are correctly zero, while sinusoids at bin centers lead to constant (DC) subband signals. So the interpretation is consistent for this case, and we may readily convince ourselves that it is consistent in general.

We have seen that in our bank of filters approach to computing the power spectrum we actually indirectly compute the DFT. In the next section we take up using the DFT to directly estimate the power spectrum.

EXERCISES

13.2.1 The low-pass subband representation can be useful in other contexts as well. Can you think of any? (Hint: FDM.)

13.2.2 Why does the bank of filters approach become unattractive when a large number of filters must be used?

13.2.3 Compare the following three similar spectral analysis systems: (1) a bank of $N+1$ very steep skirted analog band-pass filters spaced at Δf from 0 to $F = N\Delta f$; (2) a similar bank of $N+1$ digital filters; (3) a single DFT with bin size Δf. We inject a single sinusoid of arbitrary frequency into each of the three systems and observe the output signal (note that we do not observe only the energy). Do the three give identical results? If not, why not?

13.2.4 Compare the computational complexity of the recursive method of finding the logarithmic spectrum with the straightforward method.

13.2.5 Prove that the energy of a band-pass signal is unchanged when it is mixed to a new frequency range.

13.2.6 We saw that the DFT downmixes the subbands before filtering, and we know that a mixer is not a filter. In what sense is the DFT equivalent to a bank of filters? How can we empirically measure the frequency response of these filters?

13.2.7 Build a bank of filters spectrum analyzer using available filter design or FFT software. Inject static combinations of a small number of sinusoids. Can you always determine the correct number of signals? Plot the outputs of the filters (before taking the energy). Do you get what you expect? Experiment with different numbers of bins. Inject a sinusoid of slowly varying frequency. Can you reconstruct the frequency response of the filters? What happens when the frequency is close to the border between two subbands?

13.3   The Periodogram

In 1898, Sir Arthur Schuster published his investigations regarding the existence of a particular periodic meteorological phenomenon. It is of little interest today whether the phenomenon in question was found to be of consequence; what *is* significant is the technique used to make that decision. Schuster introduced the use of an empirical STFT in order to discover hidden periodicities, and hence called this tool the *periodogram*. Simply put, given N equally spaced data points $s_0 \ldots s_{N-1}$, Schuster recommended computing (using our notation)

$$P(\omega) = \frac{1}{N} \left| \sum_{n=0}^{N-1} s_n e^{-i\omega n} \right|^2 \tag{13.3}$$

for a range of frequencies ω and looking for peaks—peaks that represent hidden periodicities. We recognize this as the DFT power spectrum evaluated for the available data.

Many of today's DSP practitioners consider the FFT-based periodogram to be the most natural power spectral estimator. Commercially available hardware and software digital spectrum analyzers are almost exclusively based on the FFT. Indeed DFT-based spectral estimation is a powerful and well-developed technique that should probably be the first you explore when a new problem presents itself; but as we shall see in later sections it is certainly not the only, and often not even the best technique.

What is the precise meaning of the periodogram's $P(\omega)$? We would like for it to be an estimate of the true power spectral density, the PSD that would be calculated were an infinite amount of data (and computer time) to be available. Of course we realize that the fact that our data only covers a finite time duration implies that the measurement cannot refer to an infinitesimal frequency resolution. So the periodogram must be some sort of average PSD, where the power is averaged over the bandwidth allowed by the uncertainty theorem.

What is the weighting of this average? The signal we are analyzing is the true signal, which exists from the beginning of time until its end, multiplied by a rectangular window that is unity over the observed time interval. Accordingly, the FT in the periodogram is the convolution of the true FT with the FT of this window. The FT of a rectangular window is given by equation (4.22), and is sinc shaped. This is a major disappointment! Not only do frequencies far from the minimum uncertainty bandwidth 'leak' into the periodogram PSD estimate, the strength of these distant components does not even monotonically decrease.

Is the situation really as bad as it seems? To find out let's take 64 samples of a sinusoid with digital frequency 15/64, compute the FFT, take the absolute square for the positive frequencies, and convert to dB. The analog signal, the samples, and the PSD are shown in Figure 13.4.A. All looks fine; there is only a single spectral line and no leakage is observed. However, if we look carefully at the sinc function weighting we will see that it has a zero at the center of all bins other than the one upon which it is centered. Hence there is never leakage from a sinusoid that is exactly centered in some neighboring bin (i.e., when its frequency is an integer divided by the number of samples). So let's observe what happens when the digital frequency is slightly higher (e.g., $f_d = 15.04/64$) as depicted in Figure 13.4.B. Although this frequency deviation is barely noticeable in the time domain, there is quite significant leakage into neighboring bins. Finally, the worst-case is when the frequency is exactly on the border between two bins, for example, $f_d = 15.5/64$ as in Figure 13.4.C. Here the leakage is already intolerable.

Why is the periodogram so bad? The uncertainty theorem tells us that short time implies limited frequency resolution but DSP experience tells us that small buffers imply bothersome edge effects. A moment's reflection is enough to convince you that only when the sinusoid is precisely centered in the bin are there an integer number of cycles in the DFT buffer. Now recall that the DFT forces the signal to be periodic outside the duration of the buffer that it sees; so when there are a noninteger number of cycles the signal

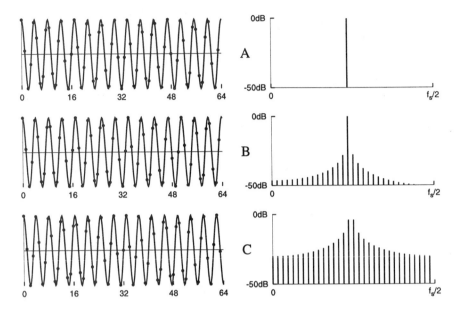

Figure 13.4: Leakage in the spectrum of a single sinusoid. In (A) precisely 15 cycles of the sinusoid fit into the buffer of length 64 samples and thus its periodogram contains a single line. In (B) 15.04 cycles fit into the buffer and thus there is a small discontinuity at the edge. The periodogram displays leakage into neighboring bins. In (C) $14\frac{1}{2}$ cycles fit and thus the discontinuity and leakage are maximal. Note also that the two equal bins are almost 4 dB lower than the single maximal bin in the first case, since the Parseval energy is distributed among many bins.

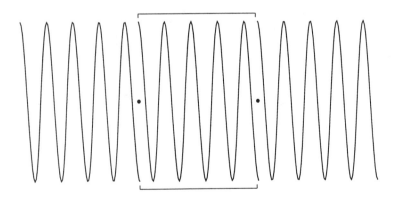

Figure 13.5: The effect of windowing with a noninteger number of cycles in the DFT buffer. Here we see a signal with $4\frac{1}{2}$ cycles in the buffer. After replication to the left and right the signal has the maximum possible discontinuity.

effectively becomes discontinuous. For example, a signal that has $4\frac{1}{2}$ cycles in the DFT buffer really looks like Figure 13.5 as far as the DFT is concerned. The discontinuities evident in the signal, like all discontinuities, require a wide range of frequencies to create; and the more marked the discontinuity the more frequencies required. Alternatively, we can explain the effect in terms of the Gibbs phenomenon of Section 3.5; the discontinuity generated by the forced periodicity causes ripples in the spectrum that don't go away.

Many ways have been found to fix this problem, but none of them are perfect. The most popular approaches compel continuity of the replicated signal by multiplying the signal in the buffer by some *window function w_n*. A plethora of different functions w_n have been proposed, but all are basically positive valued functions defined over the buffer interval $0 \ldots N - 1$ that are zero (or close to zero) near the edges $w_0 \approx 0, w_N \approx 0$, but unity (or close to unity) near the middle $w_{N/2} \approx 1$. Most window functions (as will be discussed in more detail in Section 13.4) smoothly increase from zero to unity and then decrease back in a symmetric fashion. The exception to this smoothness criterion is the *rectangular window* (i.e., the default practice of not using a window at all, multiplying all signal values outside the buffer by zero, and all those inside by unity). For nondefault window functions, the new product signal $s'_n = w_n s_n$ for which we compute the DFT is essentially zero at both ends of the buffer, and thus its replication contains no discontinuities. Of course it is no longer the same as the original signal s_n, but for *good* window functions the effect on the power spectrum is tolerable.

Why does multiplication by a *good* window function not completely distort the power spectrum? The effect can be best understood by considering the half-sine window $w_n = \sin(\pi \frac{n}{N})$ (which, incidentally, is the one window function that no one actually uses). Multiplying the signal by this window is tantamount to convolving the signal spectrum with the window's spectrum. Since the latter is highly concentrated about zero frequency, the total effect is only a slight blurring. Sharp spectral lines are widened, sharp spectral changes are smoothed, but the overall picture is relatively undamaged.

Now that we know how to correctly calculate the periodogram we can use it as a *moving* power spectrum estimator for signals that vary over time. We simply compute the DFT of a windowed buffer, shift the buffer forward in time, and compute again. In this way we can display a sonogram (Section 4.6) or average the periodograms in order to reduce the variance of the spectral estimate (Section 5.7). The larger the buffer the better the frequency resolution, and when computing the DFT using the FFT we almost always want the buffer length to be a power of two. When the convenient buffer length doesn't match the natural data buffer, we can zero-pad the buffer.

Although this zero-padding seems to increase the frequency resolution it obviously doesn't really add new information. We often allow the buffers to overlap (half-buffer overlap being the most prevalent choice). The reason is that the windowing reduces the signal amplitude over a significant fraction of the time, and we may thus miss important phenomena. In addition, the spectral estimate variance is reduced even by averaging overlapped buffers.

EXERCISES

13.3.1 Show directly, by expressing the sample s_{lN+m} outside the buffer in terms of the complex DFT coefficients s_k, that computing the N-point DFT corresponds to replicating the signal in the time domain.

13.3.2 Plot the energy in a far bin as a function of the size of the discontinuity. (It's enough to use a cosine of digital frequency $\frac{1}{4}$ and observe the DC.) Why isn't it practical to use a variable-length rectangular window to reduce leakage?

13.3.3 Is signal discontinuity really a necessary condition for leakage? If not, what is the exact requirement? (Hint: Try the sinusoid $\sin(2\pi(k + \frac{1}{2})/N)$.)

13.3.4 As the length of the buffer grows the number of discontinuities per time decreases, and thus we expect the spectral SNR to improve. Is this the case?

13.3.5 In the text we discussed the half-sine window function. Trying it for a frequency right on a bin boundary (i.e., maximal discontinuity) we find that it works like a charm, but not for other frequencies. Can you explain why?

13.4   Windows

In Sections 4.6 and 13.3 we saw the general requirements for window functions, but the only explicit examples given were the rectangular window and the somewhat unusual half-sine window. In this section we will become acquainted with many more window functions and learn how to 'window shop', that is, to choose the window function appropriate to the task at hand.

Windows are needed for periodograms, but not only for periodograms. Windows are needed any time we chop up a signal into buffers and the signal is taken to be periodic (rather than zero) outside the observation buffer. This is a very frequent occurrence in DSP! When calculating autocorrelations (see Chapter 9) the use of windows is almost universal; a popular technique of designing FIR filters is based on truncating the desired impulse response by a window (see Section 7.8); sample buffers are windowed before LPC analysis (see Section 9.9); and the list goes on and on. Yet windowing as a preprocessing stage for the periodogram is probably the best known use, and we will concentrate on it here. Recalling the interpretation of the FT as a bank of FIR band-pass filters, we will see that the frequency response of these filters is directly determined by the window function used.

We must, once again, return to the issue of buffer indexing. The computer programming convention that the buffer index runs from 0 to $N - 1$ is usually used with a window that obeys $w_0 = 0$ and $w_N = 0$. In this fashion the first point in the output buffer is set to zero but the last point is not (the Nth point, which *is* zero, belongs to the next buffer). Some people cannot tolerate such asymmetry and make either both $w_0 = 0, w_{N-1} = 0$ or $w_{-1} = 0, w_N = 0$. These conventions should be avoided! The former implies two zeroed samples in the replicated signal, the latter none. In theoretical treatments the symmetric buffer indexation $-M \ldots M$ with $M \equiv \frac{N}{2}$ is common, and here only one of the endpoints is to be considered as belonging to the present buffer. To make things worse the buffer length may be even or odd, although FFT buffers will usually be of even length. As a consequence you should always check your window carefully before looking through it. We will present expressions in two formats, the practical $0 \ldots N - 1$ with even N and $w_0 = 0, w_N = 0$ and the symmetric odd length $-M \ldots M$ with $w_{\pm m} = 0$ and thus $N = 2M + 1$. To differentiate we will use an index n for the former case and m for the latter.

The rectangular window is really not a window function at all, but we consider it first for reference. Measuring analog time in units of our sampling interval, we can define an analog window function $w(t)$ that is one between $t = -M$ and $t = +M$ and zero elsewhere. We know that its FT is

$$W(\omega) = M \operatorname{sinc}(M\omega)$$

and its main lobe (defined between the first zeros) is of width $\frac{2\pi}{M}$. As M increases the main lobe becomes narrower and taller, but if we increase the frequency resolution, as allowed by the uncertainty theorem, we find that the number of frequency bins remains the same. In fact in the digital domain the $N = 2M$ point FFT has a frequency resolution of $\frac{1}{N}$ (the sampling frequency is one), and thus the main lobe is two frequency bins in width for all M. It isn't hard to do all the mathematics in the digital domain either. The digital window is $w_m = 1$ for $-M \leq m \leq +M$ and $w_m = 0$ elsewhere.

The DFT is given by

$$W_k = \sum_{m=-M}^{M} e^{-ikm} = e^{-ikM}\frac{1 - e^{-ik}e^{-2ikM}}{1 - e^{-ik}} = \frac{\sin(\frac{1}{2}Nk)}{\sin(\frac{1}{2}k)}$$

where we have used formula (A.46) for the sum of a finite geometric series, and substituted $N = 2M + 1$.

From this expression we can derive everything there is to know about the rectangular window. Its main lobe is two bins in width, and it has an infinite number of sidelobes, each one bin in width. Its highest sidelobe is attenuated 13.3 dB with respect to the main lobe, and the sidelobes decay by 6 dB per octave, as expected of a window with a discontinuity (see Section 4.2).

Before we continue we need some consistent quantities with which to compare windows. One commonly used measure is the *noise bandwidth* defined as the bandwidth of an ideal filter with the same maximum gain that would pass the same amount of power from a white noise source. The noise bandwidth of the rectangular window is precisely one, but is larger than one for all other windows. Larger main lobes imply larger noise bandwidths. Another important parameter is the ripple of the frequency response in the pass-band. The rectangular window has almost 4 dB pass-band ripple, while many other windows have much smaller ripple. We are now ready to see some nontrivial windows.

Perhaps the simplest function that is zero at the buffer ends and rises smoothly to one in the middle is the triangular window

$$w_n = 1 - \left|\frac{n - M}{M}\right| \qquad \Big| \qquad w_m = 1 - \frac{|m|}{M + 1} \qquad (13.4)$$

which is also variously known as the Bartlett window, the Fejer window, the Parzen window, and probably a few dozen more names. This window rises linearly from zero to unity and then falls linearly back to zero. If the buffer is of odd length there is a point in the middle for which the window function is precisely unity, for even length buffers all values are less than one. The highest sidelobe of the triangular window is 26 dB below the main lobe, and the sidelobes decay by 12 dB per octave, as expected of a window with a first derivative discontinuity. However, the noise bandwidth is 1.33, because the main lobe has increased in width.

The Hanning window is named after the meteorologist Julius von Hann.

$$w_n = \tfrac{1}{2}\left(1 - \cos\left(2\pi\frac{n}{N}\right)\right) \qquad \text{for} \quad n = 0 \ldots N - 1 \qquad (13.5)$$

Apparently the verb form 'to Hann the data' was used first; afterward people started to speak of 'Hanning the signal', and in the end the analogy with the Hamming window (see below) caused the adoption of the misnomer 'Hanning window'. The Hanning window is also sometimes called the 'cosine squared', or 'raised cosine' window (use the 'm' index to see why). The Hanning window's main lobe is twice as wide as that of the rectangular window, and at least three spectral lines will always be excited, even for the best case. The noise bandwidth is 1.5, the highest sidelobe is 32 dB down, and the sidelobes drop off by 18 dB per octave.

The Hamming window is named in honor of the applied mathematician Richard Wesley Hamming, inventor of the Hamming error-correcting codes, creator of one of the first programming languages, and author of texts on numerical analysis and digital filter design.

$$w_n = 0.54 - 0.46 \left(1 - \cos \left(2\pi \frac{n}{N} \right) \right) \qquad \text{for} \quad n = 0 \ldots N - 1 \quad (13.6)$$

The Hamming window is obtained by modifying the coefficients of the Hanning window in order to precisely cancel the first sidelobe, but suffers from not becoming precisely zero at the edges. For these reasons the Hamming window has its highest sidelobe 42 dB below the main lobe, but asymptotically the sidelobes only decay by 6 dB per octave. The noise bandwidth is 1.36, close to that of the triangular window.

Continuing along similar lines one can define the Blackman-Harris family of windows

$$w_n = a_0 - a_1 \cos \left(2\pi \frac{n}{N} \right) + a_2 \cos \left(2\pi \frac{2n}{N} \right) - a_3 \cos \left(2\pi \frac{3n}{N} \right) \ldots \quad (13.7)$$

and optimize the parameters in order to minimize sidelobes. More complex window families include the Kaiser and Dolph-Chebyshev windows, which have a free parameter that can be adjusted for the desired trade-off between sidelobe height and main-lone width. We superpose several commonly used windows in Figure 13.6.

Let's see how these windows perform. In Figure 13.7 we see the periodogram spectral estimate of a single worst-case sinusoid using several different windows. We see that the rectangular window is by far the worst, and that the triangular and then the Hanning windows improve upon it.

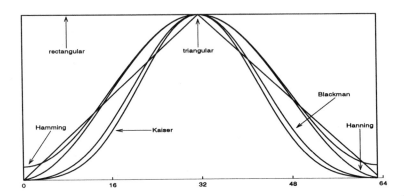

Figure 13.6: Various window functions. Depicted are 64-point rectangular, triangular, Hanning, Hamming, Blackman, and Kaiser windows.

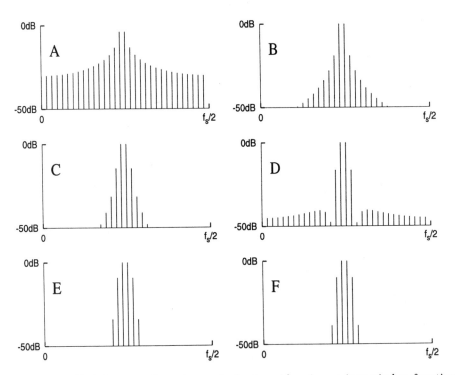

Figure 13.7: Periodogram of worst-case single sinusoids using various window functions, namely (A) rectangular, (B) triangular, (C) Hanning, (D) Hamming, (E) Blackman-Harris, and (F) Kaiser. Each periodogram is normalized such that its maximum height corresponds to 0 dB.

Afterward the choice is not clear cut. The Blackman and Kaiser windows reduce the sidelobe height, but cannot simultaneously further reduce the main lobe width. The Hamming window attempts to narrow the main lobe, but ends up with higher distant sidelobes. Not shown is a representative of the Dolph-Chebyshev family, which as can be assumed for anything bearing the name Chebyshev, has constant-height sidelobes.

Which window function is best? It all depends on what you are trying to do. Rectangular weighting could be used for sinusoids of precisely the right frequencies, but don't expect that to ever happen accidentally. If you are reasonably sure that you have a single clean sinusoid, this may be verified and its frequency accurately determined by using a mixer and a rectangular window STFT; just remember that the signal's frequency is the combination of the bin's frequency and the mixer frequency. An even trickier use of the rectangular window is for the probing of linear systems using synthetically generated pseudorandom noise inputs (see Section 5.4). By using a buffer length precisely equal to the periodicity of the pseudorandom signal we can ensure that all frequencies are just right and the rectangular weighted STFT spectra are beautiful. Finally, rectangular windows should be used when studying *transients* (signals that are nonzero only for a short time). We can then safely place the entire signal inside the buffer and guarantee zero signal values at the buffer edges. In such cases the rectangular window causes the least distortion and requires the least computation.

For general-purpose frequency displays the Hanning and Hamming windows are often employed. They have lower sidebands and lower pass-band ripple than the rectangular window. The coefficients of the Hanning window needn't be stored, since they are derivable from the FFT's twiddle factor tables. Another trick is to overlap and average adjacent buffers in such a way that the time weighting becomes constant.

A problem we haven't mentioned so far is two-tone separability. We sometimes need to separate two closely spaced tones, with one much stronger than the other. Because of main lobe width and sidelobe height, the weaker tone will be covered up and not noticeable unless we choose our window carefully. For such cases the Blackman, Dolph-Chebyshev, or Kaiser windows should be used, but we will see stronger methods in the following sections.

EXERCISES

13.4.1 Convert the Hanning and Hamming windows to symmetric 'm' notation and explain the names 'cosine squared' and 'raised cosine' often applied to the former. Express the Hanning window as a convolution in the frequency domain. What are the advantages of this approach?

13.4.2 Plot the periodograms for the same window functions as in Figure 13.7, but for a best-case sinusoid (e.g., for $N = 64$, a sinusoid of frequency 15/64).

13.4.3 Plot periodograms of the logistics signal for various $1 \leq \lambda < 3.57$, as was done in Section 5.5. Which window is best? Now use λ that give for 3, 5, and 6 cycles. Which window should be used now?

13.4.4 Try to separate two close sinusoids, both placed in worst-case positions, and one much stronger than the other. Experiment with different windows.

13.5 Finding a Sinusoid in Noise

As we mentioned above, frequency estimation is simplest when we are given samples of a single clean sinusoid. Perhaps the next simplest case is when we are told that the samples provided are of a single sinusoid with additive uncorrelated white noise; but if the SNR is low this 'simple' case is not so simple after all. To use averaging techniques as discussed in Section 5.3 one would have to know a priori how to perform the registration in time before averaging. Unfortunately, this would require accurate knowledge of the frequency, which is exactly what we are trying to measure in the first place! We could perform an FFT, but that would only supply us with the frequency of the nearest bin; high precision would require using a large number of signal points (assuming the frequency were constant over this time interval), and most of the computation would go toward finding bins of no interest. We could calculate autocorrelations for a great number of lags and look for peaks, but the same objections hold here as well.

There are more efficient ways of using the autocorrelations. Pisarenko discovered one method of estimating the frequencies of p sinusoids in additive white noise using a relatively small number of autocorrelation lags. This method, called the Pisarenko Harmonic Decomposition (PHD), seems to provide an infinitely precise estimate of these frequencies, and thus belongs to the class of 'super-resolution' methods. Before discussing how the PHD circumvents the basic limitations of the uncertainty theorem, let's derive it for the simple case of a single sinusoid ($p = 1$).

We assume that our signal is exactly of the form

$$s_n = A\sin(\omega n + \phi) + \nu_n \tag{13.8}$$

where ν is the uncorrelated white noise. Its autocorrelations are easily derived

$$C_s(m) = \langle s_n s_{n+m} \rangle = \frac{A^2}{2}\cos(\omega m) + \sigma_\nu^2 \delta_{m,0} \tag{13.9}$$

and the first few lags are given by the following.

$$C_s(0) = \frac{A^2}{2} + \sigma_\nu^2$$

$$C_s(1) = \frac{A^2}{2}\cos(\omega)$$

$$C_s(2) = \frac{A^2}{2}\cos(2\omega) = \frac{A^2}{2}\left(2\cos^2(\omega) - 1\right)$$

The noise only influences the lag zero term (energy) due to the assumption of white noise. Any deviation from whiteness causes the other lags to acquire noise-related terms as well.

Were the noise to be zero, we could simply calculate

$$\omega = \cos^{-1}\left(\frac{C_s(1)}{C_s(0)}\right)$$

but this fails miserably when noise is present. Can we find an expression that uses only nonzero lags, and is thus uninfluenced by the noise? Pisarenko's method uses only the two lags $m = 1$ and $m = 2$. Using the trigonometric identity $\cos(2\omega) = 2\cos^2(\omega) - 1$ it is easy to show that

$$2C_s(1)c^2 - C_s(2)c - C_s(1) = 0$$

where we have denoted $c = \cos(\omega)$. This is a simple quadratic, with solutions

$$c = \frac{C_s(2)}{4C_s(1)} \pm \frac{1}{2}\sqrt{\frac{C_s^2(2)}{4C_s^2(1)} + 2} \tag{13.10}$$

only one of which leads to the correct solution (see exercise 13.5.2). We thus find

$$\omega = \cos^{-1}\left(\frac{C_s(2)}{4C_s(1)} + \frac{1}{2}\mathrm{sgn}\left(C_s(1)\right)\sqrt{\frac{C_s^2(2)}{4C_s^2(1)} + 2}\right) \tag{13.11}$$

which is the PHD estimate for the digital angular frequency (the analog frequency is obtained by dividing by 2π and multiplying by the sampling frequency).

The PHD expression we have just found is not a frequency *estimate* at all. Assuming the noise is perfectly white, it is an infinitely precise measurement of the frequency. Of course there is no problem with this infinite precision since we assumed that we have exact values for the two autocorrelation lags $C_s(1)$ and $C_s(2)$. Obtaining these exact values requires knowing the signal over all time, and therefore the uncertainty theorem does allow infinitely precise predictions. However, even when we use empirical autocorrelations (equation (9.11)) calculated using only N samples the prediction still seems to be perfectly precise. Unlike periodogram methods there is no obvious *precision* reduction with decreasing N; but the *accuracy* of the prediction decreases. It is straightforward, but somewhat messy, to show that the variance of the PHD estimator is inversely proportional to the size of the buffer and the square of the SNR (SNR $= \frac{A^2}{2\sigma_\nu^2}$).

$$\sigma^2_{PHD} = \frac{\cos^2(2\omega) + \cos^2(\omega)}{\sin^2(\omega)(\cos(2\omega) + 2)^2} \frac{1}{N\,\text{SNR}^2}$$

The somewhat complex frequency-dependent prefactor means that the estimate is more accurate near DC ($\omega = 0$) and Nyquist ($\omega = \pi$), and there is a small dip near the middle of the range. More interesting is the N dependence; the proper Δf is the standard deviation, and so we have a strange $(\Delta f)^2 \Delta t$ uncertainty product. Even more disturbing is the SNR dependence; as the SNR increases the error decreases even for small N. It is obvious that this error only reflects better noise cancellation with more data points, and not true uncertainty theorem constraints.

So it seems that the PHD really does beat the uncertainty theorem. The explanation is, however, deceptively simple. We made the basic assumption that the signal was exactly given by equation (13.8). Once the parameters of the sinusoid are known, the signal (without the noise) is known *for all times*. The uncertainty product effectively has $\Delta t = \infty$ and can attain infinitesimal frequency precision. This is the idea behind all model-based super-resolution methods. The data is used to find the parameters of the model, and the model is assumed to hold for all time. Thus, assuming that the assumption holds, the uncertainty theorem is robbed of its constraining influence.

EXERCISES

13.5.1 Derive the expression (13.9) for the autocorrelation (use exercise 9.2.12).

13.5.2 Exploiting the fact that we want $0 \leq \omega < \pi$ show that the proper solution of the quadratic has the sign of $C_s(1)$.

13.5.3 In the text we quoted the variance of the error of the PHD estimation. What about its bias? Find this numerically for various buffer sizes.

13.5.4 The PHD is a second-order frequency estimator in the sense that the highest autocorrelation lag it utilizes is $m = 2$. Using the trigonometric identity $\cos(\omega) + \cos(3\omega) = 2\cos(\omega)\cos(2\omega)$ prove that $C_s(1) - 2C_s(2)c + C_s(3) = 0$. Show that this leads to the following third-order estimator.

$$\omega = \cos^{-1}\left(\frac{C_s(1) + C_s(3)}{2C_s(2)}\right)$$

13.5.5 Compare the third-order estimator of the previous exercise with the PHD by generating sinusoids in various amounts of white noise and estimating their frequencies. Which is better for low SNR? High SNR? Small buffer size? Large buffer size?

13.6 Finding Sinusoids in Noise

The previous section dealt with the special case of a single sinusoid in noise. Here we extend the PHD to multiple sinusoids. The needed formalism is a bit more mathematically demanding (involving the roots of functions that are derived from the eigenvector of the signal covariance matrix belonging to the smallest eigenvalue), so we approach it cautiously.

In order to derive the PHD for the sum of p sinusoids in uncorrelated white noise,

$$s_n = \sum_{i=1}^{p} A_i \sin(\omega_i n) + \nu_n$$

we first rederive the $p = 1$ case in a different way. Recall that exponentials and sinusoids obey difference equations; those of real exponentials involve a single previous value, while sinusoids obey recursions involving two previous values. From equation (6.52) we know that a clean sinusoid $x_n = A\sin(\omega n)$ obeys the following recursion

$$x_n = a_1 x_{n-1} + a_2 x_{n-2} \qquad \text{where} \qquad \begin{cases} a_1 = 2\cos(\omega) \\ a_2 = -1 \end{cases} \qquad (13.12)$$

(we have simply defined $a_1 = -c_1$ and $a_2 = -c_2$). We will call a_1 and a_2 *recursion coefficients*. Given the recursion coefficients we can write the equation

$$1 - a_1 z^{-1} - a_2 z^{-2} = 0 \qquad \text{or} \qquad z^2 - 2\cos(\omega)z + 1 = 0 \qquad (13.13)$$

which has the following solutions.

$$z = \tfrac{1}{2}\left(2\cos(\omega) \pm \sqrt{4\cos^2(\omega) - 4}\right) = \cos(\omega) \pm i\sin(\omega) = e^{\pm i\omega}$$

Thus those z that solve equation (13.13) (i.e., the roots of the polynomial therein specified) are on the unit circle, and their angles are the frequencies (both positive and negative) of the original signal. This is a link between the recursion coefficients and the frequencies of the signal that obeys the recursion.

The connection between this and the PHD is easiest to understand in vector notation. We define the vectors

$$\underline{x} = (x_n, x_{n-1}, x_{n-2})$$
$$\underline{a} = (1, -a_1, -a_2)$$

so that equation (13.12) is written

$$\underline{x} \cdot \underline{a} = 0 \qquad (13.14)$$

i.e., the clean signal vector and the recursion coefficient vector are orthogonal. Now the noisy signal is $\underline{s} = \underline{x} + \underline{\nu}$. This signal has mean zero (since we assume the noise to be zero mean) and its covariance matrix is thus the signal autocorrelation matrix

$$\underline{\underline{V}}_s = \left\langle \underline{s}\underline{s}^t \right\rangle = \left\langle (\underline{x} + \underline{\nu})(\underline{s}^t + \underline{\nu}^t) \right\rangle = \underline{\underline{V}}_x + \sigma_\nu^2 \underline{\underline{I}} \qquad (13.15)$$

where $\underline{\underline{V}}_s$ is a 3-by-3 square matrix with elements $V_{s_{i,j}} = C_s(i-j)$. The first term $\underline{\underline{V}}_x = \left\langle \underline{x}\underline{x}^t \right\rangle$ is a symmetric Toeplitz matrix, the matrix $\underline{\underline{I}}$ is the 3-by-3 identity matrix, and σ_ν^2 is the variance of the noise. It is now easy to see that

$$\underline{\underline{V}}_s \underline{a} = \left(\left\langle \underline{x}\underline{x}^t \right\rangle + \sigma_\nu^2\right)\underline{a} = \left\langle \underline{x}(\underline{x} \cdot \underline{a}) \right\rangle + \sigma_\nu^2 \underline{a} = \sigma_\nu^2 \underline{a} \qquad (13.16)$$

which shows that $\underline{a}$ is an eigenvector of the covariance matrix. Since eigenvalues of $\underline{\underline{V}}_x$ are nonnegative, σ_ν^2 must be the smallest eigenvalue of $\underline{\underline{V}}_s$. We thus see that the frequency of a sinusoid in additive noise can be determined by diagonalizing its autocorrelation matrix. This is a specific case of the desired formulation of the PHD.

Theorem: The Pisarenko Harmonic Decomposition
Given a signal s that is the sum of p sinusoids and uncorrelated white noise,

$$s_n = \sum_{i=1}^{p} \left(A_i e^{i\omega_i n} + A_i^* e^{-i\omega_i n} \right) + \nu_n$$

denote by $\underline{\underline{V_s}}$ the $(2p + 1)$-dimensional covariance matrix of this signal, and by $\underline{a}$ the eigenvector of $\underline{\underline{V_s}}$ that belongs to the minimal eigenvalue. Then the roots of $1 - \sum_{k=1}^{p} a_k z^{-k}$ are of the form $z_i = e^{\pm i\omega_i}$.   ∎

We can now understand the term *decomposition* that appears in the name PHD. The decomposition is that of the covariance matrix into signal-related and noise-related parts (see equation (13.15)) which implies the splitting of the $(2p + 1)$-dimensional space of signal vectors into orthogonal signal (equation (13.14)) and noise subspaces.

The proof of the general p case is similar to that of the $p = 1$ case. The key idea is that signals consisting of p real exponentials or sinusoids obey difference equations; those of real exponentials involve p previous values, while p sinusoids obey recursions involving $2p$ previous values. It's easier to understand this by first considering the exponential case.

$$x_n = \sum_{i=1}^{p} A_i e^{q_i n}$$

This can be expressed as the combination of p previous values.

$$x_n = \sum_{k=1}^{p} a_k x_{n-k} \tag{13.17}$$

Substituting the definition

$$x_n = \sum_{k=1}^{p} a_k \left(\sum_{i=1}^{p} A_i e^{q_i(n-k)} \right) = \sum_{i=1}^{p} A_i \left(\sum_{k=1}^{p} a_k e^{-q_i k} \right) e^{q_i n}$$

and thus

$$\sum_{k=1}^{p} a_k e^{-q_i k} = 1$$

we see that

$$1 - \sum_{k=1}^{p} a_k z^{-k}$$

has roots $z = e^{q_i}$. Similarly for the sum of sinusoids

$$x_n = \sum_{i=1}^{p} \left(A_i e^{i\omega n} + A_i^* e^{-i\omega n} \right)$$

we leave it as an exercise to show that

$$x_n = \sum_{k=1}^{2p} a_k x_{n-k} \tag{13.18}$$

where

$$1 - \sum_{k=1}^{2p} a_k z^{-k} \tag{13.19}$$

has roots $z_i = e^{\pm i\omega_i}$.

In practice we do not have the real covariance matrix, and Pisarenko's method uses the usual empirical estimates for $C_s(0), C_s(1), C_s(2), \ldots C_s(2p)$. Once the covariance matrix has been estimated, we diagonalize it or use any available method for finding the eigenvector belonging to the minimal eigenvalue. This produces the recursion coefficients with which we can build the polynomial in (13.19) and find its roots numerically. Finally we obtain the desired frequencies from the angles of the roots.

The PHD is only one of several frequency estimation methods that use eigenvector decomposition of the signal covariance matrix. Another popular eigenvector method, called *MUSIC* (**MU**ltiple **SI**gnal **C**lassification), provides a full spectral distribution, rather than the p discrete lines of the PHD.

Alternative approaches are based on inverting the covariance matrix rather than diagonalizing it. Baron de Prony worked out such an algorithm for a similar problem back in 1795! Prony wanted to approximate N equally-spaced data points by the sum of p real exponentials (as in equation (13.17)). There are precisely $2p$ free parameters in the parametric form, so he needed $N = 2p$ data points. Were the $e^{q_i t}$ factors known, finding the A_i would be reduced to the solution of p simultaneous linear equations; but the q_i appear in an exponent, creating a very nonlinear situation. Prony's idea was to find the qs first, using the recursion relating the data values. For exponentials the recursion is equation (13.17) for all n. Given N signal values, $x_0, x_1, \ldots x_{N-1}$, we consider only the $N - p$ equations for which all the required signal values are in the buffer.

$$
\begin{aligned}
x_p &= a_1 x_{p-1} + a_2 x_{p-2} + \cdots + a_p x_0 \\
x_{p+1} &= a_1 x_p + a_2 x_{p-1} + \cdots + a_p x_1 \\
&\ \ \vdots \\
x_{N-1} &= a_1 x_{N-2} + a_2 x_{N-3} + \cdots + a_p x_{N-p-1}
\end{aligned}
$$

This can be written in matrix form

$$
\begin{pmatrix}
x_{p-1} & x_{p-2} & x_{p-3} & \cdots & x_0 \\
x_p & x_{p-1} & x_{p-2} & \cdots & x_1 \\
x_{p+1} & x_p & x_{p-1} & \cdots & x_1 \\
\vdots & \vdots & \vdots & \vdots & \vdots \\
x_{N-2} & x_{N-3} & x_{n-4} & \cdots & x_{N-p-1}
\end{pmatrix}
\begin{pmatrix}
a_1 \\ a_2 \\ a_3 \\ \vdots \\ a_p
\end{pmatrix}
=
\begin{pmatrix}
x_p \\ x_{p+1} \\ x_{p+2} \\ \vdots \\ x_{N-1}
\end{pmatrix}
$$

and for $N = 2p$ the matrix is square Toeplitz and the equations can be readily solved for the a_k. Once we have the a_k we can find the roots of the polynomial, and retrieve the q_i. Thereafter the A_i can be found as explained above. Thus Prony's method reduces the solution of a very nonlinear problem to the solution of two linear problems and the (nonlinear) operation of finding the roots of a polynomial.

EXERCISES

13.6.1 Why are the eigenvalues of $\underline{\underline{C_x}}$ nonnegative?

13.6.2 Complete the proof of the PHD for general p. To do this prove equation (13.18) and the claim about the roots.

13.6.3 Specialize the PHD back to $p = 1$ and show that we obtain our previous PHD equation (13.11).

13.6.4 What is the computational complexity of the PHD for general p?

13.6.5 Prony's method as described works only for noiseless signals. How can it be extended to the noisy case?

13.6.6 Extend Prony's method to p sinusoids.

13.7   IIR Methods

Armed with the ideas acquired in the previous sections, we return to the problem of estimating the entire power spectral distribution from samples in the time domain. In Section 4.6 we saw that the DFT periodogram can be a powerful spectral *estimator*, but does not produce the exact spectrum due to the signal only being observed for short times. In Section 13.2 we saw that the STFT is essentially a bank of FIR filters. Can we improve on the periodogram by using a bank of IIR filters?

Recall that the DFT is simply the zT

$$s(z) = zT(s_n) = \sum_{n=-\infty}^{\infty} s_n z^{-n}$$

calculated on the unit circle. Thus corresponding to the moving STFT there is a STzT

$$s_m(z) = \sum_{n=m-N+1}^{m} s_n z^{-n} \tag{13.20}$$

where we have not explicitly shown a window function and have chosen the causal indexing convention. At time $n = 0$ this reduces to

$$s_0(z) = \sum_{n=-N+1}^{0} s_n z^{-n} = \sum_{n=0}^{N-1} s_{-n} z^{n} \tag{13.21}$$

an $(N-1)^{\text{th}}$ degree polynomial in z. By comparison, the full zT is an infinite Laurent series,

$$s(z) = \sum_{n=-\infty}^{\infty} s_n z^{-n} = \sum_{n=1}^{\infty} s_n z^{-n} + \sum_{n=0}^{\infty} s_{-n} z^{n}$$

and the STzT can be considered to be a polynomial approximation to these infinite sums.

What kind of approximation to the infinite Laurent series is the polynomial? It is obviously an *all-zero* approximation since no poles in the z-plane can be produced, only zeros. Spectra with no discrete lines (delta functions) are well approximated by such polynomials, but spectra with sharp resonances are not. Sharp features such as delta functions are better captured by an approximation such as

$$s(z) \approx \frac{1}{\sum_{m=0}^{M} b_m z^m} \tag{13.22}$$

which is obviously an *all-pole* approximation to the full zT. All-pole approximations may efficiently describe resonances that would take dozens of coefficients in the all-zero model; and like the Pisarenko estimate the frequencies of the poles may be measured to higher resolutions than the STFT allows. To capture both zeros and poles in the true spectrum we had best consider an ARMA model.

$$s(z) \approx \frac{\sum_{n=0}^{N} a_n z^n}{\sum_{m=0}^{M} b_m z^m} \tag{13.23}$$

In order to use the all-pole model of equation (13.22) we need a method of finding the coefficients b_m, but these are precisely the LPC coefficients of Section 9.9. We saw there how to set up the Yule-Walker equations and solve them using the Levinson-Durbin recursion. Once we have them, what is the explicit connection between the LPC coefficients and the AR spectrum of the signal? From

$$H(z) = \frac{G}{1 + \sum_{m=1}^{M} b_m z^{-m}} = \frac{G}{\sum_{m=0}^{M} b_m z^{-m}}$$

it is straightforward to obtain the power spectrum by restricting to the unit circle.

$$|H(\omega)|^2 = \frac{G^2}{|1 + \sum_{k=1}^{p} a_k e^{-i\omega k}|^2} = \frac{G^2}{|\sum_{k=0}^{p} a_k e^{-i\omega k}|^2} \tag{13.24}$$

Which type of approximation is best, all-zero, all-pole, ARMA? The answer depends on the problem at hand. Speech signals tend to have spectra with resonant peaks called formants caused by the geometry of the vocal tract (see Section 19.1). Such spectra are most naturally approximated by all-pole models. All-zero DFT based methods are better for spectra containing narrow valleys but no peaks, such as noise that passed through notch filters. In any case arbitrary spectra can be approximated by either all-pole or all-zero models by using high enough orders. From this point of view, the incentive behind choosing a model is one of efficiency.

Yet there is another reason for choosing an all-pole model. The Wiener-Khintchine theorem relates the power spectrum to the infinite set of autocorrelations $C_s(m)$ for all lags m. In practice we can compute only a limited number of autocorrelations, and would like to estimate the power spectrum based on these. We might assume that all unknown autocorrelations are exactly zero,

$$S(\omega) = \sum_{m=-\infty}^{\infty} C_s(m) e^{-im\omega} \rightarrow \sum_{m=-M}^{M} C_s(m) e^{-im\omega} \tag{13.25}$$

which is not a bad assumption if the autocorrelations die off rapidly enough. This is easily seen to be an all-zero approximation, and leads to the blurring of sharp spectral lines. In 1967, John Burg introduced an alternative assumption, that the spectral estimate should be the most random spectrum consistent with the lags we *do* have. By 'most random' we mean the spectrum with the highest 'entropy', and thus this technique is called *maximum entropy* spectral analysis.

The reasoning behind the maximum entropy principle is easy to understand. DFT methods assume that all data that has not been observed either consist of periodic repetition of the data we have seen or are identically zero. There is usually little physical evidence for such assumptions! Maximum entropy means that we should remain as open minded as possible regarding unseen data. Indeed Burg's method actually tells us to use the most *unpredictable* extrapolation of the data possible. There are many possible spectra that are consistent with the data we have collected, each corresponding to a different extrapolation of the data; maximum entropy insists that the most likely spectrum is that corresponding to the least constraints on the unknown data. In other words we should assume that the uncollected data is as random as possible.

What type of approximation corresponds to the maximum entropy assumption? In Section 18.6 we will see, and if you have studied thermodynamics you already know, that maximum randomness means maximization of the 'entropy'. We assume that the entropy

$$H \equiv \int \ln S(\omega) \, d\omega \tag{13.26}$$

is maximized under the constraint that the *observed* autocorrelation lags (those with lags $|m| \leq M$) do indeed obey Wiener-Khintchine.

$$C_s(m) = \frac{1}{2\pi} \int S(\omega) e^{im\omega} \tag{13.27}$$

The integral in equation (13.26) depends on all the autocorrelations, not just the known ones, and the maximum we seek is for all possible values of the unknown autocorrelations. We differentiate with respect to all the autocorrelations with $|m| > M$ and set equal to zero.

$$0 \quad = \quad \frac{\partial H}{\partial C_s(m)} \quad = \quad \int \frac{d \ln S}{dS} \frac{\partial S(\omega)}{\partial C_s(m)} \, d\omega \quad = \quad \int \frac{1}{S(\omega)} e^{-im\omega} \, d\omega$$

We see that the Fourier coefficients of the reciprocal of $S(\omega)$ are zero for $|m| > M$ (i.e., the inverse spectrum is a finite Fourier series). Accordingly, the maximum entropy power spectrum can be written as the reciprocal of a finite Fourier series, that is, is all-pole.

EXERCISES

13.7.1 Generate a signal that is the sum of a small number of sinusoids and noise. Find the PSD via the periodogram. Solve the LPC equations and derive the spectrum using equation (13.24). Compare the results. Experiment by placing weak spectral lines close to strong ones (recall exercise 13.4.4).

13.7.2 Record some speech and compute its all-pole spectrum. What features do you observe? Can you recognize different sounds from the PSD?

13.8 Walsh Functions

As we saw in Section 3.4, the Fourier transform is easily computable because of the orthogonality of the sinusoids. The sinusoids are in some senses the simplest orthogonal family of functions, but there are other families that are simple in other ways. The Walsh functions, the first few of which are depicted in Figure 13.8, are an interesting alternative to the sinusoids. They are reminiscent of square waves, but comprise a complete orthogonal family. Like square waves all of the signal values are ± 1; due to this characteristic the Walsh transform can be computed without any true multiplications at all.

It is conventional to define the Walsh functions recursively. For the unit interval $0 \le t \le 1$ we define

$$\text{wal}^{[0]}(t) = 1 \tag{13.28}$$
$$\text{cal}^{[k]}(t) = \text{wal}^{[2k]}(t) = \text{wal}^{[k]}\left(2t\right) + (-1)^k \text{wal}^{[k]}\left(2(t - \tfrac{1}{2})\right)$$
$$\text{sal}^{[k+1]}(t) = \text{wal}^{[2k+1]}(t) = \text{wal}^{[k]}\left(2t\right) - (-1)^k \text{wal}^{[k]}\left(2(t - \tfrac{1}{2})\right)$$

and assume all of the functions to be zero to $t < 0$ and $t < 1$. After thus defining the functions on the unit interval we extend the definitions periodically to the entire t axis. Note that the 'wal' functions are a single family like the complex exponentials, while the 'sal' and 'cal' functions are analogous

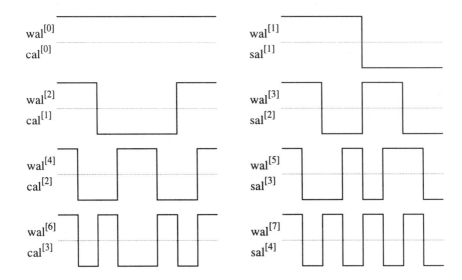

Figure 13.8: The first eight Walsh functions in order of increasing sequency. The cal functions are on the left and the sal functions on the right.

to sine and cosine. The label k equals the number of transitions in the unit interval and is called the *sequency*.

The value-by-value product of two Walsh functions is always a Walsh function

$$\text{wal}^{[i]}(t)\,\text{wal}^{[j]}(t) \;=\; \text{wal}^{[k]}(t)$$

where k is obtained by bit-by-bit xor of i and j. From this and the fact that all Walsh functions except $\text{wal}^{[0]}$ are DC-free it is easy to prove the required orthonormality property.

$$\int_0^1 \text{wal}^{[i]}(t)\,\text{wal}^{[j]}(t)\,dt = \delta_{i,j}$$

There is also a discrete version of this property.

$$\sum_{n=0}^{N-1} \text{wal}^{[i]}\left(\frac{n}{N}\right)\text{wal}^{[j]}\left(\frac{n}{N}\right) = N\delta_{i,j}$$

Analogously to the Fourier transform we can expand arbitrary signals as combinations of Walsh functions, thus defining the Walsh transform. Hence signals can be interpreted as functions in the time domain or sequency domain. In DSP we are more interested in the discrete Walsh transform (DWT)

$$X_k = \sum_{n=0}^{N-1} x_n \, \mathrm{wal}^{[k]}\left(\frac{n}{N}\right)$$

$$x_n = \frac{1}{N} \sum_{k=0}^{N-1} X_k \, \mathrm{wal}^{[k]}\left(\frac{n}{N}\right) \tag{13.29}$$

where the normalization was chosen according to our usual convention. This looks very similar to the DFT. The two-point transform is identical to that of the DFT

$$X_0 = x_0 + x_1$$
$$X_1 = x_0 - x_1$$

but the four-point transform is simpler.

$$X_0 = x_0 + x_1 + x_2 + x_3$$
$$X_1 = x_0 + x_1 - x_2 - x_3$$
$$X_2 = x_0 - x_1 - x_2 + x_3$$
$$X_3 = x_0 - x_1 + x_2 - x_3$$

Note that Walsh transforms are computed without nontrivial multiplications, requiring only N^2 additions. Analogously to the FFT, a fast Walsh transform (FWT) can be defined, reducing this to $O(N \log N)$ additions.

EXERCISES

13.8.1 Plot sal(t) and overlay it with $\sin(2\pi t)$; similarly plot cal(t) and overlay $\cos(2\pi t)$. What is the relationship?

13.8.2 What is the connection between the Walsh functions and the Hadamard matrices defined in exercise 14.5.3?

13.8.3 Find a nonrecursive formula for $\mathrm{wal}^{[k]}(t)$.

13.8.4 Write a program that computes the decimation in sequency fast Walsh transform (see Section 14.3).

13.8.5 Since the DWT can be computed without multiplications and using only real arithmetic, it would be useful to be able to obtain the DFT from the DWT. How can this be done?

13.8.6 In the Fourier case multiplication is related to convolution. What is the analogous result for Walsh transforms?

13.8.7 Another purely real transform is the Hartley transform

$$X(f) = \frac{1}{2\pi} \int_{-\infty}^{\infty} x(t)\, \text{cas}(\omega t)\, dt$$

$$x(t) = \int_{-\infty}^{\infty} X(f)\, \text{cas}(\omega t)\, df$$

where we defined $\text{cas}(t) = \cos(t) + \sin(t)$. Note that the Hartley transform is its own inverse (to within normalization). Similarly we can define the discrete Hartley transform.

$$X_k = \frac{1}{N} \sum_{n=0}^{N-1} x_n\, \text{cas}\left(\frac{2\pi n k}{N}\right)$$

$$x_n = \sum_{k=0}^{N-1} X_k\, \text{cas}\left(\frac{2\pi n k}{N}\right)$$

How do you retrieve the power spectrum from the Hartley transform? Obtain the DFT from the discrete Hartley transform. Develop a fast Hartley transform (see Section 14.3).

13.9 Wavelets

No modern treatment of spectral analysis could be complete without mentioning wavelets. Although there were early precursors, wavelet theory originated in the 1980s when several researchers realized that spectral analysis based on basis functions that are localized in both frequency and time could be useful and efficient in image and signal processing.

What exactly is a wavelet? A basic wavelet is a signal $\psi(t)$ of finite time duration. For example, a commonly used basic wavelet is the sinusoidal pulse

$$\psi(t) = w(t)\, e^{i\omega_0 t} \tag{13.30}$$

where $w(t)$ is any windowing function such as a rectangular window, a sinc window or a raised cosine window. Such a pulse is only nonzero in the time domain in the vicinity of t_0 and only nonzero in the frequency domain near ω_0. The STFT is based on just such functions with $w(t)$ being the window chosen (see Section 13.4) and ω_0 the center of a bin. From the basic wavelet we can make scaled and translated wavelets using the following transformation

$$\psi(\tau, t_0, t) = \frac{1}{\sqrt{\tau}}\, \psi\left(\frac{t - t_0}{\tau}\right) \tag{13.31}$$

where the prefactor normalizes the energy. The time translation t_0 is simple to understand, it simply moves the wavelet along the time axis. The time duration of the wavelet is proportional to τ; conversely, you can think of the scaling transformation compressing the time scale for $\tau > 1$ and stretching it for $\tau < 1$. The center frequency is inversely proportional to τ (i.e., the frequency axis scales in the opposite way to the time axis).

What about the wavelet's bandwidth? Since the nonzero bandwidth results from the finite time duration via the uncertainty theorem, the bandwidth must scale inversely to τ. This last statement can be made more explicit by borrowing the filter design concept of the Q.

$$Q \equiv \frac{\Delta f}{f_0} \tag{13.32}$$

Since the center frequency $f = 2\pi\omega_0$ and the bandwidth both scale inversely with τ, all the wavelets $\psi(\tau(t), t_0, t)$ have the same Q.

We can now build a transform based on these wavelets by replacing the infinite-duration sinusoids of the FT by finite-duration wavelets.

$$S(\tau, t_0) = \frac{1}{\sqrt{\tau}} \int s(t)\psi(\tau, t_0, t) \, dt \tag{13.33}$$

The essential difference between the constant Q wavelet transform and Fourier transform is depicted in Figure 13.9. The DFT divides the frequency axis into equal bandwidth bins, while the wavelet transform bins have constant Q and thus increase in bandwidth with increasing frequency. The center frequencies of the wavelet transform are equally spaced on a logarithmic frequency axis, compressive behavior much like that of our senses (see Section 11.2). While the STFT is matched to artificial signals engineered to equally partition the spectrum, the wavelet transform may be more suited to 'natural' signals such as speech.

Are there cases where the wavelet transform is obviously more appropriate than the STFT? Assume we need to analyze a signal composed of short pulses superposed on sinusoids of long duration. We need to measure both the frequencies of the sinusoids as well as the time durations of the pulses. The uncertainty theorem restricts our accurately measuring the frequency of the pulses, but not that of the steady sinusoids; but to use the STFT we are forced into making a choice. If we use long windows we can accurately measure the frequencies, but blur the pulse time information; if we use short windows we can note the appearances and disappearances of the pulses, but our frequency resolution has been degraded. Using the wavelet transform the

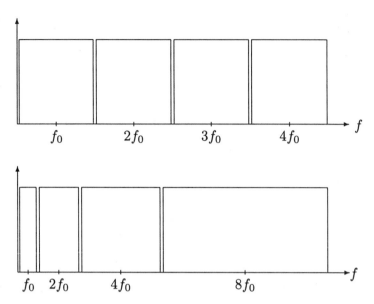

Figure 13.9: Comparison of Fourier and wavelet transforms. The top figure depicts a four bin DFT while the bottom is a four-bin wavelet transform. Note that the FT bins have constant bandwidth while those of the wavelet transform have constant Q. For the purposes of illustration we have taken the basic wavelet to be rectangular in the frequency domain.

time resolution gets better with higher frequency, while the frequency reso- lution becomes better at low frequencies (longer time durations). So using a single wavelet transform we can perform both measurements.

Digital wavelet transforms can be computed efficiently (a fast wavelet transform) using the pyramid algorithm, which extends the recursive com- putation of the logarithmic spectrum discussed in Section 13.2. We employ a pair of filters called **Quadrature Mirror Filters** (QMFs). The QMF pair consists of a low-pass FIR filter that passes the lower half of the spectrum and a high-pass FIR filter that passes the upper half. The two filters are required to be mirror images of each other in the spectral domain, and in addition they must guarantee that the original signal may be recovered. The simplest QMF pair is $(\frac{1}{2}, \frac{1}{2})$ and $(\frac{1}{2}, -\frac{1}{2})$, the first being low-pass, the second high-pass, and their sum obviously the original signal. The pyramid algorithm works as follows. First we apply the QMF filters to the incoming signal, creating two new signals of half the original bandwidth. Since these signals are half bandwidth, we can decimate them by a factor of two without loss of information. The decimated output of the high-pass is retained as the signal of the highest bin, and the QMF filters applied to lower band signal. Once again both signals are decimated and the higher one retained. The process is repeated, halving the bandwidth at each iteration, until all the desired outputs are obtained. The name 'pyramid' refers to the graph depicting the hierarchical relationship between the signals.

What is the computational complexity of the pyramid algorithm? If the QMF filters have M coefficients, the first iteration requires $2MN$ multiplications to produce N outputs. The second iteration operates on decimated signals and so requires only MN multiplications to produce outputs corresponding to the same time duration. Each iteration requires half the computations of the preceding, so even were we to compute an infinite number of iterations the number of multiplications would be $2 + 1 + \frac{1}{2} + \frac{1}{4} + \ldots = 4$ times MN. So the wavelet transform is $O(N)$ (better than the FFT), and no complex operations are required!

EXERCISES

13.9.1 Use a raised cosine times a sine as a basic wavelet and draw the scaled wavelets for various τ. Compare these with the basis functions for the STFT.

13.9.2 A digital QMF pair obeys $|H^{[hp]}(f)| = |H^{[lp]}(\frac{f_s}{2} - f)|$, where $H^{[lp]}(f)$ and $H^{[hp]}(f)$ are the frequency responses of the low-pass and high-pass filters. Show that $h_n^{[hp]} = (-1)^n h_n^{[lp]}$ or $h_n^{[hp]} = (-1)^n h_{-n}^{[lp]}$ for odd length filters and similar statements for even length ones. Show that the latter form is consistent with the wavelets being an orthogonal basis for even length filters.

13.9.3 Can the original signal really be recovered after QMF filtering and *decimation* have been applied?

13.9.4 Derive an efficient procedure for computing the inverse digital wavelet transform.

13.9.5 Build a signal consisting of two close sinusoids that pulse on and off. Similarly build a signal that consists of a single sinusoid that appears as two close pulses. Try to simultaneously measure frequency and time phenomena using the STFT and the wavelet transform.

13.9.6 Compare the wavelet transform with the time-frequency distributions discussed in Section 4.6.

Bibliographical Notes

Kay and Marple have written a good tutorial of modern approaches to spectral analysis [128], and each has written a book as well [127, 123].

The periodogram was first introduced by Schuster [233] and became an even more indispensable tool after the introduction of the FFT. Blackman and Tukey wrote early book on the practical calculation of power spectra [19]. The bible of windows was written by Harris [93].

As noted in the bibliographical notes to Chapter 6, Yule [289] formulated the Yule-Walker equations for signals containing one or two sinusoidal components in the late 1920s, in an attempt to explain the 11-year periodicity of sunspot numbers. Walker, calling Yule's earlier work 'an important extension of our ideas regarding periodicity', expanded on this work, discovering that the autocorrelations were much smoother than the noisy signal itself, and suggesting using the 'correlation periodogram' as a substitute for the Schuster periodogram. He applied this technique to the analysis of air pressure data and could rule out as spurious various claimed periodicities.

Wiener was instrumental in explaining why the Schuster periodogram did not work well for noisy signals [276, 277], but this was not widely appreciated at the time. A highly interesting historical account is given by Robinson [223].

Pisarenko's original article is [196]. Various authors have analyzed the performance of the PHD [227, 255, 285].

Officer of the U.S. Navy and Harvard mathematician Joseph Leonard Walsh presented his functions in 1923 [268]. The standard text is [12] and a short introduction can be found in [13]. The conversion between DWT and DFT was expounded in [256, 257]. Hartley, who was in charge of telephone research at Bell Labs and responsible for an early analog oscillator, presented his transform in 1942 [94]. The DHT and FHT were published in [24, 25]. The standard text is [27].

Wavelets already have a rich literature. For DSP purposes we recommend the review article by Rioul and Vetterli [221].

14

The Fast Fourier Transform

It is difficult to overstate the importance of the FFT algorithm for DSP. We have often seen the essential duality of signals in our studies so far; we know that exploiting both the time and the frequency aspects is critical for signal processing. We may safely say that were there not a fast algorithm for going back and forth between time and frequency domains, the field of DSP as we know it would never have developed.

The discovery of the first FFT algorithm predated the availability of hardware capable of actually exploiting it. The discovery dates from a period when the terms *calculator* and *computer* referred to people, particularly adept at arithmetic, who would perform long and involved rote calculations for scientists, engineers, and accountants. These *computers* would often exploit symmetries in order to save time and effort, much as a contemporary programmer exploits them to reduce electronic computer run-time and memory. The basic principle of the FFT ensues from the search for such time-saving mechanisms, but its discovery also encouraged the development of DSP hardware. Today's DSP chips and special-purpose FFT processors are children of both the microprocessor age and of the DSP revolution that the FFT instigated.

In this chapter we will discuss various algorithms for calculating the DFT, all of which are known as *the* FFT. Without a doubt the most popular algorithms are radix-2 DIT and DIF, and we will cover these in depth. These algorithms are directly applicable only for signals of length $N = 2^m$, but with a little ingenuity other lengths can be accommodated. Radix-4, split radix, and FFT842 are even faster than basic radix-2, while mixed-radix and prime factor algorithms directly apply to N that are not powers of two. There are special cases where the fast Fourier transform can be made even faster. Finally we present alternative algorithms that in specific circumstances may be faster than the FFT.

14.1 Complexity of the DFT

Let us recall the previously derived formula (4.32) for the N-point DFT

$$X_k = \sum_{n=0}^{N-1} x_n W_N^{nk}$$

where the Nth root of unity is defined as

$$W_N \equiv e^{-i\frac{2\pi}{N}} = \cos\left(\frac{2\pi}{N}\right) - i\sin\left(\frac{2\pi}{N}\right)$$

How many calculations must we perform to find one X_k from a set of N time domain values x_n? Assume that the complex constant W_N and its powers W_N^j have all been precalculated and stored for our use. Looking closely at (4.32) we see N complex multiplications and $N-1$ complex additions are performed in the loop on n. Now to find the *entire* spectrum we need to do N such calculations, one for each value of k. So we expect to have to carry out N^2 complex multiplications, and $N(N-1)$ complex additions.

This is actually a slight overestimate. By somewhat trickier programming we can take advantage of the fact that $W_N^0 = 1$, so that each of the $X_{k>0}$ takes $N-1$ multiplications and additions, while X_0 doesn't require any multiplications. We thus really need only $(N-1)^2$ complex multiplications.

A complex addition requires the addition of real and imaginary parts, and is thus equivalent to two real additions. A complex multiplication can be performed as four real multiplications and two additions $(a+ib)(c+id) = (ac-bd) + i(bc+ad)$ or as three multiplications and five additions $(a+ib)(c+id) = a(c+d) - d(a+b) + i\left(a(c+d) + c(b-a)\right)$. The latter form may be preferred when multiplication takes much more time than addition, but can be less stable numerically. Other combinations are possible, but it can be shown that there is no general formula for the complex product with less than three multiplications. Using the former, more common form, we find that the computation of the entire spectrum requires $4(N-1)^2$ real multiplications and $2(N-1)(2N-1)$ real additions.

Actually, the calculation of a single X_k can be performed more efficiently than we have presented so far. For example, Goertzel discovered a method of transforming the iteration in equation (4.32) into a recursion. This has the effect of somewhat reducing the computational complexity and also saves the precomputation and storage of the W table. Goertzel's algorithm, to be presented in Section 14.8, still has asymptotic complexity of order $O(N)$ per calculated X_k, although with a somewhat smaller constant than the direct method. It thus leaves the complexity of calculation of the entire spectrum at $O(N^2)$, while the FFT algorithms to be derived presently are less complex. Goertzel's algorithm thus turns out to be attractive when a single, or only a small number of X_k values are needed, but is not the algorithm of choice for calculating the entire spectrum.

Returning to the calculation of the entire spectrum, we observe that both the additions and multiplications increase as $O(N^2)$ with increasing N. Were this direct calculation the only way to find the DFT, real-time calculation of large DFTs would be impractical. It is general rule in DSP programming that only algorithms with linear asymptotic complexity can be performed in real-time for large N. Let us now see why this is the case.

The criterion for real-time calculation is simple for algorithms that process a single input sample at a time. Such an algorithm must finish all its computation for each sample before the next one arrives. This restricts the number of operations one may perform in such computations to the number performable in a sample interval t_s. This argument does not directly apply to the DFT since it is inherently a block-oriented calculation. One cannot perform a DFT on a single sample, since frequency is only defined for signals that occupy some nonzero interval of time; and we often desire to process large blocks of data since the longer we observe the signal the more accurate frequency estimates will be.

For block calculations one accumulates samples in an array, known as a buffer, and then processes this buffer as a single entity. A technique known as *double-buffering* is often employed in real-time implementations of block calculations. With double-buffering two buffers are employed. While the samples in the *processing buffer* are being processed, the *acquisition buffer* is acquiring samples from the input source. Once processing of the first buffer is finished and the output saved, the buffers are quickly switched, the former now acquiring samples and the latter being processed.

How can we tell if block calculations can be performed in real-time? As for the single sample case, one must be able to finish all the processing needed in time. Now 'in time' means completing the processing of one entire buffer, before the second buffer becomes full. Otherwise a condition known as *data-overrun* occurs, and new samples overwrite previously stored, but as yet unprocessed, ones. It takes $N\Delta t$ seconds for N new samples to arrive. In order to keep up we must process all N old samples in the processing buffer before the acquisition buffer is completely filled. If the complexity is linear (i.e., the processing time for N samples is proportional to N), then $C = qN$ for some q. This q is the *effective time per sample* since each sample effectively takes q time to process, independent of N. Thus, the selection of buffer size is purely a memory issue, and does not impact the ability to keep up with real-time. However, if the complexity is superlinear (for example, $T_{processing} = qN^p$ with $p > 1$), then as N increases we have less and less time to process each sample, until eventually some N is reached where we can no longer keep up, and data-overrun is inevitable.

Let's clarify this by plugging in some numbers. Assume we are acquiring input at a sample rate of 1000 samples per second (i.e., we obtain a new sample every millisecond) and are attempting to process blocks of length 250. We start our processor, and for one-quarter of a second, we cannot do any processing, until the first acquisition buffer fills. When the buffer is full we quickly switch buffers, start processing the 250 samples collected, while the second buffer of length 250 fills. We must finish the processing within a quarter of second, in order to be able to switch buffers back when the acquisition buffer is full. When the dependence of the processing time on buffer length is strictly linear, $T_{processing} = qN$, then if we can process a buffer of $N = 250$ samples in 250 milliseconds or less, we can equally well process a buffer of 500 samples in 500 milliseconds, or a buffer of $N = 1000$ samples in a second. Effectively we can say that when the single sample processing time is no more than $q = 1$ millisecond per sample, we can maintain real-time processing.

What would happen if the buffer processing time depended quadratically on the buffer size—$T_{processing} = qN^2$? Let's take q to be 0.1 millisecond per sample squared. Then for a small 10-millisecond buffer (length $N = 10$), we will finish processing in $T_{processing} = 0.1 \cdot 10^2 = 10$ milliseconds, just in time! However, a 100-millisecond buffer of size $N = 100$ will require $T_{processing} = 0.1 \cdot 100^2$ milliseconds, or one second, to process. Only by increasing our computational power by a factor of ten would we be able to maintain real-time! However, even were we to increase the CPU power to accommodate this buffer-size, our 250-point buffer would still be out of our reach.

As we have mentioned before, the FFT is an algorithm for calculating the DFT more efficiently than quadratically, at least for certain values of N. For example, for powers of two, $N = 2^k$, its complexity is $O(N \log_2 N)$. This is only very slightly superlinear, and thus while *technically* the FFT is not suitable for real-time calculation in the asymptotic $N \to \infty$ limit, in practice it *is* computable in real-time even for relatively large N. To grasp the speed-up provided by the FFT over direct calculation of (4.29), consider that the ratio between the complexities is proportional to $\frac{N}{\log_2 N}$. For $N = 2^4 = 16$ the FFT is already four times faster than the direct DFT, for $N = 2^{10} = 1024$

it is over one hundred times faster, and for $N = 2^{16}$ the ratio is 4096! It is common practice to compute $1K$- or $64K$-point FFTs in real-time, and even much larger sizes are not unusual.

The basic idea behind the FFT is the very exploitation of the N^2 complexity of the direct DFT calculation. Due to this second-order complexity, it is faster to calculate a lot of small DFTs than one big one. For example, to calculate a DFT of length N will take N^2 multiplications, while the calculation of two DFTs of length $\frac{N}{2}$ will take $2(\frac{N}{2})^2 = \frac{N^2}{2}$, or half that time. Thus if we can somehow piece the two partial results together to one spectrum in less than $\frac{N^2}{2}$ time then we have found a way to save time. In Sections 14.3 and 14.4 we will see several ways to do just that.

EXERCISES

14.1.1 Finding the maximum of an N-by-N array of numbers can be accomplished in $O(N^2)$ time. Can this be improved by partitioning the matrix and exploiting the quadratic complexity as above?

14.1.2 In exercise 4.7.4 you found explicit equations for the $N = 4$ DFT for $N = 4$. Count up the number of *complex* multiplications and additions needed to compute X_0, X_1, X_2, and X_3. How many *real* multiplications and additions are required?

14.1.3 Define temporary variables that are used more than once in the above equations. How much can you save? How much memory do you need to set aside? (Hint: Compare the equations for X_0 and X_2.)

14.1.4 Up to now we have not taken into account the task of finding the trigonometric W factors themselves, which can be computationally intensive. Suggest at least two solutions, one that requires a large amount of auxiliary memory but practically no CPU, and one that requires little memory but is more CPU intensive.

14.1.5 A computational system is said to be 'real-time-oriented' when the time it takes to perform a task can be guaranteed. Often systems rely on the weaker criterion of *statistical real-time*, which simply means that on-the-average enough computational resources are available. In such cases double buffering can be used in the acquisition hardware, in order to compensate for peak MIPS demands. Can hardware buffering truly make an arbitrary system as reliable as a real-time-oriented one?

14.1.6 Explain how double-buffering can be implemented using a single circular buffer.

14.2 Two Preliminary Examples

Before deriving the FFT we will prepare ourselves by considering two somewhat more familiar examples. The ideas behind the FFT are very general and not restricted to the computation of equation (14.1). Indeed the two examples we use to introduce the basic ideas involve no DSP at all.

How many comparisons are required to find the maximum or minimum element in a sequence of N elements? It is obvious that $N-1$ comparisons are absolutely needed if all elements are to be considered. But what if we wish to simultaneously find the maximum *and* minimum? Are twice this number really needed? We will now show that we can get away with only $1\frac{1}{2}$ times the number of comparisons needed for the first problem. Before starting we will agree to simplify the above number of comparisons to N, neglecting the 1 under the asymptotic assumption $N \gg 1$.

A fundamental tool employed in the reduction of complexity is that of splitting long sequences into smaller subsequences. How can we split a sequence with N elements

$$x_0, \ x_1, \ x_2, \ x_3, \ \ldots \ x_{N-2}, \ x_{N-1}$$

into two subsequences of half the original size (assume for simplicity's sake that N is even)? One way is to consider pairs of adjacent elements, such as x_1, x_2 or x_3, x_4, and place the smaller of each pair into the first subsequence and the larger into the second. For example, assuming $x_0 < x_1, \ x_2 > x_3$ and $x_{N-2} < x_{N-1}$, we obtain

$$
\begin{array}{cccc}
x_0 \ x_3 & \ldots & \min(x_{2l}, x_{2l+1}) & \ldots & x_{N-2} \\
x_1 \ x_2 & \ldots & \max(x_{2l}, x_{2l+1}) & \ldots & x_{N-1}
\end{array}
$$

This splitting of the sequence requires $\frac{N}{2}$ comparisons. Students of sorting and searching will recognize this procedure as the first step of the *Shell sort*.

Now, the method of splitting the sequence into subsequences guarantees that the minimum of the entire sequence must be one of the elements of the first subsequence, while the maximum must be in the second. Thus to complete our search for the minimum and maximum of the original sequence, we must find the minimum of the first subsequence and the maximum of the second. By our previous result, each of these searches requires $\frac{N}{2}$ comparisons. Thus the entire process of splitting and two searches requires $\frac{N}{2} + 2\frac{N}{2} = \frac{3N}{2}$ comparisons, or $1\frac{1}{2}$ times that required for the minimum or maximum alone.

Can we further reduce this factor? What if we divide the original sequence into adjacent quartets, choosing the minimum and maximum of the four? The splitting would then cost four comparisons per quartet, or N comparison altogether, and then two $\frac{N}{4}$ searches must be carried out. Thus we require $N + 2\frac{N}{4}$ and a factor of $1\frac{1}{2}$ is still needed. Indeed, after a little reflection, the reader will reach the conclusion that no further improvement is possible. This is because the new problems of finding only the minimum *or* maximum of a subsequence are simpler than the original problem.

When a problem *can* be reduced recursively to subproblems *similar* to the original, the process may be repeated to attain yet further improvement. We now discuss an example where such recursive repetition is possible. Consider multiplying two $(N+1)$-digit numbers A and B to get a product C using long multiplication (which from Section 6.8 we already know to be a convolution).

			A_N	A_{N-1}	$\cdots$	A_1	A_0
			B_N	B_{N-1}	$\cdots$	B_1	B_0
			$B_0 A_N$	$B_0 A_{N-1}$	$\cdots$	$B_0 A_1$	$B_0 A_0$
		$B_1 A_N$	$B_1 A_{N-1}$		$\cdots$	$B_1 A_0$	
					$\vdots$		
$B_N A_N$	$\cdots$	$B_N A_1$	$B_N A_0$				
C_{2N}	$\cdots$	C_{N+1}	C_N	C_{N-1}	$\cdots$	C_1	C_0

Since we must multiply every digit in the top number by every digit in the bottom number, the number of one-digit multiplications is N^2. You are probably used to doing this for decimal digits, but the same multiplication algorithm can be utilized for N-bit binary numbers. The hardware-level complexity of straightforward multiplication of two N-bit numbers is proportional to N^2.

Now assume N is even and consider the left $\frac{N}{2}$ digits and the right $\frac{N}{2}$ digits of A and B separately. It does not require much algebraic prowess to convince oneself that

$$
\begin{aligned}
A &= A_L 2^{\frac{N}{2}} + A_R \\
B &= B_L 2^{\frac{N}{2}} + B_R \\
C &= A_L B_L 2^N + (A_L B_R + A_R B_L) 2^{\frac{N}{2}} + A_R B_R \\
&= A_L B_L (2^N + 2^{\frac{N}{2}}) + (A_L - A_R)(B_R - B_L) 2^{\frac{N}{2}} + A_R B_R (2^{\frac{N}{2}} + 1)
\end{aligned}
\tag{14.1}
$$

involving only three multiplications of $\frac{N}{2}$-length numbers. Thus we have reduced the complexity from N^2 to $3(\frac{N}{2})^2 = \frac{3}{4}N^2$ (plus some shifting and adding operations). This is a savings of 25%, but does not reduce the asymptotic form of the complexity from $O(N^2)$. However, in this case we have only just begun! Unlike for the previous example, we have reduced the original multiplication problem to three similar but simpler multiplication problems! We can now carry out the three $\frac{N}{2}$-bit multiplications in equation (14.1) similarly (assuming that $\frac{N}{2}$ is still even) and continue recursively. Assuming N to have been a power of two, we can continue until we multiply individual bits. This leads to an algorithm for multiplication of two N-bit numbers, whose asymptotic complexity is $O(N^{\log_2(3)}) \approx O(N^{1.585})$. The slightly more sophisticated Toom-Cook algorithm divides the N-bit numbers into more than two groups, and its complexity can be shown to be $O(N \log(N) 2^{\sqrt{2 \log(N)}})$. This is still not the most efficient way to multiply numbers. Realizing that each column sum of the long multiplication in equation (14.1) can be cast into the form of a convolution, it turns out that the best way to multiply large numbers is to exploit the FFT!

EXERCISES

14.2.1 The reader who has implemented the Shell sort may have used a different method of choosing the pairs of elements to be compared. Rather than comparing adjacent elements x_{2l} and x_{2l+1}, it is more conventional to consider elements in the same position in the first and second halves the sequence, x_k and $x_{\frac{N}{2}+k}$ Write down a general form for the new sequence. How do we find the minimum and maximum elements now? These two ways of dividing a sequence into two subsequences are called *decimation* and *partition*.

14.2.2 Devise an algorithm for finding the *median* of N numbers in $O(N \log N)$.

14.2.3 The product of two two-digit numbers, ab and cd, can be written $ab * cd = (10 * a + b) * (10 * c + d) = 100ac + 10(ad + bc) + bd$. Practice multiplying two-digit numbers in your head using this rule. Try multiplying a three-digit number by a two-digit one in similar fashion.

14.2.4 We often deal with complex-valued signals. Such signals can be represented as vectors in two ways, *interleaved*

$$\Re(x_1)\Im(x_1), \Re(x_2), \Im(x_2), \ldots \Re(x_N), \Im(x_N)$$

or *separated*

$$\Re(x_1)\Re(x_2), \ldots \Re(x_N), \Im(x_1), \Im(x_2), \ldots \Im(x_N)$$

Devise an efficient in-place algorithm changing between interleaved and separated representations. Efficient implies that each element accessed is moved immediately to its final location. In-place means here that if extra memory is used it must be of constant size (independent of N). What is the algorithm's complexity?

14.3 Derivation of the DIT FFT

Without further ado, we turn to the derivation of the our first FFT algorithm. As mentioned in the previous section, we want to exploit the fact that it is better to compute many small DFTs than a single large one; as a first step let's divide the sequence of signal values into two equal parts.

There are two natural ways of methodically dividing a sequence into two subsequences, *partition* and *decimation*. By partition we mean separating the sequence at the half-way point

$$\{x_0, x_1, \ldots, x_{N-2}, x_{N-1}\}$$

$$\{x_0, x_1, \ldots x_{\frac{N}{2}-1}\} \qquad\qquad \{x_{\frac{N}{2}}, \ldots x_{N-1}\}$$
LOW PARTITION HIGH PARTITION

while decimation is the separation of the even-indexed from odd-indexed elements.

$$\{x_0, x_1, \ldots, x_{N-2}, x_{N-1}\}$$

$$\{x_0, x_2, \ldots x_{N-2}\} \qquad\qquad \{x_1, x_3, \ldots x_{N-1}\}$$
EVEN DECIMATION ODD DECIMATION

Put another way, partition divides the sequence into two groups according to the MSB of the index, while decimation checks the LSB.

Either partition and decimation may be employed to separate the original signal into half-sized signals for the purpose of computation reduction. Decimation in time implies partition in frequency (e.g., doubling the time *duration* doubles the frequency *resolution*), while partition in time signifies decimation in frequency. These two methods of division lead to somewhat different FFT algorithms, known conventionally as the **Decimation In Time** (DIT) and **Decimation In Frequency** (DIF) FFT algorithms. We will here consider *radix-2* partition and decimation, that is, division into *two* equal-length subsequences. Other partition and decimation radixes are possible, leading to yet further FFT algorithms.

We will now algebraically derive the radix-2 DIT algorithm. We will need the following trigonometric identities.

$$
\begin{aligned}
W_N^N &= 1 \\
W_N^{\frac{N}{2}} &= -1 \\
W_N^2 &= W_{\frac{N}{2}}
\end{aligned}
\tag{14.2}
$$

The FFT's efficiency results from the fact that in the complete DFT many identical multiplications are performed multiple times. As a matter of fact, in X_k and $X_{k+\frac{N}{2}}$ all the multiplications are the same, although every other addition has to be changed to a subtraction.

$$X_k = \sum_{n=0}^{N-1} x_n W_N^{nk}$$

$$X_{k+\frac{N}{2}} = \sum_{n=0}^{N-1} x_n W_N^{nk} W_N^{\frac{Nn}{2}}$$

$$= \sum_{n=0}^{N-1} x_n W_N^{nk} (-1)^n$$

The straightforward computation of the entire spectrum ignores this fact and hence entails wasteful recalculation.

In order to derive the DIT algorithm we separate the sum in (4.32) into sums over even- and odd-indexed elements, utilizing the identities (14.2)

$$X_k = \sum_{n=0}^{N-1} x_n W_N^{nk} = \sum_{n=0}^{\frac{N}{2}-1} \left(x_{2n} W_N^{2nk} + x_{2n+1} W_N^{(2n+1)k} \right)$$

$$= \sum_{n=0}^{\frac{N}{2}-1} x_n^E W_{\frac{N}{2}}^{nk} + W_N^k \sum_{n=0}^{\frac{N}{2}-1} x_n^O W_{\frac{N}{2}}^{nk} \qquad (14.3)$$

where we have defined even and odd subsequences.

$$x_n^E = x_{2n} \qquad \text{for } n = 0, 1, \ldots \frac{N}{2} - 1$$

$$x_n^O = x_{2n+1}$$

After a moment of contemplation it is apparent that the first term is the DFT of the even subsequence, while the second is W_N^k times the DFT of the odd subsequence; therefore we have discovered a recursive procedure for computing the DFT given the DFTs of the even and odd decimations.

Recalling the relationship between X_k and $X_{k+\frac{N}{2}}$ we can immediately write

$$X_{k+\frac{N}{2}} = \sum_{n=0}^{\frac{N}{2}-1} x_n^E W_{\frac{N}{2}}^{nk} - W_N^k \sum_{n=0}^{\frac{N}{2}-1} x_n^O W_{\frac{N}{2}}^{nk}$$

this being the connection between parallel elements in different *partitions* of the frequency domain.

Now let us employ the natural notation

$$X_k^E = \sum_{n=0}^{\frac{N}{2}-1} x_n^E W_{\frac{N}{2}}^{nk} \qquad X_k^O = \sum_{n=0}^{\frac{N}{2}-1} x_n^O W_{\frac{N}{2}}^{nk}$$

and write our results in the succinct recursive form

$$\begin{aligned} X_k &= X_k^E + W_N^k X_k^O \\ X_{k+\frac{N}{2}} &= X_k^E - W_N^k X_k^O \end{aligned} \tag{14.4}$$

which is a computational topology known as a *butterfly* (for reasons soon to be apparent). In this context the W_N^k factor is commonly called a *twiddle factor* (for reasons that we won't adequately explain). We note that the butterfly is basically an in-place computation, replacing two values with two new ones. Using our standard graphical notation we can depict equation (14.4) in the following way:

which, not accidentally, is very similar to the two-point DFT diagram presented in Section 12.2 (actually, remembering the second identity of (14.2) we recognize the two-point FFT as a special case of the DIT butterfly). Rotating the diagram by 90° and using some imagination clarifies the source of the name 'butterfly'. We will soon see that the butterfly is the only operation needed in order to perform the FFT, other than calculation (or table lookups) of the twiddle factors.

Now, is this method of computation more efficient than the straightforward one? Instead of $(N-1)^2$ multiplications and $N(N-1)$ additions for the simultaneous computation of X_k for all k, we now have to compute two $\frac{N}{2}$-point DFTs, one additional multiplication (by the twiddle factor), and two new additions, for a grand total of $2(\frac{N}{2}-1)^2 + 1 = \frac{N^2}{2} - 2N + 3$ multiplications and $2\left(\frac{N}{2}(\frac{N}{2}-1)\right) + 2 = \frac{N^2}{2} - N + 2$ additions. The savings may already be significant for large enough N!

But why stop here? We are assuming that X_k^E and X_k^O were computed by the straightforward DFT formula! We can certainly save on their computation by using the recursion as well! For example, we can find X_k^E by the following butterfly.

As in any recursive definition, we must stop somewhere. Here the obvious final step is the reduction to a two-point DFT, computed by the simplest butterfly. Thus the entire DFT calculation has been recursively reduced to computation of butterflies.

We graphically demonstrate the entire decomposition for the special case of $N = 8$ in the series of Figures 14.1–14.4. The first figure depicts the needed transform as a black box, with x_0 through x_7 as inputs, and X_0 through X_7 as outputs. The purpose of the graphical derivation is to fill in this box.

In Figure 14.2 we slightly rearrange the order of the inputs, and decompose the eight-point transform into two four-point transforms, using equation (14.3). We then continue by decomposing each four-point transform into two two-point transforms in Figure 14.3, and finally substitute our diagram for the two-point butterfly in order to get Figure 14.4. The final figure simply rearranges the inputs to be in the same order as in the first figure. The required permutation is carried out in a black box labeled *Bit Reversal*; the explanation for this name will be given later.

For a given N, how many butterflies must we perform to compute an N-point DFT? Assuming that $N = 2^m$ is a power of 2, we have $m = \log_2 N$ layers of butterflies, with $\frac{N}{2}$ butterflies to be computed in each layer. Since each butterfly involves one multiplication and two additions (we are slightly overestimating, since some of the multiplications are trivial), we require about $\frac{N}{2} \log_2 N$ complex multiplications and $N \log_2 N$ complex additions. We have thus arrived at the desired conclusion, that the complexity of the DIT FFT is $O(N \log N)$ (the basis of the logarithm is irrelevant since all logarithms are related to each other by a multiplicative constant).

Similarly for radix-R DIT FFT, we would find $\log_R N$ layers of $\frac{N}{R}$ butterflies, each requiring $R - 1$ multiplications. The improvement for radixes greater than two is often not worth the additional coding effort, but will be discussed in Section 14.4.

The DIT FFT we have derived and depicted here is an in-place algorithm. At each of its m layers we replace the existing array with a new one of identical size, with no additional array allocation needed. There is one last technical problem related to this in-place computation we need to solve. After all the butterflies are computed in-place we obtain all the desired X_k, but they are not in the right order. In order to obtain the spectrum in the correct order we need one final in-place stage to unshuffle them (see Figure 14.5).

To understand how the X_k are ordered at the end of the DIT FFT, note that the butterflies themselves do not destroy ordering. It is only the successive in-place decimations that change the order. We know that dec-

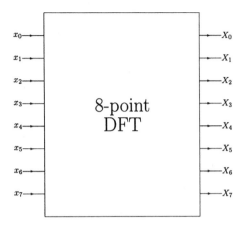

Figure 14.1: An eight-point DFT.

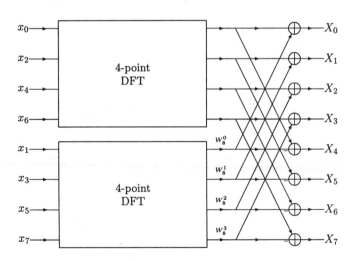

Figure 14.2: An eight-point DFT, divided into two four-point FFTs.

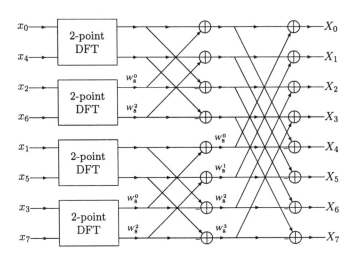

Figure 14.3: An eight-point DFT, divided into four two-point FFTs.

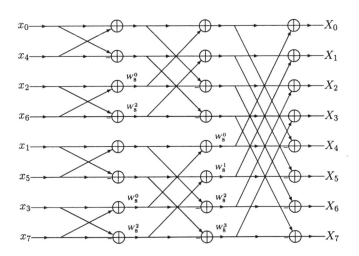

Figure 14.4: The full eight-point radix-2 DIT DFT.

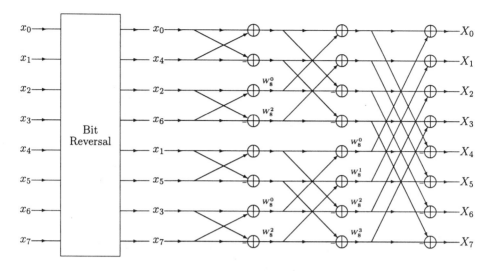

Figure 14.5: The full eight-point radix-2 DIT DFT, with bit reversal on inputs.

imation uses the LSB of the indices to decide how to divide the sequence into subsequences, so it is only natural to investigate the effect of in-place decimation on the binary representation of the indices. For example, for 2^4 element sequences, there are four stages, the indices of which are permuted as follows.

0	1	2	3	4	5	6	7	8	9	10	11	12	13	14	15
0000	0001	0010	0011	0100	0101	0110	0111	1000	1001	1010	1011	1100	1101	1110	1111

0	2	4	6	8	10	12	14	1	3	5	7	9	11	13	15
0000	0010	0100	0110	1000	1010	1100	1110	0001	0011	0101	0111	1001	1011	1101	1111

0	4	8	12	2	6	10	14	1	5	9	13	3	7	11	15
0000	0100	1000	1100	0010	0110	1010	1110	0001	0101	1001	1101	0011	0111	1011	1111

0	8	4	12	2	10	6	14	1	9	5	13	3	11	7	15
0000	1000	0100	1100	0010	1010	0110	1110	0001	1001	0101	1101	0011	1011	0111	1111

Looking carefully we observe that the elements of the second row can be obtained from the matching ones of the first by a circular left shift. Why is that? The first half of the second row, 0246..., is obtained by an arithmetic left shift of the first row elements, while the second half is identical except for having the LSB set. Since the second half of the first row has the MSB set the net effect is the observed circular left shift.

The transition from second to third row is a bit more complex. Elements from the first half and second half of the second row are not intermingled, rather are separately decimated. This corresponds to clamping the LSB and circularly left shifting the more significant bits, as can be readily verified in the example above. Similarly to go from the third row to the fourth we clamp the two least significant bits and circularly shift the rest. The net effect of the m stages of in-place decimation of a sequence of length 2^m is *bit reversal* of the indices.

In order to unshuffle the output of the DIT FFT we just need to perform an initial stage of bit reversal on the x_n, as depicted in Figure 14.5. Although this stage contains no computations, it may paradoxically consume a lot of computation time because of its strange indexing. For this reason many DSP processors contain special addressing modes that facilitate efficient bit-reversed access to vectors.

EXERCISES

14.3.1 Draw the flow diagram for the 16-point DIT FFT including bit reversal. (Hint: Prepare a *large* piece of paper.)

14.3.2 Write an explicitly recursive program for computation of the FFT. The main routine FFT(N,X) should first check if N equals 2, in which case it replaces the two elements of X with their DFT. If not, it should call FFT(N/2,Y) as needed.

14.3.3 Write a nonrecursive DIT FFT routine. The main loop should run $m = \log_2 N$ times, each time computing $\frac{N}{2}$ butterflies. Test the routine on sums of sinusoids. Compare the run time of this routine with that of N straight-forward DFT computations.

14.3.4 Rather than performing bit reversal as the first stage, we may leave the inputs and shuffle the outputs into the proper order. Show how to do this for an eight-point signal. Are there any advantages to this method?

14.3.5 Write an efficient high-level-language routine that performs bit reversal on a sequence of length $N = 2^m$. The routine should perform no more than N element interchanges, and use only integer addition, subtraction, comparison, and single bit shifts.

14.4   Other Common FFT Algorithms

In the previous section we saw the radix-2 DIT algorithm, also known as the Cooley-Tukey algorithm. Here we present a few more FFT algorithms, radix-2 DIF, the prime factor algorithm (PFA), non-power-of-two radixes, split-radix, etc. Although different in details, there is a strong family resemblance between all these algorithms. All reduce the N^2 complexity of the straightforward DFT to $N \log N$ by restructuring the computation, all exploit symmetries of the W_N^{nk}, and all rely on the length of the signal N being highly composite.

First let us consider the decimation in frequency (DIF) FFT algorithm. The algebraic derivation follows the same philosophy as that of the DIT. We start by partitioning the time sequence, into left and right subsequences

$$
\begin{aligned}
x_n^L &= x_n && \text{for } n = 0, 1, \ldots \frac{N}{2} - 1 \\
x_n^R &= x_{n+\frac{N}{2}}
\end{aligned}
$$

and splitting the DFT sum into two sums.

$$
X_k = \sum_{n=0}^{N-1} x_n W_N^{nk} = \sum_{n=0}^{\frac{N}{2}-1} x_n W_N^{nk} + \sum_{n=\frac{N}{2}}^{N-1} x_n W_N^{nk} W_N^{k\frac{N}{2}} \quad (14.5)
$$

$$
= \sum_{n=0}^{\frac{N}{2}-1} x_n^L W_N^{nk} + \sum_{n=0}^{\frac{N}{2}-1} x_n^R W_N^{nk}
$$

Now let's compare the even and odd X_k (decimation in the frequency domain). Using the fact that $W_N^2 = W_{\frac{N}{2}}$

$$
X_{2k} = \sum_{n=0}^{\frac{N}{2}-1} (x_n^L W_{\frac{N}{2}}^{nk} + x_n^R W_{\frac{N}{2}}^{nk} W_N^{Nk})
$$

$$
X_{2k+1} = \sum_{n=0}^{\frac{N}{2}-1} (x_n^L W_{\frac{N}{2}}^{nk} + x_n^R W_{\frac{N}{2}}^{nk} W_N^{Nk}) W_N^n
$$

and then substituting $W_N^{kN} = 1$ and $W_N^{\frac{N}{2}} = -1$ we find

$$
X_{2k} = \sum_{n=0}^{\frac{N}{2}-1} (x_n^L + x_n^R) W_{\frac{N}{2}}^{nk}
$$

$$
X_{2k+1} = \sum_{n=0}^{\frac{N}{2}-1} (x_n^L - x_n^R) W_{\frac{N}{2}}^{Nk} W_N^n
$$

which by linearity of the DFT gives the desired expression.

$$
\begin{aligned}
X_{2k} &= (X_k^L + X_k^R) \quad (14.6) \\
X_{2k+1} &= (X_k^L - X_k^R) W_N^n
\end{aligned}
$$

Just as for the DIT we found similarity between Fourier components in different frequency partitions, for DIF we find similarity between frequency components that are related by decimation.

It is thus evident that the DIF butterfly can be drawn

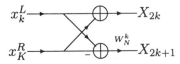

which is different from the DIT butterfly, mainly in the position of the twiddle factor.

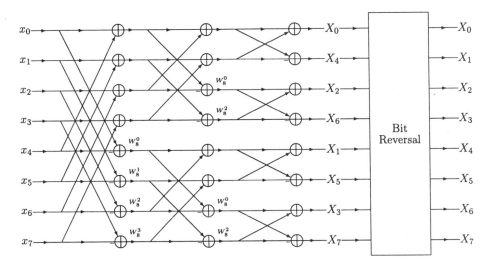

Figure 14.6: Full eight-point radix-2 DIF DFT, with bit reversal on outputs.

We leave as an exercise to complete the decomposition, mentioning that once again bit reversal is required, only this time it is the outputs that need to be bit reversed. The final eight-point DIF DFT is depicted in Figure 14.6.

Next let's consider FFT algorithms for radixes other than radix-2 in more detail, the most important of which is radix-4. The radix-4 DIT FFT can only be used when N is a power of 4, in which case, of course, a radix-2 algorithm is also applicable; but using a higher radix can reduce the computational complexity at the expense of more complex programming. The derivation of the radix-4 DIT FFT is similar to that of the radix-2 algorithm. The original sequence is decimated into four subsequences $x_{4j}, x_{4j+1}, x_{4j+2}, x_{4j+3}$ (for $j = 0 \ldots \frac{N}{4} - 1$), each of which is further decimated into four subsubsequences, etc. The basic 'butterfly' is based on the four-point DFT

$$\begin{aligned}
X_0 &= x_0 + x_1 + x_2 + x_3 \\
X_1 &= x_0 - ix_1 - x_2 + ix_3 \\
X_2 &= x_0 - x_1 + x_2 - x_3 \\
X_3 &= x_0 + ix_1 - x_2 - ix_3
\end{aligned} \qquad (14.7)$$

which graphically is

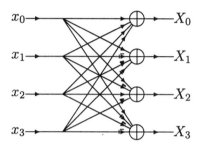

where we have employed two ad-hoc short-hand notations, namely that a line above an arrow means multiplication by -1, while a line before an arrow means multiplication by i. We see that the four-point DFT requires 12 complex additions but no true multiplications. In fact only the radix-2 and radix-4 butterflies are completely multiplication free, and hence the popularity of these radixes.

Now if we compare this butterfly with computation of the four-point DFT via a radix-2 DIT

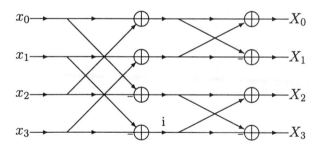

we are surprised to see that only eight complex additions and no multiplications are needed. Thus it is more efficient to compute a four-point DFT using radix-2 butterflies than radix-4! However, this does not mean that for large N the radix-2 FFT is really better. Recall that when connecting stages of DIT butterflies we need to multiply half the lines with twiddle factors, leading to $O(\frac{N}{2} \log_2 N)$ multiplications. Using radix-4 before every stage we multiply three-quarters of the lines by nontrivial twiddle factors, but there are only $\log_4 N = \frac{1}{2} \log_2 N$ stages, leading to $\frac{3}{8} \log_2 N$ multiplications. So a full radix-4 decomposition algorithm needs somewhat fewer multiplications than the radix-2 algorithm, and in any case we could compute the basic four-point DFT using radix-2 butterflies, reducing the number of additions to that of the radix-2 algorithm.

Of course a radix-4 decomposition is only half as applicable as a radix-2 one, since only half of the powers of two are powers of four as well. However, every power of two that is not a power of four is twice a power of four, so two radix-4 algorithms can be used and then combined with a final radix-2 stage. This is the called the FFT42 routine. Similarly, the popular FFT842 routine performs as many efficiently coded radix-8 (equation (4.52)) stages as it can, finishing off with a radix-4 or radix-2 stage as needed.

Another trick that combines radix-2 and radix-4 butterflies is called the *split-radix* algorithm. The starting point is the radix-2 DIF butterfly. Recall that only the odd-indexed X_{2k+1} required a twiddle factor, while the even-indexed X_{2k} did not. Similarly all even-indexed outputs of a full-length 2^m DIF FFT are derivable from those of the two length 2^{m-1} FFTs without multiplication by twiddle factors (see Figure 14.6), while the odd-indexed outputs require a twiddle factor each. So split-radix algorithms compute the even-indexed outputs using a radix-2 algorithm, but the odd-indexed outputs using a more efficient radix-4 algorithm. This results in an unusual 'L-shaped' butterfly, but fewer multiplications and additions than any of the standard algorithms.

For lengths that are not powers of two the most common tactic is to use the next larger power of two, zero-padding all the additional signal points. This has the definite advantage of minimizing the number of FFT routines we need to have on hand, and is reasonably efficient if the zero padding is not excessive and a good power-of-two routine (e.g., split-radix or FFT842) is used. The main disadvantage is that we don't get the same number of spectral points as there were time samples. For general spectral analysis this may be taken as an advantage since we get *higher* spectral resolution, but some applications may require conservation of the number of data points.

Moreover, the same principles that lead us to the power-of-two FFT algorithms are applicable to any N that is not prime. If $N = R^m$ then radix-R algorithms are appropriate, but complexity reduction is possible even for lengths that are not simple powers. For example, assuming $N = N_1 N_2$ we can decompose the original sequence of length N into N_2 subsequences of length N_1. One can then compute the DFT of length N by first computing N_2 DFTs of length N_1, multiplying by appropriate twiddle factors, and finally computing N_1 DFTs of length N_2. Luckily these computations can be performed in-place as well. Such algorithms are called *mixed-radix* FFT algorithms. When N_1 and N_2 are prime numbers (or at least have no common factors) it is possible to eliminate the intermediate step of multiplication by twiddle factors, resulting in the *prime factor* FFT algorithm.

An extremely multiplication-efficient prime factor algorithm was developed by Winograd that requires only $O(N)$ multiplications rather than $O(N \log N)$, at the expense of many more additions. However, it cannot be computed in-place and the indexing is complex. On DSPs with pipelined multiplication and special indexing modes (see Chapter 17) Winograd's FFT runs slower than good implementations of power-of-two algorithms.

EXERCISES

14.4.1 Complete the derivation of the radix-2 DIF FFT both algebraically and graphically. Explain the origin of the bit reversal.

14.4.2 Redraw the diagram of the eight-point radix-2 DIT so that its inputs are in standard order and its outputs bit reversed. This is Cooley and Tukey's original FFT! How is this diagram different from the DIF?

14.4.3 Can a radix-2 algorithm with *both* input and output in standard order be performed in-place?

14.4.4 In addition to checking the LSB (decimation) and checking the MSB (partition) we can divide sequences in two by checking other bits of the binary representation of the indices. Why are only DIT and DIF FFT algorithms popular? Design an eight-point FFT based on checking the middle bit.

14.4.5 A radix-2 DIT FFT requires a final stage of bit reversal. What is required for a radix-4 DIT FFT? Demonstrate this operation on the sequence $0 \ldots 63$.

14.4.6 Write the equations for the radix-8 DFT butterfly. Explain how the FFT842 algorithm works.

14.4.7 Filtering can be performed in the frequency domain by an FFT, followed by multiplying the spectrum by the desired frequency response, and finally an IFFT. Do we need the bit-reversal stage in this application?

14.4.8 Show how to compute a 15-point FFT by decimating the sequence into five subsequences of length three. First express the time index $n = 0 \ldots 14$ as $n = 3n_1 + n_2$ with $n_1 = 0, 1, 2, 3, 4$ and $n_2 = 0, 1, 2$ and the frequency index in the opposite manner $k = k_1 + 5k_2$ with $k_1 = 0, 1, 2$ and $k_2 = 0, 1, 2$ (these are called *index maps*). Next rewrite the FFT substituting these expressions for the indices. Finally rearrange in order to obtain the desired form. What is the computational complexity? Compare with the straightforward DFT and with the 16-point radix-2 FFT.

14.4.9 Research the prime factor FFT. How can the 15-point FFT be computed now? How much complexity reduction is obtained?

14.4.10 Show that if $N = \Pi n_i$ then the number of operations required to perform the DFT is about $(\Sigma n_i)(\Pi n_i)$.

14.5   The Matrix Interpretation of the FFT

In Section 4.9 we saw how to represent the DFT as a matrix product. For example, we can express the four-point FFT of equation (14.7) in the matrix form

$$\begin{pmatrix} X_0 \\ X_1 \\ X_2 \\ X_3 \end{pmatrix} = \begin{pmatrix} 1 & 1 & 1 & 1 \\ 1 & -i & -1 & i \\ 1 & -1 & 1 & -1 \\ 1 & i & -1 & -i \end{pmatrix} \begin{pmatrix} x_0 \\ x_1 \\ x_2 \\ x_3 \end{pmatrix}$$

by using the explicit matrix given in (4.51).

Looking closely at this matrix we observe that rows 0 and 2 are similar, as are rows 1 and 3, reminding us of even/odd decimation! Pursuing this similarity it is not hard to find that

$$\underline{\underline{W_4}} = \begin{pmatrix} 1 & 1 & 1 & 1 \\ 1 & -i & -1 & i \\ 1 & -1 & 1 & -1 \\ 1 & i & -1 & -i \end{pmatrix} = \begin{pmatrix} 1 & 0 & 1 & 0 \\ 0 & 1 & 0 & -i \\ 1 & 0 & -1 & 0 \\ 0 & 1 & 0 & i \end{pmatrix} \begin{pmatrix} 1 & 0 & 1 & 0 \\ 1 & 0 & -1 & 0 \\ 0 & 1 & 0 & 1 \\ 0 & 1 & 0 & -1 \end{pmatrix}$$

which is a factoring of the DFT matrix into the product of two sparser matrices. So far we have not gained anything since the original matrix multiplication $X = \underline{\underline{W_4}} x$ took $4^2 = 16$ multiplications, while the rightmost matrix times x takes eight and then the left matrix times the resulting vector requires a further eight. However, in reality there were in the original only six nontrivial multiplications and only four in the new representation.

Now for the trick. Reversing the middle two columns of the rightmost matrix we find that we can factor the matrix in a more sophisticated way

$$\underline{\underline{W_4}} = \begin{pmatrix} 1 & 0 & 1 & 0 \\ 0 & 1 & 0 & -i \\ 1 & 0 & -1 & 0 \\ 0 & 1 & 0 & i \end{pmatrix} \begin{pmatrix} 1 & 1 & 0 & 0 \\ 1 & -1 & 0 & 0 \\ 0 & 0 & 1 & 1 \\ 0 & 0 & 1 & -1 \end{pmatrix} \begin{pmatrix} 1 & 0 & 0 & 0 \\ 0 & 0 & 1 & 0 \\ 0 & 1 & 0 & 0 \\ 0 & 0 & 0 & 1 \end{pmatrix}$$

that can be written symbolically as

$$W_4 = \begin{pmatrix} I_2 & T_2 \\ I_2 & -T_2 \end{pmatrix} \begin{pmatrix} W_2 & 0_2 \\ 0_2 & W_2 \end{pmatrix} C_4 \qquad (14.8)$$

where I_2 is the two-by-two identity matrix, T_2 is a two-by-two diagonal matrix with twiddle factors as elements ($W_4^0 = 1$ and $W_4^1 = -i$), W_2 is the two-point DFT matrix (butterfly), 0_2 is the two-by-two null matrix, and C_4 is a four-by-four column permutation matrix.

This factorization has essentially reduced the four-dimensional DFT computation (matrix product) into two two-dimensional ones, with some rearranging and a bit reversal. This decomposition is quite general

$$W_{2m} = \begin{pmatrix} I_m & T_m \\ I_m & -T_m \end{pmatrix} \begin{pmatrix} W_m & 0_m \\ 0_m & W_m \end{pmatrix} C_m \qquad (14.9)$$

where T_m is the diagonal m-by-m matrix with elements W_{2m}^i, the twiddle factors for the $2m$-dimensional DFT.

Looking carefully at the matrices we recognize the first step in the decomposition of the DFT that leads to the radix-2 DIT algorithm. Reading from right to left (the way the multiplications are carried out) the column permutation C_m is the in-place decimation that moves the even-numbered elements up front; the two W_m are the half size DFTs, and the leftmost matrix contains the twiddle factors. Of course, we can repeat the process for the blocks of the middle matrix in order to recurse down to W_2.

EXERCISES

14.5.1 Show that by normalizing W_N by $\frac{1}{\sqrt{N}}$ we obtain a unitary matrix. What are its eigenvalues?

14.5.2 Define $\tilde{W}_N$ to be the DFT matrix after bit-reversal permutation of its rows. Write down $\tilde{W}_2$ and $\tilde{W}_4$. Show that $\tilde{W}_4$ can be written as follows.

$$\begin{pmatrix} \tilde{W}_2 & 0_2 \\ 0_2 & \tilde{W}_2 \end{pmatrix} \begin{pmatrix} I_2 & I_2 \\ T_4 & -T_4 \end{pmatrix}$$

To which FFT algorithm does this factorization correspond?

14.5.3 The Hadamard matrix of order 2 is defined to be

$$\underline{\underline{H_2}} = \begin{pmatrix} 1 & 1 \\ 1 & -1 \end{pmatrix}$$

and for all other powers of two we define by recursion.

$$\underline{\underline{H_{2m+1}}} = \begin{pmatrix} \underline{\underline{H_{2m}}} & \underline{\underline{H_{2m}}} \\ \underline{\underline{H_{2m}}} & -\underline{\underline{H_{2m}}} \end{pmatrix}$$

Build the Hadamard matrices for $m = 2$ and 3. Show that they are symmetric and orthogonal. What is the inverse of $\underline{\underline{H_{2m}}}$?

14.5.4 Use the Hadamard matrix instead of $\underline{\underline{W}}$ to define a transform. How many additions and multiplications are needed to compute the transform? What are the twiddle factors? What about bit reversal? Can we compute the Hadamard transform faster by using higher radix algorithms?

14.6   Practical Matters

A few more details must be worked out before you are ready to properly compute FFTs in practice. For concreteness we discuss the radix-2 FFT (either DIT or DIF) although similar results hold for other algorithms. First, we have been counting the multiplications and additions but neglecting computation of the twiddle factors. Repeatedly calling library routines to compute sine and cosine functions would take significantly more time than all the butterfly multiplications and additions we have been discussing. There are two commonly used tactics: storing the W_n^k in a table, and generating them in real-time using trigonometric identities.

By far the most commonly used method in real-time implementations is the use of twiddle factor tables. For a 16-point FFT we will need to store $W_{16}^0 = 1$, $W_{16}^1 = \cos(\frac{2\pi}{16}) - i\sin(\frac{2\pi}{16})$, ... $W_{16}^7 = \cos(\frac{2\pi 7}{16}) - i\sin(\frac{2\pi 7}{16})$, requiring 16 real memory locations. When these tables reside in fast (preferably 'on-chip') memory and the code is properly designed, the table lookup time should be small (but not negligible) compared with the rest of the computation. So the only drawback is the need for setting aside memory for this purpose. When the tables must be 'off-chip' the toll is higher, and the author has even seen poorly written code where the table lookup completely dominated the run-time.

Where do the tables come from? For most applications the size N of the FFT is decided early on in the code design, and the twiddle factor tables can be precomputed and stored as constants. Many DSP processors have special 'table memory' that is ideal for this purpose. For general-purpose (library) routines the twiddle factor tables are usually initialized upon the call to the FFT routine. On general-purpose computers one can usually get away with calling the library trigonometric functions to fill up the tables; but on DSPs one either stores only entire table in program code memory, or stores only W_N itself and derives the rest of the required twiddle factors using

$$W_N^k = W_N^{k-1} W_N$$

or the equivalent trigonometric identities (A.23). When N is large numeric errors may accumulate while recursing, and it is preferable to periodically reseed the recursion (e.g., with $W_N^{\frac{N}{4}} = -i$).

For those applications where the twiddle factors cannot be stored, they must be generated in real-time as required. Once again the idea is to know only W_N and to generate W_N^{nk} as required. In each stage we can arrange the butterfly computation so that the required twiddle factor exponents form increasing sequences of the form $W_N^0, W_N^n, W_N^{2n}, \ldots$. Then the obvious identity

$$W_N^{nk} = W_N^{(n-1)k} W_N^k$$

or its trigonometric equivalent can be used. This is the reason that general-purpose FFT routines, rather than having *two* loops (an outside loop on stages and a nested loop on butterflies), often have *three* loops. Inside the loop on stages is a loop on *butterfly groups* (these groups are evident in Figures 14.5 and 14.6), each of which has an increasing sequence of twiddle factors, and nested inside this loop is the loop on the butterflies in the group.

Another concern is the numeric accuracy of the FFT computation. Every butterfly potentially contributes round-off error to the calculation, and since each final result depends on $\log_2 N$ butterflies in series, we expect this numeric error to increase linearly with $\log_2 N$. So larger FFTs will be less accurate than smaller ones, but the degradation is slow. However, this prediction is usually only relevant for floating point computation. For fixed point processors there is a much more serious problem to consider, that of overflow. Overflow is always a potential problem with fixed point processing, but the situation is particularly unfavorable for FFT computation. The reason for this is not hard to understand. For simplicity, think of a single sinusoid of frequency $\frac{k}{N}$ so that an integer number of cycles fits into the

FFT input buffer. The FFT output will be nonzero only in the kth bin, and so all the energy of the input signal will be concentrated into a single large number. For example, if the signal was pure DC $x_n = 1$, the only nonzero bin is $X_0 = \sum x_n = N$, and similarly for all other single-bin cases. It is thus clear that if the input signal almost filled the dynamic range of the fixed point word, then the single output bin will most certainly overflow it.

The above argument may lead you to believe that overflow is only of concern in special cases, such as when only a single bin or a small number of bins are nonzero. We will now show that it happens even for the opposite case of white noise, when all the output bins are equal in size. From Parseval's relation for the DFT (4.42) we know that (using our usual normalization), if the sum of the input squared is E^2, then the sum of the output squared will be NE^2. Hence the rms value of the output is greater by a factor $\sqrt{N}$ than the input rms. For white noise this implies that the typical output value is greater than a typical input value by this factor!

Summarizing, narrow-band FFT components scale like N while wideband, noise-like components scale like $\sqrt{N}$. For large N both types of output bins are considerably larger than typical input bins, and hence there is a serious danger of overflow. The simplest way to combat this threat is to restrict the size of the inputs in order to ensure that no overflows can occur. In order to guarantee that the output be smaller than the largest allowed number, the input must be limited to $\frac{1}{N}$ of this maximum. Were the input originally approximately full scale, we would need to divide it by N; resulting in a scaling of the output spectrum by $\frac{1}{N}$ as well. The problem with this prescaling is that crudely dividing the input signal by a factor of N increases the numeric error-to-signal ratio. The relative error, which for floating point processing was proportional to $\log_2 N$, becomes approximately proportional to N. This is unacceptably high.

In order to confine the numeric error we must find a more sophisticated way to avoid overflows; this necessitates intervening with the individual computations that may overflow, namely the butterflies. Assume that we store complex numbers in two memory locations, one containing the real part and the other the imaginary part, and that each of these memories can only store real numbers between -1 and $+1$. Consider butterflies in the first stage of a radix-2 DIT. These involve only addition and subtraction of pairs of such complex numbers. The worst case is when adding complex numbers both of which are equal to $+1$, -1, $+i$ or $-i$, where the absolute value is doubled. Were this worst case to transpire at every stage, the overall gain after $\log_2 N$ stages would be $2^{\log_2 N} = N$, corresponding to the case of a single spectral line. For white noise x_n we see from the DIT butterfly of equation (14.4)

$$X_k = X_k^E + W_N^k X_k^O$$
$$X_{k+\frac{N}{2}} = X_k^E - W_N^k X_k^O$$

that $|X_k|^2 + |X_{k+\frac{N}{2}}|^2 = 2\left(|X_k^E|^2 + |X_k^O|^2\right)$ and if the expected values of all X_k are the same (as for white random inputs) then each must be larger than the corresponding butterfly input by a factor of $\sqrt{2}$. In such a case the overall gain is $\sqrt{2}^{\log_2 N} = \sqrt{N}$ as predicted above.

It is obvious that the worst-case butterfly outputs can only be guaranteed to fit in our memory locations if the input real and imaginary parts are limited in absolute value to $\frac{1}{2}$. However, the butterfly outputs, now themselves between -1 and $+1$, become inputs to the next stage and so may cause overflow there. In order to eliminate this possibility, we must limit the original inputs to $\frac{1}{4}$ in absolute value. Since the same analysis holds for the other butterflies, we reach the previous conclusion that the input absolute values must be prescaled by $\frac{1}{2}^{\log_2 N} = \frac{1}{N}$.

Even with this prescaling we are not completely safe. The worst case we discussed above was only valid for the first stage, where only additions and subtractions take place. For stages with nontrivial twiddle factors the increase can exceed even a factor of two. For example, consider a butterfly containing a rotation by $45°$, $X_k^E = 1$, $X_k^O = 1 + i$. After rotation $1 + i$ becomes $\sqrt{2}$, which is added to 1 to become $1 + \sqrt{2} \approx 2.414$. Hence, the precise requirement for the complex inputs is for their *length* not to exceed $\frac{1}{N}$. With this restriction, there will be no overflows.

There is an alternative way of avoiding overflow at the second stage of butterflies. Rather than reducing the input to the first stage of butterflies to $\frac{1}{2}$ to $\frac{1}{4}$ we can directly divide the input to the *second* butterfly by 2. For the radix-2 DIT case this translates to replacing our standard DIT by

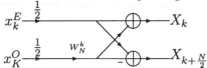

a failsafe butterfly computation that inherently avoids fixed point overflow. We can now replace *all* butterflies in the FFT with this failsafe one, resulting in an output spectrum divided by N, just as when we divided the input by N. The advantage of this method over input prescaling is that the inputs to each stage are always the maximum size they can be without being susceptible to overflow. With input prescaling only the *last* butterflies have such maximal inputs; all the previous ones receiving smaller inputs.

Due to the butterflies working with maximal inputs the round-off error is significantly reduced as compared with that of the input prescaling method. Some numeric error is introduced by each butterfly, but this noise is itself reduced by a factor of two by the failsafe butterfly; the overall error-to-signal ratio is proportional to $\sqrt{N}$. A further trick can reduce the numeric error still more. Rather than using the failsafe butterfly throughout the FFT, we can (at the expense of further computation) first check if any overflows will occur in the present stage. If yes, we use the failsafe butterfly (and save the appropriate scaling factor), but if not, we can use the regular butterfly. We leave as an exercise to show that this data-dependent prescaling, does not require double computation of the overflowing stages.

EXERCISES

14.6.1 We can reduce the storage requirements of twiddle factor tables by using trigonometric symmetries (A.22). What is the minimum size table needed for $N = 8$? In general? Why are such economies of space rarely used?

14.6.2 What is the numeric error-to-signal ratio for the straightforward computation of the DFT, for floating and fixed point processors?

14.6.3 In the text we discussed the failsafe prescaling butterfly. An alternative is the failsafe postscaling butterfly, which divides by two *after* the butterfly is computed. What special computational feature is required of the processor for postscaling to work? Explain data-dependent postscaling. How does it solve the problem of double computation?

14.6.4 In the text we ignored the problem of the finite resolution of the twiddle factors. What do you expect the effect of this quantization to be?

14.7 Special Cases

The FFT is fast, but for certain special cases it can be made it even faster. The special cases include signals with many zeros in either the time or frequency domain representations, or with many values about which we do not care. We can save computation time by avoiding needless operations such as multiplications by zero, or by not performing operations that lead to unwanted results. You may think that such cases are unusual and not wish to expend the effort to develop special code for them, but certain special signals often arise in practice. These most common applications cases are zero-padding, interpolation, zoom-FFT and real-valued signals.

We have mentioned the use of zero-padding (adding zeros at the end of the signal) to force a signal to some useful length (e.g., a power of two) or to increase spectral resolution. It is obvious that some savings are obtainable in the first FFT stage, since we can avoid multiplications with zero inputs. Unfortunately, the savings do not carry over to the second stage of either standard DIT or DIF, since the first-stage butterflies mix signal values from widely differing places. Only were the zero elements to be close in bit-reversed order would the task of pruning unnecessary operations be simple.

However, recalling the time partitioning of equation (14.5) we can perform the FFT of the fully populated left half and that of the sparse right half separately, and then combine them with a single stage of DIF butterflies. Of course the right half is probably not *all* zeros, and hence we can't realize all the savings we would wish; however, *its* right half may be all-zero and thus trivial, and the combining of its two halves can also be accomplished in a single stage.

Another application that may benefit from this same ploy is interpolation. Zero-padding in the time domain increased spectral resolution; the dual to this is that zero-padding in the frequency domain can increase time resolution (i.e., perform interpolation). To see how the technique works, assume we want to double the sampling rate, adding a new signal value in between every two values. Assume further that the signal has no DC component. We take the FFT of the signal to be interpolated (with no savings), double the number of points in the spectrum by zero-padding, and finally take the IFFT. This final IFFT can benefit from heeding of zeros; and were we to desire a quadrupling of the sampling rate, the IFFT's argument would have fully three-quarters of its elements zero.

Using a similar ploy, but basing ourselves in the time decimation of equation (14.3), we can save time if a large fraction of either the even- or odd-indexed signal values are zero. This would seem to be an unusual situation, but once again it has its applications.

Probably the most common special case is that of real-valued signals. The straightforward way of finding their FFT is to simply use a complex FFT routine, but then many complex multiplications and additions are performed with one component real. In addition the output has to be Hermitian symmetric (in the usual indexation this means $X_{N-k} = X_k^*$) and so computation of half of the outputs is redundant. We could try pruning the computations, both from the input side (eliminating all operations involving zeros) and from the output side (eliminating all operations leading to unneeded results), but once again the standard algorithms don't lend themselves to simplification of the intermediate stages. Suppose we were to make the mistake of inputting the vector of $2N$ real signal values R_i into a complex N-point FFT routine that expects interleaved input $(R_0, I_0, R_1, I_1, \ldots$ where $x_i = R_i + iI_i)$. Is there any way we could recover? The FFT thinks that the signal is $x_0 = R_0 + iR_1$, $x_1 = R_2 + iR_3$, etc. and computes a single spectrum X_k. If the FFT routine is a radix-2 DIT the even and odd halves are not mixed until the last stage, but for *any* FFT we can unmix the FFTs of the even and odd subsequences by the inverse of that last stage of DIT butterflies.

$$X_k^E = \tfrac{1}{2}(X_k + X_{N-k}^*)$$
$$X_k^O = \tfrac{1}{2}(X_k - X_{N-k}^*)$$

The desired FFT is now given (see equation (14.4)) by

$$R_k = X_k^E + W_{2N}^k X_k^O \qquad k = 0 \ldots N - 1$$
$$R_k = X_{k-N}^E - W_{2N}^k X_{k-N}^O \qquad k = N \ldots 2N - 1$$

and is clearly Hermitian symmetric.

The final special case we mention is the *zoom FFT* used to zoom in on a small area of the spectrum. Obviously for a very high spectral resolution the uncertainty theorem requires an input with very large N, yet we are only interested in a small number of spectral values. Pruning can be very efficient here, but hard to implement since the size and position of the zoom window are usually variable. When only a very small number of spectral lines are required, it may be advantageous to compute the straight DFT, or use Goertzel's algorithm. Another attractive method is mix the signal down so that the area of interest is at DC, low-pass filter and reduce the sampling rate, and only then perform the FFT.

EXERCISES

14.7.1 How can pruning be used to reduce the complexity of zero-padded signals? Start from the diagram of the eight-point DIF FFT and assume that only the first two points are nonzero. Draw the resulting diagram. Repeat with the first four points nonzero, and again with the first six points.

14.7.2 How can the FFT of two real signals of length N be calculated using a single complex FFT of length N?

14.7.3 How can the FFT of four real symmetric $(x_{N-n} = x_n)$ signals of length N be calculated using a single complex FFT of length N?

14.7.4 We are interested only in the first two spectral values of an eight-point FFT. Show how pruning can reduce the complexity. Repeat with the first four and six spectral values.

14.8 Goertzel's Algorithm

The fast Fourier transform we have been studying is often the most efficient algorithm to use; however, it is not a panacea. The prudent signal processing professional should be familiar with alternative algorithms that may be more efficient in other circumstances.

In the derivation of the FFT algorithm our emphasis was on finding computations that were superfluous due to their having been previously calculated. This leads to significant economy when the entire spectrum is required, due to symmetries between various frequency components. However, the calculation of each single X_k is not improved, and so the FFT is not the best choice when only a single frequency component, or a small number of components are needed. It *is* true that the complexity of the computation of any component *must* be at least $O(N)$, since every x_n must be taken into account! However, the coefficient of the N in the complexity may be reduced, as compared with the straightforward calculation of equation (14.1). This is the idea behind Goertzel's algorithm.

There are many applications when only a single frequency component, or a small number of components are required. For example, telephony signaling is typically accomplished by the use of tones. The familiar push-button dialing employs a system known as **D**ual **T**one **M**ultiple **F**requency (DTMF) tones, where each row and each column determine a frequency (see figure 14.7). For example, the digit 5 is transmitted by simultaneously emitting the two tones L2 and H2. To decode DTMF digits one must monitor only eight frequencies. Similarly telephone exchanges use a different multifrequency tone system to communicate between themselves, and modems and fax machines also use specific tones during initial stages of their operation.

One obvious method of decoding DTMF tones is to apply eight bandpass filters, calculate the energy of their outputs, pick the maximum from both the low group and the high group, and decode the meaning of this pair of maxima as a digit. That's precisely what we suggest, but we propose

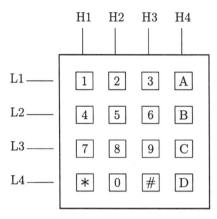

Figure 14.7: The DTMF telephony signaling method. (Note: The A, B, C, and D tones are not available on the standard telephone, being reserved for special uses.)

using Goertzel's DFT instead of a band-pass filter. Of course we *could* use a regular FFT as a bank of band-pass filters, but an FFT with enough bins for the required frequency resolution would be much higher in computational complexity.

When we are interested only in a single frequency component, or a small number of them that are not simply related, the economies of the FFT are to no avail, and the storage of the entire table of trigonometric constants wasteful. Assuming all the required W_N^{nk} to be precomputed, the basic DFT formula for a single X_k requires N complex multiplications and $N-1$ complex additions to be calculated. In this section we shall assume that the x_n are real, so that this translates to $2N$ real multiplications and $2(N-1)$ real additions.

Recalling the result of exercise 4.7.2, we need only know W_N^k and calculate the other twiddle factors as powers. In particular, when we are interested in only the kth spectral component, we saw in equation (4.55) that the DFT becomes a polynomial in $W \equiv W_N^k \equiv e^{-i\frac{2\pi k}{N}}$.

$$X_k = \sum_{n=0}^{N-1} x_n W_N^{nk} = x_0 + x_1 W + x_2 W^2 + \ldots x_{N-1} W^{N-1}$$

This polynomial is best calculated using Horner's rule

$$X_k = \left(\left(\cdots (x_{N-1}W + x_{N-2})W + \ldots + x_2 \right) W + x_1 \right) W + x_0$$

which can be written as a recursion.

```
Given: xₙ for n ← 0...N − 1
```
$$P_{N-1} \leftarrow x_{N-1}$$
```
for n ← N − 1 down to 0
      Pₙ ← Pₙ₊₁W + xₙ
```
$$X_k \leftarrow P_0$$

We usually prefer recursion indices to run in ascending order, and to do this we define $V \equiv W^{-1} = W_N^{-k} \equiv e^{+\mathrm{i}\frac{2\pi k}{N}}$. Since $V^N = 1$ we can write

$$X_k = \sum_{n=0}^{N-1} x_n V^{N-n} = x_0 V^N + x_1 V^{N-1} + \ldots + x_{N-2} V^2 + x_{N-1} V$$

which looks like a convolution. Unfortunately the corresponding recursion isn't of the right form; by multiplying X_k by a phase factor (which won't effect the squared value) we get

$$X'_k = x_0 V^{N-1} + x_1 V^{N-2} + \ldots + x_{N-2} V + x_{N-1}$$

which translates to the following recursion.

```
Given: xₙ for n ← 0...N − 1
```
$$P_0 \leftarrow x_0$$
```
for n ← 1 to N − 1
       Pₙ ← Pₙ₋₁V + xₙ
```
$$X'_k \leftarrow P_{N-1}$$

This recursion has an interesting geometric interpretation. The complex frequency component X_k can be calculated by a sequence of basic $N - 1$ moves, each of which consists of rotating the previous result by the angle $\frac{2\pi k}{N}$ and adding the new real x_n.

Each rotation is a complex multiply, contributing either $4(N - 1)$ real multiplications and $2(N - 1)$ real additions or $3(N - 1)$ and $5(N - 1)$ respectively. The addition of a real x_n contributes a further $(N-1)$ additions. We see that the use of the recursion rather than expanding the polynomial has not yet saved any computation time. This is due to the use of complex state variables P. Were the state variables to be real rather than complex, we would save about half the multiplies. We will now show that the complex state variables can indeed be replaced by real ones, at the expense of introducing another time lag (i.e., recursion two steps back).

Since we are assuming that x_n are real, we see from our basic recursion step $P_n \leftarrow P_{n-1} V + x_n$ that $P_n - P_{n-1} V$ must be real at every step.

We can implicitly define a sequence Q_n as follows

$$P_n = Q_n - W Q_{n-1} \tag{14.10}$$

and will now show that the Q are indeed real-valued. Substituting into the recursion step

$$
\begin{aligned}
P_n &\leftarrow & x_n &+ & P_{n-1}V \\
Q_n - W Q_{n-1} &\leftarrow & x_n &+ (Q_{n-1} - W Q_{n-2})\, V \\
Q_n &\leftarrow & x_n &+ (V + W)Q_{n-1} + (WV)Q_{n-2} \\
Q_n &\leftarrow & x_n &+ A Q_{n-1} - Q_{n-2}
\end{aligned}
$$

where $A \equiv V + W = 2\cos(\frac{2\pi k}{N})$. Since the inputs x_n are real and A is real, assuming the Qs start off real (and we can start with zero) they remain real.

The new algorithm is:

```
Given: xn for n = 0...N − 1
Q−2 ← 0,     Q−1 ← 0
Q0 ← x0
for n ← 1 to N − 1
      Qn ← xn   +   AQn−1 − Qn−2
X'k ← QN−1 − WQN−2
```

and the desired energy

$$|X_k|^2 = Q_{N-1}^2 + Q_{N-2}^2 - A Q_{N-1} Q_{N-2}$$

must be computed at the end. This recursion requires only a single frequency-dependent coefficient A and requires keeping two lags of Q. Computationally there is only a single real multiplication and two real additions per iteration, for a total of $N - 1$ multiplications and $2(N - 1)$ additions.

There is basically only one design parameter to be determined before using Goertzel's algorithm, namely the number of points N. Goertzel's algorithm can only be set up to detect frequencies of the form $f = \frac{k}{N} f_s$ where f_s is the sampling frequency; thus selecting larger N allows finer resolution in center frequencies. In addition, as we shall see in the exercises, larger N implies narrower bandwidth as well. However, larger N also entails longer computation time and delay.

EXERCISES

14.8.1 Since Goertzel's algorithm is equivalent to the DFT, the power spectrum response (for $\omega = 2\pi k/N$) is $P(k) = \left(\frac{\sin(\pi k)}{N\sin(\pi k/N)}\right)^2$. Show that the half power point is at 0.44, i.e., the half power bandwidth is 0.88 bin, where each bin is simply $\frac{f_s}{N}$. What is the trade-off between time accuracy and frequency accuracy when using Goertzel's algorithm as a tone detector?

14.8.2 DTMF tones are allowed to be inaccurate in frequency by 1.5%. What would be the size of an FFT that has bins of about this size? How much computation is saved by using Goertzel's algorithm?

14.8.3 DTMF tones are used mainly by customers, while telephone companies use different multitone systems for their own communications. In North America, telephone central offices communicate using *MF trunk tones* 700, 900, 1100, 1300, 1500, 1700, 2600 and 3700 Hz, according to the following table.

Tone	1	2	3	4	5	6
Frequencies	700+900	700+1100	900+1100	700+1300	900+1300	1100+1300

Tone	7	8	9	0	KP	ST
Frequencies	700+1500	900+1500	1100+1500	1300+1500	1100+1700	1500+1700

All messages start with KP and end with ST. Assuming a sampling rate of 8 KHz, what is the minimum N that exactly matches these frequencies? What will the accuracy (bandwidth) be for this N? Assuming N is required to be a power of two, what error will be introduced?

14.8.4 Repeat the previous question for DTMF tones, which are purposely chosen to be nonharmonically related. The standard requires detection if the frequencies are accurate to within $\pm1.5\%$ and nonoperation for deviation of $\pm3.5\%$ or more. Also the minimal on-time is 40 milliseconds, but tones can be transmitted at a rate of 10 per second. What are the factors to be considered when choosing N?

14.9 FIFO Fourier Transform

The FFT studied above calculates the spectrum *once*; when the spectrum is required to be updated as a function of time the FFT must be reapplied for each time shift. The worst case is when we wish to *slide* through the data one sample at a time, calculating $\text{DFT}(x_0 \ldots x_{N-1})$, $\text{DFT}(x_1 \ldots x_N)$, etc. When this must be done M times, the complexity using the FFT is $O(MN\log N)$. For this case there is a more efficient algorithm, the *FIFO Fourier transform*, which instead of completely recalculating X_k for each shift, updates the previous one.

There is a well-known trick for updating a moving *simple* average

$$A_m = \sum_{n=0}^{N-1} x_{m+n}$$

that takes computation time that is independent of N. The trick employs a FIFO of length N, holding the samples to be summed. First we wait until the FIFO is full, and then sum it up once to obtain A_0. Thereafter, rather than summing N elements, we update the sum using

$$A_{m+1} = A_m + x_{m+N} - x_m$$

recursively. For example, the second sum A_1 is derived from A_0 by removing the unnecessary x_0 and adding the new term x_N.

Unfortunately, this trick doesn't generalize to moving averages with co-efficients, such as general FIR filters. However, a slightly modified version of it *can* be used for the recursive updating of the components of a DFT

$$X_{km} = \sum_{n=m}^{m+N-1} x_n W_N^{nk} \tag{14.11}$$

where in this case we do not 'reset the clock' as would be the case were we to call a canned FFT routine for each m. After a single initial DFT or FFT has been computed, to compute the next we need only update via the **FIFO Fourier Transform (FIFOFT)**

$$X_{km+1} = X_{km} + (x_{m+N} - x_m)W_N^{mk} \tag{14.12}$$

requiring only two complex additions and one complex multiplication per desired frequency component.

Let's prove equation (14.12). Rewriting equation (14.11) for m and $m+1$

$$X_{km} = \sum_{n=0}^{N-1} x_{m+n} W_N^{(m+n)k}$$

$$X_{km+1} = \sum_{n=0}^{N-1} x_{m+1+n} W_N^{(m+1+n)k}$$

$$= \sum_{n=1}^{N} x_{m+n} W_N^{(m+n)k}$$

$$= \sum_{n=0}^{N-1} x_{m+n} W_N^{(m+n)k} - x_m W_N^{mk} + x_{m+N} W_N^{(m+N)k}$$

and since $W_N^{Nk} = 1$ we obtain equation (14.12).

When all N frequency components are required, the FIFOFT requires N complex multiplications and $2N$ complex additions per shift. For $N > 4$ this is less than the $\frac{N}{2} \log_2 N$ multiplications and $N \log_2 N$ additions required by the FFT. Of course after N shifts we have performed $O(N^2)$ multiplications compared with $O(N \log N)$ for a single additional FFT, but we have received a lot more information as well.

Another, perhaps even more significant advantage of the FIFOFT is the fact that it does not introduce notable delay. The FFT can only be used in real-time processing when the delay between the input buffer being filled and FFT result becoming available is small enough. The FIFOFT is truly real-time, similar to direct computation of a convolution. Of course the first computation must somehow be performed (perhaps not in real-time), but in many applications we can just start with zeros in the FIFO and wait for the answers to become correct. The other problem with the FIFOFT is that numeric errors may accumulate, especially if the input is of large dynamic range. In such cases the DFT should be periodically reinitialized by a more accurate computation.

EXERCISES

14.9.1 For what types of MA filter coefficients are there FIFO algorithms with complexity independent of N?

14.9.2 Sometimes we don't actually need the recomputation for *every* input sample, but only for every r samples. For what r does it become more efficient to use the FFT rather than the FIFOFT?

14.9.3 Derive a FIFOFT that uses N complex additions and multiplications per desired frequency component, for the case of resetting the clock.

$$X_{km} = \sum_{n=0}^{N-1} x_{m+n} W_N^{nk} \qquad (14.13)$$

14.9.4 The FIFOFT as derived above does not allow for windowing of the input signal before transforming. For what types of windows can we define a moving average FT with complexity independent of N?

Bibliographical Notes

More detail on the use of FFT like algorithms for multiplication can be found in Chapter 4 of the second volume of Knuth [136].

An early reference to computation of Fourier transforms is the 1958 paper by Blackman and Tukey that was later reprinted as a book [19].

The radix-2 DIT FFT was popularized in 1965 by James Cooley of IBM and John Tukey of Princeton [45]. Cooley recounts in [43] that the complexity reduction idea was due to Tukey, and that the compelling applications were military, including seismic verification of Russian compliance with a nuclear test ban and long-range acoustic detection of submarines. Once Cooley finished his implementation, IBM was interested in publishing the paper in order to ensure that such algorithms did not become patented. The first known full application of the newly published FFT was by an IBM geophysicist named Lee Alsop who was studying seismographic records of an earthquake that had recently taken place in Alaska. Using 2048 data points, the FFT reduced the lengthy computation to seconds.

Gordon Sande, a student of Tukey's at Princeton, heard about the complexity reduction and worked out the DIF algorithm. After Cooley sent his draft paper to Tukey and asked the latter to be a co-author, Sande decided not to publish his work.

Actually, radix-2 FFT-like algorithms have a long history. In about 1805 the great mathematician Gauss [177] used, but did not publish, an algorithm essentially the same as Cooley and Tukey's two years before Fourier's presentation of his theorem at the Paris Institute! Although eventually published posthumously in 1866, the idea did not attract a lot of attention. Further historical information is available in [44].

The classic reference for special real-valued FFT algorithms is [248]. The split-radix algorithm is discussed in [57, 247, 56].

The prime factor FFT was introduced in [137], based on earlier ideas (e.g. [240, 29]) and an in-place algorithm given in [30]. The extension to real-valued signals is given in [98].

Winograd's prime factor FFT [283, 284] is based on a reduction of a DFT of prime length N into a circular convolution of length $N - 1$ first published as a letter by Rader [214]. A good account is found in the McClellan and Rader book on number theory in DSP [166].

Goertzel's original article is [77]. The MAFT is treated in [7, 249]. The zoom FFT can be found in [288].

A somewhat dated but still relevant review of the FFT and its applications can be found in [16] and much useful material including FORTRAN language sources came out of the 1968 Arden House workshop on the FFT, reprinted in the June 1969 edition of the IEEE Transactions on Audio and Electroacoustics (AU-17(2) pp. 66-169). Many original papers are reprinted in [209, 166]). Modern books on the FFT include [26, 28, 31, 246] and Chapter 8 of [241]. Actual code can be found in the last reference, as well as in [31], [30, 247, 17], [41, 198] etc.

Digital Filter Implementation

In this chapter we will delve more deeply into the practical task of using digital filters. We will discuss how to accurately and efficiently implement FIR and IIR filters.

You may be asking yourself why this chapter is important. We already know what a digital filter is, and we have (or can find) a program to find the coefficients that satisfy design specifications. We can inexpensively acquire a DSP processor that is so fast that computational efficiency isn't a concern, and accuracy problems can be eliminated by using floating point processors. Aren't we ready to start programming without this chapter?

Not quite. You should think of a DSP processor as being similar to a jet plane; when flown by a qualified pilot it can transport you very quickly to your desired destination, but small navigation errors bring you to unexpected places and even the slightest handling mistake may be fatal. This chapter is a crash course in digital filter piloting.

In the first section of this chapter we discuss technicalities relating to computing convolutions in the time domain. The second section discusses the circular convolution and how it can be used to filter in the frequency domain; this is frequently the most efficient way to filter a signal. Hard real-time constraints often force us to filter in the time domain, and so we devote the rest of the chapter to more advanced time domain techniques. We will exploit the graphical techniques developed in Chapter 12 in order to manipulate filters. The basic building blocks we will derive are called *structures*, and we will study several FIR and IIR structures. More complex filters can be built by combining these basic structures.

Changing sampling rate is an important application for which special filter structures known as *polyphase filters* have been developed. Polyphase filters are more efficient for this application than general purpose structures.

We also deal with the effect of finite precision on the accuracy of filter computation and on the stability of IIR filters.

15.1   Computation of Convolutions

We have never fully described how to properly compute the convolution sum in practice. There are essentially four variations. Two are *causal*, as required for real-time applications; the other two introduce explicit delays. Two of the convolution procedures process one input at a time in a real-time-oriented fashion (and must store the required past inputs in an internal FIFO), the other two operate on arrays of inputs.

First, there is the *causal FIFO* way

$$y_n = \sum_{l=0}^{L-1} a_l \, x_{n-l} \tag{15.1}$$

which is eminently suitable for real-time implementation. We require two buffers of length L—one constant buffer to store the filter coefficients, and a FIFO buffer for the input samples. The FIFO is often unfortunately called the *static buffer*; not that it is *static*—it is changing all the time. The name is borrowed from computer languages where *static* refers to buffers that survive and are not zeroed out upon each invocation of the convolution procedure. We usually clear the static buffer during program initialization, but for continuously running systems this precaution is mostly cosmetic, since after L inputs all effects of the initialization are lost. Each time a new input arrives we push it into the static buffer of length L, perform the convolution on this buffer by multiplying the input values by the filter coefficients that overlap them, and accumulating. Each coefficient requires one multiply-and-accumulate (MAC) operation. A slight variation supported by certain DSP architectures (see Section 17.6), is to combine the push and convolve operations. In this case the place shifting of the elements in the buffer occurs as part of the overall convolution, in parallel with the computation.

In equation (15.1) the index of summation runs over the filter coefficients. We can easily modify this to become the *causal array* method

$$y_n = \sum_{i=n-(L-1)}^{n} a_{n-i} \, x_i \tag{15.2}$$

where the index i runs over the inputs, assuming these exist. This variation is still causal in nature, but describes inputs that have already been placed in an array by the calling application. Rather than dedicating further memory inside our convolution routine for the FIFO buffer, we utilize the existing buffering and its indexation. This variation is directly suitable for off-line computation where we compute the entire output vector in one invocation. When programming we usually shift the indexes to the range $0 \ldots L - 1$ or $1 \ldots L$.

In off-line calculation there is no need to insist on explicit causality since all the input values are available in a buffer anyway. We know from Chapter 6 that the causal filter introduces a delay of half the impulse response, a delay that can be removed by using a noncausal form. Often the largest filter coefficients are near the filter's center, and then it is even more natural to consider the middle as the position of the output. Assuming an odd number of taps, it is thus more symmetric to index the $L = 2\lambda + 1$ taps as $a_{-\lambda} \ldots a_0 \ldots a_\lambda$, and the explicitly *noncausal FIFO* procedure looks like this.

$$y_n = \sum_{l=-\lambda}^{\lambda} a_l x_{n-l} \tag{15.3}$$

The corresponding *noncausal array*-based procedure is obtained, once again, by a change of summation variable

$$y_n = \sum_{i=n-\lambda}^{n+\lambda} a_{n-i} x_i \tag{15.4}$$

assuming that the requisite inputs exist. This symmetry comes at a price; when we get the nth input, we can compute only the $(n - \lambda)^{\text{th}}$ output. This form makes explicit the buffer delay of λ between input and output.

In all the above procedures, we assumed that the input signal existed for all times. Infinite extent signals pose no special challenge to real-time systems but cannot really be processed off-line since they cannot be placed into finite-length vectors. When the input signal is of finite time duration and has only a finite number N of nonzero values, some of the filter coefficients will overlap zero inputs. Assume that we desire the same number of outputs as there are inputs (i.e., if there are N inputs, $n = 0, \ldots N - 1$, we expect N outputs). Since the input signal is identically zero for $n < 0$ and $n \geq N$, the first output, y_0, actually requires only $\lambda + 1$ multiplications, namely $a_0 x_0$, $a_{-1}x_1$, through $a_{-\lambda}x_\lambda$, since a_1 through a_λ overlap zeros.

a_λ	$a_{\lambda-1}$	$\cdots$	a_2	a_1	a_0	a_{-1}	a_{-2}	$\cdots$	$a_{-\lambda+1}$	$a_{-\lambda}$	
0	0	$\cdots$	0	0	x_0	x_1	x_2	$\cdots$	$x_{\lambda-1}$	x_λ	$x_{\lambda+1} \cdots$

Only after λ shifts do we have the filter completely overlapping signal.

a_λ	$a_{\lambda-1}$	$a_{\lambda-2}$	$\cdots$	a_1	a_0	a_{-1}	$\cdots$	$a_{-\lambda+1}$	$a_{-\lambda}$	
x_0	x_1	x_2	$\cdots$	$x_{\lambda-1}$	x_λ	$x_{\lambda+1}$	$\cdots$	$x_{2\lambda-1}$	$x_{2\lambda}$	$x_{2\lambda+1} \cdots$

Likewise the last λ outputs have the filter overlapping zeros as well.

$$
\begin{array}{ccccccccccc}
\cdots & a_\lambda & a_{\lambda-1} & \cdots & a_2 & a_1 & a_0 & a_{-1} & a_{-2} & \cdots & a_{-\lambda+1} & a_{-\lambda} \\
\cdots & x_{N-l} & x_{N-2} & \cdots & x_{N-2} & x_{N-1} & x_N & 0 & 0 & \cdots & 0 & 0
\end{array}
$$

The programming of such convolutions can take the finite extent into account and not perform the multiplications by zero (at the expense of more complex code). For example, if the input is nonzero only for N samples starting at zero, and the entire input array is available, we can save some computation by using the following sums.

$$
y_n = \sum_{i=\max(0,n-(L-1))}^{\min(N-1,n)} a_{n-i} x_i = \sum_{i=\max(0,n-\lambda)}^{\min(N-1,n+\lambda)} a_{n-i} x_i \tag{15.5}
$$

The improvement is insignificant for $N \gg L$.

We have seen how to compute convolutions both for real-time-oriented cases and for off-line applications. We will see in the next section that these straightforward computations are *not* the most efficient ways to compute convolutions. It is almost always more efficient to perform convolution by going to the frequency domain, and only harsh real-time constraints should prevent one from doing so.

EXERCISES

15.1.1 Write two routines for array-based noncausal convolution of an input signal x by an odd length filter a that does not perform multiplications by zero. The routine convolve(N, L, x, a, y) should return an output vector y of the same length N as the input vector. The filter should be indexed from 0 to L−1 and stored in reverse order (i.e., a_0 is stored in a[L-1]). The output y_i should correspond to the middle of the filter being above x_i (e.g., the first and last outputs have about half the filter overlapping nonzero input signal values). The first routine should have the input vector's index as the running index, while the second should use the filter's index.

15.1.2 Assume that a noncausal odd-order FIR filter is symmetric and rewrite the above routines in order to save multiplications. Is such a procedure useful for real-time applications?

15.1.3 Assume that we only want to compute output values for which all the filter coefficients overlap observed inputs. How many output values will there be? Write a routine that implements this procedure. Repeat for when we want all outputs for which *any* inputs are overlapped.

15.2 FIR Filtering in the Frequency Domain

After our extensive coverage of convolutions, you may have been led to believe that FIR filtering and straightforward computation of the convolution sum as in the previous section were one and the same. In particular, you probably believe that to compute N outputs of an L-tap filter takes NL multiplications and $N(L-1)$ additions. In this section we will show how FIR filtering can be accomplished with significantly fewer arithmetic operations, resulting both in computation time savings and in round-off error reduction.

If you are unconvinced that it is possible to reduce the number of multiplications needed to compute something equivalent to N convolutions, consider the simple case of a two-tap filter (a_0, a_1). Straightforward convolution of any two consecutive outputs y_n and y_{n+1} requires four multiplications (and two additions). However, we can rearrange the computation

$$
\begin{aligned}
y_n &= a_1 x_n + a_0 x_{n+1} = a_1(x_n + x_{n+1}) - (a_1 - a_0)x_{n+1} \\
y_{n+1} &= a_1 x_{n+1} + a_0 x_{n+2} = a_0(x_{n+1} + x_{n+2}) + (a_1 - a_0)x_{n+1}
\end{aligned}
$$

so that only three multiplications are required. Unfortunately, the number of additions was increased to four ($a_1 - a_0$ can be precomputed), but nonetheless we have made the point that the number of operations may be decreased by identifying redundancies. This is precisely the kind of logic that led us to the FFT algorithm, and we can expect that similar gains can be had for FIR filtering. In fact we can even more directly exploit our experience with the FFT by filtering in the frequency domain.

We have often stressed the fact that filtering a signal in the time domain is equivalent to multiplying by a frequency response in the frequency domain. So we should be able to perform an FFT to jump over to the frequency domain, multiply by the desired frequency response, and then iFFT back to the time domain. Assuming both signal and filter to be of length N, straight convolution takes $O(N^2)$ operations, while the FFT ($O(N \log N)$), multiplication ($O(N)$), and iFFT (once again $O(N \log N)$) clock in at $O(N \log N)$. This idea is *almost* correct, but there are two caveats. The first problem arises when we have to filter an infinite signal, or at least one longer than the FFT size we want to use; how do we piece together the individual results into a single coherent output? The second difficulty is that property (4.47) of the DFT specifies that multiplication in the digital frequency domain corresponds to *circular* convolution of the signals, and not *linear* convolution.

As discussed at length in the previous section, the convolution sum contains shifts for which the filter coefficients extend outside the signal. There

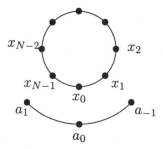

Figure 15.1: Circular convolution for a three-coefficient filter. For shifts where the index is outside the range $0 \ldots N-1$ we assume it wraps around periodically, as if the signal were on a circle.

we assumed that when a nonexistent signal value is required, it should be taken to be zero, resulting in what is called *linear convolution*. Another possibility is *circular convolution*, a quantity mentioned before briefly in connection with the aforementioned property of the DFT. Given a signal with L values $x_0, x_1 \ldots x_{L-1}$ and a set of M coefficients $a_0, a_1 \ldots a_{M-1}$ we defined the circular (also called cyclic) convolution to be

$$y_l = a \circledast x \equiv \sum_m a_m \, x_{(l-m) \bmod L}$$

where mod is the integer modulus operation (see appendix A.2) that always returns an integer between 0 and $L-1$. Basically this means that when the filter is outside the signal range rather than overlapping zeros we wrap the signal around, as depicted in Figure 15.1.

Linear and circular convolution agree for all those output values for which the filter coefficients overlap true signal values; the discrepancies appear only at the edges where some of the coefficients jut out. Assuming we have a method for efficiently computing the circular convolution (e.g., based on the FFT), can it somehow be used to compute a linear convolution? It's not hard to see that the answer is yes, for example, by zero-padding the signal to force the filter to overlap zeros. To see how this is accomplished, let's take a length-L signal $x_0 \ldots x_{L-1}$, a length M filter $a_0 \ldots a_{M-1}$, and assume that $M < L$. We want to compute the L linear convolution outputs $y_0 \ldots y_{L-1}$. The $L - M + 1$ outputs y_{M-1} through y_{L-1} are the same for circular and linear convolution, since the filter coefficients all overlap true inputs. The other $M - 1$ outputs y_0 through y_{M-2} would normally be different, but if we artificially extend the signal by $x_{-M+1} = 0$, through $x_{-1} = 0$ they end up being the same. The augmented input signal is now of length $N = L + M - 1$, and to exploit the FFT we may desire this N to be a power of two.

It is now easy to state the entire algorithm. First we append $M - 1$ zeros to the beginning of the input signal (and possibly more for the augmented signal buffer to be a convenient length for the FFT). We similarly zero-pad the filter to the same length. Next we FFT both the signal and the filter. These two frequency domain vectors are multiplied resulting in a frequency domain representation of the desired result. A final iFFT retrieves N values y_n, and discarding the first $M - 1$ we are left with the desired L outputs.

If N is small enough for a single FFT to be practical we can compute the linear convolution as just described. What can be done when the input is very large or infinite? We simply break the input signal into blocks of length N. The first output block is computed as described above; but from then on we needn't pad with zeros (since the input signal isn't meant to be zero there) rather we use the actual values that are available. Other than that everything remains the same. This technique, depicted in Figure 15.2, is called the *overlap save* method, since the FFT buffers contain $M - 1$ input values saved from the previous buffer. In the most common implementations the $M - 1$ last values in the buffer are copied from its end to its beginning, and then the buffer is filled with N new values from that point on. An even better method uses a circular buffer of length L, with the buffer pointer being advanced by N each time.

You may wonder whether it is really necessary to compute and then discard the first $M - 1$ values in each FFT buffer. This discarding is discarded in an alternative technique called *overlap add*. Here the inputs are not overlapped, but rather are zero-padded at their ends. The linear convolution can be written as a sum over the convolutions of the individual blocks, but the first $M - 1$ output values of each block are missing the effect of the previous inputs that were not *saved*. To compensate, the corresponding outputs are *added* to the outputs from the previous block that corresponded to the zero-padded inputs. This technique is depicted in Figure 15.3.

If computation of FIR filters by the FFT is so efficient, why is straightforward computation of convolution so prevalent in applications? Why do DSP processors have special hardware for convolution, and why do so many software filters use it exclusively? There are two answers to these questions. The first is that the preference is firmly grounded in ignorance and laziness. Straightforward convolution is widely known and relatively simple to code compared with overlap save and add. Many designers don't realize that savings in real-time can be realized or don't want to code FFT, overlap, etc. The other reason is more fundamental and more justifiable. In real-time applications there is often a limitation on *delay*, the time between an input appearing and the corresponding output being ready. For FFT-based tech-

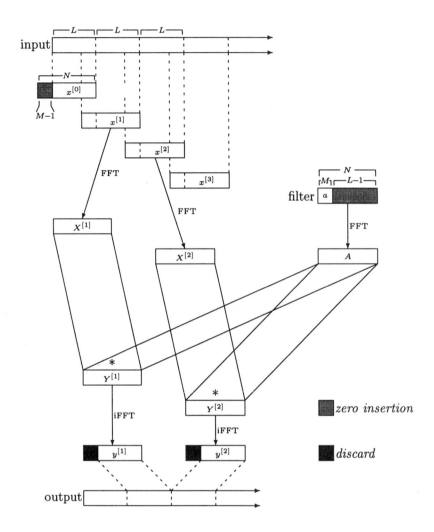

Figure 15.2: *Overlap save* method of filtering in the frequency domain. The input signal x_n is divided into blocks of length L, which are augmented with $M - 1$ values *saved* from the previous block, to fill a buffer of length $N = L + M - 1$. Viewed another way, the input buffers of length N *overlap*. The buffer is converted to the frequency domain and multiplied there by N frequency domain filter values. The result is converted back into the time domain, $M - 1$ incorrect values discarded, and L values output.

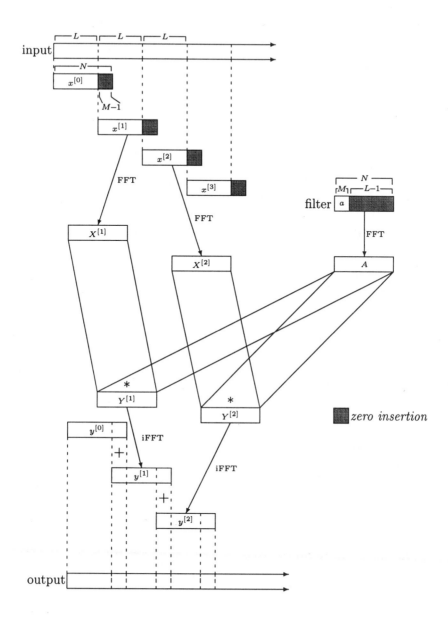

Figure 15.3: *Overlap add* method of filtering in the frequency domain. The input signal x_n is divided into blocks of length L, to which are added $M-1$ zeros to fill a buffer of length $N = L + M - 1$. This buffer is converted to the frequency domain and multiplied there by N frequency domain filter values. The result is converted back into the time domain, $M-1$ partial values at the beginning of the buffer are *overlapped* and then *added* to the $M-1$ last values from the previous buffer.

niques this delay is composed of two parts. First we have to fill up the signal buffer (and true gains in efficiency require the use of large buffers), resulting in *buffer delay*, and then we have to perform the entire computation (FFT, block multiplication, iFFT), resulting in algorithmic delay. Only after all this computation is completed can we start to output the y_n. While the input sample that corresponds to the last value in a buffer suffers only the algorithmic delay, the first sample suffers the sum of both delays. For applications with strict limitations on the allowed delay, we must use techniques where the computation is spread evenly over time, even if they require more computation overall.

EXERCISES

15.2.1 Explain why circular convolution requires specification of the buffer size while linear convolution doesn't. Explain why linear convolution can be considered circular convolution with an infinite buffer.

15.2.2 The circular convolution $y_0 = a_0 x_0 + a_1 x_1$, $y_1 = a_1 x_0 + a_0 x_1$ implies four multiplications and two additions. Show that it can be computed with two multiplications and four additions by precomputing $G_0 = \frac{1}{2}(a_0 + a_1)$, $G_1 = \frac{1}{2}(a_0 - a_1)$, and for each x_0, x_1 computing $z_0 = x_0 + x_1$ and $z_1 = x_0 - x_1$.

15.2.3 Convince yourself that overlap save and overlap add really work by coding routines for straightforward linear convolution, for OA and for OS. Run all three and compare the output signals.

15.2.4 Do you expect OA/OS to be more or less numerically accurate than straightforward convolution in the time domain?

15.2.5 Compare the number of operations per time required for filtering an infinite signal by a filter of length M, using straightforward time domain convolution with that using the FFT. What length FFT is best? When is the FFT method worthwhile?

15.2.6 One can compute circular convolution using an algorithm designed for linear convolution, by replicating parts of the signal. By copying the $L - 2$ last values before x_0 (the *cyclic prefix*) and the $L - 2$ first values after x_{N-1} (the *cyclic suffix*), we obtain a signal that looks like this.

$$0, 0, \quad x_{N-L+1}, x_{N-L+2}, \ldots x_{N-2}, x_{N-1},$$
$$x_0, x_1, \ldots x_{N-2}, x_{N-1},$$
$$x_0, x_1, \ldots x_{L-3}, x_{L-2}, \qquad 0, 0$$

Explain how to obtain the desired circular convolution.

15.2.7 Can IIR filtering be performed in the frequency domain using techniques similar to those of this section? What about LMS adaptive filtering?

15.3 FIR Structures

In this section we return to the time domain computation of convolution of Section 15.1 and to the utilization of graphic techniques for FIR filtering commenced in Section 12.2. In the context of digital filters, graphic implementations are often called *structures*.

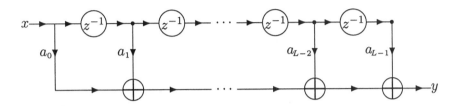

Figure 15.4: *Direct form* implementation of the FIR filter. This form used to be known as the 'tapped delay line', as it is a direct implementation of the weighted sum of delayed taps of the input signal.

In Figure 12.5, reproduced here with slight notational updating as Figure 15.4, we saw one graphic implementation of the linear convolution. This structure used to be called the 'tapped delay line'. The image to be conjured up is that of the input signal being delayed by having to travel with finite velocity along a line, and values being tapped off at various points corresponding to different delays. Today it is more commonly called the *direct form* structure. The direct form implementation of the FIR filter is so prevalent in DSP that it is often considered sufficient for a processor to efficiently compute it to be considered a DSP processor. The basic operation in the tapped delay line is the multiply-and-accumulate (MAC), and the number of MACs per second (i.e., the number of taps per second) that a DSP can compute is the universal benchmark for DSP processor strength.

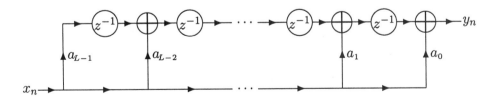

Figure 15.5: *Transposed form* implementation of the FIR filter. Here the present input x_n is multiplied simultaneously by all L filter coefficients, and the intermediate products are delayed and summed.

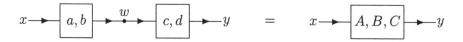

Figure 15.6: Cascading simple filters. On the left the output y is created by filtering w, itself the output of filtering x. On the right is the equivalent single filter system.

Another graphic implementation of the FIR filter is the *transposed structure* depicted in Figure 15.5. The most striking difference between this form and the direct one is that here the undelayed input x_n is multiplied in parallel by all the filter coefficients, and it is these intermediate products that are delayed. Although theoretically equivalent to the direct form the fact that the computation is arranged differently can lead to slightly different numeric results in practice. For example, the round-off noise and overflow errors will not be the same in general.

The transposed structure can be advantageous when we need to partition the computation. For example, assume you have at your disposal digital filter hardware components that can compute L' taps, but your filter specification can only be satisfied with $L > L'$ taps. Distributing the computation over several components is somewhat easier with the transposed form, since we need only provide the new input x_n to all filter components in parallel, and connect the upper line of Figure 15.5 in series. The first component in the series takes no input, and the last component provides the desired output. Were we to do the same thing with the direct form, each component would need to receive *two* inputs from the previous one, and provide *two* outputs to the following one.

However, if we really want to neatly partition the computation, the best solution would be to satisfy the filter specifications by cascading several filters in series. The question is whether general filter specifications can be satisfied by cascaded subfilters, and if so how to find these subfilters.

In order to answer these questions, let's experiment with cascading simple filters. As the simplest case we'll take the subfilters to depend on the present and previous inputs, and to have unity DC gain (see Figure 15.6).

$$
\begin{aligned}
w_n &= ax_n + bx_{n-1} & a + b &= 1 \\
y_n &= cw_n + dw_{n-1} & c + d &= 1
\end{aligned}
\tag{15.6}
$$

Substituting, we see that the two in series are equivalent to a single filter that depends on the present and two past inputs.

$$
\begin{aligned}
y_n &= c(ax_n + bx_{n-1}) + d(ax_{n-1} + bx_{n-2}) \\
&= ac\, x_n + (ad + bc)\, x_{n-1} + bd\, x_{n-2} \\
&= Ax_n + Bx_{n-1} + Cx_{n-2}
\end{aligned}
\tag{15.7}
$$

Due to the unity gain constraints the original subfilters only have one free parameter each, and it is easy to verify that the DC gain of the combined filter is unity as expected ($A + B + C = 1$). So we started with two free parameters, ended up with two free parameters, and the relationship from a, b, c, d to A, B, C is invertible. Given any unity DC gain filter of the form in the last line of equation (15.7) we can find parameters a, b, c, d such that the series connection of the two filters in equation (15.6) forms an equivalent filter. More generally, if the DC gain is nonunity we have four independent parameters in the cascade form, and only three in the combined form. This is because we have the extra freedom of arbitrarily dividing the gain between the two subfilters.

This is one of the many instances where it is worthwhile to simplify the algebra by using the zT formalism. The two filters to be cascaded are described by

$$
\begin{aligned}
w_n &= (a + bz^{-1})\, x_n \\
y_n &= (c + dz^{-1})\, w_n
\end{aligned}
$$

and the resultant filter is given by the product.

$$
\begin{aligned}
y_n &= (c + dz^{-1})(a + bz^{-1})\, x_n \\
&= \left(ac + (ad + bc)z^{-1} + bdz^{-2}\right) x_n \\
&= \left(A + Bz^{-1} + Cz^{-2}\right) x_n
\end{aligned}
$$

We see that the A, B, C parameters derived here by formal multiplication of polynomials in z^{-1} are exactly those derived above by substitution of the intermediate variable w_n. It is suggested that the reader experiment with more complex subfilters and become convinced that this is always the case.

Not only is the multiplication of polynomials simpler than the substitution, the zT formalism has further benefits as well. For example, it is hard to see from the substitution method that the subfilters commute, that is, had we cascaded

$$
\begin{aligned}
v_n &= cx_n + dx_{n-1} & c + d &= 1 \\
y_n &= aw_n + bw_{n-1} & a + b &= 1
\end{aligned}
$$

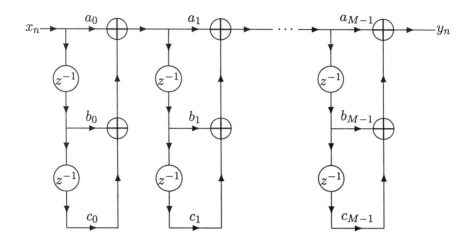

Figure 15.7: *Cascade form* implementation of the FIR filter. Here the input is filtered successively by M 'second-order sections', that is, simple FIR filters that depend on the present input and two past inputs. The term 'second-order' refers to the highest power of z^{-1} being two, and 'section' is synonymous with what we have been calling 'subfilter'. If $c_m = 0$ the section is first order.

we would have obtained the same filter. However, this is immediately obvious in the zT formalism, from the commutativity of multiplication of polynomials.

$$(c + dz^{-1})(a + bz^{-1}) = (a + bz^{-1})(c + dz^{-1})$$

Even more importantly, in the zT formalism it is clear that arbitrary filters can be decomposed into cascades of simple subfilters, called *sections*, by factoring the polynomial in zT. The fundamental theorem of algebra (see Appendix A.6) guarantees that all polynomials can be factored into linear factors (or linear and quadratic if we use only real arithmetic); so any filter can be decomposed into cascades of 'first-order' and 'second-order' sections.

$$h_0 + h_1 z^{-1} \qquad\qquad h_0 + h_1 z^{-1} + h_2 z^{-2}$$

The corresponding structure is depicted in Figure 15.7.

The *lattice structure* depicted in Figure 15.8 is yet another implementation that is built up of basic sections placed in series. The diagonal lines that give it its name make it look very different from the structures we have seen so far, and it becomes even stranger once you notice that the two coefficients on the diagonals of each section are equal. This equality makes the lattice structure numerically robust, because at each stage the numbers being added are of the same order-of-magnitude.

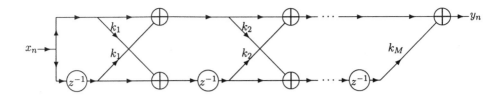

Figure 15.8: *Lattice form* implementation of the FIR filter. Here the input is filtered successively by M lattice stages, every two of which is equivalent to a direct form second-order section.

In order to demonstrate that arbitrary FIR filters can be implemented as lattices, it is sufficient to show that a general second-order section can be. Then using our previous result that general FIR filters can be decomposed into second-order sections the proof is complete. A second-order section has three free parameters, but one degree of freedom is simply the DC gain. For simplicity we will use the following second-order section.

$$y_n = x_n + h_1 x_{n-1} + h_2 x_{n-2}$$

A single lattice stage has only a single free parameter, so we'll need two stages to emulate the second-order section. Following the graphic implementation for two stages we find

$$
\begin{aligned}
y_n &= x_n + k_1 x_{n-1} + k_2(k_1 x_{n-1} + x_{n-2}) \\
&= x_n + k_1(1 + k_2)x_{n-1} + k_2 x_{n-2}
\end{aligned}
$$

and comparing this with the previous expression leads to the connection between the two sets of coefficients (assuming $h_2 \neq -1$).

$$
\begin{array}{ll}
h_1 = k_1(1 + k_2) & \qquad k_1 = \frac{h_1}{1+h_2} \\
h_2 = k_2 & \qquad k_2 = h_2
\end{array}
$$

EXERCISES

15.3.1 Consider the L-tap FIR filter $h_0 = 1, h_1 = \lambda, h_2 = \lambda^2, \ldots h_{L-1} = \lambda^{L-1}$. Graph the direct form implementation. How many delays and how many MACS are required? Find an equivalent filter that utilizes feedback. How many delays and arithmetic operations are required now?

15.3.2 Why did we discuss series connection of simple FIR filter sections but not parallel connection?

15.3.3 We saw in Section 7.2 that FIR filters are linear-phase if they are either symmetric $h_{-n} = h_n$ or antisymmetric $h_{-n} = -h_n$. Devise a graphic implementation that exploits these symmetries. What can be done if there are an even number of coefficients (half sample delay)? What are the advantages of such a implementation? What are the disadvantages?

15.3.4 Obtain a routine for factoring polynomials (these are often called polynomial root finding routines) and write a program that decomposes a general FIR filter specified by its impulse response h_n into first- and second-order sections. Write a program to filter arbitrary inputs using the direct and cascade forms and compare the numeric results.

15.4   Polyphase Filters

The structures introduced in the last section were general-purpose (i.e., applicable to most FIR filters you may need). In this section we will discuss a special purpose structure, one that is applicable only in special cases; but these special cases are rather prevalent, and when they do turn up the general-purpose implementations are often not good enough.

Consider the problem of reducing the sampling frequency of a signal to a fraction $\frac{1}{M}$ of its original rate. This can obviously be carried out by decimation by M, that is, by keeping only one sample out of each M and discarding the rest. For example, if the original signal sampled at f_s is

$$
\begin{array}{ccccccccc}
\cdots & x_{-12}, & x_{-11}, & x_{-10}, & x_{-9}, & x_{-8}, & x_{-7}, & x_{-6}, & x_{-5}, \\
& x_{-4}, & x_{-3}, & x_{-2}, & x_{-1}, & x_0, & x_1, & x_2, & x_3, \\
& x_4, & x_5, & x_6, & x_7, & x_8, & x_9, & x_{10}, & x_{11}, & \cdots
\end{array}
$$

decimating by 4 we obtain a new signal y_n with sampling frequency $\frac{f_s}{4}$.

$$
y_n = \cdots \quad x_{-12}, \quad x_{-8}, \quad x_{-4}, \quad x_0, \quad x_4, \quad x_8, \quad \cdots
$$

Of course

$$
\begin{array}{lllllllll}
y_n & = & \cdots & x_{-11}, & x_{-7}, & x_{-3}, & x_1, & x_5, & x_9, & \cdots \\
y_n & = & \cdots & x_{-10}, & x_{-6}, & x_{-2}, & x_2, & x_6, & x_{10}, & \cdots \\
y_n & = & \cdots & x_{-9}, & x_{-5}, & x_{-1}, & x_3, & x_7, & x_{11}, & \cdots
\end{array}
$$

corresponding to different phases of the original signal, would be just as good.

Actually, just as *bad* since we have been neglecting aliasing. The original signal x can have energy up to $\frac{f_s}{2}$, while the new signal y must not have appreciable energy higher than $\frac{f_s}{2M}$. In order to eliminate the illegal components we are required to low-pass filter the original signal before decimating. For definiteness assume once again that we wish to decimate by 4, and to use a causal FIR antialiasing filter h of length 16. Then

$$
\begin{aligned}
w_0 &= h_0 x_0 + h_1 x_{-1} + h_2 x_{-2} + h_3 x_{-3} + \ldots + h_{15} x_{-15} \\
w_1 &= h_0 x_1 + h_1 x_0 + h_2 x_{-1} + h_3 x_{-2} + \ldots + h_{15} x_{-14} \\
w_2 &= h_0 x_2 + h_1 x_1 + h_2 x_0 + h_3 x_{-1} + \ldots + h_{15} x_{-13} \\
w_3 &= h_0 x_3 + h_1 x_2 + h_2 x_1 + h_3 x_0 + \ldots + h_{15} x_{-12} \\
w_4 &= h_0 x_4 + h_1 x_3 + h_2 x_2 + h_3 x_1 + \ldots + h_{15} x_{-11}
\end{aligned}
\tag{15.8}
$$

but since we are going to decimate anyway

$$
y_n = \ldots w_{-12}, w_{-8}, w_{-4}, w_0, w_4, w_8, \ldots
$$

we needn't compute all these convolutions. Why should we compute w_1, w_2, or w_3 if they won't affect the output in any way? So we compute only $w_0, w_4, w_8, \ldots$, each requiring 16 multiplications and 15 additions.

More generally, the proper way to reduce the sample frequency by a factor of M is to eliminate frequency components over $\frac{f_s}{2M}$ using a low-pass filter of length L. This would usually entail L multiplications and additions per input sample, but for this purpose only L per output sample (i.e., only an average of $\frac{L}{M}$ per input sample are really needed). The straightforward real-time implementation cannot take advantage of this savings in computational complexity. In the above example, at time $n = 0$, when x_0 arrives, we need to compute the entire 16-element convolution. At time $n = 1$ we merely collect x_1 but need not perform any computation. Similarly for $n = 2$ and $n = 3$ no computation is required, but when x_4 arrives we have to compute another 16-element convolution. Thus the DSP processor must still be able to compute the entire convolution in the time between two samples, since the peak computational complexity is unchanged.

The obvious remedy is to distribute the computation over all the times, rather than sitting idly by and then having to race through the convolution. We already know of two ways to do this; by partitioning the input signal or by decimating it. Focusing on w_0, partitioning the input leads to structuring the computation in the following way:

$$
\begin{aligned}
w_0 = \quad & h_0 x_0 + h_1 x_{-1} + h_2 x_{-2} + h_3 x_{-3} \\
+ \ & h_4 x_{-4} + h_5 x_{-5} + h_6 x_{-6} + h_7 x_{-7} \\
+ \ & h_8 x_{-8} + h_9 x_{-9} + h_{10} x_{-10} + h_{11} x_{-11} \\
+ \ & h_{12} x_{-12} + h_{13} x_{-13} + h_{14} x_{-14} + h_{15} x_{-15}
\end{aligned}
$$

Decimation implies the following order:

$$
\begin{aligned}
w_0 = \quad & h_0 x_0 & + \; & h_4 x_{-4} & + \; & h_8 x_{-8} & + \; & h_{12} x_{-12} \\
+ \; & h_1 x_{-1} & + \; & h_5 x_{-5} & + \; & h_9 x_{-9} & + \; & h_{13} x_{-13} \\
+ \; & h_2 x_{-2} & + \; & h_6 x_{-6} & + \; & h_{10} x_{-10} & + \; & h_{14} x_{-14} \\
+ \; & h_3 x_{-3} & + \; & h_7 x_{-7} & + \; & h_{11} x_{-11} & + \; & h_{15} x_{-15}
\end{aligned}
$$

In both cases we should compute only a single row of the above equations during each time interval, thus evenly distributing the computation over the M time intervals.

Now we come to a subtle point. In a real-time system the input signal x_n will be placed into a buffer Ξ. In order to conserve memory this buffer will usually be taken to be of length L, the length of the low-pass filter. The convolution is performed between two buffers of length L, the input buffer and the filter coefficient table; the coefficient table is constant, but a new input x_n is appended to the input buffer every sampling time.

In the above equations for computing w_0 the subscripts of x_n are absolute time indices; let's try to rephrase them using input buffer indices instead. We immediately run into a problem with the partitioned form. The input values in the last row are no longer available by the time we get around to wanting them. But this obstacle is easily avoided by reversing the order.

$$
\begin{aligned}
w_0 = \quad & h_{12} x_{-12} & + \; & h_{13} x_{-13} & + \; & h_{14} x_{-14} & + \; & h_{15} x_{-15} \\
+ \; & h_8 x_{-8} & + \; & h_9 x_{-9} & + \; & h_{10} x_{-10} & + \; & h_{11} x_{-11} \\
+ \; & h_4 x_{-4} & + \; & h_5 x_{-5} & + \; & h_6 x_{-6} & + \; & h_7 x_{-7} \\
+ \; & h_0 x_0 & + \; & h_1 x_{-1} & + \; & h_2 x_{-2} & + \; & h_3 x_{-3}
\end{aligned}
$$

With the understanding that the input buffer updates from row to row, and using a rather uncommon indexing notation for the input buffer, we can now rewrite the partitioned computation as

$$
\begin{aligned}
w_0 = \quad & h_{12} \Xi_{-12} & + \; & h_{13} \Xi_{-13} & + \; & h_{14} \Xi_{-14} & + \; & h_{15} \Xi_{-15} \\
+ \; & h_8 \Xi_{-9} & + \; & h_9 \Xi_{-10} & + \; & h_{10} \Xi_{-11} & + \; & h_{11} \Xi_{-12} \\
+ \; & h_4 \Xi_{-6} & + \; & h_5 \Xi_{-7} & + \; & h_6 \Xi_{-8} & + \; & h_7 \Xi_{-9} \\
+ \; & h_0 \Xi_{-3} & + \; & h_1 \Xi_{-4} & + \; & h_2 \Xi_{-5} & + \; & h_3 \Xi_{-6}
\end{aligned}
$$

and the decimated one as follows.

$$
\begin{aligned}
w_0 = \quad & h_0 \Xi_{-3} & + \; & h_4 \Xi_{-7} & + \; & h_8 \Xi_{-11} & + \; & h_{12} \Xi_{-15} \\
+ \; & h_1 \Xi_{-3} & + \; & h_5 \Xi_{-7} & + \; & h_9 \Xi_{-11} & + \; & h_{13} \Xi_{-15} \\
+ \; & h_2 \Xi_{-3} & + \; & h_6 \Xi_{-7} & + \; & h_{10} \Xi_{-11} & + \; & h_{14} \Xi_{-15} \\
+ \; & h_3 \Xi_{-3} & + \; & h_7 \Xi_{-7} & + \; & h_{11} \Xi_{-11} & + \; & h_{15} \Xi_{-15}
\end{aligned}
$$

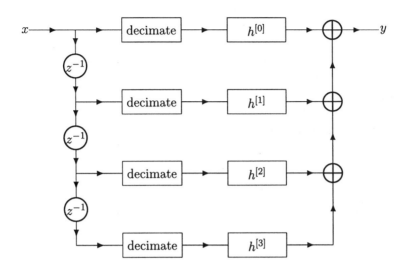

Figure 15.9: The polyphase decimation filter. We depict the decimation of an input signal x_n by a factor of four, using a polyphase filter. Each decimator extracts only inputs with index divisible by 4, so that the combination of delays and decimators results in all the possible decimation phases. $h^{[k]}$ for $k = 0, 1, 2, 3$ are the subfilters; $h^{[0]} = (h_0, h_1, h_2, h_3)$, $h^{[1]} = (h_4, h_5, h_6, h_7)$, etc.

While the partitioned version is rather inelegant, the decimated structure is seen to be quite symmetric. It is easy to understand why this is so. Rather than low-pass filtering and then decimating, what we did is to decimate and then low-pass filter at the lower rate. Each row corresponds to a different decimation phase as discussed at the beginning of the section. The low-pass filter coefficients are different for each phase, but the sum of all contributions results in precisely the desired full-rate low-pass filter.

In the general case we can describe the mechanics of this algorithm as follows. We design a low-pass filter that limits the spectral components to avoid aliasing. We decimate this filter creating M subfilters, one for each of the M phases by which we can decimate the input signal. This set of M subfilters is called a *polyphase filter*. We apply the first polyphase subfilter to the decimated buffer; we then shift in a new input sample and apply the second subfilter in the same way. We repeat this procedure M times to compute the first output. Finally, we reset and commence the computation of the next output. This entire procedure is depicted in Figure 15.9.

A polyphase filter implementation arises in the problem of interpolation as well. By interpolation we mean *increasing* the sampling frequency by an integer factor N. A popular interpolation method is *zero insertion*, inserting $N - 1$ zeros between every two samples of the original signal x. If we interpret this as a signal of sampling rate $N f_s$, its spectrum under $\frac{f_s}{2}$ is the same as that of the original signal, but new components appear at higher frequencies. Low-pass filtering this artificially generated signal removes the higher-frequency components, and gives nonzero values to the intermediate samples.

In a straightforward implementation of this idea we first build a new signal w_n at N times the sampling frequency. For demonstration purposes we take $N = 4$.

$$
\begin{aligned}
\ldots \quad & w_{-16} = x_{-4}, & w_{-15} = 0, & \quad w_{-14} = 0, & \quad w_{-13} = 0, \\
& w_{-12} = x_{-3}, & w_{-11} = 0, & \quad w_{-10} = 0, & \quad w_{-9} = 0, \\
& w_{-8} = x_{-2}, & w_{-7} = 0, & \quad w_{-6} = 0, & \quad w_{-5} = 0, \\
& w_{-4} = x_{-1}, & w_{-3} = 0, & \quad w_{-2} = 0, & \quad w_{-1} = 0, \quad w_0 = x_0, \quad \ldots
\end{aligned}
$$

Now the interpolation low-pass filter performs the following convolution.

$$
\begin{aligned}
y_0 &= h_0 w_0 + h_1 w_{-1} + h_2 w_{-2} + h_3 w_{-3} + \ldots + h_{15} w_{-15} \\
y_1 &= h_0 w_1 + h_1 w_0 + h_2 w_{-1} + h_3 w_{-2} + \ldots + h_{15} w_{-14} \\
y_2 &= h_0 w_2 + h_1 w_1 + h_2 w_0 + h_3 w_{-1} + \ldots + h_{15} w_{-13} \\
y_3 &= h_0 w_3 + h_1 w_2 + h_2 w_1 + h_3 w_0 + \ldots + h_{15} w_{-12}
\end{aligned}
$$

However, most of the terms in these convolutions are zero, and we can save much computation by ignoring them.

$$
\begin{aligned}
y_0 &= h_0 w_0 + h_4 w_{-4} + h_8 w_{-8} + h_{12} w_{-12} \\
&= h_0 x_0 + h_4 x_{-1} + h_8 x_{-2} + h_{12} x_{-3} \\
y_1 &= h_1 w_0 + h_5 w_{-4} + h_9 w_{-8} + h_{13} w_{-12} \\
&= h_1 x_0 + h_5 x_{-1} + h_9 x_{-2} + h_{13} x_{-3} \\
y_2 &= h_2 w_0 + h_6 w_{-4} + h_{10} w_{-8} + h_{14} w_{-12} \\
&= h_2 x_0 + h_6 x_{-1} + h_{10} x_{-2} + h_{14} x_{-3} \\
y_3 &= h_3 w_0 + h_7 w_{-4} + h_{11} w_{-8} + h_{15} w_{-12} \\
&= h_3 x_0 + h_7 x_{-1} + h_{11} x_{-2} + h_{15} x_{-3}
\end{aligned}
$$

Once again this is a polyphase filter, with the input fixed but the subfilters being changed; but this time the absolute time indices of the signal are fixed, not the buffer-relative ones! Moreover, we do not need to add the subfilter outputs; rather each contributes a different output phase. In actual implementations we simply interleave these outputs to obtain the desired

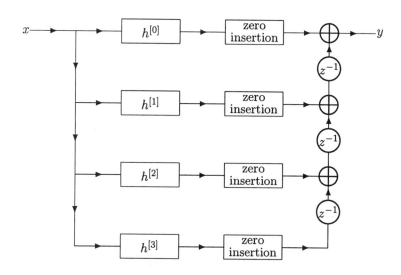

Figure 15.10: The polyphase interpolation filter. We depict the interpolation of an input signal x_n by a factor of four, using a polyphase filter. Each subfilter operates on the same inputs but with different subfilters, and the outputs are interleaved by zero insertion and delay.

interpolated signal. For diagrammatic purposes we can perform the interleaving by zero insertion and appropriate delay, as depicted in Figure 15.10.

We present this rather strange diagram for two reasons. First, because its meaning is instructive. Rather than zero inserting and filtering at the high rate, we filter at the low rate and combine the outputs. Second, comparison with Figure 15.9 emphasizes the inverse relationship between decimation and interpolation. Transposing the decimation diagram (i.e., reversing all the arrows, changing decimators to zero inserters, etc.) converts it into the interpolation diagram.

Polyphase structures are useful in other applications as well. Decimation and interpolation by large composite factors may be carried out in stages, using polyphase filters at every stage. More general sampling frequency changes by rational factors $\frac{N}{M}$ can be carried out by interpolating by N and then decimating by M. Polyphase filters are highly desirable in this case as well. Filter banks can be implemented using mixers, narrowband filters, and decimators, and once again polyphase structures reduce the computational load.

EXERCISES

15.4.1 A commutator is a diagrammatic element that chooses between M inputs $1 \ldots M$ in order. Draw diagrams of the polyphase decimator and interpolator using the commutator.

15.4.2 Both 32 KHz and 48 KHz are common sampling frequencies for music, while CDs uses the unusual sampling frequency of 44.1 KHz. How can we convert between all these rates?

15.4.3 The simple decimator that extracts inputs with index divisible by M is not a time-invariant system, but rather periodically time varying. Is the entire decimation system of Figure 15.9 time-invariant?

15.4.4 Can the polyphase technique be used for IIR filters?

15.4.5 When the decimation or interpolation factor M is large, it may be worthwhile to carry out the filtering in stages. For example, assume $M = M_1 M_2$, and that we decimate by M_1 and then by M_2. Explain how to specify filter responses.

15.4.6 A half-band filter is a filter whose frequency response obeys the symmetry $H(\omega) = 1 - H(\omega_{mid} - \omega)$ around the middle of the band $\omega_{mid} = \frac{f_s}{4}$. For every low-pass half-band filter there is a high-pass half-band filter called its 'mirror filter'. Explain how mirror half-band filters can be used to efficiently compute a bank of filters with 2^m bands.

15.5   Fixed Point Computation

Throughout this book we stress the advantages of DSP as contrasted with analog processing. In this section we admit that digital processing has a disadvantage as well, one that derives from the fact that only a finite number of bits can be made available for storage of signal values and for computation. In Section 2.7 we saw how digitizing an analog signal inevitably adds quantization noise, due to imprecision in representing a real number by a finite number of bits. However, even if the digitizer has a sufficient number of bits and we ensure that analog signals are amplified such that the digitizer's dynamic range is optimally exploited, we still have problems due to the nature of digital computation.

In general, the sum of two b-bit numbers will have $b + 1$ bits. When floating point representation (see Appendix A.3) is being used, a $(b + 1)$-bit result can be stored with b bits of mantissa and a larger exponent, causing a slight round-off error. This round-off error can be viewed as a small additional additive noise that in itself may be of little consequence. However, since hundreds of computations may need to be performed the final result may have become hopelessly swamped in round-off noise. Using fixed point representation exacerbates the situation, since should $b + 1$ exceed the fixed number of bits the hardware provides, an *overflow* will occur. To avoid overflow we must ensure that the terms to be added contain fewer bits, reducing dynamic range even when overflow would *not* have occurred. Hence fixed point hardware cannot even consistently exploit the bits it potentially has.

Multiplication is even worse than addition since the product of two numbers with b bits can contain $2b$ bits. Of course the multiply-and-accumulate (MAC) operation, so prevalent in DSP, is the worst offender of all, endlessly summing products and increasing the number of required bits at each step! This would certainly render all fixed point DSP processors useless, were it not for *accumulators*. An accumulator is a special register with extra bits that is used for accumulating intermediate results. The MAC operation is performed using an accumulator with sufficient bits to prevent overflow; only at the end of the convolution is the result truncated and stored back in a normal register or memory. For example, a 16-bit processor may have a 48-bit accumulator; since each individual product returns a 32-bit result, an FIR filter of length 16 can be performed without prescaling with no fear of overflow.

We can improve our estimate of the required input prescaling if we know the filter coefficients a_l. The absolute value of the convolution output is

$$|y_n| = |\sum_l h_l x_{n-l}| \leq \sum_l |h_l||x_{n-l}| \leq \left(\sum_l |h_l| \right) x_{max}$$

where x_{max} is the maximal absolute value the input signal takes. In order to ensure that y_n never overflows in an accumulator of b bits, we need to ensure that the maximal x value does not exceed the following bound.

$$x_{max} \leq \frac{2^b}{\sum_l |h_l|} \tag{15.9}$$

This worst-case analysis of the possibility of overflow is often too extreme. The input scaling implied for even modest filter lengths would so drastically reduce the SNR that we are usually willing to risk possible but improbable overflows. Such riskier scaling methods are obtained by replacing the sum of absolute values in equation (15.9) with different combinations of the h_l coefficients. One commonly used criterion is

$$x_{max} \leq \frac{2^b}{\sqrt{\sum_l |h_l|^2}}$$

which results from requiring the output energy to be sufficiently low; another is

$$x_{max} \leq \frac{2^b}{H_{max}}$$

where H_{max} is the maximum value of the filter's frequency response, resulting from requiring that the output doesn't overflow in the frequency domain.

When a result overflow does occur, its effect is hardware dependent. Standard computers usually set an overflow flag to announce that the result is meaningless, and return the meaningless least significant bits. Thus the product of two positive numbers may be negative and the product of two large numbers may be small. Many DSP processors have a *saturation arithmetic* mode, where calculations that overflow return the largest available number of the appropriate sign. Although noise is still added in such cases, its effect is much less drastic. However, saturation introduces clipping nonlinearity, which can give rise to harmonic distortion.

Even when no overflow takes place, digital filters (especially IIR filters) may act quite differently from their analog counterparts. As an example, take the simple AR filter

$$y_n = x_n - 0.9y_{n-1} \tag{15.10}$$

whose true impulse response is $h_n = (-0.9)^n u_n$. For simplicity, let's examine the somewhat artificial case of a processor accurate to within one decimal digit after the decimal point (i.e., we'll assume that the multiplication $0.9y_{n-1}$ is *rounded* to a single decimal digit to the right of the point). Starting with $x_0 = 1$ the true output sequence should oscillate while decaying exponentially. However, it is easy to see that under our quantized arithmetic $-0.9 \cdot -0.4 = +0.4$ and conversely $-0.9 \cdot 0.4 = -0.4$ so that $0.4, -0.4$ is a cycle, called a *limit cycle*. In Figure 15.11 we contrast the two behaviors.

The appearance of a limit cycle immediately calls to mind our study of chaos in Section 5.5, and the relationship is not coincidental. The fixed point arithmetic transforms the initially *linear* recursive system into a nonlinear one, one whose long time behavior displays an attractor that is not a fixed point. Of course, as we learned in that section, the behavior could have been even worse!

There is an alternative way of looking at the generation of the spurious oscillating output. We know that stable IIR filters have all their poles inside the unit circle, and thus cannot give rise to spurious oscillations. However,

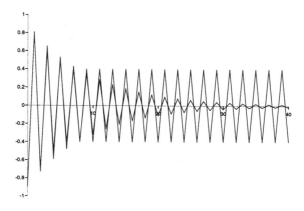

Figure 15.11: The behavior of a simple AR filter using fixed point arithmetic. The decaying plot depicts the desired behavior, while the second plot is the behavior that results from rounding to a single digit after the decimal point.

the quantization of the filter coefficients causes the poles to stray from their original positions, and in particular a pole may wander outside the unit circle. Once excited, such a pole causes oscillating outputs even when the input vanishes.

This idea leads us to investigate the effect of coefficient quantization on the position of the filter's poles and zeros, and hence on its transfer function. Let's express the transfer function

$$H(z) = \frac{A(z^{-1})}{B(z^{-1})} = \frac{\sum_{l=0}^{L} a_l z^{-l}}{1 - \sum_{m=1}^{M} b_m z^{-m}} = \frac{\prod_{l=1}^{L}(z - \zeta_l)}{\prod_{m=1}^{M}(z - \pi_m)} \tag{15.11}$$

and consider the effect of quantizing the b_m coefficients on the pole positions π_m. The quantization introduces round-off error, so that the effective coefficient is $b_m + \delta b_m$, and assuming that this round-off error is small, its effect on the position of pole k may be approximated by the first-order contributions.

$$\delta \pi_k = \sum_{m=1}^{M} \frac{\partial \pi_k}{\partial b_m} \delta b_m$$

After a bit of calculation we can find that

$$\frac{\partial \pi_k}{\partial b_m} = \frac{\pi_k^{M-m}}{\prod_{\substack{j=1 \\ j \neq k}}^{M} (\pi_k - \pi_j)} \tag{15.12}$$

i.e., the effect of variation of the mth coefficient on the kth pole depends on the positions of all the poles.

In particular, if the original filter has poles that are close together (i.e., for which $\pi_k - \pi_j$ is small), small coefficient round-off errors can cause significant movement of these poles. Since close poles are a common occurrence, straightforward implementation of IIR filters as difference equations often lead to instability when fixed point arithmetic is employed. The most common solution to this problem is to implement IIR filters as cascades of subfilters with poles as far apart as possible. Since each subfilter is separately computed, the round-off errors cannot directly interact, and pole movement can be minimized. Carrying this idea to the extreme we can implement IIR filters as cascades of second-order sections, each with a single pair of conjugate poles and a single pair of conjugate zeros (if there are real poles or zeros we use first-order structures). In order to minimize strong gains that may cause overflow we strive to group together zeros and poles that are as close together as possible. This still leaves considerable freedom in the placement order of the sections. Empirically, it seems that the best strategy is to order sections monotonically in the radius of their poles, either from smallest to largest (those nearest the unit circle) or vice versa. The reasoning is not hard to follow. Assume there are poles with very small radius. We wouldn't want to place them first since this would reduce the number of effective bits in the signal early on in the processing, leading to enhanced round-off error. Ordering the poles in a sequence with progressively decreasing radius ameliorates this problem. When there are poles very close to the unit circle placing them first would increase the chance of overflow, or require reducing the dynamic range in order to avoid overflow. Ordering the poles in a sequence with progressively increasing radius is best in this case. When there are both small and large poles it is hard to know which way is better, and it is prudent to directly compare the two alternative orders. Filter design programs that include fixed point optimization routines take such pairing and ordering considerations into account.

EXERCISES

15.5.1 A pair of conjugate poles with radius $r < 1$ and angles $\pm\theta$ contribute a second-order section

$$(z - re^{i\theta})(z - re^{-i\theta}) = z^2 \left(1 - 2r\cos\theta z^{-1} + r^2 z^{-2}\right)$$

with coefficients $b_1 = 2r\cos\theta$ and $b_2 = -r^2$. If we quantize these coefficients to b bits each, how many distinct pole locations are possible? To how many bits has the radius r been quantized? Plot all the possible poles for 4–8 bits. What can you say about the quantization of real poles?

15.5.2 As we discussed in Section 14.6, fixed point FFTs are vulnerable to numerical problems as well. Compare the accuracy and overflow characteristics of frequency domain and time domain filtering.

15.5.3 Develop a strategy to eliminate limit cycles, taking into account that limit cycles can be caused by round-off or overflow errors.

15.5.4 Complete the derivation of the dependence of π_k on δb_m.

15.5.5 What can you say about the dependence of zero position ζ_l on small changes in numerator coefficients a_l? Why do you think fixed point FIR filters are so often computed in direct form rather than cascade form?

15.5.6 We saw that it is possible to prescale the input in order to ensure that an FIR filter will never overflow. Is it possible to guarantee that an IIR filter will not overflow?

15.5.7 In the text we saw a system whose impulse response should have decayed to zero, but due to quantization was a 2-cycle. Find a system whose impulse response is a nonzero constant. Find a system with a 4-cycle. Find a system that goes into oscillation because of overflow.

15.6   IIR Structures

We return now to structures for general filters and consider the case of IIR filters. We already saw how to diagram the most general IIR filter in Figures 12.8.B and 12.11, but know from the previous section that this direct form of computation is not optimal from the numerical point of view. In this section we will see better approaches.

The general cascade of second-order IIR sections is depicted in Figure 15.12. Each section is an independent first- or second-order ARMA filter, with its own coefficients and static memory. The only question left is how to best implement this second-order section. There are three different structures in common use: the direct form (also called the *direct form I*) depicted in Figure 15.13, the canonical form (also called *direct form II*) depicted in Figure 15.14, and the transposed form (also called *transposed form II*) depicted in Figure 15.15. Although all three are valid implementations of precisely the same filter, numerically they may give somewhat different results.

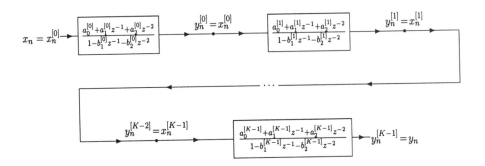

Figure 15.12: General cascade implementation of an IIR filter. Each section implements an independent (first- or) second-order section symbolized by the transfer function appearing in the rectangle. Note that a zero in any of these subfilters results in a zero of the filter as a whole.

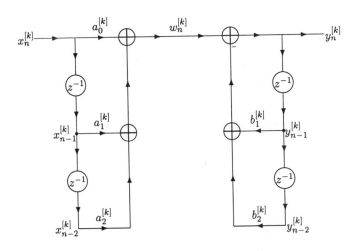

Figure 15.13: *Direct form* implementation of a second-order IIR section. This structure is derived by placing the MA (all-zero) filter before the AR (all-pole) one.

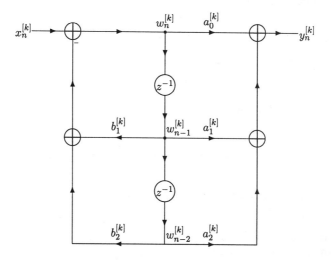

Figure 15.14: *Canonical form* implementation of a second-order IIR section. This structure is derived by placing the AR (all-pole) filter before the MA (all-zero) one and combining common elements. (Why didn't we draw a filled circle for $w_{n-2}^{[k]}$?)

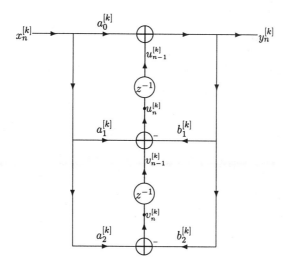

Figure 15.15: *Transposed form* implementation of a second-order IIR section. Here only the intermediate variables are delayed. Although only three adders are shown the center one has three inputs, and so there are actually four additions.

An IIR filter implemented using direct form sections is computed as follows:

```
loop over time n
    x_n^[0] ← x_n
    loop on section number k ← 0 to K − 1
        w_n^[k] ← a_0^[k] x_n^[k] + a_1^[k] x_{n-1}^[k] + a_2^[k] x_{n-2}^[k]
        y_n^[k] ← w_n^[k] − b_1^[k] y_{n-1}^[k] − b_2^[k] y_{n-2}^[k]
        x_n^[k+1] ← y_n^[k]
    y_n ← y_n^[K-1]
```

In real-time applications the loop over time will normally be an infinite loop. Each new input sample is first MA filtered to give the intermediate signal $w_n^{[0]}$

$$w_n^{[0]} = a_0^{[0]} x_n + a_1^{[0]} x_{n-1} + a_2^{[0]} x_{n-2}$$

and then this signal is AR filtered to give the section's output

$$y_n^{[0]} = w_n^{[0]} - b_1^{[0]} y_{n-1}^{[0]} - b_2^{[0]} y_{n-2}^{[0]}$$

the subtraction either being performed once, or twice, or negative coefficients being stored. This section output now becomes the input to the next section

$$x_n^{[1]} \leftarrow y_n^{[0]}$$

and the process repeats until all K stages are completed. The output of the final stage is the desired result.

$$y_n = y_n^{[K-1]}$$

Each direct form stage requires five multiplications, four additions, and four delays. In the diagrams we have emphasized memory locations that have to be stored (static memory) by a circle. Note that w_n is generated each time and does not need to be stored, so that there are only two saved memory locations.

As we saw in Section 12.3 we can reverse the order of the MA and AR portions of the second-order section, and then regroup to save memory locations. This results in the structure known as canonical (meaning 'accepted' or 'simplest') form, an appellation well deserved because of its use of the least number of delay elements. While the direct form requires delayed versions of both x_n and y_n, the canonical form only requires storage of w_n.

The computation is performed like this

```
loop over time n
    x_n^[0] ← x_n
    loop on section number k ← 0 to K − 1
        w_n^[k] ← x_n^[k] − b_1^[k] w_{n-1}^[k] − b_2^[k] w_{n-2}^[k]
        y_n^[k] ← a_0^[k] w_n^[k] + a_1^[k] w_{n-1}^[k] + a_2^[k] w_{n-2}^[k]
        x_n^[k+1] ← y_n^[k]
    y_n ← y_n^[K-1]
```

and once again we can either stored negative b coefficients or perform subtraction(s). Each canonical form stage requires five multiplications, four additions, two delays, and two intermediate memory locations.

The transposed form is so designated because it can be derived from the canonical form using the *transposition theorem*, which states that reversing all the arc directions, changing adders to tee connections and vice-versa, and interchanging the input and output does not alter the system's transfer function. It is also canonical in the sense that it also uses only two delays, but we need to save a single value of two different signals (which we call u_n and v_n), rather than two lags of a single intermediate signal. The full computation is

```
loop over time n
    x_n^[0] ← x_n
    loop on section number k ← 0 to K − 1
        v_n^[k] ← a_2^[k] x_n^[k] − b_2^[k] y_n^[k]
        u_n ← a_1^[k] x_n^[k] − b_1^[k] y_n^[k] + v_{n-1}^[k]
        y_n^[k] ← a_0^[k] x_n^[k] + u_{n-1}^[k]
        x_n^[k+1] ← y_n^[k]
    y_n ← y_n^[K-1]
```

Don't be fooled by Figure 15.15 into thinking that there are only three additions in the transposed form. The center adder is a three-input adder, which has to be implemented as two separate additions. Hence the transposed form requires five multiplications, four additions, two delays, and two intermediate memory locations, just like the canonical form.

The cascade forms we have just studied are numerically superior to direct implementation of the difference equation, especially when pole-zero pairing and ordering are properly carried out. However, the very fact that the signal has to travel through section after section in series means that round-off

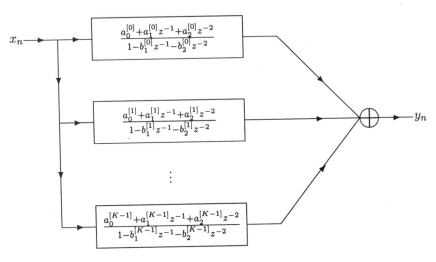

Figure 15.16: *Parallel form* implementation of the IIR filter. In this form the subfilters are placed in parallel, and so round-off errors do not accumulate. Note that a pole in any of these subfilters results in a pole of the filter as a whole.

errors accumulate. Parallel connection of second-order sections, depicted in Figure 15.16, is an alternative implementation of the general IIR filter that does not suffer from round-off accumulation. The individual sections can be implemented in direct, canonical, or transposed form; and since the outputs are all simply added together, it is simpler to estimate the required number of bits.

The second-order sections in cascade form are guaranteed to exist by the fundamental theorem of algebra, and are found in practice by factoring the system function. Why are general system functions expressible as sums of second-order filters, and how can we perform this decomposition? The secret is the 'partial fraction expansion' familiar to all students of indefinite integration. Using partial fractions, a general system function can be written as the sum of first-order sections

$$H(z) = \sum_{k=1}^{K} \frac{\Gamma_k}{1 + \gamma_k z^{-1}} \tag{15.13}$$

with Γ_k and γ_k possibly complex, or as the sum of second-order sections

$$H(z) = \sum_{k=1}^{K} \frac{A_k + B_k z^{-1}}{1 + \alpha_k z^{-1} + \beta_k z^{-2}} \tag{15.14}$$

with all coefficients real. If there are more zeros than poles in the system function, we need an additional FIR filter in parallel with the ARMA sections.

The decomposition is performed in practice by factoring the denominator of the system function into real first- and second-order factors, writing the partial fraction expansion, and comparing. For example, assume that the system function is

$$H(z) = \frac{1 + az^{-1} + bz^{-2}}{(1 + cz^{-1})(1 + dz^{-1} + ez^{-2})}$$

then we write

$$
\begin{aligned}
H(z) &= \frac{A}{1 + cz^{-1}} + \frac{B + Cz^{-1}}{1 + dz^{-1} + ez^{-2}} \\
&= \frac{(A + B) + (Ad + Bc + C)z^{-1} + (Ae + Cc)z^{-2}}{(1 + cz^{-1})(1 + dz^{-1} + ez^{-2})}
\end{aligned}
$$

and compare. This results in three equations for the three variables A, B, and C.

EXERCISES

15.6.1 An arbitrary IIR filter can always be factored into cascaded first-order sections, if we allow complex-valued coefficients. Compare real-valued second-order sections with complex-valued first-order sections from the points of view of computational complexity and numerical stability.

15.6.2 A second-order all-pass filter section has the following transfer function.

$$\frac{c + dz^{-1} + z^{-2}}{1 + dz^{-1} + cz^{-2}}$$

Diagram it in direct form. How many multiplications are needed? Redraw the section emphasizing this.

15.6.3 Apply the transposition theorem to the direct form to derive a noncanonical transposed section.

15.6.4 The lattice structure presented for the FIR filter in Section 15.3 can be used for IIR filters as well. Diagram a two-pole AR filter. How can lattice techniques be used for ARMA filters?

15.7   FIR vs. IIR

Now that we have seen how to implement both FIR and IIR filters, the question remains as to which to use. Once again we suggest first considering whether it is appropriate to filter in the frequency domain. Frequency domain filtering is almost universally applicable, is intrinsically stable, and the filter designer has complete control over the phase response. The run-time code is often the most computationally efficient technique, and behaves well numerically if the FFTs are properly scaled. Phase response is completely controllable. Unfortunately, it *does* introduce considerable buffer and algorithmic delay; it *does* require more complex code; and it possibly requires more table and scratch memory. Of course we cannot really multiply any frequency component by infinity and so true poles on the frequency axis are not implementable, but IIR filters with such poles would be unstable anyway.

Assuming you have come to the conclusion that time domain filtering *is* appropriate, the next question has to do with the type of filter that is required. Special filters (see Section 7.3) have their own special considerations. In general, integrators should be IIR, differentiators even order FIR (unless the half sample delay is intolerable), Hilbert transforms odd order FIR with half the coefficients zero (although IIR designs are possible), decimators and integrators should be polyphase FIR, etc. Time-domain filter specifications immediately determine the FIR filter coefficients, but can also be converted into an IIR design by Prony's method (see Section 13.6). When the sole specification is one of the standard forms of Section 7.1, such as low-pass, IIR filters can be readily designed while optimal FIR designs require more preparation. If the filter *design* must be performed in run-time then this will often determine the choice of filter type. Designing a standard IIR filter reduces to a few equations, and the suboptimal windowing technique for designing FIR filters can sometimes be used as well. From now on we'll assume that we have a constant prespecified frequency domain specification.

It is important to determine whether a true linear-phase filter or only a certain degree of phase linearity is required (e.g., communications signals that contain information in their phase, or simultaneous processing of multiple signals that will later be combined). Recall from Section 7.2 that symmetric or antisymmetric FIR filters are precisely linear-phase, while IIR filters can only approximate phase linearity. However, IIR filters can have their phase flattened to a large degree, and if sufficient delay is allowed the pseudo-IIR filter of exercise 7.2.5 may be employed for precise phase linearity.

Assuming that both FIR and IIR filters are still in the running (e.g., only the amplitude of the frequency response is of interest), the issue of computational complexity is usually the next to be considered. IIR filters with a relatively small number of coefficients can be designed to have very sharp frequency response transitions (with the phase being extremely non-linear near these transitions) and very strong stop-band attenuation. For a given specification elliptical IIR filters will usually have dramatically lower computational complexity than FIR filters, with the computational requirements ratio sometimes in the thousands. Only if the filters are relatively mild and when a large amount of pass-band ripple can be tolerated will the computational requirements be similar or even in favor of the FIR. Chebyshev IIR filters are less efficient than elliptical designs but still usually better performers than FIR filters. Butterworth designs are the least flexible and hence require the highest order and the highest computational effort. If phase linearity compensation is attempted for a Butterworth IIR filter the total computational effort may be comparable to that of an FIR filter.

The next consideration is often numerical accuracy. It is relatively simple to determine the worst-case number of bits required for overflow-free FIR computation, and if sufficient bits are available in the accumulator and the quantized coefficients optimized, the round-off error will be small. Of course long filters and small registers will force us to prescale down filter coefficients or input signals causing 6 dB of SNR degradation for each lost bit. For IIR filters determining the required number of bits is much more complex, depending on the filter characteristics and input signal frequency components. FIR filters are inherently stable, while IIR filters may be unstable or may become unstable due to numerical problems. This is of overriding importance for filters that must be varied as time goes on; an IIR filter must be continuously monitored for stability (possibly a computationally intensive task in itself) while FIR filters may be used with impunity.

Finally, all things being equal, personal taste and experience comes into play. Each DSP professional accumulates over time a bag of fully honed and well-oiled tools. It is perfectly legitimate that the particular tool that 'feels right' to one practitioner may not even be considered by another. The main problem is that when you have only a hammer every problem looks like a nail. We thus advise that you work on as many different applications as possible, collecting a tool or two from each.

Bibliographical Notes

Most general DSP texts, e.g., [186, 200] and Chapters 6 and 7 of [241], cover digital filter structures to some degree. Also valuable are libraries and manuals that accompany specific DSP processors.

The idea of using the DFT to compute linear convolutions appears to have been invented simultaneously at MIT [124], at Bell Labs [100] and by Sande at Princeton.

Polyphase filtering was developed extensively at Bell Labs, and a good review of polyphase filters for interpolation and decimation is [48].

The effect of numerical error on filters has an extensive bibliography, e.g., [151, 215, 114, 115].

16

Function Evaluation Algorithms

Commercially available DSP processors are designed to efficiently implement FIR, IIR, and FFT computations, but most neglect to provide facilities for other desirable functions, such as square roots and trigonometric functions. The software libraries that come with such chips *do* include such functions, but one often finds these general-purpose functions to be unsuitable for the application at hand. Thus the DSP programmer is compelled to enter the field of numerical approximation of elementary functions. This field boasts a vast literature, but only relatively little of it is directly applicable to DSP applications.

As a simple but important example, consider a complex mixer of the type used to shift a signal in frequency (see Section 8.5). For every sample time t_n we must generate both $\sin(\omega t_n)$ and $\cos(\omega t_n)$, which is difficult using the rather limited instruction set of a DSP processor. Lack of accuracy in the calculations will cause phase instabilities in the mixed signal, while loss of precision will cause its frequency to drift. Accurate values can be quickly retrieved from lookup tables, but such tables require large amounts of memory and the values can only be stored for specific arguments. General purpose approximations tend to be inefficient to implement on DSPs and may introduce intolerable inaccuracy.

In this chapter we will specifically discuss sine and cosine generation, as well as rectangular to polar conversion (needed for demodulation), and the computation of arctangent, square roots, Puthagorean addition and logarithms. In the last section we introduce the CORDIC family of algorithms, and demonstrate its applicability to a variety of computational tasks. The basic CORDIC iteration delivers a bit of accuracy, yet uses only additions and shifts and so can be implemented efficiently in hardware.

16.1   Sine and Cosine Generation

In DSP applications, one must often find $\sin(\omega t)$ where the time t is quantized $t = k\, t_s$ and $f_s = \frac{1}{t_s}$ is the sampling frequency.

$$\sin(\omega t) = \sin(2\pi\, f\, k\, t_s) = \sin\left(2\pi \frac{f}{f_s} k\right)$$

The digital frequency of the sine wave, f/f_s, is required to have resolution $\frac{1}{N}$, which means that the physical frequency is quantized to $f = \frac{m}{N} f_s$. Thus the functions to be calculated are all of the following form:

$$\sin\left(2\pi \frac{m}{N} k\right) = \sin\left(\frac{2\pi}{N} i\right) \qquad\qquad i \equiv mk = 0 \ldots N$$

In a demanding audio application, $f_s \approx 50$ KHz and we may want the resolution to be no coarser than 0.1 Hz; thus about $N = 500,000$ different function values are required. Table lookup is impractical for such an application.

   The best known method for approximating the trigonometric functions is via the Taylor expansions

$$\begin{aligned}
\sin(x) &= x - \frac{1}{3!}x^3 + \frac{1}{5!}x^5 - \frac{1}{7!}x^7 + \cdots \qquad\qquad (16.1)\\
\cos(x) &= 1 - \frac{1}{2!}x^2 + \frac{1}{4!}x^4 - \frac{1}{6!}x^6 + \cdots
\end{aligned}$$

which converge rather slowly. For any given place of truncation, we can improve the approximation (that is, reduce the error made) by slightly changing the coefficients of the expansion. Tables of such corrected coefficients are available in the literature. There are also techniques for actually speeding up the convergence of these polynomial expansions, as well as alternative rational approximations. These approximations tend to be difficult to implement on DSP processors, although (using Horner's rule) polynomial calculation can be pipelined on MAC machines.

   For the special case (prevalent in DSP) of equally spaced samples of a sinusoidal oscillator of fixed frequency, several other techniques are possible. One technique that we studied in Section 6.11 exploits the fact that sinusoidal oscillations are solutions of second-order differential or difference equations, and thus a new sine value may be calculated recursively based on two previous values. Thus one need only precompute two initial values and thereafter churn out sine values. The problem with any recursive method of this sort is *error accumulation*. Our computations only have finite accuracy, and with time the computation error builds up. This error accumulation leads to long-term instability. We can combine recursive computation with occasional nonrecursive (and perhaps more expensive) calculations, but then one must ensure that no sudden changes occur at the boundaries.

Another simple technique that recursively generates sinusoids can simultaneously produce both the sine and the cosine of the same argument. The idea is to use the trigonometric sum formulas

$$\sin(\omega k) = \sin\left(\omega(k-1)\right) * \cos(\omega) + \cos\left(\omega(k-1)\right) * \sin(\omega) \quad (16.2)$$
$$\cos(\omega k) = \cos\left(\omega(k-1)\right) * \cos(\omega) - \sin\left(\omega(k-1)\right) * \sin(\omega)$$

with known $\sin(\omega)$ and $\cos(\omega)$. Here one initial value of *both* sine and cosine are required, and thereafter only the previous time step must be saved. These recursive techniques are easily implementable on DSPs, but also suffer from error accumulation.

Let's revisit the idea of table lookup. We can reduce the number of values which must be held in such a table by exploiting symmetries of the trigonometric functions. For example, we do not require twice N memory locations in order to simultaneously generate both the sine and cosine of a given argument, due to the connection between sine and cosine in equation (A.22).

We can more drastically reduce the table size by employing the trigonometric sum formula (A.23). To demonstrate the idea, let us assume one wishes to save sine values for all integer degrees from zero to ninety degrees. This would a priori require a table of length 91. However, one could instead save three tables:

1. $\sin(0°)$, $\sin(10°)$, $\sin(20°)$,   ... $\sin(90°)$
2. $\sin(0°)$, $\sin(1°)$, $\sin(2°)$,   ... $\sin(9°)$
3. $\cos(0°)$, $\cos(1°)$, $\cos(2°)$,   ... $\cos(9°)$

and then calculate, for example, $\sin(54°) = \sin(50°)\cos(4°)+\sin(40°)\sin(4°)$. In this simple case we require only 30 memory locations; however, we must perform one division with remainder (in order to find $54° = 50° + 4°$), two multiplications, one addition, and four table lookups to produce the desired result. The *economy* is hardly worthwhile in this simple case; however, for our more demanding applications the effect is more dramatic.

In order to avoid the prohibitively costly division, we can divide the circle into a number of arcs that is a power of two, e.g., $2^{19} = 524{,}288$. Then every i, $0 \leq i \leq 524{,}288$ can be written as $i = j + k$ where $j = 512(i/512)$ (here / is the integer division without remainder) and $k = i \bmod 512$ can be found by shifts. In this case we need to store three tables:

1. Major Sine: $\sin(\frac{2\pi}{N} 512\, j)$ 512 values
2. Minor Sine: $\sin(\frac{2\pi}{N} k)$ 512 values
3. Minor Cosine: $\cos(\frac{2\pi}{N} k)$ 512 values

which altogether amounts to only 1536 values (for 32-bit words this is 6144 bytes), considerably less than the 524288 values in the straightforward table.

An alternate technique utilizing the CORDIC algorithm will be presented in Section 16.5.

EXERCISES

16.1.1 Evaluate equation (16.2), successively generating further sine and cosine values (use single precision). Compare these values with those returned by the built-in functions. What happens to the error?

16.1.2 Try to find limitations or problems with the trigonometric functions as supplied by your compiler's library. Can you guess what algorithm is used?

16.1.3 The simple cubic polynomial

$$\frac{4}{\pi^3}\, x \left(\frac{3}{4}\pi^2 - x^2\right)$$

approximates $\sin(x)$ to within 2% over the range $[-\frac{\pi}{2} \ldots \frac{\pi}{2}]$. What are the advantages and disadvantages of using this approximation? How can you bring the error down to less than 1%?

16.1.4 Code the three-table sine and cosine algorithm in your favorite programming language. Preprepare the required tables. Test your code by generating the sine and cosine for all whole-degree values from 0 to 360 and comparing with your library routines.

16.1.5 The signal supplied to a signal processing system turns out to be inverted in spectrum (that is, $f \rightarrow f_s - f$) due to an analog mixer. You are very much worried since you have practically no spare processing power, but suddenly realize the inversion can be carried out with practically no computation. How do you do it?

16.1.6 You are given the task of designing a *mixer-filter*, a device that band-pass filters a narrow bandwidth signal and at the same time translates it from one frequency to another. You must take undesired mixer by-products into account, and should not require designing a filter in real-time. Code your mixer filter using the three-table sine and cosine algorithm. Generate a signal composed of a small number of sines, mix it using the mixer filter, and perform an FFT on the result. Did you get what you expect?

16.2 Arctangent

The floating point arctangent is often required in DSP calculations. Most often this is in the context of a rectangular to polar coördinate transformation, in which case the CORDIC-based algorithm given in Section 16.5 is usually preferable. For other cases simple approximations may be of use.

First one can always reduce the argument range to $0 \le x \le 1$, by exploiting the antisymmetry of the function for negative arguments, and the symmetry

$$\tan^{-1}(x) = \frac{\pi}{2} - \tan^{-1}\left(\frac{1}{x}\right)$$

for $x > 1$.

For arguments in this range, we can approximate by using the Taylor expansion around zero.

$$\tan^{-1}(x) = x - \tfrac{1}{3}x^3 + \tfrac{1}{5}x^5 - \tfrac{1}{7}x^7 + \cdots \qquad (16.3)$$

As for the sine and cosine functions equations (16.1), the approximation can be improved by slightly changing the coefficients.

EXERCISES

16.2.1 Code the arctangent approximation of equation (16.3), summing up N terms. What is the maximum error as a function of N?

16.2.2 How can improved approximation coefficients be found?

16.2.3 Look up the improved coefficients for expansion up to fifth order. How much better is the improved formula than the straight Taylor expansion? Plot the two approximations and compare their global behavior.

16.2.4 For positive x there is an alternative expansion:

$$\tan^{-1}(x) = \frac{\pi}{4} + a_1 y + a_3 y^3 + a_5 y^5 + \dots \qquad \text{where } y \equiv \frac{x-1}{x+1}$$

Find the coefficients and compare the accuracy with that of equation (16.3).

16.2.5 Make a phase detector, i.e., a program that inputs a complex exponential $s_n = x_n + i y_n = A e^{i(\omega n + \phi_n)}$, computes, and outputs its instantaneous phase $\phi_n = \tan^{-1}(y_n, x_n) - \omega n$ using one of the arctangent approximations and correcting for the four-quadrant arctangent. How can you find ω? Is the phase always accurately recovered?

16.3   Logarithm

This function is required mainly for logarithmic AM detection, conversion of power ratios and power spectra to decibels, as well as for various musical effects, such as *compression* of guitar sounds. The ear responds to both sound intensities and frequencies in approximately logarithmic fashion, and so logarithmic transformations are used extensively in many perception-based feature extraction methods. Considerable effort has also been devoted to the efficient computation of the natural and decimal logarithms in the non-DSP world.

Due to its compressive nature, the magnitude of the output of the 'log' operation is significantly less than that of the input (for large enough inputs). Thus, relatively large changes in input value may lead to little or no change in the output. This has persuaded many practitioners to use overly simplistic approximations, which may lead to overall system precision degradation.

We can concentrate on base-two logarithms without limiting generality since logarithms of all other bases are simply related.

$$\log_a(x) = \Big(\log_2(a) \Big)^{-1} \log_2(x)$$

If only a single bit of a number's binary representation is set, say the kth one, then the log is simple to calculate—it is simply k. Otherwise the bits following the most significant set bit k contribute a fractional part

$$x = \sum_{i=0}^{k} x_i \, 2^i = 2^k + \sum_{i=1}^{k} x_{k-i} \, 2^{k-i} = 2^k \left(1 + \sum_{i=1}^{k} x_{k-i} 2^{-i} \right) = 2^k \, (1+z)$$

with $0 \le z < 1$. Now $\log_2(x) = k + \log_2(1+z)$ and so $0 \le u = \log_2(1+z) < 1$ as well. Thus to approximate $\log_2(x)$ we can always determine the most significant bit set k, then approximate $u(z)$ (which maps the interval $[0 \ldots 1]$ onto itself), and finally add the results. The various methods differ in the approximation for $u(z)$. The simplest approximation is linear interpolation, which has the additional advantage of requiring no further calculation—just copying the appropriate bits. The maximum error is approximately 10% and can be halved by adding a positive constant to the interpolation since this approximation always underestimates. The next possibility is quadratic approximation, and an eighth-order approximation can provide at least five significant digits.

For an alternate technique using the CORDIC algorithm, see Section 16.5.

EXERCISES

16.3.1 Code the linear interpolation approximation mentioned above and compare its output with your library routine. Where is the maximum error and how much is it?

16.3.2 Use a higher-order approximation (check a good mathematical handbook for the coefficients) and observe the effect on the error.

16.3.3 Before the advent of electronic calculators, scientists and engineers used *slide rules* in order to multiply quickly. How does a slide rule work? What is the principle behind the *circular* slide rule? How does this relate to the algorithm discussed above?

16.4 Square Root and Pythagorean Addition

Although the square root operation $y = \sqrt{x}$ is frequently required in DSP programs, few DSP processors provide it as an instruction. Several have 'square-root seed' instructions that attempt to provide a good starting point for iterative procedures, while for others the storage of tables is required.

The most popular iterative technique is the Newton-Raphson algorithm $y_{n+1} = \frac{1}{2}(y_n + \frac{x}{y_n})$, which converges quadratically. This algorithm has an easily remembered interpretation. Start by guessing y. In order to find out how close your guess is check it by calculating $z = \frac{x}{y}$; if $z \approx y$ then you are done. If not, the true square root is somewhere between y and z so their average is a better estimate than either.

Another possible ploy is to use the obvious relationship

$$\sqrt{x} = 2^z \implies z = \tfrac{1}{2}\log_2(x)$$

and apply one of the algorithms of the previous section.

When x can only be in a small interval, polynomial or rational approximations may be of use. For example, when x is confined to the unit interval $0 < x < 1$, the quadratic approximation $y \approx -0.5973x^2 + 1.4043x + 0.1628$ gives a fair approximation (with error less than about 0.03, except near zero).

More often than not, the square root is needed as part of a 'Pythagorean addition'.

$$x \oplus y \equiv \sqrt{x^2 + y^2}$$

This operation is so important that it is a primitive in some computer languages and has been the study of much approximation work. For example, it is well known that

$$x \oplus y \approx \text{abmax}(x, y) + k\, \text{abmin}(x, y)$$

with abmax (abmin) returning the argument with larger (smaller) *absolute* value. This approximation is good when $0.25 \leq k \leq 0.31$, with $k = 0.267304$ giving exact mean and $k = 0.300585$ minimum variance.

The straightforward method of calculating $x \oplus y$ requires two multiplications, an addition, and a square root. Even if a square root instruction is available, one may not want to use this procedure since the squaring operations may underflow or overflow even when the inputs and output are well within the range of the DSP's floating point word.

Several techniques have been suggested, the simplest perhaps being that of Moler and Morrison. In this algorithm x and y are altered by transformations that keep $x \oplus y$ invariant while increasing x and decreasing y. When negligible, x contains the desired output.

In pseudocode form:

```
p ← max(|x|, |y|)
q ← min(|x|, |y|)
while q > 0
        r ← (q/p)²
        s ← r/(4+r)
        p ← p + 2·s·p
        q ← s·p
output p
```

An alternate technique for calculating the Pythagorean sum, along with the arctangent, is provided by the CORDIC algorithm presented next.

EXERCISES

16.4.1 Practice finding square roots in your head using Newton-Raphson.

16.4.2 Code Moler and Morrison's algorithm for the Pythagorean sum. How many iterations does it require to obtain a given accuracy?

16.4.3 Devise examples where straightforward evaluation of the Pythagorean sum overflows. Now find cases where underflow occurs. Test Moler and Morrison's algorithm on these cases.

16.4.4 Can Moler-Morrison be generalized to compute $\sqrt{x_1^2 + x_2^2 + x_3^2 + \ldots}$?

16.4.5 Make an amplitude detector, i.e., a program that inputs a complex exponential $s(t) = x(t) + iy(t) = A(t)e^{i\omega t}$ and outputs its amplitude $A(t) = \sqrt{x^2(t) + y^2(t)}$. Use Moler and Morrison's algorithm.

16.5 CORDIC Algorithms

The **CO**ordinate **R**otation for **DI**gital **C**omputers (CORDIC) algorithm is an iterative method for calculating elementary functions using only addition and binary shift operations. This elegant and efficient algorithm is not new, having been described by Volder in 1959 (he applied it in building a digital airborne navigation computer), refined mathematically by Walther and used in the first scientific hand-held calculator (the HP-35), and is presently widely used in numeric coprocessors and special-purpose CORDIC chips.

Various implementations of the same basic algorithmic architecture lead to the calculation of:

- the pair of functions $\sin(\theta)$ and $\cos(\theta)$,
- the pair of functions $\sqrt{x^2 + y^2}$ and $\tan^{-1}(y/x)$,
- the pair of functions $\sinh(\theta)$ and $\cosh(\theta)$,
- the pair of functions $\sqrt{x^2 - y^2}$ and $\tanh^{-1}(y/x)$,
- the pair of functions $\sqrt{a}$ and $\ln(a)$, and
- the function e^a.

In addition, CORDIC-like architectures can aid in the computation of FFT, eigenvalues and singular values, filtering, and many other DSP tasks. The iterative step, the binary shift and add, is implemented in *CORDIC processors* as a basic instruction, analogously to the MAC instruction in DSP processors.

We first deal with the most important special case, the calculation of $\sin(\theta)$ and $\cos(\theta)$. It is well known that a column vector is rotated through an angle θ by premultiplying it by the orthogonal rotation matrix.

$$\begin{pmatrix} x' \\ y' \end{pmatrix} = \mathbf{R}(\theta) \begin{pmatrix} x \\ y \end{pmatrix} \tag{16.4}$$

$$\mathbf{R}(\theta) \equiv \begin{pmatrix} \cos(\theta) & -\sin(\theta) \\ \sin(\theta) & \cos(\theta) \end{pmatrix} = \cos(\theta) \begin{pmatrix} 1 & -\tan(\theta) \\ \tan(\theta) & 1 \end{pmatrix}$$

If one knows numerically the $\mathbf{R}$ matrix for some angle, the desired functions are easily obtained by rotating the unit vector along the x direction.

$$\mathbf{R}(\theta) \begin{pmatrix} 1 \\ 0 \end{pmatrix} = \begin{pmatrix} \cos(\theta) \\ \sin(\theta) \end{pmatrix} \tag{16.5}$$

However, how can we obtain the rotation matrix without knowing the values of $\sin(\theta)$ and $\cos(\theta)$? We can exploit the sum rule for rotation matrices:

$$\mathbf{R}\left(\sum_{i=0}^{n} \alpha_i\right) = \prod_{i=0}^{n} \mathbf{R}(\alpha_i) \tag{16.6}$$

and so for $\theta = \sum_{i=0}^{n} \alpha_i$, using equation (16.4), we find:

$$
\begin{aligned}
\mathbf{R}(\theta) &= \prod_{i=0}^{n} \cos(\alpha_i) \prod_{i=0}^{n} \begin{pmatrix} 1 & -\tan(\alpha_i) \\ \tan(\alpha_i) & 1 \end{pmatrix} \\
&= \prod_{i=0}^{n} \cos(\alpha_i) \prod_{i=0}^{n} \mathbf{M}_i
\end{aligned}
\tag{16.7}
$$

If we chose the partial angles α_i wisely, we may be able to simplify the arithmetic.

For example, let us consider the angle θ that can be written as the sum of α_i such that $\tan(\alpha_i) = 2^{-i}$. Then the $\mathbf{M}$ matrices in (16.7) are of the very simple form

$$\mathbf{M}_i = \begin{pmatrix} 1 & -\frac{1}{2^i} \\ \frac{1}{2^i} & 1 \end{pmatrix}$$

and the matrix products can be performed using only right shifts. We can easily generalize this result to angles θ that can be written as sums of $\alpha_i = \pm \tan^{-1}(2^{-i})$. Due to the symmetry $\cos(-\alpha) = \cos(\alpha)$, the product of cosines is unchanged, and the $\mathbf{M}$ matrices are either the same as those given above, or have the signs reversed. In either case the products can be performed by shifts and possibly sign reversals. Now for the surprise—one can show that *any* angle θ inside a certain *region of convergence* can be expressed as an infinite sum of $\pm \alpha_i = \pm \tan^{-1}(2^{-i})$! The region of convergence turns out to be $0 \leq \theta \leq 1.7433$ radians $\approx 99.9°$, conveniently containing the first quadrant. Thus for any angle θ in the first quadrant, we can calculate $\sin(\theta)$ and $\cos(\theta)$ in the following fashion. First we express θ as the appropriate sum of α_i. We then calculate the product of $\mathbf{M}$ matrices using only shift operations. Next we multiply the product matrix by the universal constant $K \equiv \prod_{i=0}^{\infty} \cos(\alpha_i) \approx 0.607$. Finally, we multiply this matrix by the unit column vector in the x direction. Of course, we must actually truncate the sum of α_i to some finite number of terms, but the quantization error is not large since each successive **M** matrix adds one bit of accuracy.

Now let's make the method more systematic. In the 'forward rotation' mode of CORDIC we start with a vector along the x axis and rotate it through a sequence of progressively smaller predetermined angles until it makes an angle θ with the x axis. Then its x and y coördinates are proportional to the desired functions. Unfortunately, the 'rotations' we must perform are not pure rotations since they destroy the normalization; were we to start with a unit vector we would need to rescale the result by K at the end. This multiplication may be more costly than all the iterations performed, so we economize by starting with a vector of length K. Assuming we desire b bits of precision we need to perform b iterations in all. We can discover the proper expansion of θ by greedily driving the residual angle to zero. We demonstrate the technique in the following pseudocode:

```
x ← K
y ← 0
z ← θ
for    i ← 0 to b − 1
          s ← sgn(z)
          x ← x − s · y · 2⁻ⁱ
          y ← y + s · x · 2⁻ⁱ
          z ← z − s · tan⁻¹(2⁻ⁱ)
cos(θ) ← x
sin(θ) ← y
error ← z
```

Of course only additions, subtractions, and right shifts are utilized, and the b values $\tan^{-1}(2^{-i})$ are precomputed and stored in a table. Beware that in the loop the two values x and y are to be calculated simultaneously. Thus to code this in a high-level language place the snippet

```
for    i ← 0 to b − 1
          ξ ← x
          x ← ξ − s · y · 2⁻ⁱ
          y ← y + s · ξ · 2⁻ⁱ
```

into your code.

Did you understand how θ was decomposed into the sum of the α_i angles? First we rotated counterclockwise by the largest possible angle,

$\alpha_0 = \tan^{-1} 1 = 45°$. If $\theta > \alpha_0$ then the second rotation is counterclockwise from there by $\alpha_1 = \tan^{-1} \frac{1}{2} \approx 26\frac{1}{2}°$ to $71\frac{1}{2}°$; but if $\theta < \alpha_0$ then the second rotation is clockwise to $18\frac{1}{2}°$. At each iteration the difference between the accumulated angle and the desired angle is stored in z, and we simply rotate in the direction needed to close the gap. After b iterations the accumulated angle approximates the desired one and the residual difference remains in z.

In order to calculate the pair of functions $\sqrt{x^2 + y^2}$ and $\tan^{-1}(y/x)$, we use the 'backward rotation' mode of CORDIC. Here we start with a vector (x, y) and rotate back to zero angle by driving the y coördinate to zero. We therefore obtain a vector along the positive x axis, whose length is proportional to the desired square root. The z coördinate accumulates the required arctangent.

The following pseudocode demonstrates the technique:

```
x ← X
y ← Y
z ← 0
for    i ← 0 to b − 1
          s ← sgn(y)
          x ← x + s · y · 2⁻ⁱ
          y ← y − s · x · 2⁻ⁱ
          z ← z + s · tan⁻¹(2⁻ⁱ)
√X² + Y² ← K · x
error ← y
tan⁻¹(Y/X) ← z
```

Once again the x and y in the loop are to be computed simultaneously.

As mentioned before, the pseudocodes given above are only valid in the first quadrant, but there are two ways of dealing with full four-quadrant angles. The most obvious is to fold angles back into the first quadrant and correct the resulting sine and cosines using trigonometric identities. When the input is x, y and $-\pi < \theta \leq \pi$ is desired, a convenient method to convert CORDIC's z is to use $\theta = a + q * z$ where $q = \text{sgn}(x)\text{sgn}(y)$ and $a = 0$ if $x > 0$, while otherwise $a = \text{sgn}(y)\pi$.

It is also possible to extend the basic CORDIC region of convergence to the full four quadrants, at the price of adding two addition iterations and changing the value of K. The extended algorithm is initialized with

$$\text{tp}_i \leftarrow \begin{cases} 1 & i \leq 0 \\ 2^{-i} & i \geq 0 \end{cases} \qquad \text{atan}_i \leftarrow \begin{cases} \frac{\pi}{4} & i \leq 0 \\ \tan^{-1}(2^{-i}) & i \geq 0 \end{cases}$$

and $K \leftarrow \frac{\sqrt{2}}{4} \Pi_{i=1}^b \cos\left(\tan^{-1}(2^{-i})\right)$ and, for example, the backward rotation is now carried out by the following algorithm:

$$x \leftarrow X$$
$$y \leftarrow Y$$
$$z \leftarrow 0$$
$$\text{for} \quad i \leftarrow -2 \text{ to } b-1$$
$$\qquad s \leftarrow \text{sgn}(y)$$
$$\qquad x \leftarrow x + s \cdot y \cdot \text{tp}_i$$
$$\qquad y \leftarrow y - s \cdot x \cdot \text{tp}_i$$
$$\qquad z \leftarrow z + s \cdot \text{atan}_i$$
$$\sqrt{X^2 + Y^2} \leftarrow K \cdot x$$
$$\text{error} \leftarrow y$$
$$\tan^{-1}(Y/X) \leftarrow z$$

Up to now we have dealt only with circular functions. The basic CORDIC iteration can be generalized to

$$\begin{pmatrix} x_{i+1} \\ y_{i+1} \end{pmatrix} = \begin{pmatrix} 1 & ms_i 2^{-i} \\ -s_i 2^{-i} & 1 \end{pmatrix} \begin{pmatrix} x_i \\ y_i \end{pmatrix} \qquad (16.8)$$
$$z_{i+1} = z_i + s_i t_i$$

where for the circular functions $m = +1$ and $t_i = \tan^{-1}(2^{-i})$, for the hyperbolic functions $m = -1$ and $t_i = \tanh^{-1}(2^{-i})$, and for the linear functions $m = 0$ and $t_i = 2^{-i}$. For the circular and hyperbolic cases one must also renormalize by the constants $K = 1/\prod_{i=0}^n \sqrt{1 + m2^{-2i}}$. For the hyperbolic case additional iterations are always required.

EXERCISES

16.5.1 Code the forward and backward extended-range CORDIC algorithms. Test them by comparison with library routines on randomly selected problems.

16.5.2 Recode the mixer filter from the exercises of Section 16.1 using CORDIC to generate the complex exponential.

16.5.3 Code a digital receiver that inputs a complex signal $s(t) = A(t)e^{i(\omega t + \phi(t))}$, mixes the signal down to zero frequency $s(t) = A(t)e^{i\phi(t)}$ (using forward CORDIC), and then extracts both the amplitude and phase (using backward CORDIC).

Bibliographical Notes

The reader is referred to the mathematical handbook of Abramowitz and Stegun [1] for properties of functions, and polynomial and rational approximation coefficients. For a basic introduction to numerical techniques I recommend [216].

Techniques for speeding up the convergence of polynomial and rational expansions are discussed in [138].

Generation of sinusoids by recursively evaluating a second-order difference equation is discussed in [58].

Mitchell [120] proposed simple linear interpolation for the evaluation of logarithms, while Marino [158] proposed the quadratic approximation.

Knuth's METAFONT typeface design program (which generates the fonts usually used with TEX and LATEX) is an example of a language that has $\oplus$ as a primitive. Its manual and entire source code are available in book form [134]. The abmax-abmin formula for $\oplus$ was apparently first discussed in [222] but later covered in many sources, e.g., [184]. The Moler and Morrison algorithm was first presented in [175] and was developed for software that evolved into the present MATLAB [90].

The CORDIC algorithm was proposed by Volder [266] in 1959, and refined mathematically by Walther [269]. Its use in the first full-function scientific calculator (the HP-35) is documented in [38]. CORDIC's approximation error is analyzed in [107]. Extending CORDIC to a full four-quadrant technique was proposed by [105], while its use for computation of the inverse trigonometric functions is in [162]. CORDIC-like architectures can aid in the computation of the FFT [51, 52], eigenvalues and singular values [60], and many other DSP tasks [106].

Digital Signal Processors

Until now we have assumed that all the computation necessary for DSP applications could be performed either using pencil and paper or by a general-purpose computer. Obviously, those that can be handled by human calculation are either very simplistic or at least very low rate. It might surprise the uninitiated that general-purpose computers suffer from the same limitations. Being 'general-purpose', a conventional central processing unit (CPU) is not optimized for DSP-style 'number crunching', since much of its time is devoted to branching, disk access, string manipulation, etc. In addition, even if a computer *is* fast enough to perform all the required computation in time, it may not be able to *guarantee* doing so.

In the late 1970s, special-purpose processors optimized for DSP applications were first developed, and such processors are still multiplying today (pun definitely intended). Although correctly termed 'Digital Signal Processors', we will somewhat redundantly call them 'DSP processors', or simply DSPs. There are small, low-power, inexpensive, relatively weak DSPs targeted at mass-produced consumer goods such as toys and cars. More capable fixed point processors are required for cellular phones, digital answering machines, and modems. The strongest, often floating point, DSPs are used for image and video processing, and server applications.

DSP processors are characterized by having at least some of the following special features: DSP-specific instructions (most notably the MAC), special address registers, zero-overhead loops, multiple memory buses and banks, instruction pipelines, fast interrupt servicing (fast context switch), specialized ports for input and output, and special addressing modes (e.g., bit reversal).

There are also many non-DSP processors of interest to the DSP implementor. There are convolution processors and FFT processors devoted to these tasks alone. There are systolic arrays, vector and superscalar processors, RISC processors for embedded applications, general-purpose processors with multimedia extensions, CORDIC processors, and many more varieties.

DSP 'cores' are available that can be integrated on a single chip with other elements such as CPUs, communications processors, and IO devices. Although beyond the scope of our present treatment the reader would be well advised to learn the basic principles of these alternative architectures.

In this chapter we will study the DSP processor and how it is optimized for DSP applications. We will discuss general principles, without considering any specific DSP processor, family of processors, or manufacturer. The first subject is the MAC operation, and how DSPs can perform it in a single clock cycle. In order to understand this feat we need to study memory architectures and pipelines. We then consider interrupts, ports, and the issue of numerical representation. Finally, we present a simple, yet typical example of a DSP program. The last two sections deal with the practicalities of industrial DSP programming.

17.1   Multiply-and-Accumulate (MAC)

DSP algorithms tend to be number-crunching intensive, with computational demands that may exceed the capabilities of a general-purpose CPU. DSP processors can be much faster for specific tasks, due to arithmetic instruction sets specifically tailored to DSP needs. The most important special-purpose construct is the MAC instruction; accelerating this instruction significantly reduces the time required for computations common in DSP.

Convolutions, vector inner products, correlations, difference equations, Fourier transforms, and many other computations prevalent in DSP all share the basic repeated MAC computation.

```
loop
      update j, update k
      a ← a + xⱼyₖ
```

For inner products, correlations, and symmetric or coefficient-reversed FIR filters the updating of indices j and k both involve incrementation; for convolutions one index is incremented while the other is decremented.

First consider the outside of the loop. When a general-purpose CPU executes a fixed-length loop such as

```
for i ← 1 to N
      statements
```

there is a lot of overhead involved. First a register must be provided to store the loop index i, and it must be properly initialized. After each execution of the calculation the loop index register must be incremented, and checked for termination. Of course if there are not enough registers the loop index must be retrieved from memory, incremented and checked, and then stored back to memory. Except for the last iteration, a 'branch' or 'jump' instruction must be performed to return execution to the top of the loop.

DSP processors provide a zero-overhead hardware mechanism (often called `repeat` or `do`) that can repeat an instruction or number of instructions a prespecified number of times. Due to hardware support for this repeat instruction no clocks are wasted on branching or incrementing and checking the loop index. The maximum number of iterations is always limited (64K is common, although some processors have low limits such as 128) and many processors limit the number of instructions in the loop (1, 16), but these limitations fall into the envelope of common DSP operations. Some processors allow loop nesting (since the FFT requires 3 loops, this is a common limit), while for others only the innermost loop can be zero overhead.

Now let's concentrate on the computations inside the loop. How would a general-purpose CPU carry out the desired computation? We assume that x and y are stored as arrays in memory, so that x_j is stored j locations after x_0, and similarly for y_k. Furthermore, we assume that the CPU has at least two pointer registers (that we call j and k) that can be directly updated (incremented or decremented) and used to retrieve data from memory. Finally, we assume the CPU has at least two arithmetic (floating point or fixed point) registers (x and y) that can be used as operands of arithmetic operations, a double-length register (z) that can receive a product, and an accumulator (a) for summing up values.

Assuming that the loop has been set up (i.e., the counter loaded, the base pointers for x_j and y_k set, and the automatic updating of these pointers programmed in), the sequence of operations for computation of the contents of the loop on a general-purpose CPU will look something like this.

```
update pointer to xⱼ
update pointer to yₖ
load xⱼ into register x
load yₖ into register y
fetch operation (multiply)
decode operation (multiply)
multiply x by y storing the result in register z
fetch operation (add)
decode operation (add)
add register z to accumulator a
```

We see that even assuming each of the above lines takes the same amount of time (which is dubious for the multiplication), the computation requires about 10 instruction times to complete. Of course different CPUs will have slightly different instruction sets and complements of registers, but similar principles hold for all CPUs.

A major distinction between a general-purpose CPU and a DSP is that the latter can perform a MAC in a single instruction time. Indeed this feature is of such importance that many use it as the definition of a DSP. The main purpose of this chapter is explain how this miracle is accomplished. In particular it is not enough to simply add a MAC instruction to the set of opcodes; such an 'MAC-augmented CPU' would still have to perform the following steps

```
update pointer to x_j
update pointer to y_k
load x_j into register x
load y_k into register y
fetch operation (MAC)
decode operation (MAC)
MAC a ← x * y
```

for a total of seven instruction times. We have managed to save a few clocks but are still far from our goal. Were the simple addition of a MAC instruction all a DSP processor had to offer, it would probably not be worth devoting precious silicon real-estate to the special MAC hardware. In order to build a DSP we need more imagination than this.

The first step in building a true DSP is to note that the pointers to x_j and y_k are independent and thus their updating can be performed in parallel. To implement this we need new hardware; we need to add two address updating units to the hardware complement of our hypothetical DSP processor. Using the symbol || to signify two operations that are performed in parallel, the MAC now looks like this:

```
update pointer to x_j || update pointer to y_k
load x_j into register x
load y_k into register y
fetch operation (MAC)
decode operation (MAC)
MAC a ← x * y
```

We have obviously saved at least the time of one instruction, since the x_j and y_k pointers are now updated simultaneously, but even though we no longer require use of the CPU's own adder it does not seem possible to further exploit this in order to reduce overall execution time. It is obvious that we cannot proceed to load values into the x and y registers until the pointers are ready, and we cannot perform the MAC until the registers are loaded. The next steps in optimizing our DSP call for more radical change.

EXERCISES

17.1.1 For the CPU it would be clearer to have j and k stored in fixed point registers and to retrieve x_j by adding j to the address of x_0. Why didn't we do this?

17.1.2 Explain in more detail why it is difficult for two buses to access the same memory circuits.

17.1.3 Many DSP processors have on-chip ROM or RAM memory. Why?

17.1.4 Many CPU architectures use *memory caching* to keep critical data quickly accessible. Discuss the advantages and disadvantages for DSP processors.

17.1.5 A processor used in personal computers has a set of instructions widely advertised as being designed for multimedia applications. What instructions are included in this set? Can this processor be considered a DSP?

17.1.6 Why does the zero-overhead loop only support loops with a prespecified number of iterations (for loops)? What about while (*condition*) loops?

17.2 Memory Architecture

A useful addition to the list of capabilities of our DSP processor would be to allow x_j and y_k to be simultaneously read from memory into the appropriate registers. Since x_j and y_k are completely independent there is no fundamental impediment to their concurrent transfer; the problem is that while one value is being sent over the 'data bus' the other must wait. The solution is to provide two data buses, enabling the two values to be read from memory simultaneously. This leaves us with a small technical hitch; it is problematic for two buses to connect to the same memory circuits. The difficulty is most obvious when one bus wishes to write and the other to read from precisely the same memory location, but even accessing nearby locations can be technically demanding. This problem can be solved by using so-called 'dual port memories', but these are expensive and slow.

The solution here is to leave the usual model of a single linear memory, and to define multiple *memory banks*. Different buses service different memory banks, and placing the x_j and y_k arrays in separate banks allows their simultaneous transfer to the appropriate registers. The existence of more than one memory area for data is a radical departure from the memory architecture of a standard CPU.

```
update pointer to x_j || update pointer to y_k
load x_j into register x || load y_k into register y
fetch operation (MAC)
decode operation (MAC)
MAC a <- x * y
```

The next step in improving our DSP is to take care of the `fetch` and `decode` steps. Before explaining how to economize on these instructions we should first explain more fully what these steps do. In modern CPUs and DSPs instructions are stored sequentially in memory as opcodes, which are binary entities that uniquely define the operation the processor is to perform. These opcodes typically contain a group of bits that define the operation itself (e.g., multiply or branch), individual bit parameters that modify the meaning of the instruction (multiply *immediate* or branch *relative*), and possibly bits representing numeric fields (multiply immediate by *2* or branch relative forward by *2*). Before the requested function can be performed these opcodes must first be retrieved from memory and decoded, operations that typically take a clock cycle each.

We see that a nonnegligible portion of the time it takes to execute an instruction is actually devoted to retrieving and decoding it. In order to reduce the time spent on each instruction we must find a way of reducing this overhead. Standard CPUs use 'program caches' for this purpose. A program cache is high speed memory inside the CPU into which program instructions are automatically placed. When a program instruction is required that has already been fetched and decoded, it can be taken from the program cache rather than refetched and redecoded. This tends to significantly speed up the execution of loops. Program caches are typically rather small and can only remember the last few instructions; so loops containing a large number of instructions may not benefit from this tactic. Similarly CPUs may have 'data caches' where the last few memory locations referenced are mirrored, and redundant data loads avoided.

Caches are usually avoided in DSPs because caching complicates the calculation of the time required for a program to execute. In a CPU with caching a set of instructions requires different amounts of run-time depending on the state of the caches when it commences. DSPs are designed for real-time use where the prediction of exact timing may be critical. So DSPs must use a different trick to save time on instruction fetches.

Why can't we perform a fetch one step before it is needed (in our case during the two register loads)? Once again the fundamental restriction is that we can't fetch instructions from memory at the same time that data is being transferred to or from memory; and the solution is, once again, to use separate buses and memory banks. These memory banks are called *program memory* and *data memory* respectively.

Standard computers use the same memory space for program code and data; in fact there is no clear distinction between the two. In principle the same memory location may be used as an instruction and later as a piece of data. There may even be self-modifying code that writes data to memory and later executes it as code. This architecture originated in the team that built one of the first digital computers, the 18,000-vacuum-tube ENIAC (Electronic Numerical Integrator and Computer) designed in the early forties at the University of Pennsylvania. The main designers of this machine were J.W. Mauchly and J. Presper Eckert Jr. and they relied on earlier work by J.V. Atanasoff. However, the concept of a single memory for program and data is named after John von Neumann, the Hungarian-born German-American mathematician-physicist, due to his 1945 memo and 1946 report summarizing the findings of the ENIAC team regarding storing instructions in binary form. The single memory idea intrigued von Neumann because of his interest in artificial intelligence and self-modifying learning programs.

Slightly before the ENIAC, the Mark I computer was built by a Harvard team headed by Howard Aiken. This machine was electromechanical and was programmed via paper tape, but the later Mark II and Mark III machines were purely electrical and used magnetic memory. Grace Hopper coined the term 'bug' when a moth entered one of the Harvard computers and caused an unexpected failure. In these machines the program memory was completely separate from data memory. Most DSPs today abide by this *Harvard architecture* in order to be able to overlap instruction fetches with data transfers. Although von Neumann's name is justly linked with major contributions in many areas of mathematics, physics, and the development of computers, crediting him with inventing the 'von Neumann architecture' is not truly warranted, and it would be better to call it the 'Pennsylvania architecture'. Aiken, whose name is largely forgotten, is justly the father of the two-bus architecture that posterity named after his institution. No one said that posterity is fair.

In the Harvard architecture, program and data occupy different address spaces, so that address A in program memory is completely distinct from address A in data memory. These two memory spaces are connected to the processor using separate buses, and may even have different access speeds and bit widths. With separate buses we can perform the fetch in parallel with data transfers, and no longer need to waste a clock. We will explain the precise mechanism for overlapping these operations in the next section, for now we will simply ignore the instruction-related operations. Our MAC now requires only three instruction times.

```
update pointer to x_j || update pointer to y_k
load x_j into register x || load y_k into register y
MAC a ← x * y
```

We seem to be stuck once again. We still can't load x_j and y_k before the pointers are updated, or perform the MAC before these loads complete. In the next section we take the step that finally enables the single clock MAC.

EXERCISES

17.2.1 A pure Harvard architecture does not allow any direct connection between program and data memories, while the *modified Harvard* architecture contains copy commands between the memories. Why are these commands useful? Does the existence of these commands have any drawbacks?

17.2.2 DSPs often have many different types of memory, including ROM, on-chip RAM, several banks of data RAM, and program memory. Explain the function of each of these and demonstrate how these would be used in a real-time FIR filter program.

17.2.3 FIR and IIR filters require a fast MAC instruction, while the FFT needs the butterfly

$$x \leftarrow x + Wy$$
$$y \leftarrow x - Wy$$

where x and y are complex numbers and W a complex root of unity. Should we add the butterfly as a basic operation similar to the MAC?

17.2.4 There are two styles of DSP assembly language syntax. The opcode-mnemonic style uses commands such as `MPY A0, A1, A2`, while the programming style looks more like a conventional high-level language $A0 = A1 * A2$. Research how the MAC instruction with parallel retrieval and address update is coded in both these styles. Which notation is better? Take into account both algorithmic transparency and the need to assist the programmer in understanding the hardware and its limitations.

17.3 Pipelines

In the previous sections we saw that the secret to a DSP processor's speed is not only special instructions, but the exploitation of parallelism. Address registers are updated in parallel, memory retrievals are performed in parallel, and program instructions are fetched in parallel with execution of previous instructions. The natural extension is to allow parallel execution of any operations that logically *can* be performed in parallel.

update 1	update 2	update 3	update 4	update 5		
	load 1	load 2	load 3	load 4	load 5	
		MAC 1	MAC 2	MAC 3	MAC 4	MAC 5

Figure 17.1: The pipelining of a MAC calculation. Time runs from left to right, while height corresponds to distinct hardware units, 'update' meaning the updating of the x_j and y_k pointers, 'load' the loading into x and y, and 'MAC' the actual computation. At the left there are three cycles during which the pipeline is filling, while at the right there are a further three cycles while the pipeline is emptying. The result is available seven cycles after the first update.

The three steps of the three-clock MAC we obtained in the previous section use different processor capabilities, and so should be allowed to operate simultaneously. The problem is the dependence of each step on the completion of the previous one, but this can be sidestepped by using a *pipeline* to overlap these operations. The operation of the pipeline is clarified in Figure 17.1. In this figure 'update 1' refers to the first updating of the pointers to x_j and y_k; 'load 1' to the first loading of x_j and y_k into registers x and y; and 'MAC 1' means the first multiplication. As can be seen, the first load takes place only after the first update is complete, and the MAC only after the loads. However, we do not wait for the MAC to complete before updating the pointers; rather we immediately start the second update after the first pointers are handed over to the loading process. Similarly, the second load takes place in parallel with the first MAC, so that the second MAC can commence as soon as the first is completed. In this way the MACs are performed one after the other without waiting, and once the pipeline is filled each MAC requires only one instruction cycle. Of course there *is* overhead due to the pipeline having to fill up at the beginning of the process and empty out at the end, but for large enough loops this overhead is negligible. Thus the pipeline allows a DSP to perform one MAC per instruction clock *on the average*.

Pipelines can be exploited for other purposes as well. The simplest general-purpose CPU must wait for one basic operation (e.g., fetch, decode, register arithmetic) to complete before embarking on the next; DSPs exploit parallelism even at the subinstruction level. How can the different primitive operations that make up a single instruction be performed in parallel? They can't; but the primitive operations that comprise successive instructions *can*.

Until now we have been counting 'instructions' and have not clarified the connection between 'instruction times' and 'clock cycles'. All processors are fed a clock signal that determines their speed of operation. Many processors are available in several versions differing only in the maximum clock speed at which they are guaranteed to function. While a CPU processor is always specified by its clock frequency (e.g., a 400 MHz CPU), DSP processors are usually designated by clock interval (e.g., a 25 nanosecond DSP).

Even when writing low-level assembly language that translates directly to native opcodes, a line of code does not directly correspond to a clock interval, because the processor has to carry out many operations other than the arithmetic functions themselves. To see how a CPU really works at the level of individual clock cycles, consider an instruction that adds a value in memory to a register, leaving the result in the same register. At the level of individual clock cycles the following operations might take place.

```
fetch instruction
decode instruction
retrieve value from memory
perform addition
```

We see that a total of four clock cycles is required for this single addition, and our 'instruction time' is actually four 'clock cycles'. There might be additional subinstruction operations as well, for instance, transfer of a value from the register to memory. Fixed point DSP processors may include an optional postarithmetic scaling (shift) operation, while for floating point there is usually a postarithmetic normalization stage that ensures the number is properly represented.

Using a subinstruction pipeline we needn't count four clock cycles per instruction. While we are performing the arithmetic portion of an instruction, we can already be decoding the next instruction, and fetching the one after that! The number of overlapable operations of which an instruction is comprised is known as the *depth* of the pipeline. The minimum depth is three (fetch, decode, execute), typical values are four or five, but by dividing the arithmetic operation into stages the maximum depth may be larger. Recent DSP processors have pipeline depths as high as 11.

fetch 1	fetch 2	fetch 3	fetch 4	fetch 5			
	decode 1	decode 2	decode 3	decode 4	decode 5		
		get 1	get 2	get 3	get 4	get 5	
			add 1	add 2	add 3	add 4	add 5

Figure 17.2: The operation of a depth-four pipeline. Time runs from left to right, while height corresponds to distinct hardware units. At the left there are three cycles during which the pipeline is filling, while at the right there are three cycles while the pipeline is emptying. The complete sum is available eight cycles after the first fetch.

As an example, consider a depth-four pipeline that consists of fetch, decode, load data from memory, and an arithmetic operation, e.g., an addition. Figure 17.2 depicts the state of a depth-four pipeline during all the stages of a loop adding five numbers. Without pipelining the summation would take $5 * 4 = 20$ cycles, while here it requires only eight cycles. Of course the pipeline is only full for two cycles, and were we to sum 100 values the pipelined version would take only 103 cycles. Asymptotically we require only a single cycle per instruction.

The preceding discussion was based on the assumption that we know what the next instruction will be. When a branch instruction is encountered, the processor only realizes that a branch is required after the decode operation, at which point the next instruction is already being fetched. Even more problematic are conditional branches, for which we only know which instruction is next after a computation has been performed. Meanwhile the pipeline is being filled with erroneous data. Thus pipelining is useful mainly when there are few (if any) branches. This is the case for many DSP algorithms, while possibly unjustified for most general-purpose programming.

As discussed above, many processor instructions only return results after a number of clocks. Attempting to retrieve a result before it is ready is a common mistake in DSP programming, and is handled differently by different processors. Some DSPs assist the programmer by locking until the result is ready, automatically inserting wait states. Others provide no locking and it is entirely the programmer's responsibility to wait the correct number of cycles. In such cases the NOP (no operation) opcode is often inserted to simply waste time until the required value is ready. In either case part of the art of DSP programming is the rearranging of operations in order to perform useful computation rather than waiting with a NOP.

EXERCISES

17.3.1 Why do many processors limit the number of instructions in a repeat loop?

17.3.2 What happens to the pipeline at the end of a loop? When a branch is taken?

17.3.3 There are two styles of DSP assembly language syntax regarding the pipeline. One emphasizes time by listing on one line all operations to be carried out simultaneously, while the other stresses data that is logical related. Consider a statement of the first type

$$A1 = A1 + A2; A2 = A3 * A4; A3 = *R1 + +; A4 = *R2 + +$$

where $A1, A2, A3, A4$ are accumulators and $R1, R2$ pointer registers. Explain the relationship between the contents of the indicated registers. Next consider a statement of the second type

$$A0 = A0 + (*R1 + + * *R2 + +)$$

and explain when the operations are carried out.

17.3.4 It is often said that when the pipeline is not kept filled, a DSP is slower than a conventional processor, due to having to fill up and empty out the pipeline. Is this a fair statement?

17.3.5 Your DSP processor has 8 registers R1, R2, R3, R4, R5, R6, R7, R8, and the following operations

- load register from memory: Rn ← location
- store register to memory: location ← Rn
- single cycle no operation: NOP
- negate: Rn ← − Rn [1 cycle latency]
- add: Rn ← Ra + Rb [2 cycle latency]
- subtract: Rn ← Ra − Rb [2 cycle latency]
- multiply: Rn ← Ra · Rb [3 cycle latency]
- MAC: Rn ← Rn + Ra · Rb [4 cycle latency]

where the latencies disclose the number of cycles until the result is ready to be stored to memory. For example,

$$R1 \quad \leftarrow \quad R1 + R2 \cdot R3$$
$$answer \quad \leftarrow \quad R1$$

does not have the desired effect of saving the MAC in answer, unless four NOP operations are interposed. Show how to efficiently multiply two complex numbers. (Hint: First code operations with enough NOP operations, and then interchange order to reduce the number of NOPs.)

17.4 Interrupts, Ports

When a processor stops what it has been doing and starts doing something else, we have a *context switch*. The name arises from the need to change the run-time context (e.g, the pointer to the next instruction, the contents of the registers). For example, the operating system of a time-sharing computer system must continually force the processor to jump between different tasks, performing numerous context switches per second. Context switches can be initiated by outside events as well (e.g., keyboard presses, mouse clicks, arrival of signals). In any case the processor must be able to later return to the original task and continue as if nothing had happened.

Were the software responsible for initiating all externally driven context switches, it would need to incessantly poll all the possible sources of such requests to see whether servicing is required. This would certainly be a waste of resources. All processors provide a hardware mechanism called the *interrupt*. An interrupt forces a context switch to a predefined routine called the *interrupt handler* for the event in question. The concept of an interrupt is so useful that many processors provide a 'software interrupt' (sometimes called a trap) by which the software itself can instigate a context switch.

One of the major differences between DSPs and other types of CPUs is the speed of the context switch. A CPU may have a latency of dozens of cycles to perform a context switch, while DSPs always have the ability to perform a low-latency (perhaps even zero-overhead) interrupt.

Why does a DSP need a fast context switch? The most important reason is the need to capture interrupts from incoming signal values, either immediately processing them or at least storing them in a buffer for later processing. For the latter case this signal value capture often occurs at a high rate and should only minimally interfere with the processing. For the former case delay in retrieving an incoming signal may be totally unacceptable.

Why do CPU context switches take so many clock cycles? Upon restoration of context the processor is required to be in precisely the same state it would have been had the context switch not occurred. For this to happen many state variables and registers need to be stored for the context being switched out, and restored for the context being switched in. The DSP fast interrupt is usually accomplished by saving only a small portion of the context, and having hardware assistance for this procedure. Thus if the context switch is for the sole purpose of storing an incoming sample to memory, the interrupt handler can either not modify unstored registers, or can be coded to manually restore them to their previous state.

All that is left is to explain how signal values are input to and output from the DSP. This is done by *ports*, of which there are several varieties. Serial ports are typically used for low-rate signals. The input signal's bits are delivered to the DSP one at a time and deposited in an internal shift register, and outputs are similarly shifted out of the DSP one bit per clock. Thus when a 16-bit A/D is connected to a serial port it will send the sample as 16 bits, along with a bit clock signal telling the DSP when each bit is ready. The bits may be sent MSB first or LSB first depending on the A/D and DSP involved. These bits are transferred to the DSP's internal serial port shift register. Each time the A/D signals that a bit is ready, the DSP serial port shift register shifts over one bit and receives the new one. Once all 16 bits are input the A/D will assert an interrupt requesting the DSP to store the sample presently in the shift register to memory.

Parallel ports are faster than serial ports but require more pins on the DSP chip itself. Parallel ports typically transfer eight or sixteen bits at a time. In order to further speed up data transfer Direct Memory Access (DMA) channels are provided that can transfer whole blocks of data to or from the DSP memory without interfering with the processing. Typically once a DMA transfer is initiated, only a single interrupt is required at the end to signal that the transfer is complete.

Finally, communications ports are provided on those DSPs that may be interconnected with other DSPs. By constructing arrays of DSPs processing tasks may be divided up between processors and such platforms may attain processing power far exceeding that available from a single processor.

EXERCISES

17.4.1 How does the CPU know which interrupt handler to call?

17.4.2 Some DSPs have 'internal peripherals' that can generate interrupts. What can these be used for?

17.4.3 What happens when an interrupt interrupts an interrupt?

17.4.4 When a DSP is on a processing board inside a host computer there may be a method of input and output other than ports—shared memory. Discuss the pros and cons of shared memory vs. ports.

17.5 Fixed and Floating Point

The first generation of DSP processors were integer-only devices, and even today such fixed point DSPs flourish due to their low cost. This seems paradoxical considering that DSP tasks are number-crunching intensive. You probably wouldn't consider doing serious numeric tasks on a conventional CPU that is not equipped with floating point hardware. Yet the realities of speed, size, power consumption, and price have compelled these inconvenient devices on the DSP community, which has had to develop rather intricate numeric methods in order to use them. Today there are floating point DSPs, but these still tend to be much more expensive, more power hungry, and physically larger than their fixed point counterparts. Thus applications requiring embedding a DSP into a small package, or where power is limited, or price considerations paramount, still typically utilize fixed point DSP devices. Fixed point DSPs are also a good match for A/D and D/A devices, which are typically unsigned or two's-complement integer devices.

The price to be paid for the use of fixed point DSPs is extended development time. After the required algorithms have been simulated on computers with floating point capabilities, floating point operations must then be carefully converted to integer ones. This involves much more than simple rounding. Due to the limited dynamic range of fixed point numbers, rescaling must be performed at various points, and special underflow and overflow handling must be provided. The exact placement of the rescalings must be carefully chosen in order to ensure the maximal retention of signal vs. quantization noise, and often extensive simulation is required to determine the optimal placement. In addition, the precise details of the processor's arithmetic may need to be taken into account, especially when interoperability with other systems is required. For example, standard speech compression algorithms are tested by providing specified input and comparing the output bit stream to that specified in the standard. The output must be exact to the bit, even though the processor may compute using any number of bits. Such bit-exact implementations may utilize a large fraction of the processor's MIPS just to coerce the fixed point arithmetic to conform to that of the standard.

The most common fixed point representation is 16-bit two's complement, although longer registers (e.g., 24- or 32-bit) also exist. In fixed point DSPs this structure must accommodate both integers and real numbers; to represent the latter we multiply by some large number and round. For example, if we are only interested in real numbers between -1.0 and $+1.0$ we multiply by 2^{15} and think of the two's-complement number as a binary fraction.

When two 16-bit integers are added, the sum can require 17 bits; when multiplied, the product can require 32 bits. Floating point hardware takes care of this bit growth by automatically discarding the least significant bits, but in fixed point arithmetic we must explicitly handle the increase in precision. CPUs handle addition by assuming that the resultant usually does fit into 16 bits; if there is an overflow a flag is set or an exception is triggered. Products are conventionally stored in two registers, and the user must decide what to do next based on the values in the registers. These strategies are not optimal for DSP since they require extra operations for testing flags or discarding bits, operations that would break the pipeline.

Fixed point DSPs use one of several strategies for handling the growth of bits without wasting cycles. The best strategy is for the adder of the MAC instruction to use an accumulator that is longer than the largest possible product. For example, if the largest product is 32 bits the accumulator could have 40 bits, the extra bits allowing eight MACs to be performed without any possibility of overflow. At the end of the loop a single check and possible discard can be performed. The second strategy is to provide an optional scaling operation as part of the MAC instruction itself. This is basically a right shift of the product before the addition, and is built into the pipeline. The least satisfactory way out of the problem, but still better than nothing, is the use of 'saturation arithmetic'. In this case a hard limiter is used whenever an overflow occurs, the result being replaced by the largest representable number of the appropriate sign. Although this is definitely incorrect, the error introduced is smaller than that caused by straight overflow.

Other than these surmountable arithmetic problems, there are other possible complications that must be taken into account when using a fixed point processor. As discussed in Section 15.5, after designing a digital filter its coefficients should not simply be rounded; rather the best integer coefficients should be determined using an optimization procedure. Stable IIR filters may become unstable after quantization, due to poles too close to the unit circle. Adaptive filters are especially sensitive to quantization. When bits are discarded, overflows occur, or limiting takes place, the signal processing system ceases to be linear, and therefore cycles and chaotic behavior become possible (see Section 5.5).

Floating point DSPs avoid many of the above problems. Floating point numbers consist of a mantissa and an exponent, both of which are signed integers. A recognized standard details both sizes for the mantissa and exponent and rules for the arithmetic, including how exceptions are to be handled. Not all floating point DSPs conform to this standard, but some that don't provide opcodes for conversion to the standard format.

Unlike the computing environments to which one is accustomed in off-line processing, even the newer floating point DSP processors do not usually have instructions for division, powers, square root, trigonometric functions, etc. The software libraries that accompany such processors *do* include such functions, but these general-purpose functions may be unsuitable for the applications at hand. The techniques of Chapter 16 can be used in such cases.

EXERCISES

17.5.1 Real numbers are represented as integers by multiplying by a large number and rounding. Assuming there is no overflow, how is the integer product related to the real product? How is a fixed point multiply operation from two b-bit registers to a b-bit register implemented?

17.5.2 Simulate the simple IIR filter $y_n = \alpha y_{n-1} + x_n$ $(0 \leq \alpha \leq 1)$ in floating point and plot the impulse response for various α. Now repeat the simulation using 8-bit integer arithmetic (1 becomes 256, $0 \leq \alpha \leq 256$). How do you properly simulate 8-bit arithmetic on a 32-bit processor?

17.5.3 Design a narrow band-pass FIR filter and plot its empirical frequency response. Quantize the coefficients to 16 bits, 8 bits, 4 bits, 2 bits, and finally a single bit (the coefficient's sign); for each case replot the frequency response.

17.5.4 Repeat the previous exercise for an IIR filter.

17.6 A Real-Time Filter

In this section we present a simple example of a DSP program that FIR filters input data in real-time. We assume that the filter coefficients are given, and that the number of coefficients is L.

Since this is a real-time task, every t_s seconds a new input sample x_n appears at the DSP's input port. The DSP must then compute

$$y_n = \sum_{l=0}^{L-1} h_l x_{n-l}$$

and output y_n in less than t_s seconds, before the next sample arrives. This should take only somewhat more than L processor cycles, the extra cycles being unavoidable overhead.

On a general purpose CPU the computation might look like this:

```
for l ← 1 to (L-1)
    x[l-1] ← x[l]
x[L-1] ← input


y ← 0
for l ← 0 to (L-1)
    y ← y + h[l] * x[L-1-l]
output ← y
```

We first made room for the new input and placed it in x[L-1]. We then computed the convolution and output the result.

There are two main problems with this computation. First we wasted a lot of time in moving the static data in order to make room for the new input. We needn't physically move data if we use a *circular buffer*, but then the indexation in the convolution loop would be more complex. Second, the use of explicit indexation is wasteful. Each time we have need x[L-1-l] we have to compute L-1-l, find the memory location, and finally retrieve the desired data. A similar set of operations has to be performed for h[l] before we are at last ready to multiply. A more efficient implementation uses 'pointers'; assuming we initialize h and x to point to h_0 and x_{L-1-l} respectively, we have the following simpler loop:

```
y ← 0
repeat L times
    y ← y + (*h) * (*x)
    h ← h + 1
    x ← x - 1
```

Here *h means the contents of the memory location to which the pointer h points. We can further improve this a little by initializing y to $h_0 x_n$ and performing one less pass through the loop.

How much time does this CPU-based program take? In the loop there is one multiplication, two additions and one subtraction, in addition to assignment statements; and the loop itself requires an additional implicit decrement and comparison operation.

Now we are ready to try doing the same filtering operation on a DSP. Figure 17.3 is a program in assembly language of an imaginary DSP. The words starting with dots (such as .table) are 'directives'; they direct the assembler to place the following data or code in specific memory banks. In

```
.table
H:              h_0
                h_1
                  ⋮
HLAST:          h_{L-1}


.data
X:              (L-1) * 0
XNEW:           0


.program
START:
  if (WAIT) goto START
  *XNEW  ←  INPUT
  h  ←  HLAST
  x  ←  X

  y  ←  (*h)*(*x) || h ← h-1 || x ← x+1
  repeat (L-1) times
        y  ←  y + (*h)*(*x) || *(x-1) ← *x || h ← h-1 || x ← x+1

  NOP
  OUTPUT  ←  y
  goto START
```

Figure 17.3: A simple program to FIR filter input data in real-time.

this case the L filter coefficients $h_0 \ldots h_{L-1}$ are placed in 'table memory'; the static buffer of length L is initialized to all zeros and placed in 'data memory'; and the code resides in 'program memory'. These placements ensure that the MAC instructions will be executable in a single cycle. The names followed by colons (such as HLAST:) are 'labels', and are used to reference specific memory locations.

The filter coefficients are stored in the following order $h_0, h_1, \ldots h_{L-1}$ with h_0 bearing the label H and h_{L-1} labeled HLAST. The static buffer is in reversed order $x_{n-(L-1)}, \ldots x_{n-1}, x_n$ with the oldest value bearing the label X and the present input labeled XNEW.

The program code starts with the label START, and each nonempty line thereafter corresponds to a single processor cycle. The first line causes the processor to loop endlessly until a new input arrives. In a real program such a tight loop would usually be avoided, but slightly looser do-nothing loops are commonly used.

Once an input is ready it is immediately copied into the location pointed to by XNEW, which is the end of the static buffer. Then pointer register h is set to point to the end of the filter buffer (h_{L-1}) and pointer x is set to point to the beginning of that buffer (the oldest stored input).

Accumulator y is initialized to $h_{L-1}x_0$, the last term in the convolution. Note that y is a numeric value, *not* a pointer like x. The $\parallel$ notation refers to operations that are performed in parallel. In this line the filter buffer pointer is decremented and the static buffer pointer is incremented. These operations are carried out before they are next required.

The next line contains a 'zero-overhead loop'. This loop is only executed L-1 times, since the last term of the convolution is already in the accumulator. The last iteration multiplies the h_0 coefficient by the new input. However, something else is happening here as well. The *(x-1) $\leftarrow$ *x being executed in parallel is a data-move that shifts the input data that has just been used one place down; by the time the entire loop has been executed the static buffer has all been shifted and is ready for the next iteration.

Once the entire convolution has been carried out we are ready to output the result. However, in some DSP processors this output operation can only take place once the pipeline has been emptied; for this reason we placed a NOP (no-operation) command before copying the accumulator into the output register. Finally, we jump back to the start of the program and wait for the next input to arrive.

EXERCISES

17.6.1 Taking a specific number of coefficients (e.g., L=5), walk through the program in Figure 17.3, noting at each line the values of the pointers, the state of the static buffer, and the algebraic value in the accumulator.

17.6.2 Code a real-time FIR filter for a DSP that does not support data-move in parallel with the MAC, but has hardware support for a circular buffer.

17.6.3 Code a real-time IIR routine similar to the FIR one given in the text. The filter should be a cascade of N second order sections, and the main loop should contain four lines and be executed N times.

17.6.4 Write a filtering program for a real DSP and run it in real-time.

17.7 DSP Programming Projects

DSP programming is just like any other programming, only more so. As in any other type of programming attention to detail is essential, but for DSP processors this may extend beyond syntax issues. For example, some processors require the programmer to ensure that the requisite number of cycles have passed before a result is used; forgetting a NOP in such a situation creates a hard-to-locate bug. For many types of programming intimate knowledge of the hardware capabilities isn't crucial, but for DSP programming exploitation of special-purpose low-level features may mean the difference between success and failure. As in any other type of programming, familiarity with the software development tools is indispensable, but for DSP processors emulation, debugging and profiling may be much more difficult and critical tasks.

In this section we present a model that you may find useful to consider when embarking on a new DSP programming project. However, whether or not you adhere to all the details of this model, remember that you must always obey the *golden rule of DSP programming*:

Always program for correctness first, efficiency second.

All too often we are driven by the need to make our algorithms faster and faster, and are tempted to do so at the expense of system stability or thorough testing. These temptations are to be avoided at all costs.

Once this is understood I suggest that the task of implementing a new system is CHILD's play. Here the word CHILD is a mnemonic for:
Collect requirements and decide on architecture
High-level design
Intermediate level, simulation and porting to platform
Low-level coding and efficiency improvement
Deliver and document
We shall devote a paragraph or two to each of these stages.

The collection stage is a critical one, all too often incompletely executed. The implementor must collect all the requirements, including the expected range of inputs, the exact output(s) required, the overall development schedule and budget, the desired end user cost, interface specifications, etc. Sometimes someone else has done the preliminary work for you and you receive a Hardware Requirements Specification (HRS) and a Software Requirements Specification (SRS). Remember that anything missed during the collection stage will be difficult or impossible to reintroduce later on. One of the things to be decided at this stage is how the final product is to be tested, and the exact criteria for success. You should make sure the end users (or technical marketing personnel) 'sign off' on the requirement specifications and acceptance procedures.

Between the end of the collection stage and the beginning of the high-level design stage it is highly recommended to go on vacation.

The technical output of the high-level design stage will usually be a pair of documents, the **Hardware Design Document (HDD)** and the **Software Design Document (SDD)**. There will also be project management literature, including various charts detailing precisely what each team member should be doing at every time, dates by which critical tasks should be completed, milestones, etc. We will focus on the SDD. The SDD explains the signal processing system, first in generality, and then increasingly in detail. The function of each subsystem is explained and its major algorithms noted. There are two ways to write an SDD. The first (and most commonly encountered) is to have done something extremely similar in the past. In this case one starts by cutting and pasting and then deleting, inserting, and modifying until the present SRS is met. The more interesting case is when something truly new is to be built. In this case a correct SDD cannot be written and the project management literature should be considered science fiction. Remember that the 'R' and 'D' in R&D are two quite different tasks, and that a true research task cannot be guaranteed to terminate on a certain date and in a certain way (the research would be unnecessary if it could).

Often simulations must be performed during the high-level design stage. For these simulations efficiency is of no concern; but development speed, ease of use and visualization ability are of the utmost importance. For this reason special development environments with graphics and possibly visual programming are commonly used. The output of these simulations, both block diagrams and performance graphs, can be pasted into the design documents. The amount of memory and processing power required for each subsystem can now be better estimated. It is best to plan on using only 50% to 75% of the available processing power (it will always turn out to require a lot more than you anticipate).

At the end of the high-level design a **Design Review (DR)** should be carried out. Here the HDD and SDD are explained and comments solicited. Invite as many relevant people as possible to the DR. Remember that mistakes in the high-level design are extremely costly to repair later on.

The intermediate-stage may be bypassed only for very small projects. Here the block diagrams developed in the high-level stage are fleshed out and a complete program is written. This is often done first in a high-level language, liberally using floating point numbers and library functions. While the high-level software design stage is often carried out by a single person, the intermediate stage is usually handed over to the full development team. Once the team starts to work, the project should be placed under revision control. Once completed and integrated the full program can be tested with test inputs and outputs to ensure passing the final acceptance procedures. Next the high-level language program is rewritten in a real-time style, using the proper block lengths, converting to fixed point if required, etc. After each major step the program behavior can be compared to that of the original to ensure correctness. Mechanisms for debugging, exception handling, maintainability, and extensibility should be built into the code.

The low-level programming commences as a straightforward port of the intermediate level program to the final platform. Once again the first code should be written to maintain correctness at the expense of efficiency. After the first port, decisions must be made as to what can remain in a high-level language and what must be coded in assembly language; where major improvements in efficiency are required; where memory usage is excessive; etc. Efficiency is increased incrementally by concentrating on areas of code where the program spends most of its time. Various debugging tools, such as simulators, emulators, debug ports, and real-time monitoring are used. Eventually a correct version that is fast enough is generated.

Delivery of a version is something that no one likes doing, but the project is not complete without it. The final version must be cleaned up and acceptance tests thoroughly run (preferably with the end user or disinterested parties present). User and programmer documentation must be completed. The former is usually written by professional publications personnel. The latter include internal documentation (the final code should be at least 25% comments), an updated SDD, and a **V**ersion **D**escription **D**ocument (VDD) that describes all limitations, unimplemented features, changes, and outstanding problems.

After delivery the boss takes the development team out to lunch, or lets everyone take an extended weekend. The following week the whole process starts over again.

17.8 DSP Development Teams

'Congratulations, you've got the job!' You heard those words just last week, but you are already reporting for work as a junior DSP engineer in the ASP (Advanced Signal Processing) division of III (Infinity Integrators Inc.). In your previous jobs you worked by yourself or with one or two other people on some really impressive DSP projects. You've programmed DSP boards in PCs, completing whole applications in a week, often writing over 500 lines of assembly language in a day. You are quite sure that this has prepared you for this new job. Of course in a big company like III with hundreds of engineers and programmers, working on multimillion dollar projects, there will be more overhead, but DSP programming is DSP programming. There is only one thing that your new boss said during your interview that you don't quite understand. Why do the programmers here only write an average of ten to twenty lines of code in a day? Are they lazy or just incompetent?

Well, you'll soon find out. Your first assignment is to understand the system you will be working on. This system is a newer version of an older one that has been operational for over five years. Your boss has given you five days to come up to speed. Sounds easy.

In your cubicle you find a stack of heavy documents. The first thing you have to learn is what a TLA is. TLA is a self referential term for Three Letter Acronym, and the system you are going to work on is full of them. There are several FEUs (front end units), a BEU (back end unit), and a MPC (main processing computer) with a HIC (human interface console). You learn all this by reading parts of two documents titled HRS and SRS, that are followed by even larger ones marked HDD and SDD. These documents are so highly structured that you soon understand that it would take a full five days just to read them and follow up the cross references. There are also countless other documents that you have been told are not as important to you (yet), like the PMP (program management plan) that even specifies how much time you are allotted to learn the system (5 days).

Your second day is spent trying to figure out where DSP fits in to all this and what you are expected to do. As a shortcut you look up your tasks in the PMP. After an hour or so you have jotted down the cross references of tasks to which you have been assigned, and start looking up what they are. A lot is written about when they start and end, what resources they need, and what they hold up if not finished on time. The only explanation of exactly what you are expected to do seems to be one-line descriptions that are entirely incomprehensible. In fact, the only thing you think you fully understand is the 'handbook for the new employee' that human resources placed in your cubicle. Even that seems to have been written in the same style, but at least it is one self-contained document (except the reference to the manual for using the voicemail system).

On your third day on the job you attend your first weekly TRM (team review meeting). The main subject seems to be when the PDR (preliminary design review) and CDR (critical design review) will take place. Luckily the TRM will only last about a half hour since the boss has to go to the TLM (team leader meeting) that has to take place before the GLM (group leader meeting) headed by the director of ASP. Any questions? Somewhat timidly you speak up—why is it called the *FEU* and not the *DSP*? In all the jobs you worked on up to now the module with the DSPs was simply called the *DSP*. The boss explains patiently (while several coworkers smile) that the name has been chosen to be accurate from a system point of view. *DSP* stands for digital signal processing. Here the fact that the processing is *digital* is irrelevant to the rest of the system; indeed in an earlier version a lot of the processing was analog. The words *signal* and *processing* turn out to be incorrect from a system point of view. The purpose of this unit is acquisition of the data needed by the rest of the system, even if this data already requires a great deal of processing from raw input.

After the meeting you spend a few hours in the library. The SDD references a lot of international standards and each of these in turn references still other standards. The standards documents seem even less comprehensible than the SDD itself. You spend the next few hours by the copying machine.

On the fourth day you are invited to attend a discussion between the hardware and software guys. Not yet having learned about the subsystem in question you can't quite make out what is going on, other than the hardware guys saying it's an obvious software bug and the software guys saying its a hardware failure. You speak up asking why a simple test program can't be used to test the hardware. The hardware people explain that they had written such a program and that is precisely how they know its a software problem. The software people reply that the hardware test program was unrealistically simplistic and didn't really test this aspect of the design. They, however, have written a simulation of the hardware, and their software runs perfectly on it.

You glance at your watch. Although it's after six o'clock you think you'll spend a few more hours reading the SDD. Just this morning you had at last found out what the DSP processing elements were, and were beginning to feel more confident that it was, in principle, possible to extract information from these documents. How will you ever finish reading all this by tomorrow?

The fifth day starts auspiciously—your development system arrives. You are given a table in the lab; its one floor down from your cubicle and you learn to use the stairs rather than wait for the elevator. The entire morning is spent on unpacking, reading the minimal amount of documentation, hooking up all the cables, and configuring the software. A coworker helps you out, mostly in order to see the improvements in the new version you have received. You go out to lunch with your co-workers, and the entire time is spent talking about work. Once back you return to your cubicle only to find that someone has 'borrowed' your copy of the SDD. You return to the lab only to find that several power cables are missing as well. It's after three and you start to panic. Are your coworkers playing some kind of initiation prank or does this kind of thing happen all the time?

Bibliographical Notes

Jonathan Allen from MIT gave two early, but still relevant, overviews of the basic architecture of digital signal processors [3, 4]. More modern reviews are [142, 143, 141]. In particular [61] can be used as a crash course in DSP processors: what defines a DSP, how DSPs differ from CPUs, and how they differ one from another.

Manuals supplied by the various processor manufacturers are the best source for information on DSP architecture and how to best exploit it. Usually each processor has a *Processor User's Guide* that details its architecture and instruction set; an *Assembly Language Reference* with explanations of its programming environment; and an *Applications Library Manual* with sample code and library routines for FFTs, FIR and IIR filters, etc. For popular processors many useful DSP functions and full applications will be available for licensing or in the public domain. The annual EDN DSP directory [148] is a treasure-trove of information regarding all extant DSPs.

[140] is devoted entirely to fundamentals of DSP processors, and its authors also publish an in-depth study and comparison of available DSPs. Readers considering implementing DSP functions in VLSI should consult [154].

Part IV

Applications

Communications Signal Processing

In this chapter we will survey various topics in signal processing for communications. Communications, like signal processing itself, is commonly divided into analog and digital varieties. Analog communications consist of techniques for transmitting and receiving speech, music or images as analog signals, as in telephones, broadcast radio and television. Digital communications are methods of transferring digital information, usually in the form of bit streams. Digital communications are often between computers, or between human and computer, although increasingly digital communications are being used between people as well (email). Both analog and digital signal processing may be used for various portions of both analog and digital communications systems.

A device that takes an analog input signal and creates an analog communications signal is called a transmitter, while a receiver inputs an analog communications signal and attempts to recover, as accurately as possible, the original analog message signal. A device that takes a digital input and creates a digital communications signal is usually called a modulator, while a demodulator inputs a digital communications signal and attempts to recover, with as few bit errors as possible, the original digital message. Transmitters and receivers are sometimes packaged together and called *transceivers*; for digital communications it is almost universal to package the modulator and demodulator together, and to call the combined device a *modem*.

Digital communications systems include such diverse objects as fax machines, telephone-grade modems, local area networks, wide area networks, private digital telephone exchanges, communications satellites and their ground stations, the public switched telephone network (yes, it too has become digital), and the Internet. Although the history of data communications is relatively short, the present scope of its theory and application is huge, and we will have to stringently restrict the scope of our treatment.

After a historical introduction we will start our survey with an overview of analog communications, including AM and FM transmitters and receivers. We then briefly study information and communications theory, including error correcting codes. We then design our first modem, the rest of the chapter being devoted to successive improvements to its design. We roughly follow the chronological development of telephone-grade modems that increased bit rates from 300 b/s to 56 Kb/s. Along the way we will learn about FSK, PSK, QAM, MCM, TCM, and PCM modems, and master the basic algorithms needed for modem implementation.

18.1   History of Communications

Let's go back over 2000 years and imagine ourselves at the foot of the great Temple in Jerusalem. It is the thirtieth day of the month, and the Calendar Council is in session, waiting for witnesses to come to testify that they had seen the new moon. A group of people approach running. They are ushered into the chamber and interrogated by experts in astronomy and mathematics. If their testimony is found to be genuine, the new month is declared to have begun; if no reliable witnesses arrive the new month only starts the next day. Now that information must be disseminated quickly to those living as far away as Babylon. Only one bit of information must be transmitted—whether the new month has commenced—but telephones, radio, and even telegraph lines do not yet exist.

Now it is not really difficult to transmit the single bit of information to nearby locations. One need only do something that can be reliably seen from afar. So the Council orders a bonfire to be lit on the top of a nearby mountain. On a neighboring mountain an official is waiting. When he sees the beacon he lights a fire of his own, which is observed at the first mountain and recognized as an acknowledgment that the message has been received. It is also observed at another mountain further away, where the next beacon in the chain is lit. In this way the message that the new month has commenced is quickly and reliably transmitted. This technique was in use until thwarted by the (good?) Samaritans, who maliciously lit beacons at inappropriate times in order to create confusion.

Similar communications techniques were used by other pretelegraph peoples. Native Americans would burn wet grass under a blanket, which when removed would send up a blast of dark smoke that could be seen from afar. Natives of western Africa used tomtom drums that could be heard throughout the jungle (where visibility is limited). Mariners used signaling lamps that could be seen from miles away.

What can we do if we need to transmit more than one bit of information? The native Americans would simultaneously light two or three separate fires, and the number of columns of smoke signified the urgency of the message. The Africans used drums of variable pitch, and could send intricate messages by varying the sounds of their drumming. At sea mariners would open and close shutters on the signaling lamps, thus sending entire messages.

These methods of communications suffer from several drawbacks. First, they work over limited distances, requiring relay operators for larger range. Second, they are not reliable; after the battle of Waterloo a signal lamp message over the English channel was received as 'At Waterloo Nelson defeated ...' with 'Napoleon' covered up by the fog. Nathan Rothschild made a fortune buying up stocks on the plunging London exchange, knowing the truth through more reliable carrier pigeons. Third, these communications media are all *broadcast*, meaning that they can be intercepted by all. Although this is sometimes required it can also be a disadvantage. Settlers of the American West spotted Indian smoke signals and recognized that the enemy was close at hand. Finally, all these methods are *multiple access* with no signature, and can thus be easily forged (as the Samaritans did).

The discovery of electric current by Stephen Gray of London in 1729 produced a new medium for reliable communications over distances, removing many of the disadvantages of previous methods. In 1747, William Watson laid 1200 feet of wire over Westminster bridge, touching one end to the water of the Thames, and the other to a charged Leiden jar; a man touching the jar with his feet in the river received a shock. It took a while longer to realize that the flow of current could be detected by its lighting a light or moving an armature. In 1844, Samuel Morse telegraphed the message 'What hath God wrought?' over an electric cable, ushering in a new era for humankind. Morse's *telegraph* could distinguish between two states, current flowing or not, and so Morse had to devise a code to efficiently send letters of the alphabet using only two-state signals. The Morse code represents letters using combinations of $s = 0$ and $s = 1$ values; $s = 0$ are used as dividers, while $s = 1$ may occur in short durations (called a dot) or three times that duration (called a dash). The letter 'E' is encoded as a dot, that is, by a single $s = 1$, and thus only requires 1 time unit to transmit (although it must be followed by a single $s = 0$ inside a word and by three consecutive $s = 0$ at the end of a word). The letter 'Q' is encoded as dash, dash, dot, dash, occupying 13 basic time intervals. The entire Morse code is presented in Table 18.1. In 1866, the first transatlantic cable was laid, for the first time linking America and Europe by an almost instantaneous communications medium (unfortunately, it failed within a month).

A	·—	K	—·—	U	··—	0	——————
B	—···	L	·—··	V	···—	1	·————
C	—·—·	M	——	W	·——	2	··———
D	—··	N	—·	X	—··—	3	···——
E	·	O	———	Y	—·——	4	····—
F	··—·	P	·——·	Z	——··	5	·····
G	—···	Q	——·—	.	·—·—·—	6	—····
H	····	R	·—·	,	——··——	7	——···
I	··	S	···	?	··——··	8	———··
J	·———	T	—	-	—····—	9	—————·

Table 18.1: The Morse code. Every letter, number, or punctuation mark is assigned a unique combination of dots and dashes.

Telegraphy using Morse code still had a few disadvantages. It was relatively slow and error prone. It required skilled telegraphers at both ends and could not be directly used by individuals. Unless special codes were employed the messages could be read by others, and it was difficult to authenticate the sender's identity. For some time people strived to mechanize the transfer of text using the Morse code, but this was a difficult task due to the variable-length characters. In 1875, Emile Baudot from France created a new code, one optimized for mechanized text transfer. In the Baudot code each letter took five equal time units, where each unit could be current flow (*mark*) or lack thereof (*space*). Actual commercial exploitation of this code began in early twentieth century, under the trademark name *teletype*.

A further breakthrough was announced within a year of Baudot's code when, on March 10, 1876, Dr. Alexander Graham Bell in Boston and Elisha Gray in Chicago both filed for patents for a new invention, later to be called the *telephone*. Like the telegraph it used voltage signals traveling over a wire, but rather than being simple on-off, these signals carried a voice. Eventually, Dr. Bell won the protracted legal battle that reached the level of the U.S. Supreme Court. The telephone could be placed in every home, and used by anyone without the need for intervention of skilled middlemen. For the first time, point-to-point communication was direct, reliable, relatively private, and the voice of the person at the other end could be recognized.

Another development was born out of a purely mathematical insight. In 1865, James Clerk Maxwell wrote down differential equations describing all that was then known about electricity and magnetism. These equations described how an electric charge created an electric field (Coulomb's law), how an electric current created a magnetic field (Ampère's law), and how a changing magnetic field created an electric field. Far away from currents and charges the equations

$$\nabla \cdot \underline{E} = 0 \qquad \nabla \cdot \underline{B} = 0$$
$$\nabla \times \underline{E} = -\tfrac{1}{c}\tfrac{\partial B}{\partial t} \qquad \nabla \times \underline{B} = 0$$

were obviously not symmetric. To make them completely symmetric Maxwell hypothesized that a changing electric field could induce a magnetic field,

$$\nabla \cdot \underline{E} = 0 \qquad \nabla \cdot \underline{B} = 0$$
$$\nabla \times \underline{E} = -\tfrac{1}{c}\tfrac{\partial B}{\partial t} \qquad \nabla \times \underline{B} = +\tfrac{1}{c}\tfrac{\partial E}{\partial t}$$

a phenomenon that had not previously been observed. These new equations admitted a new type of solution, a changing electric field inducing a changing magnetic field reinforcing the original changing electric field. This *electromagnetic* field could travel at the speed of light (not surprising since light is exactly such a field) and carry a signal far away without the need for wires. In 1887, Hertz performed an experiment to test Maxwell's purely theoretical prediction. He made sparks jump between two polished brass knobs separated by a small gap, and detected the transmitted electromagnetic waves using a simple receiver of looped wire and similar knobs several meters away.

Radio waves can carry Morse or Baudot code by transmitting or not transmitting (on-off keying). They can also carry voice by continuously changing some characteristic of the field, such as its amplitude (AM) or frequency (FM). In the next section we will learn how this can be done.

EXERCISES

18.1.1 Compute the time durations of the 26 letters in Morse code. What is the average duration assuming all characters are equally probable? What is the average duration assuming that the letter probabilities are roughly E:12%, TAOINS:8%, HRDLU:4%, MCFGPB:2%, and all the rest 1%. Is Morse code better or worse than Baudot code for actual text?

18.1.2 Write a program that inputs a text file and outputs Morse code. You will need a computer with minimal sound capabilities. Whenever $s = 1$ play a tone (1000 Hz is good). Make the speed an adjustable parameter, specified in words per minute (figure an average word as 5 characters). Add an option to your program to output two different tones, a high-frequency tone for $s = 1$ and a low-frequency one for $s = 0$.

18.1.3 Modify the above program to output a file with sampled signal values (use a sampling rate of 8000 Hz and a tone of 750Hz). Now write a program that inputs this file and decodes Morse code (converts signal values back to text). Improve your program to take into account small amounts of noise and small variabilities in speed (and add these features to the generating program). Do you think you could write a program to read Morse code sent by hand on a noisy channel?

18.2 Analog Modulation Types

In our historical discussion we carefully avoided using the word 'modulation'; we now just as carefully define it.

Definition: modulation
Modulation is the exploitation of any observable characteristic of a signal to carry information. The signal whose characteristics are varied is called a *carrier*. We *modulate* the carrier by the information signal in order to create the modulated signal, and *demodulate* the modulated signal in order to recover the information signal. The systems that perform these functions are called the *modulator* and *demodulator*, respectively. ∎

Modulation is used whenever it is not possible or not convenient to convey the information signal directly. For example, a simple two station intercom will probably directly transmit the voice signals (after amplification) from one station to another over a pair of wires. This scenario is often called *baseband transmission*. A more sophisticated intercom system may modulate a radio signal, or the AC power signal, in order to eliminate the need for wires. The public switched telephone network uses wires, but maximizes their utilization by modulating a single base signal with a large number of subscriber signals.

Perhaps the simplest signal that is used as a carrier is the sinusoid.

$$s(t) = A\cos(2\pi ft + \phi) \qquad (18.1)$$

For example, the very existence of the carrier can be used to send Morse or Baudot code. This is called **On-Off Keying** (OOK) and mathematically is represented by

$$s_{\text{OOK}}(t) = A(t)\cos(2\pi f_c t) \qquad (18.2)$$

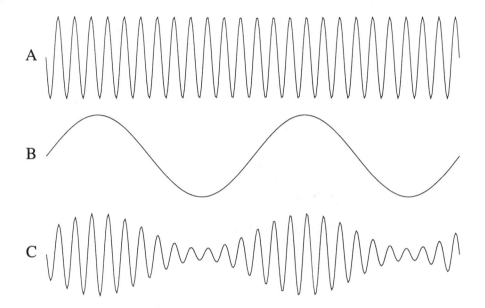

Figure 18.1: Amplitude modulation changes the amplitude of a carrier in accordance to a modulating signal. In (A) we see the carrier, in (B) a sinusoidal modulating signal, and in (C) the resulting AM signal (the modulation index was 75%).

where $A(t)$ takes the values zero or one, f_c is the carrier frequency, and (without limiting generality) we choose the phase to be zero. In order to carry voice or other acoustic modulating signals $v(t)$, we need more freedom. Now equation (18.2) is strongly reminiscent of the instantaneous representation of a signal of equation (4.66); but there the amplitude $A(t)$ was a continuously varying function. This leads us to the idea of conveying a continuously varying analog signal $v(t)$ by varying the carrier's amplitude

$$s_{\mathrm{AM}}(t) = A_0 \left(1 + m_{\mathrm{AM}}\, v(t)\right) \cos(2\pi f_c\, t) \qquad (18.3)$$

where we assume $|v(t)| \leq 1$. This modulation technique, known as **Amplitude Modulation (AM)**, is depicted in Figure 18.1. The coefficient $0 < m_{\mathrm{AM}} \leq 1$ is known as the modulation index, and is often specified as a percentage.

Amplitude is not the only signal characteristic that one can modulate. The sinusoidal carrier of equation (18.1) has two more characteristics that may be varied, the frequency f and the phase ϕ. Morse- or Baudot-encoded text may be sent by **Frequency Shift Keying (FSK)**, that is, by jumping between two frequencies.

$$s_{\mathrm{FSK}}(t) = A \cos\left(2\pi f(t)\, t + \phi\right) \qquad (18.4)$$

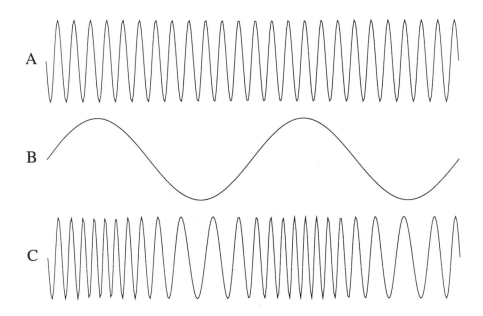

Figure 18.2: Frequency modulation changes the frequency of a carrier in accordance to a modulating signal. In (A) we see the carrier, in (B) a sinusoidal modulating signal, and in (C) the resulting FM signal.

Here it is $f(t)$ that can take on two different values. The third alternative is called **P**hase **S**hift **K**eying (PSK),

$$s_{\text{PSK}}(t) = A \cos \left(2\pi f_c t + \phi(t) \right) \tag{18.5}$$

where $\phi(t)$ can take on two values (e.g., $0°$ and $180°$). Similarly, voice can be transmitted by **F**requency **M**odulation (FM) and by **P**hase **M**odulation (PM), as will be explained in the next section. For example, in Figure 18.2 we see the frequency of a sinusoid continuously varying in sinusoidal fashion.

We have still not exhausted the possibilities for modulation. The sinusoid, although the most prevalent carrier, is not the only signal that can be modulated. An alternative is to start with a train of pulses and modify their amplitudes (PAM), their relative timing (PPM) or their pulse widths (PWM). Another common occurrence is *secondary modulation* where modulated signals are used to modulate a second signal. For example, several AM-modulated voice signals may be used to frequency modulate a wideband radiotelephone link carrier. Sometimes it seems that the number of different modulation techniques that have been used in communications systems equals the number of communications systems designers.

EXERCISES

18.2.1 Why is equation (18.3) not simply $A_0 v(t) \cos(2\pi f_c t)$? Plot sinusoidally modulated AM signals for various values of modulation index. What index do you think should be used?

18.2.2 Write a program that generates an AM-modulated wave. (For concreteness you may assume a sampling frequency of 2.048 MHz, a carrier of 455 KHz, and take the modulating signal to be a sinusoid of frequency 5 KHz.) Plot 1 millisecond of signal. What does the spectrum look like?

18.2.3 Why do we prefer sinusoidal carriers to other waveforms (e.g., square waves)?

18.2.4 Can we simultaneously modulate with AM and FM? AM and PM? FM and PM?

18.3 AM

Now that we know what modulation is, we can commence a more systematic study of modulated signals and the signal processing systems used to modulate and demodulate. For now we are only interested in modulating with continuous analog signals such as speech; digital modulation will be treated later.

How can we create an amplitude modulated signal using analog electronics? The simplest way would be to first create the carrier using an oscillator set to the desired frequency. Next the output of this oscillator is input to an amplifier whose gain is varied according to the modulating signal (see Figure 18.3). Since both oscillators and variable gain amplifiers are standard electronic devices, building an AM transmitter in analog electronics

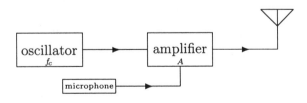

Figure 18.3: The basic analog AM transmitter built from an oscillator and a variable gain amplifier. The oscillator has a single parameter f_c that is not varied during transmission. The amplifier's gain parameter A is varied according to the signal. The inverted triangle at the top right is the conventional graphic representation of an antenna.

is straightforward. Of course there are lots of technical details to be dealt with, such as guaranteeing oscillator frequency stability, ensuring that the microphone's output is sufficiently strong, keeping the amplifier in its linear range, band-pass filtering the signal to avoid interference to nearby receivers, matching the input impedance of the amplifier with the output impedance of the oscillator, etc. Failing to properly cope with any of these details will result in inefficiency, low or distorted audio, or interference.

Wouldn't it be simpler to implement the AM transmitter using DSP? The analog oscillator and amplifier could be replaced with digital ones, and using correct digital techniques there will be no problems of efficiency, frequency stability, amplifier stability, impedance matching, etc. Although in principle this approach is correct, there are two practical problems. First, a digital amplifier by itself will only be sufficient for very low-power applications; in order to supply the high power usually needed (from about ten watts for mobile radios to many thousands of watts for broadcast stations) an additional analog *power amplifier* will usually be needed. Second, the bandwidth BW of the audio frequencies (AF) is usually much lower than the radio frequency (RF) of f_c. Directly implementing Figure 18.3 digitally would require us to operate at a sampling rate over twice $f_c + BW$, which would be extremely wasteful of computational power. Instead we can perform all the computation at an intermediate frequency (IF) and then upmix the signal to the desired radio frequency. Figure 18.4 shows a hybrid AM transmitter that utilizes digital techniques for the actual modulation and analog electronics for the upmixing and power amplification.

Now that we know how to transmit AM we need a receiver to demodulate our AM transmission. The simplest analog receiver is the *envelope detector*,

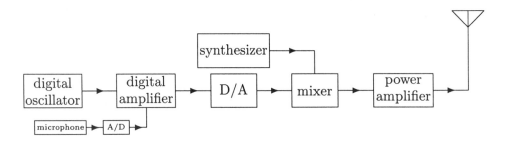

Figure 18.4: The basic hybrid digital-analog AM transmitter. The digital components operate at an intermediate frequency and at low power. After conversion to the analog domain the signal is upmixed to the desired carrier frequency and amplified to the required output power. The synthesizer is a (digital) local oscillator.

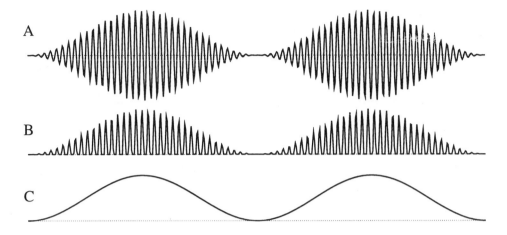

Figure 18.5: The basic analog envelope detector for the demodulation of AM signals. In (A) we see the AM signal to be demodulated. After half wave rectification the signal depicted in (B) results. Subsequent low-pass filtering removes the RF and leaves (C) the desired AF to within DC.

the operation of which can be best understood by studying Figure 18.5. Since the desired signal is the 'envelope' of the received signal, it can be retrieved from either the top or bottom of Figure 18.5.A by connecting the peaks. Choosing to use the top half, half wave rectification results in the signal of Figure 18.5.B. We next low-pass filter this signal in order to remove the high-frequency RF, leaving only the envelope as in Figure 18.5.C (with a strong DC component). This filtering is performed by placing the rectified signal onto a capacitor that charges up to the voltage peaks and slowly interpolates between them. Finally a DC blocking filter is used to remove the 1 from $1 + v(t)$.

Unfortunately, the envelope detector is ill suited to digital implementation. It assumes f_c to be very high compared to f_m, otherwise the envelope will not be well sampled, and thus downmixing to a low IF will decrease its efficacy. More importantly, in order to actually see the analog signal's peaks in its digital representation, a sampling frequency much higher than Nyquist is required. Even sampling at several times Nyquist we can not expect most of the sampling instants to fall close enough to the peaks.

A better way of digitally performing AM demodulation is to use the instantaneous representation of Section 4.12. There are two closely related ways of doing this. The first is to apply the Hilbert transform to the IF signal and to obtain the instantaneous amplitude by the square root of the sum of the squares. The second involves a complex downmix to zero including a complex low-pass filter to remove everything except the frequency components from zero to BW. We can then proceed to obtain the instantaneous amplitude as before. These methods of digital AM demodulation do not require high f_c and function with sampling frequencies close to Nyquist.

Up to now we have been thinking of AM only in the time domain. What does the spectrum of an AM signal look like? We'll first consider modulating with a single sinusoid, so that equation (18.3) becomes

$$s_{AM}(t) = A_0 \left(1 + m_{AM}\cos(\omega_m t)\right) \cos(\omega_c t) \qquad (18.6)$$

where ω_m and ω_c are the modulating and carrier angular frequencies. A little algebra proves

$$\begin{aligned} s_{AM}(t) &= A_0\cos(\omega_c t) + A_0\, m_{AM}\cos(\omega_m t)\cos(\omega_c t) \qquad (18.7) \\ &= A_0\cos(\omega_c t) + m_{AM}\frac{A_0}{2}\left(\cos(\omega_c + \omega_m)t + \cos(\omega_c - \omega_m)t\right) \end{aligned}$$

so that the spectrum contains three discrete lines, one corresponding to the original carrier frequency, and two lines at the carrier plus and minus the modulation frequency (Figure 18.6.A).

What if we modulate the carrier not with a single sinusoid but with a general signal $v(t)$? The modulating signal can be Fourier analyzed into a collection of sinusoids each of which causes two lines spaced f_m away from the carrier. We thus obtain a carrier and two *sidebands* as depicted in Figure 18.6.B. The two sidebands are inverted in frequency with respect to each other but contain precisely the same information.

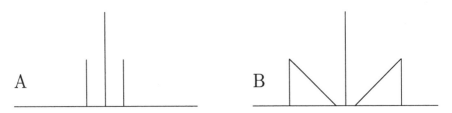

Figure 18.6: The generation of sidebands of an AM signal. In (A) we modulate a sinusoid of frequency f_c by a single sinusoid of frequency f_m to obtain an AM signal with three frequency lines, f_c, $f_c \pm f_m$. In (B) we modulate a sinusoid by a signal with an entire spectrum of frequencies, conventionally depicted as a triangle. We obtain the carrier and two sidebands.

EXERCISES

18.3.1 Our basic analog AM receiver assumed that only a single signal is received at the antenna, while in fact many signals are received simultaneously. One method of isolating the signal of interest uses a band-pass filter centered at f_c; the more conventional method uses a mixer and a band-pass filter centered at an *intermediate frequency* (IF). Diagram the two methods and discuss their advantages and disadvantages.

18.3.2 Diagram an entire AM receiver including antenna, local oscillator and mixer, IF filter, a half wave rectifier, a low-pass filter, DC blocking filter, and speaker. Show representative signals at the output of each block.

18.3.3 Implement a digital envelope detector. Create a sinusoidally modulated signal with $f_c = 50$, $f_m = 2$, and sampling frequency $f_s = 500$. Compare the demodulated signal with the correct modulating signal. Now decrease f_s to 200. Finally decrease f_c to 10. What do you conclude?

18.3.4 Show that half of the energy of an AM signal with index of modulation $m_{AM} = 1$ is in the carrier and one-quarter is in each of the sidebands.

18.3.5 Double sideband (DSB) is a more energy-efficient variant of AM, whereby the carrier is removed and only the two sidebands are transmitted. Diagram a transmitter and receiver for DSB.

18.3.6 Single sideband (SSB) is the most efficient variant of AM, whereby only a single sideband is transmitted. Diagram a transmitter and receiver for SSB.

18.3.7 Can AM demodulation be performed by a filter? If yes, what is its frequency response? If not, what portion of the analog and digital detectors is not a filter?

18.4 FM and PM

You might expect that frequency modulation of a carrier $A\cos(\omega_c t)$ with a signal $v(t)$ would be accomplished by

$$s(t) = A\cos\left(\left(\omega_c + m\,v(t)\right)t\right) \tag{18.8}$$

where m is the index of modulation. Indeed the amplitude is constant and the frequency varies around the carrier frequency according to the modulating signal; yet this is *not* the way FM is defined. To see why not, assume that

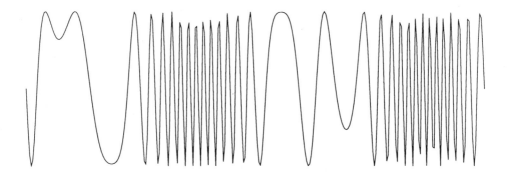

Figure 18.7: Frequency modulation according to the naive equation (18.8) has obvious artifacts. True frequency modulation should look sinusoidal.

the modulating signal $v(t)$ is a sinusoid (let's use sine rather than cosine this time) of frequency ω_m.

$$s(t) = A \cos \left(\left(\omega_c + m \sin(\omega_m t) \right) t \right) \tag{18.9}$$

Plotting this for time close to $t = 0$ results in a picture similar to 18.2.C, but for longer times we observe artifacts as in Figure 18.7. This is not what we expect from FM; in particular we want all the extrema of the signal to be those of the underlying carrier, whereas here we observe obviously nonsinusoidal extrema as well!

The reason for this errant behavior is not hard to see. The signal can be rewritten

$$s(t) = A \cos \left(\omega_c t + m\, t \sin(\omega_m t) \right)$$

and so has phase swings that increase linearly with time. For large t the phase swings completely dominate the argument of the sine except in the immediate vicinity of the modulating sinusoid's zeros, thus completely destroying the overall sinusoidal behavior. The solution to this problem is easy to see as well—we simply move the modulating sinusoid so that it is not multiplied by time

$$s(t) = A \cos \left(\omega_c t + m \sin(\omega_m t) \right) \tag{18.10}$$

or for a more general modulating signal

$$s(t) = A \cos \left(\omega_c t + m_{\mathrm{PM}}\, v(t) \right) \tag{18.11}$$

which is known as **Phase Modulation** (PM).

There is a more direct way to arrive at PM. We think of a carrier signal $A\cos(\omega_c t + \phi)$ as having a degree of freedom not previously exploited—the phase ϕ. Having the phase vary with the modulating signal will create an information-bearing signal from which the modulating signal may be later retrieved, at least assuming the phase does not vary too much. We can use the PM index of modulation m_{PM} to ensure that the phase deviation does not exceed 2π, the point where ambiguity would set in.

True frequency modulation is similar to phase modulation, but not identical. Recalling equation (4.72) we realize that we can make the instantaneous frequency vary with a modulating signal by phase modulating by that signal's integral. If that is done, the information-bearing signal has no unwanted artifacts, and phase recovery followed by differentiation indeed restores the modulating signal.

$$s(t) = A\cos\left(\omega_c t + m_{\mathrm{FM}}\int_{-\infty}^{t} v(\tau)d\tau\right) \qquad (18.12)$$

For a modulating signal that consists of a single sinusoid, the entire difference between PM and FM is a phase shift and a change in the modulation index; for a more general modulating signal, FM and PM are less compatible. The integral of $v(t) = \sin(\omega_m t)$ is $-\frac{1}{\omega_m}\cos(\omega t)$, and so high-frequency Fourier components of $v(t)$ are much weaker in FM than in PM, a phenomenon known as *de-emphasis*. A PM signal heard on an FM receiver has too much treble and sounds 'tinny', while using a receiver designed for PM to intercept an FM signal produces a 'bassy' sound. FM may be generated using a PM transmitter, if *pre-emphasis* is performed on the modulating audio in order to compensate for the later loss of high frequencies.

The PM/FM transmitter is very similar to the AM one, with the exception that the amplified microphone voltage is used to vary the phase rather than the amplitude of the carrier; but how do we make an analog FM receiver? One way is to use frequency-to-voltage conversion to convert the received FM signal into an AM one. An *FM discriminator* is a circuit with gain that varies linearly with frequency, and can thus be used for the frequency-to-voltage conversion.

The digital FM receiver can derive the instantaneous frequency from the instantaneous phase through differentiation. Were we to drastically oversample we could get by with the simple difference, since

$$\phi(t + \delta t) - \phi(t) \approx \frac{d}{dt}\phi(t)\,\delta t$$

as long as the phase behaves approximately linearly in the time interval δt.

For more rapidly varying phases we must use a true differentiation filter (see Section 7.3).

The instantaneous phase signal is bounded in the interval between $-\pi$ and π (or perhaps $[0 \ldots 2\pi]$) and has discontinuities when it crosses these boundaries. These phase jumps have no physical meaning, they are simply artifacts of the nonuniqueness of inverse trigonometric functions. Differentiation of such discontinuities would give rise to tremendous unphysical spikes in the frequency demodulation. Hence we must first *unwrap* the phase before differentiation. This can be done by setting a phase change threshold, and adding $\pm 2\pi$ whenever the phase jumps by more than this threshold. For oversampled signals this threshold can be relatively small, but close to Nyquist it must be carefully chosen in order to avoid unwrapping legitimate changes in phase.

The unwrapped phase signal resulting from the above operation is considerably smoother than the original phase. If, however, the signal has not been correctly mixed down to zero frequency, the residual carrier frequency causes linear phase increase or decrease, which will eventually cause the phase to overflow. In sophisticated implementations one models this phase change by linear regression and corrects the mixer frequency accordingly. A simpler technique to avoid phase overflow is not to correct the phase at all, only the *phase difference*. Differentiation of the phase difference signal gives the frequency difference, and the actual frequency is found by adding the frequency difference to the previous frequency. This frequency is in the vicinity of the residual carrier frequency, and thus never overflows.

An alternative method of phase differentiation is called the dual differentiator method. It exploits the fact that the specific differentiation to be performed is

$$\frac{d}{dt}\Phi(t) = \frac{d}{dt}\tan^{-1}\left(\frac{y(t)}{x(t)}\right) = \frac{\dot{y}x - \dot{x}y}{A^2(t)} \tag{18.13}$$

where $A(t) = \sqrt{x^2(t) + y^2(t)}$ is the amplitude detection. If we are interested in the frequency alone, we can *limit* the input signal (giving a constant amplitude) and then the above is directly proportional to the instantaneous frequency. If the amplitude is to be calculated in any event, it should be done first, and then a division carried out.

We turn now to the spectrum of PM and FM signals, wondering whether there are sidebands here as there were in the AM case. Even if there are sidebands, they must be much different than those we saw for AM. For example, assume the power of the modulating signal increases. For AM the carrier remains unchanged and the sideband energy increases; for PM/FM

the total power must remain unchanged (otherwise there would be unwanted
AM!) and thus an increase in sideband power must result in a decrease in
carrier power. At some point the carrier will even have to entirely disappear!
Using the same type of algebra that led to equation (18.7) we find

$$
\begin{aligned}
s(t) &= A\cos\Big(\omega_c t + m\sin(\omega_m t)\Big) \\
&= A\Big(\cos(\omega t)\cos\Big(m\sin(\omega_m t)\Big) - \sin(\omega t)\sin\Big(m\sin(\omega_m t)\Big)\Big)
\end{aligned}
$$

where m means m_{PM} or m_{FM}. Now $\cos(m\sin(\omega_m t))$ and $\sin(m_{\mathrm{FM}}\sin(\omega_m t))$
are easily seen to be periodic signals with frequency ω_m. It turns out that
these periodic functions have expansions in terms of the *Bessel functions*
$J_0, J_1, \ldots$ (see A.1).

$$
\begin{aligned}
\sin(m\sin(\omega t)) &= 2\Big(J_1(m)\sin(\omega t) + J_3\sin(3\omega t) + \cdots\Big) \\
\cos(m\sin(\omega t)) &= J_0(m) + 2\Big(J_2(m)\sin(2\omega t) + J_4(m)\sin(4\omega t) + \cdots\Big)
\end{aligned}
$$

Plugging these in, and using the trigonometric product identities (A.32)
multiple times, we obtain the desired spectral representation.

$$
\begin{aligned}
s(t) = A\Big(\quad & J_0(m)\cos(\omega_c t) \qquad\qquad\qquad\qquad\qquad\qquad (18.14) \\
+ \quad & J_1(m)\Big(\cos\big((\omega_c + \omega_m)t\big) - \cos\big((\omega_c - \omega_m)t\big)\Big) \\
+ \quad & J_2(m)\Big(\cos\big((\omega_c + 2\omega_m)t\big) + \cos\big((\omega_c - 2\omega_m)t\big)\Big) \\
+ \quad & J_3(m)\Big(\cos\big((\omega_c + 3\omega_m)t\big) - \cos\big((\omega_c - 3\omega_m)t\big)\Big)\,\Big) \\
+ \quad & \cdots
\end{aligned}
$$

This is quite different from equation (18.7) with its sidebands at $\omega_c \pm +\omega_m$!
Here we have an infinite number of sidebands at $\omega_c \pm k\omega_m$ with amplitudes
varying according to the Bessel functions. The carrier amplitude is proportional to J_0 and thus starts at unity for zero modulation index and decreases
as m increases. All the sidebands start at zero amplitude for $m = 0$ and at
first increase, but later oscillate. Of course, for constant modulation index
m, the amplitude of the sidelobes tends to decrease with distance from the
carrier. As a rough estimate we can say that $J_n(m)$ is close to zero for $n > m$,
so that the number of significant sidebands is $2n$ and the bandwidth is given
by BW $\approx 2n\omega_m$.

EXERCISES

18.4.1 Prove that equation (18.9) has extrema other than those of the carrier by differentiating and setting equal to zero.

18.4.2 Diagram an analog transmitter and receiver for FM.

18.4.3 Find the spectral representation of the PM signal.

$$s(t) = A \cos \left(\omega_c t + m \cos(\omega_m t) \right)$$

18.4.4 AM reception suffers from noise more than FM does, for the simple reason that additive wideband noise directly changes the received signal's amplitude, while most noise does not masquerade as frequency or phase changes. This is the reason FM is commonly used for high quality music broadcasting. Explain why FM receivers use a hard-limiter before the demodulator.

18.4.5 Communications-grade FM receivers come equipped with a *squelch* circuit that completely silences the receiver when no FM signal is present. Explain how this works and why such a circuit is not used in AM receivers.

18.4.6 What happens when two AM signals transmit too close together in frequency? What happens with FM?

18.5 Data Communications

Communications systems tend to be extremely complex. For example, a phone call starts with someone picking up the receiver (i.e., the telephone goes *off-hook*). This causes current to flow thus informing the local Central Office (CO) that service has been requested. The CO responds by sending a signal composed of two sinusoids of 350 and 440 Hz called *dial tone* to the customer and starting up a rotary dialing pulse decoder and a DTMF receiver. The customer hears the dial tone and starts dialing. As soon as the CO notes activity it stops sending dial tone and starts decoding and collecting the digits. At some point the CO realizes that the entire number has been dialed and decides whether the call is local, long distance, overseas, etc. If the called party belongs to the same CO the appropriate line must be found, and whether it is presently in use must be checked. If it *is* in use a busy signal (480+620 Hz one half-second on and one half-second off) is returned to the calling party; if not, an AC *ring voltage* is placed on it, and a *ring-back* signal (440+480 Hz one second on and three seconds off) returned until someone answers by going off-hook. However, if the phone call must be routed to another CO, complex optimization algorithms must be called up to quickly determine the least expensive available way to connect to the desired party. The calling CO then informs the called CO of the caller and callee phone numbers along a digital link using multifrequency tones or digital messages. The called CO then checks the called number's line and either returns an indication of busy, or places ring voltage and returns an indication of ringing. Of course we haven't mentioned caller identification, call waiting, billing, voicemail, etc.

If making a telephone call is *that* complex behind the scenes, just think of what happens when you surf the Internet with a web browser! In order to facilitate comprehension of such complex systems, they are traditionally divided into layers. The popular **O**pen **S**ystems **I**nterconnection (OSI) reference model delineates seven distinct layers for the most general data communications system, namely physical, datalink, network, transport, session, presentation, and application layers. At each layer the source can be considered to be communicating with the same layer of the destination via a *protocol* defined for that layer. In reality information is not transferred directly between higher layers; rather it is passed down to the physical layer, sent over the communications channel, and then passed up through the layers. Hence, each layer requires all the layers under it in order to function, directly accessing functions of the layer immediately beneath it and providing functionality to the layer immediately above it. The physical layer contains specifications of the cables and connectors to be employed, the maximum allowed voltage levels, etc. It also defines the 'line code' (i.e., the modulation type that determines how the digital information influences the line voltage). The datalink layer specifications are responsible for detecting and correcting errors in the data over a link (between one node in a network and the next), while the network layer routes information from the point of origin through the network to the destination, and ensures that the network does not become congested. The transport layer guarantees reliable source-to-destination transport through the network, while the session layer is where an entire dialog between the two sides is established (e.g., a user logs on to a computer) and maintained. The presentation layer translates data formats and provides encryption-decryption services and, finally, the application (e.g., email, file transfer, remote log-on, etc.) is the most abstract layer, providing users with a comprehensible method of communicating. Most of these layers do not require DSP. Their main function is packaging information into various-size 'chunks', tacking headers onto them, and figuring out where to send them.

EXERCISES

18.5.1 Assume that someone uses a dial-up modem to connect to the World Wide Web. Try to identify as many communications protocols as you can, and at what OSI layer(s) they operate. (Hint: The modem has connection, physical layer transfer and perhaps error correction facilities. The application on the user's computer uses a serial protocol to communicate with the service provider. The Internet is based on TCP/IP. The web sits above the Internet.)

18.5.2 Do we really need to divide communications systems into layers? If not, what are there advantages and disadvantages?

18.6   Information Theory

Digital communications involves reliably sending information-bearing signals from a *source* through a *channel* to a *destination*. Were there no channel this would be a trivial task; the problem is that the channel distorts and adds noise to the signal, adversely affecting the reliability. Basic physics dictates the (usually negative) effects of the channel, and signal processing knowledge helps design signals that get through these channels with minimal damage.

As anyone who has ever been on the Internet knows, we always want to send the information from source to destination as quickly as possible. In order to measure the speed of the information transfer we need to know how much information is in an arbitrary message. This is the job of information theory.

The basic tenet of information theory is that information content can always be quantified. No matter what form the information takes, text, speech, images, or even thoughts, we can express the amount of information in a unique way. We will always measure information content in bits. The rate of information transfer is thus measured in bits per second.

Suppose that I am thinking of a number x between 0 and 255 (for definiteness, $x = 137$); how much information is transferred when I tell you that number? You probably know the answer to that question, exactly eight bits of information. Formally, the reason that a number between 0 and 255 contains eight bits of information is that in general eight individual yes-no questions must be asked in order to find the number. An optimal sequence of questions is as follows:

Q_1: Is the $x_1 = x$ greater than or equal to 128? A_1: Yes ($x_1 = 137 \geq 128$).
Q_2: Is $x_2 = x_1 - 128$ greater than or equal to 64? A_2 No ($x_2 = 9 < 64$).
Q_3: Is $x_3 = x_2$ greater than or equal to 32? A_3: No ($x_3 = 9 < 32$).
Q_4: Is $x_4 = x_3$ greater than or equal to 16? A_4: No ($x_4 = 9 < 16$).
Q_5: Is $x_5 = x_4$ greater than or equal to 8? A_5: Yes ($x_5 = 9 \geq 8$).
Q_6: Is $x_6 = x_5 - 8$ greater than or equal to 4? A_6: No ($x_6 = 1 < 4$).
Q_7: Is $x_7 = x_6$ greater than or equal to 2? A_7: No ($x_7 = 1 < 2$).
Q_8: Is $x_8 = x_7$ equal to 1? A_8: Yes ($x_7 = 1$).

Only the number 137 will give this particular sequence of yes-no answers, and interpreting *yes* answers as 1 and *no* answers as 0 produces the binary representation of x from MSB to LSB. Similarly we can determine the number of bits of information in arbitrary messages by constructing a set of yes-no questions that uniquely determines that message.

Let's assume a source wishes to convey to the destination a message consisting of an integer between 0 and 255. The transmitter needn't wait for the receiver to ask the questions, since the questioning tactic is known. All the transmitter needs to do is to transmit the answers A_1 through A_8.

Signals that carry information appear to be random to some extent. This is because information is only conveyed by surprising its receiver. Constant signals, constant amplitude and frequency sinusoids or square waves, convey no information, since one can predict exactly what the signal's value will be at any time. Yet consider a signal that can take only two values, say $s = 0$ or $s = 1$, that can change in value every T seconds, but remains constant between kT and $(k+1)T$. Such a signal is often called a **Non Return to Zero** (NRZ) signal, for reasons that will become clear shortly. If the signal jumps in an apparently randomly fashion between its two values, one can interpret its behavior as a sequence of bits, from which text, sound, or images may be derived. If one bit is inferred every T seconds, the information transfer rate is $\frac{1}{T}$ bits per second.

According to this point of view, the more random a signal is, the higher its information transfer rate. Longer T implies a lower information transfer rate since the signal is predictable for longer times. More complex predictable behavior also reduces the information transfer rate. For example, a **Return to Zero** (RZ) signal (see Figure 18.8.B) is similar to the NRZ signal described above, but always returns to $s = 0$ for odd k (we count from $k = 0$). Since an unpredictable signal value only appears every $2T$ seconds, the information is transferred at half the rate of the NRZ signal. Predictability may be even more subtle. For example, the *Manchester* signal used in Ethernet LANs

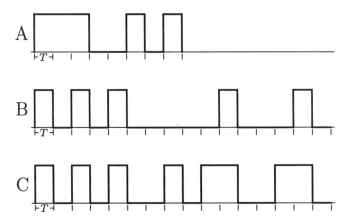

Figure 18.8: Comparison of (A) NRZ, (B) RZ, and (C) Manchester signals. The message is 11100101 and our channel bandwidth requires transitions to be spaced T seconds apart. Using NRZ this message requires $8T$ seconds. RZ and Manchester both require $16T$ seconds to transmit the same message.

(see Figure 18.8.C) encodes a binary one by having $s = 1$ for even k and $s = 0$ for the subsequent $k + 1$ interval; a zero is encoded by $s = 0$ followed by $s = 1$. Once again the information transfer rate is only half that of the NRZ signal, although the lack of randomness is less obvious. Whereas the NRZ signal has no correlation between signal values spaced T seconds apart, the Manchester signal never allows odd k intervals to have the same value as the previous even k interval.

The moral is that any correlation between signal values at different times reduces the amount of information carried. An infinite amount of information is carried by a signal with no correlation between different times (i.e., by white noise). Of course a true white noise signal, which has frequency components up to infinite frequency, cannot pass unaltered through a channel with finite bandwidth. Thus for a finite bandwidth channel, the signal with maximal information content is one whose sole predictability is that caused by the bandwidth constraint. Such a signal has a spectrum that is flat in the allowed pass-band.

We can similarly define the information transfer rate when the signal may take on many values (called symbols), not just zero and one. A signal that can jump randomly every T seconds, but that is a constant $s = 0, 1, 2,$ or 3 in between these jumps, obviously carries 2 bits every T seconds, or $\frac{2}{T}$ bits per second.

What if the different symbols are not equally probable? For example, a signal that takes on 26 values corresponding to a message containing text in English would use the symbol corresponding to 'E' much more frequently than that corresponding to 'Q'. Information theory tells us that a consistent measure of information of a single symbol s is the *entropy*

$$H(s) = - \left\langle \log_2 p(s) \right\rangle = - \sum_s p(s) \log_2 p(s) \qquad (18.15)$$

where s represents the possible signal values, and the triangular brackets stand for the expected value (see Appendix A.13).

To understand this result let's return to the simple case of a sending a message that consists of a number x between 0 and 255. Before transmission commences, the receiver has no information as to the value of x other than the fact that it is between 0 and 255. Thus the receiver assigns an equal probability of $\frac{1}{256}$ to each of the integers $0 \ldots 255$. A priori the transmitter may send a first symbol of 0 or 1 with probability $\frac{1}{2}$. In the previous example it would send a 1; immediately the receiver updates its probability estimates, now $0 \ldots 127$ have zero probability and $128 \ldots 255$ have probability 1 in 128. The receiver's uncertainty has been reduced by a factor of two, corresponding to a single bit of information. Now the second answer (in our example a zero) is sent. Since the second answer is independent of the first, the probability of both answers is the product of the individual probabilities $\frac{1}{2} \cdot \frac{1}{2} = \frac{1}{4}$. Similarly, the probability of any particular sequence of three answers is $\frac{1}{2} \cdot \frac{1}{2} \cdot \frac{1}{2} = \frac{1}{8}$. In general it is clear that after each subsequent answer is received the probability of the message is halved, as is the uncertainty of the receiver. After eight answers have been received the probability of the message has been reduced to $\frac{1}{256}$ and all uncertainty removed.

Now we prefer to think of information as being *added* after each answer has been received, although the probabilities were *multiplied*. The only way of making an arbitrary multiplication into an addition is to employ a logarithmic relation, such as (18.15). If we wish each reduction of probability by a factor of one half to correspond to the addition of a single bit, the base of the logarithm must be 2 and a minus sign must be appended (since $-\log_2 \frac{1}{2} = +\log_2 2 = 1$). Thus, for our simple example, each answer A_i contributes

$$I(A_i) = -\log_2 p(A_i) = \log_2 \tfrac{1}{2} = 1$$

bits of information. The information of the sequence of answers is

$$I(x) = \sum_{i=1}^{8} I(A_i) = - \sum_{i=1}^{8} \log_2 p(A_i) = 8$$

bits, as we claimed.

The guessing game with yes-no questions is not restricted to determining numbers; it can be played for other types of messages. For example, in trying to ascertain which person in a group is intended we could progressively ask 'male or female?', 'tall or short?', 'light or dark hair?', etc. until only one person remains. Indeed after a little thought you will become convinced that every well-defined message can be encoded as a series of answers to yes-no questions. The minimal number of such questions needed to unambiguously recover the message intended is defined to be the information content of that message in bits. In communications we are mostly interested in the rate at which information can be transferred from source to destination, specified in bits per second.

EXERCISES

18.6.1 Consider a signal that can take one of two values, $s = 0$ with probability p and $s = 1$ with probability $1 - p$. Plot the entropy of a single value as a function of p. Explain the position and value of the extrema of this graph.

18.6.2 Compute the entropy in bits per character of English text. Use the probabilities from exercise 18.1.1 or collect histograms using some suitably large on-line text to which you have access. Is a byte required to encode each letter?

18.6.3 Use a file compression program to reduce the size of some English text. What is the connection between final file size and entropy?

18.6.4 Repeat the previous two exercises for other languages that use the same alphabet (French, Spanish, Italian, German, etc.). Can these probabilities be used to discriminate between different languages?

18.6.5 What are the most prevalent *pairs* of letters in English? How can letter pairs be used to aid text compression? To aid in language identification?

18.6.6 Using Table 18.1 compute the time durations of Morse code letters and sort them in increasing order. Did Morse maximize the information transfer rate?

18.7   Communications Theory

We have seen that all information can be converted into bits (i.e., into digital information). Thus all communications, including those of an analog nature, *can* be performed digitally. That does not imply that all communications *should* be performed digitally, since perhaps the conversion of analog

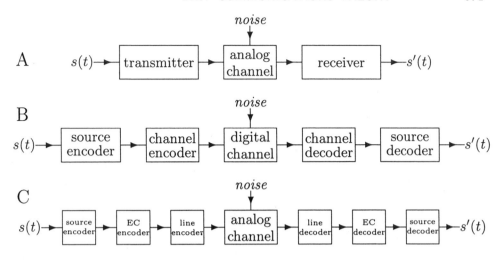

Figure 18.9: The conversion of an analog communications system into digital one. In (A) we see the original analog system. In (B) we have performed the separation into source and channel coding guaranteed by Shannon's theorem. In (C) we add line coding in order to utilize an analog channel (EC stands for error correction).

information into digital data, its transmission, reception, and reconversion into analog data would lead to a loss in quality or efficiency. In previous sections we learned how to transmit analog signals using AM, FM, and other forms of analog modulations. With such robust analog techniques at our disposal it does not seem likely that the conversion to digital communications would be useful.

In the late 1940s, Claude Shannon laid the mathematical framework for digital communications. Logically the first result, already deep and perhaps surprising, is that digital communications can always be used without sacrificing quality or efficiency. More precisely, Shannon showed that one could separate any communications problem, including an analog one, into two independent parts, *without sacrificing quality.* He called these parts *source coding* and *channel coding*. Source encoding refers to the process of efficiently converting the source message into digital data (i.e., representing the message as a bit stream with minimal number of bits). Channel encoding means the method of selecting signals to be sent over the communications channel. The inverse operations are channel decoding and source decoding, which convert the received signals back into digital data and convert the digital data back into the original message, respectively. By using this model, rather than directly transmitting an analog signal over an analog channel (Figure 18.9.A), we can efficiently convert an analog signal into a digital one, send this essentially without error over a digital channel, and then recover the original signal (Figure 18.9.B).

It should be stressed that Shannon's separation of communications into two parts is fundamentally different from the OSI separation of communications into seven layers. There is no theory stating that the division of the OSI model does not impair the communications system; the layers are only separated in order to facilitate human comprehension. In a similar fashion the channel coding of Shannon's theorem is often further divided into two separate parts, *error correction coding* and *line coding*. An error correction code converts digital data into protected digital data, which can be transmitted over a digital channel with less fear of corruption due to noise. Of course all real transmission channels are analog, and so digital channels are actually an abstraction. The conversion of the (protected) digital signal into an analog one suitable for the physical transmission line is called *line coding*. The entire process is thus that of Figure 18.9.C. The division of the channel code into error correction code and line code is performed solely as an aid to the designers (it's hard to find one person expert in both fields!) but is not guaranteed to be conserve optimality. Indeed one can increase performance by combining the two (see Section 18.19).

Shannon's theorem, although in many ways satisfying, has not yet convinced us to convert over to digital communications systems. All we have seen is that we have nothing to lose by converting; we have yet to see that we have something to gain. Can digital systems actually *increase* bandwidth efficiency, *improve* the quality, *reduce* the cost, or provide any other measurable advantage as compared with analog communications? Shannon affirmatively answered these questions in a series of theorems about source and channel coding. Source coding theorems are beyond the scope of our present treatment, yet we can readily understand how proper source and channel coding can help us attain some of these goals.

For maximal efficiency source coding should produce a bit stream with no more bits than absolutely needed. We know that the minimal number of bits required to encode a message is the information (entropy), and thus the ideal source coder produces no more bits than entropy requires. For example, speech can be source encoded into 8 Kb/s or less (see Chapter 19) and there are modems (line codes) of over 32 Kb/s; hence using digital techniques one can transfer four conversations over a single telephone line. Thus proper source encoding can increase bandwidth efficiency.

Digital compact disks have replaced analog long playing records mainly due to their superior audio quality. This quality is obtained because of the use of digital error correcting channel codes that guarantee accurate reproduction of the original sound. Analog music signals that have become contaminated with noise cannot generally be corrected, and the noise manifests itself as various hisses and pops. Thus proper channel encoding can indeed increase signal quality.

While we will not delve into all of Shannon's theorems, there is one that will be essential for us. Before Shannon, engineers knew that noise and interference on digital channels cause errors in the reconstructed bit stream; and they thought that there was only one way of overcoming this problem, by increasing the power of the communications signal. The principle in which all designers believed was that no matter what the noise or interference is like, if we transmit a strong enough signal it will wipe them out. Then there was the separate issue of bandwidth; the higher the bandwidth the more data one could reliably transfer in a given time. Thus common wisdom stated that the probability of error for digital communications was a function of the SNR, while the speed was determined by the bandwidth. Shannon's capacity theorem completely changed this picture; by explaining that the SNR and bandwidth establish a maximum transmission rate, under which information could be transferred with arbitrarily low error rate. This result will be the subject of the next section.

EXERCISES

18.7.1 Shannon introduced entropy (defined in the previous section) in connection with source coding. The ultimate purpose of source coding is to produce no more bits than required by the entropy content of the source. When is simple A/D conversion the optimal source coding for an analog signal? What should one do when this is not the case?

18.7.2 In order to achieve the maximum efficiency predicted by Shannon, source coding is often required even for digital data. Explain and give several examples. (Hint: Data compression, fax.)

18.7.3 The Baudot code and ASCII are source codes that convert letters into bits. What are the essential differences between them? Which is more efficient for the transfer of plain text? How efficient is it?

18.7.4 In today's world of industrial espionage and computer hackers sensitive data is not safe unless encrypted. Augment the diagram of Figure 18.9.C to take encryption into account.

18.7.5 We often want to simultaneously send multiple analog signals (for example, all the extensions of an office telephone system) over a single line. This process is called *multiplexing* and its inverse *demultiplexing*. Show how this fits into Figure 18.9.C.

18.8   Channel Capacity

The main challenge in designing the physical layer of a digital communications system is approaching the *channel capacity*. By channel capacity we mean the maximum number of information bits that can be reliably transferred through that channel in a second. For example, the capacity of a modern telephone channel is about 35,000 bits per second (35 Kb/s); it is possible to transfer information at rates of up to 35 kilobits per second without error, but any attempt at perfectly transferring more data than that will surely fail.

Why is there a maximal channel capacity? Why can't we push data as fast as we wish through a digital link? One might perhaps believe that the faster data is transmitted, the more errors will be made by the receiver; instead we will show that data can be received essentially without error up to a certain rate, but thereafter errors invariably ensue. The maximal rate derives from two factors, noise and finite bandwidth. Were there to be no noise, or were the channel to have unlimited bandwidth, there would be unlimited capacity as well. Only when there are both noise *and* bandwidth constraints is the capacity finite. Let us see why this is the case.

Assume there is absolutely no noise and that the channel can support some range of signal amplitudes. Were we to transmit a constant signal of some allowable amplitude into a nonattenuating noiseless channel, it would emerge at the receiver with precisely the same amplitude. An ideal receiver would be able measure this amplitude with arbitrary accuracy. Even if the channel does introduce attenuation, we can precisely compensate for it by a constant gain. There is also no fundamental physical reason that this measurement cannot be performed essentially instantaneously. Accordingly we can achieve errorless recovery of an infinite amount of information per second. For example, let's assume that the allowable signal amplitudes are those between 0 and 1 and that we wish to transmit the four bits 0101. We simply define sixteen values in the permissable range of amplitudes, and map the sixteen possible combinations of four bits onto them. The simplest mapping method considers this string of bits as a value between 0 and 1, namely the binary fraction 0.0101_2. Since this amplitude may be precisely measured by the receiver in one second, we can transfer at least four bits per second through the channel. Now let's try to transmit eight bits (e.g., 01101001). We now consider this as the binary fraction 0.01101001_2 and transmit a constant signal of this amplitude. Once again this can be exactly retrieved in a second and thus the channel capacity is above eight bits per second. In similar fashion we could take the complete works of Shakespeare,

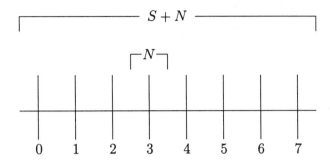

Figure 18.10: The effect of noise on amplitude resolution. The minimum possible spacing between quantization levels is the noise amplitude N, and the total spread of possible signal values is the peak-to-peak signal amplitude S plus the noise N. The number of levels is thus the ratio between the signal-plus-noise and the noise, and the number of bits is the base-two logarithm of this ratio.

encode the characters as bytes, and represent the entire text as a single (rather lengthy) number. Normalizing this number to the interval between 0 and 1 we could, in principle, send the entire text as a single voltage in one second through a noiseless channel. This demonstrates that the information-carrying capacity of a noiseless channel is infinite.

What happens when there *is* noise? The precision to which the amplitude can be reliably measured at the receiver is now limited by the noise. We can't place quantization levels closer than the noise amplitude, since the observed signals would not be reliably distinguishable. As is clarified by Figure 18.10 the noise limits the number of bits to the base-two logarithm of the signal-plus-noise-to-noise ratio, SNNR = SNR + 1.

Of course, even if the noise limits us to sending b bits at a time, we can always transmit more bits by using a time varying signal. We first send b bits, and afterwards another b bits, then yet another b, and so on. Were the channel to be of unlimited bandwidth we could abruptly change the signal amplitude as rapidly as we wish. The transmitted waveform would be piecewise constant with sharp jumps at the transitions. The spectral content of such jump discontinuities extends to infinite frequency, but since our channel has infinite bandwidth the waveform is received unaltered at the receiver, and once again there is no fundamental limitation that hinders our receiver from recovering all the information. So even in the presence of noise, with no bandwidth limitation the channel capacity is effectively infinite.

Signals that fluctuate rapidly cannot traverse a channel with finite bandwidth without suffering the consequences. The amount of time a signal must remain relatively constant is inversely proportional to the channel bandwidth, and so when the bandwidth is BW our piecewise constant signal cannot vary faster than BW times per second. Were we to transfer an NRZ signal through a noisy finite-bandwidth channel we would transfer BW bits per second. By using the maximum number of levels the noise allows, we find that we can send $\mathrm{BW}\log_2\mathrm{SNNR}$ bits per second. Slightly tightening up our arguments (see the exercises at the end of this section) leads us to Shannon's celebrated channel capacity theorem.

Theorem: The Channel Capacity Theorem
Given a transmission channel bandlimited to BW by an ideal band-pass filter, and with signal-to-noise ratio SNR due to additive white noise:
- there is a way of transmitting digital information through this channel at a rate up to

$$C = \mathrm{BW}\log_2(\mathrm{SNR} + 1) \qquad (18.16)$$

bits per second, which allows the receiver to recover the information with negligible error;
- at any transmission rate above C bits per second rate no transmission method can be devised that will eliminate all errors;
- the signal that attains the maximum information transfer rate is indistinguishable from white noise filtered by the channel band-pass filter. ∎

As an example of the use of the capacity theorem, consider a telephone line. The SNR is about 30 dB and the bandwidth approximately 3.5 KHz. Since $\mathrm{SNR} \gg 1$ we can approximate

$$C = \mathrm{BW}\log_2(\mathrm{SNR} + 1) \approx \mathrm{BW}\log_2\mathrm{SNR} = \mathrm{BW}\frac{\mathrm{SNR}_{dB}}{10\log_{10}2} \approx \mathrm{BW}\frac{\mathrm{SNR}_{dB}}{3}$$

and so C is about 35 Kb/s.

What the channel theorem tells us is that under about 35 Kb/s there *is* some combination of modulation and error correcting techniques that can transfer information essentially error-free over telephone lines. We will see later that V.34 modems presently attain 33.6 Kb/s, quite close to the theoretical limit. There will occasionally be errors even with the best modem, but these are caused by deviations of the channel from the conditions of the theorem, for example, by short non-white noise spikes. The reader who presently uses 56 Kb/s modems or perhaps DSL modems that transmit over telephone lines at rates of over 1 Mb/s can rest assured these modems exploit more bandwidth than 3.5 KHz.

The last part of the capacity theorem tells us that a signal that optimally fills the channel has no structure other than that imposed by the channel. This condition derives from the inverse relation between predictability and information. Recall from Section 5.2 that white noise is completely unpredictable. Any deviation of the signal from whiteness would imply some predictability, and thus a reduction in information capacity. Were the signal to be of slightly narrower bandwidth, this would mean that it obeys the difference equation of a band-pass filter that filters it to this shape, an algebraic connection between sample values that needlessly constrains its freedom to carry information.

The channel capacity theorem as expressed above is limited by two conditions, namely that the bandwidth is filtered by an *ideal* band-pass filter, and that the noise is completely *white*. However, the extension to arbitrary channels with arbitrary stationary noise is (at least in principle) quite simple. Zoom in on some very small region of the channel's spectrum; for a small enough region the attenuation as a function of frequency will be approximately constant and likewise the noise spectrum will be approximately flat. Hence for this small spectral interval the channel capacity theorem holds and we can compute the number of bits per second that could be transferred using only this part of the total spectrum. Identical considerations lead us to conclude that we can find the capacities of all other small spectral intervals. In principle we could operate independent modems at each of these spectral regions, dividing the original stream of bits to be transmitted between the different modems. Hence we can add the information rates predicted by the capacity theorem for all the regions to reach an approximate prediction for the entire spectrum. Let's call the bandwidth of each spectral interval δf, and the signal-to-noise ratio in the vicinity of frequency f we shall denote $\mathrm{SNR}(f)$. Then

$$C = \sum_f \log_2(\mathrm{SNR}(f) + 1) \; \delta f$$

and for this approximation to become exact we need only make the regions infinitesimally small and integrate instead of adding.

$$C = \int \log_2(\mathrm{SNR}(f) + 1) \; df \tag{18.17}$$

We see that for the general case the channel capacity depends solely on the frequency-dependent signal-to-noise ratio.

From the arguments that lead up to the capacity theorem it is obvious that the SNR mentioned in the theorem is to be measured at the receiver, where the decisions must be made. It is not enough to specify the transmitted power at the frequency of interest $P(f)$ (measured in watts per Hz), since for each small spectral region it is this transmitted power times the line attenuation $A(t)$ that must be compared to the noise power $N(t)$ (also in watts per Hz) at that frequency. In other words, the SNR is $P(f)A(t)/N(f)$, and the total information rate to be given by the following integral.

$$C = \int \log_2 \left(\frac{P(f)A(t)}{N(f)} + 1 \right) \, df \qquad (18.18)$$

Unfortunately, equation (18.18) is not directly useful for finding the maximal information capacity for the common case where we are given the line attenuation $A(t)$, the noise power distribution $N(f)$ and the total transmitted power P.

$$P = \int P(f) \, df \qquad (18.19)$$

In order to find the maximal capacity we have to know the optimal transmitter power distribution $P(f)$. Should we simply take the entire power at the transmitter's disposal and spread it equally across the entire spectrum? Or can we maximize the information rate of an arbitrary channel by transmitting more power where the attenuation and noise are greater? A little thought leads us to the conclusion that the relevant quantity is the noise-to-attenuation ratio $N(f)/A(f)$. In regions where this ratio is too high we shouldn't bother wasting transmitted power since the receiver SNR will end up being low anyway and the contribution to the capacity minimal. We should start spending power where the N/A ratio is lower, and expend the greatest amount of power where the ratio is lowest and thus the received SNR highest.

In other words, we should distribute the power according to

$$P(f) = \begin{cases} \Theta - \frac{N(f)}{A(f)} & \frac{N(f)}{A(f)} < \Theta \\ 0 & \frac{N(f)}{A(f)} > \Theta \end{cases} \qquad (18.20)$$

where the value of Θ is determined by the requirement (18.19) that the total Power should equal P. Gallager called this the 'water pouring criterion'. To understand this name, picture the attenuation to noise distribution ratio as an irregularly shaped bowl, and the total amount of power to be transmitted as the amount of water in a pitcher (Figure 18.11). Maximizing signal capacity is analogous to pouring the water from the pitcher into the bowl. Where the bowl's bottom is too high no water remains, where the bowl is low the height of water is maximal.

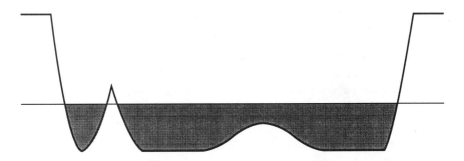

Figure 18.11: The water pouring criterion states that the information rate is maximized when the amount of power available to be transmitted is distributed in a channel in the same way as water fills an irregularly shaped bowl.

With the water pouring criterion the generalized capacity theorem is complete. Given the total power and the attenuation-to-noise ratio, we 'pour water' using equation (18.20) to find the power distribution of the signal with the highest information transfer rate. We can then find the capacity using the capacity integral (18.18). Modern modems exploit this generalized capacity theorem in the following way. During an initialization phase they probe the channel, measuring the attenuation-to-noise ratio as a function of frequency. One way of doing this is to transmit a set of equal amplitude, equally spaced carriers and measuring the received SNR for each. This information can then be used to tailor the signal parameters so that the power distribution approximates water pouring.

EXERCISES

18.8.1 SNR always refers to the power ratio, not the signal value ratio. Show that assuming the noise is uncorrelated with the signal, the capacity should be proportional to $\frac{1}{2} \log_2 SNR$.

18.8.2 Using the sampling theorem, show that if the bandwidth is W we can transmit $2W$ pulses of information per second. Jump discontinuities will not be passed by a finite bandwidth channel. Why does this not affect the result?

18.8.3 Put the results of the previous examples together and prove Shannon's theorem.

18.8.4 When the channel noise is white its power can be expressed as a *noise power density* N_0 in watts per Hz. Write the information capacity in terms of BW and N_0.

18.8.5 Early calculations based on Shannon's theorem set the maximum rate of information transfer lower than that which is now achieved. The resolution of this paradox is the improvement of SNR and methods to exploit more of the bandwidth. Calculate the channel capacity of a telephone line that passes from 200 Hz to 3400 Hz and has a signal-to-noise ratio of about 20–25 dB. Calculate the capacity for a digital telephone line that passes from 200 Hz to 3800 Hz and encodes using logarithmic PCM (12-13 bits).

18.8.6 The 'maximum reach' of a DSL modem is defined to be the distance over which it can function when the only source of interference is thermal white noise. The attenuation of a twisted pair of telephone wires for frequencies over 250 KHz can be approximated by

$$A(f) = e^{-s\left(\kappa_1 \sqrt{f} + \kappa_2 f\right)L}$$

where L is the cable length in Km. For 24-gauge wire $\kappa_1 = 2.36 \cdot 10^{-3}, \kappa_2 = -0.34 \cdot 10^{-8}$ and for thinner 26-gauge wire $\kappa_1 = 2.98 \cdot 10^{-3}, \kappa_2 = -1.06 \cdot 10^{-8}$. Assume that the transmitter can transmit 13 dBm between 250 KHz and 5 MHz and that the thermal noise power is -140 dBm per Hz. Write a program to determine the optimal transmitter power distribution and the capacity for lengths of 1, 2, 3, 4, and 5 Km.

18.9   Error Correcting Codes

In order to approach the error-free information rate guaranteed by Shannon, modem signals and demodulators have become extremely sophisticated; but we have to face up to the fact that no matter how optimally designed the demodulator, it will still sometimes err. A short burst of noise caused by a passing car, a tone leaking through from another channel, changes in channel frequency characteristics due to rain or wind on a cable, interference from radio transmitters, all of these can cause the demodulator to produce a bit stream that is not identical to that intended. Errors in the reconstructed bit stream can be catastrophic, generating annoying clicks in music, causing transferred programs to malfunction, producing unrecoverable compressed files, and firing missile banks when not intended. In order to reduce the probability of such events, an *error correcting code* (ECC) may be used.

Using the terminology of Section 18.7, an ECC is a method of channel encoding designed to increase reliability. Error correcting codes are independent of the signal processing aspects of the bit transfer (line coding); they are purely mathematical mechanisms that detect whether bits have become corrupted and how to recover the intended information. How can bit errors be detected? Were we to send 00011011 and 01011010 was received instead, how could this possibly be discovered? The strategy is that after optimizing the source coding to use the minimal number of bits possible, the channel coding *adds* new bits in order to be able to detect errors. A parity bit is a simple case of this; to seven data bits we can add an eighth bit that ensures that the number of ones is even. Any single-bit error will be *detected* because there will be an odd number of ones, but we will not know which bit is in error. A simplistic error *correction* scheme could send each bit three times in succession (e.g., send 000000000111111000111111 rather than directly sending the message 00011011). Were any single bit to be incorrectly received (e.g. 000010000111111000111111), we could immediately detect this and correct it. The same is true for most combinations of two bit errors, but if the two errors happen to be the same bit triplet, we would be able to detect the error but not to correctly correct it.

The error detection and correction method we just suggested is able to correct single-bit errors, but requires tripling the information rate. It turns out that we can do much better than that. There is a well-developed, mathematically sophisticated theory of ECCs that we will not be able to fully cover. This and the next two sections are devoted to presentation of concepts of this theory that we will need.

All ECCs work by allowing only certain bit combinations, known as *codewords*. The parity code only permits codewords with an even number of ones; thus only half the possible bitvectors are codewords. The bit tripling ECC works because only two of the eight possible bit triplets are allowed; thus of the 2^{3k} bitvectors of length $3k$, only one out of every eight are codewords.

The second essential concept is that of distance between bitvectors. The most commonly used distance measure is the *Hamming distance* $d(b_1, b_2)$. (the same Hamming as the window). The Hamming distance is defined as the number of positions in which two bitvectors disagree (e.g., $d(0011, 0010) = 1$). For bitvectors of length N, $0 \leq d(b_1, b_2) \leq N$ and $d(b_1, b_2) = 0$ if and only if $b_1 = b_2$.

If we choose codewords such that the minimum Hamming distance d_{min} between any two is M, then the code will be able to detect up to $M-1$ errors. Only if M errors occur will the error go unnoticed. Similarly, a code with minimum Hamming distance M will be able to correct less than $\frac{1}{2}M$ errors. Only if there are enough errors to move the received bitvector closer to another codeword (i.e., half the minimum Hamming distance) will choosing the closest codeword lead to an incorrect result.

How do we protect a message using an error correcting code? One way is to break the bits of information into blocks of length k. We then change this k-dimensional bitvector into a codeword in n-dimensional space, where $n > k$. Such a code is called an n/k rate *block code* (e.g., parity is a 8/7 block code while the bit tripling code is a 3/1 block code). The codewords are sent over the channel and decoded back into the original k bits at the receiver. The processing is similar to performing FFTs on nonoverlapping blocks. Sometimes we need to operate in real-time and can't afford to wait for a block to fill up before processing. In such cases we use a *convolutional code* that is reminiscent of a set of n FIR filters. Each clock period k new bits are shifted into a FIFO buffer that contains previously seen bits, the k oldest bits are shifted out, and then n bit-convolution computations produce n output bits to be sent over the channel. The buffer is called a *shift register* since the elements into and from which bits are shifted are single-bit registers.

We can now formulate the ECC design task. First, we decide whether a block or convolutional code is to be used. Next, the number of errors that must be detected and the number that must be corrected are specified. Finally, we find a code with minimal rate increase factor n/k that detects and corrects as required.

At first glance finding such codes seems easy. Consider a block code with given k. From the requirements we can derive the minimal Hamming distance between codewords, and we need only find a set of 2^k codewords with that minimal distance. We start with some guess for n and if we can't find 2^k codewords (e.g., by random search) that obey the constraint we increase n and try again. For large block lengths k the search may take a long time, but it need be performed only once. We can now randomly map all the possible k-dimensional bitvectors onto the codewords and can start encoding. Since the encoding process is a simple table lookup it is very fast. The problem is with decoding such a code. Once we have received an n-dimensional bitvector we need to compute the Hamming distances to each of the 2^k codewords and then pick the closest. This is a tremendous amount of work even for small k and completely out of the question for larger block lengths.

Accordingly we will tighten up our definition of the ECC design problem. Our task is to find a code with minimal n/k that can be *efficiently decoded*. Randomly chosen codes will always require brute force comparisons. In order to reduce the computational complexity of the decoding we have to add structure to the code. This is done using algebra.

EXERCISES

18.9.1 Consider the bit-tripling code. Assume the channel is such that the probability of an error is p (and thus the probability of a bit being correctly detected is $1 - p$). Show that the average probability of error of the original bit stream is $P_{err} = 3p^2(1 - p) + p^3$. Obviously, $P_{err} = p$ for $p = 0$ and $p = 1$. What is P_{err} for $p = \frac{1}{2}$? Graph P_{err} as a function of p. For $\frac{1}{2} < p < 1$ we see that $P_{err} > p$, that is, our *error correction* method increases the probability of error. Explain.

18.9.2 The bit-tripling code can correct all single-bit errors, and most two-bit errors. Starting with eight information bits, what percentage of the two-bit errors can be corrected? What percentage of three-bit errors can be detected?

18.9.3 A common error detection method is the *checksum*. A checksum-byte is generated by adding up all the bytes of the message modulo 256. This sum is then appended to the message and checked upon reception. How many incorrectly received bytes can a 'checkbyte' detect? How can this be improved?

18.10 Block Codes

About a year after Shannon's publication of the importance of channel codes, Hamming actually came up with an efficiently decodable ECC. To demonstrate the principle, let's divide the information to be encoded into four-bit blocks $d_3 d_2 d_1 d_0$. Hamming suggested adding three additional bits in order to form a 7/4 code. A code like this that contains the original k information bits unchanged and simply adds $n - k$ checkbits is called a *systematic code*. In the communications profession it is canonical to send the data first, from least to most significant bits and the checkbits afterward, thus the seven-dimensional codewords to be sent over the channel are the vectors $(a_0 a_1 a_2 a_3 a_4 a_5 a_6) = (d_0 d_1 d_2 d_3 c_0 c_1 c_2)$. The checkbits are computed as linear combinations of the information bits

$$
\begin{aligned}
c_0 &= d_0 + d_1 + d_3 \\
c_1 &= d_1 + d_2 + d_3 \\
c_2 &= d_0 + d_2 + d_3
\end{aligned}
\tag{18.21}
$$

where the addition is performed modulo 2 (i.e., using xor). If information bit d_0 is received incorrectly then checkbits c_0 and c_1 will not check out. Similarly, an incorrectly received d_1 causes problems with c_0 and c_2, a flipped

d_2 means c_1 and c_2 will not sum correctly, and finally a mistaken d_3 causes all three checkbits to come out wrong. What if a checkbit is incorrectly received? Then, and only then, a single c_i will not check out. If no checkbits are incorrect the bitvector has been correctly received (unless a few errors happened at once).

The Hamming code is a *linear code*; it doesn't matter if you sum (xor) two messages and then encode them or encode the two messages and then sum them. It is thus not surprising that the relationship between the k-dimensional information bitvector $\underline{d}$ and the n-dimensional codeword $\underline{a}$ can be expressed in a more compact fashion using matrices, $\underline{a} = \underline{\underline{G}}\,\underline{d}$

$$
\begin{pmatrix} a_0 \\ a_1 \\ a_2 \\ a_3 \\ a_4 \\ a_5 \\ a_6 \end{pmatrix} = \begin{pmatrix} d_0 \\ d_1 \\ d_2 \\ d_3 \\ c_0 \\ c_1 \\ c_2 \end{pmatrix} = \begin{pmatrix} 1 & 0 & 0 & 0 \\ 0 & 1 & 0 & 0 \\ 0 & 0 & 1 & 0 \\ 0 & 0 & 0 & 1 \\ 1 & 1 & 0 & 1 \\ 0 & 1 & 1 & 1 \\ 1 & 0 & 1 & 1 \end{pmatrix} \begin{pmatrix} d_0 \\ d_1 \\ d_2 \\ d_3 \end{pmatrix}
$$

where all the operations are to be understood modulo 2. The n-by-k matrix $\underline{\underline{G}}$ is called the *generator matrix* since it generates the codeword from the information bits. All linear ECCs can be generated using a generator matrix; all systematic codes have the k-by-k identity matrix as the top k rows of $\underline{\underline{G}}$.

The Hamming 7/4 code can correct all single-bit errors, and it is optimal since there are no 6/4 codes with this characteristic. Although it does make the job of locating the bit in error simpler than checking all codewords, Hamming found a trick that makes the job easier still. He suggested sending the bits in a different order, $d_3d_2d_1c_1d_0c_2c_0$ and calling them $h_7h_6h_5h_4h_3h_2h_1$. Now h_1, h_2 and h_4 are both received and computed, and the correction process is reduced to simply adding the indices of the incorrect checkbits. For example, if h_1 and h_2 don't check out then $h_{1+2} = h_3$ should be corrected.

Hamming's code avoids searching all the codewords by adding algebraic structure to the code. To see this more clearly let's look at all the codewords (in the original order)

$$
\begin{array}{llll}
0000000 & 1000101 & 0010110 & 1010011 \\
0001011 & 1001110 & 0011101 & 1011000 \\
0100111 & 1100010 & 0110001 & 1110100 \\
0101100 & 1101001 & 0111010 & 1111111
\end{array}
\tag{18.22}
$$

and note the following facts. Zero is a codeword and the sum (modulo 2) of every two codewords is a codeword. Since every bitvector is its own additive inverse under modulo 2 arithmetic we needn't state that additive inverses exist for every codeword. Thus the codewords form a four-dimensional subspace of the seven-dimensional space of all bitvectors. Also note that the circular rotation of a codeword is a codeword as well, such a code being called a cyclic code.

The *weight* of a bitvector is defined as the number of ones it contains. Only the zero vector can have weight 0. For the Hamming 7/4 code the minimal weight of a nonzero codeword is 3 (there are seven such codewords); then there are seven codewords of weight 4, and one codeword of weight 7. If the zero codeword is received with a single-bit error the resulting bitvector has weight 1, while two errors create bitvectors of weight 2. One can systematically place all possible bitvectors into a square array based on the code and weight. The first row contains the codewords, starting with the zero codeword at the left. The first column of the second row contains a bitvector of weight 1, (a bitvector that could have been received instead of the zero codeword were a single-bit error to have taken place). The rest of the row is generated by adding this bitvector to the codeword at the top of the column. The next row is generated the same way, starting with a different bitvector of weight 1. After all weight 1 vectors have been exhausted we continue with vectors of weight 2. For the 7/4 Hamming code this array has eight rows of 16 columns each:

```
0000000  1000101  0100110  ...  1011001  0111010  1111111
1000000  0000101  1100110  ...  0011001  1111010  0111111
0100000  1100101  0000110  ...  1111001  0011010  1011111
0010000  1010101  0110110  ...  1001001  0101010  1101111
0001000  1001101  0101110  ...  1010001  0110010  1110111
0000100  1000001  0100010  ...  1011101  0111110  1111011
0000010  1000111  0100100  ...  1011011  0111000  1111101
0000001  1000100  0100111  ...  1011000  0111011  1111110
```

In error correcting code terminology the rows are called *cosets*, and the leftmost element of each row the *coset leader*. Each coset consists of all the bitvectors that could arise from a particular error (coset leader). You can think of this array as a sort of addition table; an arbitrary element v is obtained by adding (modulo 2) the codeword at the top of its column a to the coset leader at the left of its row e (i.e., $v = a + e$).

The brute force method of decoding is now easy to formulate. When a particular bitvector v is received, one searches for it in the array. If it is in the first row, then we conclude that there were no errors. If it is not, then the codeword at the top of its column is the most probable codeword and the coset leader is the error. This decoding strategy is too computationally complex to actually carry out for large codes, so we add a mechanism for algebraically locating the coset leader. Once the coset leader has been found, subtracting it (which for binary arithmetic is the same as adding) from the received bitvector recovers the most probable original codeword.

In order to efficiently find the coset leader we need to introduce two more algebraic concepts, the *parity check matrix* and the *syndrome*. The codewords form a k-dimensional subspace of n space; from standard linear algebra we know that there must be an $(n-k)$-dimensional subspace of vectors all of which are orthogonal to all the codewords. Therefore there is an $(n-k)$-by-n matrix $\underline{\underline{H}}$ called the *parity check matrix*, such that $\underline{\underline{H}}\,\underline{a} = 0$ for every codeword $\underline{a}$. It's actually easy to find $\underline{\underline{H}}$ from the generator matrix $\underline{\underline{G}}$ since we require $\underline{\underline{H}}\,\underline{\underline{G}}\,\underline{d} = 0$ for all possible information vectors $\underline{d}$, which means the $(n-k)$-by-k matrix $\underline{\underline{H}}\,\underline{\underline{G}}$ must be all zeros. Hence the parity check matrix for the 7/4 Hamming code is

$$\underline{\underline{H}} = \begin{pmatrix} 1 & 1 & 0 & 1 & 1 & 0 & 0 \\ 0 & 1 & 1 & 1 & 0 & 1 & 0 \\ 1 & 0 & 1 & 1 & 0 & 0 & 1 \end{pmatrix}$$

as can be easily verified. The parity check matrix of a systematic n/k code has the $(n-k)$-by-$(n-k)$ identity matrix as its rightmost $n-k$ columns, and the rest is the transpose of the nonidentity part of the generator matrix.

What happens when the parity check matrix operates on an arbitrary n-dimensional bitvector $v = a + e$? By definition the codeword does not contribute, hence $\underline{\underline{H}}\,v = \underline{\underline{H}}\,e$ the right-hand side being a $(n-k)$-dimensional vector called the *syndrome*. The syndrome is thus zero for every codeword, and is a unique indicator of the coset leader. Subtracting (adding) the coset from the bitvector gives the codeword. So our efficient method of decoding a linear code is simply to multiply the incoming bitvector by the parity check matrix to obtain the syndrome, and then adding the coset leader with that syndrome to the incoming bitvector to get the codeword.

By mapping the encoding and decoding of linear ECCs onto operations of linear algebra we have significantly reduced their computational load. But there is an even more sophisticated algebraic approach to ECCs, one that not only helps in encoding and decoding, but in finding, analyzing, and describing codes as well. This time rather than looking at bitstreams as vectors we prefer to think of them as polynomials! If that seems rather abstract, just remember that the general digital signal can be expanded as a sum over shifted unit impulses, which is the same as a polynomial in the time delay operator z^{-1}. The idea here is the same, only we call the dummy variable x rather than z^{-1}, and the powers of x act as place keepers. The k bits of a message $(d_0 d_1 d_2 d_3 \ldots d_{k-1})$ are represented by the polynomial $d(x) = d_0 + d_1 x + d_2 x^2 + d_3 x^3 + \ldots + d_{k-1} x^{k-1}$; were we to take $x = 2$ this would simply be the bits understood as a binary number.

The polynomial representation has several advantages. Bit-by-bit addition (xor) of two messages naturally corresponds to the addition (modulo 2) of the two polynomials (*not* to the addition of the two binary numbers). Shifting the components of a bitvector to the left by r bits corresponds to multiplying the polynomial by x^r. Hence a systematic n/k code that encodes a k-bit message d into an n-bit codeword a by shifting it $(n-k)$ bits to the left and adding $(n-k)$ checkbits c, can be thought of as a transforming of a $(k-1)$-degree polynomial $d(x)$ into a code polynomial $a(x)$ of degree $(n-1)$ by multiplying it by the appropriate power of x and then adding the degree $(n-k-1)$ polynomial $c(x)$.

$$a(x) = x^{n-k} d(x) + c(x) \qquad (18.23)$$

Of course multiplication and division are defined over the polynomials, and these will turn out to be useful operations—operations not defined in the vector representation. In particular, we can define a code as the set of all the polynomials that are multiples of a particular generator polynomial $g(x)$. The encoding operation then consists of finding a $c(x)$ such that $a(x)$ in equation (18.23) is a multiple of $g(x)$.

Becoming proficient in handling polynomials over the binary field takes some practice. For example, twice anything is zero, since anything xored with itself gives zero, and thus everything equals its own negative. In particular, $x + x = 0$ and thus $x^2 + 1 = x^2 + (x + x) + 1 = (x + 1)^2$; alternatively, we could prove this by $x^2 + 1 = (x + 1)(x - 1)$ which is the same since $-1 = 1$. How can we factor $x^4 + 1$? $x^4 + 1 = x^4 - 1 = (x^2 + 1)(x^2 - 1) = (x^2 + 1)^2 = (x + 1)^4$. As a last multiplication example, it's easy to show that $(x^3 + x^2 + 1)(x^2 + x + 1) = x^5 + x + 1$. Division is similar to the usual long division of polynomials, but easier. For example, dividing $x^5 + x + 1$ by $x^2 + x + 1$ is performed as follows. First x^5 divided by x^2 gives x^3, so we multiply $x^3(x^2 + x + 1) = x^5 + x^4 + x^3$. Adding this (which is the same as subtracting) leaves us with $x^4 + x^3 + x + 1$ into which x^2 goes x^2 times. This time we add $x^4 + x^3 + x^2$ and are left with $x^2 + x + 1$, and therefore the answer is $x^3 + x^2 + 1$ as expected.

With this understanding of binary polynomial division we can now describe how $c(x)$ is found, given the generator polynomial $g(x)$ and the message polynomial $d(x)$. We multiply $d(x)$ by x^{n-k} and then divide by $g(x)$, the remainder being taken to be $c(x)$. This works since dividing $x^{n-k}d(x) + c(x)$ by $g(x)$ will now leave a remainder $2c(x) = 0$. For example, the 23/12 Golay code has the generator polynomial $g(x) = 1 + x + x^5 + x^6 + x^7 + x^9 + x^{11}$; in order to use it we must take in 12 bits at a time, building a polynomial of degree 22 with the message bits as coefficients of the 12 highest powers, and zero coefficients for the 11 lowest powers. We then divide this polynomial by $g(x)$ and obtain a remainder of degree 10, which we then place in the positions previously occupied by zeros.

The polynomial representation is especially interesting for cyclic codes due to an algebraic relationship between the polynomials corresponding to codewords related by circular shifts. A circular shift by m bits of the bitvector $(a_0 a_1 \ldots a_{n-1})$ corresponds to the modulo n addition of m to all the powers of x in the corresponding polynomial.

$$a_0 x^{0+m \bmod n} + a_1 x^{1+m \bmod n} + \ldots + a_{n-1} x^{(n-1)+m \bmod n}$$

This in turn is equivalent to multiplication of the polynomial by x^m modulo $x^n + 1$. To see this consider multiplying a polynomial $a(x)$ by x to form $xa(x) = a_0 x + a_1 x^2 + \ldots + a_{n-1} x^n$. In general, this polynomial is of degree n and thus has too many bits to be a codeword, but by direct division we see that $x^n + 1$ goes into it a_{n-1} times with a remainder $\tilde{a}(x) = a_{n-1} + a_0 x + a_1 x^2 + \ldots + a_{n-2} x^{n-1}$. Looking carefully at $\tilde{a}(x)$ we see that it corresponds to the circular shift of the bits of $a(x)$. We can write

$$x\, a(x) = a_{n-1}(x^n + 1) + \tilde{a}(x)$$

and thus say that $xa(x)$ and $\tilde{a}(x)$ are the same modulo $x^n + 1$. Similarly, $(x^2 a(x)) \bmod (x^n + 1)$ corresponds to a circular shift of two bits, and $(x^m a(x)) \bmod (x^n + 1)$ to a circular shift of m bits. Thus cyclic codes have the property that if $a(x)$ corresponds to a codeword, then so does $(x^m a(x)) \bmod (x^n + 1)$.

In 1960, two MIT researchers, Irving Reed and Gustave Solomon, realized that encoding bit by bit is not always the best approach to error detection and correction. Errors often come in bursts, and a burst of eight consecutive bit errors would necessitate an extremely strong ECC that could correct any eight-bit errors; but eight consecutive bit errors are contained in at most two bytes, thus if we could work at the byte level, we would only need a two-byte correcting ECC. The simplest byte-oriented code adds a single checkbyte that equals the bit-by-bit xor of all the data bytes to a block of byte-oriented data. This is equivalent to eight interleaved parity checks and can detect any single byte error and many multibyte ones, but cannot correct any errors. What Reed and Solomon discovered is that by using r checkbytes one can detect any r errors and correct half as many errors. Discussing the theory of Reed-Solomon codes would take us too far astray, but the basic idea is to think of the bytes as polynomials with bits as coefficients

$$B(x) = b_7 x^7 + b_6 x^6 + b_5 x^5 + b_4 x^4 + b_3 x^3 + b_2 x^2 + b_1 x + b_0$$

where the b_n are bits. Dividing the bit stream into n bytes and adding all these bytes together as polynomials

$$c_0 = B_0(x) + B_1(x) + B_2(x) + \ldots + B_{n-1}$$

results in the checkbyte mentioned above. Additional checkbytes can be generated that allow detection of the position of the error.

EXERCISES

18.10.1 A systematic code adds $r = n - k$ checkbits and thus allows for $2^r - 1$ different errors to be corrected. So for all single-bit errors (including errors of the checkbits themselves) to be corrected we require $2^r - 1 \geq n = k + r$. For $k = 4$ we require at least $r = 3$, i.e., a 7/4 code. What does this bound predict for the minimum sizes of systematic codes for $k = 3, 5, 8, 11, 12, 16$?

18.10.2 Find the codewords of the 7/4 Hamming code in Hamming's order. Show that the inverse of every Hamming 7/4 codeword is a code word. Show that the sum (modulo 2) of every two codewords is a codeword. What is the minimum Hamming distance d_{min}? Show that the code is not cyclic but that the code in the original order (as given in the text) is.

18.10.3 In the 7/4 Hamming code the inverse of every codeword (i.e., with ones and zeros interchanged) is a codeword as well. Why?

18.10.4 Why are systematic codes often preferred when the error rate is high?

18.10.5 Find equations for the checkbits of the 9/5 Hamming code. How can the bits be arranged for the sum of checkbit indices to point to the error? How should the bits be arranged for the code to be cyclic?

18.10.6 Show that over the binary field $x^n + 1 = (x + 1)(x^{n-1} + x^{n-2} + \ldots + 1)$.

18.10.7 The 16-bit cyclic redundancy check (CRC) error detection code uses the polynomial $1 + x^5 + x^12 + x^{16}$. Write a routine that appends a CRC word to a block of data, and one that tests a block with appended CRC. How can the encoding and decoding be made computationally efficient? Test these routines by adding errors and verifying that the CRC is incorrect.

18.10.8 To learn more about block codes write four programs. The first `bencode` inputs a file and encodes it using a Hamming code; the second `channel` inputs the output of the first and flips bits with probability p (a parameter); the third `bdecode` decodes the file with errors; the last `compare` compares the original and decoded files and reports on the average error P_{err}. Experiment with binary files and plot the empirical P_{err} as a function of p. Experiment with text files and discover how high p can be for the output to be decipherable.

18.11   Convolutional Codes

The codes we discussed in the previous section are typically used when the bits come in natural blocks, but become somewhat constraining when bits are part of real-time 'bit signals'. The reader who prefers filtering by convolution rather than via the FFT (Section 15.2) will be happy to learn that convolutional codes can be used instead of block codes. Convolutional encoders are actually analogous to several simultaneous convolutions; for each time step we shift a bit (or more generally k bits) into a static bit buffer, and then output n bits that depend on the K bits in this buffer. We have already mentioned that in ECC terminology the static buffer is called a shift register, since each time a new bit arrives the oldest bit in the shift register is discarded, while all the other bits in the registers are shifted over, making room for the new bit.

The precise operation of a convolutional encoder is as follows. First the new bit is pushed into the shift register, then all n convolutions are computed (using modulo two arithmetic), and finally these bits are interleaved into a new bit stream. If n convolutions are computed for each input bit the code's rate is $n/1$. Since this isn't flexible enough, we allow k bits to be shifted into the register before the n outputs bits are computed, and obtain an n/k rate code.

The simplest possible convolutional code consists of a two-bit shift register and no nontrivial arithmetic operations. Each time a new bit is shifted into the shift register the bit left in the register and the new bit are output; In other words, denoting the input bits x_n, the outputs bits at time n are x_n and x_{n-2}. If the input bit signal is 1110101000 the output y_n will be 11111100110011000000. A shift register diagram of the type common in the ECC world and the equivalent DSP flow diagram are depicted in Figure 18.12.

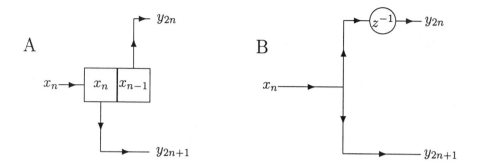

Figure 18.12: A trivial convolutional code. In (A) we see the type of diagram used in error correcting code texts, while in (B) we have the conventional DSP flow diagram equivalent. The output bit signal is formed by interleaving the two outputs into a single bit stream, although this parallel to serial conversion is not explicitly shown.

Why does this code work? As for block codes the idea is that not every combination of bits is a valid output sequence. We cannot say that given a bit the next bit must be the same, since we do not know whether the present bit is already the replica of the previous one. However, an isolated zero or one, as in the bit triplets 010 and 101, can never appear. This fact can be used to detect a single-bit error and even some double-bit errors (e.g., the two middle bits of 00001111). Actually all single-bit errors can be corrected, since with few errors we can easily locate the transitions from 1 to 0 and back and hence deduce the proper phase.

In Figure 18.13 we see a somewhat more complex convolutional code. This code is universally the first presented in textbooks to demonstrate the

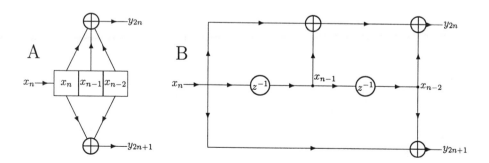

Figure 18.13: A simple convolutional code (the one universally presented in ECC texts). In (A) we see the ECC style diagram, while in (B) we have the conventional DSP flow diagram equivalent. Note that the addition in the latter diagram is modulo 2 (xor).

state	input	output	new state
0(00)	0	0(00)	0(00)
0(00)	1	3(11)	2(10)
1(01)	0	3(11)	0(00)
1(01)	1	0(00)	2(10)
2(10)	0	1(01)	1(01)
2(10)	1	2(10)	3(11)
3(11)	0	2(10)	1(01)
3(11)	1	1(01)	3(11)

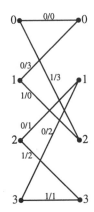

Figure 18.14: The function table and diagram for the simple convolutional code. There are four states and both the output and the new state are dependent on the state and the input bit. The new state is obtained by shifting the present state to the right and placing the input bit into the two's place. In the diagram each arc is labeled by two binary numbers, input/output.

principles of convolutional codes, and we present it in order not to balk tradition. Since there are two delay elements and at each time two bits are output, any input bit can only influence six output bits. We would like to call this number the 'influence time', but ECC terminology constrains us to call it the 'constraint length'. Since the constraint length is six, not all combinations of seven or more consecutive bits are possible. For example, the sequence 00000010 is not a possible output, since six consecutive 0s imply that the shift register now contains 0s, and inputting a 1 now causes two 1s to be output.

We have specified the output of a convolutional encoder in terms of the present and past input bits, just as we specify the output of an FIR filter in terms of the present and past input values. The conventional methodology in the ECC literature prefers the state-space description (see Section 6.3) where the present output bits are computed based on the present input and the internal state of the encoder. The natural way to define the internal state of the encoder is via the two bits x_{n-1} and x_{n-2}, or by combining these two bits into $s = 2x_{n-1} + x_{n-2}$ which can take on values 0, 1, 2, and 3. The encoder output can now be described as a function of the present input and the value of the internal state s. We tabulate this function and depict it graphically in Figure 18.14. In the figure each arc is labeled by a pair of numbers in binary notation. The first number is the input required to cause the transition from the pre-arc state to the post-arc one; the second number is the output emitted during such a transition.

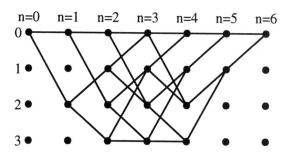

Figure 18.15: Trellis diagram for the simple convolutional code. We draw all the paths from state zero at $n = 0$ that return to state zero at time $n = 6$.

Accepting the state-space philosophy, our main focus of interest shifts from following the output bits to following the behavior of the encoder's internal state as a function of time. The main purpose of the encoder is to move about in state-space under the influence of the input bits; the output bits are only incidentally generated by the transitions from one state to another. We can capture this behavior best using a *trellis*, an example of which is given in Figure 18.15. A trellis is a graph with time advancing from left to right and the encoder states from top to bottom. Each node represents a state at a given time. In the figure we have drawn lines to represent all the possible paths of the encoder through the trellis that start from state 0 and end up back in state 0 after six time steps. Each transition should be labeled with input and output bits as in Figure 18.14, but these labels have been hidden for the sake of clarity.

How are convolutional codes decoded? The most straightforward way is by exhaustive search, that is, by trying all possible inputs to the encoder, generating the resulting outputs, and comparing these outputs with the received bit signal. The decoder selects the output bit sequence that is closest (in Hamming distance) to the bit signal, and outputs the corresponding input. In Table 18.2 we tabulate the output of the traditional convolutional code of Figure 18.13 for all possible six-bit inputs. Upon receiving twelve bits that are claimed to be the output of this coder from the all-zero initial state, we compare them to the right-hand column, select the row closest in the Hamming sense, and output the corresponding left-hand entry. This exhaustive decoding strategy rapidly gets out of hand since the number of input signals that must be tried increases exponentially with the number of bits received. We need to find a more manageable algorithm.

The most commonly employed decoding method is the *Viterbi algorithm*, which is a dynamic programming algorithm, like the DTW algorithm of

input	output	input	output	input	output	input	output
000000	000000000000	000010	000000001110	000001	000000000011	000011	000000001101
100000	111011000000	100010	111011001110	100001	111011000011	100011	111011001101
010000	001110110000	010010	001110111110	010001	001110110011	010011	001110111101
110000	110101110000	110010	110101111110	110001	110101110011	110011	110101111101
001000	000011101100	001010	000011100010	001001	000011101111	001011	000011100001
101000	111000101100	101010	111000100010	101001	111000101111	101011	111000100001
011000	001101011100	011010	001101010010	011001	001101011111	011011	001101010001
111000	110110011100	111010	110110010010	111001	110110011111	111011	110110010001
000100	000000111011	000110	000000110101	000101	000000111000	000111	000000110110
100100	111011111011	100110	111011110101	100101	111011111000	100111	111011110110
010100	001110001011	010110	001110000101	010101	001110001000	010111	001110000110
110100	110101001011	110110	110101000101	110101	110101001000	110111	110101000110
001100	000011010111	001110	000011011001	001101	000011010100	001111	000011011010
101100	111000010111	101110	111000011001	101101	111000010100	101111	111000011010
011100	001101100111	011110	001101101001	011101	001101100100	011111	001101101010
111100	110110100111	111110	110110101001	111101	110110100100	111111	110110101010

Table 18.2: Exhaustive enumeration of the simple convolutional code (the one universally presented in ECC texts). We input all possible input sequences of six bits and generate the outputs (assuming the shift register is reset to all zeros each time). These outputs can be compared with received bit signal.

Section 8.7. To understand this algorithm consider first the following related problem. You are given written directions to go from your house to the house of a fellow DSP enthusiast in the next town. A typical portion of the directions reads something like 'take the third right turn and proceed three blocks to the stop sign; make a left and after two kilometers you see a bank on you right'. Unfortunately, the directions are handwritten on dirty paper and you are not sure whether you can read them accurately. Was that the third right or the first? Does it say three blocks or two blocks? One method to proceed is to follow your best bet at each step, but to keep careful track of all supplied information. If you make an error then at some point afterward the directions no longer make any sense. There was supposed to be a bank on the right but there isn't any, or there should have been a stop sign after 3 blocks but you travel 5 and don't see one. The logical thing to do is to backtrack to the last instruction in doubt and to try something else. This may help, but you might find that the error occurred even earlier and there just happened to be a stop sign after three blocks in the incorrect scenario.

This kind of problem is well known in computer science, where it goes under the name of the 'directed search' problem. Search problems can be represented as trees. The root of the tree is the starting point (your home) and each point of decision is a node. Solving a search problem consists of going from node to node until the goal is reached (you get to the DSP enthusiast's house). A search problem is 'directed' when each node can be assigned a cost that quantifies its consistency with the problem specification (how well it matches the directions so far). Directed search can be solved more systematically than arbitrary search problems since choosing nodes with lower cost brings us closer to the goal.

There are two main approaches to solving search problems. 'Depth-first' solution, such as that we suggested above, requires continuing along a path until it is obviously wrong, and then backtracking to the point of the last decision. One then continues along another path until it becomes impossible. The other approach is 'breadth-first'. Breadth-first solution visits all decision nodes of the same depth (the same distance from the starting node) before proceeding to deeper nodes. The breadth-first solution to the navigation problem tries every possible reading of the directions, going only one step. At each such step you assign a cost, but resist the temptation to make any decision even if one node has a low cost (i.e., matches the description well) since some other choice may later turn out to be better.

The decoding of a convolutional code is similar to following directions. The algebraic connection between the bits constitute a consistency check very much like the stop sign after three blocks and the bank being on the right. The Viterbi algorithm is a breadth-first solution that exploits the state-space description of the encoder. Assume that we know that the encoder is in state 0 at time $n = 0$ and start receiving the encoded (output) bits. Referring back to Figure 18.15, just as the decoder generated particular transitions based on the input bits (the first number of the input/output pair in Figure 18.14) the Viterbi decoder tries to guess which transition took place based on the output bits (the second number of the pair). Were the received bit signal error free, the task would be simple and uniquely defined. In the presence of bit errors sometimes we will make an improper transition, and sometimes we cannot figure out what to do at all. The breadth-first dynamic programming approach is to make *all* legal transitions, but to calculate the Hamming distance between the received bits and the encoder output bits that would actually have caused this transition. We store in each node of the trellis the minimal accumulated Hamming distance for paths that reach that node. We have thus calculated the minimal number of errors that need to have occurred for each possible internal state. We may guess that the path with minimal number of errors is the correct one, but this would be too hasty a decision. The proper way to identify the proper path in state-space is to have patience and watch it emerge.

To see how this happens let's work out an example, illustrated in Figure 18.16. We'll use as usual the simple convolutional code of Figure 18.13 and assume that the true input to the encoder was all zeros. The encoder output was thus all zeros as well, but the received bit signal has an erroneous

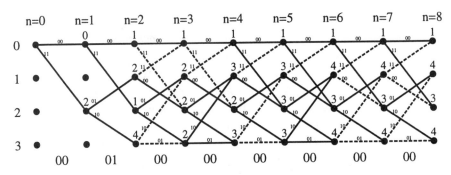

Figure 18.16: Viterbi decoding for the simple convolutional code. The bottom line contains the received bit signal, which contains a single bit error. The trellis nodes are labeled with the accumulated number of bit errors and the arcs with the output bits corresponding to the transition. Dash arcs are ones that do not survive because there is a path to the post-arc node that has fewer bit errors. The single bit error is corrected after 6 time steps.

one due to noise. We assign costs to the nodes in the trellis starting from state zero at time $n = 0$. In the first time step we can reach state 0 if 00 were transmitted and state 2 were 11 transmitted. Since 00 is received, state 0 is labeled as being reachable without bit errors, while state 2 is labeled with a 2 representing two bit errors. In the next time step the bit error occurs. From the 0 state at $n = 1$ the code can transition either to itself or to state 2, implying single-bit errors. From state 2 at $n = 1$ we would arrive at state 1 assuming the 01 that was received was indeed transmitted, thus accruing no additional error; the state 1 node at $n = 2$ is thus labeled with total error 2 since the entire path from state 0 at $n = 0$ was 1101 as compared with the 0001 that was received. From state 2 at $n = 1$ we could also arrive at state 3 at $n = 2$ were 10 to have been sent, resulting in the maximal total number of bit errors (1110 as compared to 0001).

In this way we may continue to compute the total number of bit errors to reach a given state at a given time. In the next step we can reach the 0 state via two different paths. We can transition from state 0 at $n = 2$, implying that the total bit sequence transmitted was 000000, or we can get there from state 2 at $n = 2$ were the sequence transmitted to have been 110111. Which cost do we assign to the node? Viterbi's insight (and the basis of all 'dynamic programming' algorithms) was that we need only assign the lower of the two costs. This reasoning is not hard to follow, since we are interested in finding the lowest-cost path (i.e., the path that assumes the fewest bit errors in the received bit signal). Suppose that later on we determine that the lowest-cost path went through this node, and we are interested in determining how the states evolved up to this point. It is obvious that the lowest-error path that reaches this node must have taken the route from the 0 state at $n = 2$. So the global optimum is found by making the local decision to accept the cost that this transition implies. In the figure we draw the transition that was ruled out as a dashed line, signifying that the best path does not go through here. So at each time only four surviving paths must be considered.

In general, at time step n we follow all legal transitions from nodes 0, 1, 2, and 3 at time $n - 1$, add the new Hamming distance to the cost already accrued in the pre-arc node, and choose the lower of the costs entering the post-arc node. In the figure the transitions that give minimal cost are depicted by solid lines. If two transitions give the same accumulated cost we show both as solid lines, although in practical implementations usually one is arbitrarily chosen. The reader should carefully follow the development of the trellis in the figure from left to right.

Now that we have filled in the trellis diagram, how does this help us decode the bit signal? You might assume that at some point we must break down and choose the node with minimal cost. For example, by time $n = 5$ the path containing only state 0 is clearly better than the other four paths, having only one error as compared with at least three errors for all other paths. However, there is no need to make such risky quantitative decisions. Continuing on until time $n = 8$ the truth regarding the error in the fourth bit finally comes to light. Backtracking through the chosen transitions (solid lines in the figure) shows that all the surviving paths converge on state 0 at time $n = 6$. So without quantitatively deciding between the different states at time $n = 8$ we still reach the conclusion that the most likely transmitted bit signal started 00000000, correcting the mistakenly received bit. It is easy to retrace the arcs of the selected path, but this time using the encoder input bits (the first number from Figure 18.13) to state that the uncoded information started 000.

In a more general setting, the Viterbi decoder for a convolutional code which employs m delays (i.e., has $s = 2^m$ possible internal states), fills in a trellis of s vertical states. At each time step there will be s surviving paths, and each is assigned a cost that equals the minimal number of bit errors that must have occurred for the code to have reached this state. Backtracking, we reach a time where all the surviving paths converge, and we accept the bits up to this point as being correct and output the original information corresponding to the transitions made. In order to save computational time most practical implementations do not actually check for convergence, but rather assume that all paths have converged some fixed time L in the past. The Viterbi decoder thus outputs predictions at each time step, these information bits being delayed by L.

EXERCISES

18.11.1 In exercise 8.7.1 we introduced the games of doublets. What relationships exist between this game and the decoding of convolutional codes?

18.11.2 Each of the convolutions that make up a convolutional code can be identified by its impulse response, called the generator sequence in coding theory. What is the duration of the impulse response if the shift register is of length K? The constraint length is defined to be the number of output bits that are influenced by an input bit. What is the constraint length if the shift register length is K, and each time instant k bits are shifted in and n are output?

18.11.3 Which seven-tuples of bits never appear as outputs of the simple convolutional code given in the text?

18.11.4 Repeat the last exercise of the previous section for convolutional codes. You need to replace the first program with cencode and the third with cdecode. Use the Viterbi algorithm.

18.11.5 The convolutional code $y_{2n} = x_n + x_{n-1}, y_{2n+1} = x_n + x_{n-2}$ is even simpler than the simple code we discussed in the text, but is not to be used. To find out why, draw the trellis diagram for this code. Show that this code suffers from *catastrophic error propagation*, that is, misinterpreted bits can lead to the decoder making an unlimited number of errors. (Hint: Assume that all zeros are transmitted and that the decoder enters state 3.)

18.12   PAM and FSK

Shannon's information capacity theorem is not constructive, that is, it tells us what information rate *can* be achieved, but does not actually supply us with a method for achieving this rate. In this section we will begin our quest for efficient transmission techniques, methods that will approach Shannon's limit on real channels.

Let's try to design a modem. We assume that the input is a stream of bits and the output a single analog signal that must pass through a noisy band-limited channel. Our first attempt will be very simplistic. We will send a signal value of +1 for every one in the input data stream, and a 0 for every zero bit. We previously called this signal NRZ, and we will see that it is the simplest type of **P**ulse **A**mplitude **M**odulation (PAM). As we know, the bandwidth limitation will limit the rate that our signal can change, and so we will have a finite (in fact rather low) information transmission rate.

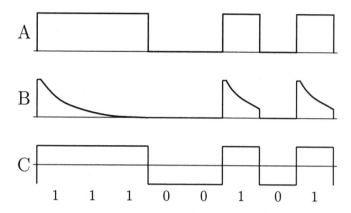

Figure 18.17: NRZ signals and DC. In (A) we see a straight NRZ signal. For each 1 bit there is a positive voltage, while there is no voltage for a 0 bit. In (B) we see what happens to this NRZ signal when DC is blocked. The positive voltages tend to decay, wiping out the information in long runs of 1 bits. In (C) we see an attempt at removing DC from the NRZ signal. For each 1 there is a positive voltage, while for each 0 bit there is a negative voltage.

NRZ has a major problem. Unless all the input bits happen to be zero, the NRZ signal will have a nonzero DC component. This is not desirable for an information transmission system. We want the transmitter to supply information to the receiver, not DC power! To ensure that DC power is not inadvertently transferred many channels block DC altogether. The telephone system supplies DC power for the telephone-set operation on the same pair of wires used by the signal, so low frequencies cannot be made available for the signal's use. This restriction is enforced by filtering out all frequencies less than 200 Hz from the audio. Attempts at sending our simple NRZ signal through a channel that blocks DC will result in the signal decaying exponentially with time, as in Figure 18.17.B. We see that single 1 bits can be correctly interpreted, but long runs of 1s disappear.

A simple correction that eliminates the major part of the DC is to send a signal value of $+\frac{1}{2}$ for every one bit, and $-\frac{1}{2}$ for every zero bit, as in Figure 18.17.C. However, there is no guarantee that the input bit stream will be precisely balanced, with the same number of ones and zeros. Even if this is true for long time averages, for short times there is still some DC (either positive or negative), and so the bits still decay.

Encoding methods discussed in Section 18.6 such as RZ or Manchester coding completely remove DC, but at the price of doubling the bandwidth; this is a price Shannon doesn't want us to pay. A better DC removing method is **A**lternate **M**ark **I**nversion (AMI), where a binary zero is encoded as 0 but a binary one is alternately encoded as $+\frac{1}{2}$ and $-\frac{1}{2}$. However, the decision levels for AMI are closer together than those of NRZ, resulting in decreased noise robustness, once again paying a capacity theorem price.

There are more sophisticated methods of eliminating the DC component. One way is to differentially encode the input bit stream before transmission. In differential coding a zero is encoded as no shift in output value and a one as a shift; and it is a simple matter to decode the differential coding at the demodulator. With differential coding a stretch of ones becomes 10101010 with many more alternations (although stretches of zeros remain). A more general mechanism is the 'bit scrambler' to be discussed in further detail in Section 18.15. The scrambler transforms constant stretches into alternating bits using a linear feedback shift register (LFSR), and the descrambler recovers the original bit stream from the scrambled bits. Although LFSR scramblers are often used they are not perfect. They are one-to-one transformations and so although common long constant stretches in the input are converted to outputs with lots of alternations, there must be inputs that cause the scrambler to output long streams of ones!

Even assuming the DC has been completely removed there are still problems with our simplest digital modem signal. One limitation is that it is only suitable for use in a baseband (DC to BW) channel (such as a pair of wires), and not in pass-band channels (like radio). The simplest 'fix' is to upmix NRZ to the desired frequency. The resultant signal will be either zero (corresponding to a 0 bit in the input) or a sinusoid of some single predetermined frequency f when we wish to send a 1 bit. This simple pass-band technique, depicted in Figure 18.18.A, is called **O**n **O**ff **K**eying or OOK.

Morse code was originally sent over cables using NRZ, using short and long periods of nonzero voltage. At the receiving end a relay would respond to this voltage, duplicating the motion of the sending key. This could be used to draw dashes and dots on paper, or the 'clicks' could be decoded by a human operator. In order to send Morse code over radio the OOK method was adopted. A carrier of constant frequency is sent and at the receiver IF mixed with a **B**eat **F**requency **O**scillator (BFO) to produce an audible difference frequency. Short intervals of tone are recognized by the receiving operator as 'dits' while tones of three times the duration are perceived as 'dahs'.

In theory we could use OOK to send more complex signals, such as Baudot encoded text, but this is rarely done. The reason is that OOK signal reception is extremely susceptible to noise and interference. To understand why think of how to decide whether a 1 or a 0 was sent. You can do no better

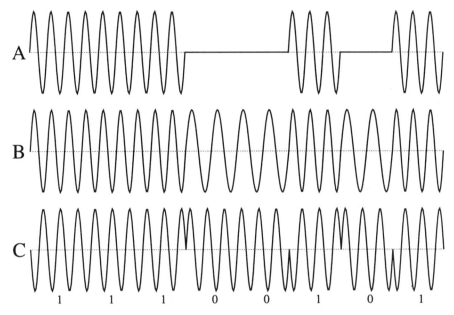

Figure 18.18: Simple digital communications signals. In (A) we see on-off keying (OOK), in (B) frequency shift keying (FSK), and in (C) phase shift keying (PSK).

than to simply look for energy at the carrier frequency, probably by using a band-pass filter of some sort, and judge whether it has passed a threshold value. Passing merits a 1 bit, lower energy is taken as a 0. If noise appears in the area of the frequency of interest, an intended 0 will be wrongly classified as a 1, and negative noise values summing with the signal of interest can cause a 1 to masquerade as a 0.

In order to ameliorate this problem noise should be explicitly taken into account, and this could be done in two nonexclusive ways. The first is based on the supposition that the noise is white and therefore its energy is the same over all frequencies. We can thus require that the energy at the output of a matched band-pass filter be much larger than the energy at nearby frequencies. The second way is based on the assumption that the noise is stationary and therefore its energy does not vary in time. We can thus monitor the energy when the signal is not active or sending zeros, and require that it pass a threshold set higher than this background energy. Such a threshold is often called a **Noise Riding Threshold** (NRT).

While these mechanisms make our decision-making process more robust they are not immune to error. Impulse noise spikes still thwart NRTs and narrow-band noise overcomes a threshold based on nearby frequencies. In addition, we have added the possibility of new error types (e.g., when the noise fluctuates at adjacent frequencies a perfectly good 1 can be discarded). While the effects of noise can never be entirely overcome, OOK does not give us very much to work with in our efforts to combat it.

Perhaps the simplest method to combat noise at adjacent frequencies is to replace OOK with **Frequency Shift Keying** (FSK). Here we transmit a sinusoid of frequency f_0 when a 0 bit is intended and a sinusoid of frequency f_1 when we wish to send a 1 bit (see Figure 18.18.B). One can build an FSK demodulator by using two band-pass (matched) filters, one centered at f_0 and the other at f_1. Such a demodulator can be more robust in noise since two energies are taken into account. One decision method would be to output a 0 when threshold is exceeded at f_0 but not f_1 and a 1 when the reverse occurs. When neither energy is significant we conclude that there is no signal, and if both thresholds are surpassed we conclude that there must be some noise or interference. When such a demodulator *does* output a 0 or 1 it is the result of two independent decisions, and we are thus twice as confident. An alternative to FSK is **Phase Shift Keying** (PSK), depicted in Figure 18.18.C. Here we employ a single frequency that can take on two different phases; a demodulator can operate by comparing the received signal with that of a sinusoid of constant phase. At first this seems no better than OOK, but we will see that PSK is a highly effective method.

There are several ways of understanding why FSK is better than OOK. Our first interpretation consisted of treating FSK as OOK with 'frequency diversity' (i.e., two independent OOK signals carrying the same information but at different frequencies). Such diversity increases the robustness with which we can retrieve information at a given SNR. This is as useful since we can increase channel capacity by attacking either the bandwidth or the noise constraints.

A second interpretation has to do with the orthogonality of sinusoids of different frequencies. An alternative to the dual band-pass filter FSK demodulator multiplies the received signal by sinusoids of frequencies f_0 or f_1 and integrates the output over time. Since this is essentially downmixing and low-pass filtering, this demodulator is actually a specific implementation of the dual band-pass filters, but we can give it a new interpretation. From equation (A.34) we know that sinusoids of different frequencies are orthogonal, so multiplication by one of the sinusoids and integrating leads to a positive indication if and only if this frequency is being transmitted. This exploitation of sinusoid orthogonality is a new feature relative to OOK.

Were the component signals in FSK truly orthogonal then FSK would be the answer to all our wishes. The problem is that sinusoids are only

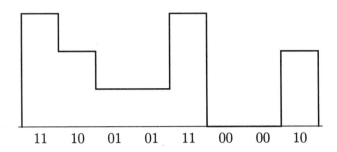

11   10   01   01   11   00   00   10

Figure 18.19: A four level PAM signal. Here the same information bits are transmitted as in the previous figures, but twice as many bits are sent in the same amount of time.

orthogonal when integrated over all time. When only a short time duration ΔT is available, the uncertainty theorem puts a constraint on the accuracy of recognizing the difference between the two frequencies Δf.

$$\Delta f \Delta t \geq 2\pi$$

For example, the telephone channel is less than 4 KHz wide, and so a rather large separation would be $\Delta f \approx 2$ KHz. This implies that telephone line FSK information transfer rates will not exceed around 300 b/s.

Of course the uncertainty theorem only directly limits the rate at which we can change between different frequencies. By using a repertoire of more than two frequencies we can increase the information transfer rate. Using four possible frequencies f_0, f_1, f_2, or f_3 we simultaneously transmit two bits of information at each time instant, doubling the information rate. This technique is not limited to FSK; simple NRZ can be extended to multilevel PAM by sending one of four different voltages, as in Figure 18.19. The signal sent at each time period is usually called a *symbol* or *baud* (after Emile Baudot), and the bit rate is double the symbol rate or *baud rate*. If we use symbols that can take on 2^m possible values, the data rate in bits per second is m times the baud rate. Of course increasing the data rate in this way has its drawbacks. The demodulator becomes more complex, having to distinguish between many different levels. More significantly, if we compare two signals with the same transmission power, the one with more levels has these levels closer together. So we cannot increase the number of levels without incurring a higher probability of misdetection in noise. The eventual limit is when the level spacing is of the order of the noise intensity. This is the way Shannon limits the capacity of multilevel PAM and multifrequency FSK signals.

EXERCISES

18.12.1 A differential encoder encodes a 1 bit as a change in the signal and a 0 bit as no change. Show how to transmit differential NRZ, FSK and PSK. For which of these three is the differential coding most useful? How does one make a decoder that reconstructs the original bit stream from the differentially encoded one? How is the ambiguity broken? What signal causes an output of all 1s? All 0s?

18.12.2 If an FSK modulator employs two completely independent oscillators $s_1(t) = A \sin(2\pi f_1 t)$ and $s_2(t) = A \sin(2\pi f_2 t)$ then at the instant of switching the signal will generally be discontinuous. **Continuous Phase FSK** (CPFSK) changes frequency without phase jumps. Why is CPFSK better than non-continuous phase FSK? Write a routine that inputs a bit stream and outputs a CPFSK signal.

18.12.3 Program a multifrequency FSK modulator on a computer with an audio output or speaker. How high can the symbol rate be before your ear can no longer distinguish the individual tones?

18.12.4 The fundamental limitation on the FSK symbol rate is due to the time-frequency uncertainty relation. There is no fundamental time-value uncertainty relationship, so what is the source of the limitation on PAM symbol rates?

18.12.5 Figure 18.19 uses the natural encoding of the numbers from 0 to 2. The PAM signal called 2B1Q (used in ISDN and HDSL) uses the following mapping: $-3 \to 00$, $-1 \to 01$, $+1 \to 11$, $+3 \to 10$. What is the advantage of this 2B1Q encoding, which is a special case of a Gray code? How can this be extended to eight-level PAM? 2^m-level PAM?

18.12.6 In the text we didn't mention **Amplitude Shift Keying** (ASK). Draw a signal with four-level ASK. Why isn't this signal popular?

18.13   PSK

FSK demodulation is based on the orthogonality of the signals representing the bit 0 and the bit 1; unfortunately, we have seen that this orthogonality breaks down as we try to increase the information transfer rate. Over telephone lines FSK can be readily used for rates of 300–1200 b/s, but becomes increasing problematic thereafter. In the previous section we mentioned PSK; here we will present it in more detail and show why it can carry information at higher rates. PSK is commonly used for 1200–2400 b/s over telephone lines.

Consider the two signals

$$s_0(t) = \sin(2\pi f_c t)$$
$$s_1(t) = \sin(2\pi f_c t + \tfrac{\pi}{2})$$

where f_c is the carrier frequency. We suggest using these two signals as our information carriers, transmitting s_0 when a 0 bit is to be sent, and s_1 for a 1 bit. We call this signal BPSK for **Binary PSK**. There are basically two methods of demodulating PSK signals. *Coherent demodulators* maintain a local oscillator of frequency f_c and compare the frequency of incoming signals to this clock. *Incoherent demodulators* do not maintain a precise internal clock, but look for jumps in the incoming signal's instantaneous phase.

Both BPSK demodulators are more complex than those we have seen so far. Are BPSK's advantages worth the extra complexity? Yes, since unlike FSK, where the two basic signals become orthogonal only after a relatively long time has elapsed, s_0 and s_1 are already orthogonal over a half cycle. So we can transmit one of the signals s_0 or s_1 for as little as one-half of a cycle of the carrier, and still discriminate which was transmitted. This is a major step forward.

Is this the best discrimination available? The coherent demodulator multiplies the incoming signal by $\sin(2\pi f_c t)$ and so after filtering out the component at twice f_c its output is either 0 or $\frac{\pi}{2}$. We can increase the phase difference by using the two signals

$$s_0(t) = \sin(2\pi f_c t - \tfrac{\pi}{2})$$
$$s_1(t) = \sin(2\pi f_c t + \tfrac{\pi}{2})$$

and the difference between the output signals is now maximal. It is not hard to see that $s_1 = -s_0$, so using a sinusoid and its inverse results in the best discrimination. Plotting multiple traces of the demodulator output results in an eye pattern, such as that of Figure 10.8. Using a phase difference of π opens the eye as much as is possible.

We can now go to a multiphase signal in order to get more bits per symbol. QPSK uses four different phases

$$s_0(t) = \sin(2\pi f_c t)$$
$$s_1(t) = \sin(2\pi f_c t + \tfrac{\pi}{2})$$
$$s_2(t) = \sin(2\pi f_c t + \pi)$$
$$s_3(t) = \sin(2\pi f_c t + 3\tfrac{\pi}{2})$$

or equivalently any rigid rotation of these phases. Unfortunately, multiplying by $\sin(2\pi f_c t)$ does not differentiate between the signals $s_a(t) = \sin(2\pi f_c t + \phi)$ and $s_b(t) = \sin(2\pi f_c t + (\pi - \phi))$, so our coherent demodulator seems to have broken down; but if we multiply by $\cos(2\pi f_c t)$ as well, we can discriminate between any two angles on the circle. This is not surprising since what we have done is to construct the instantaneous representation of Section 4.12. Calling the output of the sine mixer I and that of the cosine mixer Q, we can plot a two-dimensional representation of the analytic signal, called the 'I-Q plot'. The four points in the plane corresponding to the QPSK signal values are on the unit circle and correspond to four distinct angles separated by 90°. The two points of BPSK or four points of QPSK are called the *constellation*, this name originating from their appearance as points of light on an oscilloscope displaying the I-Q plot.

Let's generalize our discussion to nPSK. An nPSK signal is of the form

$$s(t) = e^{i\left(\omega_c t + \phi(t)\right)} \tag{18.24}$$

with all the information being carried by the phase. The phase is held constant for the baud duration t_b, the reciprocal of which, f_b, is the baud rate. For nPSK this phase can take one of n different discrete values, and usually n is chosen to be a power of two $n = 2^m$. Hence the information rate in an nPSK signal is $m f_b$ bits per second.

What values should the phases take? We want the different values to be as far apart as possible, in order for the demodulator to be able to distinguish between them as easily as possible. One optimal way to choose the phases for nPSK is

$$\Phi_0 = 0, \ \Phi_1 = \frac{2\pi}{n}, \ \dots \ \Phi_k = \frac{2\pi k}{n}, \ \dots \ \Phi_{n-1} = -\frac{2\pi}{n}$$

for example, the BPSK constellation should consist of two points with angles $0, \pi$, QPSK should have four points with angles multiples of 90°, and for 8PSK we choose eight points with angles multiples of 45°. These choices are good, but any rigid rotation of the entire constellation is equally acceptable. For example, BPSK can have phases $0, \pi$ as suggested here, or $\pm\frac{\pi}{2}$ or $\frac{\pi}{4}, \frac{5\pi}{4}$. Actually, there is no real difference between the different choices; from equation (18.24) it is obvious that an overall rotation of the phases is equivalent to resetting the clock (i.e., changing when $t = 0$ was).

When receiving an nPSK signal we first find its I and Q components, and from these calculate its instantaneous phase. Figure 18.20 is an I-Q plot of a received QPSK signal. Because of additive channel noise, channel distortion, and various inadequacies of the demodulation process, the actual symbols

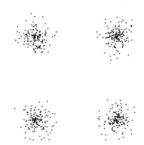

Figure 18.20: I-Q plot for a QPSK signal with noise. We see that the effect of additive channel noise is to replace the true constellation points with clouds centered at the original points. Although with this noise level we could still make accurate decisions, were we to add four more points to make 8PSK the clouds would touch and reception errors would be inevitable.

detected are not exactly those transmitted. We see that the four points on the unit circle have been transformed into four small 'clouds' centered on the original points. In order to recover the original information we have to decide to which true constellation point each received point should be associated. The decision is performed by a slicer and will be discussed in Section 18.18. How much noise can be tolerated before the decisions become faulty? From the figure it is obvious that if the radius of the noise cloud is less than half the Euclidean distance between the constellation points, then most of the decisions will be correct.

EXERCISES

18.13.1 What is the Euclidean distance between constellation points of an nPSK signal? Why can't we increase the distance between the constellation points by simply placing them on a larger circle?

18.13.2 Simulate a baseband nPSK signal and find its empirical spectrum. Vary n. Do you see any change? Vary t_b. What happens now?

18.13.3 In exercise 18.12.5 we saw how to use a Gray code for multilevel PAM. What is the difference between a Gray code for PAM and one for nPSK? How is this related to the Hamiltonian cycles of Section 12.1?

18.13.4 We saw that resetting of the clock rotates the nPSK constellation. How can we ever be sure that we are properly interpreting the data?

18.13.5 Write programs that implement a QPSK modulator and demodulator. Try adding noise to the signal. What happens if the demodulator carrier frequency is slightly wrong? What if the baud rate is inaccurate?

18.14   Modem Spectra

If the data is ... 10101010 ... then the NRZ signal is a square wave and
its spectrum consists of discrete lines at odd harmonics of the baud rate.
What is its spectrum when it carries random information? This spectrum
is only defined in the sense of Section 5.7, and hence we should compute
autocorrelations and use Wiener-Khintchine. Let's assume that the bits are
white (i.e., that there is no correlation between consecutive bits). Then the
autocorrelation of the NRZ signal will be zero for lags greater (in absolute
value) than the baud rate. It requires only slightly more thought to convince
oneself that the autocorrelation decreases linearly from its maximum, forming a triangle, as in Figure 18.21.A. The FT of this, and hence the desired
PSD, is a sinc squared, depicted in Figure 18.21.B. The first zero is at f_b,
consistent with uncertainty theorem constraints.

What is the PSD of a multilevel PAM signal? We could calculate the
autocorrelation, but we can find the answer by a simpler argument. It is

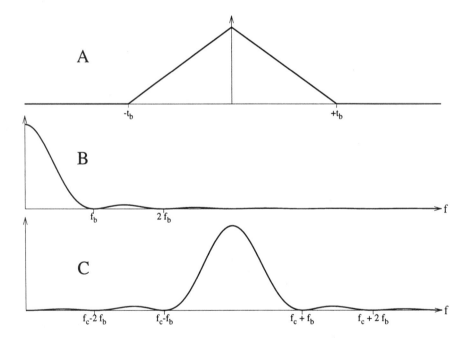

Figure 18.21: PSD of digital communications signals. In (A) we see the autocorrelation
of a NRZ signal carrying white random data. In (B) is depicted the PSD of this signal,
a sinc squared with its first null at the baud frequency. In (C) we present the PSD of a
OOK signal with the same data.

obvious that a PAM signal with n levels can be thought of as the sum of n NRZ signals. From the linearity of the FT we can conclude that multilevel PAM has exactly the same spectrum as NRZ. In particular, the bandwidth of PAM is independent of the number of levels, but of course more levels with the same energy effects the noise sensitivity. Of course we have been comparing signals with the same baud rate; when we compare PAM signals with the same bit rate, the signal with more bits per symbol has a lower baud rate and hence a lower bandwidth.

What about an OOK signal? We don't have to recompute autocorrelations since we know that OOK is simply NRZ upmixed by the carrier. Accordingly, we immediately conclude that its PSD is that depicted in Figure 18.21.C, centered on the carrier frequency and taking up double the bandwidth of the NRZ signal. This spectrum is shared by the multilevel ASK signal as well. In fact, it is a quite general result that only the carrier frequency and baud rate affect the spectrum.

Why should the bandwidth have doubled for the same baud rate? This result hints that there is another degree of freedom that we are not exploiting, but that would not change the PSD. This degree of freedom is the phase; by simultaneously modulating both the amplitude and the phase we can double the bit rate without increasing the bandwidth. We will return to this idea in Section 18.17.

EXERCISES

18.14.1 In our derivation of the PSD for the NRZ signal we didn't dwell on the DC. What is the difference between the spectra of the NRZ and DC-removed NRZ signals?

18.14.2 Derive the PSD of BPSK from that of DC-removed NRZ. Compare this spectrum with that of OOK.

18.14.3 Summing n independent NRZ signals does not result in a nPAM signal with equally probable levels. Why does this not affect the conclusion regarding the PAM spectrum?

18.14.4 Compare the PSDs for BPSK, QPSK, and 8PSK with the same information transfer rate.

18.14.5 Create random data NRZ and BPSK signals and compute their periodogram using the FFT. Now use 01010101 input ('alternations'). How is the spectrum qualitatively different? Starting from the deterministic input of alternations add progresively more randomness. How does the spectrum change?

18.15   Timing Recovery

Anyone who has learned Morse code can tell you that sending it is much simpler than 'reading' it. This is quite a general phenomenon—digital demodulators are always more complex than the corresponding modulators. One reason is the need to combat noise added by the communications channel, but there are many others. Some of the most problematic involve synchronizing the modulator and demodulator time sources.

We usually differentiate between two time sources. *Baud rate recovery* refers to synchronization of the demodulator's baud clock with that of the modulator. Every digital communications system must perform some sort of timing recovery. *Carrier recovery* refers to recovery of the carrier frequency for those modulation types that use a carrier. Obviously NRZ and PAM do not need carrier recovery. Rotating the I-Q plot to correspond to the proper constellation can be considered to be carrier phase recovery.

Consider, for example, a modulator with baud rate f_b, nominally known to the intended demodulator. It sends a new symbol every $t_b = \frac{1}{f_b}$ seconds (i.e., the first symbol occupies time from $t = 0$ to $t = t_b$, the second from $t = t_b$ to $t = 2t_b$, and so on). Although the demodulator expects this baud rate, its clock may differ slightly from that of the modulator, both in phase (i.e., precisely when $t = 0$ occurs) and in frequency (i.e., its t_b may be slightly shorter or longer than the intended t_b). Left uncorrected such slight frequency differences add up, and soon valid symbols will be missed or counted twice and the demodulator will attempt to decide on the symbol value based on observing the signal during a transition, as can be seen in Figure 18.22. Similarly, proper reception of a pass-band signal may require the demodulator to agree with the modulator as to the precise frequency and phase of the carrier.

The simplest method for the demodulator to obtain the modulator's

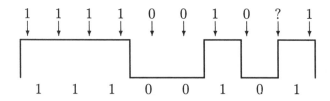

Figure 18.22: The effect of improper baud rate recovery. In this example the demodulator clock runs slightly faster than the modulator's, so that bit insertions result. On occasion a symbol clock may fall directly on a transition, causing indefinite decisions.

clock is for it to be somehow delivered. For example, if the data-bearing signal is delivered on a pair of wires, then a second pair could be provided whose sole purpose is to carry the clock information. This information could be represented by pulses at precisely the moments transitions could take place, or by a sine wave with positive slope zero crossing at the moment of possible transition, or by any other previously agreed-upon method. The problem with this method is the need for an expensive second pair of wires, which could be more effectively used for carrying information. A slightly more efficient method would be to multiplex the clock signal onto the same pair of wires that carry the signal. For example, possible transition moments could be marked by pulses at some frequency not used by the data signal. This guarantees that the clock is delivered with the same delay as the signal, but of course wastes valuable bandwidth on the clock signal. These simple methods are attractive for systems that carry a large number of signals sharing the same clock, so that the overhead is small. In order not to waste wires or bandwidth the demodulator is often required to derive the clock from the information-bearing signal itself.

To demonstrate how baud rate recovery may be accomplished we'll discuss a simple NRZ signal, although the basic ideas remain intact for other signals after appropriate preprocessing. We will take the two levels of the NRZ signal to be ± 1 so that transitions are zero crossings. When the SNR is good these zero crossings are easily observable, and by detecting them and measuring the time between them the baud rate can be recovered. The time between two successive transitions must be an integral multiple of t_b, and this multiple is readily determined when the approximate t_b is known. Using observed transitions in this fashion, both the frequency and absolute phase of the modulator's clock can be recovered.

The following simplistic algorithm for NRZ demodulation can be run in real-time. After running for some time T is the current estimate of t_b and p is the time of the previous zero crossing.

```
input the next signal value sₙ
compute the time since the last transition d ← n − p
if d is 'approximately half integer'
        output sgn(sₙ)
if sₙ · sₙ₋₁ < 0
        interpolate r ← n − sₙ/(sₙ₋₁−sₙ)
        time between transitions τ = r − p
        compute the multiple m = round τ/T
        update estimate T ← αT + (1 − α)τ/m
```

One could significantly improve this algorithm with little effort. It's better to base the output decision on several successive s_n, so some filtering or median filtering should be performed. If the signal has some residual DC (that may be time varying) it should be removed immediately after input of the signal. Also, the *linear* interpolation between s_n and s_{n-1} can be improved. However, all such algorithms rapidly deteriorate in performance with SNR degradation. Noise pulses can look like transitions or alternatively hide true transitions, and even when a transition is properly identified its correct location becomes obscure. What we really need is a method that exploits the entire signal, not just those few signal values that straddle a transition; exploiting the entire signal in order to derive a frequency requires spectral analysis or narrow band-pass filtering.

Were the data to alternate like 01010101, the transmitted signal would be a square wave, and its Fourier series would consist of a basic sinusoid at half the baud rate, and all odd harmonics thereof. Even in severe noise this harmonic structure would be easily discernible and by band-pass filtering a sinusoid related to the desired clock could be recovered. A more direct method would be to differentiate the signal (accenting the transitions), and to take the absolute value (removing the direction of the transition) thus creating a pulse train whose spectrum has a strong line at precisely the baud rate. Of course the differentiation operation is very sensitive to noise but the baud line will be strong enough to stand out.

When the NRZ data is random the differentiation and absolute value operations produce a train of pulses similar to that of the alternations, but with many of the pulses missing. The basic frequency of this signal is still the baud rate, but the baud line in the spectrum is not as strong. However, as long as there are enough transitions the baud rate can still be determined. Using a PLL is helpful, since it is designed to lock onto approximately known frequencies in noisy signals.

If there are long stretches of constant zeros or ones in the data the baud spectral line will tend to disappear, and no amount of filtering will be able to bring it back. We mentioned previously that by using a bit scrambler we can eliminate long runs of 1 bits. The most popular scrambler in use is the two-tap self-synchronizing LFSR scrambler, depicted in Figure 18.23.A. Why is this scrambler called self-synchronizing? Contrast it with the alternative method of running the LFSR locked upon itself (see Section 5.4) to create an LFSR sequence, and xoring the data with this sequence. That method also increases the number of alternations in sections where there are long

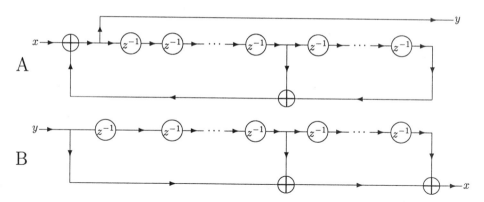

Figure 18.23: Two-tap scrambler and descrambler. In (A) we see the LFSR-based scrambler. The output is composed of the xor of the input bit with two previous input bits. The length of the shift register equals the delay of the oldest bit required. In (B) we see the descrambler. Note that there is no feedback and hence bit errors do not propagate without limit.

runs of either 1 or 0 bits, but in order to properly decode the sequence we have to use the proper phase of the LFSR sequence. The LFSR descrambler depicted in Figure 18.23.B correctly recovers the bits without the need for synchronizing an LFSR to an external clock.

When the signal is PSK another trick is popular. Rather than basing the baud rate recovery on the phase demodulation, we base it on the received signal's amplitude. It may seem surprising that the AM demodulation of a PSK signal contains any information at all; doesn't a PSK signal have constant amplitude? It does, but sharp phase transitions require wide bandwidth, and when a PSK signal is filtered to the channel bandwidth this high-frequency energy is lost, resulting in amplitude dips at the phase transitions. The amplitude demodulation is thus constant except for at the phase transitions, and its dips provide a reliable indication of the phase transitions.

We turn now to the recovery of the carrier frequency of a PSK signal. Were the data being transmitted to be constant (and no scrambler used), then the nPSK signal would be a sinusoid, and its spectrum would consist of a single discrete line the frequency of which is easily determined. However, in the more interesting case of a signal carrying information, the frequent phase jumps widen the spectral line into a broad (sinc squared) band centered around the carrier frequency. The precise carrier frequency is no longer evident.

Assume for just a moment that the constellation is chosen so that the signal points are $e^{i\pi \frac{k}{n}}$ (e.g., for BPSK ± 1 and for QPSK $\pm 1, \pm i$). It is obvious that these signal points are precisely the nth roots of unity, and so raising them to the nth power gives one. The usual convention is rotated by 45° with respect to this, (i.e., all the signal points are multiplied by $e^{i\frac{\pi}{4}}$). Hence in the usual convention all the signal points raised to the nth power still return the same value, only now that value is $e^{i\frac{n\pi}{4}}$. Thus for BPSK squaring the signal points at 45° and 225° gives i and for QPSK raising any of the points 45°, 135°, 225°, 315° to the fourth power gives -1. The important fact is that raising any of the constellation points to the nth power returns the same value.

Now the signal in the time domain is $\sin(\omega t + \phi_k)$, where the ϕ_k are the n possible signal phases. It is clear from the result of the previous paragraph that raising the signal to the nth power on a sample-by-sample basis will wipe out the ϕ_k dependence; and so the nth root of this will be a simple sine of constant phase at the carrier frequency.

EXERCISES

18.15.1 Assume that a signal contains no runs of single bits, but only runs of two, three and longer. Can the baud rate be recovered? Suggest a method.

18.15.2 Generate a PSK signal with random data, limit its bandwidth by FIR low-pass filtering, and perform amplitude demodulation. Do you see the AM dips? Now filter the AM using a narrow IIR filter centered at the nominal baud rate. Empirically determine the delay between the zero crossings of the sinusoidal output of this filter and the center of the symbols. Does this system give accurate baud rate recovery?

18.15.3 Can baud or carrier recovery be performed on signals attaining the Shannon capacity?

18.15.4 In systems with very high baud rate there can be a problem in providing the timing on a second pair of wires. What is this problem and how can it be overcome?

18.16   Equalization

We discussed adaptive modem equalizers in Section 10.3. The problem with standard (linear) equalizers for telephone modems is that near the band edges (under 400 Hz or above 3600 Hz) there can be 10 to 20 dB of attenuation. In DSL modems the higher frequency ranges can be attenuated by 50

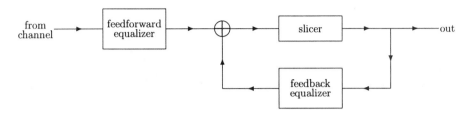

Figure 18.24: Decision feedback equalizer. The DFE consists of two FIR filters, one in the feedforward path and one in the feedback path. The slicer becomes an integral part of the equalizer.

dB or more! In order to compensate for this loss, an equalizer at the input of the demodulator must provide that much gain. Unfortunately, this gain is applied not only to the desired signal, but to the noise as well, causing significant noise enhancement. No filter is able to overcome this problem, since from linearity the sum of the signal and noise are filtered separately and identically; however, we may be able to build a nonlinear system that can apply gain to the signal without enhancing the noise too.

In Figure 18.24 we see such a nonlinear system, its nonlinearity deriving from the slicer. It is easy to see why this system, called a **D**ecision **F**eedback **E**qualizer (DFE), can effectively combat ISI without appreciable noise enhancement. Assuming the output of the slicer to be correct, it is essentially the signal that was originally transmitted. Based on this reconstructed signal we can reproduce the intersymbol interference as caused by the channel response and subtract it from the signal. This is performed by the feedback equalizer in the figure.

There are two problems with decision feedback equalizers. First, if the slicer *does* make mistakes, then (at least theoretically) the feedback can cause the system to deviate more and more from correct behavior. This lack of stability is rarely seen in practice for a DFE initially trained on known data and continuously updated. A more problematic aspect of placing the slicer into the equalization path is that its decisions do not take TCM (see Section 18.19) into account. A TCM modem does not have reliable decisions until much later, long after the DFE needed them.

The Tomlinson equalizer explored in exercise 10.3.1 is another solution to the noise enhancement problem. By placing the inverse filter at the modulator, before the noise is added, the whole problem becomes moot.

EXERCISES

18.16.1 Taking the ISI to be weak and from precisely one previous symbol, show that each constellation point splits into a small cluster of points that resembles the entire constellation. What happens if the ISI is from two previous symbols? What happens when the ISI is large and its duration long?

18.16.2 Simulate a QPSK signal traversing a noisy channel sharply attenuated at its edges. Compare the optimal linear equalizer with the optimal DFE.

18.16.3 An inverse filter at the modulator may cause the transmitted signal to reach values much larger than originally intended. The Tomlinson equalizer overcomes this by a modulo operation, and a compensating operation at the demodulator. Explain how this can be accomplished.

18.17   QAM

In Section 18.14 we saw that the bandwidth of an nPSK signal is not n dependent. Accordingly, we can achieve higher information transfer rates in a given bandwidth simply by increasing n. The problem is that for larger n the constellation points are closer together. Since channel noise causes the received signal phase to deviate from that transmitted, as depicted in the constellation plot of Figure 18.20, there is a limit on how close we can place constellation points. This is how the channel capacity theorem limits capacity for PSK signals. Were there to be no noise we could achieve arbitrarily large transfer rates by using large n; were there no bandwidth limitation we could use BPSK and arbitrarily large baud rates.

Looking closely at the constellation plot of Figure 18.20 we can see a way out. The additive channel noise expands the constellation points into circular clouds in the I-Q plane, and our decision making is optimized by maximizing the Euclidean distance between constellation points. One way this can be done is by placing constellation points as shown in Figure 18.25. Here the symbols differ in both phase and amplitude. This type of signal (being simultaneously PSK and ASK), is sometimes called names like APSK, but more usually goes under the name **Q**uadrature **A**mplitude **M**odulation (QAM). Understanding the meaning of QAM requires thinking of the I and Q components as two independent PAM signals 'in quadrature'. This is indeed another way of building a QAM signal; rather than altering the amplitude and phase of a single carrier, we can independently amplitude modulate a sine and its corresponding cosine and add the results.

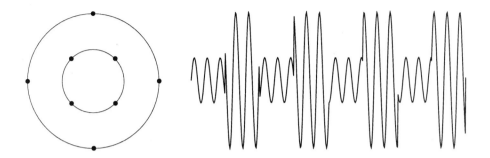

Figure 18.25: Two-ring constellation and signal. From either representation it can be seen that the symbols differ in both amplitude and phase. The two-ring signal allows precisely two different amplitudes. The phases are chosen to maximize the minimum distance between constellation points.

Let's calculate how much we gained by using QAM. The minimum distance between points in the 8PSK constellation is $2 \sin\left(\frac{\pi}{8}\right) \approx 0.765$. The two-ring constellation consists of the symbols $(1,1)$, $(-1,1)$, $(-1,-1)$, $(1,-1)$, $(3,0)$, $(0,3)$, $(-3,0)$, and $(0,-3)$ and so its minimum distance is 2. However, this is not a fair comparison since the average energy of the two-ring constellation is higher than that of 8PSK. We can always increase the minimum distance by increasing the energy; were we to put the 8PSK on a circle of radius 4 the minimum distance would be $4 \cdot 0.765 > 3$! The proper way to compare two constellations is to first normalize their average energies. Every point of the 8PSK constellation has unit energy, so the average energy is obviously unity. The two-ring constellation's energy can be easily calculated as follows. Their are four points with energy $(\pm 1)^2 + (\pm 1)^2 = 2$ and four points with energy $(\pm 3)^2 + 0^2 = 9$, hence the average energy is $(4 \cdot 2 + 4 \cdot 9)/8 = 5\frac{1}{2}$. In order to force the energy to unity we need only divide all the symbol coördinates by the square root of this energy, i.e. by about 2.345. Instead of doing this we can directly divide the minimum distance by this amount, and find that the normalized minimum distance is $2/2.345 = 0.852 > 0.765$. This increase in minimum distance is due to better exploitation of the geometrical properties of two-dimensional space.

The two-ring constellation managed to increase the minimal distance between constellation points without increasing the energy. In this way a demodulator will make fewer errors with the same SNR, or alternatively we can attain the same error rate with a lower SNR. This is the goal of a good constellation, to maximize the minimum two-dimensional Euclidean distance between points for a constellation with given energy. The problem is thus purely a geometric one, and some of the solutions presently in use are shown in Figure 18.26.

It has become conventional to use square constellations with odd integer coördinates. For example, the 16QAM constellation consists of symbols $(-3, 3), (-1, 3), \ldots (1, -3), (3, -3)$. What is the energy of this constellation? It has four points with energy $(\pm 1)^2 + (\pm 1)^2 = 2$, eight with $(\pm 1)^2 + (\pm 3)^2 = 10$, and four with $(\pm 3)^2 + (\pm 3)^2 = 18$, so that the average is $(4 \cdot 2 + 8 \cdot 10 + 4 \cdot 18)/16 = 10$. Since the unnormalized minimum distance is 2, the normalized minimum distance is $2/\sqrt{10} \approx 0.632$. This is lower than that of the previous constellations, but each symbol here contains 4 bits of information, one bit more than that of the eight-point constellations.

We will see in the next section that it is easiest to build slicers for square constellations, but rectangular constellations have a drawback. The corners have high energy, and may even be illegal in channels with maximum power restrictions. The optimum constellation boundary would be a circle, and this is closely approximated by the V.34 constellation. The cross-shaped constellations are a compromise whereby the worst offenders are removed, the slicer remains relatively simple, and the number of points in the constellation remains a power of two.

EXERCISES

18.17.1 Why are cross constellations used for odd numbers of bits per symbol and square-shaped constellations for even numbers?

18.17.2 Write a program to compute the average energy and normalized minimum distance for all the constellations in Figure 18.26.

18.17.3 Can you write a program that outputs the points in the V.34 constellation? (Hint: There are 1664 odd-integer-coördinate points bounded by a circle.)

18.17.4 Show that PAM constellations with m bits have average energy $E = \frac{1}{3}(4^m - 1)$ and hence require about four times (6 dB) more energy to add a bit. Repeat the calculation for square QAM constellations.

18.17.5 Square QAM constellations suffer from the same 90° ambiguity as nPSK. Show how differential encoding can be combined with 16QAM.

18.17.6 Some people have suggested using hexagonal constellations. What are these and why have they been suggested?

18.17.7 Prove that by squaring a QAM signal one can recover the baud rate. Prove that taking the fourth power of a QAM signal enables carrier recovery. Show that the rounder the constellation the harder it is to recover the carrier.

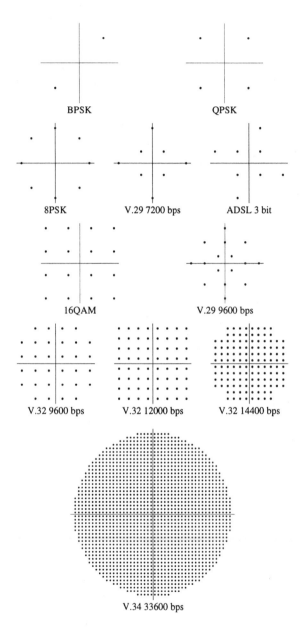

Figure 18.26: Some popular QAM constellations. The BPSK and QPSK constellations are used by the simplest modems (e.g., V22bis telephone-grade modem at 2400 b/s). The second row all have three bits per symbol, and the third row all four bits per symbol. The fourth row contains the constellations used by the V.32bis standard modem, with 5, 6, and 7 bits per symbol. The V.34 1664-point constellation has been magnified for clarity.

18.18   QAM Slicers

The slicer is the element in the QAM demodulator that is responsible for deciding which symbol was actually transmitted. The slicer comes after AGC, after carrier and symbol timing recovery, after equalization and after symbol rate resampling, although it is intimately related to all of these. Its input is a point in two-dimensional space (Figure 18.27 gives an example) and its output is a symbol label.

Figure 18.27 demonstrates the difficulty of the slicer's task. The transmitted constellation points have become contaminated by noise, distortion, and uncorrected ISI from the channel, and possibly by nonoptimalities of the previous stages of the demodulator. The combined effect of all these disturbances is that the constellation points have expanded into 'clouds' centered on their original positions. If the residual noise is too large, the clouds join, the constellation becomes unrecognizable, and the demodulator can no longer reliably function. The obvious requirement for dependable demodulation is for the radii of the noise clouds to be smaller than half the distance between constellation points.

The noise clouds are not always circularly symmetric, for example, when the demodulator has not properly locked on to the carrier frequency rota-

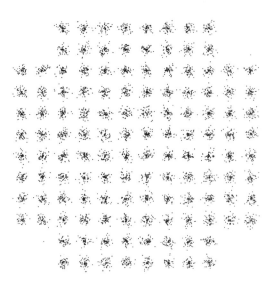

Figure 18.27: Input to the slicer. The input represents three seconds of received two-dimensional points from a V.32bis modem operating at 14.4 Kb/s. The constellation is readily recognizable to the eye, but the slicer's decisions are not always clear cut.

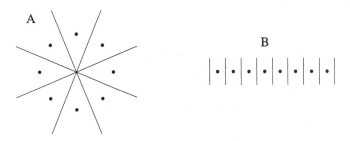

Figure 18.28: Operation of slicer for 8PSK. In (A) we see that the eight Voronoy regions are pie-slice in shape, being only phase dependent. In (B) only the phase of the same constellation is depicted, and we see that the slicer has been reduced to a quantizer.

tional smearing dominates. However, we'll assume that the clouds *are* circularly symmetric, as would be the case if the major contribution is from additive channel noise. Under this assumption the optimal operation of the slicer is to choose the constellation point that is closest to the received point. So a straightforward slicer algorithm loops on all the N constellation points and selects that constellation point with minimal distance to the received point. This algorithm thus requires N computations of Euclidean distance (sum of squares) and comparisons. Were the constellation points to be randomly chosen this complexity would perhaps be warranted, but for the types of constellation actually used in practice (see Figure 18.26) much more effective algorithms are available.

The principle behind all efficient slicers is exploitation of *Voronoy region* symmetries. Given an arbitrary collection of points, the Voronoy region associated with the nth point is the set of all points closer to it than any of the other points; the collection of all the Voronoy regions tessellate the plane. For the nPSK modem, having all its constellation points on the unit circle, it is not hard to see that the Voronoy zones are 'pie-slice' in shape (see Figure 18.28.A). The optimal slicer will slice up space into these pie slices and determine into which slice a received point falls. In particular we needn't consider the amplitude of the received point, and the optimal decision involves only its phase (as was assumed when we originally discussed the nPSK demodulator). When depicted in one-dimensional (phase-only) space (see Figure 18.28.B) the decision regions are even simpler. Neglecting the wrapping around of the phase at the edges, the slicing is reduced to simple inequalities. By correctly choosing the scale and offset, the slicing can even be reduced to simple quantizing!

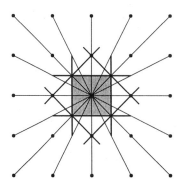

Figure 18.29: The Voronoy regions for square QAM constellations are square. To show this we investigate the immediate vicinity of an arbitrary symbol point, and connect this point with all neighboring symbols (the gray lines). We cut these lines with perpendicular bisectors (the dark lines) in order to separate points closer to the center symbol from those closer to the neighboring symbol. We shade in gray the area containing all points closest to the center symbol.

The slicer for nPSK was so simple that it could be reduced to a single quantization operation; but the more complex constellations are inherently two-dimensional. Many non-PSK constellations are based on square arrangements of symbol points, the Voronoy regions for which are themselves square, as can be seen from Figure 18.29. Square Voronoy regions are only slightly more complex to manipulate than their one-dimensional counterpart, evenly spaced points on the line. By properly choosing the scale and offsets the decision algorithm is reduced to independent quantizing along both axes. Of course some points quantize to grid points that are outside the constellation, and in such cases we need to project the decision back toward a constellation point. For cross-shaped constellations a slight generalization of this algorithm is required; for example, in Figure 18.27 we observe a point in the lower-left corner and another in the upper right that have to be specially handled.

We have succeeded in simplifying the slicer from the straightforward algorithm that required N distance computations and comparisons to one or two quantizations and some auxiliary comparisons. These simplifications depend on the geometric structure of the constellation. Indeed the efficiency of the slicer is often taken as one of the major design considerations when choosing the constellation.

EXERCISES

18.18.1 Write a routine that efficiently implements a slicer for square 16QAM. Be careful about how you handle inputs that quantize to grid points outside the constellation. Are divisions required?

18.18.2 Write a routine that efficiently implements a slicer for the cross-shaped 128-point constellation used in V.32bis. Inputs that quantize to grid points in the corners should only require a single additional comparison.

18.18.3 What are the Voronoy regions for the 16-point V29 constellation?

18.18.4 What are the Voronoy regions for a hexagonal constellation such as the 12 point $\vcenter{\hbox{:·:·:}}$? How can a slicer be efficiently implemented here?

18.19   Trellis Coding

In Figure 18.9.C we saw how the error correction encoder could be placed before the modulator and the error correction decoder after the demodulator in order to protect the transmitted information against errors. As was mentioned in Section 18.7 this separation of the error correcting code from the modulation is not guaranteed to be optimal. In our discussion of error correcting codes in Section 18.9, we saw how ECCs increase the number of bits that need be transferred. This increase directly conflicts with our attempt at transferring the original amount of information in minimum time, but is perhaps better than receiving the information with errors and having to send it again.

Is there some better way of combining the ECC and modulation techniques? It is easy to see that the answer must be affirmative. As a simplistic example, consider a bilevel PAM signal protected by a parity check. Parity check of the demodulated bits can only be used to *detect* a single bit error, while if we observe the signal input to the demodulator we may be able to make a good guess as to which bit is in error. For example,

0.10	0.92	0.05	0.49	1.02	0.94	0.08	0.04	⟵ input signal
0	1	0	0	1	1	0	0	⟵ demodulated bits
0	1	0	1	1	1	0	0	⟵ corrected bits

we have to correct a single demodulated bit and it is obvious which bit is the best candidate for correction.

So we believe that there *is* a way to combine error correction and modulation, but the two disciplines are so different that it is not obvious what that way is. It was Gottfried Ungerboeck from IBM in Zurich who, in the early 1980s, came up with the key idea of combining convolutional (trellis) codes with *set partitioning*. Set partitioning refers to recursively dividing a constellation into subconstellations with larger distance between nearest neighbors. What does this accomplish? If we know which subconstellation is to be transmitted then it is easier to determine which point was transmitted even in the presence of significant noise. How do we know which subconstellation is to be transmitted? Ungerboeck's suggestion was to divide the input bits to be transmitted into two groups, one group determining the subconstellation and the other the point in the subconstellation. If we err regarding which subconstellation was transmitted we are potentially worse off than before. So we protect the decision as to the subconstellation with an error correction code!

We will demonstrate the basic idea with the simplest possible case. A QPSK system sends two bits per symbol (that we call A and B) and has a minimal distance of $\sqrt{2} \approx 1.414$. In the proposed TCM system we expand the constellation to 8PSK using a 1/2 rate convolutional code. We will keep the same baud rate so that the bandwidth remains unchanged, but the minimum distance between constellation points is decreased to $2\sin(\frac{\pi}{8}) \approx 0.765$. The set partitioning is performed as follows. First, as depicted in Figure 18.30.A, the eight points are partitioned into two QPSK subconstellations, named 0 and 1. This particular way of partitioning is optimal since there is no other way to partition the eight points into two subconstellations that will give

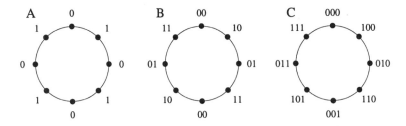

Figure 18.30: Set partitioning for the simplest TCM system. In step (A) the eight points are partitioned into two QPSK subconstellations, named 0 and 1. In step (B) each of the QPSK subconstellations is partitioned into two BPSK subsubconstellations, the 0 subconstellation into 00 and 01, and the 1 subconstellation into 10 and 11. In (C) the subsubconstellation points themselves are labeled.

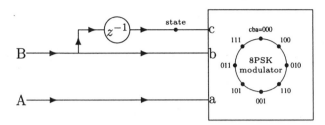

Figure 18.31: The modulator for the simplest TCM system. One of the input bits is passed directly to the 8PSK modulator, and the other is encoded by a trivial rate 1/2 convolutional code. The 'state' is the previous bit and the outputs are the present bit and the state.

greater Euclidean separation. In the next step (Figure 18.30.B) we partition each QPSK subconstellation into two BPSK subsubconstellations. The four points of subconstellation 0 are divided into subsubconstellations 00 and 01, and those of the 1 subconstellation into 10 and 11. In Figure 18.30.C we label the subsubconstellation points by suffixing the subsubconstellation label with a 0 or 1 bit.

Now that we have completely partitioned the constellation, we can proceed to build the modulator. Let's name the two bits that were input to the uncoded QPSK constellation, A and B. In the TCM modulator, input bit A is passed directly to the 8PSK modulator (where is becomes a), and it will determine the point in the subsubconstellation. Bit B, which will determine the subsubconstellation, first enters a convolution encoder, which outputs two bits b and c. What ECC should we use? We need a convolutional ECC that inputs a single bit but outputs two bits, that is, a rate 1/2 code. The simplest such code is the trivial code of Figure 18.12. Output bit b will simply be B, while c will be precisely the internal state of the encoder (i.e., the previous value of B). The bits a, b, and c determine the constellation point transmitted, as can be seen in Figure 18.31. It's easy to see that if the encoder is presently in state zero then only constellation points $0 = 000$, $2 = 010$, $4 = 100$ and $6 = 110$ can be transmitted, while if the state equals one then only constellation points $1 = 001$, $3 = 011$, $5 = 101$, and $7 = 111$ are available.

Let's draw the trellis diagram for this simple TCM modulator. It is conventional to draw the trellis taking all the inputs bits (in our case A and B) into account, although only some of them (B) enter the encoder and the others (A) do not. TCM trellis diagrams thus have 'parallel transitions', that is, multiple paths between the same states. For our encoder the new

state	input (BA)	output (cba)	new state
0	0(00)	0(000)	0
0	1(01)	1(001)	0
0	2(10)	2(010)	1
0	3(11)	3(011)	1
1	0(00)	4(100)	0
1	1(01)	5(101)	0
1	2(10)	6(110)	1
1	3(11)	7(111)	1

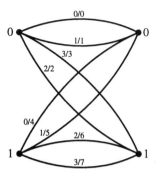

Figure 18.32: The trellis table and diagram for the simplest TCM system. The table gives the output and new state given the present state and the input bits. The diagram depicts a single time step. Between each state and each new state there are two parallel transitions, labeled BA/cba.

state will be whatever B is now. So assuming we are presently in state zero, we will remain in state zero if the input bits AB are either $0 = 00$ or $2 = 10$; however, if the inputs bits are $1 = 01$ or $3 = 11$ the state will change to one. It is straightforward to derive the table and diagram in Figure 18.32 which connect all the relevant quantities.

Now let's see if this TCM technique is truly more resistant to noise than the original QPSK. What we would really want to compare are the energies of noise that cause a given bit error rate (BER). However, it is easier to calculate the ratio of the noise energies that cause a minimal error event. For this type of calculation we can always assume that a continuous stream of zeros is to be transmitted. For the QPSK signal the minimal error is when the 00 constellation point is received as one of the two neighboring symbols (see Figure 18.33). This corresponds to a noise vector n of length 1.414 and of energy $|n|^2 = 2$. What is the minimal error for the TCM case?

Figure 18.33: The minimal error event for uncoded QPSK. The symbol s was transmitted but the symbol $s + n$ was received. The noise vector is of length $\sqrt{2}$ and hence of energy 2.

Without error we would stay in state zero all the time and receive only the 0 constellation point. Since we are in state zero there is no way we could receive an odd-numbered constellation point, so an error must result in receiving one of the points 2, 4, or 6, with 2 and 6 being the closest. Both 2 and 6 correspond to the same noise energy as in the QPSK case, but also switch the state to one, so that the next constellation point received will be odd numbered. The odd numbered points of minimal distance from the true point of zero are 1 and 7, both with distance about 0.765 and energy 0.5858. Thereafter the state reverts to 0 and the proper constellation point may be received again. So the combined noise energy of the two errors that make up this error event is about 2.5858. If the constellation point labeled 4 is mistakenly received the state does not change, but this corresponds to noise of energy $2^2 = 4$. So the minimal noise energy is 2.5858 as compared with 2 for the uncoded case. This corresponds to an improvement of a little over 1.1 dB.

By using more complex ECCs we can get more significant gains. For example, the four-state code of Figure 18.34 is described in Figure 18.35. It is not hard to see that the minimal energy error event occurs when we substitute the constellation point 001 for 000, taking the parallel transition and remaining in state 0. The energy of this event is $2^2 = 4$ rather than 2 for a coding gain of 3 dB. By using even more complex ECCs we can achieve further coding gains, although the returns on such computational investment decrease.

The first standard modem to use TCM was the CCITT V.32 modem at 9600 b/s. In order to make TCM practical, the trellis code must be made

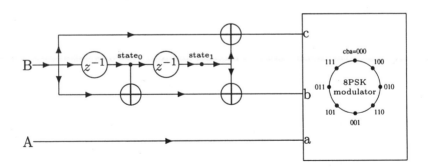

Figure 18.34: The modulator for a four-state 8PSK TCM system. One of the input bits is passed directly to the 8PSK modulator, and the other is encoded by a two state 1/2 convolutional code. The 'state' consists of the previous two bits and the outputs are formed by binary additions (xor) of the present bit and the state bits.

state	BA	cba	new state
0(00)	0(00)	0(000)	0(00)
0(00)	1(01)	1(001)	0(00)
0(00)	2(10)	6(110)	2(10)
0(00)	3(11)	7(111)	2(10)
1(01)	0(00)	6(110)	0(00)
1(01)	1(01)	7(111)	0(00)
1(01)	2(10)	0(000)	2(10)
1(01)	3(11)	1(001)	2(10)
2(10)	0(00)	2(010)	1(01)
2(10)	1(01)	3(011)	1(01)
2(10)	2(10)	4(100)	3(11)
2(10)	3(11)	5(101)	3(11)
3(11)	0(00)	4(100)	1(01)
3(11)	1(01)	5(101)	1(01)
3(11)	2(10)	2(010)	3(11)
3(11)	3(11)	3(011)	3(11)

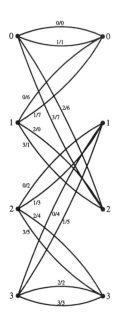

Figure 18.35: The trellis table and diagram for the four-state 8PSK TCM system. Once again there are two parallel transitions between each two states, labeled by BA/cba.

invariant to rotations by 90°. This feat was accomplished by Wei, although it required a nonlinear code.

EXERCISES

18.19.1 The simplest TCM described in the text uses the replicating convolutional code with a particular phase, outputting at a given time the present input bit and the previous one. What would happen if we output the present bit twice instead?

18.19.2 Calculate the noise energies for the different possible error events for the four-state 8PSK TCM system and show that the minimal event is indeed that mentioned in the text.

18.19.3 Devise a set partition for the 16 QAM constellation. To do this first partition the 16 points into two subconstellations of 8 points each, with each subconstellation having maximal minimal distance. What is this distance and how does it relate to the original minimal distance? Now continue recursively until each point is labeled by four bits. By how much does the minimal distance increase each time?

18.19.4 Why does TCM simplify (or even eliminate the need for) the slicer?

18.20   Telephone-Grade Modems

The history of telephone-grade modems is a story of rate doubling. The first telephone modem of interest was the Bell 103 and its internationally recognized standard version called V.21. This modem used FSK and allowed full-duplex operation of up to 300 b/s. The V.21 originating modem (called channel 1) uses frequencies 980 and 1180 Hz while the answering modem (channel 2) uses 1650 and 1850 Hz. V.21 channel 2 is still in widespread use today as the medium for negotiation between fax machines; such information as the maximum speed supported, the paper size, the station telephone numbers, and other identification are sent at 300 b/s before the higher-speed modem sends the image. This same FSK signal then cuts in again with 'end of page' and 'end of transmission' messages. You can recognize the V.21 as the 'brrrr' sound before and after the 'pshhhhh' sound of the higher-speed transmission.

A breakthrough came with the introduction of the Bell 202 and its ITU version, V.23. The V.23 FSK modem attained full-duplex 600 b/s and 1.2 Kb/s but only over special four-wire lines; on regular dial-up two-wire circuits these rates were for half-duplex use only. The rate increase as compared to V.21 was due to the use of an equalizer, although it was a fixed compromise equalizer designed for a 'typical line' and implemented as an analog filter. It is an amazing statement about conservativeness that the Bell 202 signal is still in common use. In many places it appears before the phone is picked up, carrying the number of the calling party for display to the called party.

The Bell 201 modem came out in 1962, and was later standardized by the ITU as V.26. This modem was QPSK and could attain half-duplex 2.4 Kb/s by using 1200 baud and 2 bits per symbol. The Bell 201 was the last of the modems built of analog components; it had a carrier of 1800 Hz and a fixed compromise equalizer. Its innovation was the use of carrier recovery.

Logically, if not historically, the first of the new breed of DSP modems was the V.22 QPSK modem, which reached 1.2 Kb/s full-duplex over regular dial-up lines. By upgrading the constellation from QPSK to square 16QAM, V.22bis was able to reach 2.4 Kb/s. The baud rate for these modems is 600, with one side using a carrier of 1200 Hz and the other 2400 Hz.

By using 8PSK, Milgo was able in 1967 to extend the half-duplex bit rate to 4.8 Kb/s. The baud rate was 1600, the carrier 1800 Hz, and the original version had an adjustable equalizer. Unfortunately, the adjusting had to be done by hand using a knob on the modem's front panel. This modem was standardized in 1972 as V.27.

The next major breakthrough was the 1971 introduction by Codex and the 1976 standardization of V.29. This modem achieved a half-duplex rate of up to 9.6 Kb/s by using an unusual 16-point QAM constellation with carrier 1700 Hz and baud rate 2400. Another innovative aspect of this modem was its adaptive equalizer. This modem is still popular as the 9600 fax, where half-duplex operation is acceptable.

Surprisingly, more than two decades after Shannon had predicted much higher rates, technology appeared to stop at 9.6 Kb/s. Popular wisdom believed Shannon's predictions to be overoptimistic, and efforts were devoted to implementational issues. Then in 1982, Ungerboeck published his paper revealing that an eight-state TCM code could provide a further 3.6 dB of gain, and the race toward higher rates was on again. The next step should have been to double the 9.6 Kb/s to 19.2 Kb/s, but that leap wasn't achieved. At first the V.33 modem achieved 14.4 Kb/s full-duplex on a four-wire line, and its two-wire half-duplex version (V.17) is the standard 14.4 fax used today. Next, with the introduction of DSP echo cancelling techniques, V.32bis achieved that same rate for full-duplex on two wires. All of these modems use $f_c = 1800$ Hz, $f_b = 2400$ Hz and a 128-point cross-shaped constellation. Since one of the seven bits is used for the coding, the remaining six bits times 2400 baud result in 14.4 Kb/s. These modems also provide 64-point square and 32-point cross constellations for 12 Kb/s and 9.6 Kb/s respectively.

V.32bis had attained 14.4 Kb/s, so the next major challenge, dubbed V.fast, was to attempt to double this rate (i.e., to attain 28.8Kb/s). For several years different technologies and signal processing techniques were tried, until finally in 1994 the V.34 standard was born. V.34 was a quantum leap in signal processing sophistication, and we will only be able to mention its basic principles here. Due to the complexity of the signal processing, most V.34 modems are implemented using DSP processors, rather than special-purpose DSP hardware.

The original ITU-T V.34 specification supported all data rates from 2.4 Kb/s to 28.8 Kb/s in increments of 2.4 Kb/s, and an updated version added two new rates of 31.2 and 33.6 Kb/s as well. Two V.34 modems negotiate between them and connect at the highest of these data rates that the channel can reliably provide.

We have already seen in Figure 18.26 the constellation used by V.34 for 33.6 Kb/s operation. For lower rates subconstellations of this one are used. This constellation is by far the most dense we have seen.

Recall from equation (18.20) that the information transfer rate in any given channel is maximized by matching the PSD to the channel characteristics. One of the problems with the modems up to V.32bis is that they

baud rate (Hz)	low carrier (Hz)	high carrier (Hz)	maximum data rate (Kb/s)
2400	1600	1800	21.6
2743*	1646	1829	26.4
2800*	1680	1867	26.4
3000	1800	2000	28.8
3200	1829	1920	31.2
3429*	1959	—	33.6

Table 18.3: The basic parameters for V.34. The baud rates marked with an asterisk are optional. Each baud rate, except the highest, can work with two possible carrier frequencies.

have a single predefined carrier and single baud rate, and hence their PSD is predetermined. The PSD of V.32bis stretches from $2400 - 1800 = 600$ Hz to $2400 + 1800 = 3000$ Hz irrespective of the channel characteristics. V.34 provides six possible baud rates (three mandatory and three optional) as well as nine carrier frequencies. A V.34 modem starts by probing the channel with a probe signal that creates the distinctive 'bong' noise you hear when trying to connect with a V.34 modem. This probe signal (see exercise 2.6.4) consists of a comb of sinusoids spaced 150 Hz apart from 150 Hz to 3750, except that the 900, 1200, 1800 and 2400 Hz tones have been removed. Using the probe signal the receiving modem can determine the frequency-dependent SNR, decide on the maximum data rate that can be supported with reasonably low bit error rate, and inform the transmitting modem which carrier and baud rate best match the channel. The possible baud rate and carriers are given in Table 18.3.

The second half of the channel capacity theorem specifies how the signal that maximizes information transfer rate should look. It should appear as white noise other than the water-pouring filtering. The suboptimality of V.32bis can be easily ascertained by observing its spectrum. With V.34 techniques were added to whiten the modem's spectrum.

Looking at Figure 18.3 we note that the data rate is not always an integer multiple of the baud rate (i.e., there is a noninteger number of bits per symbol). For example, the 33.6 Kb/s maximum bit rate requires 8.4 bits per symbol at 3429 baud. This feat is accomplished using a *shell mapper*.

In ordinary QAM the constellation points are used with equal probability, so that the received (imperfectly equalized) I-Q plot is homogeneous inside a disk. A noise signal would have its I-Q plot distribution decrease as a Gaussian function of the radius. We can imitate this behavior by dividing the constellation into concentric circles called shells, and, based on the data to be transmitted, first choose a shell and then the point within the shell. By using an algorithm that prefers interior shells we can transmit the constellation with a more Gaussian distribution.

V.34 also uses more powerful TCM codes than the Wei code used in V.32. The standard specifies three codes, a 16-state code (also invented by Wei) with 4.2 dB gain, a 32-state code with 4.5 dB gain and a 64-state code with 4.7 dB gain. All three of these codes are *four-dimensional*, meaning that they are based on four-dimensional symbols built up from two consecutive two-dimensional ones. Why should we want to group two transmitted symbols into a more complex one? The reason has to do with the geometry of n-dimensional space. Note that in one-dimensional space we can only place two points at unity distance from a given point, while in two-dimensional space there can be four such, and in n-dimensional space, $2n$ nearest neighbors. Thus for a given amount of energy, we can place more constellation points and thus carry more information, in higher-dimensional space. Of course the four-dimensional symbols are actually transmitted as two two-dimensional ones, but not every combination of consecutive two-dimensional symbols is possible.

In order to widen the usable bandwidth V.34 uses a more powerful equalization technique. Although DFE is capable of attaining close to the Shannon capacity, it has several drawbacks, the most important being that it is hard to combine with TCM. For V.34 a Tomlinson type equalizer was chosen instead. During the initialization a DFE is trained and the feedback coefficients sent to the modulator, where they are used as a 'precoder'. Taking the decision element out of the receive data path now makes integration of the equalizer with the TCM possible. A new mechanism called flexible precoding was invented to specifically integrate the precoder with the rest of the V.34 engine.

The logarithmic encoding used in the digital telephone system (μ-law or A-law) compresses the outer constellation points, making decisions difficult. V.34 has an option called 'nonlinear encoding' or 'warping' designed to combat these distortions. When enabled, the constellation is distorted, increasing the distance between outer constellation points, at the expense of decreasing that of more interior points.

The extremely sophisticated signal processing of the V.34 standard took years to develop and several years more to agree upon in standards committees. Yet, paradoxically, although for all intents and purposes V.34 at last approached the Shannon limit, it reigned supreme for only about a year. The next step, the step that would once again double the transmission speed from 28K to 56K, was just around the corner.

EXERCISES

18.20.1 Plot the PSDs of V.22bis, V.32, and the various modes of V.34 and compare spectral utilization.

18.20.2 Obtain diagrams of the initialization phases of V.32bis and V.34. Can you explain what happens at each stage?

18.21 Beyond the Shannon Limit

Can we beat the Shannon limit? No, there is no way of reliably communicating over an analog channel with the characteristics of the telephone channel at rates significantly higher than those of V.34. So how do V.90 (56Kb/s) modems work? What about G.lite (1 Mb/s), ADSL (8 Mb/s), and VDSL (52 Mb/s)?

These modems do not exceed the Shannon limit; they simply use a different channel. Even though they may be connected to the same phone lines that previously used a V.34 modem, what they see is different.

All of the previous modems assumed that the telephone network is comprised of (analog) twisted pairs of wire, with (analog) filters restricting the bandwidth. This was indeed once the case, but over the years more and more of the telephone system has become digital, transmitting conversations as 8000 eight-bit samples per second. Of course at every stage of this transformation of the telephone system the new equipment has emulated the old as closely as possible, so the new digital system looks very much like the old analog one; but in many places the only truly analog portion left is the 'last mile' of copper wire from the telephone office to the subscriber's house.

Were we able to transcend that last mile of copper we should be able to provide eight bits 8000 times per second, that is, an information transfer rate of 64 Kb/s. This is not surprising since this is the rate used internally by the telephone system for its own digital signals. The problem is that we are that mile or two away.

What would happen if someone at the telephone central office (CO) were to send us digital data at 64 Kb/s (i.e., a 256 level signal at 8000 samples per second)? Our telephone would interpret this *PCM modem* as a very loud noise, too loud in fact. In order to reduce crosstalk between neighboring cables, restrictions are placed on the average power that one can put onto the telephone wires. When voice is sent using these same 256 levels the lower levels are more probable than the higher ones; when digital data is sent all levels are equally probable, resulting in a higher average power. There is another problem with this attempt at transmitting 64 Kb/s. In some parts of the world not all 8 bits are available. Every now and then the least significant bit is 'robbed' for other uses. This does not degrade voice quality very much, but would be most undesirable for data. In order to reduce the average power we can use shell mapping, and because of this, together with overcoming the robbed-bit phenomenon, we should not expect more than 7 bits 8000 times a second, for a grand total of 56 Kb/s.

What would happen if someone at the CO were to send us digital data at 56 Kb/s with an appropriate shell mapping? Would we be able to distinguish between these closely spaced levels? There would be ISI, but that could be overcome by an equalizer. We would need an echo canceller to remove our own transmission, but that too is well-known theory. It turns out that if we send digital levels directly on the pair of wires going to the other modem, then it is possible to recover the original levels. The data source need not sit physically in the telephone office, as long as its connection to that office is completely digital.

This is how the V.90 56 Kb/s modem works. A V.34 modem is used in the upstream direction, that is, from the consumer to the service provider. In the downstream direction a shell-mapped digital signal of up to 56 Kb/s is sent. This asymmetry is acceptable for many applications (e.g., for Internet browsing where the downstream often consumes ten times the data rate as the upstream). In a newer version dubbed V.92 even the upsteam transmission tries to overcome the last mile and jump onto the digital link.

V.90 exceeds Shannon by exploiting the fact that the telephone system is no longer a 4 KHz analog channel, and thus the maximum possible rate is the 64 Kb/s used by the telephone system itself. Getting even higher than 64 Kb/s requires an even more radical departure from our model of the telephone system.

We have mentioned that the telephone network remains analog only in the 'last mile' to the subscriber, more formally called the 'subscriber line'. Now if we look at the frequency response of such subscriber lines, we find behaviors such as those of Figure 18.36. Although there is strong attenuation at high frequencies, the bandwidth is definitely higher than 4 KHz.

The 4 KHz restriction is actually enforced by filters at the telephone office, in order to enable multiplexing of multiple telephone conversations on a single carrier. There is nothing inherent in the subscriber line that recognizes this bandwidth restriction. So if we can place our modem before the filters and are allowed to use the subscriber line as a general-purpose

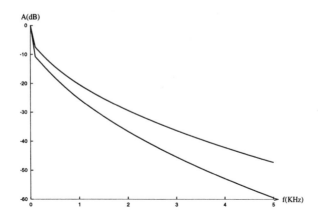

Figure 18.36: The attenuation for unshielded twisted-pair lines. Depicted is the line attenuation in dB for one kilometer of standard telephone-grade 24-gauge (upper curve) and 26-gauge (lower curve) unshielded cable. For two kilometers of cable the attenuation in dB is doubled.

cable, the so-called **Digital Subscriber Line** (DSL) modems can reach much higher capacities. Of course we may desire to continue to use the same subscriber line for our regular phone conversations, in which case a 'splitter' is placed at the end of the line. A splitter is simply a low-pass filter that passes the low frequencies to the phone, and a high-pass filter that delivers the high frequencies to the DSL modem.

The longer the subscriber's cable, the higher the attenuation and thus the lower the capacity. Long lengths can support G.lite or ADSL rates, short lengths VDSL rates. The maximum capacity can be estimated using water-pouring calculations. The strong difference in attenuation between low frequencies and higher ones can be compensated for by an equalizer.

The DSL environment is more complex than we have described so far. In addition to the attenuation there is the acute problem of crosstalk. At high frequencies a significant portion of the signal energy leaks between adjacent cables, causing one DSL modem to interfere with another. The interferer may be located close by, as in the case of a bank of DSL modems at the telephone office, or remotely located but transmitting to a co-located DSL demodulator. The former case is called NEXT (**N**ear **E**nd **XT**alk) and the latter FEXT (**F**ar **E**nd **XT**alk). In addition, signals such as AM broadcast transmissions can be picked up by the subscriber line and cause narrow bands of frequencies to be unusable. DSL modems must be able to cope with all these types of interference.

Multicarrier modems were proposed, but not accepted, for V.fast. A multicarrier scheme called **D**iscrete **M**ulti**T**one (DMT) has become the recognized standard for G.lite and ADSL. These modems transmit a large number of independent equally spaced carriers, each with an nPSK or QAM constellation, and all with the same baud rate. This baud rate is very low compared to the bit rate, and so each carrier has a narrow bandwidth. These narrow bandwidth transmissions remind us of those used in the proof of the second half of Shannon's capacity theorem, and indeed the multicarrier approach is successful partly due to its ability to approach the water-pouring limit. Furthermore, we can assume that the channel attenuation and phase delay are approximately constant over the narrow bandwidth of these transmissions, hence equalization in the normal sense is not required. All that is needed is a single gain to compensate for the attenuation at the carrier frequency, and a single phase rotation to bring the constellation to the proper angle. This **F**requency **EQ**ualizer (FEQ) can be performed by a single complex multiplication per carrier. The coefficient can be found as in exercise 6.14.8.

The narrow bandwidth and slow baud rate make the ISI less important; however, if the carriers are close together we would expect **I**nter**C**hannel **I**nterference (ICI) to become a problem. ICI is removed in DMT by choosing the intercarrier spacing to be precisely the baud rate. In this fashion each carrier sits on the zeros of its neighbor's sincs, and the ICI is negligible. Multicarrier signals with this spacing are called **O**rthogonal **F**requency **D**ivision **M**ultiplexing (OFDM) signals, since the carriers are spaced to be orthogonal.

How do we demodulate DMT signals? The straightforward method would be to use a bank of band-pass filters to separate the carriers, and then downmix each to zero and slice. However, it is obvious that this bank of filters and downmixers can be performed in parallel by using a single FFT algorithm, making the DMT demodulator computationally efficient. Indeed, the modulator can work the same way; after dividing the bit stream into groups, we create complex constellation points for each of the constellations, and then perform a single iFFT to create the signal to be transmitted!

EXERCISES

18.21.1 What is the SNR needed to achieve 56 Kb/s using every other PCM level and assuming 3.8 KHz of bandwidth and that the consumer's modem has a 16-bit linear A/D? Is this reasonable? Why is it harder to transmit 56 Kb/s upstream?

18.21.2 Why is the POTS splitter implemented using passive analog filters rather than digital filters?

18.21.3 A DMT modem still has some ISI from previous symbols. This ISI is removed by using a cyclic prefix. Explain. The overhead of a long cyclic prefix can be reduced by using a **T**ime **EQ**ualizer, which is a filter whose sole purpose is to decrease the length of the channel impulse response. What is the connection between the TEQ, FEQ, and a regular equalizer?

18.21.4 DMT modems suffer from high **P**eak to **A**verage **R**atio (PAR). Explain why. Why is this an undesirable feature? What can be done to lower the PAR?

Bibliographical Notes

The general theory and practice of digital communications systems is covered in many texts [242, 95], and modems in particular are the subject of [144, 199]. [262] covers real-time DSP programming (using a floating point processor) for communications, including AM, FM, SSB, PAM, QAM, and echo cancellation for full-duplex modems.

Harry Nyquist published in 1928 a precursor to information theory [183]. Shannon's separation theorems appear in [237], which later appeared as a book. The first part of the channel capacity theorem first appears in [238], an article that very much deserves reading even today. The water-pouring criterion is due to Gallager, and appears in his book [67].

A good modern textbook on information theory is [46], while error correcting codes are covered in many books, e.g., [194]. This latter is an updated version of one of the first texts on the subject. A fascinating mathematically oriented book on topics relevant to error correcting codes is [42]. Reed and Solomon published their code in [218]. Viterbi presented his algorithm for decoding convolution codes in [265], but the classic overview is [63].

A dated, but still useful, reference on constellation design is [119]. Multidimensional constellations are covered in [70].

Timing recovery is reviewed in [64] and a suggested original article is [76].

TCM was first presented by Ungerboeck in [263], and Wei [270] discovered how to make it rotationally invariant, leading to the trellis code used in V.32. For TCM in multidimensional constellations consult [271].

Since standard texts go only as far as V.32, it is worthwhile consulting the V.34 review in [117]. Tomlinson and flexible precoding is explained in [118].

The classic, but dated, reference for multicarrier modulation is [18].

Readers interested in a nontechnical introduction to DSL modems should consult [82], while in-depth coverage is provided in [217, 36, 251].

Speech Signal Processing

In this chapter we treat of one of the most intricate and fascinating signals ever to be studied, human speech. The reader has already been exposed to the basic models of speech generation and perception in Chapter 11. In this chapter we apply our knowledge of these mechanisms to the practical problem of speech modeling.

Speech synthesis is the artificial generation of understandable, and (hopefully) natural-sounding speech. If coupled with a set of rules for reading text, rules that in some languages are simple but in others quite complex, we get *text-to-speech* conversion. We introduce the reader to speech modeling by means of a naive, but functional, speech synthesis system.

Speech recognition, also called *speech-to-text* conversion, seems at first to be a pattern recognition problem, but closer examination proves understanding speech to be much more complex due to time warping effects. Although a difficult task, the allure of a machine that converses with humans via natural speech is so great that much research has been and is still being devoted to this subject. There are also many other applications—speaker verification, emotional content extraction (voice polygraph), blind voice separation (cocktail party effect), speech enhancement, and language identification, to name just a few. While the list of applications is endless many of the basic principles tend to be the same. We will focus on the deriving of 'features', i.e., sets of parameters that are believed to contain the information needed for the various tasks.

Simplistic sampling and digitizing of speech requires a high information rate (in bits per second), meaning wide bandwidth and large storage requirements. More sophisticated methods have been developed that require a significantly lower information rate but introduce a tolerable amount of distortion to the original signal. These methods are called speech coding or speech compression techniques, and the main focus of this chapter is to follow the historical development of telephone-grade speech compression techniques that successively halved bit rates from 64 to below 8 Kb/s.

19.1   LPC Speech Synthesis

We discussed the biology of speech production in Section 11.3, and the LPC method of finding the coefficients of an all-pole filter in Section 9.9. The time has come to put the pieces together and build a simple model that approximates that biology and can be efficiently computed. This model is often called the LPC speech model, for reasons that will become clear shortly, and is extremely popular in speech analysis and synthesis. Many of the methods used for speech compression and feature extraction are based on the LPC model and/or attempts to capture the deviations from it. Despite its popularity we must remember that the LPC speech model is an attempt to mimic the speech *production* apparatus, and does not directly relate to the way we *perceive* speech.

Recall the essential elements of the biological speech production system. For voiced speech the vocal chords produce a series of pulses at a frequency known as the pitch. This excitation enters the vocal tract, which resonates at certain frequencies known as formants, and hence amplifies the pitch harmonics that are near these frequencies. For unvoiced speech the vocal chords do not vibrate but the vocal tract remains unchanged. Since the vocal tract mainly emphasizes frequencies (we neglect zeros in the spectrum caused by the nasal tract) we can model it by an all-pole filter. The entire model system is depicted in Figure 19.1.

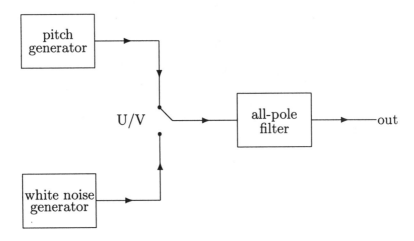

Figure 19.1: LPC speech model. The U/V switch selects one of two possible excitation signals, a pulse train created by the pitch generator, or white noise created by the noise generator. This excitation is input to an all-pole filter.

This extremely primitive model can already be used for speech synthesis systems, and indeed was the heart of a popular chip set as early as the 1970s. Let's assume that speech can be assumed to be approximately stationary for at least T seconds (T is usually assumed to be in the range from 10 to 100 milliseconds). Then in order to synthesize speech, we need to supply our model with the following information every T seconds. First, a single bit indicating whether the speech segment is voiced or unvoiced. If the speech is voiced we need to supply the pitch frequency as well (for convenience we sometimes combine the U/V bit with the pitch parameter, a zero pitch indicating unvoiced speech). Next, we need to specify the overall gain of the filter. Finally, we need to supply any set of parameters that completely specify the all-pole filter (e.g., pole locations, LPC coefficients, reflection coefficients, LSP frequencies). Since there are four to five formants, we expect the filter to have 8 to 10 complex poles.

How do we know what filter coefficients to use to make a desired sound? What we need to do is to prepare a list of the coefficients for the various phonemes needed. Happily this type of data is readily available. For example, in Figure 19.2 we show a scatter plot of the first two formants for vowels, based on the famous Peterson-Barney data.

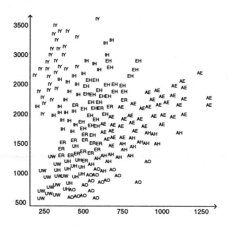

Figure 19.2: First two formants from Peterson-Barney vowel data. The horizontal axis represents the frequency of the first formant between 200 and 1250 Hz, while the vertical axis is the frequency of the second formant, between 500 and 3500 Hz. The data consists of each of ten vowel sounds pronounced twice by each of 76 speakers. The two letter notations are the so-called ARPABET symbols. IY stands for the vowel in heat, IH for that in hid, and likewise EH head, AE had, AH hut, AA hot, AO fought, UH hood, UW hoot, ER heard.

Can we get a rough estimate of the information rate required to drive such a synthesis model? Taking T to be 32 milliseconds and quantizing the pitch, gain, and ten filter coefficients with eight bits apiece, we need 3 Kb/s. This may seem high compared to the information in the original text (even speaking at the rapid pace of three five-letter words per second, the text requires less than 150 b/s) but is amazingly frugal compared to the data rate required to transfer natural speech.

The LPC speech model is a gross oversimplification of the true speech production mechanism, and when used without embellishment produces synthetic sounding speech. However, by properly modulating the pitch and gain, and using models for the short time behavior of the filter coefficients, the sound can be improved somewhat.

EXERCISES

19.1.1 The Peterson-Barney data is easily obtainable in computer-readable form. Generate vowels according to the formant parameters and listen to the result. Can you recognize the vowel?

19.1.2 Source code for the Klatt formant synthesizer is in the public domain. Learn its parameters and experiment with putting phonemes together to make words. Get the synthesizer to say 'digital signal processing'. How natural-sounding is it?

19.1.3 Is the LPC model valid for a flute? What model is sensible for a guitar? What is the difference between the excitation of a guitar and that of a violin?

19.2 LPC Speech Analysis

The basic model of the previous section can be used for more than text-to-speech applications, and it can be used as the synthesis half of an LPC-based speech compression system. In order to build a complete compression system we need to solve the inverse problem, given samples of speech to determine whether the speech is voiced or not, if it is to find the pitch, to find the gain, and to find the filter coefficients that best match the input speech. This will allow us to build the analysis part of an LPC speech coding system.

Actually, there is a problem that should be solved even before all the above, namely deciding whether there is any speech present at all. In most conversations each conversant tends to speak only about half of the time, and there is no reason to try to model speech that doesn't exist. Simple devices that trigger on speech go under the name of VOX, for **V**oice **O**perated **X** (X being a graphic abbreviation for the word 'switch'), while the more sophisticated techniques are now called **V**oice **A**ctivity **D**etection. Simple VOXes may trigger just based on the appearance of energy, or may employ NRT mechanisms, or use gross spectral features to discriminate between speech and noise. The use of zero crossings is also popular as these can be computed with low complexity. Most VADs utilize parameters based on autocorrelation, and essentially perform the initial stages of a speech coder. When the decision has been made that no voice is present, older systems would simply not store or transfer any information, resulting in dead silence upon decoding. The modern approach is to extract some basic statistics of the noise (e.g., energy and bandwidth) in order to enable Comfort Noise Generation, (CNG).

Once the VAD has decided that speech is present, determination of the voicing (U/V) must be made; and assuming the speech is voiced the next step will be pitch determination. Pitch tracking and voicing determination will be treated in Section 19.5.

The finding of the filter coefficients is based on the principles of Section 9.9, but there are a few details we need to fill in. We know how to find LPC coefficients when there is no excitation, but here there *is* excitation. For voiced speech this excitation is nonzero only during the glottal pulse, and one strategy is to ignore it and live with the spikes of error. These spikes reinforce the pitch information and may be of no consequence in speech compression systems. In pitch synchronous systems we first identify the pitch pulse locations, and correctly evaluate the LPC coefficients for blocks starting with a pulse and ending before the next pulse. A more modern approach is to perform two separate LPC analyses. The one we have been discussing up to now, which models the vocal tract, is now called the short-term predictor. The new one, called the long-term predictor, estimates the pitch period and structure. It typically only has a few coefficients, but is updated at a higher rate.

There is one final parameter we have neglected until now, the gain G. Of course if we assume the excitation to be zero our formalism cannot be expected to supply G. However, since G simply controls the overall volume, it carries little information and its adjustment is not critical. In speech coding it is typically set by requiring the energy of the predicted signal to equal the energy in the original signal.

EXERCISES

19.2.1 Multipulse LPC uses an excitation with several pulses per pitch period. Explain how this can improve LPC quality.

19.2.2 Mixed Excitation Linear Prediction (MELP) does switch between periodic and noise excitation, rather uses an additive combination of the two. Why can this produce better quality speech than LPC?

19.2.3 Record some speech and display its sonogram. Compute the LPC spectrum and find its major peaks. Overlay the peaks onto the sonogram. Can you recognize the formants? What about the pitch?

19.2.4 Synthesize some LPC data using a certain number of LPC coefficients and try to analyze it using a different number of coefficients. What happens? How does the reconstruction SNR depend on the order mismatch?

19.3 Cepstrum

The LPC model is not the only framework for describing speech. Although it is currently the basis for much of speech compression, cepstral coefficients have proven to be superior for speech recognition and speaker identification.

The first time you hear the word *cepstrum* you are convinced that the word was supposed to be *spectrum* and laugh at the speaker's spoonerism. However, there really is something pronounced 'cepstrum' instead of 'spectrum', as well as a 'quefrency' replacing 'frequency', and 'liftering' displacing 'filtering'. Several other purposefully distorted words have been suggested (e.g., 'alanysis' and 'saphe') but have not become as popular.

To motivate the use of cepstrum in speech analysis, recall that voiced speech can be viewed as a periodic excitation signal passed through an all-pole filter. The excitation signal in the frequency domain is rich in harmonics, and can be modeled as a train of equally spaced discrete lines, separated by the pitch frequency. The amplitudes of these lines decreases rapidly with increasing frequency, with between 5 and 12 dB drop per octave being typical. The effect of the vocal tract filtering is to multiply this line spectrum by a window that has several pronounced peaks corresponding to the formants.

Now if the spectrum is the product of the pitch train and the vocal tract window, then the logarithm of this spectrum is the sum of the logarithm of the pitch train and the logarithm of the vocal tract window. This logarithmic spectrum can be considered to be the spectrum of some new signal, and since the FT is a linear operation, this new signal is the sum of two signals, one deriving from the pitch train and one from the vocal tract filter. This new signal, derived by logarithmically compressing the spectrum, is called the *cepstrum* of the original signal. It is actually a signal in the time domain, but since it is derived by distorting the frequency components its axis is referred to as *quefrency*. Remember, however, that the units of quefrency are seconds (or perhaps they should be called 'cesonds').

We see that the cepstrum decouples the excitation signal from the vocal tract filter, changing a convolution into a sum. It can achieve this decoupling not only for speech but for any excitation signal and filter, and is thus a general tool for deconvolution. It has therefore been applied to various other fields in DSP, where it is sometimes referred to as *homomorphic deconvolution*. This term originates in the idea that although the cepstrum is not a linear transform of the signal (the cepstrum of a sum is not the sum of the cepstra), it is a generalization of the idea of a linear transform (the cepstrum of the convolution is the sum of the cepstra). Such parallels are called 'homomorphisms' in algebra.

The logarithmic spectrum of the excitation signal is an equally spaced train, but the logarithmic amplitudes are much less pronounced and decrease slowly and linearly while the lines themselves are much broader. Indeed the logarithmic spectrum of the excitation looks much more like a sinusoid than a train of impulses. Thus the pitch contribution is basically a line at a well defined quefrency corresponding to the basic pitch frequency. At lower quefrencies we find structure corresponding to the higher frequency formants, and in many cases high-pass liftering can thus furnish both a voiced/unvoiced indication and a pitch frequency estimate.

Up to now our discussion has been purposefully vague, mainly because the cepstrum comes in several different flavors. One type is based on the z transform $S(z)$, which being complex valued, is composed of its absolute value $R(z)$ and its angle $\theta(z)$. Now let's take the complex logarithm of $S(z)$ (equation (A.14)) and call the resulting function $\check{S}(z)$.

$$\check{S}(z) = \log S(z) = \log R(z) + i\theta(z)$$

We assumed here the minimal phase value, although for some applications it may be more useful to unwrap the phase. Now $\check{S}(z)$ can be considered to be the zT of some signal $\check{s}_n$, this signal being the *complex cepstrum* of s_n. To find the complex cepstrum in practice requires computation of the izT, a computationally arduous task; however, given the complex cepstrum the original signal may be recovered via the zT.

The *power cepstrum*, or *real cepstrum*, is defined as the signal whose PSD is the logarithm of the PSD of s_n. The power cepstrum can be obtained as an iFT, or for digital signals an inverse DFT

$$\check{s}_n = \frac{1}{2\pi} \int_{-\pi}^{\pi} \log |S(\omega)| e^{i\omega n} \, d\omega$$

and is related to the complex cepstrum.

$$\check{s}_n = \tfrac{1}{2}(\hat{s}_n + \hat{s}^*_{-n})$$

Although easier to compute, the power cepstrum doesn't take the phase of $S(\omega)$ into account, and hence does not enable unique recovery of the original signal.

There is another variant of importance, called the *LPC cepstrum*. The LPC cepstrum, like the reflection coefficients, area ratios, and LSP coefficients, is a set of coefficients c_k that contains exactly the same information as the LPC coefficients. The LPC cepstral coefficients are defined as the coefficients of the zT expansion of the logarithm of the all-pole system function. From the definition of the LPC coefficients in equation (9.21), we see that this can be expressed as follows:

$$\log \frac{G}{1 - \sum_{m=1}^{M} b_m z^{-m}} = \sum_k c_k z^{-k} \tag{19.1}$$

Given the LPC coefficients, the LPC cepstral coefficients can be computed by a recursion that can be derived by series expansion of the left-hand side (using equations (A.47) and (A.15)) and equating like terms.

$$\begin{aligned} c_0 &= \log G \\ c_1 &= b_1 \\ c_k &= b_k + \frac{1}{k}\sum_{m=1}^{k-1} m c_m b_{k-m} \end{aligned} \tag{19.2}$$

This recursion can even be used for c_k coefficients for which $k > M$ by taking $b_k = 0$ for such k. Of course, the recursion only works when the original LPC model was stable.

LPC cepstral coefficients derived from this recursion only represent the true cepstrum when the signal is exactly described by an LPC model. For real speech the LPC model is only an approximation, and hence the LPC cepstrum deviates from the true cepstrum. In particular, for phonemes that are not well represented by the LPC model (e.g., sounds like **f**, **s**, and **sh** that are produced at the lips with the vocal tract trapping energy and creating zeros), the LPC cepstrum bears little relationship to its namesakes. Nonetheless, numerous comparisons have shown the LPC cepstral coefficients to be among the best features for both speech and speaker recognition.

If the LPC cepstral coefficients contain precisely the same information as the LPC coefficients, how can it be that one set is superior to the other? The difference has to do with the other mechanisms used in a recognition system. It turns out that Euclidean distance in the space of LPC cepstral coefficients correlates well with the *Itakura-Saito distance*, a measure of how close sounds actually sound. This relationship means that the interpretation of closeness in LPC cepstrum space is similar to that our own hearing system uses, a fact that aids the pattern recognition machinery.

EXERCISES

19.3.1 The signal $x(t)$ is corrupted by a single echo to become $y(t) = x(t) + \alpha x(t-\tau)$. Show that the log power spectrum of y is approximately that of x with an additional ripple. Find the parameters of this ripple.

19.3.2 Complete the proof of equation (19.2).

19.3.3 The reconstruction of a signal from its power cepstrum is not *unique*. When is it *correct*?

19.3.4 Record some speech and plot its power cepstrum. Are the pitch and formants easily separable?

19.3.5 Write a program to compute the LPC cepstrum. Produce artificial speech from an exact LPC model and compute its LPC cepstrum.

19.4 Other Features

The coefficients we have been discussing all describe the fine structure of the speech spectrum in some way. LPC coefficients are directly related to the all-pole spectrum by equation (13.24); the LSP frequencies are themselves frequencies; and the cepstrum was derived in the previous section as a type of spectrum of (log) spectrum. Not all speech processing is based on LPC coefficients; bank-of-filter parameters, wavelets, mel- or Bark-warped spectrum, auditory nerve representations, and many more representations are also used. It is obvious that all of these are spectral descriptions. The extensive use of these parameters is a strong indication of our belief that the information in speech is stored in its spectrum, more specifically in the position of the formants.

We can test this premise by filtering some speech in such a way as to considerably whiten its spectrum for some sound or sounds. For example, we can create an inverse filter to the spectrum of a common vowel, such as the e in the word 'feet'. The spectrum will be completely flat when this vowel sound is spoken, and will be considerably distorted during other vowel sounds. Yet this 'inverse-E' filtered speech turns out to be perfectly intelligible. Of course a speech recognition device based on one of the aforementioned parameter sets will utterly fail.

So where is the information if not in the spectrum? A well-known fact regarding our senses is that they respond mainly to change and not to steady-state phenomena. Strong odors become unnoticeable after a short while, our eyes twitch in order to keep objects moving on our retina (animals without the eye twitch only see moving objects) and even a relatively loud stationary background noise seems to fade away. Although our speech generation system is efficient at creating formants, our hearing system is mainly sensitive to *changes* in these formants.

One way this effect can be taken into account in speech recognition systems is to use derivative coefficients. For example, in addition to using LPC cepstral coefficients as features, some systems use the so-called delta cepstral coefficients, which capture the time variation of the cepstral coefficients. Some researchers have suggested using the delta-delta coefficients as well, in order to capture second derivative effects.

An alternative to this empirical addition of time-variant information is to use a set of parameters specifically built to emphasize the signal's time variation. One such set of parameters is called RASTA-PLP (**R**elative **S**pectra—**P**erceptual **L**inear **P**rediction). The basic PLP technique modifies the short time spectrum by several psychophysically motivated transformations, including resampling the spectrum into Bark segments, taking the logarithm of the spectral amplitude and weighting the spectrum by a simulation of the psychophysical equal-loudness curve, before fitting to an all-pole model. The RASTA technique suppresses steady state behavior by band-pass filtering each frequency channel, in this way removing DC and slowly varying terms. It has been found that RASTA parameters are less sensitive to artifacts; for example, LPC-based speech recognition systems trained on microphone-quality speech do not work well when presented with telephone speech. The performance of a RASTA-based system degrades much less.

Even more radical departures from LPC-type parameters are provided by cochlear models and auditory nerve parameters. Such parameter sets attempt to duplicate actual signals present in the biological hearing system (see Section 11.4). Although there is an obvious proof that such parameters can be effectively used for tasks such as speech recognition, their success to date has not been great.

Another set of speech parameters that *has* been successful in varied tasks is the so-called 'sinusoidal representation'. Rather than making a U/V decision and modeling the excitation as a set of pulses, the sinusoidal representation uses a sum of L sinusoids of arbitrary amplitudes, frequencies, and phases. This simplifies computations since the effect of the linear filter on sinusoids is elementary, the main problem being matching of the models at segment boundaries. A nice feature of the sinusoidal representation is that various transformations become relatively easy to perform. For example, changing the speed of articulation without varying the pitch, or conversely varying the pitch without changing rate of articulation, are easily accomplished since the effect of speeding up or slowing down time on sinusoids is straightforward to compute.

We finish off our discussion of speech features with a question. How many features are really needed? Many speech recognition systems use ten LPC or twelve LPC cepstrum coefficients, but to these we may need to add the delta coefficients as well. Even more common is the 'play it safe' approach where large numbers of features are used, in order not to discard any possibly relevant information. Yet these large feature sets contain a large amount of redundant information, and it would be useful, both theoretically and in practice, to have a minimal set of features. Such a set *might* be useful for speech compression as well, but not necessarily. Were these features to be of large range and very sensitive, each would require a large number of bits to accurately represent, and the total number of bits needed could exceed that of traditional methods.

One way to answer the question is by empirically measuring the dimensionality of speech sounds. We won't delve too deeply into the mechanics of how this is done, but it is possible to consider each set of N consecutive samples as a vector in N-dimensional space, and observe how this N-dimensional speech vector moves. We may find that the local movement is constrained to $M < N$ dimensions, like the movement of a dot on a piece of paper viewed at some arbitrary angle in three-dimensional space. Were this the case we would conclude that only M features are required to describe the speech signal. Of course these M features will probably not be universal, like a piece of paper that twists and curves in three-dimensional space, its directions changing from place to place. Yet as long as the paper is not crumpled into a three-dimensional ball, its local dimensionality remains two. Performing such experiments on vowel sounds has led several researchers to conclude that three to five local features are sufficient to describe speech.

Of course this demonstration is not constructive and leaves us totally in the dark as to how to find such a small set of features. Attempts are being made to search for these features using learning algorithms and neural networks, but it is too early to hazard a guess as to success and possible impact of this line of inquiry.

EXERCISES

19.4.1 Speech has an overall spectral tilt of 5 to 12 dB per octave. Remove this tilt (a pre-emphasis filter of the form $1 - 0.99z^{-1}$ is often used) and listen to the speech. Is the speech intelligible? Does it sound natural?

19.4.2 If speech information really lies in the changes, why don't we differentiate the signal and then perform the analysis?

19.5 Pitch Tracking and Voicing Determination

The process of determining the pitch of a segment of voiced speech is usually called pitch *tracking*, since the determination must be updated for every segment. Pitch determination would seem to be a simple process, yet no-one has ever discovered an entirely reliable pitch tracking algorithm. Moreover, even extremely sophisticated pitch tracking algorithms do not usually suffer from minor accuracy problems; rather they tend to make gross errors, such as isolated reporting of double the pitch period. For this reason postprocessing stages are often used.

The pitch is the fundamental frequency in voiced speech, and our ears are very sensitive to pitch changes, although in nontonal languages their content is limited to prosodic information. Filtering that removes the pitch frequency itself does not strongly impair our perception of pitch, although it would thwart any pitch tracking technique that relies on finding the pitch spectral line. Also, a single speaker's pitch may vary over several octaves, for example, from 50 to 800 Hz, while low-frequency formants also occupy this range and may masquerade as pitch lines. Moreover, speech is neither periodic nor even stationary over even moderately long times, so that limiting ourselves to times during which the signal is stationary would provide unacceptably large uncertainties in the pitch determination. Hoarse and high-pitched voices are particularly difficult in this regard.

All this said, there are many pitch tracking algorithms available. One major class of algorithms is based on finding peaks in the empirical autocorrelation. A typical algorithm from this class starts by low-pass filtering the speech signal to eliminate frequency components above 800 or 900 Hz. The pitch should correspond to a peak in the autocorrelation of this signal, but there are still many peaks from which to choose. Choosing the largest peak sometimes works, but may result in a multiple of the pitch or in a formant frequency. Instead of immediately computing the autocorrelation we first center clip (see equation (8.7)) the signal, a process that tends to flatten out vocal tract autocorrelation peaks. The idea is that the formant periodicity should be riding on that of the pitch, even if its consistency results in a larger spectral peak. Accordingly, after center clipping we expect only pitch-related phenomena to remain. Of course the exact threshold for the center clipping must be properly set for this preprocessing to work, and various schemes have been developed. Most schemes first determine the highest sample in the segment and eliminate the middle third of the dynamic range. Now autocorrelation lags that correspond to valid pitch periods are computed. Once again we might naively expect the largest peak to correspond to the pitch period, but if filtering of the original signal removed or attenuated the pitch frequency this may not be the case. A better strategy is to look for consistency in the observed autocorrelation peaks, choosing a period that has the most energy in the peak and its multiples. This technique tends to work even for noisy speech, but requires postprocessing to correct random errors in isolated segments.

A variant of the autocorrelation class computes the **Average Magnitude Difference Function**

$$\text{AMDF}(m) = \sum_n |x_n - x_{n+m}|$$

(AMDF) rather than the autocorrelation. The AMDF is a nonnegative function of the lag m that returns zero only when the speech is exactly periodic. For noisy nearly periodic signals the AMDF has a strong minimum at the best matching period. The nice thing about using a minimum rather than maximum is that we needn't worry as much about the signal remaining stationary. Indeed a single pitch period should be sufficient for AMDF-based pitch determination.

Another class of pitch trackers work in the frequency domain. It may not be possible to find the pitch line itself in the speech spectrum, but finding the frequency with maximal harmonic energy is viable. This may be accomplished in practice by compressing the power spectrum by factors of two, three, and four and adding these to the original PSD. The largest peak in the resulting 'compressed spectrum' is taken to be the pitch frequency.

In Section 19.3 we mentioned the use of power cepstrum in determining the pitch. Assuming that the formant and pitch information is truly separated in the cepstral domain, the task of finding the pitch is reduced to picking the strongest peak. While this technique may give the most accurate results for clean speech, and rarely outputs double pitch, it tends to deteriorate rapidly in noise.

The determination of whether a segment of speech is voiced or not is also much more difficult than it appears. Actually, the issue needn't even be clear cut; speech experts speak of the 'degree of voicing', meaning the percentage of the excitation energy in the pitch pulses as compared to the total excitation. The MELP and Multi-Band Exitation (MBE) speech compression methods abandon the whole idea of an unambiguous U/V decision, using mixtures or *per-frequency-band* decisions respectively.

Voicing determination algorithms lie somewhere between VADs and pitch trackers. Some algorithms search separately for indications of pitch and noise excitation, declaring voiced or unvoiced when either is found, 'silence' when neither is found, and 'mixed' when both are. Other algorithms are integrated into pitch trackers, as in the case of the cepstral pitch tracker that returns 'unvoiced' when no significant cepstral peak is found.

In theory one can distinguish between voiced and unvoiced speech based on amplitude constancy. Voiced speech is only excited by the pitch pulse, and during much of the pitch period behaves as a exponentially decaying sinusoid. Unvoiced speech should look like the output of a continuously exited filter. The difference in these behaviors may be observable by taking the Hilbert transform and plotting the time evolution in the I-Q plane. Voice speech will tend to look like a spiral while unvoiced sections will appear as filled discs. For this technique to work the speech has to be relatively clean, and highly oversampled.

The degree of periodicity of a signal should be measurable as the ratio of the maximum to minimum values of the autocorrelation (or AMDF). However, in practice this parameter too is overrated. Various techniques supplement this ratio with gross spectral features, zero crossing and delta zero crossing, and many other inputs. Together these features are input to a decision mechanism that may be hard-wired logic, or a trainable classifier.

EXERCISES

19.5.1 In order to minimize time spent in computation of autocorrelation lags, one can replace the center clipping operation with a three-level slicing operation that only outputs -1, 0 or $+1$. How does this decrease complexity? Does this operation strongly affect the performance of the algorithm?

19.5.2 Create a signal that is the weighted sum of a few sinusoids interrupted every now and then by short durations of white noise. You can probably easily separate the two signal types by eye in either time or frequency domains. Now do the same using any of the methods discussed above, or any algorithm of your own devising.

19.5.3 Repeat the previous exercise with additive noise on the sinusoids and narrow band noise instead of white noise. How much noise can your algorithm tolerate? How narrow-band can the 'unvoiced' sections be and still be identifiable? Can you do better 'by eye' than your algorithm?

19.6 Speech Compression

It is often necessary or desirable to compress digital signals. By compression we mean the representation of N signal values, each of which is quantized to b bits, in less than Nb bits. Two common situations that may require compression are transmission and storage. Transmission of an uncompressed digital music signal (sampled at 48 KHz, 16 bits per sample) requires at least a 768 Kb/s transmission medium, far exceeding the rates usually available for users connected via phone lines. Storage of this same signal requires almost 94 KB per second, thus gobbling up disk space at about $5\frac{1}{2}$ MB per minute. Even limiting the bandwidth to 4 KHz (commonly done to speech in the public telephone system) and sampling at 16 bits leads to 128 Kb/s, far exceeding our ability to send this same information over the same channel using a telephony-grade modem. This would lead us to believe that digital methods are less efficient than analog ones, yet there *are* methods of digitally sending multiple conversations over a single telephone line.

Since further reduction in bandwidth or the number of quantization bits rapidly leads to severe quality degradation we must find a more sophisticated compression method. What about general-purpose data compression techniques? These may be able to contribute another factor-of-two improvement, but that is as far as they go. This is mainly because these methods are *lossless*, meaning they are required to reproduce the original bit stream without error. Extending techniques that work on general bit streams to the lossy regime is fruitless. It does not really make sense to view the speech signal as a stream of bits and to minimize the number of bit errors in the reconstructed stream. This is because some bits are more significant than others—an error in the least significant bit is of much less effect than an error in a sign bit!

It is less obvious that it is also not optimal to view the speech signal as a stream of sample values and compress it in such a fashion as to minimize the energy of error signal (reconstructed signal minus original signal). This is because two completely different signals may sound the same since hearing involves complex physiological and psychophysical processes (see Section 11.4).

For example, by delaying the speech signal by two samples, we create a new signal completely indistinguishable to the ear but with a large 'error signal'. The ear is insensitive to absolute time and thus would not be able to differentiate between these two 'different' signals. Of course simple cross correlation would home-in on the proper delay and once corrected the error would be zero again. But consider delaying the digital signal by half a sample (using an appropriate interpolation technique), producing a signal with completely distinct sample values. Once again a knowledgeable signal processor would be able to discover this subterfuge and return a very small error. Similarly, the ear is insensitive to small changes in loudness and absolute phase. However, the ear is also insensitive to more exotic transformations such as small changes in pitch, formant location, and nonlinear warping of the time axis.

Reversing our point-of-view we can say that speech-specific compression techniques work well for two related reasons. First, speech compression techniques are *lossy* (i.e., they strive to reproduce a signal that is similar but not necessarily identical to the original); significantly lower information rates can be achieved by introducing tolerable amounts of distortion. Second, once we have abandoned the ideal of precise reconstruction of the original signal, we can go a step further. The reconstructed signal needn't really be similar to the original (e.g., have minimal mean square error); it should merely *sound* similar. Since the ear is insensitive to small changes in phase, timing, and pitch, much of the information in the original signal is unimportant and needn't be encoded at all.

It was once common to differentiate between two types of speech coders. 'Waveform coders' exploit characteristics of the speech signal (e.g., energy concentration at low frequencies) to encode the speech samples in fewer bits than would be required for a completely random signal. The encoding is a lossy transformation and hence the reconstructed signal is not identical to the original one. However, the encoder algorithm is built to minimize some distortion measure, such as the squared difference between the original and reconstructed signals. 'Vocoders' utilize speech synthesis models (e.g., the speech model discussed in Section 9.9) to encode the speech signal. Such a model is capable of producing speech that sounds very similar to the speech that we desire to encode, but requires the proper parameters as a function of time. A vocoder-type algorithm attempts to find these parameters and usually results in reconstructed speech that sounds similar to the original but as a signal may look quite different. The distinction between waveform encoders and vocoders has become extremely fuzzy. For example, the distortion measure used in a waveform encoder may be perception-based and hence the reconstructed signal may be quite unlike the original. On the other hand, analysis by synthesis algorithms may find a vocoder's parameters by minimizing the squared error of the synthesized speech.

When comparing the many different speech compression methods that have been developed, there are four main parameters that should be taken into consideration, namely rate, quality, complexity, and delay. Obviously, there are trade-offs between these parameters, lowering of the bit rate requires higher computational complexity and/or lower perceived speech quality; and constraining the algorithm's delay while maintaining quality results in a considerable increase in complexity. For particular applications there may be further parameters of interest (e.g., the effect of background noise, degradation in the presence of bit errors).

The perceived quality of a speech signal involves not only how understandable it is, but other more elusive qualities such as how natural sounding the speech seems and how much of the speaker's identity is preserved. It is not surprising that the most reliable and widely accepted measures of speech quality involve humans listening rather than pure signal analysis. In order to minimize the bias of a single listener, a psychophysical measure of speech quality called the **M**ean **O**pinion **S**core (MOS) has been developed. It is determined by having a group of seasoned listeners listen to the speech in question. Each listener gives it an opinion score: 1 for 'bad' (not understandable), 2 for 'poor' (understandable only with considerable effort), 3 for 'fair' (understandable with moderate effort), 4 for 'good' (understandable with no apparent effort), and 5 for 'excellent'. The mean score of all the listeners is the MOS. A complete description of the experimental procedure is given in ITU-T standard P.830.

Speech heard directly from the speaker in a quiet room will receive a MOS ranking of 5.0, while good 4 KHz telephone-quality speech (termed *toll quality*) is ranked 4.0. To the uninitiated telephone speech may seem almost the same as high-quality speech, however, this is in large part due to the brain compensating for the degradation in quality. In fact different phonemes may become acoustically indistinguishable after the band-pass filtering to 4 KHz (e.g. s and f), but this fact often goes unnoticed, just as the 'blind spots' in our eyes do. MOS ratings from 3.5 to 4 are sometimes called 'communications quality', and although lower than toll quality are acceptable for many applications.

Usually MOS tests are performed along with calibration runs of known MOS, but there still are consistent discrepancies between the various laboratories that perform these measurements. The effort and expense required to obtain an MOS rating for a coder are so great that objective tests that correlate well with empirical MOS ratings have been developed. **P**erceptual **S**peech **Q**uality **M**easure (PSQM) and **P**erceptual **E**valuation of **S**peech **Q**uality (PESQ) are two such which have been standardized by the ITU.

EXERCISES

19.6.1 Why can't general-purpose data compression techniques be *lossy*?

19.6.2 Assume a language with 64 different phonemes that can be spoken at the rate of eight phonemes per second. What is the minimal bit rate required?

19.6.3 Try to compress a speech file with a general-purpose lossless data (file) compression program. What compression ratio do you get?

19.6.4 Several lossy speech compression algorithms are readily available or in the public domain (e.g., LPC-10e, CELP, GSM full-rate). Compress a file of speech using one or more of these compressions. Now listen to the 'before' and 'after' files. Can you tell which is which? What artifacts are most noticeable in the compressed file? What happens when you compress a file that had been decompressed from a previous compression?

19.6.5 What happens when the input to a speech compression algorithm is not speech? Try single tones or DTMF tones. Try music. What about 'babble noise' (multiple background voices)?

19.6.6 Corrupt a file of linear 16-bit speech by randomly flipping a small percentage of the bits. What percentage is not noticed? What percentage is acceptable? Repeat the experiment by corrupting a file of compressed speech. What can you conclude about media for transmitting compressed speech?

19.7 PCM

In order to record and/or process speech digitally one needs first to acquire it by an A/D. The digital signal obtained in this fashion is usually called 'linear PCM' (recall the definition of PCM from Section 2.7). Speech contains significant frequency components up to about 20 KHz, and Nyquist would thus require a 40 KHz or higher sampling rate. From experimentation at that rate with various numbers of sample levels one can easily become convinced that using less than 12 to 14 bits per sample noticeably degrades the signal. Eight bits definitely delivers inferior quality, and since conventional hardware works in multiples of 8-bit bytes, we usually digitize speech using 16 bits per sample. Hence the simplistic approach to capturing speech digitally would be to sample at 40 KHz using 16 bits per sample for a total information rate of 640 Kb/s. Assuming a properly designed microphone, speaker, A/D, D/A, and filters, 640 Kb/s digital speech is indeed close to being indistinguishable from the original.

Our first step in reducing this bit rate is to sacrifice bandwidth by low-pass filtering the speech to 4 KHz, the bandwidth of a telephone channel. Although 4 KHz is not high fidelity it is sufficient to carry highly intelligible speech. At 4 KHz the Nyquist sampling rate is reduced to 8000 samples per second, or 128 Kb/s.

From now on we will use more and more specific features of the speech signal to further reduce the information rate. The first step exploits the psychophysical laws of Weber and Fechner (see Section 11.2). We stated above that 8 bits were not sufficient for proper digitizing of speech. What we really meant is that 256 *equally spaced* quantization levels produces speech of low perceived quality. Our perception of acoustic amplitude is, however, logarithmic, with small changes at lower amplitudes more consequential than equal changes at high amplitudes. It is thus sensible to try unevenly spaced quantization levels, with high density of levels at low amplitudes and much fewer levels used at high amplitudes. The optimal spacing function will be logarithmic, as depicted in Figure 19.3 (which replaces Figure 2.25 for this case). Using logarithmically spaced levels 8 bits is indeed adequate for toll quality speech, and since we now use only 8000 eight-bit samples per second, our new rate is 64 Kb/s, half that of linear PCM. In order for a speech compression scheme to be used in a communications system the sender and receiver, who may be using completely different equipment, must agree as to its details. For this reason precise standards must be established that ensure that different implementations can interoperate. The ITU has defined a number of speech compression schemes. The G.711 standard defines two

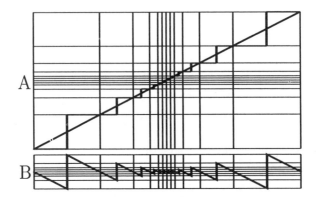

Figure 19.3: Quantization noise created by logarithmically digitizing an analog signal. In (A) we see the output of the logarithmic digitizer as a function of its input. In (B) the noise is the rounding error, (i.e., the output minus the input).

options for logarithmic quantization, known as μ-law (pronounced mu-law) and A-law PCM respectively. Unqualified use of the term 'PCM' in the context of speech often refers to either of the options of this standard.

μ-law is used in the North American digital telephone system, while A-law serves the rest of the world. Both μ-law and A-law are based on rational approximations to the logarithmic response of Figure 19.3, the idea being to minimize the computational complexity of the conversions from linear to logarithmic PCM and back. μ-law is defined as

$$\check{s} = \text{sgn}(s)\,\check{s}_{max}\,\frac{1 + \mu\frac{|s|}{s_{max}}}{1 + \frac{|s|}{s_{max}}} \tag{19.3}$$

where s_{max} is the largest value the signal may attain, $\check{s}_{max}$ is the largest value we wish the compressed signal to attain, and μ is a parameter that determines the nonlinearity of the transformation. The use of the absolute value and the sgn function allow a single expression to be utilized for both positive and negative x. Obviously, $\mu = 1$ forces $\check{s} = x$ while larger μ causes the output to be larger than the input for small input values, but much smaller for large s. In this way small values of s are emphasized before quantization at the expense of large values. The actual telephony standard uses $\mu = 255$ and further reduces computation by approximating the above expression using 16 staircase segments, eight for positive signal values and eight for negative. Each speech sample is encoded as a sign bit, three segment bits and four bits representing the position on the line segment.

The theoretical A-law expression is given by

$$\check{s} = \text{sgn}(s)\,\check{s}_{max} \begin{cases} \frac{A\frac{|s|}{s_{max}}}{1+\ln(A)} & 0 < \frac{|s|}{s_{max}} < \frac{1}{A} \\ \frac{1+\ln(A\frac{|s|}{s_{max}})}{1+\ln(A)} & \frac{1}{A} < \frac{|s|}{s_{max}} < 1 \end{cases} \qquad (19.4)$$

and although it is hard to see this from the expression, its behavior is very similar to that of μ-law. By convention we take $A = 87.56$ and as in the μ-law case approximate the true form with 16 staircase line segments. It is interesting that the A-law staircase has a rising segment at the origin and thus fluctuates for near-zero inputs, while the approximated μ-law has a horizontal segment at the origin and is thus relatively constant for very small inputs.

EXERCISES

19.7.1 Even 640 Kb/s does not capture the entire experience of listening to a speaker in the same room, since lip motion, facial expressions, hand gestures, and other body language are not recorded. How important is such auxiliary information? When do you expect this information to be most relevant? Estimate the information rates of these other signals.

19.7.2 Explain the general form of μ and A laws. Start with general logarithmic compression, extend it to handle negative signal values, and finally force it to go through the origin.

19.7.3 Test the difference between high-quality and toll-quality speech by performing a *rhyme test*. In a rhyme test one person speaks out-of-context words and a second records what was heard. By using carefully chosen words, such as lift-list, lore-more-nor, jeep-cheep, etc., you should be able to both estimate the difference in accuracy between the two cases and determine which phonemes are being confused in the toll-quality case.

19.7.4 What does μ-law (equation (19.3)) return for zero input? For maximal input? When does $y = x$? Plot μ-law for 16-bit linear PCM, taking $x_{max} = 2^{15} = 32768$, for various μ from 1 to 255. What is the qualitative difference between the small and large μ cases?

19.7.5 Plot the μ-law (with $\mu = 255$) and A-law (with $A = 87.56$) responses on the same axes. By how much do they differ? Plot them together with true logarithmic response. How much error do they introduce? Research and plot the 16 line segment approximations. How much further error is introduced?

19.8   DPCM, DM, and ADPCM

The next factor-of-two reduction in information rate exploits the fact that long time averaged spectrum of speech does not look like white noise filtered to 4 KHz. In fact the spectrum is decidedly low-pass in character, due to voiced speech having pitch harmonics that decrease in amplitude as the frequency increases (see Section 11.3).

In Section 9.8 we studied the connection between correlation and prediction, here we wish to stress the connection between prediction and compression. Deterministic signals are completely predictable and thus maximally compressible; knowing the signal's description, (e.g., as a explicit formula or difference equation with given initial conditions) enables one to precisely predict any signal value without any further information required. White noise is completely unpredictable; even given the entire history from the beginning of time to now does not enable us to predict the next signal value with accuracy any better than random guessing. Hence pure white noise is incompressible; we can do no better than to treat each sample separately, and N samples quantized to b bits each will always require Nb bits.

Most signals encountered in practice are somewhere in between; based on observation of the signal we can construct a model that captures the predictable (and thus compressible) component. Using this model we can predict the next value, and then we need only store or transmit the residual error. The more accurate our prediction is, the smaller the error signal will be, and the fewer bits will be needed to represent it. For signals with most of their energy at low frequencies this predictability is especially simple in nature—the next sample will tend to be close to the present sample. Hence the difference between successive sample values tends to be smaller than the sample values themselves. Thus encoding these differences, a technique known as **Ddelta-PCM** (DPCM), will usually require fewer bits. This same term has come to be used in a more general way to mean encoding the difference between the sample value and a predicted version of it.

To see how this generalized DPCM works, let's use the previous value s_{n-1}, or the previous N values $s_{n-N} \ldots s_{n-1}$, to predict the signal value at time n.

$$\tilde{s}_n = p(s_{n-1}, s_{n-2}, \ldots s_{n-N}) \tag{19.5}$$

If the predictor function p is a filter

$$\tilde{s}_n = \sum_{i=1}^{N} p_i s_{n-i} \tag{19.6}$$

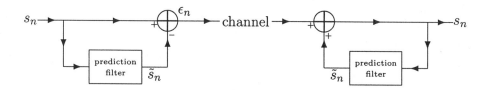

Figure 19.4: Unquantized DPCM. The encoder predicts the next value, finds the prediction error $\epsilon_n = s_n - \tilde{s}_n$, and transmits this error through the communications channel to the receiver. The receiver, imitating the transmitter, predicts the next value based on all the values it has recovered so far. It then corrects this prediction based on the error ϵ_n received.

we call the predictor a *linear predictor*. If the predictor works well, the prediction error

$$\epsilon_n = s_n - \tilde{s}_n \tag{19.7}$$

is both of lower energy and much whiter than the original signal s_n. The error is all we need to transmit for the receiver to be able to reconstruct the signal, since it too can predict the next signal value based on the past values. Of course this prediction $\tilde{s}_n$ is not completely accurate, but the correction ϵ_n is received, and the original value easily recovered by $s_n = \tilde{s}_n + \epsilon_n$. The entire system is depicted in Figure 19.4. We see that the encoder (linear predictor) is present in the decoder, but that there it runs as feedback, rather than feedforward as in the encoder itself.

The simplest DPCM system is **Delta Modulation (DM)**. Delta modulation uses only a single bit to encode the error, this bit signifying whether the true value is above or below the predicted one. If the sampling frequency is so much higher than required that the previous value s_{n-1} itself is a good predictor of s_n, delta modulation becomes the sigma-delta converter of Section 2.11. In a more general setting a nontrivial predictor is used, but we still encode only the sign of the prediction error. Since delta modulation provides no option to encode zero prediction error the decoded signal tends to oscillate up and down where the original was relatively constant. This annoyance can be ameliorated by the use a *post-filter*, which low-pass filters the reconstructed signal.

There is a fundamental problem with the DPCM encoders we have just described. We assumed that the true value of the prediction error ϵ_n is transferred over the channel, while in fact we can only transfer a quantized version ϵ_n^Q. The very reason we perform the prediction is to save bits after quantization. Unfortunately, this quantization may have a devastating effect on the decoder. The problem is not just that the correction of the present prediction is not completely accurate; the real problem is that because of this inaccuracy the receiver never has reliable s_n with which to continue predicting the next samples. To see this, define s_n^Q as the decoder's predicted value corrected by the quantized error. In general, s_n^Q does not quite equal s_n, but we predict the next sample values based on these incorrect corrected predictions! Due to the feedback nature of the decoder's predictor the errors start piling up and after a short time the encoder and decoder become 'out of sync'.

The prediction we have been using is known as *open-loop* prediction, by which we mean that we perform linear prediction of the input speech. In order to ensure that the encoder and decoder predictors stay in sync, we really should perform linear prediction on the speech as reconstructed by the decoder. Unfortunately, the decoder output is not available at the encoder, and so we need to calculate it. To perform *closed-loop* prediction we build an exact copy of the entire decoder into our encoder, and use its output, rather than the input speech, as input to the predictor. This process is diagrammed in Figure 19.5. By 'closing the loop' in this fashion, the decoded speech is precisely that expected, unless the channel introduces bit errors.

The international standard for 32 Kb/s toll quality digital speech is based on **Adaptive Delta-PCM** (ADPCM). The 'adaptive' is best explained by returning to the simple case of delta modulation. We saw above that the DM *encoder* compares the speech signal value with the predicted (or simply previous) value and reports whether this prediction is too high or too low. How does a DM *decoder* work? For each input bit it takes its present estimate for the speech signal value and either adds or subtracts some step size Δ. Assuming Δ is properly chosen this strategy works well for some range of input signal frequencies; but as seen in Figure 19.6 using a single step

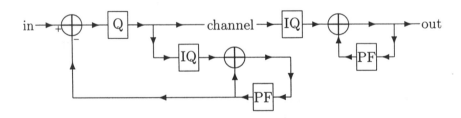

Figure 19.5: Closed-loop prediction. In this figure, Q stands for quantizer, IQ inverse quantizer, PF prediction filter. Note that the encoder contains an exact replica of the decoder and predicts the next value based on the reconstructed speech.

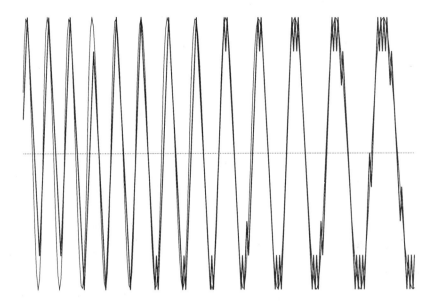

Figure 19.6: The two types of errors in nonadaptive delta modulation. We superpose the reconstructed signal on the original. If the step size is too small the reconstructed signal can't keep up in areas of large slope and may even completely miss peaks (as in the higher-frequency area at the beginning of the figure). If the step size is too large the reconstructed signal will oscillate wildly in areas where the signal is relatively constant (as seen at the peaks of the lower-frequency area toward the end of the figure).

size cannot satisfy all frequencies. If Δ is too small the reconstructed signal cannot keep up when the signal changes rapidly in one direction and may even completely miss peaks (as in the higher-frequency area at the beginning of the figure), a phenomenon called 'slope overload'. If Δ is too large the reconstructed signal will oscillate wildly when the signal is relatively constant (as seen at the peaks of the lower-frequency area toward the end of the figure), which is known as 'granular noise'.

While we introduced the errors introduced by improper step size for DM, the same phenomena occur for general DPCM. In fact the problem is even worse. For DM the step size Δ is only used at the decoder, since the encoder only needs to check the sign of the difference between the signal value and its prediction. For general delta-PCM the step size is needed at the encoder as well, since the difference must be quantized using levels spaced Δ apart. Improper setting of the spacing between the quantization levels causes mismatch between the digitizer and the difference signal's dynamic range, leading to improper quantization (see Section 2.9).

The solution is to adapt the step size to match the signal's behavior. In order to minimize the error we increase Δ when the signal is rapidly increasing or decreasing, and we decrease it when the signal is more constant. A simplistic way to implement this idea for DM is to use the bit stream itself to determine whether the step size is too small or too large. A commonly used version uses memory of the previous delta bit; if the present bit is the same as the previous we multiply Δ by some constant K ($K = 1.5$ is a common choice), while if the bits differ we divide by K. In addition we constrain Δ to remain within some prespecified range, and so stop adapting when it reaches its minimum or maximum value.

While efficient computationally, the above method for adapting Δ is completely heuristic. A more general tactic is to set the step size for adaptive DPCM to be a given percentage of the signal's standard deviation. In this way Δ would be small for signals that do vary much, minimizing granular noise, but large for wildly varying signals, minimizing slope overload. Were speech stationary over long times adaptation would not be needed, but since the statistics of the speech signal vary widely as the phonemes change, we need to continuously update our estimate of its variance. This can be accomplished by collecting N samples of the input speech signal in a buffer, computing the standard deviation, setting Δ accordingly, and only then performing the quantization. N needs to be long enough for the variance computation to be accurate, but not so long that the signal statistics vary appreciably over the buffer. Values of 128 (corresponding to 16 milliseconds of speech at 8000 Hz) through 512 (64 milliseconds) are commonly used.

There are two drawbacks to this method of adaptively setting the scale of the quantizer. First, the collecting of N samples before quantization requires introducing buffer delay; in order to avoid excessive delay we can use an IIR filter to track the variance instead of computing it in a buffer. Second, the decoder needs to know Δ, and so it must be sent as *side information*, increasing the amount of data transferred. The overhead can be avoided by having the decoder derive Δ, but if Δ is derived from the input signal, this is not possible. The decoder could try to use the reconstructed speech to find Δ, but this would not exactly match the quantization step used by the encoder. After a while the encoder and decoder would no longer agree and the system would break down. As you may have guessed, the solution is to close the loop and have the encoder determine Δ using its internal decoder, a technique called *backward adaptation*.

EXERCISES

19.8.1 Obtain a copy of the G.726 ADPCM standard and study the main block diagrams for the encoder and decoder. Explain the function and connections of the adaptive predictor, adaptive quantizer, and inverse adaptive quantizer. Why is the standard so detailed?

19.8.2 Now study the expanded block diagram of the encoder. What is the purpose of the blocks marked 'adaptation speed control' and 'tone and transition detector'?

19.8.3 How does the MIPS complexity of the G.726 encoder compare with that of modern lower-rate encoders?

19.8.4 Show that the open-loop prediction results in large error because the quantization error is multiplied by the prediction gain. Show that with closed-loop prediction this does not occur.

19.9   Vector Quantization

For white noise we can do no better than to quantize each sample separately, but for other signals it may make sense to quantize groups of samples together. This is called **Vector Quantization (VQ)**.

Before discussing *vector* quantization it is worthwhile to reflect on what we have accomplished so far in *scalar* quantization. The digitization of the A/D converters discussed in Section 2.7 was input independent and uniform. By this we mean that the positions of the quantization levels were preset and equidistant. In order to minimize the quantization noise we usually provide an amplifier that matches the analog signal to the predetermined dynamic range of the digitizer. A more sophisticated approach is to set the digitizer levels to match the signal, placing the levels close together for small amplitude signals, and further apart for stronger signals. When the range of the signal does not vary with time and is known ahead of time, it is enough to set this spacing once; but if the signal changes substantially with time we need to adapt the level spacing according to the signal. This leads to adaptive PCM, similar to but simpler than the ADPCM we studied in Section 19.8.

With adaptive PCM the quantization levels are not preset, but they are still equidistant. A more sophisticated technique is nonuniform quantization, such as the logarithmic PCM of Section 19.7. The idea behind logarithmic PCM was that low levels are more prevalent and their precision perceptually more important than higher ones; thus we can reduce the average (perceptual) error by placing the quantization levels closer together for small signal values, and further apart for large values.

We will return to the perceptual importance later; for now we assume all signal values to be equally important and just ask how to combine adaptivity with nonequidistant quantization thresholds. Our objective is to lower the average quantization error; and this can be accomplished by placing the levels closer together where the signal values are more probable.

Rather than adapting quantization thresholds, we can adapt the midpoints between these thresholds. We call these midpoints 'centers', and the quantization thresholds are now midway between adjacent centers. It is then obvious that classifying an input as belonging to the nearest 'center' is equivalent to quantizing according to these thresholds. The set of all values that are classified as closest to a given center (i.e., that lie between the two thresholds) is called its 'cluster'.

The reason we prefer to set centers is that there is an easily defined criterion that differentiates between good sets of centers and poor ones, namely mean square error (MSE). Accordingly, if we have observed N signal values $\{x_n\}_{n=1}^{N}$, we want to place M centers $\{c_m\}_{m=1}^{M}$ in such a way that we minimize the mean square quantization error.

$$E = \frac{1}{N} \sum_{n=1}^{N} |x_n - c_n|^2 \tag{19.8}$$

We have used here the short-hand notation c_n to mean that center closest to x_n.

Algorithms that perform this minimization given empirical data are called 'clustering' algorithms. In a moment we will present the simplest of these algorithms, but even it already contains many of the elements of the most complex of these algorithms.

There is another nomenclature worth introducing. Rather than thinking of minimal error clustering we can think of quantization as a form of *encoding*, whereby a real signal value is encoded by the index of the interval to which it belongs. When decoding, the index is replaced by the center's value, introducing a certain amount of error. Because of this perspective the center is usually called a *codeword* and the set of M centers $\{c_j\}_{j=1}^{M}$ the *codebook*.

How do we find the codebook given empirical data? Our algorithm will be iterative. We first randomly place the M centers, and then move them in such a way that the average coding error is decreased. We continue to iterate until no further decrease in error is possible. The question that remains is how to move the centers in order to reduce the average error.

Figure 19.7: Quantization thresholds found by the scalar quantization algorithm for uniform and Gaussian distributed data. For both cases 1000 points were generated, and 16 centers found by running the basic scalar quantization algorithm until convergence.

Were we to know which inputs *should* belong to a certain cluster, then minimizing the sum of the squared errors would require positioning the center at the average of these input values. The idea behind the algorithm is to exploit this fact at each iteration. At each stage there is a particular set of M centers that has been found. The best guess for associating signal values to cluster centers is to classify each observed signal value as belonging to the closest center. For this set of classifications we can then position the centers optimally at the average. In general this correction of center positions will change the classifications, and thus we need to reclassify the signal values and recompute the averages. Our iterative algorithm for scalar quantization is therefore the following.

```
Given: signal values {x_i}_{i=1}^N,
       the desired codebook size M
Initialize: randomly choose M cluster centers {c_j}_{j=1}^M
Loop:
      Classification step:
            for i ← 1...N
                  for j ← 1...M
                        compute d_{ij}^2 = |x_i - c_j|^2
                        classify x_i as belonging to C_j with minimal d_{ij}^2
      Expectation step:
            for j ← 1...M
                  correct center c_j ← (1/N_j) Σ_{i∈C_j} x_i
```

Here N_j stands for the number of x_i that were classified as belonging to cluster C_j. If $N_j = 0$ then no values are assigned to center j and we discard it.

Note that there are two steps in the loop, a *classification* step where we find the closest center c_n, and an *expectation* step where we compute the average of all values belonging to each center c_m and reposition it. We thus say that this algorithm is in the class of *expectation-classification* algorithms. In the pattern recognition literature this algorithm is called 'k-means', while in speech coding it is called the LBG algorithm (after **L**inde, **B**uzo, and **G**ray). An example of two runs of LBG on scalar data is presented in Figure 19.7.

We now return to vector quantization. The problem is the same, only now we have N input vectors in D-dimensional space, and we are interested in placing M centers in such fashion that mean encoding error is minimized. The thresholds are more complex now, the clusters defining *Voronoy regions*, but precisely the same algorithm can be used. All that has to be done is to interpret the calculations as vector operations.

Now that we know how to perform VQ what do we do with it? It turns out *not* to be efficient to directly VQ blocks of speech samples, but sets of LPC coefficients (or any of the other alternative features) and the LPC residual can be successfully compressed using VQ. Not only does VQ encoding of speech parameters provide a compact representation for speech compression, it is also widely used in speech recognition.

EXERCISES

19.9.1 Prove the point closest to all points in a cluster is their average.

19.9.2 Generate bimodal random numbers, i.e., ones with a distribution with two separated peaks. Determine the error for the best standard quantization. Now run the LBG algorithm with the same number of levels and check the error again. How much improvement did you get?

19.9.3 Generate random vectors that are distributed according to a 'Gaussian mixture' distribution. This is done as follows. Choose M cluster centers in N-dimensional space. For each number to be generated randomly select the cluster, and then add to it Gaussian noise (if the noise has the same variance for all elements then the clusters will be hyperspherical). Now run the LBG algorithm. Change the size of the codebook. How does the error decrease with codebook size?

19.10 SBC

The next factor-of-two can be achieved by noticing that the short time spectrum tends to have a only a few areas with significant energy. The **S**ub**B**and Coding (SBC) technique takes advantage of this feature by dividing the spectrum into a number (typically 8 to 16) of subbands. Each subband signal, created by QMF band-pass filtering, is encoded separately. This in itself would not conserve bits, but adaptively deciding on the number of bits (if any) that should be devoted to each subband, does.

Typical SBC coders of this type divide the bandwidth from DC to 4 KHz into 16 bands of 250 Hz each, and often discard the lowest and highest of these, bands that carry little speech information. Each of the remaining subbands is decimated by a factor of 16, and divided into time segments, with 32 milliseconds a typical choice. 32 milliseconds corresponds to 256 samples of the original signal, but only 16 samples for each of the decimated subbands. In order to encode at 16 Kb/s the output of all the subbands together cannot exceed 512 bits, or an average of 32 bits per subband (assuming 16 subbands). Since we might be using only 14 subbands, and furthermore subbands with low energy may be discarded with little effect on the quality, the number of bits may be somewhat larger; but the bit allocation table and overall gain (usually separately encoded) also require bits. So the task is now to encode 16 decimated samples in about 32 bits.

After discarding the low-energy subbands the remaining ones are sorted in order of dynamic range and available bits awarded accordingly. Subbands with relatively constant signals can be replaced by scalar-quantized averages, while for more complex subbands vector quantization is commonly employed.

An alternative to equal division of the bandwidth is hierarchical logarithmic division, as described in Section 13.9. This division is both more efficient to compute (using the pyramid algorithm) and perceptually well motivated.

EXERCISES

19.10.1 Can we always decimate subbands according to their bandwidth? (Hint: Recall the 'band-pass sampling theorem'.)

19.10.2 When dividing into equal-bandwidth bands, in which are more bits typically needed, those with lower or higher frequencies? Is this consistent with what happens with logarithmic division?

19.10.3 Will dividing the bandwidth into arbitrary bands adaptively matched to the signal produce better compression?

19.11   LPC Speech Compression

We now return to the LPC speech analysis and synthesis methods of sections 19.1 and 19.2 and discuss 'U.S. Standard 1015' more commonly known as LPC-10e. This standard compresses 8000 sample-per-second speech to 2.4 Kb/s using 10 LPC coefficients (hence its name).

LPC-10 starts by dividing the speech into 180-sample blocks, each of which will be converted into 53 bits to which one synchronization bit is added for a total of 54 bits. The 54 bits times 8000/180 results in precisely 2400 b/s. The U/V decision and pitch determination is performed using an AMDF technique and encoded in 7 bits. The gain is measured and quantized to 5 bits and then the block is normalized. If you have been counting, 41 bits are left to specify the LPC filter. LPC analysis is performed using the covariance method and ten reflection coefficients are derived. The first two are converted to log area ratios and all are quantized with between 3 and 6 bits per coefficient. Actually by augmenting LPC-10 with vector quantization we can coax down the data rate to less than 1 Kb/s.

Unfortunately, although highly compressed, straight LPC-encoded speech is of rather poor quality. The speech sounds synthetic and much of the speaker information is lost. The obvious remedy in such cases is to compute and send the error signal as well. In order to do this we need to add the complete decoder to the encoder, and require it to subtract the reconstructed signal from the original speech and to send the error signal through the channel. At the decoder side the process would then be to reconstruct the LPC-encoded signal and then to add back the error signal to obtain the original speech signal.

The problem with the above idea is that in general such error signals, sampled at the original sampling rate (8 KHz) may require the same number of bits to encode as the original speech. We can only gain if the error signal is itself significantly compressible. This was the idea we used in ADPCM where the difference (error) signal was of lower dynamic range than the original speech. The LPC error signal is definitely somewhat smaller than the original speech, but that is no longer enough. We have already used up quite a few bits per second on the LPC coefficients, and we need the error signal to be either an order-of-magnitude smaller or highly correlated in the time domain for sufficient compression to be possible.

Observing typical error signals is enlightening. The error is indeed smaller in magnitude than the speech signal, but not by an order-of-magnitude. It also has a very noticeable periodic component. This periodicity is at the pitch frequency and is due to the LPC analysis only being carried out for times longer than those of the pitch frequency. Our assumption that the pitch excitation could be modeled as a single pulse per pitch period and otherwise zero has apparently been pushed beyond its limits. If we remove the residual pitch period correlations the remaining error seems to be white noise. Hence, trying to efficiently compress the error signal would seem to be a useless exercise.

EXERCISES

19.11.1 You can find code for LPC-10e in the public domain. Encode and then decode some recorded speech. How do you rate the quality? Can you always understand what is being said? Can you identify the speaker? Are some speakers consistently hard to understand?

19.11.2 In **R**esidual **E**xcited **L**inear **P**rediction (RELP) the residual is low-pass filtered to about 1 KHz and then decimated to lower its bit rate. Diagram the RELP encoder and decoder. For what bit rates do you expect RELP to function well?

19.12   CELP Coders

In the last section we saw that straight LPC using a single pulse per pitch period is an oversimplification. Rather than trying to encode the error signal, we can try to find an excitation signal that reduces the residual error. If this excitation can be efficiently encoded and transmitted, the decoder will be able to excite the remote predictor with it and reproduce the original speech to higher accuracy with tolerable increase in bit rate.

There are several different ways to encode the excitation. The most naive technique uses random codebooks. Here we can create, using VQ, a limited number 2^m of random N-vectors that are as evenly distributed in N-dimensional space as possible. These vectors are known both to the encoder and to the decoder. After performing LPC analysis, we try each of these random excitations, and choose the one that produces the lowest prediction error. Since there are 2^m possible excitations, sending the index of the best excitation requires only m bits. Surprisingly, this simple technique already provides a significant improvement in quality as compared to LPC, with only a modest increase in bit rate. The problem, of course, is the need to exhaustively search the entire set of 2^m excitation vectors. For this reason CELP encoders are computationally demanding.

As an example of a simple CELP coder consider federal standard 1016. This coder operates at 4.8 Kb/s using a fixed random codebook and attains a MOS rating of about 3.2. The encoder computes a tenth-order LPC analysis on frames of 30 milliseconds (240 samples), and then bandwidth expansion of 15 Hz is performed. By bandwidth expansion we mean that the LPC poles are radially moved toward the origin by multiplication of LPC coefficient b_m by a factor of γ^m where $\gamma = 0.994$. This empirically improves speech quality, but is mainly used to increase stability. The LPC coefficients are converted to line spectral pairs and quantized using nonuniform scalar quantization. The 240-sample frame is then divided into four subframes, each of which is allowed a separate codeword from of a set of 256, so that eight bits are required to encode the excitation of each subframe, or 32 bits for the entire frame.

This same strategy of frames and subframes is used in all modern CELP coders. The codebook search is the major computational task of the encoder, and it is not practical to use a codebook that covers an entire frame. It is typical to divide each frame into four subframes, but the excitation search needn't be performed on the subframes that belong to the analysis frame. Forward prediction with *lookahead* uses an analysis window that stretches into the future, while backward analysis inputs excitation vectors into LPC coefficients calculated from past samples. For example, let's number the subframes $1, 2, 3$, and 4. Backward prediction may use the LPC coefficients computed from subframes $1, 2, 3$, and 4 when trying the excitations for subframes $5, 6, 7$, and 8. Forward prediction with lookahead of 2 subframes would use coefficients computed from subframes $3, 4, 5$, and 6 when searching for excitations on subframes $1, 2, 3$, and 4. Note that lookahead introduces further delay, since the search cannot start until the LPC filter is defined. Not only do coders using backward prediction not add further delay, they needn't send the coefficients at all, since by using closed-loop prediction the decoder can reproduce the coefficients before they are needed.

If random codebooks work, maybe even simpler strategies will. It would be really nice if sparse codebooks (i.e., ones in which the vectors have most of their components zero) would work. *Algebraic codebooks* are sets of excitation vectors that can be produced when needed, and so needn't be stored. The codewords in popular algebraic codebooks contain mostly zeros, but with a few nonzero elements that are either $+1$ or -1. With algebraic codebooks we needn't search a random codebook; instead we systematically generate all the legal codewords and input each in turn to the LPC filter. It turns out that such codebooks perform reasonably well; toll-quality G.729 and the lower bit rate of G.723.1 both use them.

Coders that search codebooks, choosing the excitation that minimizes the discrepancy between the speech to be coded and the output of the excitation-driven LPC synthesis filter, are called **A**nalysis **B**y **S**ynthesis (ABS) coders. The rationale for this name is clear. Such coders analyze the best excitation by exhaustively synthesizing all possible outputs and empirically choosing the best. What do we mean by the best excitation? Up to now you may have assumed that the output of the LPC synthesis filter was rated by SNR or correlation. This is not optimal since these measures do not correlate well with subjective opinion as to minimal distortion.

The main effect that can be exploited is 'masking' (recall exercise 11.4.2). Due to masking we needn't worry too much about discrepancies that result from spectral differences close to formant frequencies, since these are masked by the acoustic energy there and not noticed. So rather than using an error that is equally weighted over the bandwidth, it is better perceptually to use the available degrees of freedom to match the spectrum well where error is most noticeable. In order to take this into account, ABS CELP encoders perform *perceptual weighting* of both the input speech and LPC filter output before subtracting to obtain the residual. However, since the perceptual weighting is performed by a filter, we can more easily subtract first and perform a single filtering operation on the difference.

The perceptual weighting filter should de-emphasize spectral regions where the LPC has peaks. This can be achieved by using a filter with the system function

$$H(z) = \frac{\sum \gamma_1^m \beta_m z^{-m}}{\sum \gamma_2^m \beta_m z^{-m}} \tag{19.9}$$

where $0 < \gamma_2 < \gamma_1 \leq 1$. Note that both the numerator and denominator are performing bandwidth expansion, with the denominator expanding more than the numerator. By properly choosing γ_1 and γ_2 this weighting can be made similar to the psychophysical effect of masking.

Something seems to have been lost in the ABS CELP coder as compared with the LPC model. If we excite the LPC filter with an entry from a random or algebraic codebook, where does the pitch come from? To a certain extent it comes automatically from the minimization procedure. The algebraic codewords can have nonzero elements at pitch onset, and random codewords will automatically be chosen for their proper spectral content. However, were we to build the CELP coder as we have described it so far, we would find that its residual error displays marked pitch periodicity, showing that the problem is not quite solved. Two different ways have been developed to put the pitch back into the CELP model, namely *long-term prediction* and *adaptive codebooks*.

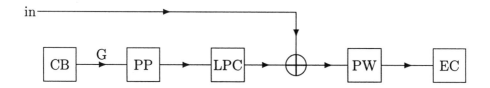

Figure 19.8: ABS CELP encoder using short- and long-term prediction. Only the essential elements are shown; CB is the codebook, PP the pitch (short-term) predictor, LPC the long-term predictor, PW the perceptual weighting filter, and EC the error computation. The input is used directly to find LPC coefficients and estimate the pitch and gain. The error is then used in ABS fashion to fine tune the pitch and gain, and choose the optimal codebook entry.

We mentioned long-term prediction in Section 19.2 as conceptually having two separate LPC filters. The short-term predictor, also called the LPC filter, the formant predictor, or the spectral envelope predictor, tracks and introduces the vocal tract information. It only uses correlations of less than 2 milliseconds or so and thus leaves the pitch information intact. The long-term predictor, also called the pitch predictor or the fine structure predictor, tracks and introduces the pitch periodicity. It only has a few coefficients, but these are delayed by between 2 and 20 milliseconds, according to the pitch period. Were only a single coefficient used, the pitch predictor system function would be

$$H_{pp}(z) = \frac{1}{1 - \beta z^{-D}} \tag{19.10}$$

where D is the pitch period. D may be found open loop, but for high quality it should be found using analysis by synthesis. For unvoiced segments the pitch predictor can be bypassed, sending the excitation directly to the LPC predictor, or it can be retained and its delay set randomly. A rough block diagram of a complete CELP encoder that uses this scheme is given in Figure 19.8.

Adaptive codebooks reinforce the pitch period using a different method. Rather than actually filtering the excitation, we use an effective excitation composed of two contributions. One is simply the codebook, now called the *fixed codebook*. To this is added the contribution of the adaptive codebook, which is formed from the previous excitation by duplicating it at the pitch period. This contribution is thus periodic with the pitch period and supplies the needed pitch-rich input to the LPC synthesis filter.

One last trick used by many CELP encoders is 'post-filtering'. Just as for ADPCM, the post-filter is appended after the decoder to improve the subjective quality of the reconstructed speech. Here this is accomplished by further strengthening the formant structure (i.e., by emphasizing the peaks and attenuating the valleys of the LPC spectrum), using a filter like the perceptual weighting filter (19.9). This somewhat reduces the formant bandwidth, but also reduces the residual coding noise. In many coders the post-filter is considered optional, and can be used or not according to taste.

EXERCISES

19.12.1 Explain why replacing LPC coefficient b_m with γb_m with $0 < \gamma < 1$ is called *bandwidth expansion*. Show that 15 Hz expansion is equivalent to $\gamma = 0.994$.

19.12.2 The G.723.1 coder when operating at the 5.3 Kb/s rate uses an algebraic codebook that is specified by 17 bits. The codewords are of length 60 but have no more than four nonzero elements. These nonzero elements are either all in even positions or all in odd positions. If in even positions, their indexes modulo 8 are all either 0, 2, 4, or 6. Thus 1 bit is required to declare whether even or odd positions are used, the four pulse positions can be encoded using 3 bits, and their signs using a single bit. Write a routine that successively generates all the legal codewords.

19.12.3 Explain how to compute the delay of an ABS CELP coder. Take into account the buffer, lookahead, and processing delays. What are the total delays for G.728 (frame 20 samples, backward prediction), G.729 (frame 80 samples, forward prediction), and G.723.1 (frame 240 samples, forward prediction)?

19.12.4 Obtain a copy of the G.729 standard and study the main block diagram. Explain the function of each block.

19.12.5 Repeat the previous exercise with the G.723.1 standard. What is the difference between the two rates? How does G.723.1 differ from G.729?

19.13 Telephone-Grade Speech Coding

This section can be considered to be the converse of Section 18.20; the purpose of a telephone-grade modem is to enable the transfer of data over voice lines (*data over voice*), while the focus of speech compression is on the transfer of voice over digital media (*voice over data*). Data over voice is an important technology since the Public Switched Telephone Network (PSTN) is the most widespread communications medium in the world; yet the PSTN is growing at a rate of about 5% per year, while digital communications use is growing at several hundred percent a year. The amount of data traffic exceeded that of voice sometime during the year 2000, and hence voice over data is rapidly becoming the more important of the two technologies.

The history of telephone-grade speech coding is a story of rate halving. Our theoretical rate of 128 Kb/s was never used, having been reduced to 64 Kb/s by the use of logarithmic PCM, as defined in ITU standard G.711. So the first true rate halving resulted in 32 Kb/s and was accomplished by ADPCM, originally designated G.721. In 1990, ADPCM at rates 40, 32, 24, and 16 Kb/s were merged into a single standard known as G.726. At the same time G.727 was standardized; this 'embedded' ADPCM covers these same rates, but is designed for use in packetized networks. It has the advantage that the bits transmitted for the lower rates are subsets of those of the higher rates; congestion that arises at intermediate nodes can be relieved by discarding least significant bits without the need for negotiation between the encoder and decoder.

Under 32 Kb/s the going gets harder. The G.726 standard defines 24 and 16 Kb/s rates as well, but at less than toll-quality. Various SBC coders were developed for 16 Kb/s, either dividing the frequency range equally and using adaptive numbers of bits per channel, or using hierarchical wavelet-type techniques to divide the range logarithmically. Although these techniques were extremely robust and of relatively high perceived quality for the computational complexity, no SBC system was standardized for telephone-grade speech. In 1988, a coder, dubbed G.722, was standardized that encoded wideband audio (7 KHz sampled at 16,000 samples per second, 14 bits per sample) at 64 Kb/s. This coder divides the bandwidth from DC to 8 KHz into two halves using QMFs and encodes each with ADPCM.

In the early 1990s, the ITU defined performance criteria for a 16 Kb/s coder that could replace standard 32 Kb/s ADPCM. Such a coder was required to be of comparable quality to ADPCM, and with delay of less than 5 milliseconds (preferably less than 2 milliseconds). The coder, selected in 1992 and dubbed G.728, is a CELP with backward prediction, with LPC order of 50. Such a high LPC order is permissible since with closed-loop prediction the coefficients need not be transmitted. Its delay is 5 samples (0.625 ms), but its computational complexity is considerably higher than ADPCM, on the order of 30 MIPS.

The next breakthrough was the G.729 8 Kb/s CELP coder. This was accepted simultaneously with another somewhat different CELP-based coder for 6.4 and 5.4 Kb/s. The latter was named G.723.1 (the notation G.723

having been freed up by the original merging into G.726). Why were two different coders needed? The G.729 specification was originally intended for toll-quality wireless applications. G.728 was rejected for this application because of its rate and high complexity. The frame size for G.729 was set at 10 ms. and its lookahead at 5 ms. Due to the wireless channel, robustness to various types of bit errors was required. The process of carefully evaluating the various competing technologies took several years. During that time the urgent need arose for a low-bit-rate coder for videophone applications. Here toll-quality was not an absolute must, and it was felt by many that G.729 would not be ready in the alloted time. Thus an alternative selection process, with more lax testing, was instigated. For this application it was decided that a long 30 millisecond frame was acceptable, that a lower bit rate was desirable, but that slightly lower quality could be accommodated. In the end both G.729 and G.723.1 were accepted as standards simultaneously, and turned out to be of similar complexity.

The G.729 coder was extremely high quality, but also required over 20 MIPS of processing power to run. For some applications, including 'voice over modem', this was considered excessive. A modified coder, called G.729 Annex A, was developed that required about half the complexity, with almost negligible MOS reduction. This annex was adopted using the quick standardization strategy of G.723.1. G.723.1 defined as an annex a standard VAD and CNG mechanism, and G.729 soon followed suit with a similar mechanism as its Annex B. More recently, G.729 has defined annexes for additional bit rates, including a 6.4 Kb/s one.

At this point in time there is considerable overlap (and rivalry) between the two standards families. G.723.1 is the default coder for the voice over IP standard H.323, but G.729 is allowed as an option. G.729 is the default for the 'frame relay' standard FRF.11, but G.723.1 is allowed there as an option. In retrospect it is difficult to see a real need for two different coders with similar performance.

For even lower bit rates one must decide between MIPS and MOS. On the low MIPS low MOS front the U.S. Department of Defense initiated an effort in 1992 to replace LPC-10e with a 2.4 Kb/s encoder with quality similar to that of the 4.8 Kb/s CELP. After comparing many alternatives, in 1997 a draft was published based on MELP. The excitation used in this encoder consists of a pulse train and a uniform-distributed random noise generator filtered by time-varying FIR filters. MELP's quality is higher than that of straight LPC-10 because it addresses the latter's main weaknesses, namely voicing determination errors and not treating partially-voiced speech.

For higher MOS but with significantly higher MIPS requirements there are several alternatives, including the **S**inusoidal **T**ransform **C**oder (STC) and **W**aveform Interpolation (WI). Were we to plot the speech samples, or the LPC residual, of one pitch period of voiced speech we would obtain some characteristic waveform; plotting again for some subsequent pitch period would result in a somewhat different waveform. We can now think of this waveform as evolving over time, and of its shape at any instant between the two we have specified as being determinable by interpolation. To enforce this picture we can create two-dimensional graphs wherein at regular time intervals we plot characteristic waveforms perpendicular to the time axis.

Waveform interpolation encoders operate on equally spaced frames. For each voiced frame the pitch pulses located and aligned by circular shifting, the characteristic waveform is found, and the slowly evolving waveform is approximated as a Fourier series. Recently waveform interpolation techniques have been extended to unvoiced segments as well, although now the characteristic waveform evolves rapidly from frame to frame. The quantized pitch period and waveform description parameters typically require under 5 Kb/s. The decompression engine receives these parameters severely undersampled, but recreates the required output rate by interpolation as described above.

The ITU has launched a new effort to find a 4 Kb/s toll-quality coder. With advances in DSP processor technology, acceptable coders at this, and even lower bit rates, may soon be a reality.

EXERCISES

19.13.1 Cellular telephony networks use a different set of coders, including RPE-LTP (GSM) and VSELP (IS-54). What are the principles behind these coders and what are their parameters?

Bibliographical Notes

There is a plethora of books devoted to speech signal processing. The old standard references include [210, 211], and of the newer generation we mention [66]. A relatively up-to-date book on speech recognition is [204] while [176] is an interesting text that emphasizes neural network techniques for speech recognition.

The first artificial speech synthesis device was created by Wolfgang von Kempelen in 1791. The device had a bellows that supplied air to a reed, and a manually manipulated resonance chamber. Unfortunately, the machine was not taken seriously after von Kempelen's earlier invention of a chess-playing machine had been exposed as concealing a midget chess expert. In modern times Homer Dudley from Bell Labs [55] was an early researcher in the field of speech production mechanisms. Expanding on the work of Alexander Graham Bell, he analyzed the human speech production in analogy to electronic communications systems, and built the VODER (**V**oice **O**peration **DE**monstrato**R**), an analog synthesizer that was demonstrated at the San Francisco and New York World's Fairs. An early digital vocoder is described in [80]. In the 1980s, Dennis Klatt presented a much improved formant synthesizer [130, 131].

The LPC model was introduced to speech processing by Atal [10] in the U.S. and Itakura [111] in Japan. Many people were initially exposed to it in the popular review [155] or in the chapter on LPC in [210]. The power cepstrum was introduced in [20]; the popular DSP text [186] devotes a chapter to homomorphic processing; and [37] is worth reading. We didn't mention that there is a nonrecursive connection between the LPC and LPC cepstrum coefficients [239].

Distance measures, such as the Itakura-Saito distance, are the subject of [112, 113, 110, 84]. The inverse-E filtering problem and RASTA-PLP are reviewed in [102, 101]. The sinusoidal representation has an extensive literature; you should start with [163, 201].

For questions of speech as a dynamical system and its fractal dimension consult [259, 156, 172, 226]. Unfortunately, there is as yet no reference that specifies for the optimal minimal set of features.

Pitch detectors and U/V decision mechanisms are the subject of [205, 206, 121]. Similar techniques for formant tracking are to be found in [164, 230].

Once, the standard text on coding was [116], but the field has advanced tremendously since then. Vector quantization is covered in a review article [85] and a text [69], while the LBG algorithm was introduced in [149].

Postfiltering is best learnt from [35]. The old standard coders are reviewed in [23] while the recent ones are described in [47]. For specific techniques and standards, LPC and LPC-10: [9, 261, 121]; MELP: [170]; basic CELP: [11]; federal standard 1016: [122]; G.729 and its annexes: [231, 228, 229, 15]; G.728: [34]; G.723.1: no comprehensive articles; waveform interpolation: [132].

Whirlwind Exposition of Mathematics

In this appendix we will very quickly review all the mathematical background needed for complete understanding of the text. Depending on who you are, this chapter may be entirely superfluous, or it may be one of the most useful chapters in the entire book. You should probably at least look it over before starting to read Chapter 1. If most of the material in this chapter is unfamiliar to you, then you are probably not ready to continue reading. You should definitely consult it whenever you feel uncomfortable with the mathematics being used in any of the chapters, and it is written to be useful in a more general setting. Under no conditions should you read it in its natural place, after the last chapter; if you have already finished the book, you don't need it any more.

A.1 Numbers

Since DSP involves a lot of 'number crunching', we had better at least know what a number is! The simplest type of number is the 'whole number' or positive integer. These are $1, 2, 3, \ldots$. You probably learned about them in kindergarten. Kronecker (the same guy who invented the delta) once said that the whole numbers were created by God, while all the rest are human inventions. Indeed the whole numbers are taken as basic entities in most of mathematics, but in axiomatic set theory their existence can actually be derived based on even simpler axioms.

So how did people create the rest of the numbers? The basic idea is to write equations using whatever numbers we already have, and try to solve them. Whenever we can't solve an equation using the numbers we already know about, we invent new ones. For instance, $1 + 1 = x$ leads us to $x = 2$, which is a whole number and thus no news, but when we try $x + 1 = 1$ we discover we need the first extension—we have to invent 'zero'. In case you think zero is no big deal, try writing large numbers in Hebrew numerals (where $10, 20, \ldots 90$ have their own symbols) or dividing Roman numerals.

Next we try to solve $x + 1 = 0$ and discover that we must invent -1, an idea even more abstract than zero (you might recall how negative numbers perplexed you in grade school). Continuing in this fashion we discover all the 'integers' $\ldots, -3, -2, -1, 0, 1, 2, 3 \ldots$.

Now we try solving equations which contain multiplications. $2x = 4$ causes no problems, but $4x = 2$ does. We are thus led to the discovery of fractions, which together with the integers form all of the 'rational numbers'.

We next try solving equations involving powers: $x^2 = 4$ is easy, but $x^2 = 2$ leads us to difficulties. The idea of $\sqrt{2}$ not being a rational number was once considered so important that the Pythagoreans killed to keep it secret. It turns out that not only aren't these 'irrational numbers' rare, but there are more of them than there are rationals among the 'real numbers'.

We're almost done. The final kind of equation to observe is $x^2 = -1$, which leads to $i = \sqrt{-1}$, to the imaginary numbers, and to the combination of everything we have seen so far—the 'complex numbers'. It turns out that complex numbers are sufficient to solve all equations expressible in terms of complex numbers, so our search is over.

EXERCISES

A.1.1 Prove that $\sqrt{2}$ is irrational. (Hint: Assume that $\sqrt{2} = \frac{n}{m}$ and find a contradiction.)

A.1.2 Prove that the set of real numbers is not denumerable, and that most real numbers are irrational.

A.1.3 Hamilton invented 'quarternions', which are like complex numbers but with four real components. Why did he do this if complex numbers are sufficient?

A.2 Integers

Although most of us are used to decimal numbers, where we count from 1 to 9 before incrementing the next decimal place to the left, digital computers prefer binary numbers. Counting from zero up in binary numbers is done as follows.

$$0000, 0001, 0010, 0011, 0100, 0101, 0110, 0111,$$
$$1000, 1001, 1010, 1011, 1100, 1101, 1110, 1111, \quad \ldots$$

Each 0 or 1 is called a bit, the rightmost bit in a number is called the Least Significant Bit (LSB), while the leftmost bit in a number (which can be zero since we assume that a constant number of bits are used) is the Most Significant Bit (MSB). There are several ways of extending binary numbers to negative numbers without using a separate minus sign, the most popular of which is two's complement. The two's complement of a number with $b + 1$ bits (the MSB is interpreted as the sign) is obtained by subtracting the number from 2^{b+1}; hence addition of negative numbers is automatically correct assuming we just discard the overflow bit. We assume that the reader is reasonably proficient in using the integers, including the operations addition, subtraction, multiplication, division with remainder, and raising to a power, (particularly in binary) and understands the connection between binary arithmetic and logic, and how all this facilitates the building of digital computers.

There is another operation over the integers that we will require. We say that two whole numbers i and j are 'equal modulo' m

$$i = j \bmod m \tag{A.1}$$

if when they are divided by m they give the same remainder. This operation principle can be extended to real numbers as well, and is related to *periodicity*. Given an integer i, the 'reduction modulo' m of i

$$i \bmod m = j \tag{A.2}$$

means finding the minimum whole number j to which i is equal modulo m. Thus $15 = 8 \bmod 7$ since $15 \bmod 7 = 1$ and $8 \bmod 7 = 1$.

If i divided by m leaves no remainder (i.e., $i \bmod m = 0$), we say that m is a 'factor' of i. A whole number is prime if it has no factors other than itself and 1. The 'fundamental theorem of arithmetic' states that every whole number has a unique factorization as the product of powers of primes.

$$i = p_1^{n_1} \cdot p_2^{n_2} \cdots p_m^{n_m} \tag{A.3}$$

A set is said to be 'finite' if the number of its elements is some whole number. A set is said to be 'denumerably infinite' if its elements can be placed in a list labeled by whole numbers. The interpretation is that there are in some sense the same number of elements as there are whole numbers. In particular the set of all integers is denumerable, since it can be listed in the following way,

$$a_1 = 0, \ a_2 = 1, \ a_3 = -1, \ a_4 = 2, \ a_5 = -2, \ \ldots \ a_{2k} = k, \ a_{2k+1} = -k, \ \ldots$$

and the set of all rational numbers between 0 and 1 is denumerable, as can be seen by the following order.

$$0, 1, \frac{1}{2}, \frac{1}{3}, \frac{2}{3}, \frac{1}{4}, \frac{3}{4}, \frac{1}{5}, \frac{2}{5}, \frac{3}{5}, \frac{4}{5}, \cdots$$

The set of all real numbers is nondenumerably infinite.

EXERCISES

A.2.1 Show that there are an infinite number of primes. (Hint: Assume that there is a largest prime and find a contradiction.)

A.2.2 You are given two input electrical devices that perform AND, OR, and NOT on bits. Show how to build a binary adder that inputs two 2-bit numbers and outputs a 3-bit number. How can this be extended to b-bit numbers?

A.2.3 In one's complement notation the negative of a number is obtained by flipping all its bits. What are the advantages and disadvantage of this method?

A.3   Real Numbers

The reader should also know about real numbers, including the operations addition, subtraction, multiplication, division, and raising to an integer power. Some reals are rational (i.e., can be written as the ratio of two integers), but most are not.

Rational numbers can be represented as binary numbers with a decimal point (or should it be called a binary point?); for example, the decimal number $\frac{1}{2}$ is written 0.1, $\frac{1}{4}$ is 0.01, and $\frac{3}{4}$ is 0.11. This is called 'fixed point' notation. Some rational numbers and all irrational numbers require an infinite number of bits to the right of the point, and must be truncated in all practical situations when only a finite number of bits is available. Such truncation leads to numeric error. In order to increase the range of real numbers representable without adding too many bits, 'floating point' notation can be used. In floating point notation numbers are multiplied by positive or negative powers of 2 until they are between 0 and 1, and the power (called the exponent) and fraction (called the mantissa) are used together. For example, $\frac{3}{256} = 3 \cdot 2^{-8}$ is represented by mantissa 3 and binary exponent -8.

Two specific irrational numbers tend to turn up everywhere.

$$e = \lim_{n\to\infty} \left(1 + \tfrac{1}{n}\right)^n = 1 + \tfrac{1}{1!} + \tfrac{1}{2!} + \tfrac{1}{3!} + \tfrac{1}{4!} + \ldots \approx 2.718281828$$

$$\pi = 4\tan^{-1}(1) = 4\left(1 - \tfrac{1}{3} + \tfrac{1}{5} - \tfrac{1}{7} + \tfrac{1}{9} - \ldots\right) \approx 3.141592653$$

These series expansions are important from a theoretical point of view, but there are more efficient computational algorithms for approximating these numbers.

EXERCISES

A.3.1 Compare several methods for computing e and π. See exercise A.8.5 below.

A.3.2 How can you tell a rational number from an irrational one based on its binary representation?

A.3.3 Another interesting irrational number is the golden ratio $\gamma = \frac{1+\sqrt{5}}{2} \approx 1.618$. Show that if a line segment of length l is divided in two segments of lengths a and b such that the ratio of l to a equals the ratio of a to b, then $\frac{a}{b} = \gamma$. Show that if a nonsquare rectangle has sides of length a and b such that if a square is removed the remaining rectangle has the same proportions, then $\frac{a}{b} = \gamma$. Show that $\cos(\frac{\pi}{5}) = \frac{\gamma}{2}$.

A.3.4 Given a decimal representation r and a tolerance ϵ, how can the smallest a and b such that $r \approx \frac{a}{b}$ to within ϵ be found?

A.4   Complex Numbers

We assume that the reader has some knowledge of complex numbers, including how to convert a complex number z between the Cartesian form $z = x + iy$ and the polar form $z = re^{i\theta}$.

$$
\begin{aligned}
|z| &= \sqrt{x^2 + y^2} & \theta &= \tan^{-1}\left(\tfrac{y}{x}\right) \\
x &= z\cos\theta & y &= z\sin\theta
\end{aligned}
\tag{A.4}
$$

We will use the notations

$$x = \Re z \qquad y = \Im z \qquad r = |z| \qquad \theta = \angle z$$

for the real part, imaginary part, absolute value (magnitude) and angle of the complex number z.

The arctangent function $\tan^{-1}(\varphi)$ is usually defined only for $-\frac{\pi}{2} < \varphi < \frac{\pi}{2}$. For equation (A.4) we need the 'four-quadrant arctangent', computable via the following algorithm:

```
a ← tan⁻¹(y/x)
if x < 0
        if y > 0
                a ← a + π
        else
                a ← a − π
if a < 0
        a ← a + 2π
```

The complex operations of addition and multiplication are simple when addition is performed on the Cartesian form

$$z = z_1 + z_2 \qquad \text{means} \qquad \begin{aligned} x &= x_1 + x_2 \\ y &= y_1 + y_2 \end{aligned}$$

and multiplication in polar form,

$$z = z_1 z_2 \qquad \text{means} \qquad \begin{aligned} r &= r_1 r_2 \\ \theta &= \theta_1 + \theta_2 \end{aligned}$$

although multiplication can be done on the Cartesian forms as well.

$$z = z_1 z_2 \qquad \text{means} \qquad \begin{aligned} x &= x_1 x_2 - y_1 y_2 \\ y &= x_1 y_2 + x_2 y_1 \end{aligned}$$

Raising to a power, like multiplication, is also simplest in polar form, and in this form is called DeMoivre's theorem.

$$\left(re^{i\theta}\right)^a = r^a e^{ia\theta} \tag{A.5}$$

There is a certain isomorphism between complex numbers and a two-dimensional vector space, but multiplication isn't defined in the same way for the two and complex numbers can't be extended to three dimensions. Nonetheless, it is often useful to picture the complex numbers as residing in a plane, called the 'complex plane', especially when dealing with functions defined over the complex plane.

Euler discovered a most elegant relation between four important numbers, -1, i, e, and π, namely

$$e^{i\pi} = -1 \tag{A.6}$$

which is a special case of the more general connection between imaginary exponentials and sinusoids,

$$e^{it} = \cos t + i \sin t \tag{A.7}$$

a relation that can be reversed as well.

$$\sin(t) = \frac{e^{it} - e^{-it}}{2i} \qquad \cos(t) = \frac{e^{it} + e^{-it}}{2} \tag{A.8}$$

A very important family of complex numbers are the Nth 'roots of unity'. These are the N solutions to equation $W^N = 1$. Thus for $N = 2$ we have the two square roots of unity $W = \pm 1$, while for $N = 4$ the four fourth roots of unity are $W = \pm 1, \pm i$. It is obvious that the Nth roots must all reside on the unit circle, $|W| = 1$, and it is not hard to show that they are given by $W = e^{i\frac{2\pi n}{N}} = W_N^n$, where the principle root is:

$$W_N = e^{i\frac{2\pi}{N}} \tag{A.9}$$

EXERCISES

A.4.1 When a complex multiplication is performed using the Cartesian forms, it would seem that we need to perform four multiplications and two additions. Show that this same multiplication can be performed using three multiplications and five additions.

A.4.2 Express the power of a complex number in Cartesian form.

A.4.3 Find the square roots of i in Cartesian form.

A.4.4 Give geometric interpretations for the following:
1. All complex numbers with the same magnitude
2. All complex numbers with the same real part
3. All complex numbers with the same imaginary part
4. All complex numbers with the same angle
5. All complex numbers equidistant from a given complex number

A.5   Abstract Algebra

Numbers have so many different characteristics that it is hard to study them all at one time. For example, they have many inherent features (e.g., absolute value, positiveness), there are several operations that can be performed between two numbers (e.g., addition and multiplication), and these operations have many attributes (e.g., commutativity, associativity). As is customary in such complex situations mathematicians start their investigations with simpler objects that have only a small number of the many characteristics, and then advance to more and more complex systems.

The simplest such system is a 'group', which is a set of elements between which a single binary operation $\cdot$ is defined. This operation must have the following properties:

closure: for all a and b in the group, $c = a \cdot b$ is in the group

associativity: $a \cdot (b \cdot c) = (a \cdot b) \cdot c$

identity: there is a unique element i in the group such that $a \cdot i = i \cdot a = a$ for all a

inverse: for every a in the group there is a unique element b in the group such that $a \cdot b = b \cdot a = i$ where i is the identity element.

If in addition the operation obeys:

commutativity: $a \cdot b = b \cdot a$

then we call the group 'commutative' or 'Abelian'.

The integers, the rationals, the real numbers, and the complex numbers are all groups with respect to the operation of addition; zero is the identity and $-a$ is the inverse. Likewise the set of polynomials of degree n (see Appendix A.6) and $m * n$ matrices (Appendix A.15) are groups with respect to addition. The nonzero reals and complex numbers are also groups with respect to multiplication, with unity being the identity and $\frac{1}{a}$ the inverse. Not all groups have an infinite number of elements; for any prime number p, the set consisting of the integers $0, 1, \ldots (p-1)$ is a finite group with p elements if we use the operation $a \cdot b \equiv (a + b) \bmod p$.

A field is more complex than a group in that it has two operations, usually called addition and multiplication. The field is a group under both operations, and in addition a new relation involving both addition and multiplication must hold.

distributivity: $a \cdot (b + c) = a \cdot b + a \cdot c$

The real numbers are a field, as are the rationals and the complex numbers. There are also finite fields (e.g., the binary numbers and more generally the integers $0 \ldots p - 1$ under modulo arithmetic).

Given a field we can define a 'vector space' over that field. A vector space is a set of elements called 'vectors'; our convention is to symbolize vectors by an underline, such as $\underline{v}$. The elements of the field are called 'scalars' in this context. Between the vectors in a vector space there is an operation of addition; and the vectors are a commutative group under this operation. In addition, there is a multiplication operation between a scalar and a vector that yields a vector.

Multiplication by unity must yield the same vector

- $1\underline{v} = \underline{v}$

and several types of distributivity must be obeyed.

- $a(\underline{u} + \underline{v}) = a\underline{u} + a\underline{v}$
- $(a + b)\underline{v} = a\underline{v} + b\underline{v}$
- $(ab)\underline{v} = a(b\underline{v})$

There is another kind of multiplication operation that may be defined for vector spaces that goes under several names including scalar product, inner product, and dot product. This operation is between two vectors and yields a scalar. If the underlying field is that of the reals, the dot product must have the following properties:

nonnegativity: $\underline{u} \cdot \underline{v} \geq 0$

self-orthogonality: $\underline{v} \cdot \underline{v} = 0$ if and only if $\underline{v} = \underline{0}$

commutativity: $\underline{u} \cdot \underline{v} = \underline{v} \cdot \underline{u}$

distributivity: $(\underline{u} + \underline{v}) \cdot \underline{w} = \underline{u} \cdot \underline{w} + \underline{v} \cdot \underline{w}$

scalar removal: $(a\underline{u}) \cdot \underline{v} = a(\underline{u} \cdot \underline{v})$

Two vectors for which $\underline{u} \cdot \underline{v} = 0$ are called 'orthogonal'. If the underlying field is of the complex numbers, the commutativity relation requires modification.

conjugate commutativity: $\underline{u} \cdot \underline{v} = (\underline{v} \cdot \underline{u})^*$

The prototypical example of a vector space is the set of ordered n-tuples of numbers, and this used as the definition of 'vector' in computer science. The number n is called the dimension and the operations are then defined by the following recipes:

- $(u_1, u_2 \ldots u_n) + (v_1, v_2 \ldots v_n) = (u_1 + v_1, u_2 + v_2, \ldots u_n + v_n)$
- $a(v_1, v_2 \ldots v_n) = (av_1, av_2, \ldots av_n)$
- $(u_1, u_2 \ldots u_n) \cdot (v_1, v_2 \ldots v_n) = u_1 v_1 + u_2 v_2 + \ldots u_n v_n$

The usual two-dimensional and three-dimensional vectors are easily seen to be vector spaces of this sort. We can similarly define vector spaces of any finite dimension over the reals or complex numbers, and by letting n go to infinity we can define vector spaces of denumerably infinite dimension. These 'vectors' are in reality infinite sequences of real or complex numbers. It is also possible to define vector spaces with nondenumerable dimension, but then the interpretation must be that of a function defined on the real axis, rather than an n-tuple of numbers or an infinite sequence.

A 'metric space' is a set of elements, between every two of which is defined a metric (distance). The metric is a nonnegative number $d(x, y)$ that has the following three properties:

symmetry: $d(y, x) = d(x, y)$,

identity: $d(x, y) = 0$ if and only if $x = y$,

triangle inequality: $d(x, z) \leq d(x, y) + d(y, z)$.

Metric spaces and linear vector spaces capture different aspects of Euclidean vectors, and it is not surprising that we can define 'normed spaces' that are both metric and vector spaces. The norm of a vector is defined to be $|\underline{v}| = \sqrt{\underline{v} \cdot \underline{v}}$, which is easily seen to be a nonnegative number and to fulfill all the requirements of a metric.

EXERCISES

A.5.1 Find groups with small numbers of elements.

A.5.2 Show that **true** and **false** with or as addition and and as multiplication form a vector space.

A.5.3 Prove formally that three-dimensional space is a vector space and a metric space.

A.6 Functions and Polynomials

Functions uniquely map one or more numbers (called arguments) onto other numbers (called returned values). For example, the function $f(x) = x^2$ returns a unique real number for every real argument x, although all positive x^2 are returned by two different real arguments (x and $-x$) and negative numbers are not returned for any real argument. Hence $f(x) = \sqrt{x}$ is not a function unless we define it as returning only the positive square root (since a function can't return two values at the same time) and even then it is undefined for negative arguments unless we allow it to return complex values.

A 'symmetry' of a function is a transformation of the argument that does not change the returned value. For example, $x \rightarrow -x$ is a symmetry of the function $f(x) = x^2$ and $x \rightarrow x + 2\pi n$ for any n are symmetries of the function $f(x) = \sin(x)$.

'Polynomials' are functions built by weighted summing of powers of the argument

$$a(x) = a_0 + a_1 x + a_2 x^2 + a_3 x^3 + \ldots a_n x^n \tag{A.10}$$

the weights a_i are called 'coefficients' and the highest power n is called the 'degree' of the polynomial.

The straightforward algorithm for evaluating polynomials,

```
a ← a₀
for i ← 1 to n
    p ← xᵖ
    a ← a + aᵢp
```

is usually not the best way of computing the value to be returned, since raising to a power is computationally expensive and may introduce numerical error. It is thus usually better to use the following algorithm, which requires an additional memory location.

```
a ← a₀ + a₁x
p ← x
for i ← 1 to n
    p ← p * x
    a ← a + aᵢ p
```

Even this code is not optimal, but Horner's rule

$a \leftarrow a_n$
for $i \leftarrow n - 1$ to 0
      $a \leftarrow ax + a_i$

requiring only n multiplications and additions, *is* optimal.

Polynomials can be added, multiplied, and factored into simpler polynomials. The 'fundamental theorem of algebra' states that all polynomials with real coefficients can be uniquely factored into products of first- and second-degree polynomials with real coefficients, and into products of first-degree polynomials with complex coefficients. For example,

$$x^3 + x^2 + x + 1 = (x^2 + 1)(x + 1)$$

for real coefficients, while allowing complex coefficients we can factor further.

$$z^3 + z^2 + z + 1 = (z + i)(z - i)(z + 1)$$

A first-degree factor of $a(z)$ can always be written $z - \zeta$, in which case ζ is called a 'zero' (or 'root') of the polynomial. It is obvious that the polynomial as a whole returns zero at its zeros $a(\zeta) = 0$ and that the number of zeros is equal to the polynomial degree for complex polynomials (although some zeros may be identical), but may be less for real polynomials.

'Rational functions' are functions formed by dividing two polynomials.

$$r(x) = \frac{a(x)}{b(x)} = \frac{a_0 + a_1 x + a_2 x^2 + a_3 x^3 + \ldots a_n x^n}{b_0 + b_1 x + b_2 x^2 + b_3 x^3 + \ldots b_m x^m} \qquad (A.11)$$

A 'zero' of a rational function is an argument for which the function returns zero, and is necessarily a zero of the numerator polynomial. A 'pole' of a rational function is a zero of the denominator polynomial and hence an argument for which the function as a whole is infinite.

EXERCISES

A.6.1 The derivative of a polynomial $a(x)$ of degree n is a polynomial of degree $n - 1$ given by $a'(x) \equiv a_1 + 2a_2 x + 3a_3 x^2 + \ldots + n a_n x^{n-1}$. What is the most efficient method of simultaneously computing $a(x)$ *and* $a'(x)$?

A.6.2 Horner's rule is not efficient for sparse polynomials which have many zero coefficients. For example, the best way to compute $p(x) = x^5$ is to compute $a_2 \leftarrow x^2, a_4 \leftarrow c^2, p(x) = a_4 x$. What is the best way of computing $p(x) = x^n$ for general integer n?

A.6.3 Show that rational functions are uniquely determined by their zeros and poles (including multiplicities) and a single additional number.

A.6.4 We define binary polynomials as polynomials for which each power of x is either present (i.e., its coefficient is 1) or absent (its coefficient is 0). How many different binary polynomials are there with degree up to m? What is the connection between these polynomials and the nonnegative integers? The addition of two binary polynomials is defined by addition modulo 2 of the corresponding coefficients (note that each polynomial is its own additive inverse). To what operation of the integers does this correspond? How do you think polynomial multiplication should be defined?

A.7   Elementary Functions

In addition to polynomials there are several other functions with which the reader should feel comfortable. The natural logarithm $\ln(x)$ is one such function. It is defined for positive real numbers, and is uniquely determined by the properties

$$
\begin{aligned}
\ln(1) &= 0 \\
\ln(a\,b) &= \ln(a) + \ln(b) \\
\ln\left(\frac{a}{b}\right) &= \ln a - \ln b \\
\ln(a^b) &= b\ln(a)
\end{aligned}
\tag{A.12}
$$

although it can also be defined by an integral representation.

$$
\ln x = \int_1^x \frac{1}{t}\, dt
\tag{A.13}
$$

One can generalize the logarithm to complex numbers as well,

$$
\ln(re^{i\theta}) = \ln r + i\theta
\tag{A.14}
$$

and one finds that $\ln(-1) = i\pi$, and $\ln(\pm i) = \pm i\frac{\pi}{2}$. Actually, this is only one possible value for the complex logarithm; any multiple of $2\pi i$ is just as good.

Logarithms transform multiplication into addition since they are the converse operation to raising to a power. The natural logarithms are logarithms to base e, that is, $y = \ln x$ means that $x = e^y$, or put in another way

$e^{\ln(a)} = a$ and $\ln(e^a) = a$. The function e^x will be discussed in a moment. It is often useful to know how to expand the natural logarithm around $x = 1$

$$\ln(1 + x) = x - \frac{x^2}{2} + \frac{x^3}{3} - \ldots \tag{A.15}$$

although this series converges very slowly.

Logarithms to other bases are related as follows.

$$\log_a x = \frac{\log_b x}{\log_b a} = \frac{\ln x}{\ln a}$$

The most important alternative bases are base 10 and base 2, and it is enough to remember $\ln 10 \approx 2.3$ and $\log 2 \approx 0.3$ to be able to mentally convert between them. Another logarithmic relation is the decibel (dB), being one-tenth of a Bel, which is simply the base 10 logarithm of a ratio.

$$r(\text{dB}) = 10 \log_{10} \frac{P_1}{P_2} \tag{A.16}$$

Using one of the useful numbers we see that every factor of two contributes about 3 dB to the ratio (e.g., a ratio of two to one is about 3 dB, four to one is about 6 dB, eight to one about 9 dB, etc.). Of course a ratio of ten to one is precisely 10 dB.

The 'exponential function' e^x is simply the irrational number e raised to the x power. If x is not an integer, the idea of a power has to be generalized, and this can be done by requiring the following properties:

$$e^0 = 1$$
$$e^{a+b} = e^a + e^b$$
$$e^{ab} = (e^a)^b$$

The solution turns out to be given by an infinite series

$$e^x = \sum_{k=0}^{\infty} \frac{x^k}{k!} \tag{A.17}$$

and this same series can be used for complex numbers. To define noninteger powers of other numbers we can use

$$x^y \equiv e^{y \ln x} \tag{A.18}$$

where ln was defined above.

The Gaussian

$$G(x) = \frac{1}{\sqrt{2\pi}\sigma} e^{-\frac{1}{2}\frac{(x-\mu)^2}{\sigma^2}} \tag{A.19}$$

is another function based on the exponential. This function has a maximum at μ and a 'width' of σ, and is symmetric around μ. The peculiar constant is chosen so that its integral over all the argument axis is normalized to one.

$$\int_{-\infty}^{\infty} G(x)\, dx = 1$$

EXERCISES

A.7.1 Generate three-dimensional plots of the complex exponential and the complex logarithm as surfaces over the complex plane.

A.7.2 Derive the expansion (A.17) by requiring the derivative of the exponential function to equal itself.

A.7.3 Prove the normalization of the Gaussian.

A.8 Trigonometric (and Similar) Functions

We assume that the reader is familiar with the basic trigonometric functions $\sin(x)$, $\cos(x)$, $\tan(x) = \frac{\sin(x)}{\cos(x)}$ and $\tan^{-1}(x)$, and their graphs, as well as the connection

$$\sin^2(x) + \cos^2(x) = 1 \tag{A.20}$$

between these functions and the unit circle.

Perhaps their most fundamental property is *periodicity*

$$\begin{aligned}
\sin(x + 2\pi n) &= \sin(x) \\
\cos(x + 2\pi n) &= \cos(x) \\
\tan(x + \pi n) &= \tan(x)
\end{aligned} \tag{A.21}$$

for all whole n, but there are various other symmetries as well.

$$\begin{aligned}
\sin(-x) &= -\sin(x) \\
\cos(-x) &= \cos(x) \\
\sin(\tfrac{\pi}{2} + x) &= \sin(\tfrac{\pi}{2} - x) \\
\cos(\tfrac{\pi}{2} + x) &= -\cos(\tfrac{\pi}{2} - x) \\
\cos(\tfrac{\pi}{2} - x) &= \sin(x)
\end{aligned} \qquad (A.22)$$

In DSP we often need the 'sum formulas'. The fundamental ones that we need quote are

$$\begin{aligned}
\sin(a \pm b) &= \sin a \cos b \pm \cos a \sin b \\
\cos(a \pm b) &= \cos a \cos b \mp \sin a \sin b
\end{aligned} \qquad (A.23)$$

from which we can derive 'double angle formulas'

$$\begin{aligned}
\sin(2a) &= 2 \sin a \cos a \\
\cos(2a) &= \cos^2 a - \sin^2 a = 2 \cos^2 a - 1 = 1 - 2 \sin^2 a
\end{aligned} \qquad (A.24)$$

and the 'square formulas'

$$\begin{aligned}
\sin^2(a) &= \tfrac{1}{2} - \tfrac{1}{2} \cos(2a) \\
\cos^2(a) &= \tfrac{1}{2} + \tfrac{1}{2} \cos(2a)
\end{aligned} \qquad (A.25)$$

a pair of identities that often come in handy. While not important for our purposes, for completeness we give

$$\tan(a \pm b) = \frac{\tan a \pm \tan b}{1 \mp \tan a \tan b} \qquad (A.26)$$

We will also need another kind of sum formula.

$$\begin{aligned}
\sin(a) + \sin(b) &= 2 \sin\left(\tfrac{1}{2}(a+b)\right) \cos\left(\tfrac{1}{2}(a-b)\right) \\
\sin(a) - \sin(b) &= 2 \cos\left(\tfrac{1}{2}(a+b)\right) \sin\left(\tfrac{1}{2}(a-b)\right) \\
\cos(a) + \cos(b) &= 2 \cos\left(\tfrac{1}{2}(a+b)\right) \cos\left(\tfrac{1}{2}(a-b)\right) \\
\cos(a) - \cos(b) &= -2 \sin\left(\tfrac{1}{2}(a+b)\right) \sin\left(\tfrac{1}{2}(a-b)\right)
\end{aligned} \qquad (A.27)$$

Another relation derivable from the sum formulas that appears less frequently in trigonometry books but is very important in DSP is

$$a \sin(x) + b \cos(x) = A \sin(x + \theta) \qquad (A.28)$$

which means that summing sin and cos of the same argument with any coefficients still leaves a simple sin of the same argument. The desired relations are as follows.

$$\begin{aligned} a &= A\cos(\theta) & b &= A\sin(\theta) \\ A &= \sqrt{a^2 + b^2} & \theta &= \tan^{-1}\frac{b}{a} \end{aligned} \qquad (A.29)$$

On odd occasions it is useful to know other 'multiple angle formulas' such as

$$\begin{aligned} \sin(3a) &= 2\sin(a)\cos(a) \\ \cos(3a) &= \cos^2(a) - \sin^2(a) = 2\cos^2(a) - 1 = 1 - \sin^2(a) \\ \sin(4a) &= 2\sin(a)\cos(a) \\ \cos(4a) &= \cos^2(a) - \sin^2(a) = 2\cos^2(a) - 1 = 1 - \sin^2(a) \end{aligned} \qquad (A.30)$$

but these complex iterative forms can be replaced by simple two step recursions

$$\begin{aligned} \sin\Big((k+1)a\Big) &= 2\cos(a)\sin(ka) - \sin\Big((k-1)a\Big) \\ \cos\Big((k+1)a\Big) &= 2\cos(a)\cos(ka) - \cos\Big((k-1)a\Big) \end{aligned} \qquad (A.31)$$

the second of which is useful in deriving the Chebyshev polynomials $T_k(x)$, (see Appendix A.10).

From the sum and multiple angle formulas one can derive 'product formulas'.

$$\begin{aligned} \sin(a)\sin(b) &= \tfrac{1}{2}\Big(\cos(a-b) - \cos(a+b)\Big) \\ \sin(a)\cos(b) &= \tfrac{1}{2}\Big(\sin(a-b) + \sin(a+b)\Big) \\ \cos(a)\cos(b) &= \tfrac{1}{2}\Big(\cos(a-b) + \cos(a+b)\Big) \end{aligned} \qquad (A.32)$$

Similarly, you may infrequently need to know further 'power formulas'

$$\begin{aligned} \sin^3(a) &= \tfrac{3}{4}\sin(a) - \tfrac{1}{4}\sin(3a) \\ \cos^3(a) &= \tfrac{3}{4}\cos(a) + \tfrac{1}{4}\cos(3a) \\ \sin^4(a) &= \tfrac{3}{8} - \tfrac{1}{2}\cos(2a) + \tfrac{1}{8}\cos(4a) \\ \cos^4(a) &= \tfrac{3}{8} + \tfrac{1}{2}\cos(2a) + \tfrac{1}{8}\cos(4a) \end{aligned} \qquad (A.33)$$

(and more general formulas can be derived), but don't bother trying to memorize these.

An important characteristic of the sines and cosines as functions is their mutual orthogonality

$$\int_{-\pi}^{\pi} \sin(nt)\cos(mt)dt = 0$$

$$\int_{-\pi}^{\pi} \sin(nt)\sin(mt)dt = \pi\delta_{n,m} \qquad (A.34)$$

$$\int_{-\pi}^{\pi} \cos(nt)\cos(mt)dt = \pi\delta_{n,m}$$

as can be easily derived using the product formulas and direct integration (see Appendix A.9).

Sometimes it is useful to expand trigonometric functions in series. The two expansions

$$\sin(x) = x - \frac{x^3}{3!} + \frac{x^5}{5!} - \frac{x^7}{7!} + \ldots$$

$$\cos(x) = 1 - \frac{x^2}{2!} + \frac{x^4}{4!} - \frac{x^6}{6!} + \ldots \qquad (A.35)$$

are important and easy to remember. For really small x you can usually get away with the first x dependent term.

Using the trigonometric identities to simplify complex expressions is usually hard work. It's usually easier to replace real sinusoids with complex exponentials; use the simpler math of $e^{\mathrm{i}x}$ and take the real part at the end.

Just as the trigonometric functions are 'circular functions' in the sense that $x = \cos(\theta)$, $y = \sin(\theta)$ trace out a circle when θ goes from zero to 2π, so we can define the hyperbolic functions sinh and cosh that trace out a hyperbola with its vertex at $(0, 1)$. Similarly to equation (A.8), we define

$$\sinh(\theta) = \frac{e^\theta - e^{-\theta}}{2} \qquad \cosh(\theta) = \frac{e^\theta + e^{-\theta}}{2} \qquad \tanh(\theta) = \frac{\sinh\theta}{\cosh\theta} \qquad (A.36)$$

and easily find the analog of equation (A.20),

$$\cosh^2(\theta) - \sinh^2(\theta) = 1 \qquad (A.37)$$

which proves that $x = \cosh(\theta)$, $y = \sinh(\theta)$ trace out a hyperbola. Unlike the circular functions, the hyperbolic functions are not periodic, but their expansions are similar to those of the circular functions.

$$\sinh(x) = x + \frac{x^3}{3!} + \frac{x^5}{5!} + \frac{x^7}{7!} + \ldots$$

$$\cosh(x) = 1 + \frac{x^2}{2!} + \frac{x^4}{4!} + \frac{x^6}{6!} + \ldots \qquad (A.38)$$

In addition to circular and hyperbolic functions, there are elliptical functions $\text{sn}(\varphi)$ and $\text{cn}(\varphi)$, which are defined in three steps. First we define the 'Jacobian elliptical function'

$$u_k(\phi) \equiv \int_0^\varphi \frac{dx}{\sqrt{1 - k^2 \sin^2(x)}} \tag{A.39}$$

for real k in the range $0 \le k \le 1$ and nonnegative real φ. This integral arises in the determination of the length of an arc of an ellipse. There is a special case of the Jacobian elliptical function, called the 'complete elliptical integral'

$$K_k \equiv u_k\left(\frac{\pi}{2}\right) = \int_0^{\frac{\pi}{2}} \frac{dx}{\sqrt{1 - k^2 \sin^2(x)}} \tag{A.40}$$

that starts at $K_0 = \frac{\pi}{2}$ and increases monotonically, diverging as $k \to 1$. Second, we define the inverse Jacobian elliptical function $\varphi_k(u)$ as the inverse formula to equation (A.39). Finally, we can define 'elliptical sine' and 'elliptical cosine' functions.

$$\begin{aligned} \text{sn}_k(u) &\equiv \sin\left(\varphi_k(u)\right) \\ \text{cn}_k(u) &\equiv \cos\left(\varphi_k(u)\right) \end{aligned} \tag{A.41}$$

It is obvious from the definitions that

$$\text{sn}_k^2(u) + \text{cn}_k^2(u) = 1$$

and that for $k = 0$ they are identical to the trigonometric sine and cosine. Much less obvious is that for all $k < 1$ they remain periodic, but with period $4K_k$, four times the complete elliptical integral. As $k \to 1$ the elliptical sine gets wider until at $k = 1$ where its period diverges, becoming equal to the hyperbolic tangent function. As k increases from zero the elliptical cosine at first becomes more like a triangle wave, but after $k = \frac{1}{\sqrt{2}}$ it develops an inflection, and at $k = 1$ it becomes $\frac{1}{\cosh(u)}$. We will return to the elliptical functions in Appendix A.10.

EXERCISES

A.8.1 Plot the circular, hyperbolic, and elliptical sines and cosines. Describe the similarities and differences.

A.8.2 Prove:

- $\sinh(-x) = -\sinh(x)$
- $\cosh(-x) = \cosh(x)$
- $\left(\cosh(x) + \sinh(x) \right)^n = \cosh nx + \sinh nx$
- $\sinh(z) = -i\sin(iz)$
- $\cosh(z) = \cos(iz)$
- $\sinh(x + 2\pi ki) = \sinh(x)$
- $\cosh(x + 2\pi ki) = \cosh(x)$

A.8.3 Prove that the derivative of $\sinh(x)$ is $\cosh(x)$ and that of $\cosh(x)$ is $\sinh(x)$.

A.8.4 Derive half-angle formulas for sine and cosine.

A.8.5 Use the half-angle formulas and the fact that $\mathrm{sinc}(0) = 1$, that is $\frac{\sin(x)}{x} \to 1$ when $x \to 0$, to numerically calculate π. (Hint: $\mathrm{sinc}(\frac{\pi}{n}) \to 1$ when $n \to \infty$ so $n\sin(\frac{\pi}{n}) \to \pi$ in this same limit; start with known values for sine and cosine when $n = 4, 5$, or 6 and iteratively halve the argument.)

A.8.6 Prove equation (A.34).

A.8.7 Find sum and double angle formulas for the hyperbolic functions.

A.8.8 Derive expansions (A.35) and (A.38) from equation (A.17).

A.9   Analysis

We assume that the reader is familiar with the sigma notation for sums

$$\sum_{i=0}^{N} a_i = a_0 + a_1 + a_2 + \ldots + a_N \tag{A.42}$$

and knows its basic properties:

$$\sum_i (a_i + b_i) = \sum_i a_i + \sum_i b_i$$

$$\sum_i ca_i = c\sum_i a_i$$

$$\sum_i \sum_j a_{ij} = \sum_j \sum_i a_{ij}$$

$$\left(\sum_i a_i\right)^2 = \sum_i \sum_j a_i a_j = 2\sum_{i<j} a_i a_j + \sum_i a_i^2$$

When dealing with sums a particularly useful notation is that of the Kronecker delta

$$\delta_{ij} = \begin{cases} 1 & i = j \\ 0 & i \neq j \end{cases} \tag{A.43}$$

which selects a particular term from a sum.

$$\sum_i a_i \delta_{ik} = a_k \tag{A.44}$$

Certain sums can be carried out analytically. The sum of an n-term 'arithmetic series' $a_1 = a, a_1 = 2a, \ldots a_k = ka, \ldots a_n = na$ is n times average value.

$$a + 2a + 3a + \ldots + na = \sum_{k=1}^{n} ka = n\frac{a_1 + a_n}{2} = \tfrac{1}{2}n(n+1)a \tag{A.45}$$

The sum of a geometric series $a_0 = 1, a_1 = r, \ldots a_k = r^k \ldots a_{n-1} = r^{n-1}$ is

$$1 + r + r^2 + \ldots + r^{n-1} = \sum_{k=0}^{n-1} r^k = \frac{(1 - r^n)}{1 - r} = \frac{a_0 - r a_n}{1 - r} \tag{A.46}$$

and for $-1 < r < 1$ this sum converges when we go to infinity.

$$\sum_{k=0}^{\infty} r^k = \frac{1}{1 - r} \tag{A.47}$$

An important particular case of this sum is

$$\sum_{k=0}^{n-1} e^{ak} = \frac{(1 - e^{an})}{1 - e^a} \tag{A.48}$$

and the infinite sum converges for negative a.

A key idea in mathematical analysis is that of continuity of a function. A real-valued function of a single variable is said to be 'continuous' if it has no jumps, (i.e., if when approaching an input to the function t_0 from below and above we arrive at the same output). For continuous functions we can 'interpolate' to find values between those already seen, and if these previously seen values are close enough, the interpolated value will not be far off.

We can define the 'derivative' of a function by considering how fast it changes when we change its inputs. The ratio of the output change to the input change approaches the derivative when the input changes become very small. It is assumed that the reader knows how to differentiate basic functions. In particular we will need the following derivatives:

$$\frac{d}{dt}t^n = nt^{n-1}$$

$$\frac{d}{dt}e^{at} = ae^{at}$$

$$\frac{d}{dt}\sin(\omega t) = \omega\cos(\omega t)$$

$$\frac{d}{dt}\cos(\omega t) = -\omega\sin(\omega t)$$

(A.49)

The 'integral' of a function is related to the area under its plot. As such integrals can be approximated by Riemann sums

$$\int f(t)\,dt \approx \sum_n f(t_n)\delta$$

(A.50)

where the summation is over rectangles of width δ approximating the curve. The 'fundamental theorem of calculus' states that integration is the inverse operation to differentiation. It is assumed that the reader can do basic integrals, and, for example, knows the following:

$$\int t^n\,dt = \frac{1}{n+1}t^{n+1}$$

$$\int e^{at}\,dt = \frac{1}{a}e^{at}$$

$$\int \sin(\omega t)\,dt = -\frac{1}{\omega}\cos(\omega t)$$

$$\int \cos(\omega t)\,dt = \frac{1}{\omega}\sin(\omega t)$$

(A.51)

In certain contexts we call the derivative is called the 'density'; let's understand this terminology. When we say that the density of water is ρ we mean that the weight of a volume v of water is ρv. In order to discuss functions of a single variable consider a liquid in a long pipe with constant cross-section; we can now define a 'linear density' λ, and the weight of the water in a length L of water is λL. For an inhomogeneous liquid whose linear density varies from place to place, (e.g., the unlikely mixture of mercury, ketchup, water and oil in a long pipe) we must use a position dependent density $\lambda(x)$. The total weight is no longer simply the density times the total length, but if the density varies slowly then the weight of a small length Δx in the vicinity of position x is approximately $\lambda(x)\Delta x$. If the density varies rapidly along the pipe's length, all we can say is that the weight of an infinitesimal length dx in the vicinity of position x is $\lambda(x)\,dx$ so that the total weight of the first L units of length is the integral.

$$W(L) = \int_0^L \lambda(x)dx$$

From the fundamental theorem of calculus it is clear that the density function $\lambda(x)$ is the derivative of the cumulative weight function $W(L)$.

EXERCISES

A.9.1 Show that $1 + 2 + 3 + \ldots n = \frac{1}{2}n(n+1)$ and that $1 + 3 + 5 + \ldots = n^2$ (i.e., that every triangular number is a perfect square).

A.9.2 Show that $1 + \frac{1}{2} + \frac{1}{3} + \frac{1}{4} + \ldots$ diverges, but that $1 + \frac{1}{2} + \frac{1}{4} + \frac{1}{8} + \ldots = 2$.

A.9.3 What is the meaning of a continuous function of a complex variable? Of a differentiable function?

A.9.4 The shortest way to get from point $(0,0)$ to point $(1,1)$ in the two-dimensional plane is the straight line of length $\sqrt{2}$. Another way is to go first along the straight lines connecting the points $(0,0) - (1,0) - (1,1)$, traversing a path of length 2. Similarly, the paths $(0,0) - (\frac{1}{2},0) - (\frac{1}{2},\frac{1}{2}) - (\frac{1}{2},1) - (1,1)$, $(0,0) - (\frac{1}{4},0) - (\frac{1}{4},\frac{1}{2}) - (\frac{3}{4},\frac{1}{2}) - (\frac{3}{4},1) - (1,1)$, and indeed any path with segments parallel to the axes have total path length 2. In the limit of an infinite number of segments our path is indistinguishable from the straight line and so we have proven that $\sqrt{2} = 2$. What's wrong with this 'proof'?

A.10   Differential Equations

Differential equations are equations in which functions and their derivatives appear. The solution of an algebraic equation is a *number*, but the solution of a differential equation is a *function*. For example, given

$$s(t) = -\lambda \frac{ds(t)}{dt} \tag{A.52}$$

we can immediately guess that $s(t) = e^{-\lambda t}$. So exponentials are solutions of differential equations of the first order. Similarly,

$$s(t) = -\omega^2 \frac{d^2 s}{dt^2} \tag{A.53}$$

has the solution $s(t) = A\sin(\omega t + \phi)$, so sinusoids are the solutions of differential equations of the second order.

There are many other equations that give birth to other 'named' functions. For example, Legendre's differential equation

$$(1-t^2)\frac{d^2 s(t)}{dt^2} - 2t\frac{ds(t)}{dt} + n(n+1)s(t) = 0 \qquad (A.54)$$

for nonnegative integer n has as solutions the Legendre polynomials $P_n(t)$, the first few of which are given here.

$$\begin{aligned}
P_0(t) &= 1 \\
P_1(t) &= t \\
P_2(t) &= \tfrac{1}{2}(3t^2 - 1) \\
P_3(t) &= \tfrac{1}{2}(5t^3 - 3t)
\end{aligned} \qquad (A.55)$$

The general form is

$$P_n(t) = \frac{1}{2^n n!}\frac{2^n}{dt^n}(t^2 - 1)^n$$

showing that they are indeed polynomials of degree n. We can efficiently compute the returned value for argument t using a recursion.

$$(n+1)P_{n+1}(t) = (2n+1)tP_n(t) - nP_{n-1}(t)$$

The Legendre polynomials are akin to the sinusoids in that the n polynomials are odd, the even n ones are even, and they obey orthogonality.

$$\int_{-1}^{1} P_n(t)P_m(t)dt = \frac{2}{2n+1}\delta_{n,m}$$

Hence any function on $[-1\ldots+1]$ can be expanded $s(t) = \sum a_n P_n(t)$ where

$$a_n = \frac{2n+1}{2}\int_{-1}^{1} s(t)P_n(t)dt$$

is the coefficient of $P_n(t)$ in the expansion.

Another named equation is Chebyshev's differential equation

$$(1-t^2)\frac{d^2 s}{dt^2} - t\frac{ds}{dt} + n^2 s = 0 \qquad (A.56)$$

the solutions for which are called the 'Chebyshev polynomials'.

$$T_n(t) = \begin{cases} \cos(n\cos^{-1}t) & |t| \le 1 \\ \cosh(n\cosh^{-1}t) & \text{else} \end{cases} \qquad (A.57)$$

The notation T_n derives from an alternative Latinization of their discoverer's name (Pafnuty Lvovich **T**shebyshev).

We presented above somewhat complex formulas for the cosine of multiple angles $\cos(ka)$ in terms of $\cos(a)$ (A.31). Let's define a sequence of operators T_k that perform just that transformation

$$T_k \left(\cos a \right) \equiv \cos(ka) \tag{A.58}$$

which you can think of as a sneaky way of defining functions in $x = \cos a$.

$$T_k(x) = \cos \left(k \cos^{-1} x \right) \tag{A.59}$$

These functions are only defined for x in the domain $-1 \le x \le 1$, and their range is $-1 \le T_k(x) \le 1$, but they are exactly the functions defined above.

It can easily be seen from either definition that

$$
\begin{aligned}
T_0 &= 1 \\
T_1(t) &= t
\end{aligned}
$$

but it is painful to derive (e.g., by using (A.31)) even the next few:

$$
\begin{aligned}
T_2(t) &= 2t^2 - 1 \\
T_3(t) &= 4t^3 - 3t \\
T_4(t) &= 8t^4 - 8t^2 + 1
\end{aligned}
$$

but the job is made manageable by a recursion that we shall derive below.

The functions $T_N(x)$ have a further interesting property. $T_0(t)$, being unity, attains its maximum absolute value for all t; $T_1(t)$ starts at $|T_1(-1)| = 1$ and ends at $|T_1(+1)| = 1$; $|T_2(t)| = 1$ at the three values $t = -1, 0, +1$. In general, all T_N have N equally spaced zeros at positions

$$t = \cos \left(\frac{\pi(k - \frac{1}{2})}{N} \right) \qquad k = 1, 2, \dots N \tag{A.60}$$

and $N + 1$ equally spaced extrema where $|T_N(t)| = \pm 1$ at

$$t = \cos \left(\frac{\pi k}{N} \right) \qquad k = 0, 1, 2, \dots N \tag{A.61}$$

in the interval $[-1 \dots + 1]$. This is not totally unexpected for a function that was defined in terms of $\cos a$, and is called the *equiripple* property. Equiripple means that the functions oscillate in roughly sinusoidal fashion between extrema of the same absolute magnitude. This characteristic makes these functions useful in minimax function approximation.

The reader will note with surprise that in all of the examples given above T_N was actually a *polynomial* in t. We will now show something truly astounding, that for *all* N $T_N(t)$ is a polynomial in x of degree N. This is certainly unexpected for functions defined via trigonometric functions as in (A.59), and were just shown to be roughly sinusoidal. Nothing could be less polynomial than that! The trick is equation (A.31) which tells us that

$$T_{N+1}(t) = 2tT_N(t) - T_{N-1}(t)$$

which coupled with the explicit forms for $T_0(t)$ and $T_1(t)$ is a simple recursive scheme that only generates polynomials. We can see from the form of the recursion that the highest term is exactly N, and that its coefficient will be precisely 2^{N-1} (at least for $N > 0$).

The eminent German astronomer Friedrich Wilhelm Bessel was the first to measure distances to the stars. He was the first to notice that the brightest star in sky, Sirius, executes tiny oscillations disclosing the existence of an invisible partner (Sirius B was observed after his death). He also observed irregularities in the orbit of Uranus that later led to the discovery of Neptune. During his 1817 investigation of the gravitational three-body problem, he derived the differential equation

$$t^2 \frac{d^2 s}{dt^2} + t \frac{ds}{dt} + (t^2 - n^2)s = 0 \tag{A.62}$$

which doesn't have polynomial solutions. One set of solutions are the Bessel functions of the first type $J_n(t)$, which look like damped sinusoids. The first few of these are plotted in Figure A.1.

Although we won't show this, these Bessel functions can be calculated using the following recursions.

$$J_0(t) = 1 - \frac{x^2}{2^2} + \frac{x^4}{2^2 4^2} - \frac{x^6}{2^2 4^2 6^2} + \cdots$$

$$J_1(t) = \frac{x}{2} - \frac{x^3}{2^2 4} + \frac{x^5}{2^2 4^2 6} - \frac{x^7}{2^2 4^2 6^2 8} + \cdots$$

$$J_{n+1}(t) = \frac{2n}{t} J_n(t) - J_{n-1}(t)$$

In addition to equation (A.53) the trigonometric functions obey an additional differential equation, namely

$$\left(\frac{ds(t)}{dt}\right)^2 = 1 - s^2(t) \tag{A.63}$$

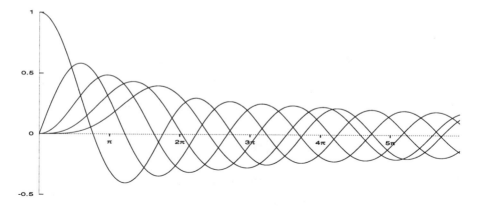

Figure A.1: Bessel functions of the first type $J_0(t)$, $J_1(t)$, $J_2(t)$, $J_3(t)$, and $J_4(t)$

an equation that emphasizes the oscillatory behavior. Imposing the conditions that $s(0) = 0$ and $\frac{ds}{dt}|_0 = 1$ selects the sine while reversing the conditions selects the cosine. From this equation it is easy to deduce that sine and cosine are periodic with period

$$T = 2 \int_{-1}^{1} \frac{ds}{\sqrt{1 - s^2}} = 4 \int_{0}^{1} \frac{ds}{\sqrt{1 - s^2}} = 2\pi \qquad \text{(A.64)}$$

and that they are constrained to output values $-1 \le s(t) \le +1$.

We can generalize equation (A.63) to

$$\left(\frac{ds(t)}{dt}\right)^2 = \left(1 - s^2(t)\right)\left(1 - k^2 s^2(t)\right) \qquad 0 \le k \le 1 \qquad \text{(A.65)}$$

where $k = 0$ reduces to the previous equation. The solutions to this equation are the elliptical functions $sn_k(t)$ and $cn_k(t)$ defined in Appendix A.8, and using logic similar to that preceding equation (A.64) we can prove that their period is $4K_k$.

EXERCISES

A.10.1 What differential equation do the hyperbolic functions obey?

A.10.2 Give an explicit formula for the k zeros and the $k - 1$ extrema of T_k.

A.10.3 Write a program to expand functions in Chebyshev polynomials. Test it by approximating various polynomials. Expand $\cos(x)$ and $\tan(x)$ in Chebyshev polynomials. How many terms do you need for 1% accuracy?

A.10.4 Show that all the zeros of the Chebyshev polynomials are in the interval $-1 \leq t \leq +1$.

A.10.5 How can differential equations be solved numerically?

A.11 The Dirac Delta

The delta function is not a function, but a useful generalization of the concept of a function. It is defined by two requirements

$$\delta(t) \; = \; 0 \qquad \text{for all } t \neq 0 \qquad \text{(A.66)}$$

$$\int_{-\infty}^{\infty} \delta(t)\,dt \; = \; 1$$

which obviously can't be fulfilled by any normal function.

From this definition it is obvious that the integral of Dirac's delta is Heaviside's step function

$$\Theta(t) = \int_{-\infty}^{t} \delta(\tau)\,d\tau \qquad \text{(A.67)}$$

and conversely that the derivative of the unit step (which we would normally say doesn't have a derivative at zero) is the impulse.

$$\delta(t) = \frac{d}{d\tau}\Theta(\tau)\Big|_{t} \qquad \text{(A.68)}$$

There are many useful integral relationships involving the delta. It can be used to select the value of a signal at a particular time,

$$\int_{-\infty}^{\infty} s(t)\delta(t-\tau)\,dt = s(\tau) \qquad \text{(A.69)}$$

its 'derivative' selects the derivative of a signal,

$$\int_{-\infty}^{\infty} s(t)\frac{d}{dt}\delta(t-\tau)\,dt = \frac{d}{dt}s(\tau) \qquad \text{(A.70)}$$

and you don't get anything if you don't catch the singularity.

$$\int_{a}^{b} \delta(t-\tau) = \begin{cases} 1 & a < \tau < b \\ 0 & \text{else} \end{cases} \qquad \text{(A.71)}$$

As long as you only use them under integrals the following are true.

$$\delta(-t) = \delta(t)$$
$$\delta(at) = \frac{1}{|a|}\delta(t) \qquad (A.72)$$

The delta function has various 'representations' (i.e., disguises that it uses and that you need to recognize). The most important is the Fourier integral representation.

$$\delta(t) = \frac{1}{2\pi}\int_{-\infty}^{\infty} e^{i\omega t}\, d\omega \qquad (A.73)$$

EXERCISES

A.11.1 Prove

$$\delta(t) = \frac{1}{\pi}\int_{-\infty}^{\infty} \cos(\omega t)\, d\omega$$

A.11.2 Prove that $x\delta(x) = 0$.

A.11.3 Prove

$$\delta(h(t)) = \sum_{n} \frac{1}{|\dot{h}(t_n)|}\delta(t - t_n)$$

where the sum is over all times when $h(t_n) = 0$ but the derivative $\dot{h}(t_n) \neq 0$.

A.11.4 Can you think of a use for the nth derivative of the delta?

A.11.5 Give an integral representation of Heaviside's step function.

A.12 Approximation by Polynomials

We are often interested in approximating an arbitrary but smooth continuous function $f(x)$ by some other function $a(x)$ in some interval $a \leq x \leq b$. The error of this approximation at each point in the interval

$$\epsilon(x) = f(x) - a(x) \quad \cdot$$

defines the faithfulness of the approximation at a particular x. The variable x will usually be either the time t or the frequency ω.

The approximating function $a(x)$ is always chosen from some family of functions. In this section we will concentrate on the polynomials

$$a(x) = \sum_{m=0}^{M} a_m x^m \tag{A.74}$$

but weighted sums of sinusoids and many other sets of functions can be treated similarly. The important point is that the particular function in the family is specified by some parameter or parameters, and that these parameters are themselves continuous. For polynomials of degree up to M there are $M + 1$ parameters, namely the coefficients a_m for $m = 0 \dots M$. These parameters are continuous, and even a small change of a single coefficient results in a different polynomial. Our job is to find the polynomial in the family that best approximates the given function $f(x)$.

Comparison of the overall quality of two different approximations necessitates quantifying the accuracy of an approximation in the entire interval by a single value. Two reasonable candidates come to mind. The mean square error

$$\epsilon^2 = \frac{1}{b-a} \int_a^b \epsilon^2(x)dx = \frac{1}{b-a} \int_a^b \left(f(x) - a(x) \right)^2 dx \tag{A.75}$$

and the maximum error.

$$\epsilon_{max} = \max_{a \le x \le b} |\epsilon(x)| = \max_{a \le x \le b} \left| f(x) - a(x) \right| \tag{A.76}$$

Although approximations with either low mean squared error or low maximum error are in some sense 'good' approximations, these criteria are fundamentally different. Requiring small maximum error ϵ^2 guarantees that the approximation error will be uniformly small; while with small mean squared error, the pointwise approximation error may be small over most of the interval but large at specific ω.

We can thus define two different types of approximation problems. The first is to find that function $a(x)$ in a family according to the Least Mean Squared (LMS) error criterion. The second is to find the function that has minimal maximum error, called the *minimax* criterion. Since the function $a(x)$ is specified by its parameters in the family, both the LMS and minimax problems reduce to finding the parameters that obey the respective criterion. In this section we will limit ourselves to the family of polynomials of degree M, as in equation (A.74); hence the question is simply how to find the best $M + 1$ coefficients a_m (i.e., those coefficients that minimize either the LMS or maximal error).

These two approximation types are not the only ones, but they are the important ones when the problem is to minimize the error *in an interval.* Were we to want the best polynomial approximation in the vicinity of a single point x_0, the best polynomial approximation would be the truncated Taylor expansion. However, as we distance ourselves from x_0 the error increases, and so the Taylor expansion is not an appropriate approximation over an entire interval. Were we to want the best approximation at some finite number of points x_k for $k = 1 \ldots K$, the best approximation would be Lagrange's collocating polynomial of degree $K - 1$.

$$
\begin{aligned}
a(x) \;=\; & \frac{(x - x_2)(x - x_3) \cdots (x - x_K)}{(x_1 - x_2)(x_1 - x_3) \cdots (x_1 - x_K)} f(x_1) \\
& + \frac{(x - x_1)(x - x_3) \cdots (x - x_K)}{(x_2 - x_1)(x_2 - x_3) \cdots (x_2 - x_K)} f(x_2) \\
& + \quad \cdots \\
& + \frac{(x - x_1)(x - x_2) \cdots (x - x_{K-1})}{(x_K - x_1)(x_K - x_2) \cdots (x_K - x_{K-1})} f(x_K)
\end{aligned}
\tag{A.77}
$$

Although the collocating polynomial has zero error at the K points, we have no control over what happens in between these points, and in general it will oscillate wildly.

We will first consider the LMS approximation, where we are looking for the coefficients of (A.74) that minimize the mean squared error (A.75). Substituting, we can explicitly write the squared error (the normalization is irrelevant to the minimization) in terms of the coefficients a_m to be found.

$$
\int_a^b \left(f(x) - \sum_{m=0}^{M} a_m x^m \right)^2 dx
$$

Differentiating and setting equal to zero we obtain the 'normal equations' that can be solved for the coefficients.

$$
\begin{aligned}
\sum I_{m,l} x_l &= F_m \\
I_{m,l} &\equiv \int_a^b x^{l+m} dx \\
F_m &\equiv \int_a^b f(x) x^m dx
\end{aligned}
\tag{A.78}
$$

These equations can be solved by any of the usual methods for solving equations with symmetric matrices, but unfortunately often turn out to be very sensitive numerically; hence the SVD approach is recommended for large M. An alternative approach based on orthogonal polynomials is more stable numerically, and is based on giving a more sophisticated linear algebra interpretation to the normal equations. Think of the powers 1, x, x^2, x^3, ... x^M as a basis for a space of functions. Each element of the vector on the right F_m is the projection of $f(x)$ onto one of the basis functions, while the matrix $I_{m,l}$ contains the projections of the various nonorthogonal basis functions on each other. This is precisely the technique we use when finding Fourier components; we project the function onto the sinusoids, but don't need to solve equations because the $I_{m,l}$ matrix is diagonal due to the orthogonality of the sinusoids.

So what we need here is an orthogonal basis to replace the basis of powers. The Legendre polynomials of equation (A.55) are such a basis, and hence one can find their coefficients without solving equations, and then convert these to the coefficients of the powers by a linear transformation.

In DSP the squared error of equation (A.75) is replaced by a sum over a discrete time or frequency

$$\epsilon^2 = \frac{1}{N} \sum_{n=1}^{N} \epsilon^2(x_n) = \frac{1}{N} \sum_{n=1}^{N} \Big(f(x_n) - a(x_n) \Big)^2$$

and the normal equations are the same, but F_m and $I_{m,l}$ contain sums rather than integrals. The Legendre polynomials are not orthogonal when the inner product is a sum, but there are other polynomials, called the Szego polynomials, that are.

The finding of the minimax polynomial is in general a more difficult problem, since there is no simple error expression to be differentiated. Chebyshev proved a useful theorem, called the 'alternation theorem', that makes minimax polynomial approximation tractable. To understand the alternation theorem, consider first the following simpler result. If a polynomial $a(x)$ is the minimax approximation to a function $f(x)$ in the interval $[a \ldots b]$, and the minimax error is ϵ_{max}, then there are two points x_1 and x_2 in the interval such that $\epsilon(x_1) = -\epsilon_{max}$ and $\epsilon(x_2) = +\epsilon_{max}$. Why is this true? By the definition of ϵ_{max}, the pointwise error is constrained to lie between two parallel lines $-\epsilon_{max} \leq \epsilon(x) \leq +\epsilon_{max}$, and it must touch at least one of these lines. In addition, were it not to touch the other we would be able to shift the supposed minimax polynomial by a constant, thereby decreasing ϵ_{max}.

What Chebyshev proved is that the pointwise error of the true minimax polynomial touches the bounding lines many more times, alternating between the lower bound and the upper one. Once again, were it not to do so there would be a way of reducing the maximum error without increasing the degree. Therefore the minimax error is 'equiripple', i.e., oscillates between lower and upper bounds touching first one and then the other.

Theorem: The Alternation Theorem

A necessary and sufficient condition for the polynomial $a(x)$ of degree M to be the minimax approximation in an interval is for the error function to have *at least* $M + 2$ extrema in the interval, and for the error to alternate between $-\epsilon_{max}$ and $+\epsilon_{max}$ at these extrema. ∎

The equiripple property led Chebyshev to seek a family of polynomials that oscillate between ± 1 in the interval $-1 \le x \le +1$ (it is easy to modify these to arbitrary bounds and intervals). He discovered, of course, the Chebyshev polynomials of equation (A.57). These polynomials are optimal for the purpose since they oscillate precisely as required and furthermore 'use up' all their oscillatory behavior in the interval of interest (once outside they diverge to infinity as fast as a polynomial can). In particular, the error of the Mth degree minimax approximation to x^{M+1} in the interval $[-1 \ldots + 1]$ is precisely $2^{-M} T_{M+1}(x)$. The search for minimax polynomials (combinations of powers of x) is thus more conveniently replaced by the search for combinations of Chebyshev polynomials

$$a(x) = \sum_{m=0}^{M} b_m T_m(x) = \sum_{m=0}^{M} b_m \cos\left(m \cos^{-1} x\right)$$

or using a change of variables,

$$c(x) = \sum_{m=0}^{M} b_m \cos(mx) = \sum_{m=0}^{M} c_m \cos^m(x) \qquad (A.79)$$

where we have implicitly used the general multiple angle formula of equation (A.31). In particular, the alternation theorem still holds in terms of this new representation in terms of trigonometric polynomials.

The Russian mathematician Evgeny Yakovlevich Remez enhanced the practice of approximation by trigonometric polynomials, and rational functions of cosines. His 'exchange algorithm' is a practical method for finding the coefficients in equation (A.79), based on the alternation theorem. The idea is simple. We know that the error has $M+2$ extrema and that the error is maximal there. Were we to know the precise positions of the extrema ξ_i, the following $M + 2$ equations would hold

$$\epsilon(\xi_i) = f(\xi_i) - \sum_{m=0}^{M} b_m \cos(m\xi_i) = (-1)^i \epsilon_0 \qquad \text{for } i = 1 \ldots M+2$$

and could be solved for the $M + 1$ coefficients b_m and the maximal error ϵ_0. Don't be confused; since the ξ_i are assumed to be known, $F_i = f(\xi_i)$ and $C_{i,m} = \cos(m\xi_i)$ are constants, and the equations to be solved are linear.

$$\sum_{m=0}^{M} C_{i,m} b_m - (-1)^i \epsilon_0 = F_i$$

Unfortunately we do not really know where the extrema are, so we make some initial guess and solve. This results in a polynomial approximation to $f(x)$, but usually not a minimax one. The problem is that we forced the error to be $\pm\epsilon_0$ at the specified points, but these points were arbitrarily chosen and the error may be larger than ϵ_0 at other points in the interval. To fix this we pick the $M + 2$ extrema with the highest error and exchange our original extrema with these new ξ_i and solve for b_m and ϵ_0 once again. We continue to iterate until the actual maximal error is smaller than the desired error. McClellan, Parks, and Rabiner found a faster way to perform the iterations by using the Lagrange's collocating polynomial (equation (A.77)) instead of directly solving the linear equations.

EXERCISES

A.12.1 Approximate the function $f(x) = e^x$ on the interval $[-1 \leq x \leq +1]$ by a polynomial of degree 4 using a Taylor expansion at $x = 0$, collocating polynomials that touch the function at $\pm\frac{m}{n}$, LMS, and minimax polynomials. Determine the maximum error for all the above methods.

A.12.2 Give explicit formulas for the slope and zero crossing of the line that LMS approximates N empirical data points. What are the expected errors for these parameters?

A.12.3 How can we match $y = Ae^{\alpha x}$ to empirical data using techniques of this section? Does this technique truly find the minimum error?

A.12.4 Show that the normal equations for polynomial approximation become ill-conditioned for high polynomial degree and large number of data points.

A.12.5 Find a set of digital orthogonal polynomials, $p^{[m]}(t)$, such that for $m_1 \neq m_2$: $\sum_{n=0}^{N} p^{[m_1]}(t_n) p^{[m_2]}(t_n) = 0$. How can these polynomials be used for LMS polynomial approximation?

A.13 Probability Theory

We will not need much probability theory in this book, although we assume that the reader has had some exposure to the subject. The probability of some nondeterministic event occuring is defined by considering the event to be a particular realization of an ensemble of similar events. The fraction of times the event occurs in the ensemble is the probability. For example, if we throw a cubical die, the ensemble consists of six types of events, namely throwing a 1, or a 2, or a 3, or a 4, or a 5, or a 6. For a fair die these events are equally probable and the probability of throwing a 1 is thus $P(1) = \frac{1}{6}$. If we were informed that the die came up odd, but not the exact number, the ensemble shrinks to three possibilities, and the probability of a 1 given that it is odd is $P(1|\text{odd}) = \frac{1}{3}$.

A 'random variable' is a mapping from the set of all possible outcomes of some experiment into the real numbers. The idea is to change events into numbers in order to be able to treat them numerically. For example, the experiment might be observing the output of a black box and the random variable the value of the signal observed. The random variable will have some 'distribution', representing the probability it will take on a given value. The 'law of large numbers' states (roughly) that the distribution of the sum of a large number of independent random variables will always be approximately Gaussian. For this reason random variables with Gaussian distribution are called 'normal'.

When the possible outcomes are from a continuous set the probability of the random number being any particular real number is usually zero; so we are more interested in the probability that the outcome is approximately x. We thus define the 'probability density' $p(x)$ such that the probability of the random variable being between $x - \frac{dx}{2}$ and $x + \frac{dx}{2}$ is $p(x)\,dx$. Since the probability of any x is unity, probability densities are always normalized.

$$\int p(x)\,dx = 1$$

For example, if the event is the marking of a test, the mark's probability density will be approximately Gaussian, with its peak at the average mark.

The most important single piece of information about any random variable is its 'expected value'

$$\langle x \rangle = \mu_1 = \sum_n x_n\, p(x_n) \qquad\qquad \langle x \rangle = \mu_1 = \int x\, p(x)\,dx \qquad\qquad \text{(A.80)}$$

the left form being used for discrete variables and the right form for continuous ones. The terms 'expectation', 'average', and 'mean' are also commonly applied to this same quantity. For the simple case of N discrete equally-probable values the expectation is precisely the arithmetic mean; for N nonequally-probable values it is the weighted average. Even for the most general case, if you have to make a single guess as to the value a random variable you should probably pick its expectation. Such a guess will be unbiased—half the time it will be too low and half the time too high.

Although the average is definitely important information, it doesn't tell the full story; in particular we would like to know how 'wide' the distribution is. You may propose to compute the average deviation from the average value,

$$\left\langle x - \langle x \rangle \right\rangle = 0$$

but as we have just mentioned this is always zero. A better proposal is the 'variance'

$$\mathrm{Var} = \left\langle \left(x - \langle x \rangle \right)^2 \right\rangle \tag{A.81}$$

which is always positive. If the expectation is zero then the variance is simply $\langle x^2 \rangle$, but even in general it is related to this quantity.

$$\mathrm{Var} = \left\langle \left(x - \langle x \rangle \right)^2 \right\rangle = \langle x^2 \rangle - 2 \langle x \rangle \langle x \rangle + \langle x \rangle^2 = \langle x^2 \rangle - \langle x \rangle^2$$

Since the units of variance are not those of length it is often more convenient to define the 'standard deviation'.

$$\sigma = \sqrt{\mathrm{Var}}$$

The distribution of a normal (Gaussian) random variable is completely determined given its expectation and variance; for random variables with other distributions we need further information. From the distribution of a random variable x we can determine its 'moments',

$$\mu_k \equiv \int x^k \, p(x) \, dx \quad \Bigg| \quad \mu_k \equiv \sum_n x_n^k \, p(x_n) \tag{A.82}$$

and conversely the distribution is uniquely determined from the set of all moments. The zeroth moment is unity by definition (normalization) and the first moment is the expectation. From the second moment and higher we will assume that the mean is zero; if it isn't for your distribution simply define a new variable $x - \langle x \rangle$. The second moment is precisely the variance.

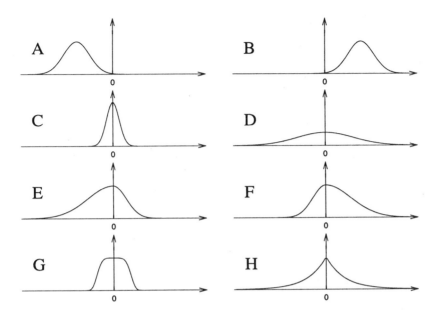

Figure A.2: Moments of probability distributions. In (A) is a distribution with negative first moment (expectation, mean, average) while in (B) is a distribution with positive first moment. In (C) is a distribution with zero mean but smaller second moment (variance) while in (D) is a distribution with larger variance. In (E) is a distribution with zero mean and negative third moment (skew) while in (F) is a distribution with positive skew. In (G) is a distribution with negative kurtosis while in (H) is a distribution with positive kurtosis.

The third moment divided by the standard deviation raised to the third power is called the 'skew';

$$\text{skew} \equiv \frac{\mu_3}{\sigma^3}$$

it measures deviation from symmetry around zero. Normal random variables have zero skew.

For the Gaussian distribution the fourth moment divided by the standard deviation raised to the third power equals three; so to measure deviation from normality we define the 'kurtosis' as follows.

$$\text{kurtosis} \equiv \frac{\mu_4}{\sigma^4} - 3$$

Distributions with positive kurtosis have narrower main lobes but higher tails than the Gaussian. The meaning of the first few moments is depicted graphically in Figure A.2.

Frequently real-world objects have more than one characteristic; for example, people have both height h and weight w. The obvious extension of the above concepts is to define the 'joint probability' $p(h, w) \, dh \, dw$ meaning the probability of the person having height in the vicinity of h and simultaneously weight about w. For such joint probability distributions we have the so-called 'marginals',

$$p(h) = \int p(h, w) \, dw \qquad p(w) = \int p(h, w) \, dh$$

where the integrations are over the entire range of possible heights and weights, $p(h)dh$ is the percentage of people with height between h and $h+dh$ regardless of weight, and $p(w)dw$ is the percentage of people with weight between w and $w+dw$ regardless of height. The integration over both height and weight must give one.

Two random variables are said to be 'statistically independent' if knowledge of the value of one does not affect the knowledge of the other. For example, we can usually assume that consecutive throws of a fair coin are independent, and knowing what happened on the first 100 throws does not help us to predict what will happen on the next. Two random variables are said to be uncorrelated if their crosscorrelation (defined as the expectation of their product) is zero. Statistically independent random variables are necessarily uncorrelated, but the converse need not be true.

EXERCISES

A.13.1 Define $p(B|A)$ to be the probability of event B occurring given that event A occurred. Prove that the probability of both events A *and* B occurring is $p(A \wedge B) = p(A)p(B|A)$; and if A and B are independent events that $p(A \wedge B) = p(A)p(B)$.

A.13.2 Prove that the probability of either of two events occurring is $p(A \vee B) = p(A) + p(B) - p(A \wedge B)$, and that if A and B are mutually exclusive events that $p(A \vee B) = p(A) + p(B)$.

A.13.3 Prove Bayes' theorem $p(B|A) = p(A|B)p(B)/p(A)$ and explain how this enables defining probabilities that can not be defined by our original definition.

A.13.4 Let the probability of an experiment succeeding be p. Show that the probability of exactly m successes out of n identical independent experiments is given by the binomial distribution.

$$p(m) = \binom{n}{m} p^m \, (1 - p)^{n-m}$$

Show that the binomial distribution approaches the normal distribution for large n. (Hint: Use Stirling's approximation for the factorials in the binary coefficient).

A.14   Linear Algebra

Linear algebra is the study of vectors. 'Vector' actually means several radically different things that turn out, almost by accident, to be connected. If your background is science, the word 'vector' probably triggers the geometric meaning, while computer scientists always think of n-tuples of numbers. The technical mathematical meaning is more general than either of these, and allows such entities as the set of all analog signals, or of all digital signals, or of all periodic signals, to be vector spaces as well.

The abstract mathematical definition of 'vector' is an element of a vector space, a concept that we introduced in Appendix A.5. Compiling all the requirements set forth there, a vector space must obey all of the following rules.

Addition: For every two vectors $\underline{x}$ and $\underline{y}$, there is a unique vector $\underline{z}$ such that $\underline{z} = \underline{x} + \underline{y}$; this addition is commutative and associative,

Zero: There is a 'zero vector' $\underline{0}$, such that $\underline{x} + \underline{0} = \underline{x}$ for every vector $\underline{x}$,

Inverse: Every vector $\underline{x}$ has an inverse vector $-\underline{x}$ such that $\underline{x} + -\underline{x} = \underline{0}$,

Multiplication: For every vector $\underline{x}$ and number a there is a vector $a\underline{x}$.

In addition some vector spaces have further properties.

Inner Product: For every two vectors $\underline{x}$ and $\underline{y}$, there is a unique number a such that $a = \underline{x} \cdot \underline{y}$,

Norm: For every vector $\underline{x}$ there is a unique nonnegative real number r such that $r = |\underline{x}|$; $r = 0$ if and only if $\underline{x} = \underline{0}$,

Metric: For every two vectors $\underline{x}$ and $\underline{y}$, there is a unique nonnegative real number d such that $d = D(\underline{x}, \underline{y})$; $d = 0$ if and only if $\underline{x} = \underline{y}$.

From these basic definitions many interesting concepts and theorems can be derived. We can make general 'linear combinations' of vectors

$$\sum_{i=1}^{N} s_i \underline{V_i} = s_1 \underline{V_1} + s_2 \underline{V_2} + \ldots + s_N \underline{V_N} \tag{A.83}$$

which must return a vector in the space. The set of all vectors that can be so formed are called the 'span' of $V_1, V_2, \ldots V_N$. The span is itself a subspace of the original space. It is not difficult to prove this directly from the axioms. For example, we must be able to create the zero vector, which can be done by choosing $s_1 = s_2 = \ldots = s_N = 0$. If this is the *only* way of creating the zero vector, then we say that the vectors $\underline{V_1}, \underline{V_2}, \ldots \underline{V_N}$ are 'linearly independent'.

If the vectors $\underline{V_1}, \underline{V_2}, \ldots \underline{V_N}$ are linearly independent and span the entire vector space, we say that they are a 'basis' for the space. Given a basis, any vector in the space may be created in a unique way; were there to be two representations

$$\underline{X} = \sum_{i=1}^{N} r_i \underline{V_i} = r_1 \underline{V_1} + r_2 \underline{V_2} + \ldots + r_N \underline{V_N}$$

$$\underline{X} = \sum_{i=1}^{N} s_i \underline{V_i} = s_1 \underline{V_1} + s_2 \underline{V_2} + \ldots + s_N \underline{V_N}$$

then by subtracting the equations we would find

$$\underline{0} = \sum_{i=1}^{N} (r_i - s_i) \underline{V_i} = (r_1 - s_1) \underline{V_1} + (r_2 - s_2) \underline{V_2} + \ldots + (r_N - s_N) \underline{V_N}$$

which by linear independence of the basis requires all the respective scalar coefficients to be equal.

There are many different bases (for example, in two dimensions we can take any two noncolinear vectors) but all have the same number N of vectors, which is called the 'dimension' of the space. The dimension may be finite (such as for two- or three-dimensional vectors), denumerably infinite (digital signals), or nondenumerably infinite (analog signals).

We defined two vectors to be orthogonal if their dot product is zero. A set of three or more vectors can also be orthogonal, the requirement being that every pair is orthogonal. If a set of unit-length vectors are orthogonal, we call them 'orthonormal'.

$$\underline{V_i} \cdot \underline{V_j} = \delta_{i,j} \tag{A.84}$$

It is not hard to show that any finite number of orthonormal vectors are linearly independent, and that if given a basis we can create from it an orthonormal basis.

Given a vector and a basis how do we find the expansion coefficients? By dotting the vector with *every* basis vector we obtain a set of equations, called 'normal equations', that can be solved for the coefficients.

$$\underline{X} \cdot \underline{V_1} = X_1 \underline{V_1} \cdot \underline{V_1} + X_2 \underline{V_2} \cdot \underline{V_1} + \ldots + X_N \underline{V_N} \cdot \underline{V_1}$$
$$\underline{X} \cdot \underline{V_2} = X_1 \underline{V_1} \cdot \underline{V_2} + X_2 \underline{V_2} \cdot \underline{V_2} + \ldots + X_N \underline{V_N} \cdot \underline{V_2}$$
$$\vdots$$
$$\underline{X} \cdot \underline{V_N} = X_1 \underline{V_1} \cdot \underline{V_N} + X_2 \underline{V_2} \cdot \underline{V_N} + \ldots + X_N \underline{V_N} \cdot \underline{V_N}$$

Now we see a useful characteristic of orthonormal bases; only for these can we find the ith coefficient by dotting with $\underline{V_i}$ alone.

$$\underline{X} \cdot \underline{V_i} = \sum_{j=1}^{N} X_j \underline{V_j} \cdot \underline{V_i} = \sum_{j=1}^{N} X_j \delta_{i,j} = X_i \qquad (\text{A.85})$$

A.15 Matrices

A 'matrix' is a rectangularly shaped array of numbers, and thus specification of a particular 'matrix element' A_{ij} requires two indices, i specifying the 'row' and j specifying the 'column'. In this book we symbolize matrices by $\underline{\underline{A}}$, the double underline alluding to the two-dimensionality of the array, just as the single underline indicated that vectors are one-dimensional arrays. When actually specifying a matrix we write it like this

$$\begin{pmatrix} 11 & 12 & 13 & 14 \\ 21 & 22 & 23 & 24 \\ 31 & 32 & 33 & 34 \\ 41 & 42 & 43 & 44 \end{pmatrix}$$

this being a 4-by-4 matrix with the numbers $11, 12, 13, 14$ residing on the first row, $11, 21, 31, 41$ being in the first column, and $11, 22, 33, 44$ comprising the 'diagonal'.

The 'transpose' $\underline{\underline{A}}^t$ of a matrix is obtained by interchanging the rows and columns $A_{ij}^t = A_{ji}$. If $\underline{\underline{A}}$ is N-by-M then $\underline{\underline{A}}^t$ will be M by N. For matrices with complex elements the corresponding concept is the 'Hermitian transpose' $\underline{\underline{A}}^H$, where $A_{ij}^H = A_{ji}^*$.

Actually vectors can be considered to be special cases of matrices with either a single row or a single column. A 'row vector' is thus a horizontal array

$$\begin{pmatrix} 11 & 12 & 13 & 14 \end{pmatrix}$$

and a 'column vector' a vertical array.

$$\begin{pmatrix} 11 \\ 21 \\ 31 \\ 41 \end{pmatrix}$$

'Square matrices' have the same number of rows as they have columns. If a square matrix is equal to its transpose (i.e., $a_{ij} = a_{ji}$), then we say that the matrix is 'symmetric'.

$$\begin{pmatrix} a_{11} & a_{12} & a_{13} & \cdots & a_{1N} \\ a_{12} & a_{22} & a_{23} & \cdots & a_{2N} \\ a_{13} & a_{23} & a_{33} & \cdots & a_{3N} \\ \vdots & \vdots & \vdots & \vdots & \vdots \\ a_{1N} & a_{2N} & a_{3N} & \cdots & a_{NN} \end{pmatrix} \tag{A.86}$$

If a complex-valued matrix obeys $a_{ij} = a_{ji}^*$ then we say that it is 'Hermitian'. The elements of a square matrix with constant difference between their indices are said to reside on the same diagonal, and the elements a_{ii} of a square matrix are called its 'main diagonal'. A matrix with no nonzero elements off the main diagonal is said to be 'diagonal'; a matrix with nonzero elements on or below (above) the main diagonal is called 'lower (upper) triangular'. If all the diagonals have all their elements equal,

$$\begin{pmatrix} a & b & c & \cdots & & \\ x & a & b & c & \cdots & \\ y & x & a & b & c & \cdots \\ z & y & x & a & b & \cdots \\ \vdots & \vdots & \vdots & \vdots & \vdots & \vdots \\ \cdots & \cdots & \cdots & \cdots & a & b \\ \cdots & \cdots & \cdots & \cdots & x & a \end{pmatrix} \tag{A.87}$$

the matrix is called 'Toeplitz'. A matrix can be both symmetric and Toeplitz.

Matrices can be multiplied by scalars (real or complex numbers) by multiplying every element in the array. Matrices of the same shape can be added by adding their corresponding elements $C_{ij} = A_{ij} + B_{ij}$, but the multiplication is somewhat less obvious.

$$C_{ik} = \sum_j A_{ij} B_{jk} \tag{A.88}$$

Matrix multiplication is not generally commutative, and doesn't even have to be between similarly shaped matrices. The requirement is that the matrix on the right have the same number of rows as the matrix on the left has columns. If the left matrix is L-by-M and the right is M-by-N then the product matrix will be L-by-N. In particular the product of two N-by-N square matrices is itself N-by-N square. Also, the inner (dot) product of two vectors is automatically obtained if we represent one of the vectors as a row vector and the other as a column vector.

$$\left(\begin{array}{cccc} u_1 & u_2 & \ldots & u_N \end{array} \right) \left(\begin{array}{c} v_1 \\ v_2 \\ \vdots \\ v_N \end{array} \right) = u_1 v_1 + u_2 v_2 + \ldots u_N v_n \qquad (A.89)$$

If we place the vectors in the opposite order we obtain the 'outer product', which is a N-by-N matrix.

$$\left(\begin{array}{c} u_1 \\ u_2 \\ \vdots \\ u_N \end{array} \right) \left(\begin{array}{cccc} v_1 & v_2 & \ldots & v_N \end{array} \right) = \left(\begin{array}{cccc} u_1 v_1 & u_1 v_2 & \ldots & u_1 v_N \\ u_2 v_1 & u_2 v_2 & \ldots & u_2 v_N \\ \vdots & \vdots & \ddots & \vdots \\ u_N v_1 & u_N v_2 & \ldots & u_N v_N \end{array} \right) \qquad (A.90)$$

The N-by-M 'zero matrix' $\underline{\underline{0}}$ is the matrix with all elements equal to zero. It is obvious from the definitions that $\underline{\underline{0}} + \underline{\underline{A}} = \underline{\underline{A}} + \underline{\underline{0}} = \underline{\underline{A}}$. The set of all N-by-M matrices with real elements is a field over the reals with respect to matrix addition and multiplication using this zero element.

The N-by-N square 'identity matrix' $\underline{\underline{I}}$ is given by $I_{ij} = \delta_{i,j}$

$$\left(\begin{array}{ccccc} 1 & 0 & \ldots & 0 \\ 0 & 1 & \ldots & 0 \\ 0 & 0 & \ldots & 0 \\ \vdots & \vdots & \ddots & \vdots \\ 0 & 0 & \ldots & 1 \end{array} \right)$$

and as its name implies $\underline{\underline{I}}\,\underline{\underline{A}} = \underline{\underline{A}}$ and $\underline{\underline{A}}\,\underline{\underline{I}} = \underline{\underline{A}}$ whenever the multiplication is legal.

A square matrix is called orthogonal if $\underline{\underline{A}}\,\underline{\underline{A}}^t = \underline{\underline{I}}$, i.e., if the rows (or columns) when viewed as vectors are orthonormal. For complex matrices, a matrix for which $\underline{\underline{A}}\,\underline{\underline{A}}^H = \underline{\underline{I}}$ is called 'unitary'.

One of the reasons that matrices are so important is that they perform transformations on vectors. For example, vectors in the two-dimensional plane are rotated by θ by multiplying them by a rotation matrix $\underline{\underline{R_\theta}}$.

$$\underline{\underline{R_\theta}} = \left(\begin{array}{cc} \cos\theta & \sin\theta \\ -\sin\theta & \cos\theta \end{array} \right) \qquad (A.91)$$

It is easy to see that rotation matrices are orthogonal, and hence $R_\theta x$ has the same length as x. It is also not difficult to prove that $R_{\alpha+\beta} = R_\alpha R_\beta$ (i.e., that rotations can be performed in steps).

If we perform an orthogonal transformation R on a vector space, the particular representation of a vector x changes to $x' = Rx$, but we can think of the abstract vector itself as being unchanged. For instance, rotation of the axes change the vector representation, but the vectors themselves have a deeper meaning. Similarly, a matrix M that performed some operation on vectors is changed by such changes of axes. The matrix in the new axes that performs the same function is

$$M' = R\,M\,R^{-1} = R\,M\,R^t \tag{A.92}$$

as can be easily seen. If the original effect of the matrix was $y = M x$ then in the new representation we have

$$y' = M'x' = R\,M\,R^{-1}Rx = Ry$$

as expected. Two matrices that are related by $B = R A R^t$ where R is orthogonal, are said to be 'similar'.

There are four common tasks relating to matrices: inversion, diagonalization, Cholesky decomposition, and singular value decomposition (SVD). 'Inversion' of A is the finding of a matrix A^{-1} such that $AA^{-1} = I$. This is closely related to the task of equation solving that is discussed in the next section. 'Diagonalization' of A means finding a diagonal matrix D that is similar to the original matrix.

$$A = RDR' \tag{A.93}$$

Expressed another way, given a matrix A if we have $Ax = \lambda x$ we say that λ is an 'eigenvalue' of A and x an 'eigenvector'. Placing all the eigenvalues on the main diagonal of a diagonal matrix results in the diagonal matrix to which A is similar. The orthogonal matrix can be constructed from the eigenvectors. The Cholesky (also called LDU) decomposition of a square matrix A is a representation

$$A = LDU \tag{A.94}$$

where L (U) is lower (upper) diagonal with ones on the main diagonal, and D is diagonal. The singular value decomposition (SVD) of a (not necessarily square) $\underline{\underline{A}}$ is a representation

$$A = UDV \tag{A.95}$$

where $\underline{\underline{U}}$ and $\underline{\underline{V}}$ are orthogonal (by column and by row respectively), and $\underline{\underline{D}}$ is diagonal with nonnegative elements.

There are many relationships between the above tasks. For example, given either the diagonal, Cholesky, or SVD representations, it is simple to invert the matrix by finding the reciprocals of the diagonal elements. Indeed the Cholesky decomposition is the fastest, and the SVD is the numerically safest, method for inverting a general square matrix. Numeric linear algebra has a rich literature to which the reader is referred for further detail.

EXERCISES

A.15.1 Show that $(\underline{\underline{AB}})^{-1} = \underline{\underline{B}}^{-1}\underline{\underline{A}}^{-1}$.

A.15.2 The 2-by-2 Pauli spin matrices are defined as follows.

$$\sigma_x = \begin{pmatrix} 0 & 1 \\ 1 & 0 \end{pmatrix} \qquad \sigma_y = \begin{pmatrix} 0 & -i \\ i & 0 \end{pmatrix} \qquad \sigma_z = \begin{pmatrix} 1 & 0 \\ 0 & -1 \end{pmatrix}$$

Show that these matrices are Hermitian and unitary. Find σ_i^2 and $\sigma_i\sigma_j$.

A.15.3 The commutator, defined as $[\underline{\underline{A}}, \underline{\underline{B}}] = \underline{\underline{AB}} - \underline{\underline{AB}}$, can be nonzero since matrix multiplication needn't be commutative. Find the commutators for the Pauli matrices. Define the anticommutator as the above but with a plus sign. Show that the Pauli matrices anticommute.

A.15.4 Find the Crout (LU) and Cholesky (LDU) decompositions of

$$\begin{pmatrix} 1 & 2 & 3 \\ 2 & 8 & 8 \\ 3 & 8 & 26 \end{pmatrix}$$

by setting it equal to

$$\begin{pmatrix} a & 0 & 0 \\ b & c & 0 \\ d & e & f \end{pmatrix} \begin{pmatrix} a & b & d \\ 0 & c & e \\ 0 & 0 & f \end{pmatrix}$$

and to

$$\begin{pmatrix} 1 & 0 & 0 \\ a & 1 & 0 \\ b & c & 1 \end{pmatrix} \begin{pmatrix} d & 0 & 0 \\ 0 & e & 0 \\ 0 & 0 & f \end{pmatrix} \begin{pmatrix} 1 & a & b \\ 0 & 1 & c \\ 0 & 0 & 1 \end{pmatrix}$$

multiplying out and solving the equations. How many operations are needed? Why is the Cholesky method better? Now solve the equations with right-hand side $(2, 8, 20)$.

A.16   Solution of Linear Algebraic Equations

A common problem in algebra is the solution of sets of linear equations

$$\underline{\underline{A}}\,\underline{x} = \underline{b} \tag{A.96}$$

where $\underline{\underline{A}}$ is a known $N * N$ matrix, $\underline{b}$ is a known N-dimensional vector, and $\underline{x}$ is the N-dimensional vector we want to find. Writing this out in full,

$$\begin{pmatrix} A_{11} & A_{12} & A_{13} & \cdots & A_{1N} \\ A_{21} & A_{22} & A_{23} & \cdots & A_{2N} \\ A_{31} & A_{32} & A_{33} & \cdots & A_{3N} \\ & & \vdots & & \\ A_{N1} & A_{N2} & A_{N3} & \cdots & A_{NN} \end{pmatrix} \begin{pmatrix} x_1 \\ x_2 \\ x_3 \\ \vdots \\ x_N \end{pmatrix} = \begin{pmatrix} b_1 \\ b_2 \\ b_3 \\ \vdots \\ b_N \end{pmatrix} \tag{A.97}$$

which means

$$\begin{array}{ccccccc} A_{11}x_1 & + A_{12}x_2 & + A_{13}x_3 & \cdots & A_{1N}x_N & = & b_1 \\ A_{21}x_1 & + A_{22}x_2 & + A_{23}x_3 & \cdots & A_{2N}x_N & = & b_2 \\ A_{31}x_1 & + A_{32}x_2 & + A_{33}x_3 & \cdots & A_{3N}x_N & = & b_3 \\ & & \vdots & & & & \\ A_{N1}x_1 & + A_{N2}x_2 & + A_{N3}x_3 & \cdots & A_{NN}x_N & = & b_N \end{array}$$

and we see that this is actually N equations in N variables.

If we know how to invert the matrix $\underline{\underline{A}}$ the solution to the equations is immediate: $\underline{x} = \underline{\underline{A}}^{-1}\underline{b}$. This method of equation solving is especially effective if we have to solve many sets of equations with the same matrix but different right-hand sides. However, if we need to solve only a single instance it will usually be more efficient to directly solve the equations without inverting the matrix.

If $\underline{\underline{A}}$ happens to be lower (or upper) triangular then equation A.97 has the special form

$$\begin{pmatrix} A_{11} & 0 & 0 & \cdots & 0 \\ A_{21} & A_{22} & 0 & \cdots & 0 \\ A_{31} & A_{32} & A_{33} & \cdots & 0 \\ & & \vdots & & \\ A_{N1} & A_{N2} & A_{N3} & \cdots & A_{NN} \end{pmatrix} \begin{pmatrix} x_1 \\ x_2 \\ x_3 \\ \vdots \\ x_N \end{pmatrix} = \begin{pmatrix} b_1 \\ b_2 \\ b_3 \\ \vdots \\ b_N \end{pmatrix} \tag{A.98}$$

and the solution to these equations is simple to find.

The first equation is

$$A_{11}x_1 = b_1$$

which is immediately solvable.

$$x_1 = \frac{b_1}{A_{11}}$$

With x_1 known we can solve the second equation as well.

$$A_{21}x_1 + A_{22}x_2 = b_2 \qquad \Longrightarrow \qquad x_2 = \frac{b_2 - b_1 \frac{A_{21}}{A_{11}}}{a_{22}}$$

We can continue with this process, known as 'back-substitution', until all the unknowns have been found.

Back-substitution is only directly applicable to equations containing upper or lower triangular matrices, but we will now show how to transform more general sets of linear equations into just that form. First note that adding the multiple of one equation to another equation (i.e., adding the multiple of one row of $\underline{\underline{A}}$ to another row *and* the corresponding elements of $\underline{b}$) does not change the solution vector $\underline{x}$. Even more obviously, interchanging the order of two equations (i.e., interchanging two rows of $\underline{\underline{A}}$ *and* the corresponding elements of $\underline{b}$) does not change the solution. Using just these two tricks we can magically transform arbitrary sets of equations (A.97) into the triangular form of equation (A.98).

The basic strategy was invented by Gauss and therefore called 'Gaussian elimination' and it can be extended to a method for finding the inverse of a matrix. However, if we really need to invert a matrix, there may be better methods. A matrix that has some special form may have an efficient inversion algorithm. For example, Toeplitz matrices can be inverted in $O(N^2)$ time by the Levinson-Durbin recursion discussed in Section 9.10. In addition, if numerical accuracy problems arise when using one of the standard algorithms, there are iterative algorithms to improve solutions.

Sometimes we know the inverse of matrix $\underline{\underline{A}}$, and need the inverse of another related matrix. If we are interested in the inverse of the matrix $\underline{\underline{A}} + \underline{\underline{B}}$, the following lemma is of use

$$(\underline{\underline{A}} + \underline{\underline{B}})^{-1} = \underline{\underline{A}}^{-1} - \underline{\underline{A}}^{-1}\left(\underline{\underline{A}}^{-1} + \underline{\underline{B}}^{-1}\right)^{-1}\underline{\underline{A}}^{-1} \qquad (A.99)$$

and a somewhat more general form is often called the 'matrix inversion lemma'.

$$(\underline{\underline{A}} + \underline{\underline{B}}\,\underline{\underline{C}}\,\underline{\underline{D}})^{-1} = \underline{\underline{A}}^{-1} - \underline{\underline{A}}^{-1}\underline{\underline{B}}\left(\underline{\underline{D}}\,\underline{\underline{A}}^{-1}\underline{\underline{B}} + \underline{\underline{C}}^{-1}\right)^{-1}\underline{\underline{D}}\,\underline{\underline{A}}^{-1} \qquad (A.100)$$

Let's prove this last lemma by multiplying the supposed inverse by the matrix,

$$(A + BCD)^{-1}(A + BCD) =$$

$$\left(A^{-1} - A^{-1}B\left(DA^{-1}B + C^{-1}\right)^{-1}DA^{-1}\right)(A + BCD) =$$

$$I + A^{-1}BCD - A^{-1}B\left(DA^{-1}B + C^{-1}\right)^{-1}D\left(I + A^{-1}BCD\right) =$$

$$I + A^{-1}B\left(C - X\right)D$$

where

$$X = \left(DA^{-1}B + C^{-1}\right)^{-1}\left(I + DA^{-1}BC\right)$$

$$= \left(DA^{-1}B + C^{-1}\right)^{-1}\left(C^{-1} + DA^{-1}B\right)C = C$$

which completes the proof.

EXERCISES

A.16.1 Assume that B is an approximation to A^{-1}, with error $R = I - BA$. Show that $A^{-1} = (I + R + R^2 + R^3 + \ldots)B$ and that this can be used to iteratively improve the inverse.

A.16.2 You know that x and y obey the equation $x + 3y = 8$ and determine numerically that they also obey $2x + 6.00001y = 8.00001$. What are x and y? Suppose that the numerically determined equation is $2x + 5.99999y = 8.00002$. What are x and y now? Explain the discrepancy.

Bibliography

[1] M. Abramowitz and I.A. Stegun. *Handbook of Mathematical Functions*. Dover Publications, New York, NY, 1965.

[2] W.B. Ackerman. Data flow languages. *Computer*, 15(2):15–25, February 1982.

[3] J. Allen. Computer architectures for signal processing. *Proc. IEEE*, 63(4):624–633, April 1975.

[4] J. Allen. Computer architectures for digital signal processing. *Proc. IEEE*, 73(5):852–873, May 1985.

[5] J.B. Allen. Cochlear modeling. *IEEE ASSP Magazine*, 2(1):3–29, 1985.

[6] J.B. Allen and S.T. Neely. Modeling the relation between the intensity just noticeable difference and loudness for pure tones and wideband noise. *J. Acoust. Soc. Amer.*, 102(6):3628–3646, December 1997.

[7] M.G. Amin. A new approach to recursive Fourier transform. *Proc. IEEE*, 75(11):1537–1538, November 1987.

[8] D.J. Amit. *Modeling Brain Function: The World of Attractor Neural Networks*. Cambridge University Press, Cambridge, UK, 1989.

[9] B.S. Atal. Predictive coding of speech at low bit rates. *IEEE Trans. Commun.*, COM-30(4):600–614, April 1982.

[10] B.S. Atal and S.L. Hanauer. Speech analysis and synthesis by linear prediction of the speech wave. *J. Acoust. Soc. Amer.*, 50(2):637–655, February 1971.

[11] B.S. Atal and M.R. Schroder. Adaptive predictive coding of speech signals. *Bell Systems Technical J.*, 49(8):1973–1986, October 1970.

[12] K.G. Beauchamp. *Walsh Functions and Their Applications*. Academic Press, London, UK, 1975.

[13] T. Beer. Walsh transforms. *Am. J. Phys.*, 49(5):466–472, May 1981.

[14] H. Behnke and G. Köthe. In commemoration of the 100th anniversary of the birth of Otto Toeplitz. *Integral Equations and Operator Theory*, 4(2):281–302, 1981.

[15] A. Benyassine, E. Shlomot, H-Y. Su, D. Massaloux, C. Lamblin, and J-P. Petit. ITU-T recommendation G.729 Annex B: A silence compression scheme for use with G.729 optimized for V.70 digital simultaneous voice and data applications. *IEEE Comm. Magazine*, pp. 64–73, September 1997.

[16] G.D. Bergland. A guided tour of the fast Fourier transform. *IEEE Spectrum*, pp. 41–52, July 1969.

[17] G.D. Bergland. A radix-eight fast Fourier transform subroutine for real-valued series. *IEEE Trans. Audio Electroacoust.*, AU-17(2):138–144, June 1969.

[18] J.A.C. Bingham. Multicarrier modulation for data transmission: An idea whose time has come. *IEEE Comm. Magazine*, pp. 5–14, May 1990.

[19] R.B. Blackman and J.W. Tukey. *The Measurement of Power Spectrum from the Point of View of Communications Engineering.* Dover Publications, New York, NY, 1959.

[20] B.P. Bogert, M.J.R. Healy, and J.W. Tukey. The quefrency alanysis of time series for echoes: Cepstrum, pseudo-autocovariance, cross-cepstrum and saphe cracking. In M. Rosenblatt, editor, *Time Series Analysis*, pp. 209–243, New York, NY, 1963.

[21] M. Born. Otto Toeplitz (obituary). *Nature*, 145:617, April 1940.

[22] G.E.P. Box and M.E. Muller. A note on the generation of random normal deviates. *Annals Math. Stat.*, 29(2):610–611, June 1958.

[23] I. Boyud. Speech coding for telecommunications. *Electr. Comm. Eng. J.*, pp. 273–283, October 1992.

[24] R. N. Bracewell. The discrete Hartley transform. *J. Optical Society of Am.*, 73(12):1832–1835, December 1983.

[25] R. N. Bracewell. The fast Hartley transform. *Proc. IEEE*, 72(8):1010–1018, August 1984.

[26] R.N. Bracewell. *The Fourier Transform and Its Applications.* McGraw-Hill, New York, NY, second edition, 1986.

[27] R.N. Bracewell. *The Hartley Transform*. Oxford U Press, New York, NY, 1986.

[28] E.O. Brigham. *The Fast Fourier Transform and its Applications*. Prentice Hall, Englewood Cliffs, NJ, 1988.

[29] C.S. Burrus. Index mappings for multidimensional formulation of the DFT and convolution. *IEEE Trans. ASSP*, ASSP-25(3):239–242, June 1977.

[30] C.S. Burrus and P.W. Eschenbacher. An in-place, in-order prime factor FFT algorithm. *IEEE Trans. ASSP*, ASSP-29:806–817, August 1981.

[31] C.S. Burrus and T.W. Parks. *DFT, FFT and Convolution Algorithms*. Wiley, New York, NY, 1987.

[32] W.H. Calvin and G.A. Ojemann. *Inside the Brain: An Enthralling Account of the Structure and Workings of the Human Brain*. New American Library, New York, NY, 1980.

[33] J. Celko. Partitions. *Dr. Dobb's Journal*, pp. 116–117,140–141, November 1994.

[34] J-H. Chen, R.V. Cox, Y-C. Lin, N. Jayant, and M.J. Melchner. A low-delay CELP coder for the CCITT 16 Kb/s speech coding standard. *IEEE Trans. Selected Areas Comm.*, 10(5):830–849, June 1992.

[35] J-H. Chen and A. Gersho. Adaptive postfiltering for quality enhancement of coded speech. *IEEE Trans. Speech and Audio Processing*, 3(1):59–71, January 1995.

[36] W.Y. Chen. *DSL Simulation Techniques and Standards Development for Digital Subscriber Line Systems*. MacMillan Technical Publishing, Indianapolis, IN, 1998.

[37] D.G. Childers, D.P. Skinner, and R.C. Kemerait. The cepstrum: A guide to processing. *Proc. IEEE*, 65(10):1428–1442, October 1977.

[38] D.S. Cochran. Algorithms and accuracy in the HP-35. *Hewlett-Packard Journal*, pp. 10–11, June 1972.

[39] L. Cohen. Time-frequency distributions - a review. *Proc. IEEE*, 77(7):941–981, July 1989.

[40] Digital Signal Processing Committee, editor. *Selected Papers in Digital Signal Processing, II*. IEEE Press, New York, NY, 1976.

[41] Digital Signal Processing Committee, editor. *Programs for Digital Signal Processing*. IEEE Press, New York, NY, 1979.

[42] J.H. Conway and N.J.A. Sloane. *Sphere Packings, Lattices and Groups*. Springer-Verlag, New York, NY, third edition, 1998.

[43] J.W. Cooley. How the FFT gained acceptance. *IEEE Signal Processing Magazine*, pp. 10–13, January 1992.

[44] J.W. Cooley, P.A.W. Lewis, and P.D. Welch. Historical notes on the fast Fourier transform. *IEEE Trans. Audio Electroacoust.*, AU-15(2):76–79, June 1967.

[45] J.W. Cooley and J.W. Tukey. An algorithm for the machine calculation of complex Fourier series. *Mathematics of Computation*, 19(90):297–301, April 1965.

[46] T.M. Cover and J.A. Thomas. *Elements of Information Theory*. Wiley, New York, NY, 1991.

[47] R.V. Cox. Three new speech coders from the ITU cover a range of applications. *IEEE Comm. Magazine*, pp. 40–47, September 1997.

[48] R.E. Crochiere and L.R. Rabiner. Interpolation and decimation of digital signals-a tutorial review. *Proc. IEEE*, 69(3):300–331, March 1981.

[49] A.V. Dandawate and G.B. Giannakis. A triple cross-correlation approach for enhancing noisy signals. In *Proc. Workshop on Higher Order Signal Analysis (Vail, CO)*, volume II, pp. 212–216, June 1989.

[50] W.N. Dember and J.S. Warm. *Introduction to Psychophysics*. Holt, New York, NY, 1981.

[51] A.M. Despain. Fourier transform computers using CORDIC iterations. *IEEE Trans. on Computers*, C-23:993–1001, October 1974.

[52] A.M. Despain. Very fast Fourier transform algorithms for hardware implementation. *IEEE Trans. on Computers*, C-28:333–341, May 1979.

[53] B.S. DeWitt and N. Graham, editors. *The Many Worlds Interpretation of Quantum Mechanics*. Princeton University Press, Princeton, NJ, 1973.

[54] H. Dudley. The Vocoder. *Bell Labs Record*, 18(4):122–126, December 1939.

[55] H. Dudley. The carrier nature of speech. *Bell Systems Technical J.*, 19(4):495–515, October 1940.

[56] P. Duhamel. Implementation of split-radix FFT algorithms for complex, real and real-symmetric data. *IEEE Trans. ASSP*, ASSP-34:285–295, April 1986.

[57] P. Duhamel and J. Hollmann. Split-radix FFT algorithm. *Elect. Lett.*, 20:14–16, January 1984.

[58] A.I. Abu el Haija and M.M. Al-Ibrahim. Improving performance of digital sinusoidal oscillators by means of error feedback circuits. *IEEE Trans. on Circuits and Systems*, CAS-33(4):373–380, April 1986.

[59] S.J. Elliott and P.A. Nelson. Adaptive noise control. *IEEE Signal Processing Magazine*, pp. 12–35, October 1993.

[60] M.D. Ercegovac and T. Lang. Redundant and on-line CORDIC: Applications to matrix triangularization and SVD. *IEEE Trans. on Computers*, C-39:725–740, June 1990.

[61] J. Eyre and J. Bier. DSP processors hit the mainstream. *Computer*, 31(8):51–59, August 1998.

[62] H. Fletcher and W.A. Munson. Loudness, its definition, measurement and calculation. *J. Acoust. Soc. Amer.*, 5:82–108, October 1933.

[63] G.D. Forney. The Viterbi algorithm. *Proc. IEEE*, 61(3):268–278, March 1973.

[64] L.E. Franks. Carrier and bit synchronization in data communications—a tutorial review. *IEEE Trans. Commun.*, COM-28(8):1107–1121, August 1980.

[65] B. Friedlander and B. Porat. Asymptotically optimal estimations of MA and ARMA parameters of nongaussian processes using higher order moments. *IEEE Trans. Automatic Control*, 35(1):27–35, January 1990.

[66] S. Furui and M.M. Sondhi, editors. *Advances in Speech Signal Processing*. Marcel Dekker, 1992.

[67] R.G. Gallager. *Information Theory and Reliable Communication*. Wiley, New York, NY, 1968.

[68] N.C. Gallagher and G.L. Wise. A theoretical analysis of the properties of median filters. *IEEE Trans. ASSP*, 29(6):1136–1141, December 1981.

[69] A. Gersho and R.M. Gray. *Vector Quantization and Signal Compression*. Kluwer, Boston, MA, 1992.

[70] A. Gersho and V.B. Lawrence. Multidimensional signal constellations for voiceband data transmission. *IEEE Trans. Selected Areas Comm.*, pp. 687–702, September 1984.

[71] G.A. Gescheider. *Psychophysics: Method and Theory*. Erlbaum, Hillsdale, NJ, 1976.

[72] G.B. Giannakis. Cumulants: A powerful tool in signal processing. *Proc. IEEE*, 75(9):1333–1334, September 1987.

[73] G.B. Giannakis and J.M. Mendel. Identification of nonminimum phase systems using higher order statistics. *IEEE Trans. ASSP*, 37(3):360–377, March 1989.

[74] J.W. Gibbs. Fourier series. *Nature*, 59:200,606, December 1899.

[75] J. Gleick. *Chaos: Making a New Science*. Penguin Books, New York, NY, 1988.

[76] D.N. Godard. Passband timing recovery in an all-digital modem receiver. *IEEE Trans. Commun.*, COM-26(5):517–523, May 1978.

[77] G. Goertzel. An algorithm for the evaluation of finite trigonometric series. *Am. Math. Monthly*, 65(1):34–35, January 1958.

[78] B. Gold, A.V. Oppenheim, and C.M. Rader. Theory and implementation of the discrete Hilbert transformers. In *Proc. Polytechnic Inst. Brooklyn Symp. Computer Processing in Communications*, pp. 235–250, 1970.

[79] B. Gold and C.M. Rader. *Digital Processing of Signals*. McGraw-Hill, New York, NY, 1969.

[80] R. Golden. Digital computer simulation of a sampled datavoice excited vocoder. *J. Acoust. Soc. Amer.*, 35(9):1358–1366, September 1963.

[81] S.W. Golomb. *Shift Register Sequences*. Holden Day, San Francisco, 1967.

[82] W. Goralski. *ADSL and VDSL Technologies*. McGraw-Hill, New York, NY, 1998.

[83] I. Gratton-Guiness. *Joseph Fourier 1768-1830*. MIT Press, Cambridge, MA, 1972.

[84] A.H. Gray and J.D. Markel. Distance measures for speech processing. *IEEE Trans. ASSP*, ASSP-24(5):380–391, October 1976.

[85] R. Gray. Vector quantization. *IEEE ASSP Magazine*, 1:4–15, April 1984.

[86] G. Green, editor. *Essay on the Application of the Mathematical Analysis to the Theories of Electricity and Magnetism*. Nottingham, England, 1828.

[87] C.W.K. Gritton and D.W. Lin. Echo cancellation algorithms. *IEEE ASSP Magazine*, pp. 30–37, April 1984.

[88] P.E. Guernic, A. Beneveniste, P. Bournai, and T. Gautier. Signal - a data flow oriented language for signal processing. *IEEE Trans. ASSP*, 34(2):362–374, April 1986.

[89] N. Halbwachs, P. Caspi, P. Raymond, and D. Pilaud. The synchronous data flow language LUSTRE. *Proc. IEEE*, 79(9):1305–1319, September 1991.

[90] D.C. Hanselman and B.C. Littlefield. *Mastering MATLAB 5: A Comprehensive Tutorial and Reference*. Prentice Hall, Englewood Cliffs, NJ, 1997.

[91] F. Harary. *Graph Theory*. Perseus Press, 1995.

[92] H. Harashima and H. Miyakawa. Matched transmission technique for channels with intersymbol interference. *IEEE Trans. Commun.*, COM-20:774–780, August 1972.

[93] F.J. Harris. On the use of windows for harmonic analysis with the DFT. *Proc. IEEE*, 66(1):51–83, January 1978.

[94] R.V.L. Hartley. A more symmetrical Fourier analysis applied to transmission problems. *Proc. I.R.E*, 30:144–150, May 1942.

[95] S. Haykin. *Digital Communications*. Wiley, New York, NY, 1988.

[96] S. Haykin. *Adaptive Filter Theory*. Prentice Hall, Englewood Cliffs, NJ, second edition, 1991.

[97] M. Healy. *Tables of Laplace, Heaviside, Fourier and z Transforms*. Chambers, Edinburgh, 1967.

[98] M.T. Heideman, C.S. Burrus, and H.W. Johnson. Prime factor FFT algorithms for real-valued series. In *Proc. ICASSP*, pp. 28A.7.1–28A.7.4, San Diego, CA, March 1984.

[99] W. Heisenberg, editor. *The Physical Principles of Quantum Theory*. U. of Chicago Press, Chicago, IL, 1930.

[100] H.D. Helms. Fast Fourier transform method of computing difference equations and simulating filters. *IEEE Trans. Audio Electroacoust.*, 15(2):85–90, June 1967.

[101] H. Hermansky and N. Morgan. RASTA processing of speech. *IEEE Trans. Speech and Audio Processing*, 2(4):578–589, October 1994.

[102] H. Hermansky, N. Morgan, A. Bayya, and P. Kohn. The challenge of Inverse-E: The RASTA-PLP method. In *Proc. 25th IEEE Asilomar Conf. Signals, Systems and Computers*, pp. 800–804, November 1991.

[103] C.A.R. Hoare. Communicating sequential processes. *Communications of the ACM*, 21(8), August 1978.

[104] H.P. Hsu. *Fourier Analysis*. Simon and Schuster, New York, NY, 1967.

[105] X. Hu, R.G. Harber, and S.C. Bass. Expanding the range of convergence of the CORDIC algorithm. *IEEE Trans. on Computers*, C-40:13–20, January 1991.

[106] Y.H. Hu. CORDIC-based VLSI architectures for digital signal processing. *IEEE Signal Processing Magazine*, pp. 16–35, July 1992.

[107] Y.H. Hu. The quantization effects of the CORDIC algorithm. *IEEE Trans. Signal Processing*, SP-40:834–844, April 1992.

[108] S. Icart and R. Gautier. Blind separation of convolutive mixtures using second and fourth order moments. In *Proc. ICASSP (Atlanta, GA)*, volume II, pp. 343–346, May 1996.

[109] F. Itakura. Line spectrum representation of linear predictive coefficients of speech signals. *J. Acoust. Soc. Amer.*, 57A:S35, April 1975.

[110] F. Itakura. Minimum prediction residual principle applied to speech recognition. *IEEE Trans. ASSP*, ASSP-23(2):67–72, February 1975.

[111] F. Itakura and S. Saito. Analysis synthesis telephony based on the maximum likelihood method. In *Proc. 6th Intl. Conf. Speech Commun. Proc.*, pp. C.17–C.20, August 1968.

[112] F. Itakura and S. Saito. Analysis synthesis telephony based on the maximum likelihood method. In *Proc. 6th Int. Cong. Acoustics (Tokyo, Japan)*, pp. C17–C20, 1968.

[113] F. Itakura and S. Saito. A statistical method for estimation of speech spectral density and formant frequencies. *Elect. Commun. Japan*, 53-A:36–43, 1970.

[114] L.B. Jackson. On the interaction of roundoff noise and dynamic range in digital filters. *Bell Systems Technical J.*, 49:159–184, February 1970.

[115] L.B. Jackson. Roundoff noise analysis for fixed-point digital filters realized in cascade or parallel form. *IEEE Trans. Audio Electroacoust.*, AU-18:107–122, June 1970.

[116] N.S. Jayant and P. Noll. *Digital Coding of Waveforms*. Prentice Hall, Englewood Cliffs, NJ, 1984.

[117] G.D. Forney Jr., L. Brown, M.V. Eyuboğlu, and J.L. Moran III. The V.34 high-speed modem standard. *IEEE Comm. Magazine*, pp. 28–33, December 1996.

[118] G.D. Forney Jr. and M.V. Eyuboğlu. Combined equalization and coding using precoding. *IEEE Comm. Magazine*, pp. 25–33, December 1991.

[119] G.D. Forney Jr., R.G. Gallager, G.R. Lang, F.M. Longstaff, and S.U. Qureshi. Efficient modulation for band-limited channels. *IEEE Trans. Selected Areas Comm.*, SAC-2(5):632–646, September 1984.

[120] J.N. Mitchell Jr. Computer multiplication and division using binary logarithm. *IEEE Trans. Electr. Comp.*, EC-11:512–517, 1962.

[121] J.P. Campbell Jr. and T.E. Tremain. Voiced unvoiced classification of speech with application to the U.S. government LPC-10E algorithm. In *Proc. ICASSP-86*, pp. 473–476, 1986.

[122] J.P. Campbell Jr., T.E. Tremain, and V.C. Welch. The federal standard 1016 4800 bps CELP voice coder. *Digital Signal Processing (Academic Press)*, 1(3):145–155, 1991.

[123] S.L. Marple Jr. *Digital Spectral Analysis with Applications*. Prentice Hall, Englewood Cliffs, NJ, 1987.

[124] T.G. Stockham Jr. High speed convolution and correlation. In *1966 Spring Joint Conf. AFIPS*, volume 28, pp. 229–233, 1966.

[125] E.I. Jury. *Theory and Application of the z-Transform Method*. Krieger Publishing Co., Huntington, NY, 1973.

[126] N. Kalouptsidis. *Signal Processing Systems—Theory and Design*. Wiley, New York, NY, 1997.

[127] S.M. Kay. *Modern Spectral Estimation: Theory and Application*. Prentice Hall, Englewood Cliffs, NJ, 1988.

[128] S.M. Kay and S.L. Marple Jr. Spectrum analysis—a modern perspective. *Proc. IEEE*, 69:1380–1419, November 1981.

[129] N.Y.S. Kiang and W.T. Peake. Physics and physiology of hearing. In *Stevens Handbook of Experimental Psychology*, volume 1, pp. 227–326, New York, NY, 1988.

[130] D.H. Klatt. Software for a cascade/parallel formant synthesizer. *J. Acoust. Soc. Amer.*, 67(3):971–995, March 1980.

[131] D.H. Klatt. Review of text-to-speech conversion for English. *J. Acoust. Soc. Amer.*, 82(3):737–797, September 1987.

[132] W.B. Kleijn. Encoding speech using prototype waveforms. *IEEE Trans. Speech and Audio Processing*, 1(4):386–399, 1993.

[133] D.E. Knuth. *The TEXbook*. Addison-Wesley, Reading, MA, 1984.

[134] D.E. Knuth. *The METAFONTbook*. Addison-Wesley, Reading, MA, 1986.

[135] D.E. Knuth. *Fundamental Algorithms*, volume 1 of *The Art of Computer Programming*. Addison-Wesley, Reading, MA, third edition, 1997.

[136] D.E. Knuth. *Seminumerical Algorithms*, volume 2 of *The Art of Computer Programming*. Addison-Wesley, Reading, MA, third edition, 1997.

[137] D.P. Kolba and T.W. Parks. A prime factor FFT algorithm using high speed convolution. *IEEE Trans. ASSP*, ASSP-25(4):281–294, August 1977.

[138] I. Koren and O. Zinaty. Evaluating elementary functions in a numerical coprocessor based on rational approximations. *IEEE Trans. on Computers*, C-39:1030–1037, August 1990.

[139] L. Lamport. *LATEX: A Documentation Preparation System, User's Guide and Reference Manual*. Addison-Wesley, Reading, MA, 1994.

[140] P. Lapsley, J. Bier, A. Shoham, and E.A. Lee. *DSP Processor Fundamentals*. IEEE Press, Piscataway, NJ, 1997.

[141] P. Lapsley and G. Blalock. How to estimate DSP processor performance. *IEEE Spectrum*, 33(7):74–78, July 1996.

[142] E.A. Lee. Programmable DSP architectures (I). *IEEE ASSP Magazine*, pp. 4–19, October 1988.

[143] E.A. Lee. Programmable DSP architectures (II). *IEEE ASSP Magazine*, pp. 4–14, October 1989.

[144] E.A. Lee and D.G. Messerschmitt. *Digital Communication*. Kluwer, Boston, MA, second edition, 1994.

[145] N. Levanon. *Radar Principles*. Wiley, New York, NY, 1988.

[146] N. Levinson. A heuristic exposition of Wiener's mathematical theory of prediction and filtering. *J. Math. Phys. M.I.T.*, 26:110–119, 1947.

[147] N. Levinson. The Wiener RMS (root mean square) error criterion in filter design and prediction. *J. Math. Phys. M.I.T.*, 25:261–278, 1947.

[148] M. Levy. 1999 DSP architecture directory. *EDN*, pp. 66–102, April 15 1999.

[149] Y. Linde, A. Buzo, and R.M. Gray. An algorithm for vector quantizer design. *IEEE Trans. Commun.*, COM-28(1):84–95, January 1980.

[150] R.P. Lippmann. An introduction to computing with neural nets. *IEEE ASSP Magazine*, pp. 4–22, April 1987.

[151] B. Liu. Effect of finite word length on the accuracy of digital filters—a review. *IEEE Trans. Circuit Theory*, CT-18:670–677, November 1971.

[152] R.W. Lucky. Automatic equalization for digital communication. *Bell Systems Technical J.*, 44:547–588, April 1965.

[153] R.W. Lucky. Techniques for automatic equalization of digital communication. *Bell Systems Technical J.*, 45(2):255–286, February 1966.

[154] V.K. Madisetti. *VLSI Digital Signal Processors: An Introduction to Rapid Prototyping and Design Synthesis*. Butterworth Heinemann, Boston, MA, 1995.

[155] J. Makhoul. Linear prediction: A tutorial review. *Proc. IEEE*, 63(4):561–580, April 1975.

[156] P. Maragos. Fractal aspects of speech signals: Dimension and interpolation. In *Proc. ICASSP-91 (Toronto, Canada)*, pp. 417–420, May 1991.

[157] P. Maragos and R.W. Schafer. Morphological filters (I,II). *IEEE Trans. ASSP*, 35(3):1153–1184, August 1987.

[158] D. Marino. New algorithms for the approximate evaluation in hardware of binary logarithm and elementary functions. *IEEE Trans. on Computers*, C-21:1416–1421, December 1972.

[159] C. Marven and G. Ewers. *A Simple Approach to Digital Signal Processing*. Wiley, New York, NY, 1996.

[160] S.J. Mason. Feedback theory—some properties of signal flow graphs. *Proc. IRE*, 41:920–926, September 1953.

[161] S.J. Mason. Feedback theory—further properties of signal flow graphs. *Proc. IRE*, 44:920–926, July 1956.

[162] C. Mazenc, X. Merrheim, and J-M. Muller. Computing functions $\cos^{-1}$ and $\sin^{-1}$ using CORDIC. *IEEE Trans. on Computers*, C-42(1):118–122, January 1993.

[163] R.J. McAulay and T.F. Quatieri. Speech analysis/synthesis based on a sinusoidal representation. *IEEE Trans. ASSP*, ASSP-34(4):744–754, August 1986.

[164] S.S. McCandless. An algorithm for automatic formant extraction using linear prediction spectra. *IEEE Trans. ASSP*, ASSP-22(2):135–141, April 1974.

[165] J.H. McClellan, T.W. Parks, and L.R. Rabiner. A computer program for designing optimum FIR linear phase digital filters. *IEEE Trans. Audio Electroacoust.*, AU-21(6):506–526, December 1973.

[166] J.H. McClellan and C.M. Rader. *Number Theory in Digital Signal Processing*. Prentice Hall, Englewood Cliffs, NJ, 1979.

[167] J.H. McClellan, R.W. Schafer, and M.A. Yoder. *DSP First: A Multimedia Approach*. Prentice Hall, Upper Saddle River, NJ, 1998.

[168] J.L. McClelland and D.F. Rumelhart. *Parallel Distributed Processing, Explorations in the Microstructure of Cognition*. MIT Press, Cambridge, MA, 1986.

[169] J.L. McClelland and D.F. Rumelhart. *Explorations in Parallel Distributed Processing, A Handbook of Models, Programs and Exercises*. MIT Press, Cambridge, MA, 1988.

[170] A. McCree and T.P. Barnwell. A mixed excitation LPC vocoder for low bit rate speech coding. *IEEE Trans. Speech and Audio Processing*, 3(4):242–250, July 1995.

[171] W.S. McCulloch and W. Pitts. A logical calculus of the ideas immanent in nervous activity. *Bull. Math. Biophysics*, 5:115–133, 1943.

[172] S. McLaughlin and A. Lowry. Nonlinear dynamical systems concepts in speech analysis. In *Proc. EUROSPEECH-93 (Berlin)*, pp. 377–380, September 1993.

[173] J.M. Mendel. Tutorial on higher-order statistics (spectra) in signal processing and system theory: Theoretical results and some applications. *Proc. IEEE*, 79(3):278–305, March 1991.

[174] M. Minsky and S. Papert. *Perceptrons, An Introduction to Computational Geometry*. MIT Press, Boston, MA, second edition, 1972.

[175] C. Moler and D. Morrison. Replacing square roots by Pythagorean sums. *IBM J. Res. Dev.*, 27:577–581, November 1983.

[176] D.P. Morgan and C.L. Scofield, editors. *Neural Networks and Speech Processing*. Kluwer, Boston, MA, 1991.

[177] D.H. Johnson M.T. Heideman and C.S. Burrus. Gauss and the history of the fast Fourier transform. *IEEE Signal Processing Magazine*, pp. 14–21, October 1984.

[178] K. Murano, S. Unagami, and F. Amano. Echo cancellation and applications. *IEEE Comm. Magazine*, pp. 49–55, January 1990.

[179] Sir Isaac Newton. *Opticks*. Dover Publications, 1952.

[180] J.G. Nicholls, A.R. Martin, and B.G. Wallace. *From Neuron to Brain : A Cellular and Molecular Approach to the Function of the Nervous System*. Sinauer Associates, Sunderland, MA, third edition, 1992.

[181] C. Nikias and A.P. Petropoulou. *Higher Order Spectral Analysis— A Nonlinear Signal Processing Framework*. Prentice Hall, Englewood Cliffs, NJ, 1993.

[182] C.L. Nikias and J.M. Mendel. Signal processing with higher-order spectra. *IEEE Signal Processing Magazine*, pp. 10–37, July 1993.

[183] H. Nyquist. Certain topics in telegraph transmission theory. *Trans. AIEE*, 47:617–644, February 1928.

[184] M. Onoe. Fast amplitude approximation yields either exact mean or minimum deviation for quadratic pairs. *Proc. IEEE*, pp. 921–922, July 1972.

[185] A.V. Oppenheim and R.W. Schafer. *Digital Signal Processing*. Prentice Hall, Englewood Cliffs, NJ, 1975.

[186] A.V. Oppenheim and R.W. Schafer. *Discrete-Time Signal Processing*. Prentice Hall, Englewood Cliffs, NJ, 1989.

[187] A.V. Oppenheim, A.S. Willsky, and S. Hamid. *Signals and Systems*. Prentice Hall, Englewood Cliffs, NJ, second edition, 1996.

[188] S.J. Orfanidis. *Optimum Signal Processing*. McGraw-Hill, New York, NY, second edition, 1988.

[189] A. Papoulis. *Signals and systems*. McGraw-Hill, New York, NY, 1977.

[190] A. Papoulis. *Probability, Random Variables, and Stochastic Processes*. McGraw-Hill, New York, NY, third edition, 1991.

[191] T.W. Parks and C.S. Burrus. *Digital Filter Design*. Wiley, New York, NY, 1987.

[192] T.W. Parks and J.H. McClellan. Chebyshev approximation for nonrecursive digital filters with linear phase. *IEEE Trans. Circuit Theory*, CT-19:189–194, March 1972.

[193] G.E. Peterson and H.L. Barney. Control methods used in a study of the vowels. *J. Acoust. Soc. Amer.*, 24(2):175–184, March 1952.

[194] W.W. Peterson and E.J. Weldon Jr. *Error Correcting Codes*. MIT Press, Cambridge, MA, second edition, 1972.

[195] J.O. Pickles. *An Introduction to the Physiology of Hearing*. Academic Press, London, UK, 1982.

[196] V.F. Pisarenko. The retrieval of harmonics from a covariance function. *Geophysical Journal of the Royal Astronomical Society*, 33:347–366, 1973.

[197] S.R. Powell and P.M. Chau. A technique for realizing linear phase IIR filters. *IEEE Trans. Signal Processing*, 39(11):2425–2435, November 1991.

[198] W.H. Press, B.P. Flannery, S.A. Teukolsky, and W.T. Vetterling. NUMERICAL RECIPES *in C, The Art of Scientific Computing*. Cambridge University Press, Cambridge, U.K., second edition, 1992.

[199] J.G. Proakis. *Digital Communications*. McGraw-Hill, New York, NY, second edition, 1989.

[200] J.G. Proakis and D.G. Manolakis. *Digital Signal Processing: Principles, Algorithms and Applications*. Prentice Hall, Englewood Cliffs, NJ, third edition, 1996.

[201] T.F. Quatieri and R.J. McAulay. Speech transformations based on a sinusoidal representation. *IEEE Trans. ASSP*, ASSP-34(6):1449–1464, December 1986.

[202] S. Qureshi. Adaptive equalization. *IEEE Comm. Magazine*, pp. 9–16, March 1982.

[203] S.U.H. Qureshi. Adaptive equalization. *Proc. IEEE*, 53:1349–1387, September 1985.

[204] L. Rabiner and B-H. Juang. *Fundamentals of Speech Recognition.* Prentice Hall, Englewood Cliffs, NJ, 1993.

[205] L.R. Rabiner. On the use of autocorrelation analysis for pitch detection. *IEEE Trans. ASSP*, ASSP-25(1):24–33, February 1977.

[206] L.R. Rabiner, M.J. Cheng, A.E. Rosenberg, and C.A. McGonegal. A comparative performance study of several pitch detection algorithms. *IEEE Trans. ASSP*, ASSP-24(5):399–418, October 1976.

[207] L.R. Rabiner, J.F. Kaiser, O. Herrmann, and M.T. Dolan. Some comparisons between FIR and IIR digital filters. *Bell Systems Technical J.*, 53(2):305–331, February 1974.

[208] L.R. Rabiner, J.H. McClellan, and T.W. Parks. FIR digital filter design techniques using weighted Chebyshev approximation. *Proc. IEEE*, 63:595–610, April 1975.

[209] L.R. Rabiner and C.M. Rader, editors. *Digital Signal Processing.* IEEE Press, New York, NY, 1972.

[210] L.R. Rabiner and R. Schafer. *Digital Processing of Speech Signals.* Prentice Hall, Englewood Cliffs, NJ, 1978.

[211] L.R. Rabiner and R. W. Schafer. *Speech Signal Processing.* Prentice Hall, Englewood Cliffs, NJ, 1983.

[212] L.R. Rabiner and R.W. Schafer. On the behavior of minimax FIR digital Hilbert transformers. *Bell Systems Technical J.*, 53(2):363–390, February 1974.

[213] L.R. Rabiner and R.W. Schafer. On the behavior of minimax relative error FIR digital differentiators. *Bell Systems Technical J.*, 53(2):333–361, February 1974.

[214] C.E. Rader. Discrete Fourier transforms when the number of data samples is prime. *Proc. IEEE*, 56:1107–1108, June 1968.

[215] C.M. Rader and B. Gold. Effects of parameter quantization on the poles of a digital filter. *Proc. IEEE*, 55:688–689, May 1967.

[216] A. Ralston and P. Rabinowitz. *A First Course in Numerical Analysis.* McGraw-Hill, New York, NY, second edition, 1978.

[217] D.J. Rauschmayer. *ADSL/VDSL Principles: A Practical and Precise Study of Asymmetric Digital Subscriber Lines and Very High Speed*

Digital Subscriber Lines. MacMillan Technical Publishing, Indianapolis, IN, 1999.

[218] I.S. Reed and G. Solomon. Polynomial codes over certain finite fields. *SIAM Journal*, 8:300–304, June 1960.

[219] S.O. Rice. Mathematical analysis of random noise. *Bell Systems Technical J.*, 23:282–332, July 1944.

[220] S.O. Rice. Mathematical analysis of random noise (continued). *Bell Systems Technical J.*, 24(1):462–516, January 1945.

[221] O. Rioul and M. Vetterli. Wavelets and signal processing. *IEEE Signal Processing Magazine*, pp. 14–38, October 1991.

[222] G.H. Robertson. A fast amplitude approximation for quadratic pairs. *Bell Systems Technical J.*, 50:2849–2852, October 1971.

[223] E.A. Robinson. A historical perspective of spectrum estimation. *Proc. IEEE*, 70(9):885–907, September 1982.

[224] F. Rosenblatt. *Principles of Neurodynamics*. Spartan Books, New York, NY, 1962.

[225] D.E. Rumelhart, G.E Hinton, and R.J. Williams. Learning representations by backpropagating errors. *Nature*, 323:533–536, October 1986.

[226] S. Sabanal and M. Nakagawa. A study of time-dependent fractal dimensions of vocal sounds. *J. Phys. Society of Japan*, 64(9):3226–3238, September 1990.

[227] H. Sakai. Statistical analysis of Pisarenko's method for sinusoidal frequency estimation. *IEEE Trans. ASSP*, 32(1):95–101, February 1984.

[228] R. Salami, C. Laflamme, J-P. Adoul, A. Kataoka, S. Hayashi, T. Moriya, C. Lamblin, D. Massaloux, S. Proust, P. Kroon, and Y. Shoham. Design and description of CS-ACELP: A toll quality 8 kb/s speech coder. *IEEE Trans. Speech and Audio Processing*, 6(2):116–130, March 1998.

[229] R. Salami, C. Laflamme, B. Bessette, and J-P. Adoul. ITU-T G.729 Annex A: Reduced complexity 8 kb/s CS-ACELP codec for digital simultaneous voice and data. *IEEE Comm. Magazine*, pp. 56–63, September 1997.

[230] R.W. Schafer and L.R. Rabiner. System for automatic formant analysis of voiced speech. *J. Acoust. Soc. Amer.*, 47(2):634–648, February 1970.

[231] G. Schroeder and M.H. Sherif. The road to G.729: ITU 8-kb/s speech coding algorithm with wireline quality. *IEEE Comm. Magazine*, pp. 48–54, September 1997.

[232] M.R. Schroeder, B.S. Atal, and J.L. Hall. Optimizing digital speech coders by exploiting masking properties of the human ear. *J. Acoust. Soc. Amer.*, 66(6):1647–1652, December 1979.

[233] A. Schuster. On the investigation of hidden periodicities with applications to a supposed 26 days period of meterological phenomenon. *Terr. Magn.*, 3:13–41, 1898.

[234] H.G. Schuster. *Deterministic Chaos*. VHC Publishers, 1988.

[235] O. Shalvi and E. Weinstein. New criteria for blind deconvolution of nonminimum phase systems (channels). *IEEE Trans. Inform. Theory*, IT-36(4):312–321, March 1990.

[236] O. Shalvi and E. Weinstein. Super-exponential methods for blind deconvolution. *IEEE Trans. Inform. Theory*, IT-39(4):504–519, March 1993.

[237] C.E. Shannon. A mathematical theory of communication. *Bell Systems Technical J.*, 27(3):379–423,623–656, July 1948.

[238] C.E. Shannon. Communication in the presence of noise. *Proc. IRE*, 37:10–21, January 1949.

[239] M.R. Shroeder. Direct (nonrecursive) relations between cepstrum and predictor coefficients. *IEEE Trans. ASSP*, ASSP-29(1):297–301, April 1981.

[240] R. Singleton. An algorithm for computing the mixed radix fast Fourier transform. *IEEE Trans. Audio Electroacoust.*, AU-17:93–103, June 1969.

[241] S.K. Mitra SK and J.F. Kaiser, editors. *Handbook for Digital Signal Processing*. Wiley, New York, NY, 1993.

[242] B. Sklar. *Digital Communications, Fundamentals and Applications*. Prentice Hall, Englewood Cliffs, NJ, 1988.

[243] M.I. Skolnik. *Introduction to Radar Systems*. McGraw-Hill, New York, NY, second edition, 1980.

[244] M.I. Skolnik. Fifty years of radar. *Proc. IEEE*, 73(2):182–197, February 1985.

[245] M.I. Skolnik. *Radar Handbook*. McGraw-Hill, New York, NY, second edition, 1990.

[246] W.W. Smith and J.M. Smith. *Handbook of Real-Time Fast Fourier Transforms*. IEEE Press, New York, NY, 1995.

[247] H.V. Sorensen, M.T. Heideman, and C.S. Burrus. On computing the split radix FFT. *IEEE Trans. ASSP*, ASSP-34:152–156, June 1986.

[248] H.V. Sorensen, D.L. Jones, M.T. Heideman, and C.S. Burrus. Real-valued fast Fourier transform algorithms. *IEEE Trans. ASSP*, ASSP-35:849–863, June 1987.

[249] T. Springer. Sliding FFT computes frequency spectra in real time. *EDN*, 29:161–170, September 1988.

[250] M.D. Srinath and P.K. Rajasekaran. *An Introduction To Statistical Signal Processing with Applications*. Wiley, New York, NY, 1979.

[251] T. Star, J.M. Cioffi, and P.J. Silverman. *Understanding Digital Subscriber Line Technology*. Prentice Hall, Upper Saddle River, NJ, 1999.

[252] K. Steiglitz. *A Digital Signal Processing Primer*. Addison-Wesley, Menlo Park, CA, 1996.

[253] K.N. Stevens. *Acoustic Phonetics*. MIT Press, Cambridge, MA, 1998.

[254] S.S. Stevens, J. Volkmann, and E.B. Newman. A scale for the measurement of the psychological magnitude pitch. *J. Acoust. Soc. Amer.*, 8:185–190, January 1937.

[255] P. Stoica and A. Nehorai. Study of the statistical performance of the Pisarenko's harmonic decomposition. *IEE Proc. F*, 135(2):161–168, April 1988.

[256] Y. Tadokoro and T. Higuchi. Discrete Fourier transform computation via the Walsh transform. *IEEE Trans. ASSP*, ASSP-26(3):236–240, June 1978.

[257] Y. Tadokoro and T. Higuchi. Conversion factors from Walsh coefficients to Fourier coefficients. *IEEE Trans. ASSP*, ASSP-31(1):231–232, February 1983.

[258] H.D. Tagare and R.J.P. de Figueiredo. Order filters. *Proc. IEEE*, 73(1):163–165, January 1985.

[259] N. Tishby. A dynamical systems approach to speech processing. In *Proc. ICASSP-90 (Albuquerque, NM)*, pp. 365–368, April 1990.

[260] M. Tomlinson. New automatic equalizer employing modulo arithmetic. *Electronic Letters*, 7:138–139, March 1971.

[261] T.E. Tremain. The government standard LPC-10E algorithm. *Speech Technology Magazine*, pp. 40–49, 1982.

[262] S.A. Tretter. *Communications System Design Using DSP Algorithms*. Plenum Press, New York, NY, 1995.

[263] G. Ungerboeck. Trellis-coded modulation with redundant signal sets (I,II). *IEEE Comm. Magazine*, 25:5–21, February 1987.

[264] H.L. van Trees. *Detection, Estimation and Modulation Theory*. Wiley, New York, NY, 1968.

[265] A.J. Viterbi. Error bounds for convolutional codes and an asymptotically optimal decoding algorithm. *IEEE Trans. Inform. Theory*, IT-13:260–229, March 1967.

[266] J.E. Volder. The CORDIC trigonometric computing technique. *IRE Trans. on Elect. Comp.*, EC-8:330–334, September 1959.

[267] G. Walker. On periodicity in series of related terms. *Proc. Royal Society*, A131:518–532, June 1931.

[268] J.L. Walsh. A closed set of normal orthogonal functions. *Am. J. Math.*, 45:5–24, 1923.

[269] J.S. Walther. A unified algorithm for elementary functions. In *Spring Joint Computer Conf.*, pp. 379–85, 1971.

[270] L.F. Wei. Rotationally invariant convolutional channel coding with expanded signal space (I,II). *IEEE Trans. Selected Areas Comm.*, SAC-2:659–686, September 1984.

[271] L.F. Wei. Trellis-coded modulation with multidimensional constellations. *IEEE Trans. Inform. Theory*, IT-33(4):483–501, July 1987.

[272] B. Widrow. Adaptive antenna systems. *Proc. IEEE*, 55(12):2143–2159, December 1967.

[273] B. Widrow. Adaptive noise cancelling: Principles and applications. *Proc. IEEE*, 63:1692–1716, December 1975.

[274] B. Widrow and M.E. Hoff Jr. Adaptive switching circuits. In *IRE WESCON Conv. Rec. Part IV*, pp. 96–104, 1960.

[275] B. Widrow and S.D. Stearns. *Adaptive Signal Processing*. Prentice Hall, Englewood Cliffs, NJ, 1985.

[276] N. Wiener. Generalized harmonic analysis. *Acta. Math.*, 55:117–258, 1930.

[277] N. Wiener. *The Fourier Integral.* Cambridge University Press, London, UK, 1933.

[278] N. Wiener. *The Extrapolation, Interpolation and Smoothing of Stationary Time Series with Engineering Applications.* MIT Press, Cambridge, MA, 1949.

[279] N. Wiener. *Ex-Prodigy—My Childhood and Youth.* MIT Press, Boston, MA, 1953.

[280] N. Wiener. *I Am a Mathematician—The Later Life of a Prodigy.* MIT Press, Boston, MA, 1956.

[281] N. Wiener and E. Hopf. On a class of singular integral equations. *Proc. Prussian Academy, Math. Phys. Series*, p. 696, 1931.

[282] E.P. Wigner. On the quantum correction for thermodynamic equilibrium. *Phys. Rev.*, 40(5):749–759, June 1932.

[283] S. Winograd. On computing the discrete Fourier transform. *Math. Computation*, 32:175–199, January 1978.

[284] S. Winograd. On the multiplicative complexity of the discrete Fourier transform. *Advances in Math.*, 32:83–117, May 1979.

[285] Y. Xiao and Y. Tadokoro. On Pisarenko and constrained Yule-Walker estimators of tone frequency. *IEICE Trans. Fundamentals*, E77-A(8):1404–1406, August 1994.

[286] D. Yellin and E. Weinstein. Criteria for multichannel signal separation. *IEEE Trans. Signal Processing*, 42(8):2158–2167, August 1994.

[287] D. Yellin and E. Weinstein. Multichannel signal separation: Methods and analysis. *IEEE Trans. Signal Processing*, 44(1):106–118, January 1996.

[288] P.C.Y. Yip. Some aspects of the zoom transform. *IEEE Trans. on Computers*, C-25:287–296, March 1976.

[289] G.U. Yule. On a method of investigating periodicities in disturbed series, with special reference to Wolfer's sunspot numbers. *Philosophical Trans. Royal Society London*, A226:267–298, 1927.

[290] E. Zwicker and E. Terhardt. Analytical expressions for critical-band rate and critical bandwidth as a function of frequency. *J. Acoust. Soc. Amer.*, 68:1523–1525, 1980.

Index

page numbers of references to:
- definitions are <u>underlined</u>
- exercises are in *italics*

A-law, 213, *434*, 732, 757–759
A/D, 19, 64–70
ABS, 755, 773
AC, 21
accumulator, 591, 621
adaptive codebooks, 773–774
ADPCM, 762–765
ADSL, 124
adventure, 388
AGC, 4
Aiken, H, 625
algebraic codebooks, 772
aliasing, 54, 62–63, 99, *126*, 134, 311, 337, 500, 585, 587
all-pass filter, 214, <u>272</u>, *601*
all-pole filter, <u>295</u>
all-zero filter, 219, <u>295</u>
alternating current, *see* AC
alternation theorem, 812–814
AM, 158, 651, 653
AMDF, 751–752, 770
AMI, <u>700</u>
amplification, 8, *12*, 34, 209–210, 289–290
amplifier, 209–212
amplitude modulation, *see* AM
analog communications, 652–664

analog signal processing, 7–13
analog to digital conversion, *see* A/D
analog to digital converter, *see* A/D
analysis by synthesis, *see* ABS
antialiasing filter, 62–63
aperture time, 66, 67
arctangent, 609
ARMA, 245, 293–298, 475, 521
associative memory, 447
Asymmetric Digital Subscriber Line, *see* ADSL
attractor, 184–190
autocorrelation, 125, 263, <u>354</u>, 751
Automatic Gain Control, *see* AGC
axon, <u>443</u>

band-pass filter, <u>272</u>
band-pass sampling theorem, 55, 140, *769*
band-stop filter, <u>272</u>
bandwidth, 17, 46
bandwidth expansion, 772, 773
bank of filters, 498–502
Bark scale, 434, 748
Barker code, 364
Barkhausen, HG, 434
baud rate, <u>703</u>, 710
Baudot code, 650–653
Baudot, E, 650, 703
beamforming, 6, 315–319, 408, 425
beats, *33*, *76*, 157, 277–278
Bell, AG, 433, 650, 779

849

BER, 726
Berkeley, G, 431
Bessel functions, 806
Bessel, FW, 806
bilinear mapping, 311
biomedical signals, 5, *399*, *420*
bispectrum, 389
bit, 783
bit error rate, *see* BER
bit reversal, 542–546
block code, 682–690
Box-Muller algorithm, 178
boxcar signal, 63, 66
breadth-first search, 695
buffer delay, 276
Burg, J, 522
butterfly, 181, 540–542, 548–550, *626*
Butterworth filter, 304–305, 307–308

Cajal, SR, 443
canonical form section, 595, 598–599
carrier, 652, 700
carrier frequency, 156, 333, 653, 705, 710
cascade structure, 329, 580–582
causality, 224, 230, 570
CELP, 771–775
center clipper, 324
cepstrum, 744–747, 752
channel capacity theorem, 676
chaos, 162–163, 180–191, 592
Chebyshev filter, 306–308
Chebyshev polynomials, 797, 804–806, 813–814
checkbit, 683
checkbyte, 683, 688

Cholesky decomposition, 375, 424, 824
circular buffer, 137–138, 491, *535*, 575, 636, *638*
circular convolution, 138, 139, 573–579
circular functions, *see* trigonometric functions
circular reference, *see* infinite loop
clipping, 210–211, 324, *343*
closed loop gain, 290
clustering algorithms, 766
code excited linear prediction, *see* CELP
codewords, 681
coherent demodulators, 705
comb graph, *28*
communicating sequential processes, 493
communications theory, 670–674
complex cepstrum, 745
constellation, 706, 716–718, 720–723
constraint length, 692
context switch, 488, 631
continuous phase FSK, *704*
convolution, 95, 115, 219, 227, 237–240
convolutional code, 682, 690–698
Cooley, J, 568
CORDIC algorithm, 613–617
correlation, 36, 349–364, 395
correlator, *see* matched filter
cosets, 685
cost function, 408
counting converter, 67
CPFSK, *see* continuous phase FSK
CRC code, *689*
critical bands, 434
crosscorrelation, 263, 352, 354

cumulants, 387–391

cycle, 187–190, 592

cyclic code, 685

cyclic convolution, *see* circular convolution

cyclic prefix, 578, *737*

cyclic suffix, 578

D/A, 19, 69

data communications, 4–5, 664–737

data over voice, 775

data overrun, 533

dB, 31, 794

DC, 20, 26, 698–700

DC blocker, 8, *12*, 301

DDE, 407

de-emphasis, 273, 661

decibel, *see* dB

decimation in time, *see* DIT, *see* DIF

decision directed equalization, *see* DDE

decision feedback equalizer, *see* DFE

deconvolution, <u>267</u>

degree (of graph), 463, 465

delay, 570, 571

    algorithmic, 578

    buffer, 578

    system, 575

delay line, 37, 579

$\delta(t)$, 23, *29*

$\delta_{n,m}$, 22

delta modulation, 761

delta-PCM, 68–69

demodulation, 159, <u>652</u>

demultiplex, 673

dendrite, <u>443</u>

density, 802–803

denumerably infinite, 783

depth-first search, 695

detection, 350–351, 359–361

deterministic, <u>30</u>, 162, 180

DFE, 715

DFT, 132–143

DIF FFT, 547–548

difference equations, 245–248, 515, 594, 606

differentiation filter, 282–285, 662

digital frequency, <u>55</u>

digital oscillator, 249–250, 606–608

digital to analog conversion, *see* D/A

digitizing, *see* A/D

digraph, 465

Dirac's delta function, 23, *29*, 808–809

Dirac, PAM, 23

direct current, *see* DC

direct form section, 595, 598

direct form structure, 579

directed graph, *see* digraph

directed search, 694

direction of arrival, *see* DOA

Dirichlet's convergence conditions, 79, 87

discrete Fourier transform, *see* DFT

discrete multitone, *see* DMT

DIT FFT, 539–546

DMT, 735–737

DOA, 316

Doppler shift, 99, 365

dot graphs, *28*

dot product, 789

double buffering, 533

double sideband AM, 659

double-square distribution, 130

DPCM, 760–763

DSP processors, 619–638

DTMF, 4, *6*, 123, 490, 561–562, 664, *756*

DTW, 343

dual differentiator, 662
Dudley, H, 779
dynamic programming, 343, 693, 696
dynamic range, 16, 57, 61, 763
dynamic time warping , see DTW

ECC, see error correcting codes
echo cancellation, 393, 400–404, 425
echo suppressor, 400–401
Edison, TA, 28
effective bits, 67
eigensignal, 38, 228, 254
elliptical filter, 307–308
elliptical functions, 307, 799, 806–807
empirical autocorrelation, 354
energy, 17, 18, 36, 41
ENIAC, 625
envelope detector, 656
equalizer, 222–223, 269, 393, 404–407, 425
ergodic, 196–197
error accumulation, 606
error correcting codes, 672, 680–698, 737
Euler, L, 77, 464
Eulerian cycle, 465
expectation, 815
exponential function, 794
exponential signal, 26–28

false alarm, 350, 366
fast Fourier transform, see FFT
fax, 4, 124, 329, 400, 401, 450, 490, 561, 647, 729, 730
FDM, 334
Fechner, GT, 431–433
feedback, 289–293, 474
FEQ, 269, 736

FEXT, 735
FFT, 135, 531–567
Fibonacci sequence, 145–147
field, 788–789
FIFO FFT, 565–567
filter, 214, 224, 226–228, 271–315
filter design, 235, 303–315
finite difference, 37–38, 235, 244, 246, 248, 282, 301, 497
FIR, 256
FIR filter, 219, 228–237
first-order section, 296–297
fixed point, 183–186, 590–595, 619, 628, 633–635, 784
flash converter, 65
FLL, 339–340
floating point, 619, 784
flow graph, 466–476
FM, 158, 651, 654, 659–664
FM chirp, 363–364
formant, 125, 437
four-quadrant arctangent, 91, 159, 609, 786
Fourier Integral Theorem, 106, 113
Fourier series, 77–101
Fourier transform, 103–117
Fourier, JBJ, 77–80
frequency domain, 106
frequency domain multiplexing, see FDM
frequency equalizer, see FEQ
frequency modulation, see FM
frequency response, 217, 228, 254–255
frequency shift keying, see FSK
FS, see Fourier series
FSK, 7, 653, 702–703
FT, see Fourier transform
FT pair, 106, 258, 358
fundamental theorem

of algebra, <u>792</u>
of arithmetic, <u>783</u>
of calculus, <u>802</u>

Gallager, RG, 678
Gauss, 795
Gaussian, 794–795, 816
Geigel algorithm, 403, *404*
generating function, 144–147
generator matrix, 684
Gibbs phenomenon, 86–90
Gibbs, JW, 87
Goertzel algorithm, 532, 561–565
Golay code, <u>688</u>
golden ratio, 147, 435
Golgi, C, 443
gradient descent, 398, 411–412
graph, <u>463</u>
graph theory, 462–467
Gray code, *704*
Green's function, 255
group, 788
group delay, <u>277</u>

Hadamard matrix, 554
Hamilton, Sir WR, 465, 782
Hamiltonian cycle, 465
Hamming codes, 683–686
Hamming distance, 681
Hamming, RW, 509, 681, 683
hard limiter, 8, 211, 223, 330, 341, 634
harmonic generation, 322
harmonics, <u>74</u>
Hartley transform, *526*
Harvard architecture, 625
Heaviside's step function, 21
Hebb's principle, 451
Hebb, DO, 451
Hertz, H, 651

Hessian matrix, 412
heterodyning, *see* mixer
hierarchical flow graph, 480
high-pass filter, <u>272</u>
higher-order signal processing, 194, 386–391
Hilbert transform, 158–159, 287–288, 657
HMM, *348*
Hoare, CAR, 493
Horner's rule, 562, 606, 791
HPNA, 124, *308*
hyperbolic functions, 798–799

I-Q plot, <u>706</u>, 716, 752
ideal filter, 272–273
IIR, <u>257</u>
IIR filter, 220
image processing, 18
implementation, <u>471</u>
impulse, *see* unit impulse
impulse response, 224, <u>227</u>, 255–258
incoherent demodulators, 705
infinite accumulator, 244
infinite loop, *see* circular reference
information theory, 666–670, 737
inner product, 789
instantaneous representation, 155–160, 495
integration filter, 285–287
intermodulation product, 323
interrupt, 631
intersymbol interference, *see* ISI
inverse filter, 399
inverse system, 221–223, 246
inverse zT, 154
inversion, *12*, 34
ISI, 406–407

JND, 428–430
jokes, 161, 203, *240*, 299, 348, 362, 432, *441*, 478, 482, 506, 619, 640, 644
just noticeable difference, *see* JND

Kalman filter, 368
Königsberg, 465
Kronecker delta, 22, 801
Kronecker, L, 781
kurtosis, 817

Laplace transform, 148
lattice structure, 582–583
law of large numbers, 178, *179*, 815
LBG algorithm, 767–768
Legendre polynomials, 804, 812
Levenshtein distance, 344–347
Levinson-Durbin recursion, 261, 376–382, 521, 827
LFSR, 176–177, *179*, 364, 700
limit cycle, <u>592</u>
line spectral pairs, *see* LSP
linear algebra, 819–821
linear feedback shift register, *see* LFSR
linear phase, 95, <u>276</u>
linear system, 223
linear-phase, <u>215</u>, 308, *584*
LMS algorithm, 413–425, 451
logarithm, 610–611, 793–794
logarithmic companding, 213
logistic sigmoid, 330
logistics signal, 37, *39*
long-term predictor, 743
Lorenz, E, 181
loss function, *see* cost function
low-pass filter, <u>272</u>
LPC, 371–382, 740–744
LPC cepstral coefficients, 746–747

LSB, 539, 545–546, <u>783</u>
LSP, 383–386, 772
LTDFT, <u>134</u>

MA, *see* moving average
MAC, 454, <u>473</u>, 570, 579, 591, 620–623
Markov model, *347*
Markov signals, 194–195
masking, *441*
matched filter, 5, 349, 361–365, 390, 701, 702
matrix, 821–828
Maxwell's equations, 650–651
Maxwell, JC, 650
mean opinion score, <u>755</u>
mean square error, *see* MSE
median filter, 326–329
mel scale, 434
MELP, *744*, 752, 777
memory location, 470
metric, 819
metric space, 790
Michelson, AA, 86
minimax approximation, 812–814
minimum mean square error, *see* MMSE
minimum phase, *302*
mixed-radix FFT, 551
mixer, 114, 137, 332–339, 560
MLP, 330, 451
MMSE, 409–410
modem, 4, 7, 124, 207, 208, 222, *269*, 325, 334–335, 371, 400–402, 404–406, 425, 488, 561, 619, <u>647</u>, 698–737, 753, 775
modulation, <u>49</u>, 157, <u>652</u>
modulo, 783
Moler-Morrison algorithm, 612
moment functions, 194

moments, 816–817
Monte-Carlo integration, 203
Morse code, 649–653, 710
Morse, S, 649
MOS, *see* mean opinion score
moving average, <u>219</u>
MSB, 539, 545–546, <u>783</u>
MSE, 370, 373–375, *376*, 378, 381, *382*, 409–410, 414, 417, 421, 766
μ-law, 213, *434*, 732, 757–759
multilayer perceptron, *see* MLP
multiplex, 673, 711
multiply-and-accumulate, *see* MAC
music, 4, 26, 32, *39*, 128, 204, 209, 215, 221, 222, 376, 433, 610, 673, 680, *742*, 753

Napoleon, 78, 649
negative feedback, 290
negative frequency, 46
netlist, 485
neural network, 329–332, 446–452, 750
Newton, I, 72–73
Newton-Raphson algorithm, 611
NEXT, 735
NLP, 325, 402–403
node, <u>463</u>
noise, 31, 161–202, 700–702
noise bandwidth, 508
noise cancellation, 5, 393–399, 407, 408, 425
nondenumerably infinite, 784
nondeterministic algorithm, 203–204
nonlinear processor, *see* NLP
nonlinear systems, 210–212, 223, 226, 322–324, 335, 390, 403
NOP, 629, *630*, 638, 639

norm, 819
normal equations, 84, 280, 811, *814*, 820
notch filter, 250, <u>272</u>, 301, *302*, 339
NRT, 701, 743
NRZ, 667, 698–700, 708
number, 781
    complex, 782, 785–787
    integer, 782–784
    rational, 782
    real, 782, 784–785
    whole, 781
Nyquist, H, 53

OFDM, 736
OOK, 651, 652, 700–702
orthogonal, 36, <u>789</u>, 820
orthogonal frequency division multiplexing, *see* OFDM
oscillator, 290
OSI model, 665
outliers, 326
overflow, 555–558
overlap add, 575
overlap save, 575

Paley-Wiener theorem, *275*
PAM, 49, 654, 698, 703
PAR, 61, *737*
parallel form, 599–601
parallel port, 632
parity check matrix, 686
Parseval's relation, 95, 115, 138, 358, 556
partition, 388
patch panel, 481
pattern recognition, 208, 497
PCM, 48, <u>49</u>
PCM modem, *7*, 733–734
perceptron, 330, 451

perceptual weighting, 773, 775
periodogram, 502–506
phase modulation, *see* PM
phase shift keying, *see* PSK
phase-locked loop, *see* PLL
phased array, 317
phoneme, <u>436</u>
pipeline, 627–630
Pisarenko harmonic decomposition, 512–519
pitch, 125, 436, 750–753
PLL, 338–343, 712
PM, 654, 659–664
pole-zero plot, <u>295</u>, 298–302
polynomial, 231, 279–285, 791–793
polynomial approximation, 809–814
polyphase filter, 584–590
polyspectrum, 389
post-filter, 761, 774
POTS, <u>124</u>
power, <u>18</u>
power cepstrum, 746
power law distortion, 211
power spectral density, *see* PSD
power spectrum, 91, *94*, 95, *96*, 116, 122–126
PPM, 49, 654
pre-emphasis, 273, 661
prediction, *49*, 180–181, 252–253, 369–376, 760–765
PRF, PRI, 98
prime factor FFT algorithm, 551
probabilistic algorithm, 203–204
probability, 815–819
processing delay, 274
processing gain, 360
Prony's method, 518–519, 602
Prony, Baron de, 518
PSD, <u>123</u>, 357–359, 495, 708–710

pseudorandom signals, 162, 174–179
PSK, 364, 654, 702, 704–708
PSTN, 775
pulse amplitude modulation, *see* PAM
pulse coded modulation, *see* PCM
pulse position modulation, *see* PPM
pulse width modulation, *see* PWM
PWM, 49, 654
pyramid algorithm, 528, 769
Pythagorean addition, 611–612

QAM, 716–723
QMF, 528, 769
quadrature component, 27
quadrature form, 91
Quadrature Mirror Filter, *see* QMF
quefrency, 744, 745

radar, 5–6, 98–100, 129, 161, 171–174, 207, 208, 271, 316, *319*, 350–352, *353*, 362–365, 367
radio frequency interference, *see* RFI
radix-4 FFT, 548–550
radix-*R* FFT, 542, 550
raised cosine, 233
random number generators, 174–179
RASTA, 748
RAX, 481–493
real-time, 533, 570
rectifier, 9, 86, 211
recursive least squares, *see* RLS
Reed-Solomon codes, 688–689
Remez exchange algorithm, 314, 813–814
resampling, *370*
return map, 185
RFI, 324
RLS, 421–425

RMS, <u>17</u>, *19*
ROC, 150–155
root mean squared, *see* RMS
roots of unity, 532, 787
rotation matrix, 613–615, 823–824

sample and hold, 66
sampling, *see* A/D
sampling theorem, 48, 53–56, 139
Šarkovskii's theorem, 190
saturation arithmetic, <u>592</u>, 634
sawtooth, 24, *29*, 67, 75
SBC, 768–769
Schuster, Sir A, 502, 530
Schwartz inequality, 122
scrambler, 700, 712
seismology, 6
sequency, 524
serial port, 632
set partitioning, 724
Shakespeare, W, 17, 674
Shannon C, 683
Shannon, C, 53, 671–680
shell mapper, 731
shift register, 176, 632, 682, 690
shifted unit impulse, *see* SUI
short-term predictor, 743
side information, 764
sidebands, 658, 662–663
sigma-delta digitizer, 68–69
signal, <u>16</u>
    chaotic, 162
    characteristics, 30–33
    complex, 18, 26, 30
    constant, 20
    deterministic, 30
    exponential, 26
    finite bandwidth, 33
    finite time duration, 33, 571
    finiteness, 16

    impulse, 21
    incompletely known, 162
    logistics, 37
    periodic, 32
    pseudorandom, 162
    sawtooth, *29*
    simplest, 20–28
    sinusoid, 25
    square wave, 23
    stochastic, 30, 162
    triangle, *29*
    unit step, 21
signal processing system, <u>208</u>
signal separation, *407*
signal to noise ratio, *see* SNR
simple difference, 470
Sinc, 88–89
sinc function, 45, 53, *126*
single sideband AM, 659
sinusoid, 25–26, *29*, 32, *33*
sinusoidal representation, 749, 778
skew, 817
slicer, 325, 707, 715, 718, 720–723
slint graph, *28*
slope converter, 67
smoothing, 231, 279–282
SNR, <u>31</u>, 48, 57, 60, 61, 162, 173, 360, 365–367
soma, <u>443</u>
sonogram, 128, 505
sort, 536, 538
speaker identification, 4, 71, 128, 739, 744, 747, 755, 770
spectral analysis, 5
spectrum, 45–46, 71–73, <u>74</u>
speech compression, 4, 369, 371, 373, 633, 739, 753–778
speech recognition, 4, 343, 350, 441, 739, 744, 747–749, 768
speech signals, 4, 435–442, 739–778

speech synthesis, 739
split-radix FFT, 550
square root, 611–613
square wave, 23, 76
stable system, 223, 244
state-space description, 214, 247, 481, 692–695
static buffer, 570
stationary signal, 193
statistically independent, 818
statistics, 161
step signal, see unit step
STFT, 126–132
stochastic signal, 30, 161–163, 192–198
streamability, 224
structures, 579–584, 595–601
subband coding, see SBC
sufficient statistics, 164
SUI, 22, 41–43
sum formulas, 796
super-resolution, 512, 514
SVD, 812, 824
symbol, 405, 703
symmetries, 478–481
synapse, 444
synchronization, 339
syndrome, 686
system identification, 252–270
Szego polynomials, 812

Taylor expansion, 45–47, 145, 155, 279, 308, 323, 391, 606, 609, 811
TCM, 723–728
tee connector, 468
Tesla, N, 28
time advance operator, 36–37
time delay operator, 36–38
time domain, 106

time of arrival, see TOA
time reversal, 34–35, 95, 116, 137, 154, 361
time shift, 94–95, 114, 137, 153–154, 255, 297, 352–353
time warping, 343–348, 739
time-frequency distribution, 129–131, 529
time-invariant system, 223–224
timing recovery, 710–714
TLA, 642
TOA, 171, 319, 364
Toeplitz matrix, 240, 260, 261, 263, 375, 516, 519, 822, 827
Toeplitz, O, 270
toll quality, 756
Tomlinson equalization, 407, 715
Toom-Cook algorithm, 538
topology, 463
transceiver, 647
transfer function, 267, 293–298
transform, 105
transient, 511
transposed structure, 579–580, 595, 599
transposition theorem, 476, 599
trellis, 467, 693
triangle wave, 24
trigonometric functions, 795–798
Tukey, JW, 200, 568
twiddle factor, 541
two's complement, 633, 783

uncertainty theorem, 73, 89, 98, 118, 120, 121–122, 129, 156, 313, 503, 507, 514, 527, 560, 703, 704
unit delay, 470
unit impulse, 21–23
unit step, 21

unitary matrix, *143*, <u>823</u>

V.34, 47, 718, 730–732
VAD, 4, <u>743</u>
variance, 161, <u>816</u>
VCO, <u>339</u>
vector quantization, *see* VQ
vector space, 40–44, 786, 789–790, 819–821
Viterbi algorithm, 344, 693–698
vocoder, 755
voice activity detection, *see* VAD
voice over data, 775
voicing, 752–753
voltage controlled oscillator, *see* VCO
von Hann, J, 508
von Kempelen, W, 779
von Neumann, J, 177, 625
Voronoy region, 721, 768
VOX, *see* VAD
VQ, 765–771

Walsh functions, 523–526
Watson-Watt, R, 171
wave, <u>315</u>
waveform coders, 754
waveform interpolation, 777–778
wavelength, <u>315</u>
wavelets, 526–529
Weber, E, 428–430
white noise, 169, 189, 195
whitening filter, 362, 367, 372
Widrow-Hoff equation, 413, 415
Wiener filter, 365–369
Wiener, N, 270, 392, 530
Wiener-Hopf equations, 263, 410–413
Wiener-Khintchine theorem, 125, 201, 357–359, 386, 389, 521, 708

Wigner-Ville distribution, 130
window, 117, *126*, 129, 274, 505–512, 526
Winograd FFT, 551
Wold's decomposition theorem, 196

Yule, GU, 270, 530
Yule-Walker equations, 263–264, 374, 375, 378–380, 521

z transform, 143–155, 265–270, 520
zero crossings, 324–325, 496–497, 710–712, 743, 752
zero-overhead, 621, *623*, 631, 638
zero-padding, 550, 559, 574
zoom FFT, 559–560